## Handbook of Urban Insects and Arachnids

This account provides the first comprehensive coverage of the insect and other arthropod pests in the urban environment worldwide. Presented is a brief description, biology, and detailed information on the development, habits, and distribution of urban and public health pests. There are 570 illustrations to accompany some of the major pest species. The format is designed to serve as a ready-reference and to provide basic information on orders, families, and species. The species coverage is international and based on distribution in domestic and peridomestic habitats. The references are extensive and international, and cover key papers on species and groups. The introductory chapters overview the urban ecosystem and its key ecological components, and review the pests' status and modern control strategies. The book will serve as a student textbook, professional training manual, and handbook for pest control professionals, regulatory officials, and urban entomologists. It is organized alphabetically throughout.

WILLIAM H ROBINSON is a major figure in the field of urban entomology. He works extensively on urban pest control strategies worldwide.

William H Robinson

# Handbook of Urban Insects and Arachnids

CAMBRIDGE
UNIVERSITY PRESS

CAMBRIDGE UNIVERSITY PRESS
Cambridge, New York, Melbourne, Madrid, Cape Town, Singapore, São Paulo

Cambridge University Press
The Edinburgh Building, Cambridge CB2 2RU, UK

www.cambridge.org
Information on this title: www.cambridge.org/9780521812534



First published 2005

Printed in the United Kingdom at the University Press, Cambridge

*A catalog record for this book is available from the British Library*

*Library of Congress Cataloging in Publication data*
Robinson, W. H (William H), 1943–
Urban insects and arachnids / William H Robinson.
p. cm.
ISBN 0 521 81253 4 (alk. paper)
1. Urban entomology – Handbooks, manuals, etc. 2. Insect pests – Handbooks, manuals, etc.
3. Arthropod pests – Handbooks, manuals, etc. I. Title.
QL472.7.R632 2005
595.7165′09173′2 – dc22 2004054537

ISBN-13 978-0-521-81253-4 hardback
ISBN-10 0-521-81253-4 hardback

# Contents

# Preface

Hugo Hartnac, Arnold Mallis, James R. Busvine, Walter Ebeling, John Gerozisis, Phillip Hadington, Kazuo Yasutomi, and Kenji Umeya.

In their time and in their part of the world, these entomologists brought together in textbooks and handbooks information on the biology and control of household, structural, and public health pests. Their efforts to collect and summarize these data, and to provide their observations and practical experience on insects and other arthropods, have served entomologists and pest control professionals around the world, and are sincerely appreciated.

The goal of this book was to build on the foundation provided by these authors, then to expand the format and provide international coverage. The discipline of urban entomology is changing; it has grown from research and information exchange on a regional basis to international research and cooperation on pests. The modern student and research entomologist needs access to information and a basic understanding of a variety of insects and other arthropods, since distribution and pest status are much less static features of pest species. The objective of this text is to provide a concise and usable reference text on urban and public health insects and other arthropods around the world. In part, this is a global census. Listed here are the invertebrates known to occur, regardless of pest status, in domestic and peridomestic habitats in the urban environment. It provides a boundary for the discipline of urban entomology, and shows the overlap with public health and medical entomology, and stored-food entomology, and the arthropods considered a part of ornamental and turfgrass entomology.

All authors know the limitations of their work. We all know very well where and to what extent our product strayed or failed from the original intent, and what more time and talent would have done for the finished product. That is certainly true for this book. I intended to provide international coverage, including peridomestic and domestic habitats, rural and urban location, and across the boundary of pest, nuisance, and occasional invader. I have certainly missed species or included some of limited importance; for some I failed to find biological data, or the data presented are incomplete, or worse. There are no excuses for the failings, but I resolve to improve what is here with the help of those using this book as a resource. Urban entomology and professional pest control can grow from shared knowledge and experience, and this work can benefit from such cooperation.

This text was prepared and organized in the format of a resource book, and a primary consideration was given to utility. The alphabetical arrangement of the orders (for the most part), families, and species used here removes phylogenetic relationships, and often sets apart related or natural taxonomic groups. The format may be very useful for some users, and may seem near-heresy for others – I apologize to the latter. There are other features that may cause problems. This text was not intended or written to be an organized whole that would be read from beginning to end. Rather it is a source book for the retrieval of information and perhaps a helpful illustration; there is some level of repetition. Morphological and biological information on groups and individual species is presented with key words (such as egg, larva, adult) as guides, with the minimum use of headings and bold type. Included are common or vernacular names for some household pests that occur around the world, and there are other names that are used regionally, locally, and sometimes only temporarily. When common names were available and appropriate, they were included; I may have missed some.

For some pest species or groups, we know perhaps too much. The pure weight of the published facts is daunting. In some groups the depth and volume of information can impede attempts to arrange and present it in a useful and meaningful manner. It becomes a decision of at what level to stop and defer to the published information. The published data on termites, ants, and some species of cockroaches are large, and would be overwhelming without the help and order provided by the authors of bibliographies, books, and subject-matter reviews. Those individuals have provided a great service to this discipline. The reference literature provided here relies on the works that have collected and cataloged scientific papers, and reviewed urban pest concepts.

Control methods and materials are not included with the biological information or in the bibliography. This is often a subjective topic and to cover properly, it must be accompanied by a large amount of published data. Chemical control methods and application equipment are becoming more standardized around the world as manufacturers adopt a global approach to pest management. However, this aspect of urban and public health entomology will always be more dynamic than static, and difficult to put between the covers of a book.

This book could not have been prepared without the investigations, research, and careful observations of pest control professionals, urban entomologists, their students, and technicians around the world. At the local or regional level these entomologists collected and published data on the arthropods that are a small and large part of the urban ecosystem. This book is a collective of those published works. I sincerely appreciate their work and have attempted to share it with other professionals in urban entomology. The majority of the illustrations used here were adapted from various US Department of Agriculture publications. Urban entomologists around the world have provided help with illustrations and translations, and I greatly appreciate their contributions.

Part one

# Urban entomology

# 1 Introduction

## Introduction

The urban environment is a complex of habitats developed by humans from natural sites or agricultural land. Houses, villages, towns, cities, buildings, roads, and other features that characterize the urban environment have gradually and irrecoverably changed the landscape of natural and agricultural areas. As a part of this change, some habitats and their associated plant and animal communities were eliminated, while others were expanded and new ones were created. Many of the new habitats were intentional – parks, waterways, street trees, turfgrass, food stores – but some were consequential – standing water in roadside ditches, garbage and landfill sites near residential neighborhoods, the underground sewer and storm drain network in urban and suburban areas. They all provided habitats for a select group of insects and other arthropods, some of which attained pest status.

Local conditions, climate, and available resources determine the distribution of some arthropods in the urban environment, and for some species their abundance is limited. Other species are broadly adapted to the resources and harborages in and around buildings, and these are cosmopolitan in their distribution and pest status. Stable habitats with resources and conditions suitable for long-term survival support reservoir populations of pest species, and from these habitats individuals or groups move or are transported to establish infestations in unstable or temporary habitats.

## Peridomestic and domestic habitats

Within and around buildings, houses, and other urban structures are habitats that support individuals or populations of plants and animals. Peridomestic habitats are outside, around the perimeter of structures. They include the external surfaces of buildings, the ornamental trees, shrubs, and turfgrass that characterize the urban and suburban landscape. Domestic habitats are indoors, and include the plant- and animal-based materials in this controlled, anthropogenic environment.

### Peridomestic

Harborage substrates, food resources, and environmental conditions of urban landscapes around the world generally support a large number of different species, if not individual species in large numbers. The soil-inhabiting and -nesting arthropods in this environment include ants that forage indoors and termites that damage structural wood, ground-nest bees and wasps, and occasional or nuisance pests such as clover mites, millipedes, centipedes, and springtails. Plant-feeding insects utilize the cultivated urban and suburban trees and shrubs, and many are aesthetic pests. Blood-feeding mites (chiggers), ticks, mosquitoes and other biting flies are associated with domestic and feral vertebrates. Species utilizing building surfaces or perimeter substrates include the umbrella wasps, hornets, yellowjackets, spiders, and scorpions. Underground sewer and storm drainage pipes provide some cockroach and rodent species access to urban and suburban neighborhoods. The garbage disposal network of collection, sorting, and landfill provide harborage and food for cockroaches, flies, rodents, and pest birds.

Reservoir populations for many of the pest species established in peridomestic habitats are in nearby natural or undisturbed areas. Woodland tracts and other small or large patches of greenspace can support populations of biting flies, wasps and hornets, ticks, and spiders. Here are the populations that provide the individuals or groups that establish or replenish infestations in less stable habitats, or re-establish populations lost to habitat destruction.

### Domestic

Environmental conditions indoors are generally stable, and the harborages and food resources are somewhat limited. There may be few species, but those adapted to specialized resources often occur in large numbers. Stored food, including packaged whole food and vegetables, organic fabrics, and other materials are the most common harborages and food resources in the domestic habitat. Directly or indirectly associated with these are dermestid beetles, flour beetles and moths, flies, and cockroaches. The global distribution of domestic products and similar storage environments across cultures has contributed to the cosmopolitan pest status of many of these insects, in both residential and commercial sites. Blood- and skin-feeding species that breed indoors are limited, but lice, fleas, bed bugs, and mites are medically important pests for more than one socioeconomic level of society. Insects and other arthropods in the living space are nuisance pests when they are few and their presence brief, but are not tolerated when they pose a health threat or persist in large numbers.

Natural habitats and populations for some domestic species, especially those infesting stored food, have been lost. Only populations in the urban environment represent many of these species, or they survive only through their link to humans (bed bugs, lice). Other indoor pests have reservoir populations in peridomestic and natural areas. Many of the common species occur in the nests of bird and rodents and from there have access to indoor habitats.

## Pest status and control

In the agroecosystem, pest status and the decision to apply control measures for arthropods are based primarily on economics. Pests can be measured by their damage and reduction in animal weight or crop yield, and controls are applied to prevent or minimize predictable loss. Pest status for insects and other arthropods in the urban environment may or may not be based on a measurable feature. The damage caused to structural wood by termites or other wood-infesting insects can be measured, and the control and repair costs determined. The health threat or medical importance, such as from stinging insects, can be measured by medical costs. A decision to apply control measures may be based on potential damage or personal injury, or solely or in part on emotion. The control decision is no less appropriate when it is based on emotion. Arthropods in the living space are generally unwanted and unwelcome, whether their numbers are few or many.

Pest status is generally based on persistence or recurrence of an arthropod indoors or outdoors, due to the failure of control methods, or the ability to reinfest from reservoir populations. The persistence of many species in the urban environment is based on a network of reservoir populations, from which individuals or groups move to infest or reinfest domestic or peridomestic habitats. Undisturbed woodlands may support populations of yellowjackets, subterranean termites, and carpenter ants, and serve as a reservoir for colonies and infestations in adjacent and distant residential areas. Sewer pipes often provide conditions suitable for American cockroach populations, and from this habit, adults and nymphs infest and reinfest buildings.

For pest control or management programs to be successful, reservoir populations and habitats must be identified and reduced. The only functional reservoir populations for some peridomestic and domestic species are in secondary habitats in the urban environment. Pests whose abundance is based on the limited availability of artificial habitats and resources are vulnerable to effective chemical and nonchemical control methods, and may be eliminated.

## Pest dispersal and distribution

International transportation, economic exchange, and globalization have brought a degree of uniformity to the urban area around the world, and increased the movement and exchange of arthropods. The majority of household and stored-food pests, including fruit flies, cockroaches, flour beetles, moths, and mites, have moved with infested commercial goods and now have cosmopolitan distribution. Peridomestic mosquitoes, subterranean termites, and wood-infesting beetles share the same potential for widespread distribution. Current distribution records for many household and structural pests are subject to change with increased movement of people and materials around the world.

Information on pest identification, biology, and habits, compiled on an international basis, is appropriate for the urban environment. A global census indicates that nearly 2300 insects and other arthropods have some level of pest status around the world. Some are only occasional invaders of houses and other buildings, some are closely associated with the foods, fabrics, and other aspects of dwellings, and others are linked to plants and animals in domestic and peridomestic habitats. Many of these species are capable of adapting to the soil conditions, climate, and building construction in other regions of the world, and becoming established in pest populations. Regional conditions may alter some behaviors, but morphological features and the basic life cycle will remain unchanged, and control strategies are usually transferable from region to region.

## Urbanization

The quality of life for most people in the future will be determined by the quality of cities. In 1950, 60% of the world's population lived in villages and small towns in countryside. By the year 2030, 60% of the world's people will be living in metropolitan areas anchored by a large city. Those cities will be bigger than ever and dominate the landscape: most of these cities will be in developing countries. Explosive growth in urban populations and the steady stream of migration of people from the countryside put great strains on city services and the quality of life. The housing, health care, water, and sanitation systems must keep pace with the growth, and the threat of disease. Despite the conditions, migration to cities continues, and that is good news. Cities provide development and growth, and generally a better life than in rural areas. The future of many developed countries is linked to their cities. Urban growth is inevitable: the challenge is how to address the consequences and improve the quality of life from city center to the unplanned housing at the perimeter of the metropolis.

Insects and other arthropods that carry and transmit disease organisms present a threat to the cities and densely populated urban areas of the world. In these areas, crowded living conditions and poor sanitation support vector populations, and the concentration of human hosts can maintain common diseases and rapidly spread new ones. Pest management and control strategies will be based on pest identification and life-cycle information, an understanding of reservoir habitats, and effective chemical and nonchemical control materials.

## Bibliography

Bornkamm, R., J. A. Lee, and M. R. D. Seward. *Urban Ecology: Second European Ecological Symposium*. London: Blackwell, 1982.

Boyden, S. *An Integrated Ecological Approach to the Study of Human Settlements*. MAB technical notes 12. Paris: UNESCO, 1979.

Boyden, S., S. Miller, K. Newcombe, and B. O'Neill. *The Ecology of a City and its People: The Case of Hong Kong*. Canberra: Australian National University Press, 1981.

Bronfenbrenner, U. *The Ecology of Human Development*. Cambridge, MA: Harvard University Press, 1979.

Chinery, M. *Collin's Guide to the Insects of Britain and Western Europe*. London: Collins, 1986.

Ebeling, W. *Urban Entomology*. Berkeley, CA: University of California Press, 1975.

Frankie, G. W. and L. E. Ehler. Ecology of insects in urban environments. *Annu. Rev. Entomol.*, **23** (1978), 367–87.

Frankie, G. W. and C. S. Koehler (eds.) *Perspectives in Urban Entomology*. New York: Academic Press, 1978.

*Urban Entomology: Interdisciplinary Perspectives*. New York: Praeger, 1983.

*The Ecology of Urban Insects*. London: Chapman and Hall, 1989.

Gerozisis, J. and P. Hadington. *Urban Pest Control in Australia*, 4th edn. Sydney: University of New South Wales Press, 2001.

Gold, R. and S. C. Jones (eds.) *Handbook of Household and Structural Insect Pests*. Lanham, MD: Entomological Society of America, 2000.

Hartnack, H. *202 Common Household Pests in North America*. Chicago, IL: Hartnack Publications, 1939.

*Unbidden House Guests*. Hartnack Publishing: Tacoma, WA: 1943.

Hedges, S. and D. Moreland (eds.) *Handbook of Pest Control: The Behavior, Life History and Control of Household Pests*. Cleveland, OH: GIE Media, 2004.

Lee, C.-Y., H. H. Yap, N. L. Chong, and Z. Jaal (eds.) *Urban Pest Control: A Malaysian Perspective*. Penang, Malaysia: School of Biological Sciences, University of Sains, 1999.

Lynch, K. *The Image of the City*. Cambridge, MA: MIT Press, 1960.

McIntyre, N. Ecology of urban arthropods: a review and a call to action. *Ann. Entomol. Soc. Am.*, **93** (2000), 825–35.

Odum, E. P. The strategy of ecosystem development. *Science* **164** (1969), 262–70.

Phillips, D. Urbanization and human health. *Parasitology* **106** (1993), 93–107.

Pisarski, B. and M. Kulesza. Characteristics of animal species colonizing urban habitats. *Memorabilia Zool.*, **37** (1982), 71–7.

Robinson, W. H. *Urban Entomology: Insect and Mite Pests in the Urban Environment*. London: Chapman and Hall, 1996.

Stearns, F. W. and T. Montag. *The Urban Ecosystem: A Holostic Approach*. Stroudsburg, PA: Dowden, Hutchinson and Ross, 1974.

Story, K. (ed.). *Handbook of Pest Control: The Behavior, Life History and Control of Household Pests by Arnold Mallis*, 7th edn. Cleveland, OH: Franzak and Foster, 1990.

Tasutomi, K. and K. Umeya. *Household Pests*. Tokyo: Zenkoku Noson Kyoiku, 1995.

Tischler, W. Ecology of arthropod fauna in man-made habitats: the problems of synanthropy. *Zool. Anz.*, **191** (1973), 157–61.

Yanitsky, O. N. Towards an eco-city: problems on integrating knowledge with practice. *Int. Soc. Sci. J.*, **34** (1982), 469–80.

# 2 The urban ecosystem

## Introduction

Major ecosystems can be broadly classified as natural, agricultural, and urban. Natural ecosystems are primitive sites where the interacting plant and animal communities have not been altered by human activity. There are few, if any, of these in the world today, and a more practical definition of natural ecosystems might be undisturbed habitats that have had limited human influence, and retain a portion of their original flora and fauna. An important feature of these habitats is the populations of native plants and animals. These are the reservoir populations of many species that have adapted to agricultural and urban conditions. Agricultural and urban ecosystems are defined by their use and the degree to which their biotic and abiotic features have been altered by human activity. These ecosystems contain few of the features that characterize their natural origins; many of the features were built or brought there, or designed by humans. The degree of change and land use can be used to subdivide these two cultural ecosystems.

Agroecosystem A is the least developed form of agriculture. It consists of small farms with a mix of domesticated animals and crop plants; it is generally expected to provide food and fiber for family groups or communities. Agroecosystem B is the most developed form of agriculture. It is characterized as mechanized farming of a single crop (soybeans, maize, wheat) or single-animal species (swine, cattle, poultry). Modern definitions of this ecosystem would include use of genetically improved or engineered crops.

The term urban is often used synonymously with city, but when used in the context of the urban environment it extends to plant and animal communities in cities and surrounding suburbs. There is a continuum of inhabited sites and human activity from the primitive farmhouse to metropolitan office building, and the division between urban, suburban, and rural is indistinct. The urban environment has levels of modification and changes in the physical landscape and biotic communities similar to those found in the agricultural ecosystem. Urban-ecosystem A is the rural–suburban landscape, and includes natural and undisturbed sites, such as small wood lots or agricultural fields. Urban-ecosystem B is the cityscape of commercial and residential neighborhoods, with a limited amount of planned greenspace and undisturbed areas. As in agriculture, these divisions are based on human interaction and intervention with the landscape and associated plant and animal communities.

## Urban ecosystems

Development of what is known as suburbia began in the 1800s with people from the upper and middle classes moving to the perimeter of the industrial cities. The crowded living and poor sanitary conditions in the early cities was an incentive to move to the rural conditions of the periphery. Movement to the suburbs continued in the 1920s and 1930s, and it increased worldwide after 1945 with improvements in transportation and highways systems. By the 1960s, major cities in the industrial countries had a distinct suburban perimeter (urban-ecosystem A), and a commercial core (urban-ecosystem B). Urbanization continues around the world through urban sprawl; this is a process in which the suburban residential and commercial land use spreads into peripheral farmland and natural areas. The outward spread and fusing-together of adjacent towns has led in many places to the formation of conurbations. The traditional concept of the city as a clearly defined entity has given way to terms that better describe its size, such as megalopolis, or ecumenopolis.

The outlines of such large urban areas can be discerned in the Great Lakes area and the northeastern seaboard of the USA, along the highways and transportation systems that link Tokyo and Osaka, Japan, and in Europe, in the zone of intense urban development that extends from London through Rotterdam to the Ruhr in Germany.

Urban-ecosystem A is typically 60% greenspace and 40% built landscape, and has a range of soil types, drainage systems, ground cover, and plant and animal species. It is a mix of land use: undisturbed areas, planned and unplanned greenspace, and commercial and residential buildings. Greenspace varies in size and use, and includes golf courses, tracts of recreation and parkland, lakes, and waterways, and the ornamental shrubs, trees, and turfgrass associated with the gardens and yards of residential housing. Undisturbed areas may be plots of trees or secondary vegetation on land bordering residential or commercial sites. The interface of suburbia with small-scale agriculture may be abrupt and with little space (often a roadway) between them. The spread of suburbia often brings residential areas close to established livestock and poultry farms, landfills for household waste, dumps for automobile tires, or industrial refuse sites. These operations have insects and other animals that become pests in adjacent areas. Interface with the city may be a gradual decrease in greenspace and increase in residential and commercial built areas.

At the periphery of cities in developing countries are zones of dense, unplanned, and impoverished housing. These shantytowns vary from country to country, but they are an established feature of major cities, and represent 20–30% of the new urban housing in the world. Most new housing in developing countries is built on unclaimed land by squatters, without consideration for local or government regulations. This housing is considered the *septic fringe*; it is composed of crowded conditions, substandard housing, and with limited access to clean water and waste removal. Here are habitats suitable for populations of vertebrate and invertebrate disease vectors with a flight and foraging range to bring them into contact with a large portion of the city's population.

Suburbia is composed of planned communities and structured greenspace; some of these peripheral areas are considered the *affluent fringe*. Houses and other buildings are surrounded by ornamental shrubs, trees, and turfgrass, and the landscape includes flower gardens, and in some neighborhoods there may be water fountains and swimming pools. As in the septic fringe, there are habitats in the affluent fringe suitable for insect vectors of disease, and successful populations of rodents and wildlife species. Planned development and improved living conditions often mean reduced diligence and less compliance with insect control programs. In these neighborhoods there may be more rather than fewer breeding sites for pests, such as mosquitoes, black flies, wasps, and beetles. The mix of vegetation and the availability of food and harborage often provide an abundance of vertebrate hosts and arthropod vectors of pathogenic organisms.

Urban-ecosystem B is the most developed ecosystem, with about 60% of the surface area consisting of hard-surface and built structures. It is the built landscape of the city and characterized by an uneven distribution of exposed soil and sparse vegetation. It is dominated by the hard surfaces of roads, sidewalks, parking lots, and permanent structures. Here the land surface has been radically altered, and the existing plants and animals selected and maintained by human activity. This ecosystem typically interfaces with a suburban landscape. The mixed-density landscape of houses, low- and high-rise residential buildings, and single-family homes at the edge of a metropolis is sometimes considered semiurban or the inner suburbs; perhaps it is a transition zone between urban-ecosystems A and B.

### Agriculture interface

The urban interface with agriculture often occurs when suburban sprawl, bringing with it residential and commercial land use, is developed close to animal farms. Dairy cattle, livestock (swine and beef), and poultry operations are often encroached upon as the suburban ring of cities spreads. The flies typically associated with the manure at these operations can disperse several kilometers and create a nuisance during nearly all months of the year. Dairy cattle herds may have 50–200 cows; in temperate regions they are housed in barns or buildings for part of the year; in warm regions they are outside most of the time. Poultry egg production is usually in 100 000-bird buildings in groups of 10 or more, and they function year-round. Manure produced at these operations can support large populations of house fly (*Musca domestica*) and stable fly (*Stomoxys calcitrans*), and in some regions *M. sorbens*. An average 1.8-kg laying hen produces about 113 g of wet manure daily; this is 11 300 kg per day or 4139 metric tons per year for each 100 000-bird poultry building. Feedlots may have 1000–3500 cattle at one time, and each feedlot cow produces about 23 000 g of wet manure daily. Stable fly and house fly maggots require about 2 g of manure to complete development, thus the potential for livestock and poultry operations to produce flies and problems is significant. Other feedlot operations, such as those for turkey and chicken production, have accumulations of manure. Accompanying the accumulation of animal and poultry manure is a concentrated manure odor, and this is also a nuisance during most of the year. Adult flies can travel 20 km or more from breeding sites, or be carried by prevailing winds to nonfarm sites and be a nuisance and sometimes a health hazard. In some countries, right-to-farm laws provide some protection to farmers, but fly control is an important feature of modern agriculture.

Other sites in or around urban areas may have accumulations of animal dung and associated flies feeding on this resource. Zoos, kennels for dogs and cats, stables for riding horses, and large recreation theme parks have large animals, and manure disposal at these sites can be difficult since it may not be easily spread on adjacent farmland. Fly populations at these sites may be seasonal in temperate regions, but small numbers of adults will be present during winter months. Other insects are associated with the manure and the fly populations, including yellowjackets, carabids, and dung beetles.

### Natural area interface

The urban interface with undisturbed or natural areas occurs when suburban sprawl brings residential housing developments close to or at the edge of land set aside or preserved as a natural site. Wilderness or relatively undisturbed areas may provide reservoir populations for domestic and peridomestic pest species, including yellowjackets and carpenter bees, carpenter ants, subterranean termites, and some species of ticks and mites (chiggers). Large- and small-animal populations would increase the potential of arthropod-borne diseases, such as Lyme disease, Rocky Mountain spotted fever, West Nile virus, and plague.

Many mountainous or wilderness areas used for recreation in the western USA contain large populations of plague-positive rodents. A large number of plague cases in the USA have been contracted during recreational pursuits, or in suburban areas adjacent to wilderness land. Increased urban growth has resulted in large numbers of people living in or near areas with rodent populations that harbor plague. The peridomestic habitats created in residential neighborhoods provide harborage and food for adaptable rodent species, such as ground squirrels and rock squirrels, chipmunks, and prairie dogs. These species have increased in density, and their fleas are efficient vectors of plague to humans and other animals, such as domestic cats. Most cats acquire plague by ingesting infected rodents, and they spread plague by a scratch or bite, or by aerosolized droplets in the case of pneumonic plague. The number of confirmed cases of plague in the USA directly transmitted by domestic cats is increasing, and is usually associated with residential areas.

## Urban habitats

The structural complexity of cities includes features that provide harborage and food for arthropods and other animals. Parks, recreation areas, and other greenspace have natural habitats for vertebrates and invertebrates; the system of storm water and sewer pipes provides artificial habitats for other animals. Garbage collection points and landfills are consistent features of urban environments around the world, and these sites provide habitats for arthropods, rodents, and pest birds. Livestock agriculture in the form of poultry egg and meat production, feedlots for swine, and beef cattle often interface with residential and commercial land.

### Parks, greenspace, and gardens

Many cities have been designed to include space for large and small parks, peripheral green belts, or forested areas along small streams and rivers. These areas break the monotony of residential and commercial buildings, influence local temperature and humidity, and provide neighborhoods with an open recreation site. Early in the development of cities in the USA and Europe large tracts of land were set aside for parks: New York's Central and Prospect Park, and Hyde Park in London are examples of this planned and dedicated space. Once established and integrated into the landscape and seasonal activities, they become an important part of the urban environment.

Cities can have two classes of open areas or greenspace: those that have been intentionally established as parks or recreation plots, and the unplanned sites of vacant lots and roadways. In the former, the diversity of plants and animals may be limited, and these sites are somewhat influenced by use patterns of people and domestic pets. Vacant lots, backyards, roadway median strips, and the rights-of-way of railroads and other roads may have a great variety of plants and animals. Modern highway and expressway systems that enter or circle urban areas often have broad medians and shoulders, and these may be planted with turfgrass, wildflowers, trees and shrubs. These narrow strips of land often have a large and diverse invertebrate fauna.

Accompanying the recent phenomenon of urban sprawl and expanding suburbs has been the increase in household flower gardens. Despite the conditions of urban high-rise buildings and a concrete and asphalt substrate, urban gardens are flourishing in many regions. Although gardens have been a feature in European cities since the 1760s, the availability of potted plants and exotic species have made it a personal pastime with psychological and economic benefits. An urban or suburban landscape of trees, shrubs, or flowers adds economic value to property: in some cases an increase of 12–30% can be achieved. However, the widespread popularity of household and public gardens can also be accompanied by some health hazards. Whether native or exotic plant species are used, urban gardens may provide food, habitat, or harborage for invertebrate disease-vectors and their vertebrate hosts. Urban wildlife, such as rabbits (*Sylvilagus*), deer (*Odocoileus*), chipmunks (*Tamias*), mice (*Peromyscus*), and voles (*Microtus*), feed on a variety of garden plants and seeds, and populations often become large and difficult to control or even manage. Their pest status is based on damage to garden plants, nesting habits, and serving as hosts for ticks and other blood-sucking insect vectors. Increases in Lyme disease and Rocky Mountain spotted fever in eastern USA may be attributed to deer and rabbit populations.

### Sanitary sewers and storm sewers

An essential urban infrastructure is the network of underground pipes that remove waste water from toilets and kitchens, and storm water runoff. Many of the urban sewers and storm drains constructed in the 1700s and 1800s are still in use, and in some cities they have been extended or connected to recently developed networks. This elaborate drainage system is hidden from view, and perhaps from the realization that it often provides food and harborage for mosquitoes, cockroaches, rats, and other invertebrate and vertebrate pests. The availability of resources and uniform environmental conditions often results in year-round pest populations in these underground pipes.

Urban areas may have different systems for handling household waste water and for removing surface or storm water. A combined system brings together household waste and surface runoff water into one network of pipes and delivers the combined discharge to a centralized sewage treatment facility. Some cities have a system which diverts household waste and storm water to separate pipes. Those pipes carrying only surface water discharge at various points into natural watercourses, and the waste water is directed to a sewage-treatment facility. The separate system diverts the majority of surface water to storm sewers, but some of it may be combined with sewage and treated before being released. While both methods can provide harborage and other resources for pests, the combined waste water system is most likely to support pest populations, because of the food contained in the toilet and kitchen refuse.

The storm water drainage system of pipes carries away large amounts of water that may otherwise accumulate on roads and streets following excessive rain or snow. Water from streets and sidewalks flows into the underground network of pipes through inlets and catch basins positioned along the curb and street corners. Inlets are covered by a grate and connected to a catch basin before leading to a drainpipe. A catch basin is usually a rectangular storage box located under the street. It is designed to trap street debris before it enters and obstructs the flow of water into drainpipes. Not until water reaches a certain height in the catch basin does it flow into the major storm drain. Because of their construction and underground location, catch basins often retain water for long periods. The combination of organic matter and standing water in a dark and protected location provides a breeding site for several species of mosquitoes. These sites also provide a source of food for cockroaches and rats. Similar conditions are present in some of the underground mass-transit systems and shopping areas in major cities of the world.

### Solid waste disposal and landfills

Collection and disposal of solid waste is important to human health and the daily operation of a city. Waste produced by households and commercial sources is collected and transferred to a landfill, a site dedicated and specifically managed for waste disposal. It may be close to the city or carried to a distant location. Municipal solid waste originates from daily activity in households, hotels, hospitals and health care facilities, and restaurants, and it contains 10–50% wet and putrescible organic material. The high organic content is a potential food resource and harborage for insects, pest birds, and rodents. The utility this material has to these pests is influenced by the techniques used for collection, and the short- and long-term disposal.

Open refuse sites may be the primary method for collecting the garbage from small communities or neighborhoods in some parts of the world. These sites are usually exposed, three-walled bins, large metal containers, or simply a vacant plot of land. Depending on the size of the areas served, there may be one or more of them in a neighborhood. Although this method leaves organic refuse vulnerable to pest infestation, concentrating household refuse in designated sites enables efficient removal and is better than uncollected garbage in the street. Depending on climate and seasonal temperatures, frequency of collection, and the organic content, open public refuse sites support large infestations of flies and rodents, and often attract birds, dogs, cats, goats, and other animals. Rodents and flies may establish long-term populations at these sites, and move from there to forage in or infest surrounding buildings. Fly maggots within the garbage at the time of collection may be removed from the population; full-grown larvae leave the refuse to pupate and avoid collection, and remain to reinfest. Hot and dry weather can reduce the attractiveness of refuse piles to flies, and hot and wet weather may extend it.

Galvanized steel or plastic containers with lids are typically used to hold household dry and putrescible material. Ideally, the garbage is secure until emptied into a collection vehicle, but lids on garbage containers may not completely prevent entry of rodents, flies, wasps, and other insects. Various species of flies can infest these containers: fruit flies access openings that are 1–2 mm wide and adult blow flies are capable of moving through openings 3.2 mm wide. Holes or cracks in the bottom of containers allow full-grown maggots to leave, or large blow fly maggots may climb the inside surface of metal containers to find a suitable pupation site outside. In some cities, 60% of the garbage containers may be infested with fly larvae. Rodents gnaw small holes in the bottom and sides of plastic containers, and leave them accessible to further attack. The lids of garbage containers are often not used and garbage is exposed. Daily or weekly garbage collection is partly a function of climate and the local authorities. Long collection intervals, combined with putrescible waste, loose-fitting lids, and damaged containers often result in pest problems.

Many of the large cities of the world rely on a local landfill to take their daily garbage; these sites are usually originally established at the periphery of the city. Landfill sites must be easily accessible and large to accommodate the quantity of solid waste and other material a city produces in the course of 10–15 years. For disposal in most large metropolitan landfills, garbage is first taken to a transfer site where it is emptied from the collection vehicle and loaded into a compactor or incinerator to reduce the volume. It is then transported to the landfill, which may be local or a long distance away. Key to the successful operation of transfer stations is the rapid processing of refuse. Regardless of their efficiency, transfer stations often attract flies, rodents, and pest birds, and their presence can cause problems in surrounding neighborhoods.

Compacted or loose garbage at the landfill is usually covered to reduce odor and the attraction it has to various pests. Soil is commonly used for cover, and the thickness of the layer is important to fly control. Cover soil that is less than about 150 mm is not sufficient to prevent fly emergence completely. House fly adults are capable of moving to the surface from beneath 250 mm of soil, and blow flies and flesh flies are known to emerge from feeding sites 450 mm within compacted refuse. When soil is unavailable or the costs for it are high, other materials, such as paper pulp, fragmented plastic, sand, woven geotextiles, and plastic sheets may be used. In direct sunlight plastic sheets create in the underlying refuse a microclimate with temperatures high enough to prevent fly development. However, these sheets may interfere with rainwater percolation and natural compaction, and trap landfill gases.

The house fly and local species of blow flies are the most common insects at urban landfills around the world. At landfills, these flies may breed continuously through the year, but with decreased numbers in the cold months. Crickets and cockroaches, including the German cockroach, can become established at landfills, depending on local conditions. Infestations of cockroaches have been linked to buried lots of household material that came to the landfill infested. Once at the site, populations were maintained by the available food and only limited compaction to provide harborage. The pest bird species varies according to location, but the most common are gulls, crows, starlings, and kites. They rarely nest at the site, but usually include the landfill within their foraging territory. The brown rat is common in landfills around the world. Large vertebrates, such as foxes, feral dogs, and goats also regularly occur.

There may be few stable habitats directly on the landfill to support vertebrate populations; most pest species only move to the landfill for feeding and have established nests offsite. Although there is a continuous source of garbage, the working face for dumping changes and there is regular (day and night) disturbance by workers and vehicles. Sudden disturbance of house fly, cricket, and cockroach populations can result in the dispersal of large numbers to areas surrounding the landfill. House flies and blow flies are capable of traveling 1–3 km from infested sites, and cockroaches can move across a varied landscape to building perimeters. Large numbers of seagulls at landfills can disrupt the operation of compaction and earth-moving equipment and spread disease. Feces from gulls at landfill sites have been shown to contain human pathogenic bacteria, such as *Escherichia coli* O157. Landfill gulls have the potential of transporting such bacteria to farm and urban sites.

## Urban environmental features

Urbanization has pronounced effects on the abiotic components of the environment. Concentrations of heat-absorbing surfaces of streets, highways, parking lots, the limited amounts of greenspace and open soil, and large amounts of pollution and particulate matter in the air result in cities having a climate different from the surrounding countryside. Climatic changes can occur in the form of seasonal temperature highs and lows, in intensity and direction of the windfields around buildings, and in amount of rainfall and runoff conditions. Climate is the net combination of temperature, water vapor in the air, precipitation, solar radiation, and speed of the wind. Meteorological variables that are usually distinctly different between cities and open country include day and night temperature and relative humidity, rainfall, and fog. The most recognized city-climate phenomena are persistent smog, early blooming or leafing of flowering plants, and longer frost-free periods in north temperate regions.

### Urban substrates

Up to 33% of the land surface in cities is occupied by hard surfaces in the form of roads, sidewalks, and parking lots. A nearly equal proportion is taken up by buildings and other built structures, with the result that 60–70% of urban areas in modern cities consists of surfaces formed from nonporous materials. Only the remaining third of urban surface can be considered porous for water circulation and water vapor exchange, but these may be covered with refuse and other debris. Hard surfaces of cities generally accept more heat energy in less time than an equal amount of soil; by the end of the day, brick or concrete surfaces will have stored more heat than an equal surface of soil. However, hard surfaces of buildings and pavement release or conduct heat about three times as fast as it is released by moist, sandy soil. The variety of light- and dark-colored building and sidewalk surface, the reflection and absorption of sunlight, and conduction of absorbed heat energy are linked to city–countryside climate differences. Urban buildings have a breaking effect on wind, and this may reduce the amount of heat that is carried away.

Buildings and other features add to the three-dimensional complexity of cities. The result is a rise in the mean temperature, forming what is called an *urban heat island*. This island results from the reduced amount of evaporative cooling, heat retained by surfaces, and heat produced by vehicles and machines. One feature of the heat island is the limited range of daily high and low temperatures. Despite the large amount of (sunlight) heat absorbed and heat radiated by structures, shading by buildings and narrow streets keeps sunlight from many urban surfaces, thus lowering the maximum daily temperature. Summer nights in the suburbs may be cool, but in the city temperatures may be only a few degrees lower at midnight than at sundown. The physical mass of the city acts as a buffer, damping temperature extremes. Since air is primarily heated more by contact with warm surfaces than it is by direct radiation, city surfaces (buildings, roads, and pavements) are capable of heating large volumes of air. The dome of warm air that is regularly over large cities forces moisture-laden clouds upward into colder air, which initiates rain. Solid, liquid, and gaseous contaminants characterize the air of most modern cities, some more than others. About 80% of the solid contaminants are particles small enough to remain suspended for long periods. These particles directly influence rainfall and air temperature in cities. Particulate matter provides nuclei for the condensation of atmospheric moisture into rain. The general rule is, as cities increase in size, air pollution increases, and rainfall increases.

Measurable rainfall in cities is shed from hard surfaces and quickly removed through drainpipes, street gutters, and storm sewers. The urban landscape was developed from agricultural or natural land; construction usually involves removing native vegetation along with upper layers of soil (topsoil), and reshaping the existing topography. One of the outcomes of these changes is altering the natural routes of rainfall runoff. Once an urban center has been developed, flood peaks in streams and rivers that are a part of the habitat often increase two to four times in comparison with preurbanization flow rates. The increases are due to pavement and roadways that cover a large percentage of the surface in suburban areas, and nearly all the surface in business and industrial areas. This reduces the amount of rainwater that infiltrates soil, and increases runoff and sediment in streams and rivers. Pollution from increased runoff affects plant and animal communities in and along the bands of these waterways.

Prevailing winds are usually rapidly decelerated over towns and cities compared with the open countryside. Wind velocity may be half what it is in the open countryside, and at the edge of the urban area wind velocity may be reduced by a third. One reason for this is the increased surface texture caused by the mixture of short and tall buildings. Cities have reduced average wind velocity in direct proportion to their size and density. Along roads and highways parallel to the wind direction, wind velocity increases and may be disruptive to people and flying insects. Trees along these wind routes, and trees in greenspace and parks can help to reduce urban wind speeds. However, the presence of large patches of vegetation and blocks of urban trees can contribute harborage and breeding sites for pests, such as birds, rodents, and other wildlife. Some insects that naturally occur in suburban or rural areas are easily moved by winds, and may be carried into the edges of the city. Cloudless skies at night and the horizontal temperature gradient across the urban/rural boundary can be sufficient to create a low-level breeze from the rural area into the city. This flow of air from suburban or agricultural areas into the city can aid and direct the movement of small, dusk- or night-flying insects, such as mosquitoes.

### Street lights

Street lights and commercial outdoor lighting have contributed to the presence, pest status, and probably the geographic distribution of some arthropods in the urban environment. A variety of flying insects, including flies, beetles, plant bugs, and moths, are attracted to bright lights at night. This behavior often results in insects indoors and outdoors at windows and on screens, and dead and dying insects near the light source. Factors that influence whether insects fly to outdoor lights include brightness (wattage), their ultraviolet light output, the heat produced, and competition from other lights in the immediate area. The number of insects attracted to the early streetlights on urban and suburban streets may have been small because of the low intensity of these lights, and their limited use. As lighting technology improved and intensity increased, the number in use increased, along with the insects. The pest status of several species of beetles, flies, and bugs is based on their occurrence at outdoor lights; June beetles, crane flies, and giant water bugs (*Belostoma*, Hemiptera) are consistently at these sites. Artificial lights may also be a contributing factor to the decrease in abundance of some populations of nocturnally active insects. Insects attracted to lights may remain there and be easily preyed upon by vertebrate scavengers, such as toads, opossums, and raccoons.

Insects respond primarily to the intensity and the color spectrum of light. Generally, they react to a light spectrum that extends from the near ultraviolet, which is 300–400 nm, up to orange, at a maximum of 600–650 nm. However, attraction is not the same throughout the spectrum. Many insects have two peaks of maximum sensitivity: one in the near ultraviolet, which is at about 350 nm, and a second in the blue-green, at about 500 nm. Perception of this color range occurs in the Hymenoptera, Diptera, Coleoptera, Lepidoptera, Neuroptera, Hemiptera, Homoptera, and Orthoptera. Sensitivity to the ultraviolet portion of the spectrum has been used to attract and trap some insects, while the blue-green component of incandescent light bulbs attracts a wide range of species at night. The light spectrum of incandescent bulbs is through the visible spectrum to the middle of the ultraviolet, which is why these lights often attract insects at night.

Mercury vapor lamps are often used for outdoor lighting. These bulbs heat mercury until it vaporizes, then an electrical discharge is passed through the vapor to produce a bright light with the blue tint. There is a strong ultraviolet and blue light content to these bulbs, and they provide a strong attraction to insects at night. Sodium vapor lamps are an economical and ecological alternative to mercury vapor lamps because they produce the most illumination for the amount of electricity used, and attract few insects. They have a distinct yellow light because they produce almost entirely one wavelength of yellow light, very little of which is below 550 nm, and only a small amount of ultraviolet light. Insects are less attracted to these and other commercial orange or yellow lights because of the light spectrum produced.

## Bibliography


Audy, J. R. The localization of disease with special reference to the zoonoses. *Trans. R. Soc. Trop. Med. Hyg.*, **52** (1958), 308–28.

Bishop, J. A. An experimental study of the cline of industrial melanism in *Biston betularia* (L.) between urban Liverpool and rural North Wales. *J. Anim. Ecol.*, **41** (1972), 209–43.

Boyden, S. and S. Millar. Human ecology and the quality of life. *Urban Ecol.*, **3** (1976), 263–87.

Brady, R. F., T. Tobias, P. F. J. Eagles *et al.* A typology for the urban ecosystem and its relationship to larger biogeographical landscape units. *Urban Ecol.*, **4** (1979), 11–28.

Brandenburg, R. and M. G. Villani (eds.) *Handbook of Turfgrass Insects.* College Park, MD: Entomological Society of America, 1995.

Bruce-Chwatt, L. J. Endemic diseases, demography and socioeconomic development of tropical Africa. *Can. J. Pub. Health*, **66** (1975), 31–7.

Davies, D. M. Seasonal variation of tabanids (Diptera) in Algonquin Park, Ontario. *Can. Ent.*, **91** (1959), 548–53.

Davis, B. N. K. Urbanisation and the diversity of insects. *Biol. Conserv.*, **10** (1978), 249–91.

Dreistadt, S. H., D. L. Dahlsten, and G. W. Frankie. Urban forests and insect ecology. *BioScience*, **40** (1990), 192–198.

Ishii, M., M. Yamada, T. Hirowatari, and T. Yasuda. Diversity of butterfly communities in urban parks in Osaka prefecture (in Japanese). *Jpn J. Environ. Entomol. Zool.*, **3** (1991), 183–95.

Johnson, W. T. and H. H. Lyon. *Insects that feed on Trees and Shrubs.* New York: Comstoack, 1991.

Kimura, Y. Defoliate insect pests in urban green zones (in Japanese). *Jpn J. Environ. Entomol. Zool.*, **3** (1991), 217–24.

Lutz, F. E. *A Lot of Insects (Entomology in a Suburban Garden).* New York, NY: Putnam Sons, 1941.

Minar, J. Synanthropisation and spreading of Dermestidae (Insecta: Coleoptera). In Robinson, W. H., F. Rettich, and G. W. Rambo (eds.) *Proceedings of the 3rd International Conference on Urban Pests*, p. 657. Hronov, Czech Republic: Grafické Závody, 1999.

Naveh, Z. Landscape ecology as an emerging branch of human ecosystem science. *Adv. Ecological Res.*, **12** (1982), 189–237.

Povolony, D. Synanthropy. In Greenberg, B. (ed.) *Flies and Disease*, pp. 17–54. Princeton, NJ: Princeton University Press, 1971.


#### Habitats and environmental features


Anginao, E. E., L. M. Magnuson, and G. F. Stewart. Effects of urbanisation on storm water runoff quality: a limited experiment, Naismith Ditch, Lawrence, Kansas. *Water Resources Res.*, **8** (1972), 135–40.

Davis, B. N. K. Urbanization and the diversity of insects. *Biol. Conserv.*, **10** (1978), 249–91.

Detwyler, T. R. and M. G. Marcus. *Urbanization and Environment: The Physical Geography of the City*. Belmont, CA: Duxbury Press, 1972.

Duckworth, F. S. and J. S. Sandberg. The effect of cities on horizontal and vertical temperature gradients. *Bull. Am. Meterol. Soc.*, **35** (1954), 198–207.

Extence, C. A. The effect of motorway construction on an urban stream. *Environ. Pollut.*, **17** (1978), 245–52.

Faeth, S. H. and T. C. Kane. Urban biogeography: city parks as islands for Diptera and Coleoptera. *Oecologia*, **32** (1962), 127–33.

Falk, J. H. Energetics of a suburban lawn ecosystem. *Ecology*, **57** (1976), 141–50.

Feldman, B. M. The problem of urban dogs. *Science*, **185** (1974), 903.

Gill, D. and P. Bonnett. *Nature in the Urban Landscape. A Study of City Ecosystems*. Baltimore, MD: York, 1973.

Hogg, I. D. and R. H. Norris. Effects of run-off from land clearing and urban development on the distribution and abundance of macroinvertebrates in pool areas of a river. *Aust. J. Mar. Freshwater Biol.*, **42** (1991), 507–18.

Landsberg, H. E. Climates and urban planning. In *Urban Climates*. World Meterological Association no. 254. Technical paper 141, note 108, pp. 364–74. Geneva: World Meterological Association, 1970.

Legner, E. F. and G. S. Olton. Distribution and relative abundance of dipterous pupae and their parasitoids in accumulations of domestic animal manure in the southwestern United States. *Hilgardia*, **40** (1971), 505–55.

Lussenhop, J. The soil arthropod community of a Chicago expressway margin. Ecology **54** (1973), 1124–37.

Newsome, E. M. Arthropod problems in recreation areas. *Annu. Rev. Entomol.*, **22** (1977), 333–53.

Owen, J. and D. F. Owen. Suburban gardens: England's most important nature reserve. *Environ. Conserv.*, **2** (1975), 53–9.

Stearns, F. Urban ecology today. *Science* **170** (1970), 1006–7.

Streu, H. T. The turfgrass ecosystem: impact of pesticides. *Bull. Entomol. Soc. Am.*, **19** (1973), 89–91.

Surtees, G. Urbanization and the epidemiology of mosquito-borne diseases. *Abstr. Hyg.*, **46** (1971), 121–34.

Tischler, W. Biozönotische Untersuchungen an Ruderalstellen. *Zool. Jahrb. Syst.*, **81** (1952), 122–74.

Untersuchungen über das Hypolithion einer Hausterrasse. *Pedobiologia*, **6** (1966), 12–36.

Ecology of arthropod fauna in man-made habitats: the problem of synanthropy. *Zool. Anz.*, **191** (1973), 157–61.

Wilton, D. P. Dog excrement as a factor in community fly problems. *Proc. Hawaii. Entomol. Soc.*, **18** (1963), 311–17.

Woodroffe, G. E. The biological origin of our domestic insect pests. *Biol. Hum. Aff.*, **18** (1952), 1–5.

Zuska, J. and P. Lastovka. Species-composition of the dipterous fauna in various types of food-processing plants in Czechoslovakia. *Acta Entomol. Bohemoslov.*, **66** (1969), 201–21.

**Landfills**

Back, E. A. *Gryllus domesticus* L. and city dumps. *J. Econ. Entomol.*, **29** (1936), 198–202.

Bose, C. J. Trends in urban refuse disposal: a pest's perspective. In Robinson, W. H., F. Rettich, and G. W. Rambo (eds.) *Proceedings of the 3rd International Conference on Urban Pests*, pp. 83–99. Hronov, Czech Republic: Grafické Závody, 1999.

Bowerman, A. G. and E. F. Redente. Biointrusion of protective barriers at hazardous waste sites. *J. Environ. Qual.*, **27** (1988), 625–32.

Calisir, B. and E. Polat. An investigation into the fly fauna of five refuse tips in Istanbul. *Turkiye Parazitoliji Dergisi*, **17** (1993), 119–29.

Campbell, E. and R. J. Black. The problems of migration of mature fly larvae from refuse containers and its implication on the frequency of refuse collection. *Calif. Vec. Views*, **7** (1960), 9–16.

Crawford, R. L. Autumn populations of spiders and other arthropods in an urban landfill. *Northwest Sci.*, **53** (1979), 51–3.

Darlington, A. *Ecology of Refuse Tips*. London: Heineman Educational Books, 1969.

Deonier, C. C. Insect pests breeding in vegetable refuse in Arizona. *J. Econ. Entomol.*, **35** (1972), 457–8.

Dirlbek, K. Species, daily frequency and succession of Diptera on refuse depositions of communal waste in Prague. In Kluzak, Z. (ed.) *Dipterologica Bohemoslovaka IV*. Sbornik referatuz VIII celostatniho dipterologiceho seminare v Ceskych Budejovicich, pp. 109–11. Ceske Budejovice, Czechoslovakia: Jihoceske Muzeum, 1986.

Feachem, R. G., D. J. Bradley, H. Garelick, and D. Duncan Mara (eds.) *Sanitation and Disease: Health Aspects of Excreta and Wastewater Management*. Chichester: John Wiley, 1983.

Ilgaz, A., Y. Ozgur, S. Ak, N. Turan, and H. Gun. Isolation and identification of bacteria from flies collected from garbage in Istanbul, and their effect on human health. *Turk. J. Infect.*, **9** (1995), 131–6.

Imai, C. Population dynamics of houseflies, *Musca domestica*, on experimentally accumulated refuse. *Res. Popul. Ecol.*, **26** (1984), 353–62.

Kohn, M. Influence of the refuse dump biotopes on ecology of some gamasoid mites and ticks. Modern acarology. In Dusbabek, F. and V. Bukva (eds.), vol. I. *Proceedings of the VIII International Congress of Acarology*, vol. I. The Hague, Netherlands: Academic, 1991.

Magy, H. I. and R. J. Black. An evaluation of the migration of fly larvae from garbage cans in Pasadena, California. *Calif. Vector Views*, **9** (1962), 55–9.

Nuorteva, P., T. Kotimaa, L. Pohjolainen, and T. Räsänen. Blowflies (Dipt., Calliphoridae) on the refuse depot of the city of Kuopio in Central Finland. *Ann. Entomol. Fenn.*, **30** (1964), 94–104.

Nuorteva, P., K. M. Kolehmainen, K. M. Korhonen, *et al.* The dimensions of nuisance caused by city garbage dumps to people living in their vicinity. *Ymparisto ja Terveys*, **11** (1980), 33–7.

Quarterman, K. D., W. C. Baker, and J. A. Jensen. The importance of sanitation in municipal fly control. *Am. J. Trop. Med.*, **29** (1949), 973–82.

Stein, W. and H. Haschemi. Dispersal and emigration of the house cricket, *Acheta domesticus* (L.) (Ensifera, Gryllidae) and the German cockroach, *Blattella germanica* (L.) (Blattodea, Blattellidae), of a rubbish tip. *Z. Ang. Zoo.*, **80** (1994), 249–58.

Strazdine, V. Anthropogenic impact on insect larvae. Complexes of insect larvae and their succession in composted municipal solid garbage. *Ekologiya*, **2** (1996), 48–58.

Süss, L., S. Cassani, B. Serra, and M. Caimi. Integrated pest management for control of the house fly *Musca domestica* (L.) (Diptera: Muscidae) in an urban solid waste treatment plant. In Robinson, W. H., F. Rettich, and G. W. Rambo (eds.) *Proceedings of the 3rd International Conference on Urban Pests*, pp. 261–9. Hronov, Czech Republic: Grafické Závody, 1999.

Toyama, G. M. A preliminary survey of fly breeding at sanitary landfills in Hawaii with an evaluation of landfill practices and their effect on fly breeding. *Proc. Hawaiian Entomol. Soc.*, **28** (1988), 49–56.

Walden, B. H. Abundance of the German roach in a city dump, *Blattella germanica* Linn. *Conn. Agr. Stat. Bull.*, **234** (1922), 188–9.

Williams, P. T. *Waste Treatment and Disposal*. London: John Wiley, 1998.

Wilton, D. P. Refuse containers as a source of flies in Honolulu and nearby communities. *Proc. Hawaiian Entom. Soc.*, **17** (1961), 477–81.

### Septic fringe and affluent fringe

Abrams, C. *Man's Struggle for Shelter in an Urbanizing World*. Cambridge, MA: MIT, 1964.

Back, K. W. *Slums, Projects and People, Puerto Rico*. Durham, NC: Duke University Press, 1962.

Kleevens, J. W. L. Housing, urbanization and health in developing (tropical) countries. *Trans. R. Soc. Trop. Med. Hyg. Suppl.*, **77** (1971), 60–72.

Majzlan, O. and M. Holecova. Anthropocoenoses of an orchard ecosystem in urban agglomerations. *Ekologia (Bratislava)* **12** (1993), 121–9.

Ragheb, I. Patterns of urban growth in the Middle East. In Breese, G. (ed.) *The City in Developing Countries: Readings on Urbanism and Urbanization*. Englewood, NJ: Prentice Hall, 1969.

Samaj, B. S. *Slums of Old Delhi*. Delhi: Atma Ram, 1958.

Turner, J. C. *Limas Barriadas and Corralones: Suburbs vs. Slums. Ekistics*, vol. 19, no. 112. Greece: 1965.

### Agriculture interface

Anderson, J. R. The behavior and ecology of various flies associated with poultry ranches in northern California. *Proc. Calif. Mosquito Control Assoc.*, **32** (1964), 30–4.

Celedova, C., N. Prokesova, B. Havlik, and O. Muller. Demonstration of migration of *Musca domestica* from pig sheds into human dwellings. *J. Hyg. Epidemiol. Microb. Immun.*, **7** (1963), 360–70.

Dipeolu, O. O. The biting flies in the zoo of the University of Ibadan. *E. Afr. Wildl. J.*, **14** (1976), 229–32.

Ek-bom, B., M. E. Erwin, and Y. Robert. *Interchanges of Insects Between Agricultural and Surrounding Landscapes*. Boston: Kluwer Academic, 2000.

Green, A. A. The control of blowflies infesting slaughter houses. I. Field observations of the habits of blowflies. *Ann. Appl. Biol.*, **38** (1951), 475–94.

Greenberg, B. and A. A. Bornstein. Fly dispersion from a rural Mexican slaughter house. *Am. J. Trop. Med. Hyg.*, **13** (1964), 881–6.

Hall, R. D., G. D. Thomas, and C. E. Morgan. Stable fly, *Stomoxys calcitrans* (L.) breeding in large round hay bales: initial associations (Diptera: Muscidae). *J. Kans. Entomol. Soc.*, **55** (1982), 617–20.

Hanec, W. A study of the environmental factors affecting the dispersion of house flies (*Musca domestica* L.) in a dairy community near Fort Whyte, Manitoba. *Can. Entomol.*, **88** (1956), 270–2.

Hulley, P. E. Factors affecting numbers of *Musca domestica* Linneaus (Diptera: Muscidae) and some other flies breeding in poultry manure. *J. Entomol. Soc. South Afr.*, **49** (1986), 19–27.

Schoof, H. F., G. A. Mail, and E. P. Savage. Fly production sources in urban communities. *J. Econ. Entomol.*, **47** (1954), 245–53.

Skoda, S. R., G. D. Thomas, and J. B. Campbell. Developmental sites and relative abundance of immature stages of the stable fly (Diptera: Muscidae) in beef cattle feedlot pens in eastern Nebraska. *J. Econ. Entomol.*, **84** (1991), 191–7.

Stafford, K. C. and D. E. Bay. Dispersion pattern and association of house fly, *Musca domestica* (Diptera: Muscidae), larvae and both sexes of *Macrocheles muscaedomesticae* (Acari: Macrochelidae) in response to poultry manure moisture, temperature, and accumulation. *Environ. Entomol.*, **16** (1987), 159–64.

# 3 Pest status and pest control

## Introduction

Pest status for arthropods in the urban environment is based, in part, on the continued presence of a species in or around the workplace and living space. Contributing to this is the potential medical or psychological reaction and economic loss linked to their occurrence. The continued presence of these animals is due, in part, to the relative ineffectiveness of control measures, and the existence of reservoir habitats and populations that provide for reinfestation. Long-term persistence and pest status of domestic and peridomestic arthropods are based on a network of small infestations in relatively unstable habitats, and large reservoir populations in relatively stable habitats. Reservoir habitats provide individuals or groups that can replenish local infestations and establish new ones. Without their reservoir populations, most of the common pest species would not sustain the abundance necessary for pest status.

Pest status is usually associated with a real or perceived medical threat, a persistent nuisance, or on economic loss. The majority of arthropods in this environment qualify for one or more of these categories. Pest status may change with the abundance of the pest species. It may begin as a nuisance by the presence of small numbers of individuals, then become a health threat by the presence of large numbers, and eventually an economic level is reached when control and repair are required. Peridomestic pests, such as umbrella wasps or yellowjackets nesting under the eaves of houses or subterranean termites damaging structural wood, may present a threat to human health or damage the physical structure of a building. Domestic pests damage food, fabric, and other materials, but also intrude on personal space to cause psychological stress. Pest status may be based solely on an aesthetic or emotional reaction to the presence of an insect or other arthropod, such as spiders and centipedes. The economic and medical basis for pest status is measurable, but may be applicable to only a select group of pests. Pest control actions based on aesthetic–emotional reasons are much less measurable and predictable, but are no less important and probably the basis of many control decisions in the urban environment.

## Aesthetic injury

The pest status of some arthropods in domestic and peridomestic habitats is based solely on an intolerance of their presence. For many people there is a psychological or emotional sensitivity to the presence of an insect or other arthropod. The living space is a personal and sacred place, the presence of insects or other animals may directly affect the quality of life there, and their presence is usually considered unacceptable. Tolerance for animals in this space is usually low, and control is based on an emotional or aesthetic threshold.

Food contaminated with foreign matter is unacceptable on aesthetic and general health basis. However, insects, mites, and other arthropods are so ubiquitous and so numerous that few, if any, food can be free of at least a small amount of damage or contamination by them. In general, government agencies have established maximum levels for natural or unavoidable defects in food for human use that present no health hazard. The assumption that these defects, which are usually in the form of live or dead insects, body fragments, and other organic material, are harmless is based more on experience than on experiment. It is expected that, if any risk to human health were identified to be associated with these allowable defects, the tolerance levels would be revised in favor of human health. The average consumer may understand and accept that pure food, such that it is free of all contamination, may be difficult to achieve in a consistent manner, but that excessive contamination by insects or other material is unacceptable, at least on an aesthetic basis.

An aesthetic injury level is a decision threshold for a pest control action that is similar to the economic threshold applied to agricultural pests. The economic threshold is a measured pest density at which control actions should be taken to prevent a pest population from reaching the economic injury level. In the urban environment, aesthetic considerations rather than economic ones are often critical in initiating control actions. Aesthetic injury may be associated with a specific number of individual pests, such as sighting one to two cockroaches within 24 h indoors, having three to four mosquito bites outdoors in 4 h, or sighting two to four wasps outdoors in the vicinity of the house. Indoors, the most common arthropods that lead to a control action at a low density are cockroaches, silverfish, moth flies, and carpet beetles (adults and larvae). Tolerance for seasonal pests, such as ants, fruit flies, cluster flies, and fungus gnats may be somewhat higher, perhaps due to their regular occurrence. Outdoors, aggregations of insects often lead to control actions; common pests in this category are boxelder bug, ladybird beetle, elm leaf beetle, and cricket. Large numbers of chironomids, winged ants, and mayflies may be a nuisance, but control measures are usually not practical.

## Medical injury

Most orders of insects and other arthropods contain species that have medical importance, either because they bite, sting, suck blood, transmit parasites and pathogens, or because they induce allergies, delusional parasitosis, or entomophobia. No medically important pest has an exclusively urban distribution; all occur in urban and natural habitats, to a greater or lesser degree. However, when these pests occur in or around the living space or workplace, their importance increases and control actions are more common. Arthropods with the highest pest status are those that inflict a painful bite, sting, or suck blood (whether painfully or not). Although they may present only a limited health risk, their presence is not tolerated. The most common of these worldwide include head louse, scabies mite, bed bugs, and spiders.

### Bites, stings, and blood-sucking arthropods

Bed bugs, scabies, and lice occur naturally in the human population, and at all socioeconomic levels around the world. People differ in their reaction to these arthropods: some are little affected, but if feeding continues or populations increase, sensitization occurs. The abundance of scabies and lice appears to be cyclic in some industrialized countries, but is more common and less cyclic in developing countries. They are commonly found on elementary schoolchildren, and there is often a social stigma associated with their presence. Lice and scabies are also common during wartime and famine when there are large numbers of refugees, poor sanitary conditions, and crowded living conditions. Bed bugs are similarly linked to humans. These blood-feeding parasites are distributed worldwide, and periodically they become numerous and infestations increase in residential and commercial buildings. Favorable indoor conditions, rapid movement of people and materials around the world, and decreased insecticide use indoors may have contributed to the re-emergence of these domiciliary pests. Regardless of the conditions or the physiological response, people dislike these ectoparasites because of their presence, and their impact on the quality of life. The pest status of lice, scabies, and bed bugs may be based on the unsightly condition of the infected skin, and the itching and discomfort caused by their feeding.

The pest status of spiders is primarily aesthetic since the majority of those found indoors are not likely to bite or be a health threat. There are a few species that have a painful bite, sometimes with severe outcomes. Nearly all spiders are poisonous, at least with regard to their normal prey, but only about 20 of the approximately 30 000 described species are dangerously poisonous for humans. The most important species are: the aggressive house spider (*Tegenaria agrestis*), which often bites people without provocation; yellow sac spiders (*Cheiracanthium* spp.), which occur indoors around the world; and species of recluse (*Loxosceles*) and widow spiders (*Latrodectus*). The bite of these spiders is generally painful and the venom may be locally or systemically toxic.

### Transmission of parasites and pathogens

Mosquitoes, reduviids (conenose bugs), and ticks transmit the major arthropod-borne diseases in the urban environment. Most of the vectors occur primarily in domestic and peridomestic habitats, or readily move to these habitats from reservoir populations outside urban areas. Their success and worldwide distribution are based in part on their ability to adapt to new hosts or substitute their natural breeding sites for those available in or around human dwellings.

Species of *Aedes*, *Anopheles*, and *Culex* mosquitoes occur in urban habitats. Many salt marsh and floodwater species of *Aedes*, such as *Ae. dorsalis*, *Ae. sollicitans*, *Ae. squamiger*, *Ae. taeniorhynchus*, and *Ae. vexans*, have flight distances from 6.4 to 64 km, which brings them within range of urban habitats. Worldwide distribution of *Ae. aegypti* is linked to its adaptation to human habitats, such as its ability to breed in artificial containers and to travel with humans. Around human dwellings are various containers that hold water and easily substitute for the ancestral tree hole conditions of this species. The ability of the eggs to survive desiccation provides for long-distance transport to new areas. This species survives best where there is open-water storage and artificial containers. Because of its adaptation abilities and occurrence in urban environments around the world, it has the potential of transmitting new arboviruses that may develop into regional epidemics. *Ae. albopictus* is another species that has substituted its natural breeding site of plant cavities for household containers and automobile tires in the urban environment. It has expanded its distribution out of Asia, and is now a major pest in urban and suburban areas, and an important urban vector of dengue.

*Culex tarsalis*, the vector for western equine encephalitis virus, and *Cx. tritaeniorhynchus*, the vector of Japanese encephalitis virus, are rural species but enter urban habitats after heavy rainfall and flooding. They breed in structures that hold water. *Cx. pipiens* is one of the most common nuisance species in urban environments, and it transmits several arboviruses. The subspecies *Cx. pipiens quinquefasciatus* (= *Cx. pipiens fatigans*) is the major mosquito vector in urban environments throughout the world. It breeds in ground pools and in water that collects in household containers, and readily enters houses. This species is well-adapted to urban and industrial conditions, and it is a dominant species in the septic fringe in developing countries. In the USA, *Cx. pipiens quinquefasciatus* breeds in pools at the ends of culverts and street drain catch basins, and it is the vector of the urban cycle of St. Louis encephalitis virus. The decrease in Japanese encephalitis in urban Japan has been attributed to people staying indoors in air-conditioned houses in the evening and watching television, instead of sitting outside exposed to urban mosquito vectors.

*Anopheles stephensi* feeds and rests indoors, and breeds in wells, cisterns, roof gutters, fountain basins, garden tanks, and discarded tins in India and the Middle East. In India, *An. culicifacies* normally breeds in natural waters, but will reproduce in flooded burrow pits and pools in urban areas. Flooding and heavy rainfall provide breeding sites for *An. atroparvus*, *An. messae*, *An. sacharovi*, and *An. superpictus* in cities in Europe. The flight range of these species, and their dispersal by the prevailing wind at the edge of cities, has influenced the urban occurrence of these and other anopheline mosquitoes. Populations of *An. gambiae* occur in urban areas in sub-Saharan Africa, where it breeds in underground cisterns and catch basins of storm drains in cities.

Species of the reduviids *Triatoma*, *Rhodnius*, and *Panstrongylus* have adapted to urban habitats. Most species occur in the western hemisphere. *Triatoma* species are often associated with rodents in natural areas, but are attracted to lights and may enter houses in suburbs and shantytowns. Because of houses built in the chaparal on the edges of cities, there has been an increase in the occurrence of conenose bugs and Chagas disease, which is caused by a *Trypanosoma* transmitted by these bugs. *Trypanosoma cruzi* is maintained in the urban environment in the domestic and peridomestic populations of cats, dogs, opossums, armadillos, squirrels, and several species of rats and mice. The primary vectors are *Triatoma* spp., *Rhodnius prolixus*, and *Panstrongylus megistus*. The focus of this disease is the poor household conditions in rural areas and septic fringe of cities.

*Dermacentor*, *Rhipicephalus*, and *Ixodes* ticks find suitable conditions and hosts in the greenspace and peridomestic habitats. In suburban areas (urban-ecosystem A) there are cases of Rocky Mountain spotted fever (RMSF) in eastern USA, and cases of Lyme disease in the USA and other countries. An increase in RMSF is associated with the success of *Dermacentor variabilis* in suburban vegetation and perhaps the abundance of domestic dogs and other host animals. The increased incidence of tick paralysis follows the abundance of *D. variabilis* in suburban areas in eastern USA, and with the abundance of *Ixodes holocyclus* in Australia. In the suburbs of Sydney, *I. holocyclus* is abundant because the mix of natural vegetation in peridomestic habitats provides suitable conditions for the bandicoot (*Parmeles* spp.), the primary host for this tick. Lyme disease is one of the most common arthropod-borne diseases in suburban areas around the world. *I. scapularis* is the principal vector in northeastern USA. The immature stages of this tick feed on numerous birds, mammals, and humans; the white-footed mouse is the primary reservoir and vector to humans, and the white-tailed deer is the primary overwintering site and host for the adult tick. Adults do not move from host to host and do not transmit the disease. The distribution range of *I. scapularis* is expanding in suburban areas, along with the incidence of Lyme disease, due to the proliferation of deer in these habitats.

*Rhipicephalus sanguineus* completes its development on domestic and feral dogs, and it has adapted to urban environments in many parts of the world. This species originated in Africa, but has been introduced into the Americas, Europe, Asia, and Australia, where it is well-established. This tick requires relatively high temperatures to complete development. In temperate countries it is associated with dogs indoors; in warm climates it occurs outdoors in suburban areas and is a vector of RMSF in the USA and a vector of boutonneuse fever in the Mediterranean region of Europe.

## Allergies

Allergic disease is a common disorder affecting about 40% of the world population. The allergen proteins that induce allergic reactions may be inhaled, ingested, and absorbed through the skin, or mucous membranes. Typical allergic reactions include swelling, itchy and watery eyes and nose, difficulty breathing, headaches, skin rash, and itching. Many species of arthropods are the sources of allergens that sensitize and cause allergic reactions in humans. These allergens are proteins and the physiological response to exposure is the same as it is for other allergen sources, such as plant pollen, molds, and some foods. Arthropods in the urban environment that induce allergic reactions in humans include flies, fleas, beetles, and moths in stored food, and stinging insects such as bees, wasps, and ants. However, the prevalence of cockroaches and dust mites in the living space and their potent allergens make these two very important sources of allergic reactions.

Cockroaches common in and around human dwellings are an important source of allergenic proteins. Sensitivity to cockroaches is worldwide and ranges from 23 to 60% of the population; it is evident as respiratory asthma and dermatitis. In some inner-city neighborhoods in the USA 37% of children may be allergic to cockroach allergens. The cockroaches known to be the sources of allergen include species of *Blattella*, *Blatta*, and *Periplaneta*. However, *Blattella germanica* and *P. americana* are the prevalent indoor pests, and contribute the most to health problems. The allergens from these insects are found in the fecal material, oral secretions, exoskeleton fragments, and dead bodies. Particles bearing cockroach allergen are mainly carried on particles less than 10 μm diameter; these particles settle quickly and reduce exposure.

Dust mites are in stored food products and inhabit the living and working space worldwide, and they are the source of allergens. Sensitivity to mite allergens is well-known; in the USA and Europe 20–35% of allergic individuals are sensitive to dust mites. Most homes and work environments inhabited by dust mites contain several species, including *Dermatophagoides farinae*, *D. pteronyssinus*, *D. microceras*, and *Euroglyphus maynei*. Dust mite populations require a source of protein-rich food and environmental conditions of 10–30 °C and at least 50% relative humidity (RH). The 0.5–1.0 g of skin scales humans shed every day provide sufficient food, and carpets and textile materials on beds, furniture, and clothing provide harborage and breeding sites for these mites. Allergens of *Dermatophagoides* species are produced in the posterior midgut and hindgut as digestive enzymes, excreted fecal pellets (10–40 $\mu$m diameter), and in cast skins of mites. These allergenic particles are relatively large and rapidly fall in undisturbed air. However, excrement pellets become dry and fragment, and small particles may become airborne.

The mites associated with stored foods and fungi include *Acaris siro*, *Glycyphagus domesticus*, *Lepidoglyphus destructor*, and *Tyrophagus putrescentive*. They feed on mold and fungi that grow on household foods, but are also found on textiles and on wall and ceiling surfaces. Most species require 70–98% humidity for development. Exposure to stored-food mite allergens can be by ingestion or by inhalation, and sensitization to these mites has been reported in many developed countries. The confused flour beetle, *Tribolium confusum*, is probably the most common contaminant of flour, cereal, prepared flour mixes, dried fruits and nuts, and various spices. In these materials, there may be fragments of all the life stages of this beetle, and for some infested material, there may be live adults and larvae. Persons who are sensitive to insect allergens may have an allergic response when ingesting contaminated flour products.

## Entomophobia and delusional parasitosis

Most people do not like having insects and spiders in their living space, and some may be fearful of their presence. Fear is a natural extension of human experience, and a reasonable and appropriate response to situations that involve potential danger. It has some value in protecting the individual from potentially harmful situations. However, irrational anxiety in situations of limited danger or health threat is a phobia. For some individuals, the presence of insects or spiders in their immediate surroundings produces an unreasonable level of fear, and this is considered entomophobia. The general symptoms of a phobia are characterized as sudden and intense feeling of anxiety, shortness of breath and increased heart rate, shaking, and sweaty palms. An important component of any phobia is avoidance and people who are extremely fearful of insects and spiders avoid them. Another component is the generalization of the fear to include other organisms, such as spiders and spider webs, or to all insects that make a buzzing sound.

The emotional condition in which individuals believe that live organisms are present on or in their skin, or periodically biting them, is called delusional parasitosis. In the late 1800s, delusional parasitosis was described by Georges Thibierge in the French literature as acarophobia. This condition was called presenile Dermatozoenwahn by Ekbom in the 1930s. He was a Swedish physician who described several cases, and for whom the condition is named. Ekbom's syndrome has been variously called dermatophobia, parasitophobia, and, more recently, monosymptomatic hypochrondriacal psychosis. About 25% of the reported cases exhibited *folie à deux* involving a family member or close associate (see below); thus this conviction of cutaneous infestation may be regarded as a contagious mental state.

It is defined as a false belief (delusions) held in spite of no evidence that there are external or internal organisms biting or stinging the skin. The apparent cause of the skin irritations is tiny, almost invisible insects or mites. This emotional state may develop quickly and persist for months or even years. It is believed that delusions of infestation are more common with advancing age, and gender (primarily females), but often patients less than 50 years of age are males. Victims are able to provide a detailed description of the supposed parasite. Individuals typically characterize the supposed parasites as black or white bugs; the bugs crawl on the skin for short periods. The supposed parasites sometimes tunnel in the skin, or jump on and off the person during various times of the day or in specific locations, which are usually indoors. The origin of the bugs can be almost any household material, including furniture and paper. The bugs may infest any portion of the body, including hair, arms, legs, and genital region; commonly the infestation will be centered in areas that are within reach of their hands. The bugs bite or sting, and often cause intense localized pain on the skin. Sometimes skin irritations develop in response to the supposed bite or sting, and the typical response to the pain or itch is intense scratching. The infestation can be so severe that the person leaves the house seeking relief, but the bugs usually reappear in the new location after a few hours or days. Others living in the household, including family members, may be convinced of the presence of these biting animals and share in the delusion (*folie à deux*).

An itch on the skin is a sensation which is sometimes described as a mild form of pain. The sensation of itching is apparently a result of chemical or physical stimulation of receptors on the skin. However, itching may not be accompanied by a stimulus to the skin. The causes of itches are many and range from medical conditions, such as diabetes, to mild irritants, such as laundry detergents, fabric sizing and conditioners, and dry skin in winter. Persons suffering from the sensation of itching of the skin often have the idea that they have mites, fleas, or some other microscopic animal. The supposed mites may be called cable mites or paper mites, and they are assumed to be associated with the wires that supply electricity to office or manufacturing equipment, or with paper that accumulates in offices or storerooms. Fleas are often suspected because these insects are usually associated with bites and intense itching. Cable mites or paper mites are often reported by groups of people performing routine and repetitive tasks, such as secretarial and bookkeeping personnel, or assembly-line workers. Cable or paper mite dermatitis is a delusional parasitosis in the sense that the victims may believe they are being exposed to the mites, although these mites do not exist. There are no such animals as paper or cable mites, and fleas are not generally present in office or manufacturing environments.

Delusions of cleptoparasitosis is an extension of the concept of delusional parasitosis, but the patients imagine arthropod infestations in their house or living area. This condition is not accompanied by the sensations of insects or mites on the body; instead the apparent infestation may be imagined in household items. Wood-infesting insects are sometimes implicated as the cause of the problem.

## Economic

The economic importance of almost any pest can be expressed in the money spent on control and prevention measures, the repair and replacement costs, and lost production or revenue due to infestations. This information is useful to manufacturers and government agencies, and for determining research and development programs. Many pests in this ecosystem are defined in economic terms of their medical importance. Malaria kills about 1 million people a year worldwide, and this mosquito-borne disease costs African countries $12 billion in lost production.

The cost of controlling the major household pests in the urban environment is unevenly divided between consumer-use products and professional service. Consumer products are generally available for controlling cockroaches, fleas, flies, ants, wasps, mosquitoes, termites, and nuisance pests. Purchase of pest control products is a monetary measure of persistence and importance of pests. The willingness to pay for control of a household pest is an emotional measure of the importance of a pest. It is expressed in terms of the money individuals would be willing to spend (if not actually spending) to eliminate a pest from their living space.

Control and repair costs are appropriate for expressing the economic importance of wood-infesting insects, such as termites, carpenter ants and bees, and wood-boring beetles. Pest control professionals usually provide control of these pests; repairs may be done by homeowners. Monetary figures are often available for termite damage in temperate regions. In the USA, conservative estimates for prevention, control, and repair of damaged wood range from about $500 to $753 million annually, but other figures are $3.4 billion annually. In the oriental region more than 550 termite species are known, and in some countries 43–54% of all buildings are infested. In Australia, *Coptotermes acinaciformis* is responsible for the majority of the damage done by termites. This is due to the extensive range of this species, to the damage capable by mature colonies, and to its success in adapting to urban habitats. This species causes most of the termite damage to buildings in Australia. Control measures include more than $4 million in control chemicals for household infestations, but damage to utility poles, estimated at about $300 million, and forest and agricultural trees significantly increases the economic importance of these insects.

Carpenter ant control and repair in the USA and Europe is estimated at $100 million. Pest prevention costs may be applied to flea and tick control on domestic pets, stinging Hymenoptera, mosquitoes, and some turfgrass insect pests. Consumers generally provide and assume the costs for prevention and control of most of these pests; mosquito control is provided and paid for, usually with tax funds, by local governments. Replacement costs for stored-food and fabric pests are borne by the homemaker, and these data are limited. Homeowners use aerosol insecticides to control common domestic and peridomestic nuisance pests.

## Pest control strategies

Control of household and structural pests includes the use of a variety of chemical and nonchemical methods. The overall size of urban and suburban areas and the importance of the domestic and peridomestic pests provide the economic incentive for manufacturers to formulate insecticides and design control programs for common pest species. Consideration for adverse exposure to humans and nontarget species has improved the delivery of modern insecticides, while maintaining or increasing efficacy.

The concept of integrated pest management (IPM) began in the agroecosystem with the need to provide an economic base for decisions to use chemical control. The objective of IPM was to provide an effective and economically efficient approach to pest control. The modern concept of pest management is the integration of biological, chemical, and other control methods into a program that restricts pest density to levels below those causing economic injury. The important feature of agricultural IPM is suppression of populations and not eradication of pest populations. Pest density is linked to an economic injury level, which is considered the lowest population density that will cause economic damage. Damage may occur below this level, but it is considered acceptable because it does not affect yield or the value of the commodity.

The concept of reducing the use of pesticides and managing pest populations has been considered for the urban ecosystem. In the peridomestic environment, pesticides are primarily used in response to seasonal pests that cause aesthetic damage. Programs that maintain these pests at low levels may be appropriate, ecologically beneficial, and economically efficient. However, applying the traditional IPM philosophy of pest management (not elimination) to arthropods in the domestic environment may be difficult or unacceptable. Control actions that have the objective of maintaining pest populations below an economic or aesthetic level may not be appropriate for insects that are a medical threat to people or pets or that may cause structural damage. Some pests in the domestic environment may not have a financial or health-related level of pest status; nevertheless, their presence in the home provides little room for tolerance and sufficient cause for their elimination. There is little evidence that the concept of a level of tolerance can be applied to household pests. Urban residents typically adjust their attitude toward the presence of pests in the home; their level of pest tolerance declines as the level of infestation decreases.

Other components of agricultural IPM programs, such as the use of biological, cultural, and mechanical methods, and monitoring pests to determine decision-making levels, may have limited application in the domestic environment. The use of predators, parasites, or other biological control strategies for arthropods indoors must consider the attitudes of the target audience toward the presence of additional organisms indoors. Use of parasites as a control strategy for indoor pests may not be acceptable for residents if these control agents are as unwelcome as the target pest. In many cases, the parasite or predator population must be significantly increased to achieve an acceptable level of pest suppression. Releasing large numbers of any insect, beneficial or not, into the living space may not be acceptable to typical homemakers.

## Microbial agents

Microbial pathogens of invertebrates include viruses, bacteria, and fungi. Their potential as pesticides is based on the feature of specificity and toxicity. Many are species-specific and are effective against selected pests, and most are noninfective to vertebrates, which provides a level of safety in the human environment. Some microbial pathogens can be mass-produced, and suitable for commercial use. Current applications include control programs for soil-infesting insects, such as termites and turfgrass pests.

The mode of action of microbials usually depends on ingestion by the target pest during normal feeding. Protoxin fragments of the virus or bacteria disrupt the cell wall lining of the midgut and they enter the body cavity; the insect dies soon after the pathogen spreads throughout the body. Efficacy depends on the toxin produced by the pathogen, rather than multiplication within the infected host. However, products based on *Bacillus thuringiensis* usually contain viable spores of the bacterium, and these may contribute to their efficacy. Environmental conditions influence the effectiveness and may limit the use of microbials in urban habitats. The virulence of bacteria and fungi decreases when temperatures drop below 18 °C, and optimal development of some viruses is at 21–29 °C. Viruses and bacteria are usually killed after prolonged exposure to the ultraviolet portion of sunlight. Pathogenic fungi survive best in soil that has a high organic content, and a suitable soil pH; for example, acid conditions are unfavorable for *Paenibacillus* (= *Bacillus*) *popilliae* spores, which control Japanese beetle grubs in turfgrass. Mode of entry into the arthropod is usually by ingestion or through damaged areas in the integument. Fungi usually enter through the spiracles; bacteria may be consumed during feeding, and some viruses can be passed from adult females to their eggs.

Viruses most commonly used for insect control include the nuclear polyhidrosis (NPV), cytoplasmic polyhidrosis (CPV), and granulosis virus (GV). Baculoviruses are specific to a few pest Lepidoptera. The infectious virus particles are embedded in a proteinaceous matrix, called the polyhedral inclusion body. The insect midgut is the route of entry; from there, they enter the body cavity and the insect soon dies. Most viruses are genus- or species-specific, but the immature stages of Lepidoptera and Hymenoptera (sawflies) are particularly affected. One limitation in using viruses to control insects is the slow control time. Usually the insect continues to feed and cause damage until the viral infection has spread from the midgut to other parts of the body. An advantage is that the body of the infected dead insect can spread the pathogen to uninfected individuals.

Bacteria used for insect control are from the genera *Bacillus*, *Paenibacillus*, and *Serratia*, but the number of species is limited. *B. thuringiensis israelensis* (BTI) and *B. sphaericus* are effective in controlling some pest species of flies, such as *Anopheles* and *Culex* mosquitoes and black flies (Simuliidae), *B. thuringiensis kurstaki* and *B. thuringiensis entomocidus* controls Lepidoptera caterpillars, *B. thuringiensis tenebrionis* controls some beetle species, and *B. japanensis* is toxic to Japanese beetle larvae. Other pathogenic bacteria include *Paenibacillus popilliae*, *P. lentimorbus*, and *Serratia entomophila*. The causal agents for the fatal milky disease of the Japanese beetle (*Popillia japonica*) are *Paenibacillus popilliae* and *P. lentimorbus*. Grubs ingest the bacteria spores along with soil and roots, the spores germinate in the gut, and vegetative cells invade the body cavity and kill the insect. Proliferation of spore bodies during the final stages of infection gives the haemocoel a milky-white color. The host-specific bacterium *Serratia entomophila* has been used effectively to control a soil-dwelling scarab pest, *Costelytra zealandica*.

Avermectins are macrocyclic lactone glycosides that are the natural products of fermentation by the soil microorganism *Streptomyces avermitilis*. The mode of action of avermectin (Avermectin B1a)-type compounds (such as ivermectin) is to increase the effect of glutamate on the chloride ion channel of nerves of the voluntary muscle system. High concentrations cause irreversible opening of the channel, which blocks any activity of muscles innervated by affected nerves. The mode of action is basically the same for vertebrates and invertebrates. These compounds undergo rapid photolysis, and their half-life in sunlight is 4–6 h.

Spinosyns are a naturally derived group of chemicals produced from a species of bacteria, *Saccharopolyspora spinosa*. Spinosad is a mixture of spinosyn compounds, and it affects species in the orders Coleoptera, Diptera, Hymenoptera, Isoptera, Lepidoptera, and Siphonaptera. The site of action is the nervous system, but the actual influence on the nerves is not completely known. As with the other bacteria-based microorganisms, this material degrades rapidly when exposed to sunlight and other environmental conditions.

Fungi infect almost all insects and other arthropods. Entomophthorales includes several species that are lethal to soil-inhabiting beetles and termites. Soil conditions provide a stable environment for these organisms; however, the upper 2–5 cm of many soils reach temperatures that are lethal to the vegetative stages of most pathogenic fungi. *Metarhizium*, *Beauveria*, and *Verticillium* have species that are pathogenic to insects. The most commonly investigated species include *M. anisopliae*, *B. bassiana*, and *V. lecanii*, all of which have a wide host range. Strains of *M. anisopliae* have been used to control subterranean termites and indoor cockroach pests. Species of the fungus genus *Cordycepioideus* have been recorded from termite nests in Mexico and Kenya.

## Nematodes

Nematode species in the families Steinernematidae (*Steinernema* spp., *Neosteinernema* spp.), Heterorhabditidae (*Heterorhabditis* spp.), and Mermithidae parasitize insects. Larvae of Steinernematidae carry the pathogenic bacteria, *Xenorhabdus nematophilus* and *X. luminescens*. *Photorhabdus* bacteria are carried by roundworms in the family Heterorhabditidae. These bacteria can enter the insect host and kill it within 24–48 h. The nematode larvae are free-swimming in water and infect the insect host by entering the mouth, anus, or spiracles. In the body cavity, the larvae release the bacteria. The nematode feeds on the bacteria cells and the decomposing tissues. When the substrate is exhausted, they leave the cadaver and seek a new host. Nematodes are environmentally safe and acceptable, and they are easily applied with standard spray nozzles. Limitations of the nematode–bacterium control strategy are that free water is necessary for their host-seeking behavior, and these animals are difficult to rear in mass cultures. The nematode *Deladenus siricidicola* sterilizes the wood wasp, *Sirix noctilio*, and has been successfully used to reduce wood wasp populations in pine (*Pinus radiata*) tree plantations in Australia.

## Botanical insecticides

Plant products for insect control have been used as attractants, repellents, as solvents, and carriers of insecticides. However, the primary use of plant compounds is as toxicants. Nicotine, extracted from the plants *Nicotiana tabacum* and *N. rustica*, has been used for hundreds of years to kill insects. The mode of action is to affect the central nervous system of the target animal directly. Limonene is extracted from citrus peels. It is effective against some external parasites, such as fleas, lice, mites, and ticks; it is nontoxic to warm-blooded animals. This botanical affects the sensory nerves of the peripheral nervous system; it is not a cholinesterase inhibitor.

Pyrethrum is a mixture of six active compounds (known collectively as pyrethrins): pyrethrins I, II; cinerin I, II, and jasmolin I, II. It has a broad range of insecticidal activity, and is effective against nearly all insects. These chemicals are extracted from flowers of several species, including *Tanacetum cinerariifolium* (= *Chrysanthemum*), *T. coccineum*, and *T. carneum*. This mode of action is to attack the peripheral nervous system, and the immediate effect is a rapid excitement and knockdown. At low dosages some recovery can occur, due to metabolism (oxidation) of the pyrethrins. They are usually formulated with synergists, such as piperonyl butoxide, to reduce the oxidation process and increase effectiveness. This group of insecticides is nontoxic to mammals.

Rotenone is found in the roots of leguminous plants *Lonchocarpus* (cube) and *Derris elliptica* (derris). These plants occur in the Amazon river basin of South America. In insects rotenone is a respiratory enzyme inhibitor; poisoning results in slow heartbeat, depressed respiratory movement, and reduction in oxygen consumption. Rotenone is highly toxic to most insects and fish, which convert rotenone to toxic metabolites, but it is nontoxic to mammals, which produce nontoxic metabolites.

Neem, *Azadirachta indica*, is a well-known member of the mahogany family (Meliaceae). Seeds and leaves of this plant contain liminoid compounds with insecticidal, fungicidal, as well as antiseptic, antiviral, and antifungal activity. These compounds include azadirachtin, which has insecticidal activity, and salannin and meliantriol, which are feeding deterrents. Extracts are obtained by crushing neem leaves or seeds and steeping them in water, alcohol, or other organic solvents.

Extracts of seaweed *Caulerpa scalpelliformis*, *Dictyota dichotoma*, and the root of the mangrove plant, *Rhizophora apiculata*, have larvicidal activity against *Aedes aegypti* and *Culex quinquefasciatus*. Ethanolic extract of the herb *Descurania sophia* kills *Cx. pipiens* larvae. Solvent extracts of *Euphorbia helioscopia* (Euphorbiaceae), *Calendula micrantha* (Compositae), and *Azadriachta indica* (Meliaceae) are toxic to *Culex pipiens* larvae.

## Inorganic insecticides

Inorganic insecticides have a long history of use in household and stored-food pest control: this group includes arsenic, boron, mercury, and sulfur. They are stable compounds and toxic to a broad range of animals. Sulfur is one of the oldest pesticides, and is toxic as a contact or stomach poison to mites, spiders, and scale insects, and as a stomach poison for some caterpillars. Sulfur dusts and sprays are also fungicidal.

Arsenic-based pesticides are stomach poisons and are typically used for insect and rodent control. The most common compounds are lead and calcium arsenate; calcium is the most toxic to insects and mammals. Arsenic trioxide dust is used for termite control. Arsenic compounds have a complex mode of action. They uncouple oxidative phosphorylation (by substitution of the arsenite ion for the phosphorus), which is a major energy-producing process of the cell. The arsenate ion inhibits certain enzymes that contain sulfhydryl groups and both the arsenite and arsenate ions coagulate proteins.

Boron compounds have a long history in household insect control as nonselective insecticides. The mode of action is a stomach poison when a lethal dose is ingested. The target sites may be the mid- and hindgut; in the hindgut borates may disrupt water regulation. Borates are also used as a contact poison because borate dusts absorb insect cuticle wax. Surface applications of boric acid and water dilutions of borates act as antiphagostimulants for insects such as wood-infesting beetles and termites. Boron is toxic to animals and humans, and the tolerable daily intake is 24 mg/day. It is a nonmetallic element that is naturally combined with oxygen and other elements, such as sodium and calcium. Boric acid is a commonly used boron compound. It is a white crystalline solid with a maximum solubility in water of 6%, and it is nonvolatile with a long residual activity. It is primarily a slow-acting stomach poison, but it is capable of penetrating insect cuticle. Borax is a naturally occurring sodium tetraborate and as an insecticide it is used in powder form or dissolved in water. Water-soluble polyborates, such as disodium octaborate tetrahydrate, are effective in protecting wood from some wood-infesting insects, such as termites, carpenter ants, and beetles.

Amorphous silicon dioxide hydrate (silica gels or silica aerogels) is made from a geological deposit composed of fossilized skeletons of siliceous marine and freshwater animals, particularly diatoms and other algae. These skeletons are made of hydrated amorphous silica and, when crushed, they break up into fine, talc-like fragments. This material has a large specific surface area of 300 $m^2/g$, and individual particles are 0.01–0.05 $\mu m$ diameter. Silica gels that have low bulk density and high porosity are called aerogels. The mode of action is based primarily on the ability of this material to adsorb waxes from insect cuticle, which permits excess water loss from the body. Silica may be abrasive to the cuticle, which further increases water loss. Mortality is due to desiccation, which is a result of abrasion and damage to the protective wax layer in the cuticle. Insect movement and other activities gradually decline until death occurs. These compounds are also repellent to some insect species. To increase efficacy and reduce the disadvantage of the extremely light weight of silica aerogels, the dusts are typically formulated with an isoparaffinic petroleum oil (solvent), pyrethrins, and a synergist. Diatomaceous earth is primarily silica; it acts as an abrasive and is slightly sorptive. It has only limited ability to adsorb the solid wax from an insect cuticle. Insects vary widely in their response to dust desiccants. Some species have a protective (cement) layer in the cuticle, which is secreted by cells in the epidermis; this appears to provide some protection against dust desiccants. It is well developed in the cuticle of many species of beetles.

## Organic insecticides

The majority of organic insecticides exert a toxic effect on parts of the insect nervous system. The nervous system of insects and mammals is dependent on the transmission of nerve impulses along the axon, from the cell body, across intermediate synapses to the nerve ending in the muscle. At the nerve ending a transmitter substance, gamma-aminobutyric acid (GABA), is released, which results in muscle contraction. Impulses pass along the axon because of changes in the electrical potential, involving sodium and potassium ions, across the outer membrane of the axon. Impulses travel along the axon and eventually reach a gap between two nerve endings, the synapse. Transmission of an impulse across the synapse is mediated by the chemical acetylcholine, which is released at the surface of the axon membrane. Acetylcholine moves across the gap, is picked up by receptors on the other side, and a fresh impulse is then generated in the opposing axon. To prevent accumulation of acetylcholine in the synapse (and repetitive impulses in the opposing axon), acetylcholinesterase is released. It is broken down to choline and acetic acid. Some of the commonly used insecticides block production of the esterase. This prevents the passage of successive messages in the nerve, and this may lead to malfunction of the nervous system and death.

Cyclodiene and gamma-HCH insecticides have played an important part in household and structural pest control, especially in controlling wood-infesting insects such as termites and beetles. These are very stable compounds when placed in the soil or applied to structural wood. The well-known compounds in this group are aldrin, chlordane, dieldrin, endrin, heptachlor, and lindane. These compounds are neurotoxicants and produce spontaneous and repetitive discharges at the synapse, which result in tremors, convulsions, and paralysis of the target insect.

Organophosphate and carbamate insecticides were originally made for agriculture, but many have been used for household and structural insect control. Organophosphates kill insects and vertebrates by binding with acetylcholinesterase in synaptic junctions of the nervous system. This results in a continuous flow of electrical–chemical signals along the length of the nerve, which results in repeated muscle contraction. A large number of agricultural and household and disease-vector insecticides have been developed in this class, including malathion, chlorpyrifos, fenthion, and diazinon. Carbamate insecticides are derivatives of carbamic acid. They have a mode of action and residual activity similar to that of the organophosphate insecticides; they affect the nervous system at the synapse. The important qualities of these insecticides include low mammalian toxicity and broad spectrum of insect control. Several carbamates are water-soluble and are used as plant systemics in agriculture. Carbamates, such as sevin and propoxur, and bendiocarb have been used to control cockroaches and other household insect pests around the world.

Pyrethroid insecticides are effective for contact control of pests. They usually provide immediate knockdown, kill, and usually some residual effectiveness. These chemicals are synthetic analogs of natural pyrethrins, they have low odor, and are effective at low dosages. Pyrethroids are generally biodegradable at varying rates, but many are relatively stable when exposed to light. They first affect the peripheral nervous system, which provides a quick knockdown; the primary target site is the ganglia of the central nervous system. Pyrethroids are a large group of chemicals and include allethrin, bifenthrin, bioallethrin, bioresmethrin, cypermethrin, cyfluthrin, deltamethrin, fenvalerate, lambda-cyhalothrin, permethrin, phenothrin, tetramethrin, and others used to control a variety of household, structural, and medically important pests.

Fluorosulfonates (fluroaliphatic sulfones) are stomach poisons, and with delayed-action toxicity. The mode of action is depressed rates of oxygen consumption and the inhibition of cellular respiration. Sulfluramid is effective against termites, cockroaches, ants, and other household and structural insect pests.

Phenyl pyrazoles were discovered in the 1980s, and they reached their full development in the form of the active ingredient fipronil. These insecticides are effective at very low concentrations. Their mode of action is as an antagonist of the GABA-gated chloride channel of nerve membranes of the central nervous system. Compounds in this class of insecticide are effective against a broad range of insects. They have been formulated for application as dusts, liquids, and in baits. A closely related family of chemicals is the pyrroles, and it is represented by chlorfenapyr. Pyrroles are effective against a range of insects; their mode of action is as a mitochondrial poison and not as a GABA antagonist.

Hydramethylnon is a fluorinated hydrocarbon insecticide in the amidinohydrazone class of insecticides. These insecticides are toxic when ingested, and the result is decreased feeding and general lethargy in the target pest. The mode of action is depressed rates of oxygen consumption and the inhibition of cellular respiration; it is an inhibitor of electron transport in the mitochrondria. This compound is typically used as a bait toxicant. Hydramethylnon is poorly metabolized in the insect body, and it can occur in the feces of individuals that have fed on treated bait. These feces are toxic to other individuals when consumed. This insecticide can be transferred to other individuals through grooming, trophallaxis, and other physical contact, and is effective in baits for ants, cockroaches, and termites.

Chloronicotinyl insecticides, such as imidacloprid, were first used in agriculture as systemic and contact insecticides. Later, products were developed for control of urban insect pests, such as ants, cockroaches, and termites. Imidacloprid was developed for termite control in Japan in 1994, and has been used in the USA and other parts of the world since 1996. In general, the neonicotinoid compounds are nonrepellent; they are effective at low rates, they have a long residual activity, and bind to organic matter. The mode of action is linked to the nervous system. Nicotinergic acetylcholine is a neurotransmitter in the synaptic junction of the cholinergic system of insects. Imidacloprid blocks the binding of this neurotransmitter to its postsynaptic receptor, and the result is a toxic reaction at the synaptic junction, which is fatal to the insect.

Thianicotinyl insecticides are second-generation neonicotinoid insecticides. The mode of action of these insecticides is primarily by stopping feeding. They are effective against a variety of household and structural pests, including ants, cockroaches, and fleas. Thiamethoxam is a member of this class of insecticides, and has been developed for subterranean termite control.

Insect growth regulators (IGRs) include compounds that mimic the juvenile hormones that regulate development and molting of immature insects. The mode of action is to disrupt biochemical and physiological processes that lead to normal development. The effects are complex and vary between chemicals and target insect. Their activities include: prolongation of larval or nymph stages with the result that development to the adult stage is prevented, increased melanization or coloration, disrupted regeneration of appendages, anomalies in reproductive organs and other structures, and morphological anomalies in sensory organs. These compounds may also affect pheromone production in adults or produce unusual morphs or castes, or influence embryonic development. They have low toxicity to mammals, birds, and fish; they rapidly degrade outdoors, but they are relatively persistent indoors. IGRs are generally limited to specific sites and pests, but are effective against Blattaria, Coleoptera, Diptera, Homoptera, Lepidoptera, and Siphonaptera. IGR compounds such as hydroprene and methoprene are modeled on natural insect juvenile hormones. These compounds are generally species-specific, volatile, and susceptible to breakdown in ultraviolet light. Fenoxycarb, which is a phenoxy-ethyl-carbamate, and pyriproxyfen display juvenile hormone activity toward nearly all insects, and they are nonvolatile and photostable.

Chitin synthesis inhibitors (CSIs) are effective because they disrupt normal development and molting by interfering with the enzyme chitin synthetase. Benzoylphenyl urea compounds, such as cryomazin, diflubenzuron, flufenoxuron, flufenuron, hexaflumuron, noviflumuron, and triflumuron, interfere with chitin deposition, and prevent proper formation of the new exoskeleton and the shedding of the old one. These compounds may possess ovicidal activity by disrupting cuticle formation in developing embryos and causing failure to hatch. This mode of action has been exploited for the control of some urban pests, such as fleas, ants, muscid flies, cockroaches, and termites. Diflubenzuron was the first of these chemicals; it was discovered by chemists at Solvay Duphar in 1972. Diflubenzuron and other CSIs are nontoxic to birds, fish, bees, and earthworms. They are typically nonsoluble in water and do not leach or wash into surface or ground water.

Synergists have little insecticidal activity, but they are typically combined with an insecticide to increase efficacy. They are usually combined with pyrethrum and pyrethroid insecticides. Piperonyl butoxide is the most common pyrethrum synergist; it increases the speed of knockdown and mortality. The mode of action is interference with detoxifying mechanisms and prevention of repair of damaged nerve cells. Other synergists for pyrethrins include sesame oil extracts, sulfoxide, and synergist 264 (MGK-264).

## Repellents

Repellents prevent a pest from reaching a food source or harborage, or move it away once it is there. They are generally considered as nontoxic to pests, and nontoxic, nonirritating, and nonallergenic to humans and domestic animals. Most of the earlier repellent substances had strong, detectable odors; the modern, synthetic repellents are nearly odorless. Although repelling pest insects such as body lice, biting flies, and carpet beetles has a long history, the modern application of this concept takes advantage of safe use and application features, and usually low cost. Protection is usually short-term, and may be effective for a small number of species.

Creosote and other oils have been used to protect structural wood from termites, wood-infesting beetles, and decay fungi. The mode of action of these materials includes toxicity to the target pest, and masking the natural insect attractants in the wood. Pyrethrin and some pyrethroid insecticides are considered repellent, and at standard or low concentrations cause insects to become active and move from harborages. Chemical bird repellents are either olfactory (odor), tactile, or gustatory (taste). The tactile irritants affect the skin, and include combinations of castor oil, polybutane, diphenylamine, pentachlorophenol, zinc oxide, and aromatic solvents. When applied to roosting or nesting sites they prevent birds from remaining on treated surfaces. The avicide, 4-aminopyridine, is used as a repellent; the effects result from a distress call made by birds that eat the treated grain. Some bird species are killed after ingesting small amounts of this chemical.

Effective insect repellents include N, N-diethyl-*m*-toluamide (Deet), ethyl butylacetyl-aminopropionate, and 1-piperidinecarboxylic acid, 2-(2-hydroxyorthyl)-1-methylpropylester. These compounds are used for skin application against mosquitoes and other flies, as well as fleas, ticks, and mites. Deet has been used to treat fabric for mosquito protection. Antimosquito coils that are ignited and smolder to produce an insecticidal smoke are the most widely used control for domiciliary mosquitoes. Most coils contain pyrethrins or a pyrethroid such as permethrin or bioallethrin, and these chemicals are either repellent or lethal. The coils are designed to burn for 8–10 h and are typically used during the night in bedrooms.

Camphor (*Cinnamomum camphora*) wood and oil has historically been used to protect wool fabric and clothing from clothes moths (*Tineola bisselliella*) and other pests. A measurable knockdown effect on adult moths can be obtained using purified oil, but larvae and adults survive long exposure in camphor-saturated atmospheres. Camphor combined with menthol has been used to repel some outdoor pests, such as the Asian ladybird beetle, *Harmonia axyridis*. Camphor is a major component of the essential oil extracted from *Ocimum kilimandscharicum* (Labiatae). *Ocimum* plants grow widely in India and many parts of eastern and southern Africa (Kenya), and are traditionally used as mosquito repellents. The camphor extract from *O. kilimandscharicum* is effective against some grain-infesting beetles. Wood and oil from species of aromatic cedars (*Cedrus*, *Chamaecyparis*, *Juniperus*, *Thuja*), such as aromatic eastern red cedar (*Juniperus virginiana*) and northern white cedar (*Thuja occidentalis*), and Lebanon cedar (*Cedrus libani*), provide some repellent action against insects. Chests and closets made of this wood have been used to protect woolens from species of clothes moths and dermestid beetles. These woods inhibit egg hatch of eggs laid in cedar-lined chests, but not eggs introduced into the chest. Toxicity of cedar chests to beetle and moth larvae declines after 16–20 months of aging. Milled wood of *Juniperus virginiana* is somewhat repellent to the German cockroach, *Blattella germanica*, but not repellent to *Periplaneta americana* or *P. fuliginosa*; it is somewhat repellent to foraging workers of the Argentine ant. Oils of Japanese mint and Scotch spearmint and bay leaves are somewhat repellent to domiciliary cockroaches.

Paradichlorobenzene (PDB) is a white crystalline substance; when exposed to air it volatizes slowly into a gas 5.1 times heavier than air. Naphthalene is a white crystalline substance that is typically formulated as flakes or in mothballs. This chemical also volatizes when exposed to air. Both chemicals are used for the protection of fabric from clothes moths and dermestid beetles.

## Attractants

The sensory mechanism involved in searching for food, oviposition sites, and mates is stimulated and controlled by chemicals. Those chemicals that deliver behavioral messages or induce a response are termed semiochemicals, and pheromones are the semiochemicals used for intraspecific communication between individuals of the same species. There are several different types of pheromones, such as alarm, dispersal, and aggregation pheromone, but it is the group of sex pheromones that is most widely used in pest management programs. The principal uses of pheromones in control and management programs are: male trapping, which reduces the reproductive potential of a pest population; mating disruption, which disrupts the mating search of males; and in the detection and monitoring of adults to determine population abundance and location of infested material.

The function of pheromones in the biology of many stored-product and fabric pests follows two general patterns. These are sex pheromones for the species that have short-lived adults, and aggregation pheromones for the species that have long-lived adults. The short-lived adults usually do not feed, and mating and oviposition are the chief activities of the adults. Soon after emergence, females of these species usually produce a strong sex pheromone to lure males for mating. The long-lived adult males and females feed, and males generally produce an aggregation pheromone to attract other males or both males and females. Females of these species often produce sex pheromones.

Polyene hydrocarbons and epoxides are used as pheromone components and sex attractants by the microlepidopteran families, Geometridae, Noctuidae, Arctiidae, and Lymantriidae. They are different from the 10–18-carbon acetates, aldehydes, and alcohols commonly produced in other species, and constitute a second major class of lepidopteran pheromones. These are biosynthesized and characterized by 17–23-carbon straight chain, and are used in pheromone blends and converted to many of the known pheromone compounds. Another group of semiochemicals are parapheromones, which are synthesized compounds structurally related to natural pheromone components. Parapheromones show a large variety of effects, and have been used as agonists and inhibitors. They can replace pheromones when these are costly to prepare or unstable in field conditions.

## Insecticidal gas

Methyl bromide is a fumigant insecticide that rapidly kills insects, mites, and nematodes. It penetrates substrates including soil and wood, it usually does not stain or taint commodities, and is noncorrosive and nonflammable. It has a boiling point of 3.6 °C and is colorless and odorless at concentrations used for fumigation. Chloropicrin is sometimes added at 2% as a warning indicator when this methyl bromide is used in structures. The mode of action is damage to nerve cell membranes, and it reacts with sulflhydryl groups in proteins. Insects usually die within 24 h of exposure, but mortality may be delayed 1–2 days. In 1992 it was listed as an ozone-depleting substance under the Montreal Protocol on Substances that Deplete the Ozone Layer, and all developed countries are scheduled to eliminate the majority of their use of this chemical by 2005.

Sulfuryl fluoride is a fumigant gas used to control household, structural, and stored-product pests. This chemical is not combustible. It has a vapor pressure of 13 442 mmHg at 25 °C, and a boiling point of −5.4 °C at 760 mmHg. The critical route of exposure is through inhalation and the threshold limit value is 5 ppm. Under practical conditions, sulfuryl fluoride is fully oxidized in the atmosphere and does not interact with ozone. It readily penetrates most materials, and has no adverse affects on metals. Sulfuryl fluoride is odorless and colorless, and a small quantity of chloropicrin is used with it as a warning agent. Mode of action is by disrupting the glycolysis cycle, thereby depriving the animal's body of metabolic energy. Mortality may be delayed for several days, depending on the animal species.

Phosphine (hydrogen phosphine) is the common name for the active ingredient released from the metal phosphides, aluminum phosphide, and magnesium phosphide. This fumigant is highly toxic, and a concentration of 400 ppm is lethal to insects, humans, and other forms of life. It will corrode metals and may ignite in air at concentrations above its flammable limit of 1.8%. Phosphine has a detectable odor for humans at concentrations as low as 0.018 ppm; normally the gas can be detected before it can cause serious effects. The mode of action includes the nervous system, paralysis of the spiracular muscles to prevent respiration in insects, and the enzyme cytochrome oxidase system is attacked. Exposure periods of 1–5 days are necessary to control most insect pests.

## Physical modifications

Physical alteration of urban structures or other features of the habitat can reduce or prevent access by pest arthropods, or limit harborage and breeding sites. Methods such as the use of screens, caulking, removing moisture, limiting wood–soil contact, and other traditional methods are effective. Screening prevents flying insects and some soil-inhabiting insects, such as subterranean termites, from entering buildings. Screen specifications for excluding house flies and similar-sized species are: mesh #10, aperture length 2.27 mm excludes house flies; mesh #16, aperture length 1.30 mm excludes most mosquitoes; and, mesh #20, aperture length 0.853 excludes ceratopogonid (Ceratopogonidae) flies. Traps based on light, ultraviolet light, carbon dioxide, pheromones, and other chemical scents can be used for local and area-wide insect control.

Ultraviolet (UV) light traps for flies and other insects utilize their sensitivity to this portion of the light spectrum. UV light is classified as light that has a wavelength between 100 and 400 nm. Blue light has a wavelength of 450–500 nm, green light 500–560 nm, orange 600–650 nm, and red 650–700 nm. The UV light bulbs used in insect light traps have an internal coating which gives off ultraviolet light when the tube is lit. The coating breaks down over time and eventually the UV light generated is not sufficient to attract insects. The tube, however, continues to give off normal, visible light. UV light is usually divided into three categories: UVA, which has light frequency of 315–400 nm; UVB, which has a frequency of 280–315 nm; and UVC, which has a frequency of 100–280 nm. UVC light is frequently used for its germicidal properties, and UVB is the sun-tanning light emitted by the sun. The UVA wavelengths are used in insect light traps, and are harmless to humans. The optimum range for attracting insects is 350–370 nm, but some insects are attracted to wavelengths near 500 nm. Some species of midges (Chironomidae) are attracted to light in the near-UV region of 300–390 nm. Many species, representative of most orders, are sensitive to UV light, and some significant behavioral responses are initiated by it. Some insects are negatively phototactic to UV light; for example, when given a choice, ants will congregate in a region not illuminated by UV.

Sound (wingbeat sounds) has been used as a component of insect traps, typically for mosquitoes and chironomids. Wavelengths are somewhat species-specific and may be combined with UV radiation to increase effectiveness in traps. Sinusoidal sounds 210–300 Hz are effective in attracting male *Chironomus plumosus*, a common chironomid pest around the world. Frequencies between 240 and 270 Hz are attractive to *C. dissidens*, and 150–180 Hz was attractive to males of *Propsilocerus akamusi* (Chironomidae).

Air currents have a long history of use as a barrier in preventing the entry of flying insects into buildings or other confined spaces. The house fly is the primary target, and velocities effective for this insect are generally effective for others. Effectiveness is achieved when air is discharged at a velocity in the range of 457–670 m/min, at a 15° angle. For the house fly, 92% exclusion can be obtained when air is discharged at 546.5 m/min, and 80% exclusion is achieved with 529 m/min.

Volatile oils and other chemicals in personal-use mosquito repellents function as a chemical barrier to host-seeking females. Bednets treated with (pyrethroid) insecticides are an effective barrier between humans and the mosquito vectors of various diseases. Other barriers for mosquito control include the use of polystyrene beads in potable water supplies to reduce the potential breeding of *Culex quinquefasciatus* and *Cx. pipiens*.

Physical barriers can limit or prevent subterranean termites from entering structures from soil nests. Barriers consisting of soil particles of specific sizes can be used to prevent species of subterranean termites from tunneling through the material and gaining access to structural wood. Termites are unable to move particles larger than about 1 mm diameter; as particle size increases, so does the size of the space between the particles. Particles about 3 mm diameter provide interspaces large enough to allow workers to crawl through. Effective termite-barrier sand has particles 1–3 mm diameter, or no larger and no smaller than that able to pass through a 16-mesh screen. Sand smaller than 16-mesh can be carried away by workers, and larger particles can support tunnel construction. For *Reticulitermes hesperus* the effective sand particle size is 1.6–2.5 mm; for *R. flavipes* the effective aquarium sand particle size is 1.4–3.35 mm; and for *Coptotermes formosanus* sand blast particles 1.4–2.36 mm are effective in establishing a barrier.

Stainless-steel screen, with a mesh of 35-mesh material with an aperture size of 0.66 × 0.45 mm, in large continuous sheets and placed over building foundations, prevents movement of termites from soil to above-ground wood. To be effective the screen must be flexible to be molded around all potential entry points; a high-quality 316 marine-grade stainless steel is used. Similarly, insecticide-impregnated plastic sheeting that covers the subslab soil, or as a fitting around pipes and other building-construction features, forms a barrier to subterranean termites. It is placed as a continuous sheet beneath the foundation.

## Modified environments

Heat or cold can be used to eliminate or at least slow the increase of populations of stored-product insect pests. Ideal conditions for stored-product insects are 25–32 °C and 65–75% RH. Above and below this range insect growth and fitness are reduced, and in extreme conditions insects die. Most stored-product insects are killed when exposed to 40 °C for 24 h, 45 °C for 12 h, 50 °C for 5 min, 55 °C for 1 min, or 60 °C for 30 s. Among the beetles, *Lasioderma serricorne* and *Rhyzopertha dominica* are highly tolerant of heat, while *Sitophilus* spp. and *Tribolium castaneum* are moderately tolerant. Acclimation to heat can occur. Brief exposures to 35–40 °C can increase survival of insects to subsequent exposure to higher temperatures, but above 55 °C there is little difference between acclimated and unacclimated individuals.

A temperature of −18 °C kills most stored product insects within 2–3 min. However, *Sitophilus granarius* can reproduce at 15 °C, and the rusty grain beetle, *Cryptolestes ferrugineus*, can survive exposures to −12 °C after a 4-week acclimation at 15 °C. Exposure of *Anobium* eggs to −14 °C produces 99% mortality. Continuous exposure at −20 °C for 2–3 weeks is lethal to all stages of clothes moth species, carpet beetles, and other dermestids.

Exposure to −15 °C for 10 h is lethal to *Pediculus* spp. eggs, and exposure to −15 °C for 2 h is lethal to adult lice; exposure to −17 °C for 2 h is lethal to *Cimex* spp. adults. As temperatures approach 0 °C, the time required to kill many species increases to about 50 days. During a short exposure to a high temperature, some insects, especially those with a large body, are somewhat cooled by the water evaporating from their body. Hot and moist air reduces the amount of cooling by evaporation and is the most effective method of using heat. Hot air is lethal for *Pediculus* spp. eggs exposed to 50 °C for 0.5 h, and lethal to adults exposed to 46 °C for 1 h. It is lethal to *Cimex* spp. eggs exposed to 45 °C for 1 h, and to adults exposed to 44 °C for 1 h. Cold temperatures have a similar effect on these two species.

Modified atmosphere generally refers to alteration of the gaseous environment in which an insect lives. Typically, it is produced artificially and maintained by enveloping an object or structure with a gas such as carbon dioxide or nitrogen. The source of the gas is usually a pressurized container, and it is important to maintain a nearly stable concentration. Application of modified atmospheres has been to control stored-food pests, and to remove insects from museum specimens, archival and library material. A high percentage of carbon dioxide coupled with limiting the oxygen concentration in the air space to 10–20% will kill insects in stored grain without damaging the product.

Temperature coupled with low levels of oxygen can control some insects. Food pests die within 30 days when exposed to low amounts of oxygen at 15 °C, and they die within 2–3 days at 30 °C. As exposure temperature increases from 32 to 43 °C in 99% nitrogen (low oxygen), the time required to kill all stages of the cigarette beetle, *Lasioderma serricorne*, decreases from 96 to 24 h. An oxygen level of 0.3% is lethal to *Anthrenus museorum* in about 32 h, and lethal to *Attagenus woodroffei* and *A. smirnovi* in about 88 h. Lethal time for larvae of *Anthrenus verbasci* and *Reesa vespulae* is about 44 h, and for *Trogoderma angustum* it is about 57 h. Exposure of 7–14-days at 0.4% oxygen kills *Tineola bisselliella*, *Lasioderma serricorne*, *Anthrenus vorax*, and *Stegobium paniceum*. For structural pests, an atmosphere of 1% oxygen kills old house borer and powderpost beetle adults within 20 days; however, they are not killed at 80% carbon dioxide. The powderpost beetle, *Lyctus brunneus*, is killed after a 7–14-day exposure at 0.4% oxygen. Exposure time necessary to produce death is generally decreased by raising the temperature, by adding 5% carbon dioxide, or by decreasing relative humidity (RH). In general, lethal time increases with increasing RH in atmospheres with a low percentage of oxygen. Exposure for 48 h in 0.32% oxygen and 33% RH provides 94% mortality of cigarette beetle larvae, but only 25% mortality at 75% RH. Increasing exposure temperature generally decreases lethal exposure time. The minimum exposure time of 45 min is required to kill the drywood termite *Incisitermes minor* and carpenter ant *Camponotus vicinus* at 48.9 °C and 49% RH.

## Bibliography


Burn, A. J., T. H. Coaker, and P. C. Jepson (eds.) *Integrated Pest Management*. New York: Academic Press, 1988.

Carson, R. L. *Silent Spring*. Boston, MA: Houghton Mifflin, 1962.

Gorham, R. J. The significance for human health of insects in food. *Annu. Rev. Entomol.*, **24** (1979), 209–24.

Kangas, E. The impact of human culture on insects, and especially on insect pests. *Fennia*, **85** (1961), 100–5.

Kogan, M. (ed.) *Ecological Theory and Integrated Pest Management Practice*. New York: John Wiley, 1986.

Krieger, R., J. Doull, D. Ecobichon *et al. Handbook of Pesticide Toxicology*. New York, NY: Academic Press, 2001.

Metcalf, R. L. and R. A. Luckman (eds.) *Introduction to Insect Pest Management*. New York: John Wiley, 1994.

Olkowski, W., H. Olkowski, R. van den Bosch, and R. Hom. Ecosystem management: a framework for urban pest control. *BioScience*, **26** (1976), 384–9.

Pedigo, L. P. *Entomology and Pest Management*. New York: Macmillan, 1989.

Raupp, M. J., C. S. Koehler, and J. A. Davidson. Advances in implementing integrated pest management for woody landscape plants. *Annu. Rev. Entomol.*, **37** (1992), 561–85.

Robinson, W. H and N. Bao. The pest status of *Periplaneta fuliginosa* in China. *Proc. Wash. Entomol. Soc.*, **90** (1988), 401–6.

Rust, M. K. Managing household pests. In Bennett, G. W. and J. M. Owens (eds.) *Advances in Urban Pest Management*, pp. 335–68. New York: Van Nostrand Reinold, 1986.

Sawyer, A. J. and R. A. Casagrande. Urban pest management: a conceptual framework. *Urban Ecol.*, **7** (1983), 145–57.

Smith, D. C. and M. J. Raupp. Economic and environmental assessment of an integrated pest management program for community-owned landscape plants. *J. Econ. Entomol.*, **79** (1986), 162–5.

Stern, V. M. and R. F. Smith. The integrated control concept. *Hilgardia*, **29** (1959), 81–101.

Takahashi, F. A concept of pest management in urban green zones. *Jpn J. Environ. Entomol. Zool.*, **3** (1991), 210–16.

### Allergies, bites, and stings

Alexander, J. O. D. Mites and skin disease. *Clin. Med.*, **79** (1972), 14–19.

Arlian, L. G. Arthropod allergens and human health. *Annu. Rev. Entomol.*, **47** (2002), 395–433.

Armstrong, R. K. and J. L. Winfield. Staphylinidae dermatitis on Okinawa. *J. Med. Entomol.*, **5** (1968), 362.

Beard, R. L. Insect toxins and venoms. *Annu. Rev. Entomol.*, **8** (1960), 1–18.

Bellas, T. E. *Insects as a Cause of Inhalent Allergies. A Bibliography*, 2nd edn. CSIRO Australia Division of Entomology report no. 25. Canberra: CSIRO, 1982.

*Insects as a Cause of Inhalent Allergies. A Bibliography 1900–1987*. Canberra: Division of Entomology, Commonwealth Scientific and Industrial Research Organization, 1989.

Bernton, H. S. and H. Brown. Insect allergy – preliminary studies of the cockroach. *J. Allergy*, **35** (1964), 506–13.

Insect allergy: the allergenicity of the excrement of the cockroach. *Ann. Allergy*, **28** (1970), 543–7.

Bettini, S. (ed.) *Arthropod Venoms*. Berlin: Springer-Verlag, 1978.

Biliotti, G., A. Passaleva, S. Romagnani, and M. Ricci. Mites and house dust allergy. I. Comparison between house dust and mite (*Dermatophagoides pteronyssinus* and *D. farinae*) skin reactivity. *Clin. Allergy*, **2** (1972), 109–13.

Brenner, R. J., K. C. Barnes, R. M. Helm, and L. W. Williams. Modernized society and allergies to arthropods. *Am. Entomol.*, **37** (1991), 143–55.

Brown, L. L. Fire ant allergy. *South. Med. J.*, **65** (1972), 273–7.

Bücherl, W. and E. E. Buckley (eds.). *Venomous Animals and Their Venoms*, vol. 3, *Venomous Invertebrates*. New York: Academic Press, 1972.

Choovivathanavanich, P. Insect allergy: antigenicity of cockroach and its excrement. *J. Med. Assoc. Thai.*, **57** (1974), 237–41.

Cloudsley-Thompson, J. On being bitten and stung. *Antenna (Lond.)*, **19** (1995), 177–80.

Cormia, F. E. Carpet beetle dermatitis. *J. A. M. A.*, **200** (1967), 799.

Feingold, B. F., E. Benjamini, and D. Michaeli. The allergic responses to insect bites. *Annu. Rev. Entomol.*, **13** (1968), 137–58.

Frankland, A. W. Bee sting allergy. *Bee World*, **57** (1976), 145–50.

Habermann, E. Bee and wasp venoms. *Science*, **177** (1972), 314–22.

Herman, J. and R. Cukerman. Contact dermatitis due to ladybirds. *Practitioner*, **226** (1982), 311.

Hewitt, M., G. I. Barrow, D. C. Miller, F. Turk, and S. Turk. Mites in the personal environment and their role in skin disorders. *Br. J. Dermatol.*, **89** (1973), 401–9.

Kang, B., D. Vellody, H. Homburger, and J. W. Yunginger. Cockroach cause of allergic asthma. Its specificity and immunologic profile. *J. Allergy Clin. Immunol.*, **63** (1979), 80–6.

Kawakami, T., C. Suto, T. Yagura, and N. Kumuda. Studies on cockroach allergy I. Allergenicity of common domestic cockroaches in Japan. *Jpn J. Sanit. Zool.*, **33** (1982), 233–8.

Kay, A. B., M. O. Gad El Rab, J. Stewart, and H. H. Erwa. Widespread IgE-mediated hypersensitivity in Northern Sudan to the chironomid, *Cladotanaytarsus lewisi* (green nimitti). *Clin. Exp. Immunol.*, **34** (1978), 106–10.

Keegan, H. L. and W. V. McFarlane. *Venomous and Poisonous Animals and Noxious Plants of the Pacific Region*. New York: Macmillan, 1963.

Kino, T. and S. Oshima. Allergy to insects in Japan. I. The reaginic sensitivity to moth and butterfly in patients with bronchial asthma. *J. Allergy Clinical Immun.*, **61** (1978), 10–16.

Langolis, C., S. Schulman, and C. E. Arbesman. Immunological studies of caddis fly. *J. Allergy*, **34** (1963), 385–94.

Marcondes, C. B. *Entomologia Médica e Veterinária (in Portuguese)*. São Paulo: Editora Atheneu, 2001.

Piek, T. (ed.) *Venoms of the Hymenoptera. Biochemical, Pharmacological and Behavioural Aspects*. London: Academic Press, 1986.

Ratcliffe, B. C. A case of tarantula-induced popular dermatitis. *J. Med. Entomol.*, **13** (1977), 745–7.

Richman, P. G., H. A. Kahn, P. C. Turkeltaub, F. J. Malveaux, and H. Baer. The important sources of German cockroach allergens as determined by RAST analyses. *J. Allergy. Clin. Immunol.*, **73** (1984), 590–5.

Schmidt, J. O. Biochemistry of insect venom. *Annu. Rev. Entomol.*, **27** (1982), 339–68.

Truit, G. W. The mushroom fly as a cause of bronchial asthma. *Annu. Allergy*, **9** (1951), 513–16.

Tu, A. T. Venoms: chemistry and molecular biology. *New York*: John Wiley, 1977.

Van Bronswijk, J. E. M. H. *Dermatophagoides pteronyssinus* (Trouessart, 1897) in mattress and floor dust in a temperate climate (Acari: Pyroglyphidae). *J. Med. Entomol.*, **10** (1973), 63–70.

Van Bronswijk, J. E. M. H. and R. N. Sinha. Pyroglyphid mites (Acari) and house dust allergy. *J. Allergy*, **47** (1971), 31–52.

Viraben, R. Papular urticaria. A cutaneous sensitivity reaction to environmental arthropods. *Ann. Dermatol. Vénéréol.*, **123** (1996), 751–6.

Voorhorst, R., M. I. A. Spieksma-Boezeman, and F. Th. M. Spieksma. Is a mite (*Dermatophagoides* sp.) the producer of the house dust allergen? *Allergic Asthma*, **10** (1964), 329.

Voorhorst, R. and F. Th. M. Spieksma. Recent progress in the house dust mite problem. *Acta Allergol.*, **24** (1969), 115–23.

Wharton, G. W. House dust mites. *J. Med. Entomol.*, **12** (1976), 577–621.

Wirtz, R. A. Occupational allergies to arthropods: documentation and prevention. *Bull. Entomol. Soc. Am.*, **26** (1980), 356–60.

Zschunke, E. Contact urticaria, contact dermatitis, and asthma from cockroaches. *Arch. Dermatol.*, **114** (1978), 1715–16.

### Biological control

Croft, B. A. *Arthropod Biological Control Agents and Pesticides*. New York: John Wiley, 1990.

DeBach, P. *Biological Control by Natural Enemies*. London: Cambridge University Press, 1974.

Ferron, P. Biological control of insect pests by entomogenous fungi. *Annu. Rev. Entomol.*, **23** (1978), 409–42.

Gaugler, R. and H. K. Kaya. *Entomopathogenic Nematodes in Biological Control*. Boca Raton, FL: CRC Press, 1990.

Ridgway, R. L. and S. B. Vinson (eds.) *Biological Control by Augmentation of Natural Enemies*. New York: Plenum, 1977.

### Botanical, microbial, and genetical pest control

Ara, I., S. Bina, S. Siddiqui, S. Shaleen, and S. Salimuzzamen. Tricyclic diterpenes from the stem bark of *Azadriachra indica*. *Planta Med.*, **56** (1990), 84–6.

Becker, N. and J. Margalit. Control of Dipteran pests by *Bacillus thuringiensis*. In Entwistle, P., M. J. Bailey, J. Cory, and S. Higgs (eds.) Bacillus thuringiensis: *Its Uses and Future as a Biological Insecticide*. London: John Wiley, 1993.

Burges, H. D. *Microbial Control of Pests and Plant Diseases, 1970–1980*. New York: Academic Press, 1981.

Burges, H. D. and J. A. Hurst. Ecology of *Bacillus thuringiensis* in storage moths. *J. Invert. Pathol.*, **30** (1977), 131–9.

Dharmshaktu, N. S., P. K. Prabhakaran, and P. K. M. Menon. Laboratory study on the mosquito larvicidal properties of leaf and seed extract of the plant *Agave americana*. *J. Trop. Med. Hyg.*, **90** (1987), 79–82.

Margalit, J., N. Becker, C. Back, and A. Zaritsky. *Bacillus thuringiensis* subsp. *israelensis* as a biological control agent of mosquitoes and black flies. In Bacillus thuringiensis *Biotechnology and Environmental Benefits*, vol. 1, pp. 521–56.

Mohsen, Z., H. Abdel-Latif, M. Jawad, M. Al-Saadi, and A. Al-Naib. Mosquito larvicidal and ovipositional activity of *Descurania sophia* extract. *Int. J. Crude Drug Res.*, **28** (1990), 77–80.

Nakanishi, K. Recent studies on bioactive compounds from plants. *J. Nat. Prod.*, **45** (1982), 15–26.

Ragoonanansingh, R. N., K. J. Njunwa, C. F. Curtis, and N. Becker. A field study of *Bacillus sphaericus* for the control of culicine and anopheline mosquito larvae in Tanzania. *Bull. Soc. Vect. Ecol.*, **17** (1992), 45–50.

Whitten, M. J. and G. G. Foster. Genetical methods of pest control. *Annu. Rev. Entomol.*, **20** (1975), 461–76.

Yuan Fang-Yu, Zhang Ji-Bin, Xu Ba-Zhao, and N. Becker. Large-scale evaluation of *Bacillus sphaericus* C3–41 local products for controlling *Culex quinquefaciatus* in urban areas. *Chinese J. Parasitic Disease Contr.*, **7** (1994), 123–7.

### Delusional parasitosis, entomophobia

Bebbington, P. Monosymptomatic hypochrondriasis, abnormal illness behavior and suicide. *Br. J. Psychiatry*, **128** (1976), 475–8.

Berrios, G. E. Tactile hallucinations: conceptual and historical aspects. *J. Neurol. Neurosurg. Psych.*, **45** (1982), 285–93.

Delusional parasitosis and physical disease. *Compr. Psychiatry*, **26** (1985), 395–403.

Bourgeois, M. L., P. Duhamel, and H. Verdoux. Delusional parasitosis: *folie à deux* and an attempted murder of a family doctor. *Br. J. Psychiatry*, **161** (1992), 709–11.

Byrne, D. N., E. H. Carpenter, E. M. Thoms, and S. T. Cotty. Public attitudes toward urban arthropods. *Bull. Entomol. Soc. Am.*, **30** (1984), 40–4.

Ekbom, K. A. *Om des k Parasitofobierna*. Stockholm: Svenska Psykiatriska Föreningens Förhandingar, 1937.

Hinkle, N. C. Delusory parasitosis. *Am. Entomol.*, **46** (2000), 17–25.

Hopkinson, G. Delusions of infestation. *Acta Psychiatr. Scand.*, **46** (1970), 111–19.

Lyell, A. Delusional parasitosis. *Br. J. Dermatol.*, **108** (1983), 485–99.

Marks, I. M. Classification of phobic disorders. *Br. J. Psychiatry*, **116** (1970), 377–86.

Mester, H. Induzierter 'Dermatozoenwahn'. *Psychiatr. Praxis*, **8** (1975), 261–3.

Morris, M. Delusional infestation. *Br. J. Psychiatry*, **159** (suppl. 14) (1991), 83–7.

Morris, M., G. Moss, and D. Jolley. When delusions of infestations afflict the elderly. *Geriatr. Medi.*, **18** (1988), 57–60.

Mumford, J. Entomophobia: the fear of arthropods. *Antennae*, **6** (1982), 156–7.

Musalek, M. and E. Kutzer. The frequency of shared delusions in delusions of infestations. *Eur. Arch. Neurol. Sci.*, **239** (1990), 263–6.

Paulson, M. J. and E. P. Petrus. Delusions of parasitosis: a psychological study. *Psychosomatics*, **10** (1969), 111–20.

Perrin, L. Des névrodermies parasitophobiques. *Ann. Dermatol. Syphiligr. (Paris)*, **7** (1896), 129–38.

Schrut, A. H. and W. G. Waldron. Psychiatric and entomological aspects of delusory parasitosis. *J. A. M. A.*, **186** (1963), 429–30.

Thirbierge, G. Les acrophobes. *Ann. Dermatol. Syphiligr. (Paris)*, **5** (1894), 730–1.

Zahner, G. E. P., S. V. Kasl, M. White, and J. C. Will. Psychological consequences of infestation of the dwelling unit. *Am. J. Publ. Health*, **75** (1985), 1303–7.

### Insecticides, repellents

Alexander, P., J. A. Kitchener, and H. V. A. Briscoe. Inert dust insecticides. Parts I, II, III. *Ann. Appl. Biol.*, **31** (1944), 143–59.

Batth, S. Evaluation of vinyl garment bags as chambers for paradichlorobenzene fumigation of fabric-insect pests. *J. Econ. Entomol.*, **65** (1972), 1074–80.

Campbell, W. C. (ed.). *Invermectin and Abamectin*. New York: Springer-Verlag, 1989.

Coats, J. R. Risks from natural versus synthetic insecticides. *Annu. Rev. Entomol.*, **39** (1994), 489–515.

Denholm, I. and M. W. Rowland. Tactics for managing pesticide resistance in arthropods: theory and practice. *Annu. Rev. Entomol.*, **37** (1992), 91–112.

Ebeling, W. Physiochemical mechanisms for the removal of insect wax by means of finely divided powders. *Hilgardia*, **30** (1961), 531–64.

Elbert, A., R. Nauen, and W. Leicht. Imidacloprid, a novel chloronicotinyl insecticide: biological activity and agricultural importance. In Ishaaya, I. and D. Deghele (eds.) *Insecticides with Novel Modes of Action, Mechanisms and Application*. New York: Springer, 1998.

Fields, P. G. and N. D. G. White. Alternatives to methyl bromide treatments for stored-product and quarantine insects. *Annu. Rev. Entomol.*, **47** (2002), 331–59.

Hayward, F. W. and R. B. Seymour. Determination of the major constituents of cedar oil vapor in cedar chests. *Anal. Chem.*, **20** (1948), 572–4.

Jepson, P. C. (ed.). *Pesticides and Non-Target Invertebrates*. New York: VCH, 1990.

Karlson, P. and M. Lüscher. Pheromones, a new term for a class of biologically active substances. *Nature*, **183** (1959), 55–6.

Lasota, J. A. and R. F. A. Dybas. Avermectins, a novel class of compounds: implications for use in arthropod pest control. *Annu. Rev. Entomol.*, **36** (1991), 91–117.

Laudani, H. and P. H. Clark. The effects of red, white, and South American cedar chests and the various stages of the webbing clothes moth and black carpet beetle. *J. Econ. Entomol.*, **47** (1954), 1107–11.

McEwen, F. L. and G. R. Stephenson. *The Use and Significance of Pesticides in the Environment*. New York: Wiley-Interscience, 1979.

Menn, J. J. and M. Beroza (eds.) *Insect Juvenile Hormones: Chemistry and Action*. New York: Academic Press, 1972.

Millar, J. G. Polyene hydrocarbons and epoxides: a second major class of lepidopteran sex attractant pheromones. *Annu. Rev. Entomol.*, **45** (2000), 575–604.

Morgan, E. D. and N. B. Mandava. *Insect Growth Regulators*. Boca Raton, FL: CRC Press, 1987.

Narahashi, T. and J. E. Chambers (eds.) *Insecticide Action: From Molecule to Organism*. New York: Plenum, 1989.

Renou, M. and A. Guerrero. Insect parapheromones in olfaction research and semiochemical-based pest control strategies. *Annu. Rev. Entomol.*, **45** (2000), 605–30.

Staal, G. B. Insect growth regulators with juvenile hormone activity. *Annu. Rev. Entomol.*, **20** (1975), 417–60.

Ware, G. W. *The Pesticide Book*, 3rd edn. Fresno, CA: Thompson, 1989.

Yamamoto, I. Mode of action of pyrethroids, nicotinoids, and rotenoids. *Annu. Rev. Entomol.*, **15** (1970), 257–68.

### Light, sound, air, physical barriers, and modified atmospheres

Bergh, J.-E., L. Stengård Hansen, K.-M. Vagn Jensen, and P. V. Nielsen. The effect of anoxic treatment on the larvae of six species of dermestids (Coleoptera). *J. Appl. Entomol.*, **127** (2003), 317–21.

Ebeling, W. and R. J. Pence. Relation of particle size to the penetration of subterranean termites through barriers of sand or cinders. *J. Econ. Entomol.*, **50** (1957), 690–2.

Fields, P. G. The control of stored-product insects and mites with extreme temperatures. *J. Stored Prod.*, **28** (1992), 89–118.

Hirabayashi, K., R. Nakazato, A. Ohara, and T. Okino. A study on phototaxis for adult Chironomidae (Diptera) by artificial light in Lake Suwa. 2. Effect of the light trap intensity and interval of electric-killing insect traps using near-ultraviolet radiation. *Jpn J. Sanit. Zool.*, **44** (1993), 299–306.

Hollingsworth, J. P., A. W. Hartsack, and D. A. Lindquist. Influence of near ultraviolet output of attractant lamps on catches of insects by light traps. *J. Econ. Entomol.*, **61** (1968), 515–21.

Ikeshoji, T. Attractive sounds for autochemosterilization of the male mosquitoes. *Jpn J. Sanit. Zool.*, **33** (1982), 41–9.

Lenz, M. and S. Rumko. Protection of buildings, other structures and materials in ground contact from attack by subterranean termites (Isoptera) with a physical barrier: a fine mesh of high-grade stainless steel. *Sociobiology*, **24** (1994), 1–16.

Mathis, W., E. A. Smith, and H. F. Schoof. Use of air barriers to prevent entrance of house flies. *J. Econ. Entomol.*, **63** (1970), 29–31.

Ogawa, K. Field trapping of male midge *Rheotantytarsus kyotoensis* (Diptera: Chironomidae) by sounds. *Jpn J. Sanit. Zool.*, **43** (1992), 77–80.

Pickens, L. G. and R. W. Thimijan. Design parameters that affect the performance of UV-emitting traps in attracting house flies (Diptera: Muscidae). *J. Econ. Entomol.*, **79** (1986), 1003–9.

Rawat, B. S. Physical barriers: non-toxic and eco-friendly alternatives to hazardous termiticides for buildings. *Int. Pest Control*, **44** (2002), 182–7.

Rust, M. K., E. O. Paine, and D. E. Reierson. Evaluation of freezing to control wood-destroying insects (Isoptera, Coleoptera). *J. Econ. Entomol.*, **90** (1997), 1215–21.

Smith, L. E. Effects of cold acclimation on supercooling and survival of the rusty grain beetle, *Cryptolestes ferrugineus* (Stephens) (Coleoptera: Cucujidae) at subzero temperatures. *Can. J. Zool.*, **48** (1970), 853–8.

Soderstrom, E. L. Effectiveness of green electroluminescent lamps for attracting stored-product insects. *J. Econ. Entomol.*, **63** (1970), 726–31.

Tamashiro, M., J. R. Yates, and R. H. Ebesu. The Formosan subterranean termite in Hawaii: problems and control. In Tamashiro M. and N.-Y. Su (eds.) *Biology and Control of the Formosan Subterranean Termite*, pp. 15–22. Honolulu: College of Tropical Agriculture and Human Resources, University Hawaii, 1987.

Tamashiro, M., J. R. Yates, R. T. Yamamoto, and R. H. Ebesu. Tunneling behavior of Formosan subterranean termite and basalt barriers. *Sociobiology*, **19** (1991), 163–70.

Thimijan, R. W. and L. G. Pickens. A method for predicting house fly attraction of electromagnetic radiant energy. *J. Econ. Entomol.*, **66** (1972), 95–100.

Thimijan, R. W., L. G. Pickens, and N. O. Morgan. Responses of the house fly, stable fly and face fly to electromagnetic radiant energy. *J. Econ. Entomol.*, **66** (1973), 1260–70.

### Pest reservoirs

Cherret, J. M. and G. R. Sager (eds.) *Origins of Pest, Parasite, Disease and Waste Problems*. Oxford: Blackwell, 1977.

Fraenkel, G. and M. Blewett. The natural foods and the food requirements of several species of stored food product insects. *Trans. R. Entomol. Soc. Lond.*, **93** (1943), 457–90.

Freeman, J. A. Methods of spread of stored products insects and origin of infestation in stored products. In *Proceedings of the Eighth International Congress of Entomology*, Stockholm, Sweden, 1950, pp. 815–25.

Haeseler, V. Anthropogene Biotype (Kahlschlag, Kiesgrub, Stadgärten) als Refugian fur Insekte, Untersucht am Beispiel der Hymenoptera Aculeata. *Zool. Jahrb. Syst. Bd.*, **99** (1972), 133–212.

Hatch, M. H. The origin and evolution of household insects. *Biologist*, **35** (1953), 57–66.

Hickin, N. E. The common furniture beetle, *Anobium punctatum* (DeG.) (Coleoptera, Anobiidae): some notes on its outdoor occurrence. *Entomologist*, **86** (1953), 216–17.

Hinton, H. E. Natural reservoirs of beetles of the family Dermestidae known to infest stored products, with notes on those found in spiders' webs. *Proc. R. Entomol. Soc. Lond. Ser.* **A 18** (1943), 33–42.

Linsley, E. G. Insect food caches as reservoirs and original sources of some stored product pests. *J. Econ. Entomol.*, **35** (1942), 434–9.

Robinson, W. H. Role of reservoir habitats and populations in the urban environment. In Jones, S., J. Zhai, and W. Robinson (eds.) *Proc. of the 4th International Conference on Urban Pests.*, pp. 217–224. Blacksburg, VA: Pocahontas Press, 2002.

Schmitz, G. Urban ruderal sites as secondary habitats for phytophagous insects. *Verh. Ges. Okol.*, **26** (1996), 581–5.

Schoof, H. F., G. A. Mail, and E. Savage. Fly production sources in urban communities. *J. Econ. Entomol.*, **47** (1954), 245–53.

Part two

# Insects in the urban environment

# 4 Blattaria

## Introduction

Cockroaches are primarily tropical and subtropical insects, but they occur in nearly all of the climatic regions of the world. Most of the approximately 4000 described species live in small populations in forest habitats, some little-known species are semi-aquatic, some are cave-dwelling, and a small number of species are associated with the nests of other insects. They are predominantly nocturnal and vegetarian, but many adjust their habits and food preferences to fit environmental conditions and available food resources. Cockroaches as a group are one of the most primitive winged insects, and seem to have remained in primitive habitats and undergone little morphological change since the Carboniferous geological period, about 250 million years ago. It is their association with decaying organic matter and humid conditions that maintains some species in the urban environment. Some have adapted to living indoors, but in this habitat they generally select sites that provide the temperature and humidity features of their natural habitat.

Adults are 10–50 mm long, brown to blackish brown or black, generally oval, and dorsoventrally flattened. Eyes are large and there are two ocelli, antennae are filiform and usually as long as or longer than the body. The head is usually concealed from above by a large pronotum, and chewing mouthparts are directed downwards. Legs have strong spines and setae, and the tarsi are five-segmented. Wings are present in most adults; the front wings are usually thickened and they overlap when closed over the abdomen. The hind wings may be large and fan-like when extended. Winged species are usually capable of directed and sustained flight or gliding. Some cockroaches are brachypterous, and females of many species have shorter wings than males. Nymphs are similar to adults except for their size and the absence of wings. Eggs of all cockroaches are enclosed within a covering, the eggcase or ootheca. This is a hard and protective shell, or it is reduced in thickness and unseen because it is within the female, or it encloses only part of the eggs. The number of eggs per ootheca ranges from 12 to 40, but not all the eggs will hatch to produce live nymphs. When incubation is complete the nymphs swallow air to increase their body size, and the collective expansion of their bodies splits the ootheca along a dorsal seam or ridge. Nymphs emerge enclosed in an embryonic cuticle, but this is shed immediately. There are 5–12 nymph stages, depending on species and sex. Adults are usually long-lived, some for up to 2.5 years.

A relatively small number, considering the large number of species, of cockroaches have adapted to peridomestic and domestic habitats. Their synanthropic association and success may be due to egg-laying and feeding habits, and some physiological features. The habitat preferences of species in tropical Africa and Asia probably brought them into close contact with human dwellings. Some *Periplaneta* and *Blattella* species occur in caves and feed on animal waste or other organic matter. They may have become associated with humans using caves as temporary living or storage space. *Periplaneta* eggcases or gravid *B. germanica* females could have easily moved with household materials from cave dwellings to outdoor shelters.

The behavioral responses of cockroaches to disturbances influence their ability to tolerate transportation. Those species with evasive behavior patterns in response to slight disturbance would be less likely to be transported by commerce. Physiological preadaptations to household conditions would have included temperature and humidity tolerances, and viability on diets that were limited to human food and food scraps. However, cockroach pest species are not tolerant of all the variations in the living space. *P. americana* and *P. fuliginosa* are restricted to warm and humid habitats because of their temperature-dependent development, and their relatively high cuticular permeability, which makes them susceptible to water loss. The permeability of their cuticle severely limits their tolerance to dry conditions. *B. germanica* is relatively restricted to humid locations indoors, such as kitchens and bathrooms, and is generally not known to breed outdoors where it occurs as a domiciliary pest.

Food includes a variety of plant and animal materials in the immediate habitat. Although they are considered omnivorous, a balance of carbohydrates, proteins, and fats is ingested. Food location and selection involve chemoreceptors that are located on the antennae and the mouthparts. In general, cockroaches prefer food in which the carbohydrate content is higher than fat and protein. For German and oriental cockroaches, rapid development and low nymph mortality occur on a diet of 22–24% protein, and for the American cockroach, 49–79% protein provided favorable development. Female *B. germanica* compensate for low dietary nitrogen (5% protein) by increasing consumption rates, but elevated dietary nitrogen (65% protein) results in delayed mating and smaller than normal oothecae. In this species, the relatively small size of the female in proportion to the ootheca produced requires that about 90% of her food reserves be incorporated into the ootheca. Food availability often controls the reproductive cycle, and in *B. germanica* females, peaks in feeding and drinking occur during egg maturation, and reduced food and water foraging when they carry the ootheca. Food availability influences the growth and molting of nymphs. Bacterial symbionts in specialized cells or bacteriocytes within the fat body increase the metabolic capabilities of cockroaches. Without these microorganisms cockroach development and reproduction are prolonged or disrupted, and mortality is increased. The symbionts have a role in uric acid metabolism and the formation of amino acid precursors for hemolymph proteins essential for egg development. The hindgut contains various symbiotic and parasitic microorganisms.

### Pest status

Pest status is primarily limited to the domiciliary and a few of the peridomestic species. Those cockroaches that primarily live outdoors may sometimes occur indoors. Some of the indoors species rarely move outdoors, except in their original habitat. These characteristics may indirectly indicate the adaptability of some species, and their original environment. Species of *Periplaneta* and *Blatta* typically occur in outdoor populations, but they may become established indoors, or include indoor habitats when foraging from outdoors. The survival of these species is not linked to human activity. They are members of reservoir populations away from buildings that are sustained habitats with adequate food and harborage. These and other peridomestic species are characterized by their relatively large size, lengthy development period, low reproductive potential, and mobility. *Blattella* and *Supella* species in the urban environment are domestic and only find suitable habitats indoors. They are closely associated with food, harborage, and conditions provided by humans. These species are typically small, have a relatively short development time from egg to adult, a high reproductive potential, and only limited natural mobility.

Pathogenic organisms that naturally infect domiciliary cockroaches include viruses and species of protozoa, bacteria, fungi, and helminths. Feeding and movement habits of cockroaches in the urban environment bring them into contact with a wide range of decaying organic matter, and provide the potential of spreading organisms that are encountered. Mechanical carrying and transferring organisms are facilitated by the spines and setae on legs and tarsi of adult and nymph cockroaches. Grooming their legs is an important behavior, and potentially transfers pathogenic organisms from the tarsi to the mouthparts and from there to the foregut. While feeding, cockroaches often regurgitate digestive fluid that contains organisms ingested in a previous meal. The habit of defecating as they rest, move about the habitat, and while feeding spreads pathogens to surfaces in the living space, including areas of food preparation and storage. Domiciliary cockroaches may not be the prime means of spread and cause of specific disease outbreaks, but they have habits that give them the potential of being chronic carriers of various pathogenic organisms.

A large number of microbes that are pathogenic to humans have been isolated from field-collected cockroaches. *B. germanica*, *P. americana*, and other species that feed on decaying organic matter, feces, or inhabit sewers have the potential of acquiring and carrying organisms pathogenic to humans. In some cases, cockroaches may be involved as a vector of disease pathogens. Bacteria associated with cockroaches include:

*Alcaligens faecalis*
*Bacillus subtilis*
*Bacillus cereus*
*Campylobacter jejuni*
*Clostridium novii*
*Clostridium perfringens*
*Enterobacter aerogenes*
*Escherichia coli*
*Klebsiella pneumoniae*
*Mycobacterium leprae*
*Proteus morgani*
*Proteus mirabilis*
*Proteus rettgeri*
*Proteus vulgaris*
*Salmonella bareilly*
*Salmonella bovis-morbificans*
*Salmonella bredeny*
*Salmonella newport*
*Salmonella oranienburg*
*Salmonella panama*
*Salmonella paratyphi-B*
*Salmonella typhi*
*Salmonella typhimurium*
*Serratia marcescens*
*Shigella dysenteriae*
*Staphylococcus aureus*
*Streptococcus faecalis*
*Streptococcus pyogenes*
*Vibrio* spp.
*Yersinia pestis*

Fungi and molds isolated from field-collected cockroaches include:

*Alternaria* spp.
*Aspergillus niger*
*Aspergillus flavus*
*Aspergillus fumigatis*
*Candida krusei*
*Candida paraspilosis*
*Candida tropicalis*
*Cladosporium* spp.
*Fusarium* spp.
*Geotrichum candidum*
*Mucor* spp.
*Penicillum* spp.
*Rhizopus* spp.
*Rhodotorula rubra*
*Trichoderma viride*
*Trichosporon cutaneum*

Helminth species isolated from field-collected cockroaches include *Ancylostomo duodenale*, *Ascaris lumbricoides*, *Enterobium vermicularis*, *Hymenolopsis* spp., *Necator americanus*, and *Trichuris trichiura*; protozoans include *Entamoeba histolytica* and *Giardia* spp.; and viruses include hepatitus and poliomyelitis.

At least 11 proteins from German and American cockroaches can cause allergic reactions and contribute to respiratory asthma in humans. Allergenic proteins from cockroaches are contained in cast skins of nymphs, fragments of antennae, legs and wings, excrement, pieces of partially consumed food, and living cockroaches. These allergens are heat-stable and persistent in the living space. Hypersensitization to these proteins occurs by inhalation of airborne allergens, such as dry feces, or through dermal contact with allergen-contaminated surfaces, or possibly by ingesting food contaminated with cockroach body fragments or feces. If cockroaches are present in the living or workspace, allergens are almost certainly present, and these proteins are produced throughout the life of the insect. Some will persist in infested harborages after the death of the insect, and long after the cockroach infestation has been controlled or eliminated.

Human encounters with cockroaches include incidences of biting. There are reports of cockroaches that gnawed the callused portions of skin, fingernails and toenails, and eyelashes of people, sometimes children, while they were sleeping. *Periplaneta* species are often implicated in gnawing on calluses, and causing small wounds on soft skin on the face of children. The bites and wounds were probably accidental and inflicted while feeding on small amounts of food that remained around the mouth and chin of these individuals. Cockroach behavior does not include aggressive biting of humans or other animals. The cockroach mite *Pimeliaphilus cunliffei* (Pterogosomatidae) is an obligatory parasite of cockroaches. It feeds on live individuals, and cannot survive on cast skins or dead cockroaches. This mite has been linked to bites of humans living in households with cockroach infestations.

## Development and distribution

Growth is relatively slow. Small-sized species usually mature more rapidly and have shorter lives than large species. The first few molts tend to occur at regular intervals, whereas later ones are often irregular and dependent on habitat and food. Duration of early instars is usually shorter than late instars, and depending on conditions the late stages can be extended many months. Under adverse conditions, such as low temperatures or limited food and water, nymphs nearly stop development. For most species, the molts, especially the last, occur in a secluded harborage, and the tanning and hardening of the cuticle are completed in a few hours. Adults and large nymphs often cannibalize other individuals in the harborage during tanning.

Egg-laying habits of cockroaches can be considered in an evolutionary series. It begins with the species that deposit thick-walled oothecae, which are unattended by the female, and extends to thin-walled and slightly sclerotized ootheca retained within the female's body until the developing nymphs are prepared to hatch. Females of many species deposit the ootheca soon after it is formed, while others carry it until the eggs complete development, and the females of other species retract the formed ootheca into a brood sac and eggs complete development there. Oviparous species deposit the recently formed ootheca, or turn the ootheca on its side and retain it in the female's genital chamber prior to deposition. Oothecae that are deposited soon after formation, such as those of the American and oriental cockroach, are tanned and hardened capsules that protect the developing eggs. Oothecae that are turned on their side and retained by the female are permeable to water provided by the female's body. In oviparous species, the dorsal edge of the ootheca is formed into a series of respiratory chambers, usually one per egg, which admit air to the developing embryo (Fig. 4.1). Nymphs hatch from the eggcase by increasing their body size with a small amount of air; this spreads apart the dorsal edge of the eggcase for the nymphs to exit.

Ovoviviparous species, such as the Madeira and lobster cockroach, retract the developed ootheca into a specialized brood sac in the female's abdomen. Hatching occurs within the brood sac, either while the ootheca is being extruded, or shortly after it is deposited by the female. Oothecae of ovoviviparous species are thin-walled to allow for the transfer of water to the developing embryos. Viviparous development occurs in *Diploptera* species, the most common of which is *D. punctata*. The ootheca of this species protrudes from the female's abdomen for a short time before it is turned on its side and then withdrawn into the

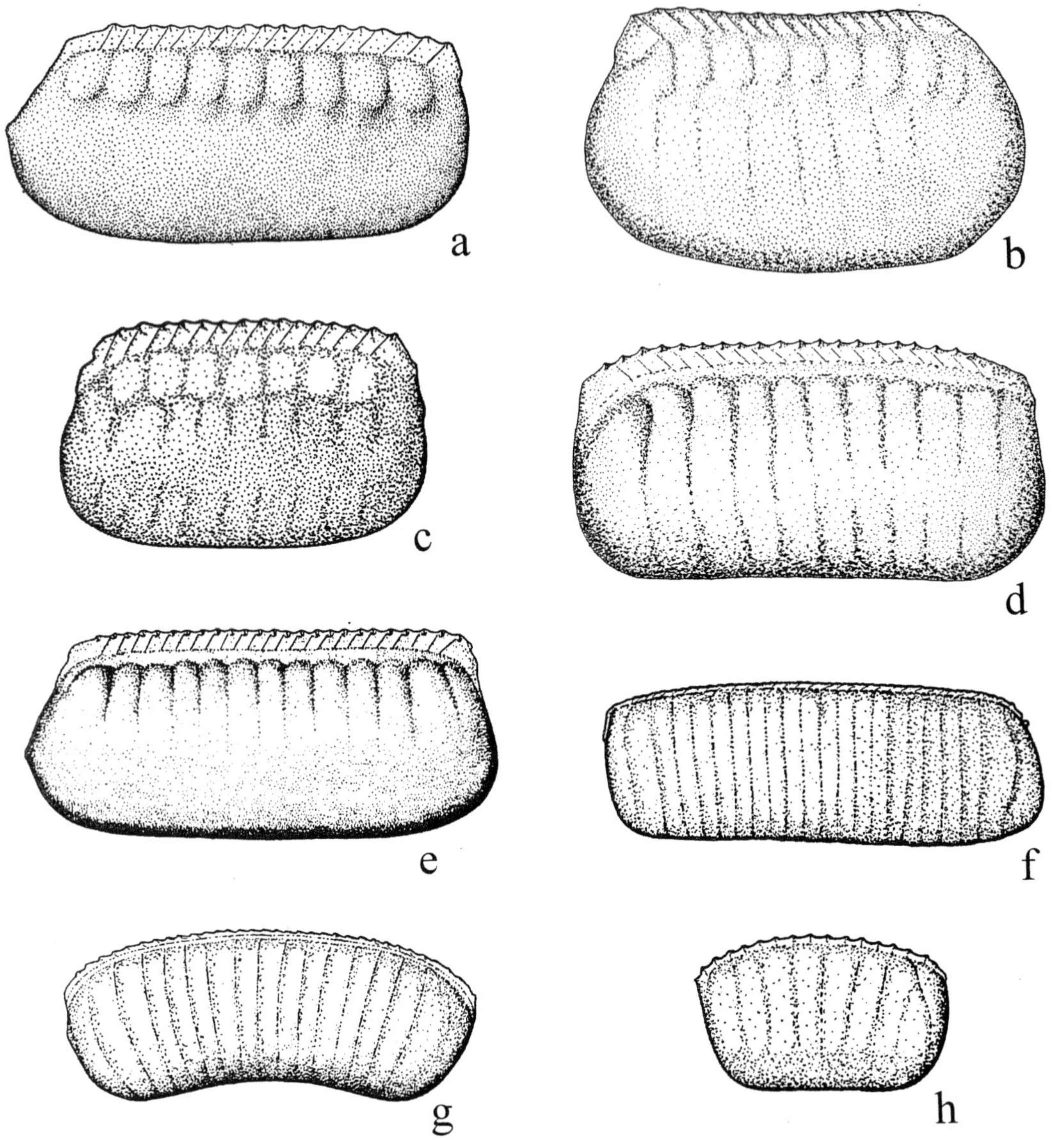

**Figure 4.1** *Blattaria oothecae.* (a) *Blatta lateralis*; (b) *B. orientalis*; (c) *Periplaneta americana*; (d) *P. brunnea*; (e) *P. fuliginosa*; (f) *B. germanica*; (g) *Parcoblatta* spp.; (h) *Supella longipalpa*.

brood sac and absorbed. The developing embryos are provided water and nutrients, and the nymphs hatch from the female's abdomen.

Early-stage nymphs resemble adult males in having abdominal sterna 8 and 9 visible, and 9 with styles. In females, these sterna gradually disappear from view and they are incorporated into the genital atrium. The number of segments of the antenna and cerci usually increase from stage to stage. First- and last-stage nymphs of some domiciliary cockroaches are distinctly colored or banded, and these characteristics are useful for species identification. In winged species, the wing buds appear late in development; usually this condition distinguishes late-stage nymphs, and wings appear on adults. Adults of brachypterous species, and short-winged females of some winged species are sometimes difficult to distinguish from large nymphs.

Aggregation is an important behavior of most cockroaches. First-stage nymphs remain close to the ootheca or the ovipositing female, as in ovoviviparous species *Rhyparobia maderae* and *Nauphoeta cinera*. Adult *Blattella germanica* and *Periplaneta americana* secrete an aggregation pheromone in the feces. Aggregation pheromone in *P. americana* attracts at low concentrations, and acts to inhibit movement at high concentrations. Pheromones probably function to mark suitable harborages or successful foraging routes, oviposition sites, or act to reduce crowding in favorable harborage sites. The odor of cockroach feces detected by humans is not a necessary component of the aggregation pheromone. Purified extracts that are perceived as odorless by humans are as attractive to cockroaches as the odorous material.

Mating in cockroaches is usually preceded by a simple courtship behavior, and typically it involves volatile and contact pheromones. These chemicals stimulate males to sexual excitement, which is usually characterized by wing fluttering and attempts at copulation, with males as well as females. *P. americana* males begin searching for a female within 20 min of being exposed to a sex pheromone. The adult male oriental cockroach responds to the American cockroach sex pheromone, and the male American cockroach responds to high concentrations of the oriental cockroach pheromone. In *B. germanica* there seem to be several short-range and contact pheromones involved in mating. Chemically, these pheromones are some of the largest among all known sex pheromones. During cockroach mating, the male and female are linked end to end, and copulation lasts from several minutes to several hours.

Males form a spermatophore and it is transferred during copulation to the female genital atrium. The empty spermatophore is discarded after a few hours or days, and before the female mates again. Subsequent ingestion of the spermatophore, into which the male has incorporated urates, by female *B. germanica* provides required nitrogen-based nutrients important for oothecae production. Parthenogenesis occurs regularly in some cockroach species or strains, such as in *Pycnoscelus surinamensis* strains in temperate regions. Parthenogenesis is known for some *Periplaneta* species, but the resulting adult cockroaches are abnormal and unable to reproduce.

Distribution of the principal domiciliary cockroaches is not indicative of their origin. Most pest species are native to northern tropical regions of Africa and southern Asia, and their present geographic range is indicative of their association with humans and their ability to adapt to alternative environments. Scientific and common names of species such as *Periplaneta americana*, *P. australasiae*, *Blattella germanica*, and *Blatta orientalis* are misleading, and not linked to their origin or predominant distribution. Commercial shipping of people, possessions, and food material was probably the major dispersal mechanism for these and the other domestic and peridomestic cockroach species. Spice, dye, aromatics, and fabric traders moved their material by ship from ports in southern Asia. These goods moved to the Mediterranean and into central Europe through a network of food warehouses that provided conditions suitable for survival of cockroaches and other six- and four-legged pests. Seaport storage in Asia and the eastern Mediterranean often combined warehouse and lodging. Under this arrangement, cockroaches that flourished in ships that came from tropical regions might find indoor habitats with temperatures and humidity suitable for survival, and the opportunity to spread further with household goods. Domiciliary cockroaches moved then and now from warehouses to markets and from markets to households, following a network of habitats and reservoir populations.

## Predators and parasites

Natural enemies of cockroaches include a variety of predators and parasites. Vertebrate predators include frogs, fish, reptiles, birds, and mammals. Among arthropods, the predators include dragonflies, mantids, reduviid bugs, carabid beetles, wasps, ants, scorpions, house centipedes, and spiders. Spiders found associated with cockroaches include the theridiid, *Steatoda grossa*, and the sparassid, *Heteropoda venatoria* in Hawaii. German cockroach populations often sustain the protozoan parasite, *Nephridiophaga blattellae* and the parasitic nematode, *Blattacola blattae*. The mite, *Pimeliaphilus cunliffei* (Pterygosomatidae) is an obligatory parasite of several species of cockroaches.

Hymenoptera egg parasites attack several cockroach species. Encyritidae: *Comperia merceti* is apparently limited to parasitizing the brownbanded cockroach, *Supella longipalpa*. It has been successful in management programs for this cockroach. Eupelmidae: *Anastatus blattidarium* is a parasite of the brownbanded cockroach. Eulophidae: *Aprostocetus hagenowii* attacks eggs of *Blattella germanica*, *Blatta orientalis*, *Periplaneta americana*, *P. fuliginosa*, and *P. australasiae*. Evaniidae: ensign wasps *Evania appendigaster* and *E. punctata* parasitize *Periplaneta* spp. and oriental cockroach. Hymenoptera predators of cockroaches are not limited to species occurring indoors, but attack inside and outside structures. Ampulicidae: *Ampulex compressa* is a cockroach-hunting wasp that is endemic in India and south Asia and extends into Africa and China. These wasps attack *Periplaneta americana* and *P. australasiae*, and often enter houses in search of their prey. They do not form nests but, after the female has stung the prey, it is dragged away to a hole or crevice. *Ampulex* and a related genus, *Dolichurus*, occur in central Europe. In Australia, species of *Aphelotoma* are predators of cockroaches in outdoor habitats. Sphecidae: *Tachyspex lativalvis* and others in this widespread genus attack cockroaches.

Cannibalism is known to occur in several cockroach species. Adult *Periplaneta* feed on eggcases in their habitat, and adults and nymphs of *Blattella* and *Blatta* feed on injured or weak individuals in crowded harborages, presumably where these individuals have little room to escape. Adult German cockroaches, and to a lesser degree nymphs, feed on molting individuals, but usually only nymphs older than the third-instar nymph. Late-stage nymphs are the subjects of attack, and molting adults usually suffer the greatest attack and mortality. There is apparently no correlation between population density and cannibalism in *Blattella germanica*.

### Classification

Current classification of cockroaches is based on adult morphology and egg-laying characteristics. The features used to establish the relationships are female genitalia and its musculature, the external male genitalia, structural aspects of the proventriculus (gizzard), and female egg-laying behavior. The order Blatteria is subdivided into two superfamilies, Blattoidea and Blaberoidea, which are further subdivided into five families and 20 superfamilies. Cockroach evolution seems to have followed two divergent lines, which are represented by the two superfamilies. The small superfamily, Blattoidea, includes many pest species: *Blatta orientalis*, *Eurycotis floridana*, *Neostylopyga rhombifolia*, *Periplaneta americana*, *P. australasiae*, *P. brunnea*, *P. fuliginosa*, and *P. japonica*. Females of these and other Blattoidea species do not turn the ootheca in the female genital chamber after it is formed. One of the most primitive living cockroaches is *Cryptocercus punctulatus*. It is a wood-eating species in the Blattoidea family Cryptocercidae. Blaberoidea is a large superfamily with many genera and species distributed worldwide. Females in the majority of Blattellidae families turn the ootheca after it is formed (Plectopterinae, represented by *Supella longipalpa*, are the exception), and female Blaberidae turn and retract the ootheca into a genital chamber, and eggs are incubated internally.

Common names for cockroaches often indicate their habits, location of their natural habitat, or the suspected location of their origin. Because they are common household pests around the world, common names are often regional and difficult to chronicle. The common names used here are considered the most frequently used.

## Blaberidae

These cockroaches are ovoviviparous; they turn and retract the formed ootheca and incubate it internally. In general, the reproductive potential of females is limited and nymphs develop slowly; some species are parthenogenetic. The legs are usually short and the femora of the front leg has one or more, short, robust spines. Many species have a hard and pitted integument and are adapted for burrowing.

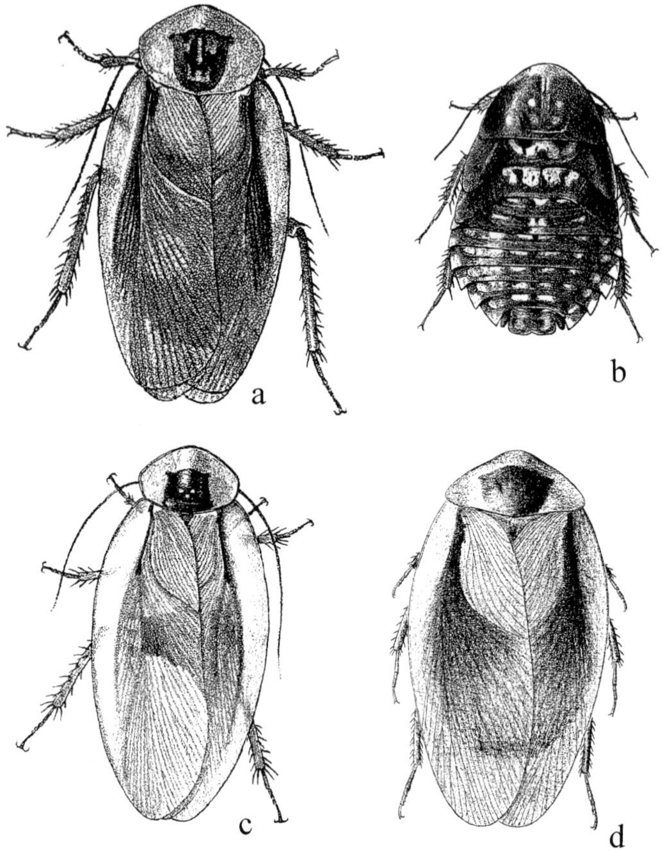


**Figure 4.2** Blattaria: Blaberidae. (a) *Blaberus craniifer* male; (b) *B. craniifer* nymph; (c) *B. discoidalis* male; (d) *B. giganteus* male.

**Death's-head cockroach, *Blaberus craniifer* (Fig. 4.2a, b)** Adult males are 43–50 mm long and females are 55–60 mm long; the body is brown to blackish brown. Wings extend by about one-third beyond the tip of the abdomen; front wings are dark brown to blackish brown, except for a pale brown region adjacent to the pronotum. The pronotum is large and has a wide, pale brown margin; the center is dark brown with yellowish-brown markings. Femora 1 has robust spines on the ventral margin; femora 2 and 3 have spines on the ventral margin. Pronotum markings and cuticle formation somewhat resemble human eyes, nose, and mouth. This feature is the origin of the common name. Nymphs have a large pronotum; yellowish orange spots are on the pronotum, thorax, and abdominal tergites. The ootheca is about 20 mm long and lightly sclerotized; it is pale brown to light brown and indentations show the position of the eggs; a distinct keel is absent. The ootheca contains about 34 eggs, and when formed it is retracted into the brood chamber. Development is completed in 257–277 days at 30–36 °C and there are 9–11 instars. Adults live 420–480 days. Natural habitats include ground vegetation and under rotting logs in wet forested areas, and in caves; they feed on plant material. In the urban environment, they occur in restaurants, food stores, and in houses. *B. craniifer* secretes an aggregation pheromone from mandibular glands, and this may promote large infestations in favorable harborages. This species occurs in the American tropics. It is established in southern Florida, where it occurs in firewood piles, refuse, and around buildings. This species is a household pest in Santiago, Chile and Havana, Cuba; it occurs in Mexico, Belize, and Dominican Republic. Adults are attracted to lights at night in Brazil.

During courtship the male of *B. craniifer* stands near a female and slightly raises his abdomen on his legs and makes trembling movements with the abdomen. Sometimes the male butts the female with his head or pronotum. The male moves his abdomen under the female, so that the female straddles the male abdomen. Once the male genitalia connect with the females, the pair move in opposite directions. They remain joined for about 4 h.

**Other *Blaberus*** There are several species that occur in the urban or agricultural environments. In the natural environment, *B. discoidalis* (Fig. 4.2c) has been taken under bark of a dead tree; in the urban environment it occurs in restaurants and food stores in Ecuador, in houses in West Indies and Puerto Rico. It is distributed throughout Central America. This species is also known from Cuba and Jamaica, and may be established in the USA, in Key West, Florida. *B. giganteus* (Fig. 4.2d) adults are 70–80 mm long; the front wings are pale brown and nearly translucent. Nymph development is 140–200 days at 30 °C and 60% relative humidity (RH); adults live as long as 600 days. *B. atropos* is native to South America, and has the death's-head markings on the mesonotum and metanotum; along with *B. boliviensis* it has been found associated with bananas.

***Calolampra irrorata*** Adult males are about 24 mm long and speckled brown; the head is not completely concealed by the pronotum; the mid and hind femora have spines. The wings extend beyond the tip of the abdomen. The female is about 20 mm long and 15 mm wide, and wingless. The females are usually in soil, but the males are active flyers, and are attracted to lights at night. This species occurs in Australia, and it was first collected in 1770 during Cook's first voyage to Australia.

**Hissing cockroach, *Gromphadorina laevigata*** Adults are about 6 cm long and the body is covered with scaly granulations. Males and females are wingless and somewhat slow-moving. Males are distinguished by a large round tubercle on each side of the pronotum, and a pair of styles at the end of the abdomen. The thoracic tubercles are used to batter other males as a part of courtship behavior. Adults and nymphs have strong spines on their legs. Disturbed adults and nymphs or males encountering other males make a hissing sound. This is produced when air is forcibly squeezed out of the spiracles on abdominal segment 2. Hissing sounds may also serve to attract females. The ootheca is 30–32 mm long and contains 32–48 eggs. The female produces an ootheca in 2–3 h, after which it extends about 25 mm from the tip of her abdomen. Then it is slowly drawn back into the internal brood sac, which is large and extends into the metathorax. Eggs are incubated internally for about 70 days, and nymphs hatch and exit the female's body over a period of about 2 days. Newly emerged nymphs usually remain grouped around and under the female for a short period. Natural habitats for this cockroach include wet or moist leaf litter; it is not known to be a household pest. This species is native to Madagascar.

**Madagascan hissing cockroach, *Gromphadorina portentosa*** Adults are about 10 cm long and blackish brown. Males and females are wingless. Similar to *G. laevigata*, when disturbed the adults and nymphs make a hissing sound. This is produced when air is forcibly squeezed out of the spiracles on abdominal segment 2. This species provides a form of maternal provisioning of newly hatched nymphs. Shortly after expelling the hatching ootheca, the female exudes from her abdominal tip a pale white substance. The newly hatched nymphs actively feed on this material. Integumentary gland cells lining the brood sac are the most likely source of the secretion. This species is sometimes sold commercially as a household pet. This species is native to Madagascar.

**Other *Gromphadorina*** This genus is native to Madagascar, and species are not known to occur naturally outside this location. Several species are used for physiological research and various educational exhibits and demonstrations, including *G. coquereliana*, and *G. portentosa*.

**Lobster cockroach, *Nauphoeta cinerea* (Fig. 4.3a, b)** Adults are 25–29 mm long and light gray. Wings are mottled or speckled, and do not cover the abdomen. The pronotum is pale brown and has irregular brown markings and a submarginal black band on each side; the posterior margin is rounded. The pronotum usually has a lobster-like color pattern, which is the origin of its common name. Nymphs have pale markings on

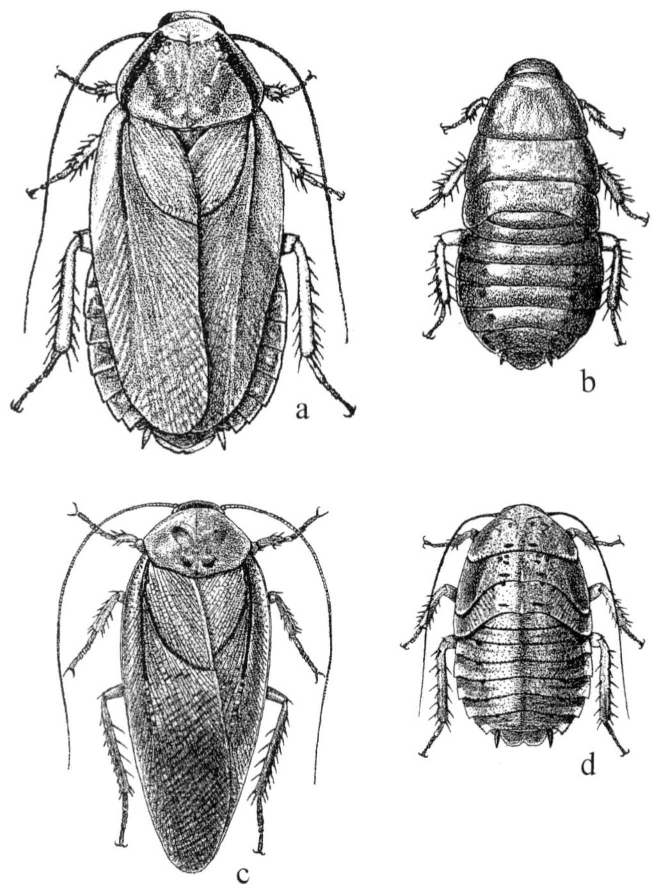


**Figure 4.3** Blattaria: Blaberidae. (a) *Nauphoeta cinera* female; (b) *N. cinera* nymph; (c) *Rhyparobia maderae* male; (d) *R. maderae* nymph.

the anterolateral margins of the abdominal tergites. The ootheca is 15–18 mm long, pale brown and curved; there are distinct indentations showing the position of the eggs. The ootheca contains 26–40 eggs and it is carried internally 36 days until the eggs hatch; fecundity is 1–6 broods. Nymphs hatch as the female extrudes the ootheca from the brood sac; they shed their embryonic membrane as they hatch, and eat both it and the ootheca. In the laboratory, the number of nymphs hatching increases from about 20 for the first eggcase to about 30 from the third, and then decreases. Newly hatched nymphs usually crawl beneath the female and remain there for about 1 h after hatching. Nymph development in isolation in the laboratory at 30–36 °C includes seven instars and takes about 73 days for males, and eight instars and about 94 days for females. Development in groups of nymphs takes about 72 days for males and about 85 days for females. Adult life span is about 344 days for females and about 365 days for males. The interval between the molt to adult and the production of the first ootheca is about 13 days. Males stridulate when courting nonreceptive females. The sound is produced by ridges laterally on the hind margin of the pronotum, which move against the costal vein at the base of the wings.

Natural habitats for this species are outdoor sites where it feeds on plant and animal materials. It is known to kill and eat individuals of the Cypress cockroach, *Diploptera punctata*. In the urban environment, it lives primarily around the outside of buildings, but occurs in houses and hospitals in Australia, and household huts in Sudan. Infesting populations occur in mills processing animal feeds, and in feed-storage rooms of commercial poultry operations. This species seems to prefer food containing fish oil. It is occasionally introduced into temperate regions with food materials and has been recorded in the UK and Germany in heated buildings. *N. cinerea* is probably native to East Africa, below the equator. It likely spread to other regions through trading ships and commerce. It was introduced into Brazil from Africa, and into Mexico from the Philippines. Current distribution is tropical regions of the world, including South America (Brazil), Cuba, West Indies, Mexico, Galapagos Islands, Hawaii, Australia, Philippines, Indonesia, Malaysia, Singapore, Mauritius, and Madagascar. It is established in the USA, around Tampa, Florida.

**Cuban cockroach, banana cockroach, green cockroach, *Panchlora nivea* (Fig. 4.4e, f)** Adults are 15–20 mm long. The body is somewhat flattened, and light or pale green to yellowish green (specimens in alcohol do not show the pale green color); integument is slightly translucent. Cerci are short and blunt, and the supraanal plate (epiproct) is distinctly bilobed. Wings extend beyond the abdomen in both sexes; cerci are short and broad. Nymphs are pale brown and somewhat flattened. The ootheca is 3–4 mm long and curved; it has indentations showing the position of the eggs. The ootheca contains about 46 eggs, and it is carried internally in the female until the eggs hatch in about 48 days at 24 °C. The interval between the oothecae production is about 63 days. Development in the laboratory at 24 °C takes about 144 days for males and about 181 days for females. Females live about 153 days at 24 °C. Courtship behavior is limited: the male moves his body sideways for several minutes, moves behind and backs up to the female to engage her genitalia, and then transfers a spermatophore.

Natural habitats include under the rotting bark of palm and coconut trees, and in moist shaded areas with abundant leaf litter; it is commonly found with bananas. In the urban environment, it occurs in peridomestic habitats, and it is attracted to lights indoors and outdoors. It does not infest structures. This species has been introduced to various parts of the world with shipments of bananas and other fruits. It is distributed in the western hemisphere tropics, and has been recorded

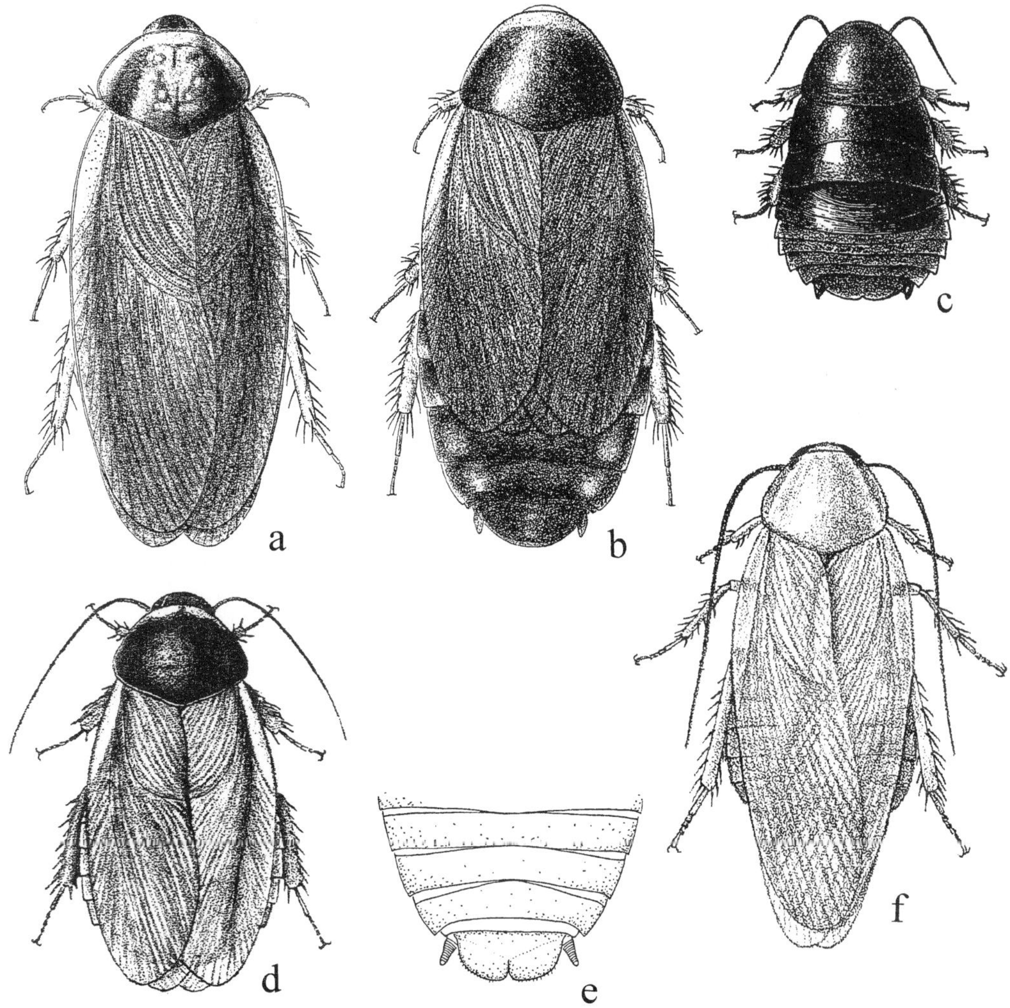

**Figure 4.4** Blattaria: Blaberidae. (a) *Pycnoscelus indicus* male; (b) *P. indicus* female; (c) *P. surinamensis* nymph; (d) *P. surinamensis* male; (e) *Panchlora nivea* male, posterior abdominal segments and supraanal plate; (f) *P. nivea* male.

from Mexico, Cuba, Puerto Rico, Trinidad, Greater Antilles, and Bahamas. It is widely distributed in Central America and northern South America. It is established along the Gulf Coast of the USA, from Texas to Florida.

***Panchlora irrorata*** Adult males are about 14 mm long and females are about 20 mm long. The body is flattened and light green to yellowish green; the integument is slightly translucent. The species name means sprinkled with dew, and probably refers to the integument. Wings extend beyond the abdomen in both sexes; antennae are not longer than the body. Last-instar nymphs are about 15 mm long and brown. The ootheca is 3–4 mm long, thin-walled and colorless; it is carried internally until the eggs hatch. The ootheca contains about 12–48 eggs, the number of undeveloped eggs ranges from 1–24; hatching occurs in about 63 days. Development in the laboratory takes about 196 days for males and about 207 days for females; adult males live about 108 days and females for about 130 days. Male courting behavior begins with rocking his body sideways; he then moves behind the female. When the male and female are in a linear position, end to end, he moves backwards, towards the female to make genital contact. Copulation lasts 20–25 min. This species is distributed in the Central American tropics, but it is carried to other regions on shipments of bananas or other fruits.

**Surinam cockroach, *Pycnoscelus surinamensis* (Fig. 4.4c, d)** Adults are 18–25 mm long. The body is dark and shiny brown to blackish brown. The pronotum is blackish brown, except for the anterior margin, which is pale brown. Wings are light brown, which gives the insect a bicolored appearance. Wings extend to the tip of the abdomen in both sexes; antennae are about one-third the length of the body. Thorax and abdominal segments 1–3 are glossy, but the dorsum of the terminal abdominal segments is dull and granulated. Nymphs are uniformly dark brown. The ootheca is 12–15 mm long and pale yellow to pale brown and membranous; it lacks a distinct keel and it is slightly curved; there are indentations showing the position of the eggs. The ootheca contains 14–48 eggs and it is carried internally until the eggs hatch in about 35 days at 18–34 °C; fecundity is 1–5 oothecae. Nymphs hatch as the female extrudes the ootheca from the brood sac. Development is through 8–10 instars and lasts 127–184 days at 18–24 °C. Adult females live about 307 days. In parthenogenetic populations, the first ootheca is formed about 7 days after the final molt. There are parthenogenetic and bisexual regional populations of *P. surinamensis*. Males are not known to occur in populations in the UK, continental Europe, and mainland USA. However, Hawaiian and some Asian populations are apparently bisexual.

In natural habitats adults and nymphs occur under stones, and all stages burrow in loose litter and in the top layer of soil; they occur in nests of wood rats in Texas. They feed on plants and cause economic damage to potato tubers in Haiti, pineapple roots in Hawaii, and tobacco plants in Sumatra. In the urban environment, this cockroach is common around buildings in warm climates. It occurs in greenhouses in parts of the USA, where it cannot live outdoors; it is occasionally found around potted plants in shopping malls, hotel lobbies, and similar locations. Adults and nymphs penetrate and find harborage in loose soil to a depth of 8–10 cm. *P. surinamensis* can be carried in nursery stock and house plants to regions outside its normal range. This species is probably native to the islands of the Malay Archipelago, now it is widely distributed in tropical and temperate regions. It is known from England, Scotland, and Germany; in North America it occurs in Texas, Florida, Louisiana, and South Carolina. It is established in Hawaii, and there is an isolated population in an urban area near Palm Springs, California. Bisexual populations are sometimes designated as *P. indicus* (Fig. 4.4a, b).

**Madeira cockroach, *Rhyparobia maderae* (= *Leucophaea*) (Fig. 4.3c, d)** Adults are 40–50 mm long and pale brown to light green, and slightly mottled. The posterior two-thirds of the front wings have a fishnet appearance. Wings cover the abdomen and there are two dark brown, curved lines at the base of the front wings. The pronotum and front wing margins are translucent. The male has a specialized organ on abdominal segment 2. Femur 1 lacks strong spines on the ventral margin, only a row of fine setae. Adults are slow-moving but readily fly; they emit an offensive odor. Adults stridulate by moving the posterior margin of the pronotum over the mesonotum. Nymphs have short spines along the dorsal posterior margin of each segment. Oothecae are 16–19 mm long, yellowish brown, and there are slight indentations showing the position of the eggs. The ootheca contains about 34 eggs and it is carried internally for 58–60 days. Nymph development in isolation in the laboratory at 30–36 °C includes seven instars and takes about 127 days for males, and eight instars and about 163 days for females. Development in groups of nymphs is complete in about 121 days for males, and about 150 days for females. Adult life span may be 2.5 years. Courtship begins with the male moving next to the female and rapidly moving his body up and down. In some instances the male raises the anterior portion of his body and strikes his abdomen against the substrate to produce a tapping sound. Adults mate 10–14 days after maturation, and the first ootheca appears in about 20 days.

In natural habitats in tropical regions this species occurs in large populations and feeds on plant material, including fruits such as banana and grapes. Banana is apparently a favorite food, and *R. maderae* is transported to other regions with shipments of this fruit. In the urban environment, it occurs indoors in Madeira, Windward Islands, Philippines, and Trinidad; in Puerto Rico, it is found in fruit stores, warehouses, and food markets. It is established in some heated buildings in New York City. It was probably introduced accidentally in household materials carried by immigrants from Puerto Rico. This species was first described from the island of Madeira, which is the origin of its common name. It is probably native to West Africa, south of the Sahara, where there are several other species of this genus. Current distribution of this species is most of the tropical regions of the world. This species occurs along the western coast of Africa, and it is established in Morocco, Spain, and Corsica. It is established in West Indies, Cuba, Jamaica, Puerto Rico, and Greater and Lesser Antilles. In South America, it occurs in coastal Brazil. It is known from Philippines, Indonesia, and Hawaii.

**Other Blaberidae** Most of the species in this family are minor pests and occur only occasionally indoors. *Epilampyra maya* has been reported from aquatic and moist habitats in Central America, and it occurs in and around houses in Acadia, Florida. *Phoetalia pallida* is widely distributed in the tropics, and it occurs in and around structures in the West Indies and the USA (Florida).

## Blattellidae

These cockroaches carry the eggcase externally. The majority of species turn the eggcase on its side after formation. This position is more streamlined and probably maintains their ability to move into narrow harborages. These cockroaches are characterized by legs that are long, slender, and with small and large spines; the ventral surface of the tarsi has spines. Many species have fully developed wings, but some have reduced wings in one or both sexes. Sternum 7 of the female is broad and rounded, not divided into lobes or valves. Tergum 10 of both sexes is usually triangular, and with long, tapering cerci.

**Asiablatta kyotensis** Adult males are 14–18 mm long and females are 16–18 mm long. The body is reddish brown; the legs and antennae are brown. Male antennae are about equal to body length; female antennae are slightly longer than the body. The ootheca contains about 32 eggs and hatching occurs in about 41 days at 25 °C. The ootheca surface has about 30 distinct longitudinal ridges extending from end to end. These ridges may be responsive to environmental humidity, and have a role in providing water to developing embryos. Nymph development is about 150 days, but extends to 2 years in dry habitats. Adult males live about 92 days and females live about 109 days. Natural habitats include moist or damp locations. This species was first described from Kyoto, Japan, but it is distributed in several areas of Japan and Korea.

**Asian cockroach, *Blattella asahinai*** Adults are 12–16 mm long; the body is light brown or yellowish brown and females are slightly darker than males. Wings cover the abdomen of the female and all except the tip of the male abdomen. Male front wings are about 3.3 mm wide and 11.6 mm long; female front wings are about 3.4 mm wide and 12 mm long. Males and females are capable of directed and sustained flight. This species closely resembles *B. germanica*, but is distinguished by its flight capabilities and morphological features. The pronotum of adults and nymphs has two longitudinal, parallel black bands. The ootheca is 6–8 mm long and brown; there are distinct indentations outlining each egg. Nymphs are blackish brown to black; the margins of abdominal segments are pale brown. Adults mate soon after maturation; the first ootheca appears in about 14 days, and it is carried until just prior to hatching. The initial preoviposition period is about 13 days and successive preoviposition periods are about 8 days. The ootheca contains 38–44 eggs; incubation is about 20 days at about 25 °C. Fecundity in the laboratory is 5–6 oothecae; in the field it is 1–2 oothecae. Nymph development at 25 °C is about 66 days for males and 68 days for females. Adult males live about 45 days and females about 103 days.

Natural habitats include grassy areas and leaf litter; they feed on dead insects, pollen and nectar in plants, and on honeydew. In the urban environment, they occur outdoors in vegetation. Adults fly to reflected light, such as illuminated areas and light-colored walls adjacent to a light source. They are capable of moving 10–200 m or more in controlled flight from wooded or grassy areas toward illuminated buildings. Flights occur at sunset if the temperature is 21 °C or above, and winds are light. Once indoors, they usually occur in moist locations, such as in kitchens and bathrooms, and in potted plants. *B. asahinai* was described in 1981 from sugarcane fields in Okinawa Island, Japan. It is widely distributed in Asia, including China, Malaysia, Myanmar, and India. It was first recorded in the western hemisphere in 1985–1986 near Lakeland, Florida. The original distribution area extended about 50 km, from the port of Tampa northeast to Lakeland. Habitats in three counties were initially infested, but by early 1989 there were 10 counties infested. Field populations of this species range from 30 000 to nearly 300 000 individuals per 0.4 ha.

**German cockroach, *Blattella germanica* (Fig. 4.1f; 4.5a, b, f)** Adults are 13–16 mm long, light brown to yellowish brown, and females are slightly darker than males. The pronotum has two reddish brown to black longitudinal stripes; the extent of the stripes is variable. The wings cover the abdomen of the female and all except the tip of the male abdomen; male front wings are about 2.8 mm wide and 10.6 mm long, while female front wings are about 3.1 mm wide and 11.2 mm long. The female abdomen is darker brown and much more round than that of the male. Powered flight is rare, but gliding flight is possible, especially for males. Male cerci are 11-segmented, and female cerci are 12-segmented. Nymphs are generally black; the margins of abdomen are brown and with a light stripe to the mid-dorsum. First-stage nymphs are 2–3 mm long and the body is dark gray or nearly black. Thoracic segments 2 and 3 are pale brown, and this band characterizes the newly emerged nymph. In late-stage nymphs the pale brown band narrows and extends in both directions to become a median longitudinal stripe.

The ootheca is 7–9 mm long, and with indentations showing the position of the eggs. The ootheca contains 35–48 eggs, and hatching is in about 14 days at 35 °C; fecundity is 4–8 oothecae, but 9 may be produced in some field strains. Females produced 27.9 nymphs per ootheca in summer and 9.7 in winter. After ootheca 4, the number of eggs per ootheca declines, and by ootheca 7 and 8 the eggs per ootheca is about 75% of the initial number. Once formed, the ootheca is turned so the keel is directed sideways, to the right or left, and it is carried until egg development is complete. Eggs become desiccated if the ootheca is removed from the female prematurely. The end of the ootheca that is held by the female is not as sclerotized as the other end, and it is permeable to water. The ootheca is provided

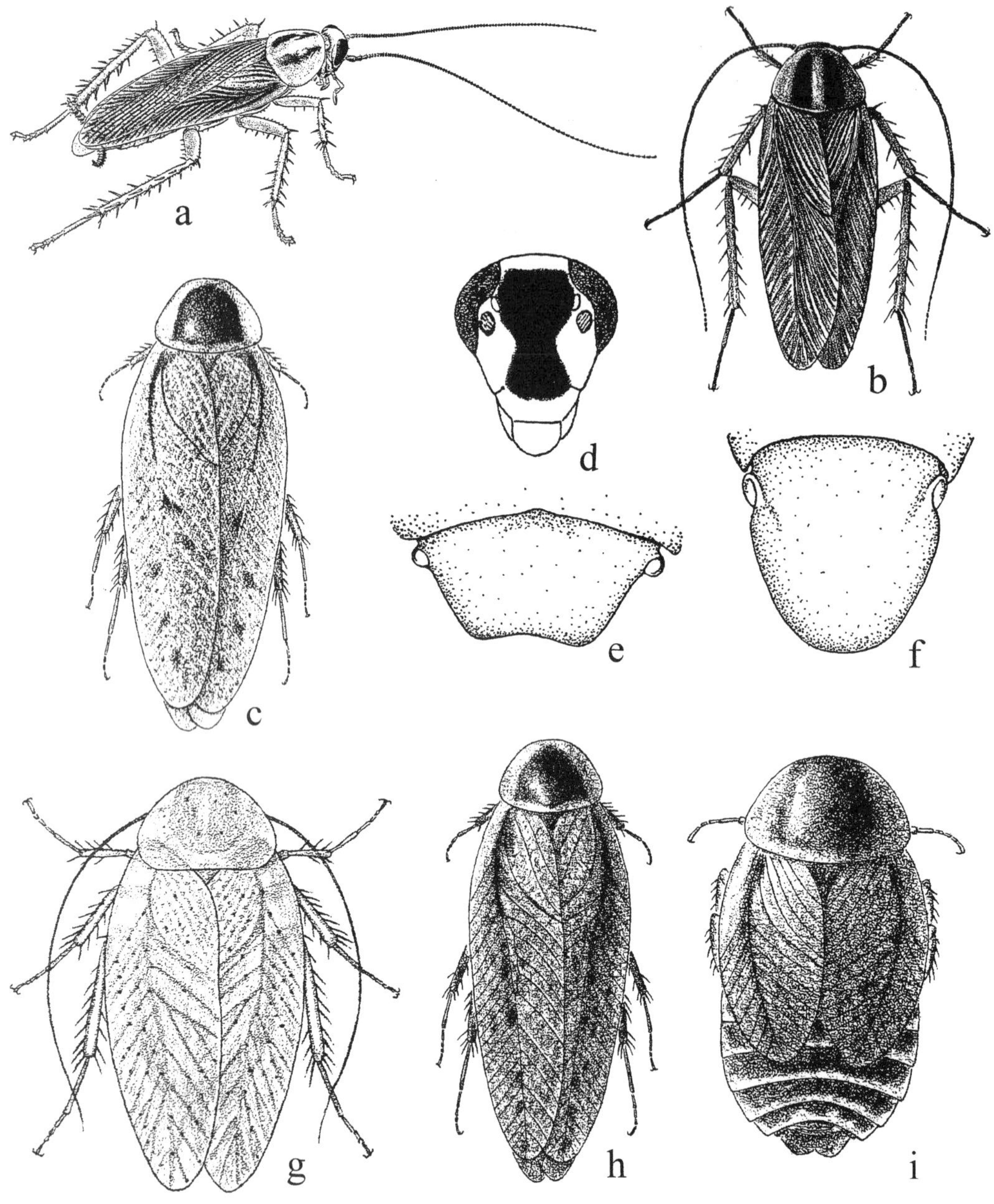


**Figure 4.5** Blattaria: Blattidae. (a) *Blattella germanica* male; (b) *B. germanica* male; (c) *Ectobius sylvestris* male; (d) *B. vega*, head anterior; (e) *B. vega* male, supraanal plate; (f) *B. germanica* male, supraanal plate; (g) *E. pallidus* male; (h) *E. lapponicus* male; (i) *E. lapponicus* female.

with water by the female and this is critical for development. A few days before hatching a green band of dark spots of condensed yolk develops along each side of the ootheca. The band intensifies as the time of hatching approaches. At that time, the female deposits the eggcase, and hatching occurs within 24 h; hatching may occur while the female is carrying the eggcase. Development is 54–215 days at 24 °C, and 60 days for males and 65 days for females at 35 °C and 90–95% RH. Gregarious behavior of nymphs, which is encouraged by an aggregation pheromone, favors their growth and development. The number of nymph stages is 5–7 in males and 6–7 in females. Nymphs with relatively small bodies undergo six molts and those with large bodies have five molts. Adults mate 7–10 days after maturation. Males mate often; females in field populations mate once or twice. Adults live about 200 days at 21 °C. Survival without food or water at 27 °C is about 8 days for males and 12 days for females; survival with water is 10 days for males and 42 days for females.

Natural habitats for this species are probably moist leaf litter in forested areas; however there are not known natural populations. Related species in southern Asia occur in caves and in ground debris. In the urban environment this species lives only indoors. In tropical climates, it prefers indoor sites with high humidity, food, and harborage. Adults can move about in space 1.6 mm in width or depth, but they seem to prefer harborages that are 4.8 mm. The spatial distribution of adults and nymphs is influenced by the aggregation behavior induced by a pheromone. Occupied harborages are marked with the pheromones contained in the feces, and the behavior-manipulating substance remains active for at least the period of one generation of *B. germanica*. The concentration of aggregation pheromone may regulate the density of individuals in local harborages. A repellent reaction occurs when the concentration of aggregation increases above a certain level, and individuals disperse rather than aggregate. Short migrations of a large number of individuals have been reported, but this behavior is not common. Distances covered by migrations range from 500 m to 1 km, and the migrating groups consist mainly of adult females.

Sexually mature male German cockroaches are generally capable of distinguishing between adult males and females, probably based on volatile and contact sex pheromones produced by the female. When a male *B. germanica* encounters another German cockroach, he investigates it with his antennae. If the other is a mature female he touches the female's body with his antennae. They then rub each other's antennae in a vibrating manner while in a head-to-head position. After a sideways motion of the abdomen, the male turns around and positions the tip of his abdomen close to the head of the female. At this time he extends his wings upward to nearly a right angle to its body. The factor responsible for eliciting the wing-raising behavior in males is a sex pheromone contained in the cuticle wax of fully developed virgin females. The raised wings expose the male abdomen, and the female palpates with her mouthparts the small glands on abdominal tergites 7 and 8, and she imbibes the material excreted from these tergal glands. After this the male moves his abdomen under the female, and extends his genitalia to contact the female's genitalia. Once joined, the male moves out from under the female, and they remain attached in a linear position for about 90 min.

The success of this cosmopolitan species as a domestic pest is linked to features of its biology and habits, and to some of the physical and environmental features common to human dwellings around the world. Adults and nymphs are relatively small and able to utilize the cracks and crevices common in households. Female fecundity and the number of eggs per ootheca are high, and the limited activity of gravid females increases the potential success of each ootheca. *B. germanica* has the unique capacity among cockroaches to develop physiological and behavioral resistance to insecticides. It seems to have the ability to acquire resistance to representatives of nearly all the major classes of chemical insecticides. The resistance-related mechanisms utilized by this cockroach include enhanced metabolic degradation, reduced rates of cuticular penetration, diminished target site sensitivity, and behavioral changes that enable avoidance of pesticide residues. In many cases, several of these resistance factors exist concurrently. High levels of resistance in this cockroach may be short-lived, but low-level resistance may persist in field populations.

***Blattella nipponica*** Adults are 12–14 mm long and the body is brown to dark brown. It closely resembles *B. germanica*, except that the adults fly readily. The preferred habitat is under leaves in forested areas, and under decayed vegetation in cultivated areas in central Japan. The ootheca is about 6 mm long and contains 12–43 eggs. Hatching occurs in 17 (June) to 42 (September) days; fecundity is about five oothecae. Adults emerge in early June, and first-stage nymphs occur in July and September. There is one generation per year, and the adults live 5–6 months.

**Field cockroach, *Blattella vega* (Fig. 4.5d, e)** Adults are 10–12 mm long. The body is yellowish brown, and with a dark brown to black region between the eyes and extending to the mouthparts. Longitudinal stripes on the pronotum are blackish brown, widely separated and sharply defined. Nymphs are pale yellow, and have dark brown cerci; large nymphs are yellowish orange. The ootheca is 5–6 mm long, light brown, and with distinct indications of the egg compartments. The first three oothecae contain about 28 eggs, but the number of eggs per ootheca drops in the succeeding oothecae. The interval between oothecae is about 24 days. Hatching is in about 20 days; fecundity is 1–8 oothecae. Development takes 45–56 days at 30–36 °C; at 30 °C males develop in about 60 days and females in about 54 days. Newly hatched and small nymphs sometimes crawl on the back of females. Adults live 100–150 days. Adults mate 3–5 days after maturation, and an ootheca appears 2–3 days later.

Natural habitats include under stones, and in moist leaf litter where it feeds on decaying vegetation; it has been found in desert locations. Recent records from eastern Texas and southwestern Louisiana are from salt marsh or coastal habitats. In the agricultural environment, *B. vega* occurs in irrigated fields. In the urban environment, it occurs around the perimeter of buildings and moves indoors during dry weather. It is active during the day, but also occurs at streetlights at night. This species is probably native to semiarid regions of southern Asia; it occurs in Afghanistan, India, Pakistan, Sri Lanka, Mexico, and in southwestern USA. It is frequently found in irrigated fields in southern Arizona and adjacent areas in southern California.

**Other *Blattella*** There are more than 40 species in this genus worldwide. Species are distributed from southern Asia, Japan, various Indo-Malaysian and Pacific islands, and Africa; and including several species from Africa, now placed in the genus *Symploce*. Most of these species occur outdoors and are not domiciliary pests. *B. lituricollis* occurs in Myanmar, China, Taiwan, Philippines, and on various islands north to Japan and east to Hawaii. It is occasionally a household pest, though it occurs mostly outdoors. It occurs in Hawaiian agricultural fields, and sometimes enters buildings. In Japan, it occurs around the outside of buildings, and indoors in kitchens.

**Tawny cockroach, spotted Mediterranean cockroach, *Ectobius pallidus* (Fig. 4.5g)** Adults are 8–9 mm long and uniformly pale brown. Wings extend to the tip of abdomen in both sexes; males and females are capable of flying. Front wings are wide and extend over the edges of the thorax and abdomen; there is a triangular area at the tip of the hind wings without veins and cross-veins. Nymphs are pale brown to dark brown. The ootheca is 2–3 mm long and dark brown; the ventral surface usually has a distinct longitudinal concavity (visible in ventral view) where the sides join. The keel has rounded teeth. The ootheca is turned on its side and carried by the female for 1–2 days. Oothecae are produced in fall and deposited in leaf litter, and hatching occurs the following spring. Nymphs develop through 5–6 instars during the spring and summer; nymphs overwinter and become adults the following spring. Natural habitats include wooded sites, but also grassy areas, heaths, and sand dunes. This species does not usually live in the urban environment, but they fly to lights at night. It occurs in western, central, and southern Europe. In the USA, it has been found in Massachusetts and Michigan.

**Other *Ectobius*** These cockroaches live outdoors and only occasionally enter structures. *E. lapponicus* (Fig. 4.5h, i) adults are about 10 mm long; the males are dark brown and the females are pale brown. Wings of the female do not cover the abdomen. The male has long wings, is capable of flight, and may occur indoors. This species occurs in Europe, including southern UK. *E. panzeri* adults are 5–7 mm long and pale brown to dark brown. The male wings extend to the tip of the abdomen; the female has short wings. This species occurs primarily in coastal regions of Europe. *E. sylvestris* (Fig. 4.5c), the forest cockroach, is closely related to *E. pallidus* and may occur indoors.

**Nicaraguan cockroach, *Ischnoptera bergrothi*** Adults are 15–18 mm long. The body is reddish brown dorsally and black ventrally; the pronotum has two distinct spots. Wings cover the body; both sexes are strong flyers. Nymphs are similar to adults. This species is native to Central America, from Panama to Nicaragua, but also occurs in outdoor populations along the Gulf and lower Atlantic coastal plain from Texas to Georgia and Florida. Indoor infestations are not common, but they fly to lights at night and occur around the outside of structures. A related species, *I. nox*, is primarily tropical and is currently known from Panama and Florida. In Fort Lauderdale, Florida, populations occur on extensively landscaped grounds of resorts.

**Pennsylvania woods cockroach, *Parcoblatta pennsylvanica* (Fig. 4.6b, c)** Adult males are 22–30 mm and females are 13–20 mm long. Males and females are light brown, and the thorax and front wings have pale brown margins. Wings of the male extend to the tip of the abdomen; female wings are usually small pads and of variable length; only the males fly. Nymphs are dark brown. The ootheca is 10–12 mm long and yellowish brown, it has a distinct keel, and there are indications of the egg compartments. The ootheca contains 32–36 eggs, and about 26 eggs hatch in 32–36 days at 24–27 °C. Oothecae are produced every 5–9 days and they are carried for 1–3 days; fecundity is about 30 oothecae. Nymphs hatch in summer and usually overwinter and complete development in the spring of the following year. Under unfavorable conditions, development is completed later in the second year.

Natural habitats include moist leaf litter and under the bark of decaying logs in hardwood and pine forests of eastern USA, and in undisturbed areas. In the urban environment this species occurs around the perimeter of buildings, in woodpiles and accumulated forest debris. They live around outbuildings in suburban areas. Males have long wings and are capable of flights over 30 m. This species is attracted to lights at night. Males occur around the outside of structures and indoors in May and June; females are abundant in July. Females and nymphs may be carried indoors with firewood or other material. These cockroaches rarely persist indoors, perhaps because of their high humidity requirements and food preferences. Nymphs remain active throughout the year, even at 0 °C.

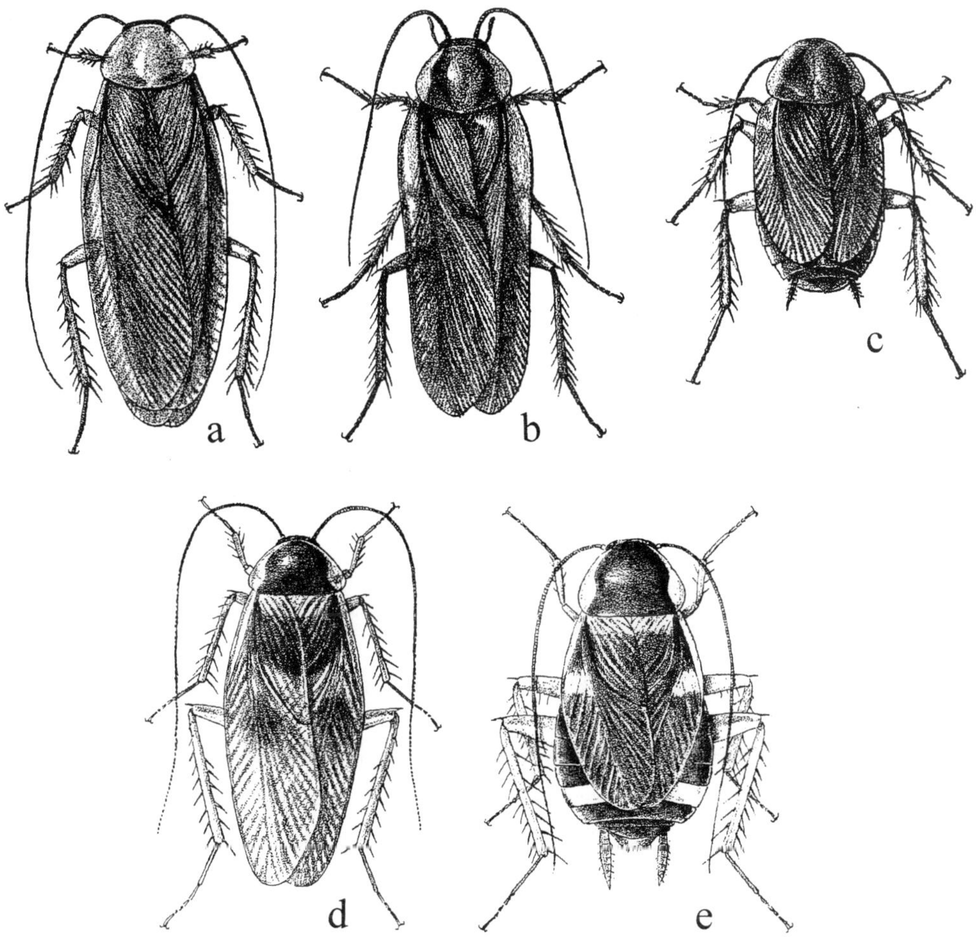


**Figure 4.6** Blattaria: Blattellidae. (a) *Parcoblatta virginica* male; (b) *P. pennsylvanica* male; (c) *P. pennsylvanica* female; (d) *Supella longipalpa* male; (e) *S. longipalpa* female.

*P. pennsylvanica* is widely distributed in the eastern and southern USA, and it occurs in southeastern Canada.

**Other *Parcoblatta*** Species in this genus are common in wooded areas in North America. They are found in decaying logs, under the bark of dead or down trees, and in woodpiles. These cockroaches are often attracted to lights at night. Adults and oothecae (Fig. 4.1g) can be accidentally carried indoors with firewood. There are 12 described *Parcoblatta* species, but aside from *P. pennsylvanica*, only a few are known to be associated with buildings or occur indoors. *P. americana* occurs in Mexico and southwestern USA and it is occasionally reported near buildings and at lights. *P. divisa* occurs in southeastern USA and west to Texas and Kansas. It is sometimes reported invading houses built in wooded areas. *P. lata* and *P. virginica* (Fig. 4.6a) occur in eastern USA and have been reported indoors, at lights, and under wooden signs on trees.

**New Zealand black cockroach, kokoroihi, *Platyzosteria novae-seelandiae* (= *P. novae-zealandiae*)** This species is uniformly blackish brown, and brachypterous. Adults and nymphs occur under loose bark of trees and down logs, or among stones on the ground; they are sometimes carried indoors. They are predaceous on other insects. This species occurs in New Zealand.

**Australian woods cockroach, *Shawella couloniana*** Adult males are 15–18 mm long and females are 20–22 mm long; both sexes are shiny and uniformly dark brown. Wings of the male cover about half of the abdominal segments, while wings of the female cover the basal 2–3 abdominal segments. Nymphs are shining, dark brown. The ootheca is 6–8 mm long, brown, and contains about 20 eggs. This species is limited to sites in Australia and New Zealand. It is an outdoor species, but occasionally occurs indoors.

**Brownbanded cockroach, *Supella longipalpa* (Fig. 4.1h; 4.6d, e)** Adult males are 13–14.5 mm long and females are 11–12 mm long; the body is brown to yellowish brown, and with distinct pale brown banding. Nymphs and adults have long palps. Wings of the male cover the abdomen, but rarely reach the tip of the abdomen in the female. Adult males fly readily when disturbed, but females do not fly. Pronotum lateral edges are translucent, the remainder is dark; the pronotum often has a pale area in the center. The common name is derived from two dark brown, transverse bands on the mesonotum and abdominal tergites. Nymphs are banded light and dark brown; one band is across the posterior margin of the mesonotum, the other across abdominal segment one, and extends dorsolaterally along the next few segments.

The ootheca is about 4 mm long, brown to reddish brown, and has a prominent keel; it is curved and there are indentations showing the position of the eggs. The ootheca contains 14–18 eggs; hatching occurs in about 96 days at 23 °C, 74 days at 25 °C, 43 days at 27.5 °C, and 37 days at 30 °C. The number of eggs per ootheca remains nearly unchanged during the life of the female. Fecundity is 10–20 oothecae. The ootheca is deposited 24–36 h after development, and the female glues it to the substrate. Many females often deposit oothecae at the same location. In large, household infestations, there may be clusters of large numbers of oothecae. Nymph development in isolation in the laboratory at 30 °C includes eight instars and lasts about 114 days for males, and six instars and about 69 days for females. Development in groups of nymphs is about 54 days for males and about 56 days for females. Nymph development is about 161 days for males and 162 days for females at 23 °C, and at 29 °C it is 90 days for males and 95 days for females. Optimum development is above 27 °C. Adult life span is about 115 days for males and 90 days for females. Adults mate 3–5 days after maturation, and the first ootheca appears in 7–10 days. Survival without food or water at 27 °C is about 9 days for males and 12 days for females; survival with water is 10 days for males and 14 days for females.

Natural habitats are probably leaf litter in nonforested regions of Africa north of the Equator. In the urban environment, it occurs outdoors and indoors; in warm climates, it is abundant year-round. In infested buildings and houses it often inhabits all rooms, and is common on furniture and in drawers of chests, and this habit has earned it the common name of furniture cockroach (*supellex*, Latin for household goods and furniture). It seems to prefer locations high on the walls of heated rooms, and favored oviposition sites include behind picture frames. Males often fly around lights in houses, and they have been observed flying outdoors during the day.

This cockroach was described as *Blatta longipalpa* in 1798 from eastern India. It was described again in 1838 as *Blatta supellectilum* from specimens collected in Mauritius. Evidence from related species and early records of *B. longipalpa* indicates that it is native to Africa. It was in Cuba in 1862, and introduced into North America (Florida) in about 1903; it spread to nearly all the contiguous states by 1967. It was first recorded in Australia in 1924, and in the UK after 1960.

***Symploce pallens* (= *Symploce lita*)** Adults are 13–16 mm long and uniformly brown. Wings of the male extend to the tip of the abdomen, and female wings extend to a third of the abdomen. Male abdominal segment 6 has a setose and slightly sculptured area. This species is distributed in the tropical regions of the world, and is a pest in Malaysia and other southeastern Asian countries; it is also known from the USA. It occurs in structures in Key West, Florida.

**Other Blattellidae** *Cariblatta lutea* is known from Cuba and southeastern USA. It has been found around houses and disturbed areas, particularly in lawns and in leaf litter. *Euthlastoblatta gemma* occurs in the Bahamas and southeastern USA. It occurs in arboreal habitats, such as behind wooden signs on trees, and in and around houses in southern Georgia. *Latiblattella rehni* occurs in the Bahamas, Cuba, and northern Florida, where it has been found in house attics. *Lupparia vilis* adults are 10–12 mm long, brown, and distinctly flattened. It occurs in Iwo Jima, China, Thailand, Malaysia, and Indonesia; it has been reported as a household pest in Iwo Jima.

## Blattidae

These cockroaches carry the ootheca upright and without turning after it is formed. Species in this family are characterized by the presence of spines on the anterior margin of the front femur; the mid and hind femora have spines on the anterior and posterior margin. Sternum 7 of the female is large; its posterior edge is modified to form two lobes or valves, and these lobes are partly separated from the remainder of the sternum by a transverse suture. Tergum 10 of both sexes is rectangular, rounded, or triangular; the cerci are often short and flattened.

**Turkestan cockroach, *Blatta lateralis* (= *Blatta tartara*, *Shelfordella tartara*) (Fig. 4.1a; 4.7b, c)** Adult males are 19–23 mm long and females are 22–25 mm long. The male is light brown to yellowish brown; the pronotum is reddish brown centrally, and has wide margins of transparent yellow.

Male supraanal plate (epiproct) has the apex broadly notched. Wings extend beyond the abdomen; the outer margins of the wings are pale yellow. Female is blackish brown to black and brachypterous. Pronotum has pale yellow wing stripes and scattered yellow markings. The front wings of the female are separated by a distance less than the width of the wing. Both sexes are capable of flying. Large nymphs have the anterior half of the body yellowish red; the posterior half is dark brown. The ootheca is 9–12 mm long, brown to blackish brown, rounded at one end, and dorsally truncated at the other end. Nymphs develop in 118–137 days at 30–35 °C. Adults live about 1 year. Natural populations of this species occur in desert regions in North Africa, from Libya eastward to Asia. In the urban environment, it occurs around the outside of buildings and in underground pipes of sewer systems. It is distributed in Egypt, Saudi Arabia, the Sudan, Israel, and Iran, Iraq, and Afghanistan. It may have been introduced to the USA on military equipment from the Middle East. It is known to occur indoors and outdoors in Texas, and other sites in southwestern USA.

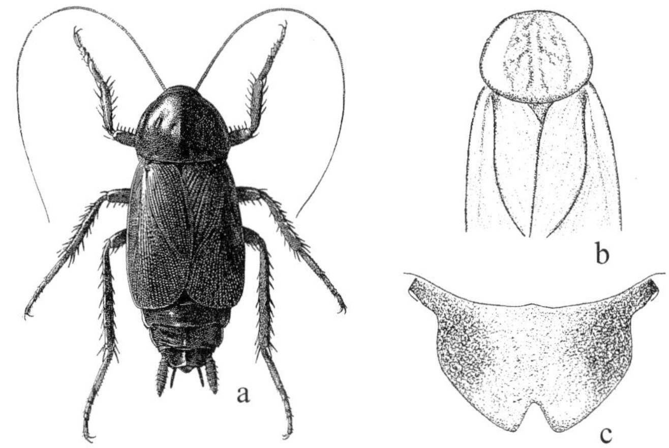


**Figure 4.7** Blattaria: Blattidae. (a) *Blatta orientalis* male; (b) *B. lateralis* adult, pronotum and elytra portion; (c) *B. lateralis* male, supraanal plate.

**Oriental cockroach, waterbug, *Blatta orientalis* (Fig. 4.1b; 4.7a)** Adult males are 17.5–28.7 mm long and females are 20–27 mm long. The body is shiny, and reddish brown to blackish brown. Wings of the male cover two-thirds of the abdomen, while female wings extend slightly past the thorax; neither sex is capable of flying. Tarsi of adult females and nymphs have a reduced arolium. Without this cushion-like pad between the claws, they have difficulty climbing smooth surfaces. The arolium is of variable size in adult males. Nymphs are reddish brown; first-instar nymphs are about 6 mm long and pale brown. The ootheca is 10–12 mm long and blackish brown; egg compartments are not distinct. The ootheca contains 16–18 eggs, and hatching occurs in 42 days at 29.5 °C and 81 days at 21 °C. Females do not produce oothecae in environments where the temperature is 15 °C or below. Oothecae lose viability when exposed to temperatures of 0 °C or below. Oothecae are produced at intervals of 1–2 weeks; fecundity is 6–8 oothecae. Females carry the ootheca for about 30 h after it is formed. Then it is deposited or attached to the substrate in a protected location. Nymph development is through 7–10 instars; at 22 °C it takes 515 days for males and 542 days for females; at 28 °C it is 288 days for males and 310 days for females; and at 30 °C it is 164 days for males and 282 days for females. Females live 35–190 days, males 112–160 days. Adults appear in May or June, and die in July or August. Parthenogenesis occurs, and some of the eggs in these oothecae develop and hatch; the resulting nymphs are females. A small number of these nymphs reach maturity. Survival without food or water at 27 °C is about 11 days for males and 13 days for females; survival with water is 20 days for males and 32 days for females.

During courtship the male moves around the female with his abdomen extended. When the male is in front of the female, he backs under her. As the female's head advances along the male's abdomen, he raises and expands both pairs of wings, and the female's mouthparts contact the male's abdominal segments. Then he continues to extend his abdomen under the female and brings his genitalia to contact hers, and if successful, he turns from under the female and the joined pair face in opposite directions. The mating process lasts about 1 h. When the ootheca first emerges from the tip of the female's abdomen it is yellowish white, and then turns reddish brown. It is not turned and the keel remains upright. Formation of the ootheca is completed in 24 h, and it is carried usually 1–2 days, and up to 7 days before it is deposited. It is placed in a sheltered place or the female attaches it to a substrate with an oral secretion. Oothecae may be partially or completely covered with small particles of debris from the substrate.

Natural habitats for this species probably include leaf litter and debris in areas with warm summer temperatures and a moderate winter. Natural populations occur on the Crimean Peninsula and in the region around the Black Sea and Caspian Sea. In the urban environment, it occurs indoors and outdoors. Large populations are typically in basements, cellars, crawl spaces, and in underground sewer pipes, and urban landfills. An aggregation pheromone in the feces promotes large populations and crowding in suitable harborages, and probably reduces the movement of adults and large nymphs. The association with cool and damp basements is probably the origin of the common name, waterbug. The preferred temperature range for *B. orientalis* is 20–29 °C, and this species is most common in north temperate regions, and less in tropical regions. Adults and large nymphs are active in early spring when daytime temperatures are 10–15 °C. Activity and pest status of this cockroach usually peak in the spring, typically from May to early July, when adults and nymphs move indoors.

**Gisborne cockroach, *Cutilia semivitta* (= *Drymaplanta*)** Adults are about 40 mm long and 12 mm wide; they are wingless and somewhat flattened. Body is shiny, dark reddish brown to black and with a pale yellow to white band on the sides of the thorax. Nymphs are light brown dorsally and yellowish brown ventrally. This species lives outdoors in cracks and crevices in rotting wood in natural areas; in the urban environment it is found in and around poultry houses, dog kennels, and under debris around buildings. It was introduced to New Zealand from Australia, and first found in Gisborne. It is now spread over the North Island and in environs of Nelson, South Island. It may occur indoors during winter months.

**Stinking cockroach, Florida wood cockroach, *Eurycotis floridana*** Adults are 30–40 mm long. The body is dark brown to blackish brown; recently molted individuals are reddish brown. Front wings extend slightly past the mesonotum, while hind wings are absent; the adults do not fly. The first segment of the hind tarsus is shorter than segments 2–5 combined, and the pulvilli of segments 2 and 3 are large. The male has short cerci and the supraanal plate (epiproct) is slightly notched. Medium and large nymphs have yellow margins on the meso- and metathorax. Adults mate about 18 days after maturation, and the first ootheca appears in about 55 days. The ootheca is 14–16 mm long, dark brown, and has distinct indentations that show the position of the eggs. The ootheca contains 21–23 eggs, and oothecae are produced in about 8-day intervals; hatching occurs in about 50 days at 30–36 °C. In natural habitats, oothecae are often buried in the soil or in decaying logs. Parthenogenesis occurs in this species, but the nymphs do not develop to adults. Nymph development is 286–302 days at 27 °C. Adults, but not the nymphs, emit an extremely foul odorous secretion when disturbed. This secretion is produced by glandular cells and stored as a yellow liquid in a bilobed sac; it is ejected through a pore in the sternal, intersegmental membrane of abdominal segments 6 and 7. A liquid spray can be ejected 2–3 cm to about 1 m. The spray irritates the eyes of humans, and it is toxic to the cockroach if it is confined in a small container.

During courtship the wingless male stands near the female and repeatedly moves his body from side to side. While doing this, he extends his abdomen slightly to expose the intersegmental membrane between abdominal segments 6 and 7. The female applies her mouthparts to the male's dorsum starting near the end of the abdomen, and progressing up to abdominal segment 1, on which is located a small glandular area bearing a patch of setae. While the female is palpating the glandular area, the male couples his genitalia with hers.

Natural habitats for this species include the cavities under the bark and the decaying holes in dead trees, stumps, and woodpiles, and sometimes in leaf litter. In the urban environment, it occasionally enters buildings, but it does not become established indoors. *Eurycotus* species occur in West Indies and tropical regions of Central and South America. *E. floridana* is recorded from USA, specifically Florida, Georgia, Alabama, and Mississippi.

**Vagabond cockroach, *Pelmatosilpha larifuga*** Adults are 28–31 mm long and they are dark reddish brown to black. Lateral margins of the pronotum and the front wings have a pale yellow band. Both sexes are brachypterous and have the posterior margins of the abdomen exposed behind the wings. Species in this genus occur in Central and South America, and the West Indies. Adults and nymphs of *P. larifuga* are often found on bananas shipped from West Indies to other regions of the world, and they occur in warehouses and food stores.

**American cockroach, *Periplaneta americana* (Fig. 4.1c; 4.8a, f)** Adult males are 34–53 mm long and female are 29–36 mm long. The body is shiny, reddish brown to brown; pronotum has a yellowish white margin with dark brown interior. Wings extend 4–8 mm beyond the abdomen in males, and as long as the abdomen in females. Adults readily fly when the temperature is above 21 °C; they usually travel short distances, but sustained flight is possible and they fly to lights at night. The male has a pair of styli and a pair of cerci at the end of the abdomen. Cerci are 18- or 19-segmented, and tapered at the tip. The male supraanal plate (epiproct) is translucent, apically rounded and deeply notched. Females lack styli, and their cerci are 13- or 14-segmented. The first-instar nymph is uniformly pale brown, and the tips of the maxillary and labial palps are white; antennae are as long as the body and pale brown. Wing pads develop during the third- or fourth-instar nymph, and the sixth-instar nymph has pale brown patches on the pronotum; wing pads on the last nymph instar are about 7 mm long. The sex of the early

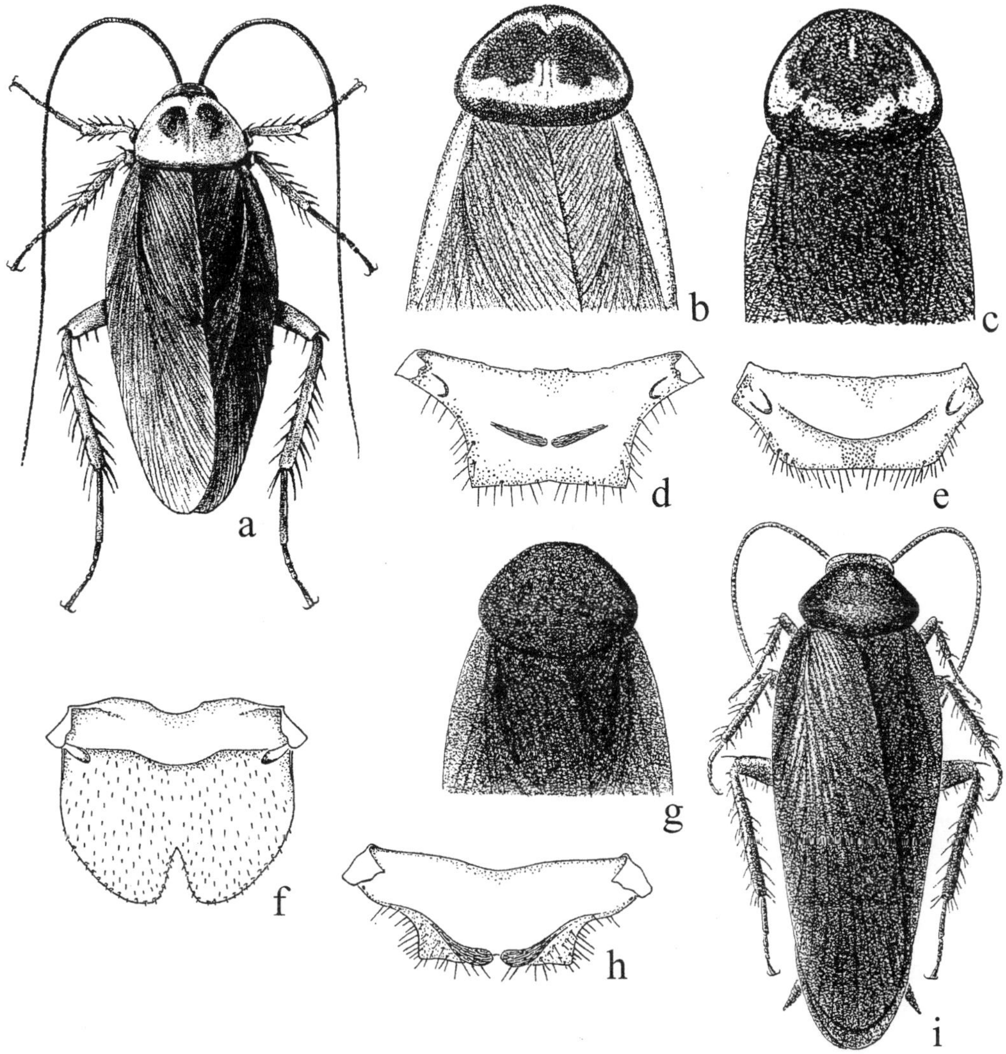


**Figure 4.8** Blattaria: Blattidae. (a) *Periplaneta americana* male; (b) *P. australasiae* adult, pronotum and elytra portion; (c) *P. brunnea* adult, pronotum and elytra portion; (d) *P. australasiae* male, supraanal plate; (e) *P. brunnea* male, supraanal plate; (f) *P. americana* male, supraanal plate; (g) *P. fuliginosa* male, pronotum and elytra portion; (h) *P. japonica* male; (i) *P. fuliginosa* male, supraanal plate.

instar nymphs can be distinguished by the posterior margin of sternite 9: it has a median notch in the female and it is only slightly indented in the male.

The ootheca is 8–10 mm long, dark brown to blackish brown; egg compartments are indistinct; there are 16 teeth on the keel. The ootheca contains 14–16 eggs. Hatching occurs in 57 days at 24 °C, 32 days at 30 °C, and 24–38 days at 30–36 °C. Females carry the ootheca for about 24 h, but it may be retained for up to 6 days before being deposited. Fecundity is 15–90 oothecae, typically 10–15 within 10 months; the interval between oothecae formation is 5–12 days. The ootheca may be cemented to the substrate by an oral secretion from the female; sometimes they are partially covered by pieces of debris from the substrate. Females often eat other *P. americana* oothecae that are encountered in the habitat. Parthenogenesis occurs but is infrequent. About 50% of the parthenogenetic nymphs hatch, and only 33% of those that hatch reach maturity. Nymph development is 7–13 instars during 5–15 months at 25–30 °C. Variations in development occur among cockroaches hatched from the same ootheca. First-instar nymphs consume their embryonic skin immediately after hatching, and the cast skin is usually eaten by other stages except the last. Preferred temperature for adults and nymphs is about 28 °C, but they remain active at 21 °C. Adult life span at 29 °C is 90–706 days for females, and 90–362 days for males. Survival without food or water at 27 °C is about 29 days for males and 42 days for females; survival with water is 43 days for males and 90 days for females.

During courtship, the male and female employ volatile pheromones and tactile stimulation to achieve copulation. It may begin with the female releasing a volatile sex pheromone, which is composed of the sesquiterpenoids, periplanone A and B. This pheromone is detected by the male antennae and directs him toward the female. The female assumes a stance with the abdomen slightly lowered and with the abdominal terga 8 and 9 spread so that the underlying tissues are exposed. When close to the female the male usually waves and contacts the female with his antennae, and raises his wings. The male then turns around and flutters his wings, and moves backwards. He pushes his abdomen under her. If the male succeeds in engaging his genitalia, he turns from under her or the female turns so that the pair is in an opposed position. They remain joined for 1 h or more while the male transfers a spermatophore to the female. The ootheca is not turned after it is formed and the keel remains upright. Fresh mating is not necessary for the production of each ootheca, and females can produce nearly their full potential of ootheca from a single mating.

Natural habitats include generally moist areas in leaf litter, under bark or the bracts of palm trees in forested or undisturbed areas with dense vegetation, but also including caves and burrows. Feral populations of this species in its native region of tropical Africa may not exist. The genus *Periplaneta* includes 47 described species, but none are endemic to the Americas. The specific name *P. americana* is misleading because of the African origin of this species. *P. americana* probably spread by trading and slave ships, and commerce to many regions of the world by the time it was described by Linnaeus in 1758. Evidence of this species was recovered from a Spanish ship sunk off the coast of Bermuda in 1625. Most *Periplaneta* species have restricted distribution and they are not closely associated with humans. Worldwide distribution and pest status characterize only the few species that are capable of adapting to domestic or peridomestic habitats.

In the urban environment, *P. americana* occurs outdoors and indoors in a wide range of habitats, from urban landfills and wastewater treatment plants to the underground sewer systems of the major cities of the world. Large populations of *P. americana* can develop in favorable habitats, such as in the holds of ships, and urban landfills. A feces-secreted aggregation pheromone, which operates during the day when individuals are inactive, promotes concentrations of adults and nymphs. Crowding and limited food may cause mass migrations of adults and nymphs from preferred habitats. Indoors, this species is common in basements and cellars, as well as on upper floors of large buildings. It is a successful and sometimes dominant species in urban environments around the world, from tropical to temperate climates.

**Australian cockroach, *Periplaneta australasiae* (Fig. 4.8b, d)**
Adult males are 27–34 mm long and females are 28–32 mm long. The body is dark brown, and the pronotum has pale brown margins and a dark brown interior. Wings extend to or beyond the abdomen in males and females; the pale margin on the front wing extends to about one-third the length. Males have cerci and styli, but females lack styli. The male supraanal plate (epiproct) has thickened transverse ridges on the ventral surface. First-instar nymphs are predominantly brown, the mesonotum is lighter in color than the remainder of the body, and there are two white spots near the anterior margin of the metanotum. Late-instar nymphs are dark brown, with pale yellow spots on lateral margins of thorax and abdomen; the tip of the antenna is white.

The ootheca is about 10 mm long, blackish brown, with slight indication of the egg compartments. The ootheca contains about 24 eggs, and hatching occurs in about 40 days. Oothecae are produced at about 10-day intervals; fecundity is 20–30 oothecae. The ootheca is not turned on its side after it is formed and the keel remains upright; it is deposited within about 24 h after formation. Oothecae are cemented to the substrate by an oral secretion from the female, and small pieces of the substrate are usually attached to the surface. The female may gnaw a small hole in wood, cardboard, or similar material, then deposit the ootheca. Debris from the substrate is usually attached to the ootheca. Nymph development is through 9–12 instars over 6–12 months, depending on conditions and resources. When nymphs are reared alone, females have 11 or 12 instars and the development time is 306–365 days. Females reared in groups develop in 213 days, and males in 198 days. Adults live about 12 months, but 2 years is possible. Adults mate within 5 days of maturation, and females mate often.

Natural habitats include moist leaf litter in forested areas, beneath logs and stones, and in caves. Natural populations of this species are not known. In the urban environment, it occurs around the perimeter of buildings, and indoors in kitchens and other sites with available food, moisture, and harborage. It occupies similar habitats to *P. americana*, but does not occur in underground sewers. It was originally described in 1775 as *Blatta australasiae*: the specific name referred to its known distribution at the time, which was in southern Asia and not Australia. It is probably native to the tropics or subtropics of Africa, but spread through ships and commerce to warm climates around the world. Among the *Periplaneta*, it is second in importance to *P. americana* as an outdoor and indoor pest. *P. australasiae* seems to prefers moist habitats with a somewhat higher temperature than that required for the development of *P. americana*. *P. australasiae* occurs in tropical and subtropical regions of the world, and is often introduced to and survives in sites in temperate regions, such as in greenhouses and other heated structures. In the USA is occurs in Florida and southeastern Texas, but there are outdoor and indoor populations in other southeastern states.

**Brown cockroach, *Periplaneta brunnea* (Fig. 4.1d; 4.8c, e)** Adults are 31–37 mm long. The body is dark brown to reddish brown; the markings on the pronotum are pale brown. Wings cover the tip of the abdomen in both sexes, and the cerci are short with a blunt tip. The male has a short, truncate supraanal plate (epiproct). The first eight and the last four antennal segments of the first-stage nymph are pale yellow to white; the intermediate segments are brown. The mesothorax has a clear medial area, and there are pale spots on the lateral margins of the second abdominal segment. This species closely resembles *P. americana*, but it is distinguished by characters of the supraanal plate (epiproct) and cerci on the adults. In *P. brunnea*, the supraanal plate is short and the cerci are short and have a blunt tip. In *P. americana* the supraanal plate (epiproct) is long and notched, and the cerci are long and tapered at the tip.

The ootheca is 12–16 mm long and brown; the egg indentations are not distinct. Oothecae of *P. brunnea* are about 2.6 mm longer than those of *P. australasiae*, and 5.2 mm longer than *P. americana*. The ootheca contains about 24 eggs at 27 °C and 65–75% RH; about 20 eggs per ootheca hatch in 35 days at 27 °C, and 61–63 days at 24 °C. Fecundity is about 30 oothecae, but many are not viable. The ootheca is deposited 20–24 h after formation, and the next is produced in 5–6 days. The female secretes from her mouthparts a frothy white substance, which is applied to the substrate where the ootheca will be placed. After deposition, the ootheca may be covered with additional frothy secretion. This substance hardens to become strong cement. Oothecae are often partially covered with pieces of debris from the substrate. Parthenogenesis occurs in *P. brunnea*, but only about 9% of the eggs hatch and complete development to adult female. Nymph development at 24 °C is about 263 days for females, and 268 days for males. Adults live about 8 months or up to 20 months, depending on environmental conditions. Adults mate within 7 days of maturation, and the first ootheca appears in about 14 days.

This species is probably native to Africa and is now distributed in tropical regions around the world. It is carried by commerce and shipments of food to temperate regions, and it survives in heated environments in urban areas. Populations are known from the UK and parts of Europe. It was first reported in the USA in 1907 in Illinois, and is now well established from Florida west to Texas; it has been introduced into California. *P. brunnea* occurs primarily indoors, but also lives outdoors around trees and in sewers.

**Smokybrown cockroach, palmettobug, *Periplaneta fuliginosa* (Fig. 4.8g, i)** Adults are 25–38 mm long and the body is uniformly dark brown to blackish brown. The pronotum is shiny, blackish brown, and without stripes or other markings. Wings are fully developed in both sexes. They are strong flyers, even females carrying an ootheca. Male abdominal segment 1 has a broad, shallow depression bearing a tuft of setae; the supraanal plate (epiproct) is apically truncate and not deeply notched, and with thickened structures on the ventral surface near the apex. First-instar nymphs are brown to dark brown. Antenna has the first 3–6 segments and last 4 or 5 segments pale brown, the intermediate segments are dark brown; tips of maxillary and labial palps are yellowish white. The mesonotum is lighter in color than the pronotum, the metanotum is uniformly dark brown, and abdominal segments 1 and 2 are brownish white ventrally. Young nymphs often move about with their abdomen elevated, and slightly bent anteriorly. Late-instar nymphs are uniformly reddish brown.

The ootheca is 8–14 mm long and has distinct indentations to show the position of the eggs. The ootheca contains 20–28 eggs, and hatching occurs in 100 days at 20 °C, 70 days at 23 °C, 56 days at 25.5 °C, and 37 days at 30 °C. It is deposited about 24 h after formation, and the next ootheca is produced in about 10 days; fecundity is 15–20 oothecae. Nymph development is 274–439 days at 21.5 °C, and 179–191 days at 30–36 °C. The number of instars is 10–12 for females, and 9–11 for males; duration of female instars ranges from 8.5 days for first-instar to 165 days for twelfth-instar. Adults live 18–24 months, depending on environmental conditions and resources. Adults and nymphs are cold-hardy and overwinter in protected sites outdoors in temperate regions. Females produce a volatile sex pheromone that initiates male courtship behavior. When the male locates the female he turns, raises his wings, and attempts to back under her abdomen. If receptive, the female mounts the back of the male and feeds on exudates from the tergal gland. Tergal stimulation by the female then initiates copulation from the male. Adults mate 7–10 days after maturation, and the first ootheca appears in about 14 days. A female may prepare an oviposition site by gnawing a hole in a soft substrate; the ootheca is then deposited and partially covered with debris.

Outdoor habitats for this species are moist, shaded sites with abundant vegetation. In natural areas these include sites such as in tree holes, under the bark of trees, leaf litter, or the bracts of palm trees, such as the fan palm (*Sabal palmetto*), which is the derivation of the common name, palmettobug. The cuticle of *P. fuliginosa* is very permeable to water, and it has the highest rate of cuticular water loss of all the cockroach pest species. Because of its high water loss rate, *P. fuliginosa* generally requires moist habitats with available water. Much of the habitat selection and survival of this species is linked to maintaining water balance. In the urban environment, it occurs primarily outdoors among landscaping vegetation and moist ground cover around the perimeter of buildings. Infestations may be completely within structures, but usually there is some exchange of adults or nymphs with outdoor habitats and populations. In temperate regions, it occurs indoors and in heated locations such as greenhouses. This species is native to temperate regions of eastern Asia, including Japan, Korea, and northern China. In the USA breeding populations are distributed in all of the southeastern states from Texas to eastern Virginia. In central and southern Florida, it is uncommon in urban areas. Isolated populations occur in urban areas of Orange County and Sacramento, California.

***Periplaneta japonica* (Fig. 4.8h)** Adults are 25–35 mm long. The body is shiny and uniformly blackish brown to black; the tarsi and labial and maxillary palps are brown. Wings of the male extend beyond the tip of the abdomen; wings of the female extend to about one-half the length of the abdomen. Male abdominal segment 1 has a broad, shallow depression bearing a tuft of setae. Male supraanal plate (epiproct) is sclerotized, the sides are parallel, and it has a deep notch on the posterior margin. First-instar nymph is uniformly dark brown; the tips of the maxillary and labial palps are white to brownish white. This species is cold-hardy. Nymphs and adults are capable of surviving 120 days and 90 days, respectively, at about 5.5 °C, following preliminary exposure of 28 days to 15 °C. There are nine instars at 27 °C, the final instar is prolonged at low temperatures, and oothecae are deposited at 15 °C but do not hatch. Incubation is about 27 days at 27 °C, 63 days at 20 °C, and hatching does not occur below 20 °C. It is distributed in Japan, China, and Russia. This species occurs primarily outdoors, but some populations have adapted to houses. It is a pest primarily in central and northern Japan.

***Methana marginalis* (Fig. 4.9e)** Adults are 25–29 mm long and the body is brown with a pale white margin on the anterior and lateral edges of the pronotum. Several species of *Methana* occur in Australia, where they live in natural habitats under the bark of trees or logs. This species occurs in Queensland and Western Australia; it sometimes enters houses.

**Harlequin cockroach, *Neostylopyga rhombifolia* (Fig. 4.9f)** Adults are 20–25 mm long. The body is shiny, blackish brown; there are yellowish brown patches over most of the body. Front wings are short and the hind wings are absent in both sexes. Large nymphs are similar to adults in color pattern; small nymphs have poorly defined markings. The ootheca is 12–15 mm long and yellowish brown; it contains about 22 eggs. Nymph development is 286–302 days at 27 °C. Adult females live about 156 days at 24 °C. Parthenogenesis has been reported for this species, but the eggs do not hatch. *N. rhombifolia* is probably native to southern Asia. It is now distributed in tropical regions around the world, including Philippines, Ryukyu and Hawaiian Islands, and Madagascar, Mauritius, Seychelles, and Madeira. In Africa, it is in the countries along the east coast. In Central America, it is known from Costa Rica, and Mexico. In South America it is known from Venezuela, Argentina, and Brazil. In North America it has been found in Baja and Sonora, California. It is primarily an outdoor pest around buildings.

***Other Blattidae*** *Pseudomops septemtrionalis* occurs infrequently indoors; it is known from Louisiana, Texas, and Oklahoma. The South African species, *Deropeltis erythrocephala* (Fig. 4.9d), occurs in wooded sites; it flies to lights at night and often enters houses.

## Cryptocercidae

These are moderate-sized, wingless cockroaches with 1–3 spines ventrally at the tip of the front femur, and without ocelli. None of the species in this family occurs indoors. There are four described species, two from China and Manchuria, and two from the USA. There is limited information on the biology and habits of these cockroaches. Except for a brief dispersal stage as late-stage nymphs or young adults, these cockroaches live their entire life cycle within galleries chewed in rotted logs.

These cockroaches are sometimes linked to termites. The association is based on their intestinal protozoa, which have the ability to digest cellulose, and the behavior of nymphs and adults remaining together in small groups. These groups consist of the breeding pair and their offspring from one reproductive season. Other cockroach species have adapted to feeding on wood. *Panesthia* in the family Blaberidae occur in Australia and they live in and are capable of feeding on decaying wood. The two common species, *P. laevicollis* and *P. australis*, are found

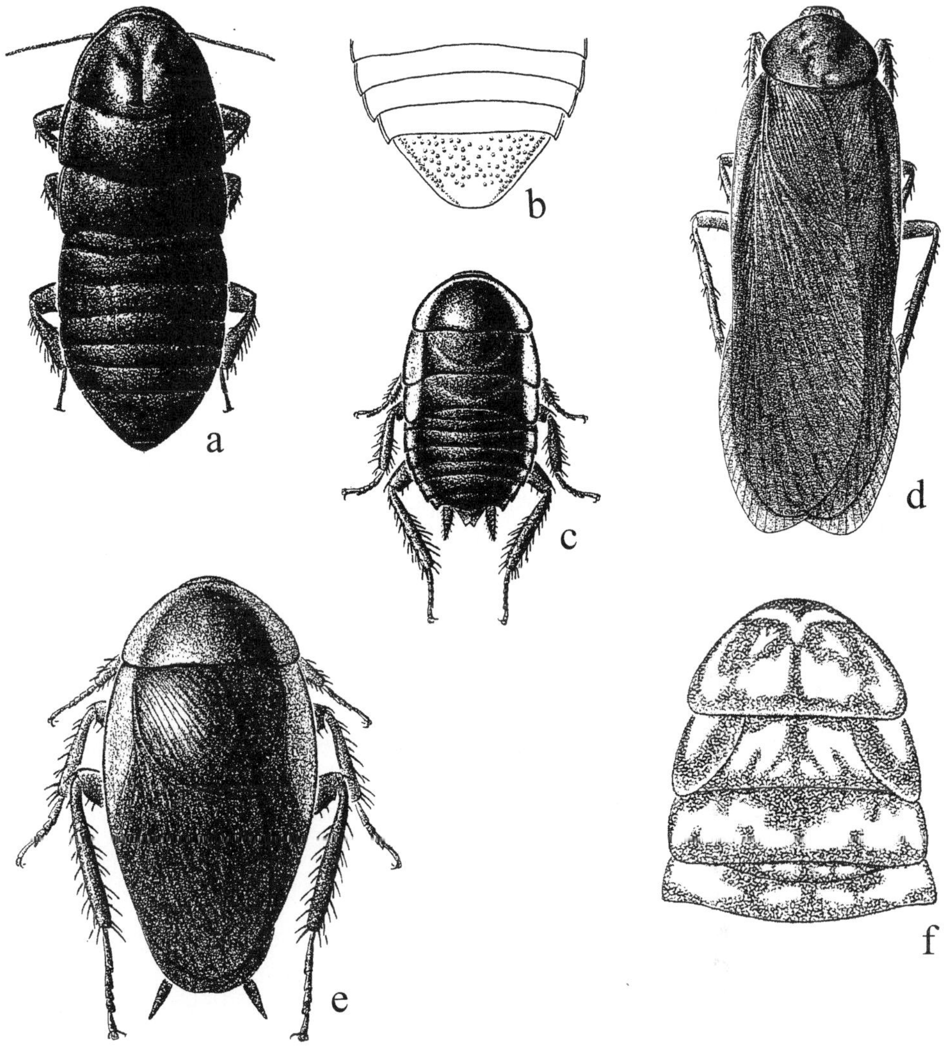


**Figure 4.9** Blattaria: Blattidae. (a) *Cryptocercus punctulatus* adult; (b) *C. punctulatus*, terminal abdominal tergum; (c) *C. punctulatus* nymph; (d) *Deropeltis erythrocephala* male; (e) *Methana marginalis* male; (f) *Neostylopyga rhombifolia* adult, pronotum and thorax.

in rotting logs. Both sexes of several *Panesthia* species have well-developed wings, but these are broken off close to the base soon after they become adults.

**Cryptocercus punctulatus (Fig. 4.9a–c)** Adults are 28–33 mm long. The body is shiny and blackish brown to black. Both sexes are completely wingless. Eyes are small and there are no ocelli. The pronotum is thickened anteriorly, and it has a wide central furrow. Anterior margin of the pronotum is raised above the head like a hood (which is the origin of the common name, brown-hooded cockroach). Legs have large spines; the front femora has three spines on the ventral margin at the tip. Cerci are surrounded and concealed by the terminal tergum and sternum. Nymphs are pale brown to brown, and the cerci protrude beyond the margins of the terminal abdominal sclerites. The ootheca is 8–10 mm long and 2.3 mm wide; it is well sclerotized and the egg chambers are distinct; the keel is well developed. The ootheca contains about 17 eggs, and it is partially concealed with debris from the substrate by the female after deposition.

Natural habitats include the sapwood of wet and decayed logs in heavily forested areas. It does not usually occur in the urban environment. The typical habitat is wet and decayed hardwood and softwood logs where individuals are found feeding in the decayed sapwood portion of the wood. Partially decayed or apparently sound wood is also infested. Adults and nymphs feed on moist and decayed wood; zooflagellate protozoa digest cellulose in the hindgut. The protozoa fauna is transferred to newly hatched nymphs through fecal pellets produced by other nymphs. When older nymphs prepare to molt, they cease feeding and no feces are passed for 2–3 days. During this time the cellulose-digesting protozoa in the hindgut become inactive and encyst. After molting, the first fecal pellets of older nymphs contain large amounts of the protozoa. Fecal pellets are eaten by first-stage nymphs. This provides them a sufficient inoculation of the cellulose-digesting protozoa. This cockroach occurs as populations in the Appalachian Mountains in the eastern USA, from New York to northern Georgia, including Ohio and Tennessee.

The Zoomastigophora in the hindgut of *C. punctulatus* include species of *Barbanympha*, *Leptospironympha*, and *Trichonympha*. The latter is one of the several symbiotic flagellates inhabiting the gut of some species of termites. Species of *Trichonympha* can be successfully exchanged between *Cryptocercus* cockroaches and *Zootermopsis* termites. Cellulose-digesting symbionts and behavior that involves parental care have led to the suggestions that *Cryptocercus* represents a stage in the development of termite-like social behavior. At one time *Cryptocercus* was considered a primitive cockroach and perhaps an ancestor of termites. However, morphological and molecular sequence studies have shown that *Cryptocercus* is distant from termites.

***Cryptocercus clevelandi*** Adults are about 32 mm long. The body is reddish brown to nearly black. Both sexes are completely wingless. The pronotum is about 8.5 mm long and 10 mm wide; nearly black, lateral margins are slightly upturned. Mesonotum and metanotum are not as dark as the pronotum, smooth, and finely punctate; lateral margins are slightly upturned. Abdominal terga 1–6 are dark reddish brown, mostly smooth, with scattered fine punctations. Abdominal sterna 2–6 are reddish brown and punctate. The punctations on the abdominal sterna are dense, particularly sternum 7. Oothecae contain about 43 eggs. *C. clevelandi* lives in biparental family groups and adults exhibit extended care of offspring. Social groups include pairs with first- and second-stage nymphs, pairs with offspring hatched the previous year, and females with nymphs hatched 2 years previously. Nymphal development takes 5–7 years to complete. Males and females pair during the year they mature, but do not reproduce until the following year (summer). This species occurs in rotting wood in upland temperate forest areas from southern Washington to northern California.

**Other *Cryptocercus*** The other extant species are wingless and have habits similar to the North American species. *C. primarius* occurs in Sichuan Province, China, and *C. relictus* in the Maritime region of the eastern Russian Federation.

## Polyphagidae

Oothecae of these cockroaches have a distinct keel, which is often strongly toothed, such as in *Polyphaga aegyptiaca*. There is usually a flange or short extension at the anterior end of the ootheca, by which it is held in the female's genitalia. In some species, including *P. aegyptiaca*, the ootheca is held in the original position and not turned after it is formed. Spines may be absent on the anterior and posterior margins of the mid and hind femora. Sternum 7 of the female is sometimes bivalvular. *Polyphaga* species are distributed nearly worldwide in shipments of vegetables. In the continental USA, they occur primarily in the southwestern states and in Florida. None of the species that occur in Central and North America are household pests. In northern Africa and southern Asia, a few species, such as *P. aegyptiaca* and *P. saussurei*, are occasionally peridomestic or domestic pests.

**Mediterranean cockroach, *Polyphaga aegyptiaca*** Adult males are 25–32 mm long and females are 25–34 mm long and wingless. The female is blackish brown and the male is dark brown. Wings are light brown and broadly rounded at the distal end; wing veins are prominent. Antennae are shorter than the body; the pronotum has a pale brown anterior margin in both sexes. Cerci are short and unsegmented in the female. The ootheca is about 12 mm long and pointed at one end, and there is a row of 12–14 teeth along the keel. The ootheca contains 7–13 eggs. This species is distributed in the region surrounding the Mediterranean sea and eastward to Iran and the Caspian sea. It is an occasional indoor pest.

**South Asian cockroach, *Polyphaga saussurei*** Adult males are 32–37 mm long and females are 35–44 mm long; the body of both sexes is strongly convex. Males are uniformly dark brown and the antennae and cerci are short. Wings are light brown and extend beyond the tip of the abdomen; wings are broadly rounded at distal end. Females are dark brown and wingless. Adults mate 10–14 days after maturation. The ootheca is 12–20 mm long and contains 34–36 eggs; it is held internally during the incubation period of about 60 days. Nymph development is 121–150 days at 30–36 °C, but under lower temperatures it is extended to 7–18 months. Adults live 2.5 years in laboratory conditions. Parthenogenesis has been reported for this cockroach. During courtship the male stands near the female and rapidly moves his body up and down. In some instances the male raises the anterior portion of his body and strikes his abdomen against the substrate to produce a prolonged tapping sound. Adults give off an offensive odor when disturbed. It is distributed in central Asia, from the Caspian sea eastward through Kazakstan, Uzbekistan, Tajikistan, and Azerbaijan. Other locations include Afghanistan, Iran, and India.

## Bibliography


Alcamo, I. E. and A. M. Frishman. The microbial flora of field-collected cockroaches and other arthropods. *J. Environ. Health*, **42** (1980), 263–6.

Appel, A. G. Water relations and thermal sensitivity of several cockroach species (Dictyoptera: Blattidae and Blaberidae). *Comp. Bioc. A.*, **100A** (1991), 353–6.

Appel, A. G., D. A. Reierson, and M. K. Rust. Comparative water relations and temperature sensitivity of cockroaches. *Comp. Bioc. A.*, **74A** (1983), 357–61.

Arnold, J. W. Adaptive features on the tarsi of cockroaches (Insecta: Dictyoptera). *Int. J. Insect Morphol. Embryol.*, **3** (1974), 317–34.

Asahina, S. A revised list of the Japanese cockroaches of sanitary importance (Insecta, Blattaria). *Jpn. J. Med. Sci. Biol.*, **14** (1961), 147–56.

Taxonomic notes on Japanese Blattaria. III. On the species of the genus *Onychostylus* Bolivar. *Jpn. J. Sanit. Zool.*, **16** (1964), 6–15.

The Blattaria from Iran and Afghanistan. *Results Kyoto Univ. Sci. Exp. Karakoram and Hindukush*, **8** (1966), 155–8.

Notes on some cockroach species in the genera *Megamareta* and *Onchostylus* (or Lupparia) from Pacific area. *Jpn. J. Sanit. Zool.*, **23** (1973), 195–200.

*Blattaria of Japan*. Tokyo: Nakayama-shoten. 1991.

Atkinson, T. H., P. G. Koehler, and R. S. Patterson. Catalog and atlas of the cockroaches (Dictyoptera) of North America north of Mexico. *Misc. Publ. Entomol. Soc. Am.*, **78** (1991), 1–86.

Beier, M. *Blatteriae (Schaben). Handbuch der Zoologie*, vol. IV, H.2, Teil 2/13, Lief. 22. Leipzig: de Gryter, 1974.

Bell, W. J., C. Parsons, and E. A. Martinko. Cockroach aggregation pheromones: analysis of aggregation tendency and species specificity. *J. Kansas Entomol. Soc.*, **45** (1972), 414–20.

Bernton, H. S. and H. Brown. Insect allergy – preliminary studies of the cockroach. *J. Allergy*, **35** (1964), 506.

Cockroach allergy II: the relation of infestation to sensitization. *South. Med. J.*, **60** (1967), 852–5.

Insect allergy: the allergenic potential of the cockroach. *South. Med. J.*, 62 (1969), 1207–10.

Bey-Bienko, G. Investigations of the Blattoidea of southern China. *Trudy Zool. Inst. Akad. Nauk. SSSR*, **15** (1954), 5–26.

Brenner, R. J. Focality and mobility of some peridomestic cockroaches in Florida (Dictyoptera: Blattaria). *Ann. Entomol. Soc. Am.*, **81** (1988), 581–92.

Brenner, R. J. and R. R. Pierce. Seasonality of peridomestic cockroaches (Blattoidea: Blattidae): mobility, winter reduction, and effect of traps and baits. *J. Econ. Entomol.*, **84** (1991), 1735–45.

Bruujning, C. F. A. Studies on Malayan Blattidae. *Zool. Meded. Leiden*, **29** (1948), 1–174.

Bunting, W. Orthoptera imported into Britain with bananas from Dominica (Leeward Isles). *Entomol. Mon. Mag.*, **91** (1955), 134.

Bunting, W. M. Preliminary notes on some Orthoptera imported into Britain with bananas from Dominica. *Entomol. Mon. Mag.*, **92** (1956), 284–6.

Cleveland, L. D., S. R. Hall, E. P. Sanders, and J. Collier. The wood-feeding roach, *Cryptocercus*, its protozoa, and the symbiosis between protozoa and roach. *Mem. Am. Acad. Arts Sci.*, **17** (1934), 185–342.

Cole, M. M., G. C. LaBrecque, and G. S. Burden. Effects of gamma radiation on some insects affecting man. *J. Econ. Entomol.*, **52** (1959), 448–50.

Cloarec, A. and C. Rivault. Age-related changes in foraging in the German cockroach (Dictyoptera: Blattellidae). *J. Insect Behav.*, **4** (1991), 661–73.

Coler, R. R., R. G. Van Driesche, and J. S. Elkinton. Effect of an oothecal parasitoid, *Comparia merceti* (Compere), in a population of brownbanded cockroaches. *Environ. Entomol.*, **13** (1984), 603–6.

Cornwell, P. B. *The Cockroach*, vol. I. *A Laboratory Insect and Industrial Pest*. London: Hutchinson, 1968.

*The Cockroach*, vol. II. *Insecticides and Cockroach Control*. New York: St. Martin's Press, 1976.

Eades, R. B., F. J. von Zuben, S. E. Bennett, and O. L. Walker. Studies on cockroaches in a municipal sewerage system. *Am. J. Trop. Med. Hyg.*, **3** (1954), 1092–8.

Faulde, M., M. E. A. Fuchs, and W. Nagl. Dispersionsauslosende proteine im speichel mehrerer Schaben Arten (Blattodea: Blattellidae, Blattidae, Blaberidae). *Entomol. Gener.*, **14** (1989), 203–10.

Gorton, R. E., Jr. A comparative ecological study of the wood cockroaches in northeastern Kansas. *Univ. Kansas Sci. Bull.*, **52** (1980), 21–30.

Gould, G. E. and H. O. Deay. The biology of six species of cockroaches which inhabit buildings. *Purdue Univ. Ag. Exp. Stn. Bull.* **451** (1940), 1–31.

Gunn, D. L. The temperature and humidity relations of the cockroach. *Nature*, **128** (1931), 186–7.

Gunn, D. L. and C. A. Cosway. The temperature and humidity relations of the cockroach. V. Humidity preference. *J. Exp. Biol.*, **15** (1938), 555–63.

Gunn, D. L. and F. B. Notley. The temperature and humidity relations of the cockroach. IV. Thermal death-point. *J. Exp. Biol.*, **13** (1936), 28–34.

Gurney, A. B. Studies in certain genera of American Blattidae (Orthoptera). *Proc. Entomol. Soc. Wash.*, **39** (1937), 101–12.

Notes on the Cuban cockroach, *Panchlora nivea* (L.). *Proc. Entomol. Soc. Wash.*, **57** (1955), 285–6.

Guthrie, D. M. and A. R. Tindall. *The Biology of the Cockroach*. New York: St. Martin's Press, 1968.

Hagenbuch, B. E., P. G. Koehler, R. S. Patterson, and R. J. Brenner. Peridomestic cockroaches (Orthoptera: Blattidae) of Florida: their species composition and suppression. *J. Med. Entomol.*, **25** (1988), 377–80.

Hanitsch, R. Malayan Blattidae. *J. R. Asiatic Soc., Straits Branch No.*, **69** (1915), 17–178.

Hanitsch, R. On a collection of Blattidae from southern Annam. *J. Siam Soc. Nat. Hist. Suppl.*, **7** (1927), 7–48.

Hebard, M. The Blattidae of North America north of the Mexican boundary. *Mem. Am. Entomol. Soc.*, **2** (1917), 1–248.

Studies in Malayan Blattidae (Orthoptera). *Proc. Acad. Nat. Sci. Philadelphia*, **81** (1929), 1–109.

Helm, R. M., A. W. Burks, L. W. Williams, R. J. Brenner. Identification of major cockroach aeroallergens from living cultures of German or American cockroaches. *J. Allergy Clin.*, **87** (1991), 189.

Huang, L. X. A survey of domestic cockroaches in Jianyang District of North Fujian. *Wuyi Sci. J.*, **6** (1986), 107–9.

Illingworth, J. F. An outbreak of cockroaches, *Nauphoeta cinerea* (Olivier), in Honolulu. *Proc. Hawaiian Ent. Soc.*, **11** (1942), 169–70.

Jiang, X. F. Analysis of the results in surveillance of cockroach density in five cities. *Chin. J. Vector Biol. Control*, **2** (1991), 20–2.

Johns, P. M. The cockroaches of New Zealand. *Rec. Canterbury Museum*, **8** (1966), 93–136.

Kanbhampti, S. A phylogeny of cockroaches and related insects based on DNA sequence of mitochondrial ribosomal RNA genes. *Proc. Natl Acad. Sci. USA*, **14** (1995), 2017–20.

Kang, B. Cockroach allergy. *Clin. R. Allergy*, **8** (1990), 87–98.

Kevan, D. K. McE. and L. Chopard. Blattodea from Northern Kenya and Jubaland. *Ann. Mag. Nat. Hist.*, **7** (1954), 166–87.

Lan, J-L., D-T. Lee, C-H. Wu, C-P. Ching, and C-L. Yeh. Cockroach hypersensitivity: preliminary study of allergic cockroach asthma in Taiwan. *J. Allergy Clin.* **82** (1988), 736–40.

Lawson, F. A. Structural features of the oothecae of certain species of cockroaches. *Ann. Entomol. Soc. Am.*, **44** (1951), 269–85.

Lawson, F. Structural features of cockroach egg capsules IV. The ootheca of *Parcoblatta uhleriana* (Orthoptera: Blattidae). *J. Kansas Entomol. Soc.*, **27** (1954), 14–20.

Lee, D. K. Distribution and seasonal abundance of cockroaches (Blattellidae and Blattidae, Blattaria) in urban general hospitals. *Korean J. Entomol.*, **25** (1995), 57–67.

Marquis, D. *Archy and Mehitabel*. New York: Garden City, 1931.

McKittrick, F. A. Evolutionary studies of cockroaches. *Cornell Univ. Agric. Exp. Stn. Mem.* (1964), 389.

Narasimham, U. A. and T. Sankaran. Domiciliary cockroaches and their oothecal parasites in India. *Entomophaga*, **24** (1979), 273–80.

Ogata, K., I. Tanaka, and T. Ogawa. Studies on the establishing factors of domiciliary cockroaches. 1. Field surveys of the distribution of domiciliary cockroaches in Tokyo and Kawasaki. *Jpn. J. Sanit. Zool.*, **26** (1975), 241–5.

Peters, H. Die synanthropen Schaben Mittleuropa (Gattungen *Blatta*, *Blattella*, *Periplaneta* und *Supella*). Mekblatter uber angewandt Parasitenkunde und Schlingsbekampfung, Merkblatt No. 3. *Z. angew. Parasit.*, **2** (1961), 1–15.

Pope, P. Studies on the life histories of some Queensland Blattidae (Orthoptera). Part 1. The domestic species. *Proc. R. Soc. Queensland*, **63** (1953), 23–46.

Studies on the life histories of some Queensland Blattidae (Orthoptera). Part 2. Some native species. *Proc. R. Soc. Queensland*, **63** (1953), 47–59.

Princis, K. Zur Systematik der Blattarien. *Eos*, **36** (1960), 429–49.

*Orthopterorum Catalogus*. Pars 3 (1962); Pars 4 (1963); Pars 6 (1964); Pars 7 (1965); Pars 8 (1966); Pars 13 (1969); Pars 14 (1971). The Hague: Junk, 1962–71.

Pul'ver, K. Yu. Dissemination of synanthropic cockroaches and their migration in some districts of a city. *Med. Parazitol. Parazit. Bolezni.*, **42** (1973), 103–4.

Rau, P. Cockroaches: the forerunner of termites (Orthoptera: Blattidae): Isoptera. *Ent. News*, **52** (1941), 156–259.

How the cockroach deposits its egg-case; a study in insect behavior. *Ann. Entomol. Soc. Am.*, **36** (1943), 221–6.

Rehn, J. A. G. African and Malagasy Blattidae (Orthoptera). Part I. *Proc. Acad. Natl Sci. Philadelphia*, **83** (1931), 305–87.

African and Malagasy Blattidae (Orthoptera). Part II. *Proc. Acad. Natl Sci. Philadelphia* **84** (1931), 405–511.

African and Malagasy Blattidae (Orthoptera). Part III. *Proc. Acad. Natl Sci. Philadelphia* **89** (1931), 17–123.

Man's uninvited fellow-traveler – the cockroach. *Sci. Monthly*, **61** (1945), 265–76.

Rehn, J. W. A key to the Genera of North American Blattaria, including established adventives. *Entomol. News*, **61** (1950), 64–7.

Roth, L. M. and E. R. Willis. The reproduction of cockroaches. *Smithsonian Misc. Coll.*, **122** (1954), 1–49.

Parthenogenesis in cockroaches. *Ann. Entomol. Soc. Am.* **49** (1956), 195–204.

The medical and veterinary importance of cockroaches. *Smith. Inst. Misc. Coll.*, **134** (1957), 1–147.

The biotic associations of cockroaches. *Smith. Inst. Misc. Coll.*, **141** (1960), 1–470.

Schal, C. and R. L. Hamilton. Integrated suppression of synanthropic cockroaches. *Annu. Rev. Entomol.*, **35** (1990), 521–51.

Schal, C., J.-Y. Gautier, and W. J. Bell. Behavioral ecology of cockroaches. *Biol. Rev.*, **59** (1984), 209–54.

Shin, Y. H., I. B. Yoon, and J. L. Kim. Studies on the cockroaches in Korea. *Entomol. Res. Bull. Korea*, **5** (1973), 1–53.

Stejskal, V. and P. H. Verner. Long-term changes of cockroach infestations in Czech and Slovak food-processing plants. *Med. Vet. Entomol.*, **10** (1996), 103–4.

Takahashi, R. Life-history of Blattidae. *Dobutsugaka Zasshi (Zool. Mag.), Tokyo*, **36** (1924), 215–30.

Willis, E. R., G. R. Riser, and L. M. Roth. Observations on reproduction and development in cockroaches. *Ann. Entomol. Soc. Am.*, **51** (1958), 53–69.

Wood, F. E., W. H. Robinson, S. K. Kraft, and P. Zungoli. Survey of attitudes and knowledge of public housing residents towards cockroaches. *Bull. Entomol. Soc. Am.*, **27** (1981), 9–13.

Zhai, J. Habitat preference of cockroaches in urban environments in Shanghai, China. *Jpn. J. Sanit. Zool.*, **41** (1990), 353–8.

### Blaberus

Brossut, R., P. Dubois, and J. Rigaud. Le gregarisme chez *Blaberus craniifer*: isolem et identification de la pheromone. *J. Insect. Physiol.*, **20** (1974), 529–43.

Hebard, M. Critical notes on certain species of *Blaberus* (Orthoptera, Blattidae). *Ent. News*, **27** (1916), 289–96.

Nutting, W. L. Observations on the reproduction of the giant cockroach, *Blaberus craniifera* Burm. *Psyche*, **60** (1953), 6–14.

Piquett, P. G. and J. H. Fales. Life-history of *Blaberus giganteus* (L.). *J. Econ. Entomol.*, **46** (1954), 1089–90.

Stewart, A. M. *Blaberus cubensis* (Orthoptera) and its ootheca. *Entomologist*, **58** (1925), 57–8.

### *Blatta*

Alexander, J. B., J. Newton, and G. A. Crowe. Distribution of Oriental and German cockroaches, *Blatta orientalis* and *Blattella germanica* (Dictyoptera), in the United Kingdom. *Med. Vet. Entomol.*, **5** (1991), 395–402.

Bao, N. and W. H. Robinson. Morphology and mating configuration of genitalia of the oriental cockroach, *Blatta orientalis* L. (Blattodea: Blattidae). *Proc. Entomol. Soc. Wash.*, **92** (1990), 416–21.

Gunn, D. L. The temperature and humidity relations of the cockroach (*Blatta orientalis*): I. Desiccation. *J. Exp. Biol.*, **10** (1933), 274–85.

The temperature and humidity relations of the cockroach (*Blatta orientalis*): II. Temperature preference. *Z. Vgl. Physiol.*, **20** (1934), 617–25.

The daily rhythm of activity of the cockroach, *Blatta orientalis*. *J. Exp. Biol.*, **17** (1940), 267–77.

Le Patourel, G. N. J. Environmental aspects of the survival and reproduction of Oriental cockroaches (*Blatta orientalis* L.). pp. 31–34. In Wildey, K. W. and W. H. Robinson (eds.) *Proceedings of the 1st International Conference on Insect Pests and the Urban Environment*, pp. 31–4. Exeter: BPCC Wheatoris, 1993.

Princis, K. Wo ist die Urheimat von *Blatta orientalis* L. zu suchen? *Opuscula Entomol.*, **19** (1954), 202–4.

Quadri, M. A. H. The life-history and growth of the cockroach *Blatta orientalis* Linn. *Bull. Ent. Res.*, **29** (1938), 263–76.

Rau, P. Food preferences of the cockroach, *Blatta orientalis* Linn. *Ent. News*, **56** (1954), 276–8.

### *Blattella*

Akers, R. C. and W. H. Robinson. Comparison of movement behavior of three strains of German cockroach, *Blattella germanica*. *Ent. Exp. et Appl.*, **34** (1983), 143–7.

Asahina, S. Taxonomic notes on Japanese *Blattaria*. II. On the occurrence of *Blattella lituricollis* in Japan. *Jpn. J. Sanit. Zool.*, **15** (1964), 61–7.

Atkinson, T. H., P. G. Koehler, and R. S. Patterson. Reproduction and development of *Blattella asahinai* (Dictyoptera: Blattellidae) *J. Econ. Entomol.*, **84** (1991), 1251–6.

Bell, W. J. and C. Schal. Patterns of turning in courtship orientation of male German cockroach. *Animal Behav.*, **28** (1980), 86–94.

Berthold, R. Jr. Behavior of the German cockroach, *Blattella germanica* (L.), in response to surface textures. *J. N. Y. Entomol. Soc.* **75** (1967), 148–53.

Berthold, R. Jr. and B. R. Wilson. Resting behavior of the German cockroach, *Blattella germanica*. *Ann. Entomol. Soc. Am.*, **60** (1967), 347–51.

Boase, C. The German cockroach – bowing out at last? *Int. Pest Control*, **44** (2002), 194–7, 202.

Breed, M. D., C. M. Hinkle, and W. J. Bell. Agonostic behavior in the German cockroach, *Blattella germanica*. *Z. Tierpysch.*, **39** (1975), 24–32.

Brenner, R. J., R. S. Patterson, and P. G. Koehler. Ecology, behavior, and distribution of *Blattella asahinai* (Orthoptera: Blattellidae) in central Florida. *Ann. Entomol. Soc. Am.*, **81** (1988), 432–6.

Buxton, G. M. and T. J. Freeman. Positive separation of *Blattella vega* and *Blattella germanica* (Orthoptera: Blattidae). *Pan-Pacific Entomol.*, **44** (1968), 168–9.

Carlson, D. A. and R. J. Brenner. Hydrocarbon-based discrimination of three North American *Blattella* cockroach species (Orthoptera: Blattellidae) using gas chromatography. *Ann. Entomol. Soc. Am.*, **81** (1988), 711–23.

Cochran, D. G. Food and water consumption during the reproductive cycle of female German cockroaches. *Ent. Exp. Appl.*, **34** (1983), 51–7.

Darchen, R. Sur l'activité exploratrice de *Blattella germanica*. *Z. Tierpsych.*, **9** (1952), 362–72.

Denzer, D. J., M. E. A. Fuchs, and G. Stein. Zum Verhalten von *Blattella germanica* L.: Aktionsradius und Refugientreue. *J. Appl. Entomol.*, **105** (1988), 330–3.

Zur Tagesrhythmik von *Blattella germanica* L. (Orthoptera: Blattellidae) 2. Weibchen. *J. Appl. Entomol.*, **105** (1988), 174–81.

Zur Tagesrhythmik von *Blattella germanica* L. (Orthoptera: Blattellidae) 3. Mischpopulation. *J. Appl. Entomol.*, **105** (1988), 262–9.

Durbin, E. J. and D. G. Cochran. Food and water deprivation effects on reproduction in female *Blattella germanica*. *Entomol. Exp. Appl.*, **37** (1985), 77–82.

Gordon, H. T. Minimal nutritional requirements of the German roach, *Blattella germanica* (L.). *Ann. N. Y. Acad. Sci.*, **77** (1959), 290–351.

Hamilton, R. and C. Schal. Effects of dietary protein levels on reproduction and food consumption in the German cockroach (Dictyoptera: Blattellidae). *Ann. Entomol. Soc. Am.*, **81** (1988), 969–76.

House, H. L. Nutritional studies with *Blattella germanica* (L.) reared under aseptic conditions. II. A chemically defined diet. *Can. Entomol.*, **81** (1949), 105–12.

Ishii, S. An aggregation pheromone of the German cockroach, *Blattella germanica* (L.). II. Species specificity of the pheromone. *Appl. Entomol. Zool.*, **5** (1970), 33–41.

Izutsu, M., U. Shuichi, and S. Ishii. Aggregation effects on the growth of the German cockroach, *Blattella germanica* (L.) (Blattaria: Blattellidae). *Appl. Entomol. Zool.*, **5** (1970), 159–71.

Koehler, P. G., R. S. Patterson, and R. J. Brenner. Spectral sensitivity and behavioral response to light quality in the German cockroach (Dictyoptera: Blattellidae). *Ann. Entomol. Soc. Am.*, **80** (1987), 820–822.

Koehler, P. G., C. A. Strong, and R. S. Patterson. Harborage width preferences of German cockroach (Dictyoptera: Blattellidae) adults and nymphs. *J. Econ. Entomol.*, **87** (1994), 699–704.

Kitamura, C., S. H. Koh, and S. Ishii. Possible role of feces for directional orientation of the German cockroach, *Blattella germanica* L. (Orthoptera: Blattellidae). *Appl. Entomol. Zool.*, **9** (1974), 271–2.

Kunkel, J. G. Development and the availability of food in the German cockroach, *Blattella germanica* (L.). *J. Insect Physiol.*, **12** (1966), 227–35.

Metzger, R. and J. Hanisch. Größe und Verteilung einer Population von *Blattella germanica*. *Z. Angew. Parasitol.*, **20** (1979), 193–202.

Metzger, R. and K. H. Trier. Zur Bedeutung der Aggregations pheromone von *Blattella germanica* und *Blatta orientalis*. *Z. Angew Parasitol.*, **16** (1975), 16–26.

Miller, D. M., P. G. Koehler, and R. S. Patterson. Influence of German cockroach (Dictyoptera: Blattellidae) fecal extract on food consumption and harborage choice. *J. Econ. Entomol.*, **89** (1996), 668–72.

Mizukubo, T. A revision of the genus *Blattella* (Blattaria: Blattellidae) of Japan. I. Terminology of the male genitalia and description of a new species from Okinawa Island. *Esakia*, **17** (1981), 149–59.

Nishida, R. and H. Fukami. Female sex pheromone of the German cockroach, *Blattella germanica*. *Mem. Coll. Agric. Kyoto Univ.*, **122** (1983), 1–24.

Nishida, R., H. Fukami, and S. Ishii. Sex pheromone of the German cockroach (*Blattella germanica* L.) responsible for male wing-raising: 3,11-dimethyl-2-nonacosanone. *Experimentia* **30** (1974), 978–9.

Noland, J. L. and C. A. Baumann. Protein requirements of the cockroach *Blattella germanica* (L.). *Ann. Entomol. Soc. Am.*, **44** (1951), 184–8.

Rivault, C. Distribution dynamics of *Blattella germanica* in closed urban environments. *Ent. Exp. Appl.* **57** (1990), 85–91.

Ross, H. H. The life history of the German cockroach, *Blattella germanica*. *Trans. Ill. State Acad. Sci.*, **21** (1928), 84–93.

Roth, L. M. A taxonomic revision of genus *Blattella* Caudell (Dictyoptera: Blattaria: Blattellidae). *Entomol. Scand.*, **22** (suppl.): (1985), 1–221.

*Blattella asahinai* introduced into Florida (Blattaria: Blattellidae). *Psyche*, **93** (1986), 371–4.

Roth, L. M. and S. Cohen. Aggregation in *Blattaria*. *Ann. Entomol. Soc. Am.*, **66** (1973), 1315–23.

Runstrom, E. S. and G. W. Bennett. Movement of German cockroaches (Orthoptera: Blattellidae) as influenced by structural features of low-income apartments. *J. Econ. Entomol.*, **77** (1984), 407–11.

Rust, M. K., J. Owens, and D. A. Reierson (eds.) *Understanding and Controlling the German Cockroach*. New York: Oxford University Press, 1995.

Samouelle, G. *The Entomological Cabinet; Being a Natural History of British Insects*, 2nd edn., pp. 1–3. London: Longman, 1841.

Schal, C., E. L. Burns, R. A. Jurenka, and G. J. Blomquist. A new component of the female sex pheromone of *Blattella germanica* (L.) (Dictyoptera: Blattellidae) and interaction with other pheromone components. *J. Chem. Ecol.*, **16** (1990), 1997–2008.

Silverman, J. and D. N. Bieman. Glucose aversion in the German cockroach, *Blattella germanica*. *J. Insect Physiol.*, **39** (1993), 925–33.

Sommer, S. H. Zum Aggregationsverhaltem bei *Blattella germanica* *Z. Angew. Parasitol.*, **16** (1975), 135–41.

Stein, W. and H. Haschemi. Untersuchungen uber Biologie, Ausbreitungs- und Wanderverhalten der Deutschen Schabe *Blattella germanica* (L.) (Blattodea, Blattellidae) in einer Mülldeponie. *Z. Angew. Zool.*, **74** (1987), 13–34.

Der Einfluß von Ausenfaktoren auf das wanderverhalten der Deutschen Schabe, *Blattella germanica* (L.) (Blattodea, Blattellidae) in einer Mülldeponie. *Anz. Schad. Pflanz. Umwelt.*, **64** (1991), 65–70.

Tsuji, H. The life cycle of *Blattella nipponica* Asahina in Koyto. *Kontyû, Tokyo*, **53** (1985), 42–8.

Zhai, J. and W. H. Robinson. Antennal and leg grooming in two strains of the German cockroach (Orthoptera: Blattellidae). *Med. Entomol. Zool.*, **47** (1996), 281–5.

### *Cryptocercus, Ectobius, Gromphadorhina*

Barth, R. H. Jr. The mating behavior of *Gromphadorhina portentosa* (Schaum): an anomalous pattern for a cockroach. *Psyche*, **75** (1968), 124–31.

Blair, K. G. A note on the British species of *Ectobius*. *Ent. Mon. Mag.*, **70** (1934), 157–9.

Brown, E. B. Observations on the life-history of the cockroach, *Ectobius panzeri* Stephens (Orth., Blattidae). *Ent. Mon. Mag.*, **88** (1952), 209–12.

Brown, V. K. The overwintering stages of *Ectobius lapponicus* (L.) (Dictyoptera: Blattidae). *J. Entomol. (A)*, **48** (1973), 11–24.

A key to the nymphal instars of the British species of *Ectobius* Stephens (Dictyoptera: Blattidae). *Entomologist*, **106** (1973), 202–9.

Grandcolas, P. Systematics, endosymbiosis, and biogeography of *Cryptocercus clevelandi* and *C. punctulatus* (Blattaria: Polyphagidae) from North America: a phylogenetic perspective. *Ann. Entomol. Soc. Am.*, **92** (1999), 285–91.

Gurney, A. B. The spotted Mediterranean cockroach, *Ectobius pallidus* (Oliver) in the United States. *U.S.D.A. Coop. Econ. Inst. Rep.*, **18** (1968), 684–6.

Nalepa, C. A. and C. Bandi. Phylogenetic status, distribution, and biogeography of *Cryptocercus* (Dictyoptera: Cryptocercidae). *Ann. Entomol. Soc. Am.*, **92** (1999), 292–302.

Nalepa, C. A., G. W. Byers, C. Bandi, and M. Sironi. Description of *Cryptocerus clevelandi* (Dictyoptera: Cryptocercidae) from the northwestern United States, molecular analysis of bacterial symbionts in its fat body, and notes on biology, distribution, and biogeography. *Ann. Entomol. Soc. Am.*, **90** (1997), 416–24.

Roth, L. M. and E. R. Willis. Observations on the biology of *Ectobius pallidus* (Oliver) (Blattaria, Blattidae). *Trans. Am. Entomol. Soc.*, **83** (1957), 31–7.

### *Panchlora, Parcoblatta*

Gurney, A. B. Notes on the Cuban cockroach, *Panchlora nivea* (L.). *Proc. Entomol. Soc. Wash.*, **57** (1955), 285–6.

Lawson, F. A. Structural features of cockroach egg capsules IV. The ootheca of *Parcoblatta uhleriana* (Orthoptera: Blattidae). *J. Kansas Entomol. Soc.*, **27** (1954), 14–20.

Ecological and collecting notes on eight species of *Parcoblatta* (Orthoptera: Blattidae) and certain other cockroaches. *J. Kansas Entomol. Soc.*, **40** (1967), 267–9.

Rau, P. The life history of the wood-roach, *Parcoblatta pennsylvania* DeGeer (Orthoptera: Blattidae). *Entomol. News*, **51** (1940), 4–9, 33–5.

Roth, L. M. and E. R. Willis. The biology of *Panchlora nivea*, with observations on the eggs of other Blattaria. *Trans. Am. Entomol. Soc.*, **83** (1957), 195–207.

Willis, E. R. Biology and behavior of *Panchlora irrorata*, a cockroach adventive on bananas (Blattaria: Blaberidae). *Ann. Entomol. Soc. Am.*, **59** (1966), 514–16.

### *Periplaneta*

Appel, A. G. Field and laboratory studies on American cockroach activity and distribution. *J. Ala. Acad. Sci.*, **57** (1986), 57–64.

Appel, A. G. and M. K. Rust. Outdoor activity and distribution of the smokybrown cockroach, *Periplaneta fuliginosa* (Serville). *Environ. Entomol.*, **14** (1985), 669–73.

A bibliography of the smokybrown cockroach (Dictyoptera: Blattidae), an urban and suburban pest. *J. Entomol. Sci.*, **22** (1987), 175–87.

Appel, A. G. and L. M. Smith II. Harborage preference of American and smokybrown cockroaches (Dictyoptera: Blattidae) for common landscape materials. *Environ. Entomol.*, **25** (1996), 817–24.

Biology and management of the smokybrown cockroach. *Annu. Rev. Entomol.*, **47** (2002), 33–55.

Barth, R. H. Jr. The mating behavior of *Periplaneta americana* (Linneaus) and *Blatta orientalis* Linneaus (Blattaria: Blattinae), with notes on three additional species of *Periplaneta* and interspecific action of female sex pheromone. *Z. Tierpsychol.*, **27** (1970), 722–48.

Bell, W. J. and K. G. Adiyodi. *The American Cockroach*. London: Chapman and Hall, 1981.

Benson, E. P. Harborage preference by *Periplaneta americana* (L.) and *Periplaneta fuliginosa* (Serville) (Dictyoptera: Blattidae) in a home in Clemson, South Carolina. *J. Entomol. Sci.*, **22** (1987), 39–44.

Benson, E. P., P. A. Zungoli, and L. M. Smith II. Comparison of developmental rates of two separate populations of *Periplaneta fuliginosa* (Dictyoptera: Blattidae) and equations describing development, preoviposition, and oviposition. *Environ. Entomol.*, **23** (1994), 979–86.

Camhi, J. M. and W. Tom. The escape system of the cockroach *Periplaneta americana*. I. Turning responses to wind puffs. *J. Comp. Physiol.*, **128** (1978), 193–201.

Camhi, J. M., W. Tom, and S. Volman. The escape system of the cockroach *Periplaneta americana*. II. Detection of natural predators by air displacement. *J. Comp. Physiol.*, **128** (1978), 203–12.

Coler, R. R., J. S. Elkenton, and R. G. Van Driesche. Density estimates and movement patterns of a population of *Periplaneta americana*. *J. Kansas Entomol. Soc.*, **60** (1987), 389–96.

Edmunds, L. R. Observations on the biology and life history of the brown cockroach, *Periplaneta brunnea* Burmeister. *Proc. Entomol. Soc. Wash.*, **59** (1957), 283–6.

Fleet, R. R., G. L. Piper, and G. W. Frankie. Studies on the population ecology of the smoky-brown cockroach, *Periplaneta fuliginosa*, in a Texas outdoor environment. *Environ. Entomol.*, **7** (1978), 807–14.

Geier, H. T. Growth rate in the cockroach *Periplaneta americana* (Linn.). *Ann. Ent. Soc. Am.*, **40** (1947), 303–17.

Gordon, J. M., P. A. Zungoli, and L. W. Grimes. Effects of relative humidity on ovipositional behavior in *Periplaneta fuliginosa* (Blattodea: Blattidae). *Environ. Entomol.*, **23** (1994), 299–303.

Griffiths, J. T. and O. E. Tauber. The nymphal development for the roach, *Periplaneta americana* L. *J. N.Y. Entomol. Soc.*, **50** (1942), 263–72.

Gupta, P. O. On the structure and formation of spermatophore in the cockroach, *Periplaneta americana* (Linn.). *Ind. J. Entomol.*, **8** (1947), 79–84.

Jackson, W. B. and P. P. Maier. Dispersion of marked American cockroaches from sewer manholes in Phoenix, Arizona. *Am. J. Trop. Med. Hyg.*, **4** (1955), 141–6.

Additional studies on dispersion patterns of American cockroaches from sewer manholes in Phoenix, Arizona. *Ohio J. Sci.*, **61** (1961), 220–6.

Jacobson, M., M. Beroza, and Y. T. Yamatoto. Isolation and identification of the sex attractant of the American cockroach. *Science*, **139** (1963), 48–9.

Komiyama, M. and K. Ogata. Observations on the life history of the Japanese cockroach, *Periplaneta japonica* (Karney), under outdoor conditions. *Jpn. J. Sanit. Zool.*, **32** (1981), 111–15.

Liang, T. L., C. C. Hsu, H. C. Tang, and A. C. Fan. Observation on the life history of *Periplaneta fuliginosa*. *Acta Entomol. Sinica*, **26** (1983), 352–5.

Nigam, L. N. The life-history of a common cockroach (*Periplaneta americana* Linneaus). *Ind. J. Agric. Sci.*, **3** (1933), 530–43.

Powell, P. K. and W. H. Robinson. Descriptions and keys to the first instar nymphs of five *Periplaneta* species (Dictyoptera: Blattidae). *Proc. Entomol. Soc. Wash.*, **82** (1980), 212–28.

Rau, P. The life history of the American cockroach, *Periplaneta americana*. Linn. (Orthopt.: Blattidae). *Ent. News*, **51** (1940), 121–4, 151–5, 186–8, 223–7, 273–8.

Schoof, H. F. and R. E. Siverly. The occurrence and movement of *Periplaneta americana* within an urban sewerage system. *Am. J. Trop. Med. Hyg.*, **3** (1954), 367–71.

Shiraki, Y. Studies on the cockroach, *Periplaneta japonica* (1) (in Japanese, English summary). *Jpn. J. Sanit. Zool.*, **15** (1964), 138–44.

Studies on the cockroach, *Periplaneta japonica* (2) (in Japanese, English summary). *Jpn. J. Sanit. Zool.*, **17** (1966), 113–19.

Simon, D. and R. H. Barth. Sexual behavior in the cockroach genera *Periplaneta* and *Blatta* I. Sex pheromones and behavioral responses. *Z. Tierpsychol.*, **44** (1977), 162–7.

Sonan, H. Observations upon *Periplaneta americana* Linnaeus and *Periplaneta australasiae* Fabricius (in Japanese). *Trans. Nat. Hist. Soc. Formosa*, **14** (1924), 4–21.

Spencer, G. J. On the oviposition habits of the Australian cockroach, *Periplaneta australasiae* (Fab.). *Proc. Ent. Soc. British Columbia*, **40** (1943), 29–30.

Tabaru, Y. and A. Kobayashi. Outdoor hibernation of *Periplaneta japonica* (Blattaria: Blattidae) in snow area (in Japanese, English summary). *Jpn. J. Sanit. Zool.*, **22** (1971), 76–7.

Takagi, M. Ecological studies on the smokybrown cockroach, *Periplaneta fuliginosa* S. I. Seasonal change of the numbers and the age structure of a natural population, and the distribution in a house. *Jpn. J. Sanit. Zool.*, **25** (1974), 27–34.

Takahashi, S., K. Watanabe, S. Saito, and Y. Nomura. Isolation and biological activity of the sex pheromone of the smokybrown cockroach, *Periplaneta fuliginosa* Serville (Dictyoptera: Blattidae). *Appl. Entomol. Zool.*, **30** (1995), 357–60.

Tsuji, H. Development of the smokybrown cockroach, *Periplaneta fuliginosa*, in relation to resistance to cold. *Jpn. J. Sanit. Zool.*, **26** (1975), 1–6.

Wei-Nung, L. Preliminary notes on the seasonal fluctuations of three species of cockroaches (*Periplaneta americana*, *P. emarginata*, *Eupolyphyga sinensis*). *Acta Entomol. Sinica*, **12** (1963), 664–9.

Wharton, D. R. A., A. Lola, and M. L. Wharton. Growth factors and population density in the American cockroach, *Periplaneta americana*. *J. Insect Physiol.*, **14** (1968), 637–53.

Woo, F. C. Species of the genus *Periplaneta* Burmeister from China, with reference to their bionomics and economic importance (Blattaria: Blattidae). *Acta Entomol. Sinica*, **25** (1982), 416–22.

Wright, C. G. Life-history of the smokybrown cockroach. *J. Ga. Entomol. Soc.*, **14** (1979), 69–75.

### *Pycnoscelus*

Caudel, A. N. *Pycnoscelus surinamensis* Linneaus (Orthoptera); on its nymphs and the damage it does to rose bushes. *Proc. Entomol. Soc. Wash.*, **27** (1925), 154–7.

Roth, L. M. Sexual isolation in parthenogenetic *Pycnocelus surinamensis* and application of the name *Pycnocelus indicus* to its bisexual relative (Dictyoptera: Blattaria: Blaberidae: Pycnoscelidae). *Ann. Entomol. Soc. Am.*, **60** (1967), 774–9.

Reproductive potential of bisexual *Pycnocelus indicus* and clones of its parthenogenetic relative *Pycnocelus surinamensis*. *Ann. Entomol. Soc. Am.*, **67** (1974), 215–23.

Cave dwelling cockroaches from Sarawak, with one new species. *System. Entomol.*, **5** (1980), 97–104.

Saupe, R. Zur Kenntnis der Lebensweise der Riesenschabe *Blabera fusca* Brunner und der Gewachshausschabe *Pycnoscelus surinamensis* L. *Z. Angew. Entomol.*, **14** (1928), 461–500.

Schwabe, C. W. Observations on the life history of *Pycnoscelus surinamensis* (Linn.), the intermediate host of the chicken eyeworm in Hawaii. *Proc. Hawaiian Entomol. Soc.*, **13** (1949), 433–6.

Zappe, M. P. Life history and development of the greenhouse cockroach, *Pycnoscelus surinamensis* Linn. *Conn. Agr. Exp. Station Bull.*, **211** (1918), 311–13.

### *Shelfordella, Supella*

Artyukhina, I. N. and M. N. Sukhova. The ecology of the Turkestan cockroach (*Shelfordella tartara* Sauss) in different types of human habitation in the Uzbek SSR (in Russian). *Med. Parazitol. Parazit. Bolezni.*, **41** (1972), 49–53.

Back, E. A. The increasing importance of the cockroach, *Supella supellectillium* Serv. as a pest in the United States. *Proc. Entomol. Soc. Wash.*, **39** (1937), 205–13.

Benson, E. P. and I. Huber. Oviposition behavior and site preference of the brownbanded cockroach, *Supella longipalpa* (F.) (Dictyoptera: Blattellidae). *J. Entomol. Sci.*, **24** (1988), 84–91.

Hafez, M. and A. M. Afifi. Biological studies on the furniture cockroach *Supella supellectillum* Serv. in Egypt (Orthoptera: Blattidae). *Bull. Soc. Entomol. Egypt*, **40** (1956), 365–96.

Kapanadze, E. I. The biology of a Turkestan cockroach, *Shelfordella tartara* Sauss., in the insectary (in Russian). *Med. Parazitol. Parazit. Bolezni.*, **40** (1971), 595–600.

Shaw, E. *Supella supellectillum* Serv. a cockroach not before recorded from Australia. *Queensland Nat.*, **4** (1924), 115.

Whelan, D. B. *Supella supellectillum* (Serville) as a household pest in Nebraska. *J. Econ. Entomol.*, **22** (1929), 421.

# 5 Coleoptera

## Introduction

Beetles comprise the largest order of insects, with over a quarter of a million described species. They have four wings, with the front pair thickened, leathery, or hardened, and usually covering the hind wings. Their chewing mouthparts are well developed in the adults and larvae. Mandibles range from slender and sharp use for predaceous habits, to large and toothed, used to crush plant and animal material, or gnawing wood. In weevils, the front of the head is produced into an elongated snout, with the mandibles at the end. Development in this order includes egg, larva, pupa, and adult stages. Larvae vary considerably in form, but most are elongate and flattened or grub-like. The life cycle varies in length to provide several generations a year to one generation in several years. Overwintering generally occurs as a larva or pupa in protected sites, but some species overwinter as adults in large aggregations.

Wood is a source of food and harborage for several species of beetles. However, cellulose and lignin that give wood strength and durability are not easily converted to food. Only a small number of insects actually derive nutrition from wood. Others, such as carpenter ants and carpenter bees, have mouthparts strong enough to tunnel into sound or slightly decayed wood, but their gut is not equipped with the enzymes or microorganisms necessary to convert it into food components. Utilization of wood as food or shelter requires strong mouthparts to bite off small pieces, and a digestive system capable of breaking down cellulose. The larvae of wood-infesting insects have these essential features, and this stage creates the powder-fill tunnels characteristic of infested wood.

To utilize wood as a harborage and food, beetle adults and larvae have evolved elaborate morphological and physiological features, and specific behavioral patterns. Adult body shape is generally cylindrical, short, rounded, and sometimes with a spiny prothorax. This body form is geometrically efficient for boring, and may explain the similarity between adults of different wood-boring species. Ridges, spikes, and horns on the head, prothorax, and elytra often occur at sites where they do not impede boring or are located on the prothorax where they probably serve as an anchor or hold-fast to assist the mandibles. Larval forms are soft-bodied and capable of stretching, contracting, and pressing against the walls of feeding tunnels. Movement in narrow tunnels and galleries is assisted by setae, spines arranged on swollen segments, and legs of various length and form. Mandibles of wood-infesting beetles are usually modified to cut and tear at the substrate, or to plane it smooth with sharp surfaces. Species that feed on dry (8–12% moisture) structural wood have short broad mandibles, and species that feed on live trees, recently felled green wood, or wood with a high moisture content have long and thin mandibles. Various features of the digestive system of some species seem specifically suited to eating wood. Larvae of *Anobium punctatum* have a grinding gizzard in their gut, which is linked to their feeding on wood. The gizzard is a spherical enlargement of the foregut that has a dense layer of spines and teeth-like structures. It functions to grind and mix the food in preparation for digestion. Mouthparts work in conjunction with the gizzard to provide a continuous supply of uniform-size wood particles. Those too small or too large fall away and are swept to the back by the front legs of the larva. A gizzard is absent from anobiids that do not utilize cellulose as a source of food.

Pest status of beetles in the urban environment is based primarily on the feeding habits of the larval stage, and the overwintering behavior of the adults of a few species. Few species transmit pathogenic organisms, but some, such as dermestids or blister beetles, cause skin irritations. About 600 beetle species in 34 families are associated with stored grain or food materials made from grains. However, stored fruits and vegetables are also attacked. Some beetles are associated with stored food because they feed on the mold and fungi growing on these substrates, or they are predators of insects or arthropods that are the primary invaders. The original habitats of food-infesting beetles were probably dry, sheltered sites where plant and animal debris collected. Animals building nests or storing food in sheltered places created some of these sites. Likely habitats include bird nests built close to human dwellings, bat roosts, and the dens of rats and mice. Species that now attack grain seeds, such as wheat, maize, and other cereals, may not have fed on the seeds of their ancestral grasses. These seeds would have been very small and not likely infested by the larvae of these beetles. Development of large varieties of primitive grains would have provided beetle larvae sufficient food to complete development. Grain borers probably evolved from attacking the structural features of storage bins to the seeds. They probably fed on larger wild seeds such as acorns, which were often scattered, and moved from these to grain, which was stored in quantity and protected.

## Anobiidae

Anobiid adults are 1.5–8 mm long, reddish brown to black, and covered with fine setae or pubescence. The head is usually hidden by the pronotum, and the 11-segmented antennae are inserted on the sides of the head, in front of the eyes. Full-grown larvae are about 8 mm long, grub-shaped, and yellowish white. Larval thoracic segments are enlarged dorsally, and the segments have a dorsal patch of large setae. Adults of many species are nocturnal and are rarely seen during the day. Feeding habits include infesting dry vegetable material, stored food materials, spices, and tobacco, and several species that feed on seasoned wood. The majority of anobiids feed in dry (8–12% moisture) hardwood and softwoods; their larvae are capable of digesting cellulose. Food-infesting species invade a broad range of substrates, and there are biotypes adapted to local conditions and food sources. The anobiid pests of stored food have worldwide distribution.

**Furniture beetle, *Anobium punctatum*** Adults are 2.5–6 mm long, cylindrical, and reddish brown. Elytra have punctures in longitudinal rows; the last three antennal segments are longer than the others. Exit holes are 1.6–3 mm diameter, and the frass has a gritty feel when rubbed between the fingers. Full-grown larvae are about 6 mm long, yellowish white with dark brown mandibles; there are scattered pale yellow setae on the body, and a double row of spines on the dorsum. Eggs are visible to the unaided eye and are smooth, white, and slightly ellipsoidal. Eggs are laid singly or in batches of 2–4. They are often cemented on to rough surfaces of lateral and end grain wood, in cracks, deposited into spaces between pieces of wood, or in emergence holes, and pupation chambers from which the female emerged. Fecundity is 20–80 eggs; hatching occurs in 3–6 weeks at 65% relative humidity (RH); no eggs hatch at or below 45% RH. First-stage larva burrows into wood through the base of the egg. As it moves out of the egg the larva ingests yeast deposited on the egg by the female, and these yeasts become established within the larval gut. Yeasts are killed at 26 °C, which restricts the initial development or reinfestations of timbers in some roofs.

Larval development takes about 2 years at 22–24 °C and 70% RH, but can extend to 3–5 years. In natural habitats, such as branches of dead trees, the life cycle is completed in about 1 year. Larvae produce a network of 1–2-mm diameter galleries filled with frass and wood fragments. Pine and spruce are most susceptible to attack; there are few infestations in Douglas fir and western hemlock. Most hardwoods in temperate regions are attacked, but rarely tropical hardwoods. The outer portion of sapwood, which is usually high in nitrogen, is the preferred larval feeding site, but heartwood of beech, birch, and poplar may be utilized. The limited amount of nitrogen in old wood results in extensive larval galleries and slow development. Larvae feeding on Scots pine digest 26–29% of the wood eaten, and utilize 7–8% of the lignin, 40% of the cellulose and hemicellulose, and most of the protein in the wood. Some types of plywood have been attacked, and beetles emerged after the wood was in place. Full-grown larvae bore close to the wood surface to form a pupal chamber. Pupal stage is 3–8 weeks, and adults bite through the outer surface of the wood to emerge and mate.

Adult beetles live 2–4 weeks and readily fly during the warm season. Adults emerge in May or June, but in northern latitudes they emerge in June and July. Mating males and females retreat into exit holes and crevices in the timber; males disengage from females after about 1 h and return to the wood surface. Females remain in the exit hole to lay eggs, and return to the surface to mate again. Males are attracted to a range of concentrations of the female sex pheromone, which is 2,3,5-trimethyl-6-(1-methyl-2-oxobutyl)-$^4$H-pyran-4-one (Stegobinone). This chemical is the same as the sex pheromone for *Stegobium paniceum*, the drugstore beetle.

*A. punctatum* is widespread in North America and Europe, especially along coastal regions where ambient air humidity is high. It occurs inside buildings and outside in natural populations in dead wood. In spite of its common name, it damages structural wood rather than furniture. However, it enters houses in infested furniture and spreads to exposed, unfinished wood. Attack occurs in structural softwood or hardwood that is free of decay fungi. Temperature, humidity, and the condition of wood limit larval development. Adults produced in wood in dry environments are generally small, and less successful in mating and egg-laying. These adults produce fewer eggs and therefore decrease population viability. This combination of factors may explain why infestations decline when wood moisture levels remain low.

There are several parasites of the furniture beetle. The mite *Pymotes tritici* (Pymetodidae) parasitizes several species of wood-boring insect larvae, including *Anobium* larvae. *P. tritici* can bite humans, resulting in what is called woodworm bite, but it is unable to burrow into human skin. The ant-like pteromalid wasp, *Theocolax formiciformis*, is also a parasite of *Anobium* larvae. The adults have rudimentary wings and do not fly. The adult wasp moves through the tunnels in the wood to locate a beetle larva, or it oviposits into the larva through the wood surface.

**Queensland pine beetle, *Calymmaderus incisus*** Adults are about 3 mm long, brown and shiny. Attack in structures usually begins in the dark, subfloor areas and spreads from there to other wood members. It has not been reported from other areas, but the damage is similar to that of *A. punctatum* and it could have been overlooked. This beetle is distributed in Queensland, Australia where it attacks softwood timber. Infestations are predominantly in hoop pine, bunya pine, and New Zealand white pine.

**Anobiid bark beetle, *Ernobius mollis* (Fig. 5.1a)** Adults are 3.5–5.5 mm long and reddish brown. Full-grown larvae are about 7 mm long, curved and pale yellow; the head is distinct and the mandibles are dark brown. Exit hole is round and about 1.5–2 mm diameter. Eggs are laid in crevices in the bark of trees; hatching is in 10–21 days. First-stage larvae are capable of crawling on the surface of the bark for a few days before finding a route to the sapwood. Larvae feed at the interface of the bark and sapwood, and frass produced from this feeding is a mixture of brown and white particles. The brown pieces are from the bark, and yellow pieces are from the sapwood. Frass is tightly packed in galleries but is dislodged and falls free when the wood dries. Larvae pupate in an enlarged portion of their gallery and the pupal stage lasts about 10 days. Larval development is completed in 1–2 years. Adults remain in the pupal chamber for 6–12 days before biting their way to the wood surface. Emergence is during summer and extends into fall. Mating takes place within 1–2 h of emergence; copulation lasts about 6 h. Females mate 2–3 times, and eggs are laid at night a few days after mating. Adults live about 30 days, but sometimes they overwinter.

This beetle infests softwood timber, especially wood that has small sections of bark remaining. The damage caused by *E. mollis* is often confused with damage caused by some bark beetles (Scolytidae). *E. mollis* restricts its feeding to the outer sapwood beneath the bark, and does not reinfest or cause structural damage to wood in use. This species has been introduced into the eastern USA and lives outdoors in various conifer trees. Related species, *E. tenuicornis* and *E. granulatus*, occur in eastern USA, but they do not infest wood in use.

**Anobiid powderpost beetle, *Euvrilletta peltata* (= *Xyletinus peltatus*) (Fig. 5.1e)** Adults are 3.4–6.3 mm long, reddish brown to black, and with fine, yellow pubescence covering the body. Eyes of male are separated by two times the width of the eye; female eyes are separated by three times the width. Antennae are 11-segmented, moderately serrate, and the last three segments are slightly enlarged. Exit holes are about 3 mm diameter; the frass from softwoods (pines, spruce, and fir) has a gritty texture when rubbed between the fingers. Full-grown larvae are about 8 mm long and yellowish white. Eggs are laid in groups of 2–3 on the surface of wood, females prefer wood that is 2–5 years old, and rough surfaces; fecundity is 20–60 eggs, but females lay 120. Hatching is in about 8 days, and is not significantly reduced by 11–12% wood moisture or low (54%) RH. First-stage larvae bore into the wood a short distance, then turn and tunnel in the direction of the wood grain. They feed first on springwood and later in the outer sapwood; galleries are packed with fecal pellets and wood fragments. Few first-stage larvae survive in wood with moisture content below 13%, but wood moisture as low as 12% does not reduce the survival of larvae that are 3–6 months old. Larvae cease or reduce feeding in response to low (winter) temperatures and low wood moisture content. They do the most feeding at about 24 °C; feeding is reduced or stops at 10 °C and at 32 °C. Larvae resume feeding in spring, but those not full-grown continue to feed for a year or more. Full-grown larvae tunnel close to the wood surface and prepare a pupal chamber. The pupal period is about 14 days, and adults cut the emergence holes in the wood surface. The life cycle is 1–5 years, depending on the quality of wood infested, fluctuations in temperature, and wood moisture.

Pine timber may be reinfested and infestations may continue until nearly all the sapwood portion of the infested wood piece has been consumed. Infestations rarely develop uniformly over

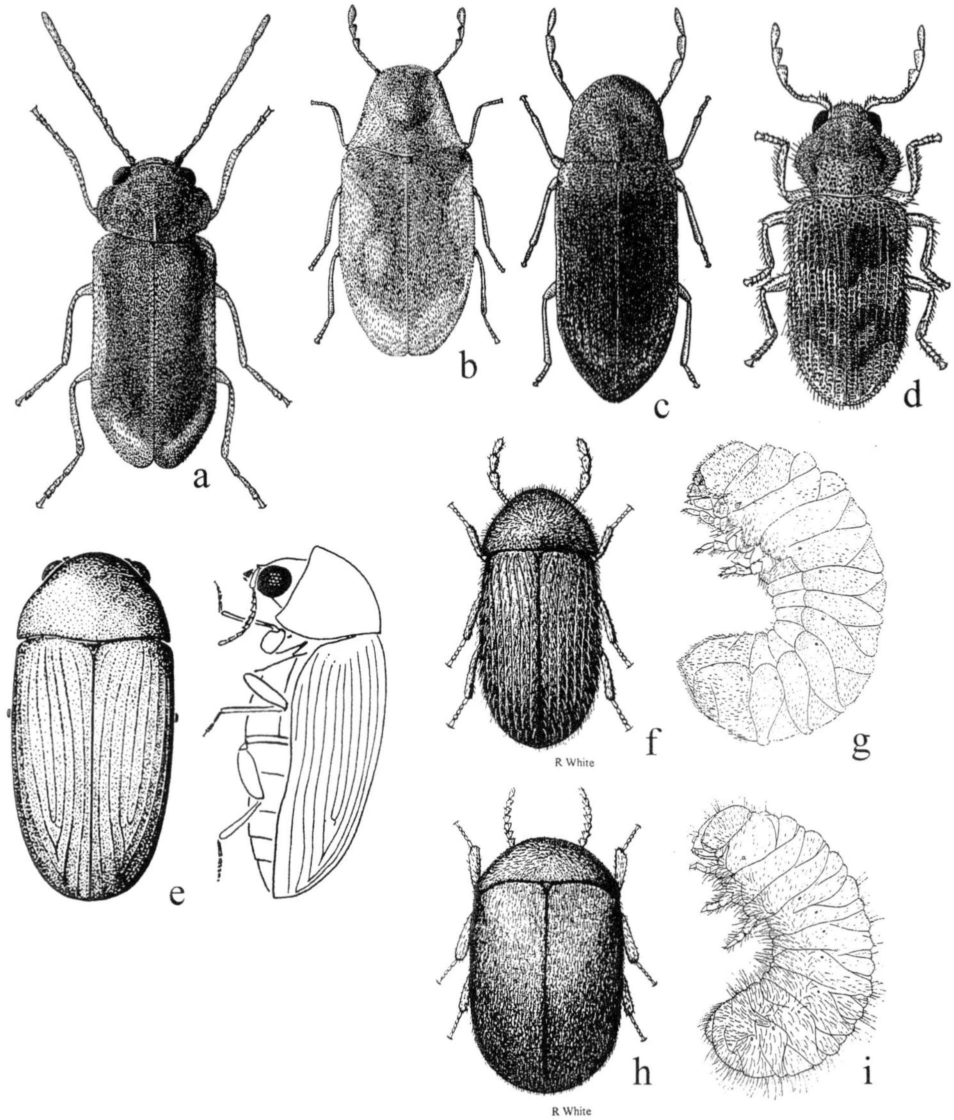


**Figure 5.1** Coleoptera: Anobiidae. (a) *Ernobius mollis*; (b) *Falsogastrallus sauteri*; (c) *Gastrallus immarginatus*; (d) *Nicobium hirtum*; (e) *Euvrilletta peltata*; (f) *Stegobium paniceum*; (g) *S. paniceum* larva; (h) *Lasioderma serricorne*; (i) *L. serricorne* larva.

an area, because of parasites, wood properties, and the size of the initial infestation. Adult beetles begin emerging in spring and continue through mid-summer; more than 90% of the total emergence occurs within 4–6 weeks. Adults do not feed and are nocturnal. This species is distributed in eastern USA, but it has probably spread to other countries. It occurs naturally in dead branches and downed logs from northern New York to Michigan, southeastward to Florida and southwestward to Arkansas. It infests dead and seasoned parts of hardwood and softwood trees, including heartwood and sapwood. A closely related species, *E. lugubris*, occurs in natural habitats, and it tunnels and feeds in dead oak twigs.

Braconid parasites of *E. peltata* in North America include *Heterospilus longicauda*. The female oviposits through the outer surface of the wood on to beetle larvae feeding close to the surface; one parasite develops on each beetle larva. Beetle larvae feeding deep in the wood are usually beyond the reach of the parasite's ovipositor. After feeding on beetle larvae and becoming adults, the *H. longicauda* wasps make an exit hole in the wood surface, which is smaller than the powderpost beetle hole. The parasitic wasp, *Theocolax formiciformis* (family Pteromalidae) attacks larvae by burrowing through its tunnels. The adults of this parasite often exit through existing beetle emergence holes.

**Decayed-wood anobiid, dampwood borer, *Hadrobregmus australiensis*** Adults are 7–8 mm long and blackish brown. Emergence holes are round and about 3 mm diameter; pupal chambers are usually close to the wood surface and stained black inside. Frass is powdery, but because of the moisture content and wood decay in the substrate, it usually sticks together in tunnels. This beetle is limited to attacking decayed and moisture-damaged wood. It infests soft- and hardwoods, but only when conditions are conducive for wood-decay fungi. It is distributed in Australia.

**Eastern death watch beetle, *Hemicoelus carinatus* (= *Hadrobregmus*)** Adults are 4–6 mm long, reddish brown to dark brown, and have 10-segmented antennae. It naturally infests ash, basswood, maple, beech, and elm. Household infestations include structural beams, roof and floor joists, sills, and flooring, both new and old wood. Damage resembles that of lyctids in hardwoods, but the emergence holes of *H. carinatus* are larger and the frass is not a fine powder. This species is distributed primarily in eastern North America.

**Western death watch beetle, *Hemicoelus gibbicollis* (= *Hadrobregmus*)** Adults are 2.5–5 mm long and pale brown to brown. The dorsum of the thorax is pointed, and the antennae are 11-segmented. Full-grown larvae are about 4 mm long and pale brown to brown anteriorly. Eggs are laid singly or in batches of 3–32 in cracks or in the end grain of the wood; rarely are eggs laid in emergence holes. Fecundity is about 200 eggs; hatching occurs in about 22 days, and with 84–88% success. First-stage larvae usually crawl about 1 cm from the egg before penetrating the wood. First-stage larvae tunnel close to the wood surface, and in the direction of the wood grain. Wood moisture between 14 and 17% is optimal for larval development; larvae do not survive longer than 18 months when wood moisture is 11–12%. Larval development takes 3–6 years, and the pupal period is 15–22 days. Adults emerge from June through August, but in warm regions of its distribution (California) emergence begins earlier and extends later in the fall. Mating occurs soon after emergence and females oviposit after an inactivity period of 18–36 h. Natural infestations occur in a variety of hardwoods and softwoods, including maple, oak, western white pine, western red cedar, and redwood. This species is a pest of building timbers in coastal regions of western North America, including Washington, Oregon, and California. Douglas fir is attacked, especially in buildings that are 20 years old or older.

**Cigarette beetle, *Lasioderma serricorne* (Fig. 5.1h, i)** Adults are about 3 mm long and light brown. The head and prothorax are bent downward to give a humpbacked appearance, the elytra are smooth, and the antennae are serrate. Full-grown larvae are about 4 mm long, scarabeiform, and setose. Larvae feed on seeds, nuts, beans, spices, yeast, dried insects, fish and vegetables, flour, meal, and tobacco. Eggs are laid singly in crevices in the larval food material; hatching is in 20–22 days at 20 °C and 5–6 days at 35 °C. Minimum temperature for egg hatch is about 19 °C and the maximum is about 39 °C. Low humidity inhibits hatching, but at 30 °C some eggs will hatch at 20% RH. First-stage larvae move away from light in search of food; they can enter small openings in the seams of food packages and infest a variety of material. Late-stage larvae are C-shaped and they are not able to penetrate smooth surfaces or cracks and crevices. Optimal humidity for development of the 4–6 larval stages is between 70 and 80% RH; temperature thresholds are 19 °C and 39 °C. Larvae are inactive at about 17.5 °C and enter dormancy at temperatures below this; they can remain dormant for several months at low temperatures. Larvae construct silk chambers for pupation, and they are often covered with fragments of the substrate. Pupal development is 4 days at 34 °C, 12 days at 20 °C, and it is not affected by humidity. Adults remain in the pupal chamber 4–12 days before emerging; they feed on the same food as the larvae. Development on various stored foods ranges from 106 to 135 days, and with a 2–15% survival rate. Development on wholemeal flour and peanuts ranges from 27 to 34 days, with a survival rate of 100–50%. Males live 21–43 days, and females live 18–46 days. This species is widely distributed in stored foods, and is nearly cosmopolitan.

**Common book beetles, *Neogastrallus librinocens*, *Nicobium hirtum* (Fig. 5.1d)** Adults are about 2.4 mm long and reddish brown. Full-grown larvae are about 3 mm long and pale white. Several anobiids are associated with the habitat conditions and mold fungi that commonly occur with books and stored grain, primarily in tropical or humid environments. *N. librinocens* infestations occur in both old and new books, especially those in storage. Larval feeding often results in book binding being cut and pages with numerous small holes. A related species, *Gastralis immarginatus* (Fig. 5.1c) is also known to infest books. This species is 2.2–3 mm long and uniformly dark brown.

**Broadheaded anobiid, *Platybregmus canadensis*** Adults are 4–6 mm long and the body is reddish brown to dark brown. Antennae are 11-segmented and the last three segments are distinctly larger than preceding segments. This species attacks maple and basswood flooring and elm timbers in buildings.

It is distributed from southern Ontario, Canada, south to New Jersey.

**Silky anobiid, *Priobium sericeum* (= *Trypopitys sericeus*)** Adults are 4.9–6.2 mm long and reddish brown. They are covered with yellowish white setae, and the head is deeply retracted within the prothorax. Antennae are 11-segmented: segments 3–7 are distinctly serrate, 8–10 somewhat serrate, and 11 is narrow and elongate. In natural habitats, this species lives in dead tree branches of oak, hickory, and cherry. Indoors it infests flooring and furniture in buildings. A related species, *Priobium punctatum* (= *Trypopitys*), infests Monterey cypress, pine, and in oak door casings and flooring, and maple wainscoting. It occurs in eastern North America, from Nova Scotia and southward to Florida, Texas, and Arizona.

**Hardwood anobiid, *Ptilinus pectinicornis*** Adults are 3–6 mm long, dark brown, and slightly cylindrical. The prothorax is globular; the antenna of the male is comb-like, and the female has toothed or serrate antennae. It infests beech, ash, maple, and sycamore. Frass is powdery, but not gritty when rubbed between fingers. This species is a common pest of structural hardwood timber, and it is often found with *A. punctatum* in mixed infestations.

***Ptilinus ruficornis*** Adults are 2.8–4.8 mm long. Males are either completely black or black with brown elytra, and have reddish-brown to reddish-yellow appendages; females are reddish brown. Female antennae are serrate; those of the male are branched, with each branch on segments 5–10 nearly as long as the combined length of segments 1–10. This species infests beech, maple, oak, sycamore, and other hardwoods, and is a common pest of woodwork in buildings. This species occurs in North America.

**Drugstore beetle, *Stegobium paniceum* (Fig. 5.1f, g)** Adults are 2.2–3.5 mm long, reddish brown, and somewhat cylindrical. Head and prothorax are bent downward, and there are fine setae arranged in rows on the elytra. It is distinguished from the cigarette beetle, *Lasioderma serricorne*, by the striate elytra and the enlarged terminal antennal segments. Full-grown larvae are about 3 mm long and slightly setose. Eggs are laid singly during 3–4 weeks; fecundity is 75–100 eggs; hatching is in 12–37 days. First-stage larvae are about 0.5 mm long and 0.125 mm wide. Their small size enables them to enter narrow openings, especially in packaged foods. They can survive for about 8 days without food. Larval development is 4–5 months and includes 4–6 instars; temperature limits are 15 °C and 34 °C, and a minimum of 35% RH. The full-grown larva produces a cocoon covered with particles from the substrate, but cocoons are absent in some populations. Pupal period is 12–18 days. Adults live about 85 days at 17.5 °C and 50–70% RH, and about 13 days at 32.5 °C and 70% RH. There are three or four generations per year. This species is cosmopolitan and infests a variety of food materials. The species name, paniceum, refers to feeding on bread, but it has a much broader food preference. Infestations are known from a variety of foods and other materials, including black and red pepper, medical drugs, grain, spices, tobacco, leather, wood, and textiles. This species is cosmopolitan.

***Trichodesma gibbosa*** Adults are 4.5–6.8 mm long and covered with grayish-white setae. Infestations have been found in joists and structural timbers made of sweet gum in historic buildings (coastal Virginia). It occurs in eastern North America, from New Hampshire to western Ontario and southward to Florida and Texas.

**Mexican book beetle, *Tricorynus herbarius*** Adults are about 3 mm long and reddish brown. It is typically associated with humid conditions and with mold fungi in books and similar material. Larvae feed in galleries in the substrate. This species is closely related to *T. confusum* and *T. tobaci*, which are similar in appearance. Another *Tricorynus* species attacks stored pinecones in southern USA.

**Deathwatch beetle, *Xestobium rufovillosum*** Adults are 6–7.5 mm long, reddish brown to dark brown, and have a pattern of yellowish-gray setae on the region behind the head and the wing covers. Antennae are 11-segmented with the last three segments somewhat enlarged. Adults probably do not fly, but they can raise their wing covers and flutter their hind wings. Full-grown larvae are about 11 mm long, pale yellow, and with black mandibles. Eyes of the larva consist of two small dark spots on each side of the head. Eggs are white, sticky, and somewhat pointed at one end; they are laid in batches of three or four in cracks and crevices or in nearby emergence holes. Fecundity is 40–60 eggs, but 200 has been recorded. Hatching occurs in 20–35 days and is successful between 23 and 41% RH; hatching does not occur above 30 °C. First-stage larvae are capable of crawling on the surface of the wood for a short time, then penetrate to the outer sapwood. They usually tunnel between the ray cells. Development takes 2–3 years, but can extend to 10 years. In the laboratory, the life cycle is completed in 1 year. Full-grown larvae tunnel close to the wood surface to form a pupal chamber. Pupal period is 3–4 weeks, and the adult overwinters in the chamber and emerges the following spring. Females emerge after the males and are immediately ready for copulation. Mating usually takes place in the afternoon and lasts about 1 h. Adult beetles do not feed and live about 10 weeks. Females begin laying eggs 5–11 days after mating.

This species is distributed in eastern North America, the UK, and continental Europe. It occurs primarily in oak and especially wood that has been previously attacked by decay fungi. Outdoor habitats are in sections of dead wood or branches of hardwood trees where fungal decay occurs. A common outdoor site for natural populations in the UK is the dead wood in the trunk of pollard willows. Indoors it usually infests wood that has been exposed to rainwater or excessive moisture; recently grown oak is not usually attacked unless fungal decay is present. European oak does not become susceptible to attack for about 60 years, which is sufficient time for exposure to moisture and establishment of decay fungi. American oak is slightly more susceptible to decay fungi and to *X. rufovillosum* attack. Infestations are in sapwood or heartwood, but sapwood has a higher nutrient content than heartwood and is the most susceptible to attack. Larvae generally restrict their feeding to the sapwood and avoid the heartwood unless there is fungal decay there. The presence of moisture in the wood and the resulting wood decay fungi determine the extent of heartwood infestation. Fungi associated with *X. rufovillosum* include *Fistulina hepatica*, *Coniophora cerebella*, *Phellinus cryptarum*, and *Polysticus versicolor*.

The common name of this anobiid is derived from its behavior of tapping on wood to attract a mate. Males and females call each other by striking the surface of wood. The sound of tapping in a quiet room of someone seriously ill was thought by superstitious people to be the knocking of the spirit of death at the door. To make the sound the male or female beetle moves its body forward rapidly to strike its head against a wood surface. There are 11–13 taps rapidly delivered at one time, but the human ear may hear only 6–8 taps. If another beetle detects tapping through vibrations in the wood, it answers by tapping in a similar manner. Tapping is usually heard in spring when adults emerge from infested wood and prepare for mating. Anobiid beetles prefer to mate in cracks and crevices; in the wood-infesting species they often retreat to the tunnels that open on to the wood surface. Entering an emergence hole while in copula is difficult, because the diameter of the hole and tunnel correspond to the height of a single beetle. After copulation has begun, the male moves off the female and extends backwards so that he is on his back. The resulting end-to-end position allows the female to enter an emergence hole while maintaining copulation with the male.

The predacious clerid beetle, *Korynetes caeruieus*, is often present in wood infested by deathwatch beetles, particularly at sites that have been infested for a long period. The adults are about 6 mm long and metallic blue. Larvae attack the deathwatch beetle larvae in their tunnels, and emerge from the same holes as the adult beetles.

**Other *Anobiidae*** *Ptilinus basalis* attacks California laurel (*Umbellularia californica*) and causes extensive damage. This wood is often used for rustic furniture. The small species, *Falsogastrallus sauteri* (Fig. 5.1b), is known to infest structural timber.

## Anthicidae

These beetles are 3–12 mm long and have a large head and narrow thorax; they are sometimes considered ant-like in appearance. The majority of species occur on flowers or in foliage or debris. A few species of *Anthicus* are associated with stored-food products, moldy foods, and dead insects indoors. Common species include *A. cervinus*, *A. ephippium*, *A. flivicans*, and *A. floralis*.

**Narrownecked grain beetle, *Anthicus floralis* (= *Omonadus*) (Fig. 5.2a)** Adults are 2.7–3.5 mm long, uniformly reddish brown. The head is broad and is wider than the thorax; eyes are oval and the antennae are clavate. The body is ant-like, with the pronotum somewhat oval in dorsal view. The full-grown larva is yellowish white and the head is light brown to dark yellow; the terminal segment is slightly sclerotized and forked. This beetle infests grain and dried fruit, and is cosmopolitan.

## Anthribidae

Adults are about 10 mm long, elongate and slightly oval; they have a short beak and straight antennae. These beetles are often refereed to as fungus weevils. There are many species in the tropics where adults and larvae are in fungi, and under the bark of dead trees. Adults feed on pollen, fungi, and dead wood, and larvae eat live plant material and seeds.

**Coffee bean weevil, nutmeg weevil, *Araecerus fasciculatus*** Adults are 3.0–4.3 mm long, brown to dark brown, and with scattered pale spots on the elytra. Antennae are longer than the head and thorax, and the last three segments are larger than

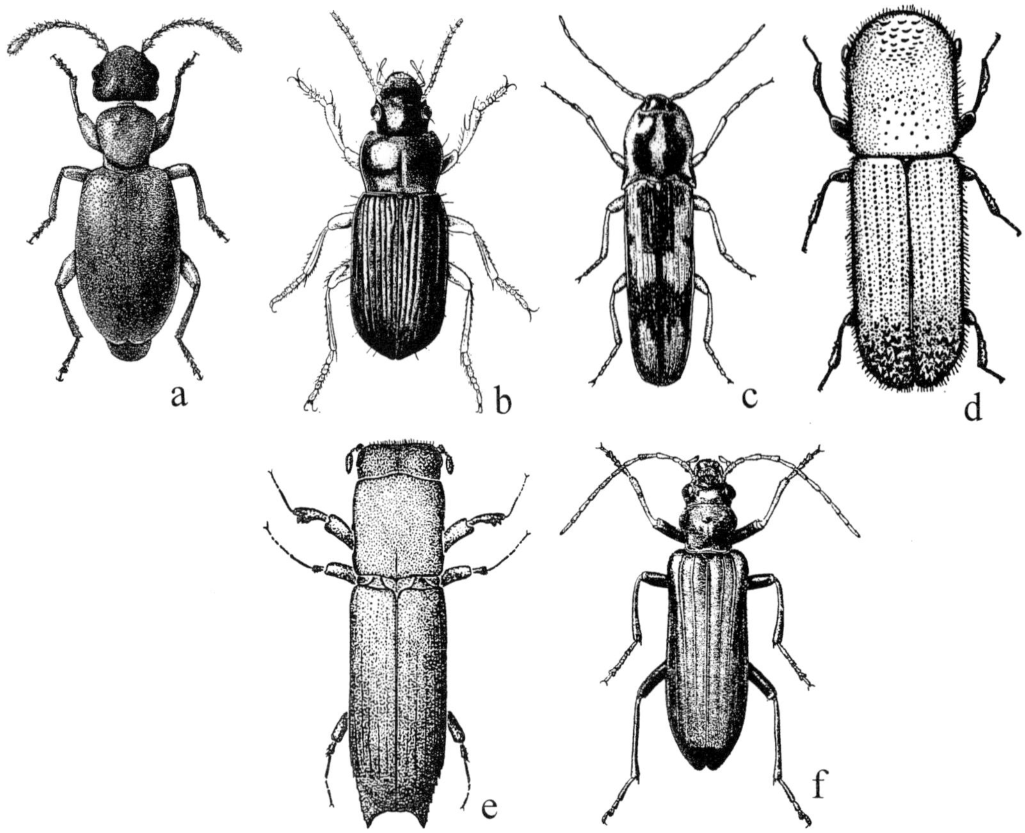


**Figure 5.2** Coleoptera: Anthicidae, Carabidae, Eleteridae, Platypodidae, Oedemeridae. (a) *Anthicus floralis* (Anthicidae); (b) *Harpalus rufipes* (Carabidae); (c) *Monocrepidius vespertinus* (Eleteridae); (d) *Xylebocus saxeseni* (Platypodidae); (e) *Platypus flavicornus*; (f) *Nacerdes melanura* (Oedemeridae).

preceding ones. Eggs are deposited into food material by the long ovipositor of the female; fecundity is about 50 eggs; hatching is in 5–8 days. Larval development takes about 30 days; the pupal period is 6–7 days at 27 °C and 50–100% RH. Larvae burrow within the seed and there is one larva per seed. On maize, adults live for 27 days at 50% RH, and live 86–134 days at 90% RH. Males are sexually mature in 3 days and females in 6 days after emergence. The infestation susceptibility of coffee, *Coffea canephora* and *C. arabica*, varies with humidity, but both are susceptible at 100% RH. Larval development takes 30–35 days at 30 °C, and 40–60 days at 25 °C. Development is completed in 65–90 days on *C. arabica* at 25 °C and 80% RH, but larvae do not develop on *C. canephora* under these conditions.

This species is distributed in tropical and subtropical countries where it attacks a variety of food crops in the field and in storage. Populations in unheated buildings in temperate regions usually die during winter. Larvae infest coffee, maize, Brazil nuts, dried fruit, ginger, palm kernels, cacao, cassava, peanuts, monkeypod, sweet potato, yams, avocado, and sorghum. Larvae also infest some dried-food material. Adults are capable flyers, and will fly from hillsides adjacent to residential buildings and gather at windows in large numbers (Hong Kong). They prefer food with high moisture content and usually infest slightly decayed material.

**Other *Araecerus*** There are several species that infest stored foods, including *A. crassicornis* in legume pods in Indonesia; and *A. levipennis*, Koa Haole seed beetle, in Hawaii.

## Apionidae

Adults are 1.5–4.5 mm long and the body is somewhat pear-shaped and blackish brown to black. They are characterized by having clubbed antennae that are not geniculate. A few species are important agricultural pests of field crops and some are carried into produce stores in infested material. The genera commonly associated with infested seeds and tubers are *Apion* and *Cylas*. Species of *Apion* are primarily pests of legumes, and the larvae complete development within the seeds. Adults occur indoors. There are several species of *Cylas* that bore into the tubers or the stems and climbers of sweet potato (*Ipomoea*).

**Sweet potato weevil, *Cylas formicarius elegantulus*** Adults are 5–6 mm long, and the body is slender, elongate, and antlike. The pronotum is reddish brown and the elytra are bluish black. Full-grown larvae are about 7 mm long and yellowish white; the head is dark brown; the body tapers posteriorly and is slightly C-shaped. Eggs are laid in the soil or in the infested tuber. Larvae bore into the stems and tubers of sweet potatoes, and pupate in the tunnel. Adults are found in the larval tunnels. Development continues in sweet potatoes that are harvested and stored. Adults emerge in storage buildings or in produce markets and stores. This species is common in tropical and subtropical areas, and it has been introduced into southern USA.

## Bostrichidae

Bostrichids are 3–6 mm long, and reddish-brown to black. They have a tuberculate and rasplike pronotum, and the body is truncate posteriorly. The head is directed down and not visible from above, the thorax is large and gives the adult a humpbacked appearance. Larvae are curved, the head is greatly reduced and the thorax enlarged. These beetles are distributed in the tropics and subtropics; many species in the family feed on wood, while others attack grain in storage. They feed in the sapwood of hardwoods, but a few attack softwoods. Larvae of some species attack freshly cut and partially seasoned wood with bark on the surface; others infest seasoned wood. Trees attacked include hickory, persimmon, and pecan, oak, beech, elm, and chestnut. Stored-food infestation is limited to a few species. Grain-feeding habits of these beetles may have been acquired recently, after the development of production agriculture and the practice of storing large quantities of grain. The large mandibles of bostrichids enable them to attack grain and structural wood. Oil seeds are not usually attacked.

Females bore circular, 2–3-mm-diameter entry holes through the bark into the sapwood; then they construct an egg tunnel across the grain. Eggs are laid at intervals in the tunnel and developing larvae feed in the sapwood. Larval feeding tunnels contain powdery frass. Feeding continues until the wood is dry. Adults can emerge after the wood has been processed into furniture or other items. Female bamboo borers deposit eggs in natural divisions and breaks in the wood, often at cut ends. Bostrichids usually have one generation per year, but the larvae take up to 6 years to develop in seasoned wood. Mating for some species is confined to the area immediately around the emergence hole. Males of *Bostrychus capucinus*, *Heterobostrychus brunneus*, and *Bostrychoplites cylincricus* approach the female backwards after first drumming on the wood surface with their front legs. This mating strategy makes it possible to copulate in the narrow tunnels below the wood surface.

**Large auger beetle, *Bostrychopis jesuita*** Adults are 8–10 mm long, dark brown, and the body is rounded. The prothorax has a patch of setae anteriorly; the posterior margins of the forewings are rounded and without lobes. Emergence holes are about 5 mm diameter. This species is distributed in Australia where it attacks hardwoods, including eucalyptus.

**Bamboo powderpost beetle, ghoon borer, *Dinoderus minutus* (Fig. 5.3b, e)** Adults are 2.5–3.5 mm long, brown and cylindrical. Eggs are laid on the surface or in crevices, and hatching occurs in 3–7 days. Larval development takes about 6 weeks and the pupal period lasts about 8 days. Adults remain in the pupal chamber for 2–3 days before chewing through the wood to the surface. Severity of the infestation is associated with the starch content of the food material. Bamboo with less than 18% moisture content is usually not attacked; smoke-dried bamboo is usually not attacked, because heating reduces the moisture content to about 5%. Larvae also feed in drugs, spices, cacao, maize, rice, stored grain, dried food, and flour. This species is native to Asia but it occurs in bamboo products around the world. Several other *Dinoderus* species attack dried cassava, bamboo, and wooden structures in Asia.

**Oriental bostrichid, oriental wood borer, *Heterobostrychus aequalis* (Fig. 5.3d, f)** Adults are 10–15 mm long, reddish brown to black and shiny. The pronotum is strongly convex and on the anterior half there are five or six broad, toothlike, marginal projections. Elytra are nearly tubular in shape; in males they end in two hook-like projections. Full-grown larvae are 6–13 mm long and yellowish white. Exit holes are about 3 mm diameter, and the frass is a fine powder. Eggs are laid on rough surfaces of lumber, logs, or in cracks or holes made by the female; hatching is in about 7 days. Larvae feed in 38-cm-long and 6-mm-diameter tunnels in the wood. Larval development takes 120–150 days; the pupal period is about 14 days. The life cycle can be completed in 1 year, but it can extend to 6 years. This large bostrichid is distributed in Asia, and occurs in Malaysia, Solomon Islands, New Caledonia, and Madagascar. It infests the sapwood of commercial woods, including *Alstonia* spp. (Pulai), *Anisoptera* spp. (Mersawa), *Artocarpus* spp. (Terap), *Dillenia* spp. (Simpoh), *Gluta* spp. (Rengas), *Hevea brasiliensis* (Rubberwood), *Intsia palembanica* (Merbau), *Mangifera* spp. (Machang), *Myristica* spp. (Penarah), and *Shorea* spp. (Yellow Meranti, Yellow Seraya).

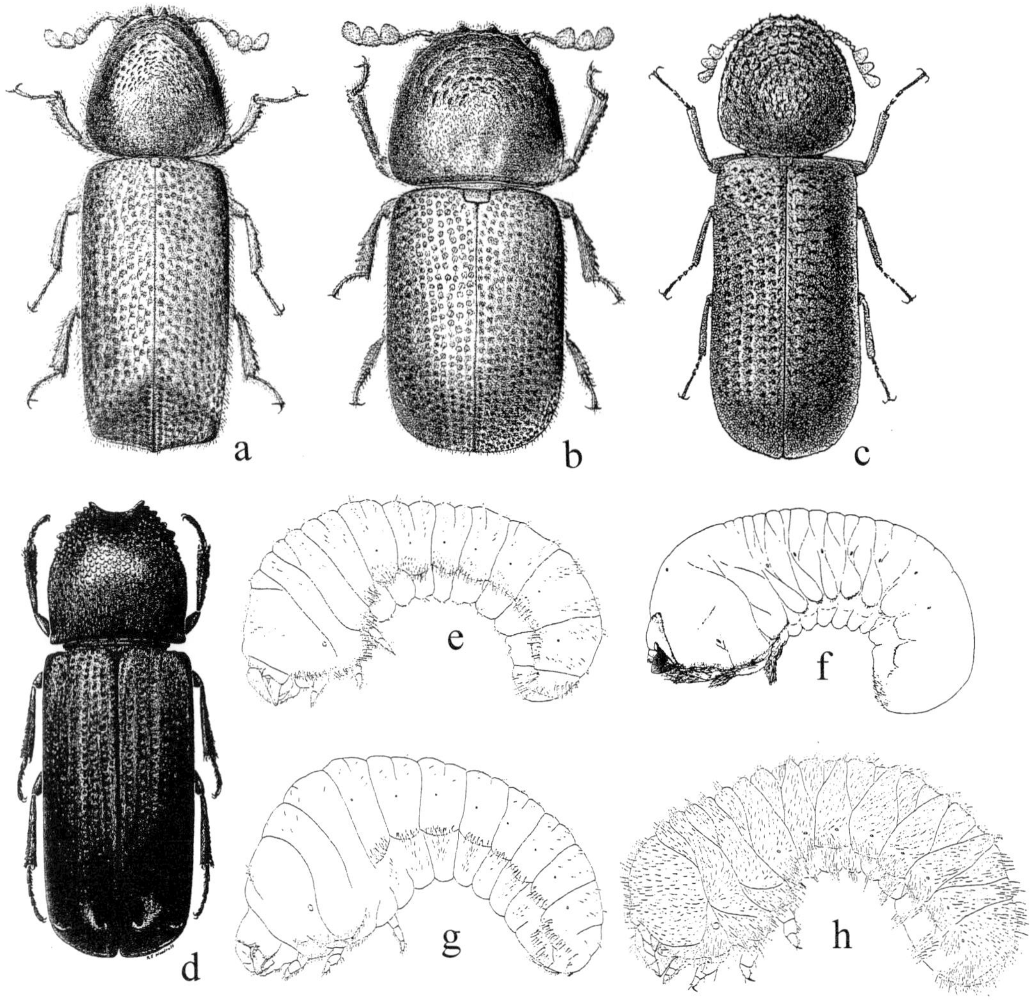

**Figure 5.3** Coleoptera: Bostrichidae. (a) *Prostephanus truncatus*; (b) *Dinoderus minutus*; (c) *Rhyzopertha dominica*; (d) *Heterobostrychus aequalis*; (e) *D. minutus* larva; (f) *H. aequalis* larva; (g) *P. truncatus* larva; (h) *R. dominica* larva.

**Boxwood borer, *Heterobostrychus brunneus*** Adults are 6–13 mm long, black, and there are fine setae covering the body. Full-grown larvae are about 7 mm long and pale yellow. Exit holes in the wood surface are 3–6 mm diameter. The life cycle is completed in about 1 year, but can extend to 3 years. This species is common in Africa where it is often a pest in debarked logs and in timber used in floors and roof construction.

**Colored auger beetle, *Mesoxylion collaris*** Adults are 4–6 mm long. This species infests new hardwoods in Australia, and infestations are usually found during the first year after construction. Emergence holes are round and 2–3 mm diameter. This species does not reinfest the original wood.

**Black polycaon, *Polycaon stouti*** Adults are 11–22 mm long, black and cylindrical; the head projects forward and the mandibles are large. Adults fly at night and seek oviposition sites. Wood attacked includes seasoned hardwoods, such as eucalyptus, maple, oak, and sycamore, plywood panels used for furniture, and buildings in mountainous areas. Larval development takes 2–3 years to complete; there are records of adults emerging from wood after 20 years. This species is distributed in western USA.

**Larger grain borer, *Prostephanus truncatus* (Fig. 5.3a, g)** Adults are 8–10 mm long. The body is elongate and cylindrical, dark brown and has a smooth polished surface. Eggs are laid in batches or singly on the grain; fecundity is 300–500 eggs. Development from egg to adult is about 27 days at 32 °C and 80% RH, and 78 days at 22 °C and 50% RH. At 70% RH the development period increases from 30 days at 30 °C to 40 days at 35 °C, and to 61 days at 22 °C. It often attacks maize before harvest and continues to infest during storage. It is capable of feeding on wheat, rice, cowpeas, beans (*Phaseolus*), peanuts, cocoa beans, and coffee beans. Although the immature stages do not feed on wood, adults produced from larvae that bored into wood live longer than those adults deprived of wood as a substrate for pupation. *P. truncatus* can survive for long periods in wooden bins and in cobs and stalks of maize. This species is established in most tropical regions and is a pest in Central America and Africa.

**Lesser grain borer, *Rhyzopertha dominica* (Fig. 5.3c, h)** Adults are about 5 cm long. The body is dark brown or black, and has a somewhat roughened surface. Eggs are laid in batches on the kernels of grain or singly in the loose frass produced by the feeding adults and larvae. Fecundity is 300–400 eggs; hatching is in about 32 days at 18 °C, and about 5 days at 36 °C. First-stage larvae cannot bore into undamaged seeds. Second-stage larvae are capable of directed movement, but the third and successive stages are C-shaped and incapable of movement on a flat surface. Larvae develop in whole grain and remain within the kernel through four or five instars; on wholemeal there are 2–7 instars. There is a reduction in the number of instars and an increase in development time at about 28 °C. Larval development on wholemeal at 70% RH requires 29 days at 28 °C and 46 days at 25 °C. Pupation usually takes place in an enlarged cell in the larval feeding tunnel. At 70% RH the pupal period is 5 days at 28 °C and 8 days at 25 °C. The adult remains in the cell for 3–5 days before emerging. Development from egg to adult takes about 25 days at 34 °C, and with 22% larval mortality; and 84 days at 22 °C with 53% larval mortality; and 36 days at 38 °C with 86% larval mortality. Adults readily fly but their flight is not well oriented, and air currents carry them. This species feeds primarily on cereals and products made of these and other grains. It commonly infests flour, macaroni, beans, chickpeas, seeds, edible bulbs, grains, and sometimes wood. Damage from *R. dominica* can be considerable because adults and larvae feed on the surface and interior of grains. It can bore into wooden objects. This species is widely distributed.

**Lead-cable borer, *Scobicia declivis*** Adults are 5–6 mm long, reddish brown, and the body is somewhat cylindrical. The pronotum has rasplike teeth or projections at the front; antennae are eight-segmented, and there is a three-segmented club. Females bore a small tunnel and deposit eggs inside it. Larval development requires about 9 months, and the pupal period lasts about 2 weeks. Adult beetles remain in the pupal chamber for several weeks before emerging to the wood surface. There is one generation per year. It normally infests seasoned oak, but also acacia, eucalyptus, maple, elm, chestnut, and other hardwoods. It has been reported attacking newly painted houses, and damaging wine casks, hardwood paneling, and flooring. The common name is derived from the ability of the adults to bore through the lead sheathing of aerial telephone cables, making holes about 2.5 mm wide. Most activity occurs from June to August, when beetles emerge in large numbers. Damage usually occurs later, when moisture enters the cables, causing short circuits. Adults are stimulated by heat, and sometimes bore into roofing materials because of high temperatures. Attacks on asphalt roofing and lead cables may be linked to local forest fires, particularly in stands of oak, where the beetle occurs in large numbers. This species is common along the coast of California and southern Oregon.

**Cylindrical auger beetle, *Xylion cylindricus*** Adults are 5–7 mm long, dark brown to blackish brown. The prothorax has small lobes at the base of the antennae; the antennal club is three-segmented, and the posterior margins of the forewings are produced into short, rounded lobes. Emergence holes are round and about 3.5 mm diameter. This species infests the sapwood portion of hardwoods. It often attacks wooden wine casks, and wine is lost through holes bored by the adults and larvae. It is distributed in Australia.

**Red-shouldered shothole borer, *Xylobiops basilaris*** Adults are about 5 mm long and black; the basal third of elytra is reddish brown. The body is cylindrical and the pronotum has rasp-like projections. The elytra posterior third is concave and there are three spines on each side. Eggs are laid singly in holes bored through the bark into the sapwood. Eggs hatch in about 10 days and the C-shaped larvae feed in the sapwood and sometimes the heartwood portion of infested wood. Galleries extend with the wood grain and they are filled with fine, powder-like frass. Development from egg to adult takes about 2 years, and adults emerge after the wood has been processed into furniture or other products. The adult attacks recently felled or dying hardwood trees and logs. The larvae feed in the sapwood portion and can cause considerable damage. Hickory, persimmon, and pecan are most frequently infested, but other hardwoods in use, such as rustic furniture, fences, small tools, and tool handles, are infested. It has been found in lumber shipped to the UK. Larvae of a related species, *X. quadrispinosus*, are frequently found in hardwoods in southwestern USA. This species is common in eastern and southern USA.

**Other Bostrichidae** *Stephanopachys rugosus* often infests bark-covered floor joists, and freshly cut or recently milled pine. Larvae complete development in 1 year, but take up to 5 years in seasoned wood. A related species, *S. substriatus*, attacks pine, hemlock, and fir; infestations occur in furniture and in construction lumber. *Scobicia bidentata* feeds in freshly cut wood and lumber of several hardwood species. Tropical and subtropical Bostrichidae are wood borers, and often encountered in food and produce stores in structural wood; the adults are found boring into stored foods. These include *Apate* spp., black borers, which are distributed in Africa, Israel, and tropical South America, and *Bostrychopis parallela*, black bamboo borer, which occurs in Southeast Asia.

## Bruchidae

Adults are 2–4 mm long and they are robust beetles that appear square at the posterior and narrow at the front. These beetles are often called weevils, because their head is slightly narrow and produced forward. Elytra do not cover the tip of the abdomen, and the antennae are serrate. The species that are pests of stored foods attack only pulses (the edible seeds of leguminous plants), such as *Phaseolus vulgaris*, *P. lunatus* (beans), *Vigna subterrana* (groundnut), *V. radiatea* (mung bean), and *V. unguiculata* (cowpea). The life histories of bruchids infesting stored food are very similar. Eggs are usually attached to the pods or the seeds, and the first-stage larvae bore through the egg into the host plant. Four larval stages occur within the seed, and pupation occurs in the tunnel of the larva within the seed. Adults are short-lived and apparently do not feed. Bruchids have evolved as pests in different parts of the world, but now the pest species are widely distributed throughout the world.

**Bean weevil, bean bruchid, *Acanthoscelides obtectus* (Fig. 5.4a)** Adults are 2.0–3.7 mm long. The femur of the hind leg has a large tooth on its lower margin, followed by 3–4 small teeth. The last exposed abdominal tergum has yellowish-white setae. Eggs are deposited singly among stored seeds or in cracks in growing pods in the field; oviposition occurs within 4 days of adult emergence; fecundity is 40–60 eggs, but may be 200. Hatching is in 5–20 days, and first-stage larva tunnels into the seed; several larvae can develop in a single seed. Full-grown larvae remain in the seed to pupate, and the adult beetle cuts a round emergence hole. Adults do not feed; they fly actively during the day when temperatures are 21 °C or more. Larval development is 23 days at 29 °C, 92 days at 18 °C, and 36 days at 32 °C. The temperature range for development is 15–35 °C, there is no development at 11 °C, and development is inhibited by very high and low humidity. Pupation occurs in a small cell inside the infested seed, behind a thin window; the pupal period is 8–25 days. On haricot beans (*Vigna vulgaris*) development is about 28 days at 30 °C and 80% RH. Mortality of immature stages is high; the minimum mortality at 25 °C is about 58%. This species feeds on *Phaseolus* spp. and *V. vulgaris*, but is capable of infesting broad beans, groundnut, and other seeds. It is capable of completing development on fungi growing on stored grain. It is native to the western hemisphere, but now it is distributed throughout warm regions of Europe and Africa, and in some regions of Asia, but not in Australia. Pteromalidae (Hymenoptera) parasitizing larvae of *Acanthoscelides* include species of *Anisopteromalus* and *Dinarmus*.

**Cowpea weevil, *Callosobruchus maculatus*** Adults are 2–3 mm long. Elytra are reddish brown and have two large red spots; the thorax is covered with fine white setae. Females prefer to oviposit on cowpeas. Eggs are laid singly on the developing pod in the field or on the seeds in storage; they are usually glued to the surface; fecundity is 100–130 eggs. Eggs are generally tolerant of low temperatures. Larval development is through five instars and on cowpea takes 21 days at 30 °C and 70% RH. Minimum temperature for development is 19 °C and the maximum is 37 °C. Pupation takes place inside the seed in a chamber covered by a thin window; the pupal period is about 7 days. This species occurs throughout tropical and subtropical regions and is generally associated with *Vigna unguiculata* (cowpea), *V. radiata* (mung bean), *V. subterranea* (groundnut), *Cicer arietinum* (chickpea), and other legumes.

There are morphologically distinct forms of *C. maculatus*: a flightless form, which infests stored seeds, and a flying form that infests growing seeds. The flying form oviposits in the field on developing beans, and larvae are brought into storage after harvest. Adults from these larvae are flightless and remain in the storage area to mate and lay eggs. After one or more generations, adults of the flying form are produced, and these move from the storage area to the field. Production of the flying form is governed by factors that operate during larval development, and include crowding, high temperature, low moisture content, and limited light. Adult longevity is about twice as long in the flying form as in the flightless form. Adults of the flightless form mate and oviposit soon after emergence, but females of the flying form emerge with undeveloped ovaries so that mating and egg laying are delayed 3–4 days. The flying form lays

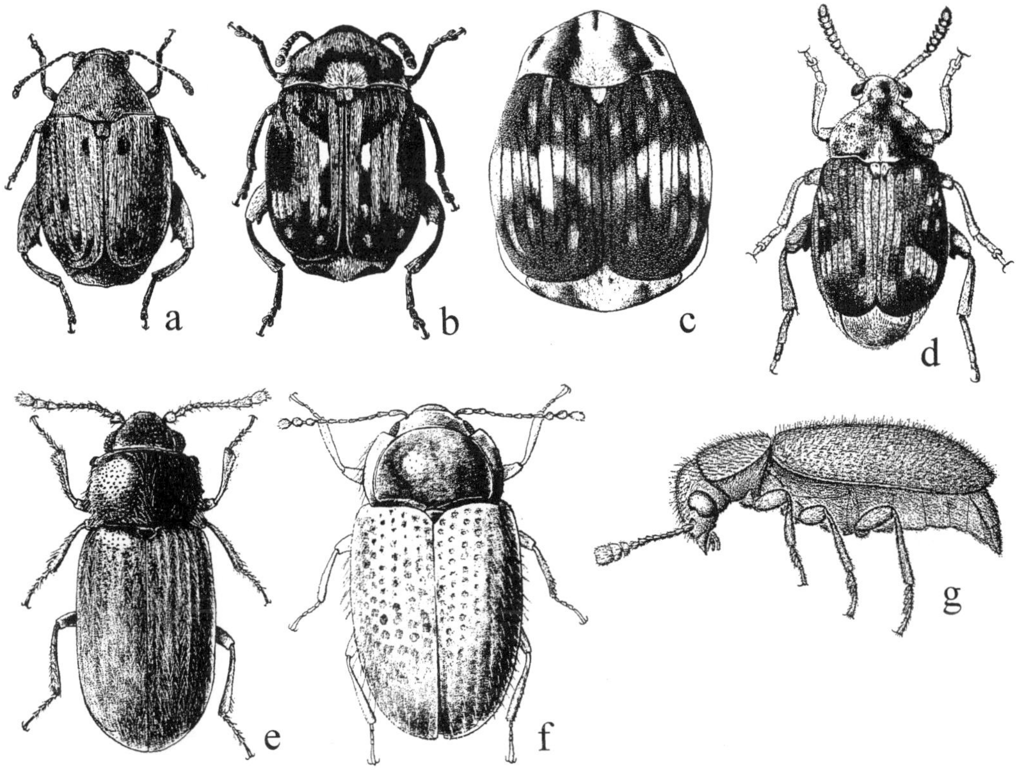


**Figure 5.4** Coleoptera: Bruchidae, Cleridae, Cryptophagidae, Endomychidae. (a) *Acanthoscelides obtectus* (Bruchidae); (b) *Bruchus pisorum* (Bruchidae); (c) *Bruchus ervi* (Bruchidae); (d) *Cryptophagus cellaris* (Cryptophagidae); (e) *Mycetaea subterranea* (Endomychidae); (f) *Necrobia rufipes* (Cleridae).

fewer eggs, but is better able to withstand low temperatures than the flightless form.

**Other *Callosobruchus*** There are several species that are pests of stored pulses around the world; these include: *C. phaseoli* in Africa and parts of Asia, *C. rhodesianus* in Africa on cowpeas, and *C. theobromae* in India on pigeon peas.

**Mexican bean weevil, *Zabrotes subfasciatus*** Adults are 1.5–2 mm long. The body is pale to dark brown or black; there is an interrupted white line across the middle of the wing covers, and a median white spot at the posterior of the prothorax. Eggs are deposited singly or in batches of 2–5 on pods in the field or on seeds in storage; fecundity is about 40 eggs. First-stage larval mortality is low when eggs are laid in batches rather than singly. Moisture produced by aggregated eggs and recently emerged larvae softens the outer shell of the bean and facilitates entrance by the larvae. Larval development limits are 20–37 °C, but few adults are produced at 35 °C when humidity is low. Optimal conditions for development on haricot beans are 32.5 °C and 70% RH; under these conditions the life cycle is completed in about 24 days. This species is widely distributed and is a pest of beans, mung bean, cowpea, and groundnuts. It is distributed in Central and South America, but also occurs in Africa, India, and the Mediterranean region.

**Other Bruchidae** Several species of *Bruchus* attack pulse crops in the field in temperate regions. Infested material may be carried indoors and adult beetles emerge there. They do not reinfest seeds indoors. The most common species include *B. chinensis* on pulses in China, *B. ervi* (Fig. 5.4c) and *B. lentis* on lentils in the Mediterranean region, *B. pisorum* (Fig. 5.4b) on peas worldwide, and *B. rufimanus* on field beans in Europe and Asia.

## Buprestidae

Adult buprestids or flatheaded borers (sometimes called jewel beetles) are 3–35 mm long, and are usually shiny and brightly colored. Adults are slightly flattened and oval-shaped, with the head strongly deflexed and inserted into the prothorax to the edge of the eyes. Antennae are serrate, 11-segmented, and inserted on the front of the head. Larvae are referred to as flat-headed borers because the body region behind the small head is enlarged and flattened. They are also referred to as cobra-headed larvae because of the snake-like shape of their head.

Larvae are distinguished by well-developed plates on the upper and lower surfaces of the first segment behind the head, by the presence of a central groove or V-shaped suture in the upper plate, and by the absence of legs. The head is small, retracted into the first thoracic segment, and scarcely visible. Tunnels made by larvae have the general shape as the thorax of the larvae, which is about three times broader than high. This flattened oval shape distinguishes buprestid tunnels from the rounded oval tunnels of cerambycids. Eggs are deposited singly or in groups in crevices that develop in freshly cut softwood and hardwood logs. First-stage larvae bore directly into the sapwood portion of the wood and feed close to the surface; older larvae tunnel deep in the wood but do not feed on heartwood. These beetles prefer recently cut rather than seasoned wood, but larvae can occur in seasoned lumber. Flatheaded borers complete development from egg to adult in 1 or 2 years, but development extends to 3 or 4 years in wood with low moisture content.

Buprestids have several natural enemies, including clerid beetles. Larvae and adults of the white-banded clerid, *Paratillus carus*, prey on the larvae of buprestids; larvae of the yellow-banded clerid beetle, *Trogodendron fasciculatum*, attack buprestid and longhorn beetle larvae. Both of these clerids occur in Australia.

**Golden buprestid, Oregon jewel beetle, *Buprestis aurulenta*** Adults are about 20 mm long. The body is golden green or blue-green, and with the median suture and margins copper-colored. Eggs are laid on the bark or sometimes in cracks and crevices of exposed wood. Eggs hatch in 7–10 days. Larvae feed in the wood and their tunnels gradually increase in size during the 2–4 years of development. Larval tunnels are oval, about 10 mm wide and 1–5 m long, and they are tightly packed with frass. Larval development takes 2–4 years, and pupation takes place in enlarged chambers near the wood surface. In spring or early summer the adult chews through the thin covering to the surface, leaving the typical oval buprestid exit hole. Emergence occurs between late fall and early spring indoors. Adults live for 2–3 months. Mating and egg-laying occur in bright sunlight. Adults prefer to lay eggs on Douglas fir, but also pine, spruce, fir, and red cedar. They oviposit on trees or down logs with the bark remaining, or on recently sawed lumber, and less frequently on seasoned structural wood. This species is distributed in the USA, Canada, and Europe (Denmark); in Australia it occurs in Douglas fir shipped from North America.

**Green buprestid, *Buprestis langii*** Adults are about 30 mm long and are bright green, blue, or coppery-bronze. This species attacks Douglas fir and other softwood species. Damage occurs in structural wood, pilings, poles, house siding, and flooring. This species is distributed in North America.

**Lined buprestid, *Buprestis lineata*** Adults are 12–20 mm long and have dark red to yellowish brown markings on the elytra. Eggs are laid on down trees, logs or round wood soon after it is cut, or on logs in storage yards. Larval development is completed in 1–3 years, and adult beetles can emerge from infested logs after building construction. This species is a common pest of modern log buildings in eastern and southeastern USA.

**Oriental metallic beetle, *Catoxantha opulenta*** Adults are about 5.7 mm long. The body is dark green to brownish green, and has an orange horizontal stripe on the wings, slightly posterior of the middle. Eggs are laid on the bark of standing trees or down logs. Larvae feed for a short time on the bark, but spend the majority of their time feeding in the cambium and sapwood. Galleries in infested wood are oval and 0.5–1.0 cm wide. Frass is sawdust-like and tightly packed in the galleries. This species is distributed in Asia.

**Pine buprestid, *Chalcophora japonica*** Adults are about 35 mm long and have distinct black longitudinal lines on the thorax and abdomen. Eggs are laid on the bark of *Pinus* spp. and other conifers. Larvae complete development in 1 or 2 years in natural habitats, but in structural wood, development takes 2 or 3 years. This species is distributed in Europe and Asia. Closely related species, *C. virginiensis*, *C. liberta*, and *C. georgiana*, infest pine trees and stumps in the USA, but there are no records of infestations persisting in structural wood.

**Sculptured pine borer, *Chalcophora virginiensis* (= *C. angulicollis*)** Adults are 25–31 mm long. The body is shining black, and the elytra are dull black to gray. Larvae infest pine, fir, and other softwood species. It occurs in southern Canada and throughout the USA, and it has been introduced into Europe.

***Melanophila consputa*** Adults are 8–13 mm long and black, and there are 12 small yellow spots on the elytra. Eggs are laid on dead lodgepole pine, Monterey pine, and species of yellow pine. Larvae mine the inner bark and outer portion of wood of injured, dying, and dead trees. The adults fly from April to October and bite people working in logging operations or others in the area of infested trees. This species is distributed in western USA, and occurs in Arizona, California, and Oregon.

**Hoop pine jewel beetle, *Prospheres aurantiopictus*** Adults are about 25 mm long, and the elytra have pale, parallel lines. Eggs are deposited primarily on hoop pine (*Araucaria* spp.). Larval development is completed in 1–2 years in natural habitats, but may extend to 20 years in structural timbers. This species is distributed in Australia.

**Other Buprestidae** Pest species commonly encountered in Australia are the cypress jewel beetle, *Diadoxus scalaris*, and the small cypress jewel beetle, *D. erythrurus*. They both oviposit on standing or down logs of fire-damaged and drought-affected cypress pine (*Callitris* spp.). Adult beetles emerge from flooring and weather boards soon after milling.

## Cantharidae

Adult soldier beetles are about 15 mm long, and somewhat resemble fireflies (Lampyridae), except the head is not covered by the prothorax and they lack light-producing organs. They are usually black or dark brown, but sometimes yellow, orange, or red with black markings. Adults are found on goldenrod and milkweed flowers, while larvae are predators of other insects under tree bark and leaf litter. Adults of many species fly to outdoor lights at night.

**Goldenrod soldier beetle, *Chauliognathus pennsylvanicus*** Adults are 9–12 mm long. The head and ventral surface are black, the thorax is yellow with a black spot; the elytra are yellow with an elongate black spot on the apical third. Adults are common in the fall on goldenrod, and sometimes fly to lights at night. This species is distributed in eastern North America.

## Carabidae

Adult ground beetles range from 1 to 40 mm long; they are dark, shiny, and somewhat flattened. Elytra usually have regular, parallel ridges, and the antennae are inserted at the sides of the head, between the eyes and mandibles. Nearly all species are predaceous on other insects. Adults and larvae are commonly found on the ground, under logs, leaves, and debris, where they search for prey and complete their life cycle. Carabids associated with turfgrass, such as species of *Aphodius* and *Harpalus*, are abundant during summer and a peak in the population can result in large numbers of adults flying to indoor and outdoor lights. These beetles are abundant in urban and rural areas, and adults will occasionally invade houses and other buildings. Instances in which large numbers occur in buildings is probably the result of a population peak in nearby habitats.

**Tule beetle, *Agonum maculicolle*** Adults are 8–10 mm long; the pronotum and elytra are dark brown with tan margins. Wings are probably dimorphic. This species typically breeds in the tule or bulrush (*Scirpus*) marshes along rivers in western USA. It is common in the Sacramento and San Joaquin valleys of California, but its range extends to Oregon. At dusk, and usually after rain in the fall, large numbers of adults enter houses close to the larval breeding sites. Adults can produce a disagreeable odor, which is characteristic of many carabids.

**Other *Agonum* species** Several species in this genus are attracted to lights at night, including *A. lutulentum*, *A. mutatum*, *A. placidum*, *A. picticorne*, and *A. albicrus*.

**Murky ground beetle, *Harpalus caliginosus*** Adults are about 23 mm long, reddish black, and slightly shiny. The thorax is about the same width as the elytra, which are deeply grooved. Full-grown larvae are about 20 mm long, black, and with a brownish-red head and with two tail-like cerci. This species is widely distributed in USA, and is often found at lights indoors and outdoors.

**Red ground beetle, *Harpalus rufipes* (Fig. 5.2b)** Adults are 10–18 mm long, reddish black, and shiny. Larvae live in the soil and are predators of other insects, especially the immature stages of other beetles and lepidopterans. This species is attracted to lights at night and is a common invader of households and commercial warehouses during the summer months. Several *Harpalus* species in the USA and Europe fly to lights at night in houses and commercial buildings.

**Caterpillar hunter, *Calosoma scrutator*** Adults are 25–36 mm and shiny black; elytra are metallic green or violet. This large beetle occurs in fields and gardens, and it often occurs at lights at night. It is considered beneficial because larvae and adults are predaceous on a variety of invertebrates, and contribute to the control of caterpillars, sawflies, and other pests in forests. Adults are active flyers and, during summer nights, they are attracted to outdoor and indoor lights. This species occurs in eastern USA.

**Seed-corn beetle, *Stenolophus leconti*** Adults are 5.5–8 mm long and the body is reddish brown; the pronotum is pale brown and sometimes has a dark center. Elytra have a spot, which is prolonged along the suture to the base. Larvae feed on planted seed corn, and the damage often prevents sprouting. Adults are often attracted to lights in large numbers in the spring. A related species, *S. lineola*, is a common ground beetle throughout the USA, and it sometimes flies to indoor lights.

**Other Carabidae** Species found indoors in the UK include *Pristonychus terricola* – the adults are 12–15 mm long, and bluish black – and *Sphodrus leucopthalmus* – adults are about 22 mm long, and the body is black. The stink beetle, *Nomius pygmaeus*, occasionally invades homes in western USA and Canada. It can produce an offensive odor, and articles of clothing that have been in contact with this beetle retain the odor for several weeks. Its presence in an area is often associated with nearby forest fires. The adults are apparently driven out of forested areas by the fire and smoke. *Pterostichus scitulus* is commonly found at lights in USA. *Carabus aeruginosus* is a common species in urban areas in southwest Siberia, Russia.

## Cerambycidae

Adult longhorned beetles are 3–60 mm long and distinguished by their elongate and cylindrical bodies, bright colors or markings, and antennae that are usually longer than the body. Larvae are yellowish white, cylindrical, and usually legless. The larval body tapers posteriorly, and the segments near the head are never enlarged or flattened. Some species attack living trees, but the majority of cerambycid larvae feed in wood of dead and down trees. Both hardwood and softwood trees are attacked. A few species attack trees soon after felling and can be pests in modern log houses, rustic furniture, or other material made of unseasoned wood. Larvae of the old house borer feed in seasoned softwoods (pine, spruce, and fir). Adult cerambycids often feed on flowers and are active during the day; some are nocturnal, and others may not feed. Most adults stridulate with organs on the thorax, and the sound is usually audible. Eggs are laid in crevices in bark or exposed wood, and all stages of larvae bore into the wood to feed. Some species are general feeders, but most are specific for the host or the condition, such as moisture content or decay fungi, of the wood attacked. Larval tunnels in wood are oval or round, and filled with fibrous, granular, or powdery frass, depending on the species of longhorned beetle. The full-grown larva tunnels to the wood surface and usually makes the exit hole. The pupal chamber is below the frass-plugged exit hole, and the adult pushes through the frass to emerge. Life cycles are 1 or 2 years, but can be extended when environmental and wood conditions are unfavorable.

Larvae of several species of cerambycid are referred to as wood borers or powderpost borers because they feed for several years in saw logs or in structural lumber, and their galleries are filled with powdery frass. Certain species attack only recently cut logs with moisture content high enough for them to complete development in 1 or 2 years. If the infested wood is stored in dry conditions, the larval development period is prolonged for several years, or the larvae die because of the low moisture content of the wood. Damage to wood from cerambycids that do not reinfest structural timber may be confused or misidentified with damage cause by the old house borer, *Hylotrupes bajulus*. This cerambycid infests seasoned lumber and the life cycle extends for several years, depending on environmental conditions.

**Oriental longhorned beetle, *Batocera rufomaculata*** Adults are about 5 cm long and the body is light brown; antennae and legs are dark brown. Elytra have a white spot medially at the base and a large white spot in the middle, and various small spots on the basal third of the wings. The prothorax has two reddish-brown spots medially. Galleries in wood are oval and packed with fibrous and some powdery frass. Eggs are laid in crevices of tree bark, and larvae feed under the bark on the cambium layer of wood before extending their galleries into the sapwood. Larval development is completed in about 1 year. This species is common in tropical regions of southern Asia, including Malaysia. It is a pest of commercial timber as standing live trees or down logs. The wood species attacked include *Acacia mangium*, *Artocarpus lanceifolius* (Keledang), *Calophyllum* spp. (Bintangor), *Cratoxylon arborescens* (Geronggang), *Heritiera* spp. (Mengkulang), *Lophopetalum* spp. (Mata Ulat), *Palaquium* spp. (Nayatoh), *Pometia* spp. (Kasai), and *Shorea* spp. (Yellow Meranti).

**Bamboo longhorn beetle, *Chlorophorus annularis*** Adults are 10–16 mm long and slender; the body has yellow and black markings. Elytra have a yellow, X-shaped mark; they are deeply notched at the distal end, and terminate in two spines. Larvae feed by boring through the stems and internode walls of bamboo. Galleries are packed with fine, powdery frass. Larval development takes about 1 year. Adults emerge indoors from bamboo stems and other pieces made into furniture; sometimes emergence is from material that is more than 1 year old. This species occurs in southern Asia, from India to Japan, including southern China and Malaysia.

**Old house borer, *Hylotrupes bajulus* (Fig. 5.5a, d, e)** Adults are 15–25 mm long and slightly oval and flattened; the body is black to brownish-black. The prothorax is rounded and has two shiny areas on each side. Elytra are completely black or with patches of grayish white that form bands medially. The

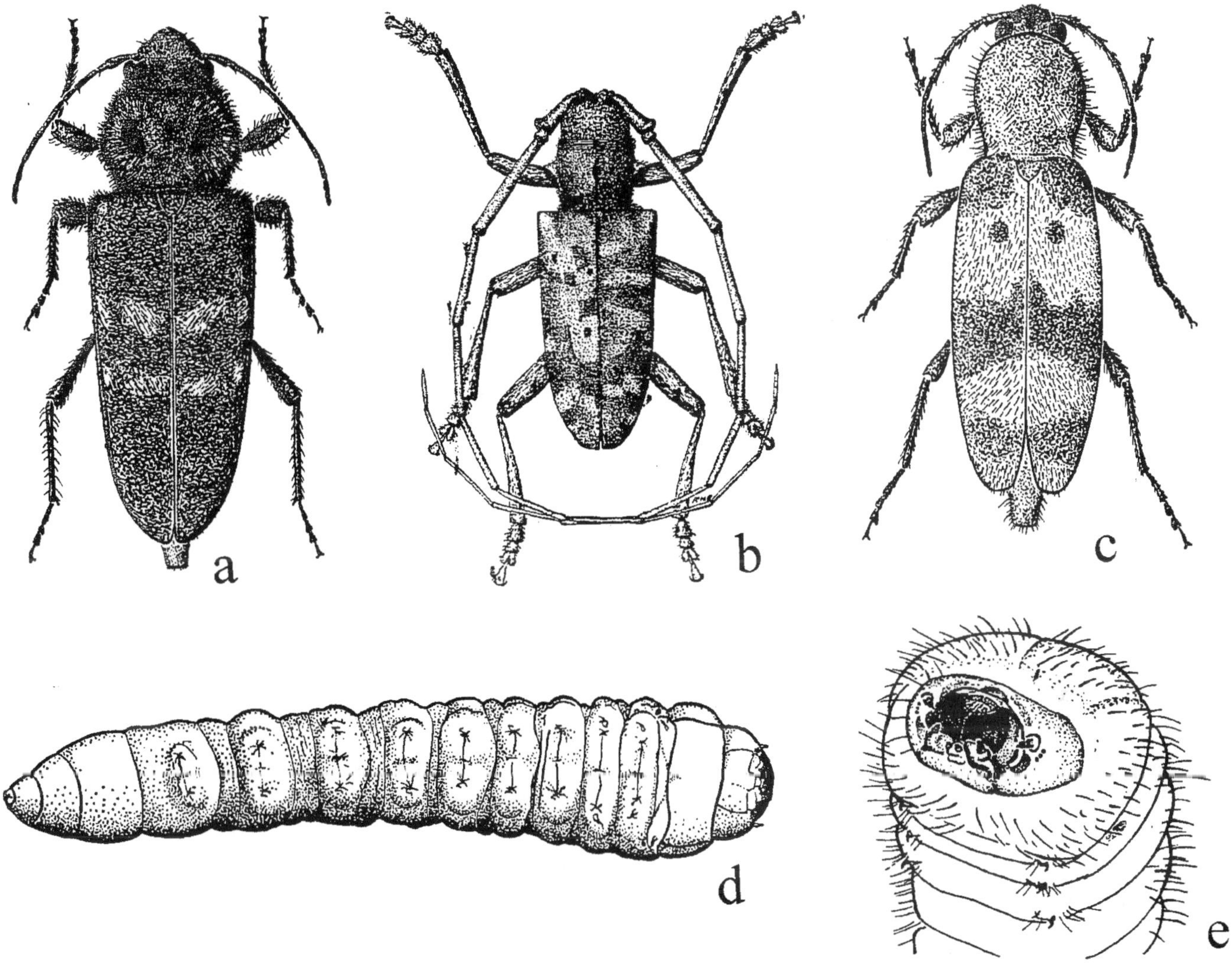


**Figure 5.5** Coleoptera: Cerambycidae. (a) *Hylotrupes bajulus*; (b) *Monochamus maculosus*; (c) *Xylotrechus colonus*; (d) *H. bajulus* larva; (e) *H. bajulus*, larva head anterior.

abdomen of the female extends slightly beyond the wing tips. Full-grown larvae weigh 150–200 mg; they are about 31 mm long, yellowish white, and have dark brown mandibles. The larval head has three dark ocelli on each side. Larvae of other cerambycids that occur in structural wood, such as *Callidium* spp., *Eburia* spp., *Neoclytus* spp., and *Monochamus* spp., have one ocellus on each side of the head. Eggs are laid in two to six batches and with 30–50 eggs per batch; they are in placed in cracks and crevices. Oviposition lasts about 5 days. Fecundity is about 165 eggs, but can be 400 or more. The ovipositor is telescoped into crevices for egg-laying, and when fully extended it is as long as or longer than the body. Hatching occurs in about 9 days; egg development is not delayed or prolonged because of environmental conditions. Females mated only once lay their full potential of eggs; the number of eggs laid and successful hatching are not significantly different between single- and multiple-mated females. Males are often aggressive during mating, and they often bite and amputate the legs and antennae of females. First-stage larvae bore immediately into the wood a short distance and begin feeding, but they can remain alive on the surface for several days before entering. Larval development ranges from 2 to 10 years, and occurs most rapidly at 20–31 °C and 80–90% RH, and when the wood moisture is above 10%. Larvae continue to feed and develop until they attain a weight of approximately 200 mg. Full-grown larvae tunnel to the surface of the wood and cut an oval exit hole. The larva retreats into a pupal chamber excavated below the exit hole, and closes the hole with fibrous pieces of wood and a small amount of powdery frass. The pupal period is about 20 days, and the adult remains in the pupal chamber for several days before emerging to the wood surface. Development is reduced at temperature extremes of 14 °C and 34 °C. Development time ranges from 2 to 10 years, but in structural wood development is usually completed in 3–5 years. After emerging from the wood, adult females live about 10 days, and males live about 15 days.

Larvae of *H. bajulus* can digest cellulose. They feed on wood with moisture content of 10–20%, and they prefer the sapwood portions. Larvae usually do not survive in wood that has less than 10% moisture. They do not feed on decayed wood, and the oils and resins of heartwood make it undesirable. Tunnels made by small larvae are usually long, parallel to the wood grain, and close to the surface. Large larvae also feed parallel to the grain, and large chambers are often produced in sections of wood favorable for larval development. The ocelli may detect small amounts of light near the wood surface, and may help the larvae to avoid piercing the wood surface, though they feed very near to it. Late-stage larvae penetrate into the inner layers of wood, but generally remain in the sapwood. The nutritional value of pine wood to *H. bajulus* larvae decreases from the periphery towards the center, and this probably influences the region of the wood favored for larval feeding. Growth of larvae is dependent on the protein content of the wood, as the concentration of nitrogen compounds is highest in the outermost layers where larvae normally feed. Frass is composed of barrel-shaped pellets of digested wood and irregular-shaped, undigested particles. Frass of other cerambycids, such as *Eburia* spp. and *Neoclytus* spp., lack the barrel-shaped pellets, and the frass of *Monochamus* spp. is fibrous and not powdery. The feeding galleries of other longhorned beetles are smooth and not sculptured with ridges as in *H. bajulus*.

*H. bajulus* is distributed from central Scandinavia to North Africa, and from Portugal to Siberia and the Near East. It has limited distribution in the UK and the USA; it is absent on the coastlines of the Northern Sea. It has been introduced and established in many other countries including Australia, New Zealand, South Africa, and parts of South America. Distribution outside continental Europe and Eurasia is almost always with household infestations. In the USA, the old house borer occurs primarily in houses less than 10 years old; in Europe it is common in houses older than 10 years, especially when wood from old houses is used in construction of new buildings. In western Europe and South Africa, it occurs in tree stumps and dead branches. In the USA, adults and larvae have been collected from lumber stored outdoors in commercial lumberyards, but not from natural habitats.

Feeding of early-stage larvae is not audible, but the sound made by the mandibles of medium and large larvae can be heard at a distance of 3–5 m. There is a periodicity cycle of larval feeding behavior: the active phase is 23–32 days, and alternated with an 8–14-day inactive phase. During feeding, the larva presses against the gallery with the thorax, and inserts a mandible into the wood as an anchor, and the other mandible scrapes the wood, bringing small pieces into the mouth. Abrasion is greater on the chewing mandible than on the stationary one. At times during the boring process, the larva pauses and turns along its longitudinal axis, and resumes feeding in the same direction. After rotation, the stationary mandible takes the function of chewing. This alternation equalizes the abrasion on both mandibles. Larvae produce pheromones that are excreted with the frass. When few larvae are feeding and a small amount of frass is produced, the low concentration of pheromones can stimulate females to oviposit in the wood. When several larvae are feeding in the wood large amounts of frass are produced, and the concentration of pheromones can be repellent to ovipositing females. Adults are most active in the daytime when temperatures are between 29 and 35 °C, and they fly when temperatures are above 30 °C. Males can be attracted to regions of infested wood containing females that have not emerged.

There are several parasites and predators of *H. bajulus*. In Europe the clerid, *Opilo domesticus*, is an important predator of larvae and of larvae of several species of wood-infesting insects. Hymenoptera parasites are less frequently encountered, and include the ichneumonid, *Ephialtes manifestor*, the braconids, *Doryctes leucogaster* and *Rhoptrocentrus piceus*, and the bethylid, *Sclerodermus domesticus*.

**Two-toothed longicorn, *Ambeodontus tristis*** Adults are 14–32 mm long and uniformly brown to dark brown. Full-grown larvae are about 30 mm long and yellowish white. Eggs are laid under bark or in cavities of dead wood, and on the rough surface of stacked lumber. Larvae tunnel in the wood in the direction of the grain, and the tunnels have fine transverse ridges. Galleries are packed with powdery frass. Larval development is completed in 2 or 3 years, and full-grown larvae tunnel close to the wood surface and create a pupal chamber. Adults emerge through oval exit holes, and they are active from February to June. This species occurs in New Zealand, but has been found in imported timber in Australia, and it may have spread to other regions. It typically attacks New Zealand rimu pine (*Dacrydium intermedium*) and is found in New Zealand houses built in the 1920s and 1930s. It is relatively uncommon now. It has been found in floor joists in buildings in Leicestershire, UK.

***Anoplodera* spp.** Adults are large and somewhat triangular-shaped beetles, and they are often brightly colored. Eggs are laid in crevices in the bark and not on bare wood. Early-stage larvae penetrate the bark and feed at the interface of the bark and outer layer of sapwood. Larval development takes 2 or 3 years. Larvae excavate oval galleries that are packed with fibrous frass. Infested wood is often reinfested until it is severely damaged. These beetles attack a variety of hardwoods, pine, spruce, and fir trees. Poles and ornamental wood members in contact with the ground are often infested. They are distributed in eastern and southeastern USA.

**Asian longhorn beetle, *Anoplophora glabripennis*** Adults are 2.0–3.8 cm long, shiny black, and with white marking on the elytra, legs, and the long antennae. Full-grown larvae are about 2.5 cm long, subcylindrical and yellowish white; the mandibles are dark brown. Adult exit holes in tree trunks, limbs, and branches are round and about 12 mm diameter. Eggs are laid singly in shallow holes chewed by females in the bark of the host tree; they are covered with a glue-like secretion from the female. Fecundity is about 50 eggs. Hatching occurs in 10–15 days, and early-stage larvae penetrate the bark and begin feeding on the outer cambium of the live tree. Late-stage larvae tunnel into the sapwood and heartwood. Larval development is completed in 12–18 months, but can be extended to 2 years or more. The full-grown larva tunnels close to the surface of the bark, and then it retreats to form a pupal chamber. The adult chews the exit hole to emerge to the surface. Adults feed on tree leaves for a few days before mating. Adults live about 60 days. The dispersal distance for males and females is within a radius of about 560 m, and the dispersal potential over a single season for males is 1029 m, and 1442 m for gravid females. This cerambycid often kills the tree after reinfesting for 2–4 years. Lumber made from infested wood can sustain living larvae, and extend the life cycle. This species is native to China, Japan, and Korea, but has recently been introduced into the USA. First introductions to North America were in packing crates from Hong Kong. It was first discovered in New York City and Long Island in 1996, and discovered in Chicago, Illinois, in 1998. In the USA, it is known to attack living maple, horse chestnut, and elm trees, but other trees are infested and killed by repeated infestations.

**New house borer, *Arhopalus productus*** Adults are 12–25 mm long, black, and with a narrow body. Full-grown larvae are about 38 mm long and yellowish white. Emergence holes are oval and about 6 mm diameter. Eggs are laid in crevices of the bark remaining on trees and round wood. Early-stage larvae feed at the junction of the bark and sapwood; later they tunnel in the sapwood and sometimes enter the heartwood. While feeding, larvae tunnel to the wood surface, but then plug the opening and continue tunneling in the interior. The full-grown larva makes a pupal chamber close to the wood surface and the adult beetle cuts its way out to emerge. The life cycle is completed within 2 years, but adults sometimes appear indoors within a few months after construction with fire-damage timber. Larvae feed in dead or dying pine and fir trees, including Douglas fir. This species is especially common in forest fire areas, and is probably attracted to the trees during the first summer after the fire. When lumber from such trees is used for building construction the beetle infestation becomes evident. House subflooring of Douglas fir is a common site of infestation. Exit holes may be cut in hardwood floors, linoleum, rugs, plasterboard, and roof coverings adjacent to infested wood. This species is distributed in eastern and central North America.

**Black-horned pine borer, *Callidium antennatum*** Adults are 9–14 mm long. The body is slightly flattened and metallic blue or bluish black; the antennae of the male are somewhat shorter than the body. The thorax is rounded and has depressions on each side of the middle; legs are black and the femora enlarged. Eggs are laid beneath bark scales, and larvae feed in the outer sapwood. Larvae make broad, meandering tunnels in the wood, and push fibrous frass through holes in the bark or wood surface. Pupation is in an enlarged portion of a gallery that is plugged with frass. Adults emerge in early spring, and there is one generation per year. This species occurs throughout the USA and develops in dead or recently felled conifers.

**Wasp beetle, *Clytus arietis*** Adults are about 12 mm long and have long legs. It has the black and yellow-banded appearance of a yellowjacket wasp (Vespidae). It has wasp-like behavior patterns, such as rapid movement of the antennae. It is known to emerge from finished beech and oak furniture, and rustic furniture in the UK.

**Ivory-marked beetle, American oak longhorn beetle, *Eburia quadrigeminata*** Adults are about 24 mm long and light brown. Elytra have yellowish-white spots at the base and middle. Full-grown larvae are about 26 mm long, somewhat wedge-shaped and shiny. Larvae feed in the dry heartwood of various hardwoods, including oak, hickory, ash, maple, elm and chestnut. Wood is infested while drying in commercial lumberyards. Larval development takes about 2 years, but is extended when the wood becomes dry after attack. Adults may emerge from flooring, doorframes, and furniture 15 years after it is placed in use. This species is distributed in eastern North America.

**Banded hickory borer, *Knulliana cincta*** Adults are about 22 mm long and dark brown; the body is covered with fine grayish-white setae. The thorax has a sharp spine on each lateral margin; elytra have yellow spots near the base and two small spines near the tip. Eggs are deposited in summer under the bark of standing trees or directly on the wood of a recently felled, dying, or dead tree. Larvae feed beneath the bark in the cambium portion of the wood, and create extensive galleries. They expel large quantities of granular frass from small openings in the surface. Larval development takes about 2 years; full-grown larvae tunnel to the surface and pupate in fibrous frass at the end of the gallery. Adults emerge in spring. It infests a variety of hardwood species. Firewood, down logs, posts, and rustic furniture are often infested. Adults occur indoors and they are attracted to lights at night. This species is distributed in eastern North America.

**Basket beetles, *Leptideela brevipennis*, *Gracilia minuta*** These two small longhorn beetles are known to infest baskets and other articles made of unbarked willow and hazel twigs. Emergence holes are 3–4 mm in diameter, and the galleries have powdery frass.

**Painted hickory borer, *Megacyllene caryae*** Adults are about 20 mm long and the body is black with yellow bands. The prosternum is wider than long, and the antennae of the male are longer than the body. Adults closely resemble the locust borer, *M. robiniae*. Their large size, tapered elytra, and long antennae distinguish the adult *M. caryae*. Eggs are laid in crevices in the bark, and hatching occurs in about 10 days. Early-stage larvae feed for 2–3 weeks between the bark and the sapwood; late-stage larvae tunnel deep into the sapwood and later into the heartwood. Full-grown larvae tunnel to the surface and cut an exit hole, then retreat to a pupal chamber plugged with fibrous frass. They overwinter as pupae and adults emerge in the spring. Adults emerge from firewood stored indoors in winter. This species develops in freshly cut logs of many hardwood species, including hickory, oak, and ash. It is distributed in eastern USA.

**Locust borer, *Megacyllene robiniae*** Adults are about 18 mm long. The body is black with bright yellow bands across the thorax and elytra; the third band on the elytra is W-shaped. Full-grown larvae are about 25 mm long. Eggs are deposited in the fall on the bark and in fresh wounds of living locust trees, and hatching occurs in about 10 days. Early-stage larvae tunnel into the wood and make an overwintering gallery in the inner bark. In spring, larvae resume feeding and bore into the sapwood. The full-grown larva tunnels to the surface of the wood and cuts an exit hole, then retreats to a pupal chamber sealed with frass. Adults emerge in late summer and early fall. The adults are active flyers and are often found feeding on goldenrod blossoms. They occur indoors when adults emerge from firewood logs. This beetle occurs in eastern Canada and nearly throughout the USA.

**White-spotted sawyer, *Monochamus scutellatus*** Adults are 18–25 mm long. The male is shiny black except for a small, round white spot at the base of the elytra. Females are marked the same or have the elytra mottled with white spots. Emergence holes are circular. Eggs are deposited in slits chewed in the bark, or in depressed areas of logs or pulpwood. Larval development is completed in 2 years in the northern region of its distribution, and 1 year in the southern region. Full-grown larvae tunnel to the surface and cut an emergence hole in the bark. This species occurs in northern USA and Canada where it attacks balsam fir, white, black, and red spruces, and larch. This species causes damage to sawlogs, and the galleries are evident in structural lumber.

**Southern pine sawyer, *Monochamus titillator*** Adults are 18–30 mm long. The body is mottled gray and brown; male antennae are two to three times as long as the body. The thorax has a strong spine on each side; elytra sutures are prolonged into sharp spines. Full-grown larvae are about 60 mm long, yellowish white, and with dark brown mandibles. Eggs are laid in crevices in the bark. Early-stage larvae feed beneath the bark in the sapwood, while late-stage larvae feed deep in the sapwood. Larval development is completed in about 1 year, but in the southern regions of the distribution range there are two generations per year. Adults emerge in April and May, and are active throughout the warm season. This beetle develops in freshly cut, recently felled, dying, or recently dead trees. It occurs in eastern and southeastern USA. A related species, *M. carolinensis*, develops in dead and dying pines in southeastern USA. Adults are about 17 mm long, and similar in appearance to *M. titillator*. Other species in this genus reported from damaged timber include *M. maculosus* (Fig. 5.5b), *M. obtusus*, *M. oregonensis*, and *M. notatus*.

Larvae are called sawyers because of the sawing-wood sound they make while feeding. These cerambycid species live in unseasoned wood, have a short life cycle, and do not re-infest the original wood. Damage characteristics include oval or round exit holes, tunnels up to 18 mm diameter in the interior of wood, and coarse, fibrous frass in the galleries. Wood that is infested before milling may have galleries that contain frass, broadly oval holes in the wood surface, and exposed galleries at the surface. These features give the false impression of an active infestation.

**Ash borers, *Neoclytus acuminatus*, *N. caprea*** Adult red-headed ash borers, *N. acuminatus*, are about 15 mm long. The body is light brown, and the head and thorax are reddish brown. Elytra are marked with four bands of yellow setae. Adult banded ash borers, *N. caprea*, are 12–25 mm long, dark brown to blackish brown. The thorax has a line of white or yellowish white hairs; elytra have four cross-bands, and the first two bands meet to form a near-circle. Eggs are deposited under the bark of dead, unseasoned wood. Early-stage larvae feed beneath the bark on the outer layer of sapwood; late-stage larvae tunnel deep into the sapwood. Pupation occurs in the fall, and adults emerge in spring. Larval development takes about 1 year, but extends to 2 or 3 years if infested logs are stored indoors or in a dry location. Several species of ash borers infest unseasoned wood. They are active in early spring, and emerge from firewood stored indoors, or are attracted to lights at night.

**Brown prionid, *Orthosoma brunneum*** Adults are 25–50 mm long, somewhat flattened, and light brown. The pronotum is narrower than the elytra and has three sharp spines on each side; elytra has six, raised longitudinal lines. Full-grown larvae are about 50 mm long, somewhat shiny and yellowish white. Eggs are laid on or in decaying wood or in the soil, and oviposition occurs from June to late summer. Larvae tunnel in wood and the galleries are packed with coarse, fibrous frass. Adults emerge in May and June, but are active for several months. This species develops in decaying coniferous and hardwood logs in southeastern Canada, and eastern USA. It damages structural timbers and other wood in use that is in contact with the ground and exposed to moisture, especially wood that has decay fungi.

**Tanbark borer, oak longhorn beetle, *Phymatodes testaceus*** Adults are about 14 mm long, elongate, and somewhat flattened. Adults have two color forms: thorax brownish yellow or dark brown with light brown elytra, and, thorax, abdomen, and legs reddish yellow, and bluish-black elytra. Eggs are laid on the surface and in cracks in the bark of dead oak trees, and occasionally in hemlock bark. Larvae feed in tunnels within the bark and in the sapwood. Larval development is completed in 1 or 2 years, and full-grown larvae tunnel to the outer layer of sapwood and pupate in an enlarged chamber. Development can be extended when oak wood is dried and used in house construction. This species infests oak flooring, and adults are known to emerge from wood many years after installation. This species is distributed in eastern North America.

**Huhu beetle, *Prionoplus reticularis*** Adults are about 36 mm long; the brown front wings have a reticulated pattern of pale lines. Full-grown larvae are about 50 mm long and yellowish white. Adults are nocturnal and make a distinct buzzing sound while in flight. Eggs are laid under the bark of trees or in the galleries of old insect burrows; hatching occurs in about 21 days. In the urban environment, eggs are deposited on the surface of lumber and in building materials where there is moisture and decay. Larvae excavate large cavities in the sapwood and heartwood, and the galleries are packed with fibrous frass. Development is complete in 2 or 3 years. The full-grown larva tunnels to the surface of the wood and excavates an oval pupal chamber, which is lined with shredded wood and frass. Pupation is in the spring and the pupal period is about 14 days. Adults emerge in November through February through an oval exit hole they cut in the wood surface. This species occurs in New Zealand.

**Tile-horned prionus, *Prionus imbricornis*** Adults are 24–50 mm long, dark brown, and shiny. The male antennae have 18–22 segments, and the segments overlap. Full-grown larvae are about 60 mm long and yellowish white. Eggs are laid in the soil. Larvae feed in the roots of living trees and shrubs. Larval development is completed in about 3 years. This species is distributed throughout eastern USA. Adults are attracted to outdoor lights at night in mid to late summer.

**Broad-necked root borer, *Prionus laticollis*** Adults are 22–45 mm long, dark brown, and shiny. The head is depressed between the eyes, and the antennae of the male are shorter than the body. Full-grown larvae are about 75 mm long, and yellowish white. Eggs are deposited in groups in the soil. Early-stage larvae feed on small tree roots; late-stage larvae feed on the surface of roots, but eventually enter the root. Larval development requires up to 3 years. Full-grown larvae leave infested roots and move close to the soil surface to pupate in oval, compact cells. This species feeds in the roots of trees and shrubs in eastern USA. Larvae are in decaying logs and stumps of hardwood trees and in buried wood. Adults occur at lights at night.

**Pine-stump prionus, *Prionus pocularis*** Adults are about 42 mm long. The body is light brown and shiny; the elytra are densely punctured. Larvae feed in decaying coniferous logs and stumps. Distribution of this species is central, Atlantic coastal, and southern USA. Like other *Prionus* species, this one occurs at lights at night.

**Slender Texas longhorn, *Psyrassa texana*** Adults are 11–13 mm long. The body is uniformly brown, slender, and somewhat tapered. The antenna of the male is longer than the body; the head, thorax, and elytra are punctate, and with long setae. Adults are attracted to lights at night. This species is distributed in southeastern North America.

**Flat oak borer, *Smodicum cucujiforme*** Adults are 7–10 mm long, yellowish brown, and shiny; the body is elongate and slightly depressed. Full-grown larvae are about 12 mm long, and have small legs and a white, triangular arch on the underside of the first segment of the thorax. Eggs are laid in cracks and crevices in exposed wood. Larvae excavate extensive galleries in the heartwood of seasoned oak and hickory. Larval development is completed in about 1 year, but in seasoned wood, it extends to 2 or 3 years. The full-grown larva forms a pupal chamber close to the surface, and the adult cuts the exit hole to emerge. Adults are active in July and August. Galleries are about 3 mm diameter, and they are tightly packed with fine granular frass. Stored lumber is frequently infested, and larvae continue to feed after the wood is in use. This species is distributed throughout eastern and central North America.

**Rustic borer, *Xylotrechus colonus* (Fig. 5.5c)** Adults are about 14 mm long, light brown to dark brown; the thorax is rounded and uniformly brown. Elytra are marked with irregular bands of yellow and gray. Antennae are not as long as the body, and femora are enlarged.

Eggs are deposited in cracks and irregularities in the bark. Early- and late-stage larvae feed within the bark; there is little feeding on the outer layer of sapwood. Larval development is completed in 1 or 2 years. Adults emerge from firewood stored indoors. This species infests recently killed hardwood trees. It is distributed in southern Canada and eastern USA.

**Other cerambycids** There are numerous longhorned beetles reported infesting unseasoned structural wood or rustic furniture. Damage from *Megacyllene antennata* has been reported from structural wood. *Hesperophanes cinerus* in Europe and *H. campestris* in Asia infest dry timber. *Phymatodes dimidiatus* has been recorded in structural timbers made of cedar. In eastern Australia, especially along the coast, the yellow longhorned beetles, *Phoracantha recurva* and *P. semipunctata*, are common under the bark of eucalyptus and pines. Damage from these beetles, and sometimes live larvae, occurs in structural wood. In India, the banded cerambycid, *Chlorophorus strobilicola*, bores into pine cones and reduces seed production. This species has been recorded in the USA in scented, decorative pinecones used in holiday decorations.

## Cerylonidae

These beetles are about 3 mm long, oval and shiny. The antennae are 10-segmented, and there is a distinct three-segmented club. The majority of species are found under bark or in decaying organic matter. Some species are known from grain storage warehouses.

**Minute beetle, *Murmidius ovalis*** Adults are about 1.5 mm long, and shiny reddish brown. Larvae are pale yellow, flattened oval and with the head hidden below, and with bristly appendages around the margin and on the back. Larval development is complete in about 2 months under humid conditions; under cool conditions, development can extend to 7 months. The full-grown larva pupates in a hemispherical cocoon, which is fastened to the substrate. Adults and larvae are sometimes found in stored grain or cereal products.

## Chrysomelidae

Adults are 6–11 mm long and nearly oval. Legs are short, and in some species the hind femora are enlarged. Larvae are soft-bodied and frequently have a highly pigmented integument. All members of the family feed on the foliage of plants.

**Fruit tree leaf beetle, *Pyrrhalta cavicollis*** Adults are about 5 mm long, shining red, and coarsely punctured. Full-grown larvae are about 6 mm long, dark brown, and with black and yellow spots. This species occurs in southern Canada and in eastern USA south to North Carolina and west to the Rocky Mountains. There is one generation per year. The preferred hosts for this species are fruit trees, including cherry, peach, and plum. Adults are often found around buildings.

**Elm leaf beetle, *Xanthogaleruca luteola*** Adults are about 6 mm long and yellowish green to dull green. Elytra have a black stripe along the sides, and a small dark spot at the base. Full-grown larvae are about 12 mm long; head, legs, and tubercles are black, and there is a broad yellow stripe in the middle of the dorsum. Pupae are bright orange-yellow and with scattered black setae. Eggs are laid in clusters of 15–20 in the soil at the base of trees or on the trunk near the base; fecundity is 400–800 eggs. Hatching occurs in about 2 weeks. Early-stage larvae climb the tree trunk and move out to the branches where they feed on the underside of the leaves. Full-grown larvae crawl into openings in the trunk of the tree, in limb crotches, or move down the tree trunk to pupate in the bark near the ground or the soil around the tree. Larval development takes 2–3 weeks; there are usually two complete generations per year, and a partial third generation in some regions. The first generation is the largest and does the most damage to the tree. Second-generation adults select hibernation sites in late summer or early fall. Overwintering locations include leaf litter around the foundations of buildings, and in the attics, eaves, and interior rooms of houses. Adults usually remain active during the winter, but do not feed. They leave overwintering sites with increasing day length in early spring; they feed between the veins of developing leaves in the spring. Egg-laying begins about 1 month after adult emergence, and when the buds on elm trees in the region begin to unfold. Elm trees close to buildings are the primary source of overwintering aggregations. This species occurs in Europe, North America, and the Middle East (Iran), and hosts include all species of elms (*Ulmus*), including American, Chinese, and European elm. Shade trees are often defoliated by large numbers of larvae; trees growing in forests are usually not seriously infested.

There are native and introduced parasites and pathogens of the pupae of the elm leaf beetle. Species of eulophid wasps, *Oomyzus*, provide limited control in some regions. The eulophid, *Aprostocetus brevistigmata*, is an internal parasite of the pupae. The tachinid fly, *Erynniopis antennata*, is an internal larval parasite that emerges from the elm leaf beetle pupa or adult.

## Cleridae

Clerids are 3–24 mm long, elongate, brightly colored, and usually covered with fine setae. They have 11-segmented antennae which are generally serrate, with open or compact clubs. The pronotum is narrower than the base of the elytra. Larvae are 9–13 mm long, soft-bodied, and pale white to purple. Most adults and larvae are predaceous on other insects, and several are predators of beetles infesting structural wood. Females are attracted to freshly cut wood or to heavily infested locations of wood. Eggs are laid in entrances to beetle tunnels, and clerid larvae actively move through the tunnels of wood-infesting beetles and consume the larvae encountered. Pupation usually occurs within the wood.

Clerid species *Korynetes caeruleus* and *Opilo mollis* are associated with deathwatch beetles in the UK. In North America, *Chariessa pilosa*, *Tarsostenus univittatus*, and *Monophyla terminata* feed on several species of borers in hardwoods, and *Cymatodera bicolor* attacks buprestid and longhorned beetle larvae in hardwoods. Clerids in which tarsal segment 4 is strongly reduced are sometimes placed in the family Corynetidae. Pest species in this group are the red-legged ham beetle, *Necrobia rufipes*, the red-shouldered ham beetle, *N. ruficollis*, and a related species, *N. violacea*.

**Red-shouldered ham beetle, *Necrobia ruficollis*** Adults are about 6 mm long, and with the front of the head and the apical three-fourths of the elytra metallic blue. The head ventral surface, the meso- and metasternum, and the legs are brownish red; the antennae and abdomen are dark brown. Adults and larvae are often associated with skin and bones of dead animals, and on fishmeal.

**Red-legged ham beetle, copra beetle, *Necrobia rufipes* (Fig. 5.4f)** Adults are 3.5–7 mm long. They are shiny green or greenish blue and have reddish-brown legs. Full-grown larvae are about 1 cm long; the head is dark brown and the body light brown; the dorsal sclerites are distinct and dark-brown. Eggs are deposited in batches of up to 28 per day, and the time between batches is 2 or 3 days or 6 weeks. Fecundity is 137 eggs in 5 months when reared on ham, and 906 eggs in 9 months when reared on larvae of the cheese skipper (*Piophila casei*); fecundity is 400–2100 eggs. Eggs are often cemented together when placed in cracks and crevices. Hatching is in 4–6 days at 21–32 °C. Larval development is complete in 17–30 days; minimum conditions for development are 20 °C and 50% RH. Larvae infesting smoked meat migrate to dry locations to pupate. They produce a white frothy oral secretion that hardens to form a chamber for pupation. The pupal period is 6–21 days, and adults remain in the cell for a few days and emerge by chewing a hole in the chamber wall. Adults live as long as 14 months. There are two or three generations per year, and in cold climates overwintering is in the larval stage. Larvae that do not construct a pupal chamber are subject to predation; adult *N. rufipes* will consume exposed pupae and break into pupal chambers to feed on pupae. Adults emit a strong odor when disturbed.

Adults fly readily in warm conditions, but typically walk about on infested material. This species infests drying meats during long storage or prolonged smoking. These beetles are animal decomposers, and they will eat the dried-flesh parts of most animals, including humans. They also feed on cheese, bacon, fish and salt fish, dried egg yolk, bones and bone meal, drying carrion, dried figs, palm nut kernels, dried coconut (copra), and guano. Adults and larvae are predatory and cannibalistic, and they will feed on other insects found on the food, including maggots of the cheese skipper, eggs of *Dermestes maculatus*, and eggs, larvae, and adults of their own species.

**White-banded clerid, *Paratillus carus*** Adult are 5–7 mm long and have a white band across the dark brown elytra. Full-grown larvae are about 13 mm long and with paired processes at the posterior end. This species is a predator of buprestid beetle larvae. *P. carus* larvae and adults search for prey in the galleries, and often dislodge frass from infested galleries during their search. This species is distributed in Australia.

**Yellow-banded clerid beetle, *Trogodendron fasciculatum*** Adults are about 20 mm long, and the antennae are yellow. Full-grown larvae are 20–25 mm long, and have paired processes at the end of the abdomen. This species occurs in Australia and is a predator of buprestid and longhorn beetle larvae.

***Other Cleridae*** *Necrobia violacea* adults are about 5 mm long and metallic dark blue or dark green, and with dark brown or black antennae and legs. It is a predator of *Dermestes* larvae in warehouses. *Thaneroclerus buqueti* adults are 5.0–6.5 mm long, and elongate and brown; the antennal club is three-segmented. It is associated with stored foods in the UK.

## Coccinellidae

Ladybird beetles are a well-known group of small and brightly colored beetles. Adults are 4–10 mm long, and the head is concealed from above. They are distinguished from chrysomelid beetles by three tarsal segments; in chrysomelids there appear to be four segments. Larvae are elongate, somewhat flattened, and covered with tubercles and spines. Most of the species are predaceous as both larvae and adults, and feed primarily on aphids. Many species in temperate regions hibernate as adults, frequently in large aggregations in protected locations. Characteristics of the ladybird species that form aggregations include: they are usually predators of aphids, they exhibit long dormancy or diapause periods, and mating occurs mostly at the aggregation site, shortly before the beetles disperse after winter dormancy.

**Two-spotted ladybird beetle, *Adalia bipunctata*** Adults are 3–5 mm long. The prothorax has an irregularly white margin, and the red elytra have two round median spots. This species is common in Europe and North America, especially along the Pacific coast from California to British Columbia. Adults occur in large numbers and seasonally invade residential and commercial buildings.

**Asian ladybird beetle, multicolored Asian lady beetle, *Harmonia axyridis*** Adults are about 6 mm long. Body color pattern varies from yellowish orange to nearly red, and the elytra can have none to more than 20 spots. Adults live several years and their color may change during this time. Larvae are elongate, somewhat flattened, and covered with small tubercles and spines. Eggs are deposited in batches of about 20 on the underside of tree leaves; hatching occurs in 3–5 days. Larval development is completed in 12–14 days. The pupa is attached to the leaf surface, and adults emerge in 5–6 days. There are two or three generations per year, and adults live 2–3 years. In the first generation, following the overwintering of adults, all females are reproductive; in the second generation a few females enter diapause; in the third and possibly fourth generation most females are in diapause. Adult beetles use visual and olfactory cues to locate a suitable overwintering site. These locations are usually in the sun and warmer than adjacent sites. Once at the overwintering site, beetles apparently use chemical cues to locate a harborage. The source of chemical cues might be beetle feces from the previous year, the odor of beetles that died at the site the previous year, or an aggregation pheromone. Adults are capable flyers and are attracted to natural and ultraviolet light indoors.

This beetle has been released in many regions of the world as a biological control agent for pest aphids. It was introduced from eastern Asia to eastern North America by the US Department of Agriculture in 1977, 1978, 1980, and 1981. The adults and larvae inhabit various trees, including maples, walnut, willow, and oak. They feed on aphids, scale insects, and other soft-bodied insects. It has been effective in reducing pest populations of aphids in pecan orchards.

Several species of flies are known to parasitize *H. axyridis*, including the phorid fly, *Phalacrotophora philaxyridis*. This pupal parasite has been recorded from Japan, and is probably distributed wherever the Asian ladybird beetle is found.

## Cryptophagidae

Adult fungus beetles are 1–2 mm long; the body has a smooth surface, but is sometimes covered with fine setae. Antennae are 11-segmented and usually have a three-segmented club; some species have a two- or four-segmented club. Tarsi are five-segmented, but four-segmented in males of some species. These beetles feed on microscopic fungi, mold, and decaying vegetation. Indoors they can be numerous when conditions favor the growth of mold on walls or exposed wood. They transmit mold spores from one location to another. Several species of *Cryptophagus* occur in bird and wasp nests. Although only small numbers of beetles are found in animal and insect nests close to buildings, these habitats form a network of reservoir populations for some of the most common pest species.

**Angular fungus beetle, *Cryptophagus acutangulus*** Adults are 1.9–3.0 mm long and dark brown. Setae on the wings are nearly equal in length and lie flat on the surface. Full-grown larvae are 2.8–3.0 mm long, yellowish white, and they have distinct urogomphi. This species feeds on the dry rot fungus *Serpula lacrymans* (= *Merulius*) indoors.

**Cellar beetle, *Cryptophagus cellaris* (Fig. 5.4d)** Adult are 2.2–2.7 mm long and uniformly dark brown. The pronotum margin has a distinct tooth; fine setae on the wings are unequal in length and only some lay flat on the surface. This species is cosmopolitan and associated with stored foods in cellars and basements.

**Other *Cryptophagus*** Adults of *C. varus* are about 3.0 mm long and brown; the teeth on the pronotum are reduced. This species is distributed in northern USA and Canada, and has been recorded in stored-food products and rodent nests. *C. valens* adults are about 3 mm long and brown; the margin of the pronotum has an indistinct tooth medially. This species is cosmopolitan and is usually found in stored foods.

**California fungus beetle, *Henoticus californicus*** Adults are 3–9 mm long and uniformly pale brown; the body is moderately hairy, and the antennae have a three-segmented club. It has been distributed to various parts of the world through commercial shipping. Larvae and adults feed on the molds and fungi that grow on stored-food products.

## Cucujidae

Cucujids are 1.4–6.0 mm long, flattened and reddish brown or yellowish brown. They are found under the bark of hardwood trees, and are predators of mites and insects, including wood-infesting beetles. A few species feed on stored grain or meal. Grain feeding may be a recent adaptation for cucujids. Originally, these species probably lived under the bark of wood used to build grain storage facilities, or they were associated with mold under bark or on grain. *Cryptolestes* is sometimes placed in the family Laemophloeidae.

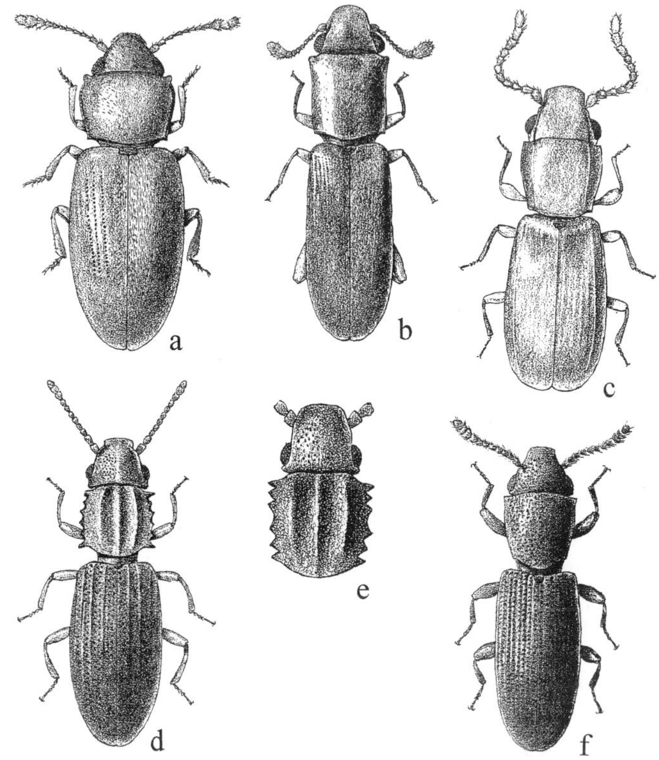


**Figure 5.6** Coleoptera: Cucujidae. (a) *Ahasverus advena*; (b) *Cathartus quadricollis*; (c) *Cryptolestes pusillus*; (d) *Oryzaephilus surinamensis*; (e) *O. mercator*; (f) *Silvanus planatus*.

**Foreign grain beetle, *Ahasverus advena* (Fig. 5.6a)** Adults are 1.9–2.5 mm long and brown to dark brown. Anterior angles of the pronotum have distinct protuberances extending anteriorly. Eggs are laid singly or in clusters of two or three. Females begin ovipositing 3–4 days after emergence; for most females there are 20–30-day periods of egg-laying, and 5–23-day periods of no egg-laying. Larval development is 11–19 days and there are four or five instars. Survival from larva to adult on rolled oats and yeast is at least 95% at 66–92% RH; no larvae survive at 58% RH. The full-grown larva constructs a pupal chamber of food particles cemented together and then attaches itself to the substrate with an anal secretion. Adults emerge in 3–5 days. Development from egg to adult on rolled oats and yeast is about 22 days at 27 °C and 75% RH. Mated males live 159 days and females live 208 days; unmated males live about 275 days, and unmated females about 301 days. This species is cosmopolitan and it has been recorded from a variety of food products, grain and cereal, copra, cocoa, peanuts, dried fruits, herbs, spices, and roots. It usually occurs in large numbers when the product is moldy, and it is capable of living in cultures of molds found on stored grain. The presence of mold, yeast, and wheat germ provides essential nutrients for development for both adults and larvae.

**Square-necked grain beetle, *Cathartus quadricollis* (Fig. 5.6b)** Adults are about 2 mm long, and the oval body is shiny, reddish brown, and somewhat flattened. This species is similar in appearance to the sawtoothed grain beetle, *Oryzaephilus surinamenis*, but lacks the six projections on the lateral margins. Larval development at 80% RH is 24–26 days at 25 °C, and 20 days or less at 30 °C. This species is cosmopolitan and has been found in wheat, rolled barley, rice, dried fruits, cacao, tobacco, and oil palms. In the USA it attacks maize in the field, and is also found in natural populations in the seed pods of a variety of plants.

**Flat grain beetle, *Cryptolestes pusillus* (Fig. 5.6c)** Adults are 1.4–1.9 mm long, brown, and slightly flattened. Antennae of male are not as long as the body; female antennae are about half as long as the body. Elytra have five parallel ridges. Adults can fly and they will jump. Full-grown larvae are about 2 mm long, and pale white with a dark head; they have a pair of brownish black projections at the posterior end. Eggs are deposited in crevices in the grain or in loose material; they hatch in 8–10 days at 25.5 °C. Larval development is 26–45 days and the pupal stage is 6–9 days at 25.5 °C. Development from egg to adult usually takes 5–9 weeks. The full-grown larva forms a pupal chamber of a gelatinous substance; food particles adhere to the surface of the chamber. The adult female lives about 1 year. This species is cosmopolitan, and it is one of the most common insect pests of stored grain around the world. It does not attack sound or uninjured kernels. It often appears after the attack by other grain pests, and it is frequently found in large numbers with the rice weevil. Larvae will also feed on dead insects.

**Rusty grain beetle, *Cryptolestes ferrugineus*** Adults are 1.6–2.2 mm long; the body is brown to dark brown and somewhat depressed. The antenna of the male is not more than half as long as the body. The pronotum is broad, and the posterior margin of the head is broad; there are distinct ridges on the lateral margins of the head and pronotum; there are four rows of setae on the second elytral interval (seen at magnification). Eggs are deposited singly in crevices or furrows of kernels or grain, or in debris. Hatching is in 4–5 days at 27 °C; fecundity is about 200 eggs. Repeated mating often results in higher egg production. Larvae burrow into kernels of grain, and leave the original kernel to seek other food. Development from egg to adult takes 69–103 days at 27 °C; optimum development is at 32–38 °C and 75% RH. The pupal chamber is a silken envelope covered with food particles and frass, or a loose group of silk strands and food material. Late-stage larvae eat the pupae and prepupae they encounter in infested material. This species is cosmopolitan and found in a variety of food materials. Adults and larvae are usually found in stored wheat, maize, rice, barley, oil seeds, copra, coffee, cassava root, and other foodstuffs, but they also occur under the bark of trees. It can also be found in stored fruits, and is usually the dominant species when the fruit is moldy.

**Flour mill beetle, *Cryptolestes turcicus*** Adults are about 2 mm long, and brown to reddish brown; the body is somewhat depressed. Males have antennae extending to about four-fifths the length of the body, and four-segmented hind tarsi. Females have antennae that extend to about half the length of the body, and five-segmented hind tarsi. There are three rows of setae on the second elytral interval (seen at magnification). This species is a pest of flour and feed mills in temperate regions of the world.

**Other *Cryptolestes*** Several species in this genus infest flour and food material. *C. pusilloides* occurs in eastern Africa, Australia, and South America and feeds on wheat and wheat products, sorghum, corn, rice, barley, almonds, and kibbled locust beans. *C. capensis* occurs in Europe and South Africa in milling flour; *C. ugandae* is limited to central Africa, and occurs in maize, millet, peanuts, cassava, cowpeas, and cottonseed.

**Sawtoothed grain beetle, *Oryzaephilus surinamensis* (Fig. 5.6d)** Adults are 1.7–3.2 mm long. The body is brown to reddish brown; the pronotum has six projections or teeth on the lateral margin, and it closely resembles *O. mercator*. The male has a tooth on the hind femur. *O. surinamensis* is distinguished from *O. mercator* by having the vertical diameter of the eye smaller than the region of the head directly behind the eye, and the head is rectangular-shaped. Full-grown larvae are about 3 mm long and yellowish white. Eggs are laid singly or in small clusters; hatching occurs at temperatures between 17.5 and 40 °C, but mortality is high at the two limits; fecundity is 45–285 eggs. Hatching is in 3 days at 40 °C and about 16 days at 17.5 °C. Oviposition usually begins during the first week of adult life and reaches a maximum during week 2 and 3. Larval development is through 2–4 instars and takes about 12 days at 30–35 °C and 75% RH. The upper limit for development is 37–40 °C. All life stages die when exposed to −18 to −15 °C for 24 h, 52 °C for 1 h. At 70% RH, development from egg to adult is about 20 days at 32.5 °C and about 80 days at 20 °C. Larvae construct a pupal chamber of particles of food material; the pupa is usually attached to the substrate. Pupal period is 4 days at 37.5 °C and 16 days at 20 °C. Adult females live for about 19 weeks at 74% RH, but adults live 3 years. There are six or seven generations per year. This species is a cosmopolitan pest of stored grain, cereal products, and dried fruit. *O. surinamensis* does not attack unbroken grain, but can utilize small lesions on the surface to gain entrance. It often feeds only on the germ, and it can complete development on only the endosperm. Larvae feed on the eggs and dead adults of stored-food moths. Adults have not been seen in flight, but they have been collected at light traps.

**Merchant grain beetle, *Oryzaephilus mercator* (Fig. 5.6e)** Adults are 2.2–3.1 mm long. The body is reddish brown to dark brown; the pronotum has six large projections or teeth on lateral margin. It closely resembles *O. surinamensis*, and the two were once considered synonymous. *O. mercator* is distinguished from *O. surinamensis* by having the vertical diameter of the eye larger than the region of the head directly behind the eye, and it has a triangular-shaped head. Eggs are laid singly on food material, and hatching is in 3–5 days. Fecundity is 20–90 eggs. Larvae complete development in about 25 days at 32.5 °C, and about 100 days at 20 °C; development is optimum between 30 and 32.5 °C, and only a few larvae complete development at 17.5 °C. The generic name *Oryzaephilus* means rice-loving, but the merchant grain beetle is usually associated with oilseeds and less with cereal grains. It is nearly cosmopolitan and in most regions *O. mercator* is a pest of processed cereals, especially those with a high oil content. It also feeds on seed-borne fungi.

**Other Cucujidae** *Silvanus planatus* (Fig. 5.6f) infests stored maize and is distributed in North America. Adults are 2.7–2.9 mm long and brown to dark brown. This species is distinguished by the pronotum, which is strongly narrowed posteriorly, with the lateral edges serrate, and with a tooth at the anterior corners.

## Curculionidae

Weevils range in size from 1 to 35 mm and the front of the head is bent downward and produced into a snout of varying length. The antennae are near the middle and at the end of the snout are the mandibles. In some species the snout is short and indistinct; in some of the nut weevils it is as long as the body or longer. Larvae are usually legless, cylindrical, and C-shaped. Weevils have diverse feeding habits, but except for those in ant nests, they are all plant feeders. This is the largest family of animals, with over 40 000 described species in the world.

There are several species of broad-nosed bark weevils in the subfamily Cossoninae that live in the sapwood of hardwoods and conifers killed by bark beetles, and some infest structural wood. The larvae cut meandering galleries across the grain of the wood and pack the galleries with granular frass. *Hexarthrum ulkei* damages damp wood in buildings, *Pselactus spadix* is known to damage wood beneath buildings and saltwater pilings above the high-water mark, and *Tomolips quercicola* damages seasoned coniferous wood. There are other weevil species that attack decayed wood in use. The wood-infesting weevils are not common worldwide, but in the UK, they may be 10% of the active infestations in buildings. Moisture and decay requirements for infestation are variable, but these beetles survive on 20% wood moisture. Wood-boring weevils attack wood that is already decayed by fungi.

The granary weevil, *Sitophilus granarius*, has infested stored grain for centuries, and is believed to be the Curculio of the Romans. The name *Sitophilus* means food-loving, and describes the habit of feeding on grain. The rice weevil, *S. oryzae* (*oryza*, Greek for rice), probably originated in southwestern Asia, perhaps in India along with its host plant. Acorns in northern India were probably a host for the granary weevil before agriculture; northern India has more described species of *Sitophilus* that any other part of the world. The closely related species, *S. sculpturatus* and *S. glandium*, are large weevils, and associated with oaks in northern India. *S. granarius* was probably a large-body species before adapting to feeding on grains. The strains of small-sized *S. oryzae* might also be adaptations to feeding on grains. The three species of *Sitophilus* are parasitized by *Anisopteromalus caladrae*, *Lariophagus distinguendus*, and *Chaetospila elegans* (Pteromalidae).

**Imported long-horned weevil, *Calomycterus setarius*** Adults are about 6 mm long and blackish brown; there are white scales and short, blunt setae on the body. It is native to Japan, but is established in the northeast and midwestern USA. The wingless and parthenogenic adults emerge from soil in June and are abundant in July and August. They feed on a variety of plants, including red clover and alfalfa. When populations peak in summer, they enter vehicles and buildings.

**Broadnosed grain weevil, *Caulophilus oryzae* (= *C. latinasus*) (Fig. 5.7a)** Adults are about 3 mm long, and dark brown. The snout is short and stout, with the dorsal margin curved. Antennae are inserted near the middle of the snout, far in front of the eye; antennae are seven-segmented, and the basal segment of

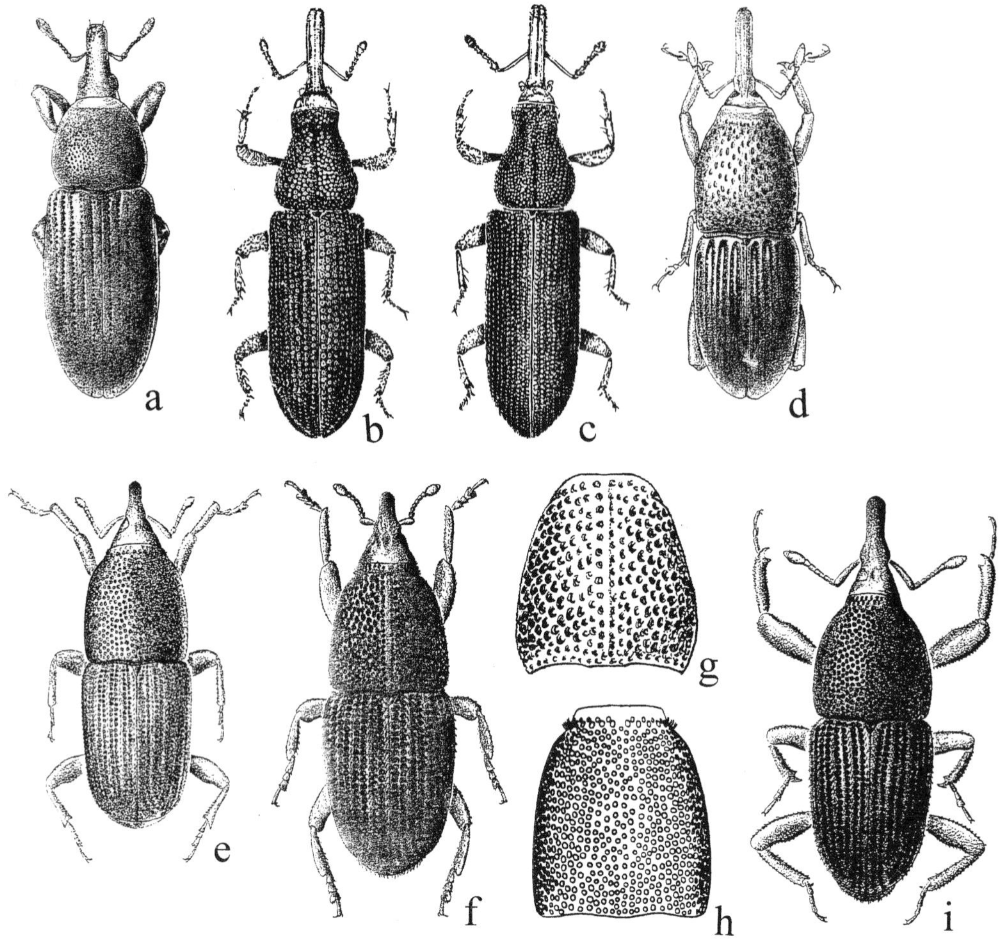


**Figure 5.7** Coleoptera: Curculionidae. (a) *Caulophilus oryzae*; (b) *Euophryum confine*; (c) *Pentarthrum huttoni*; (d) *Sitophilus granarius*; (e) *S. linearis*; (f) *S. oryzae*; (g) *S. oryzae* pronotum; (h) *S. zeamais* pronotum; (i) *S. zeamais*.

the club is not shiny. The adult is a strong flyer. Eggs are laid singly on broken portions of grain, and fecundity is 200–300 eggs; hatching occurs in about 3 days. Larvae feed on the soft portion of the grain and complete development in about 4 days; the pupal period is 2–3 days. Adults live about 5 months. This species is distributed in southern North America and Europe, where it is found infesting stored corn, dried peas, seeds, and cereals. It attacks soft and injured grain, and is often associated with infestations of the rice weevil. It is unable to develop in dry, hard, uninjured grain.

**Asiatic oak weevil, *Crytepistomus castaneus*** Adults are about 6 mm long, black to reddish brown, and with small, green scales on the body. Wings are well developed and the adults readily fly. Eggs are laid in the soil from July to September. Larvae feed on the roots of trees, usually seedlings, at a depth of about 15 cm. Overwintering is in the adult and larval stage, and adults emerge in the spring and feed on the leaves of sapling oaks and other trees. They chew from the leaf margins toward the midrib, and eat everything but the large veins. Later in the season, adults will fly from tree to tree to feed. This species is parthenogeneic and has a 1-year life cycle. The adults invade houses in large numbers in the fall, perhaps seeking an overwintering site. Oaks and chestnuts are the preferred hosts, but other hardwoods and azalea are also attacked. It occurs in eastern USA.

**New Zealand wood weevil, *Euophryum confine* (Fig. 5.7b)** Adults are about 3 mm long and reddish brown to blackish brown. Exit holes are about 1.5 mm diameter, and have a jagged edge. Frass is similar to that produced by anobiid beetles, but the pieces are less rounded and not powdery as in lyctids. This species is common in woodlands where it is found in standing trees and fallen logs. It is native to New Zealand but it has become established in the UK. *E. confine* is a secondary pest of building timbers since it is found exclusively in timbers undergoing fungal decay. As wood dries the surface hardness of the substrate increases and inhibits weevil tunneling and survival. Dispersal flight is limited by conditions of less than 30% RH and 20 °C. Softwood and hardwood species are vulnerable to attack, although most infestations are in Scots pine (*Pinus sylvestris*), *Coniophora puteana* (cellar rot), *Serpula lacrymans* (dry rot), and *Antrodia* (*Fibroporia*) *vaillantii* (white rot). All three species of fungi, which damage both softwoods and hardwoods, are found in association with *E. confine*. The mechanism by which adult weevils detect wood of suitable condition for oviposition and infestation is the odor plumes from fungi when wood decays.

**Eastern wood weevil, *Hexarthrum ulkei*** Adults are 2.5–3 mm long, shiny brownish red to black, and subcylindrical. The snout is slightly longer than the head, the antennae are 10-segmented, and the pronotum is coarsely punctate. Elytra punctures are in distinct rows. This species is distributed in southeastern Canada, eastern USA west to Indiana, and in Idaho.

**Egyptian alfalfa weevil, *Hypera brunneipennis*** Adults are about 5 mm long. The body is grayish brown to black, and covered with short gray setae. The snout is long and slender and distinctly bends downward. This species is a pest in alfalfa and clover. When populations become large in agricultural fields the adults migrate to the margins and often move into buildings.

**Tulip tree weevil, *Odontopus calceatus*** Adults are 2.5–4 mm long and black; the front legs are strongly incurved. Eggs are laid in the midrib on the underside of leaves in May and June. The midrib usually breaks at the oviposition site. Larvae bore into the leaf and mine the interior. Pupation occurs in spherical silken cocoons in enlarged portions of the mine. Newly emerged adults feed on the surrounding foliage. Adults stop feeding by mid-July, and are inactive until the following spring. They spend the winter in leaf litter on the ground, and occasionally enter houses at this time. Adults emerge in the spring when the leaf buds begin to swell. Adults and larvae feed on leaves of magnolia, sassafras, and tulip poplar trees. They are common in July and August and are attracted to lights in buildings. Distribution of this species is eastern USA.

**Black vine weevil, *Otiorhynchus sulcatus*** Adults are 9–12 mm long and blackish brown. The thorax is covered with tubercles, and the elytra are often speckled with white. The hind wings are reduced to pads, and these beetles do not fly. Eggs are laid in the ground, under loose bark, or among leaves of their host tree. Larvae enter the soil and feed on the roots of various plants. Pupation occurs in cells formed near the surface of the ground. Winter is spent as larvae or pupae, and sometimes as adults, at 15–25 cm in the soil. Adults emerge from May to July, and are present during the remainder of the warm season. They feed at night on plant and tree leaves. Adults are attracted to lights at night, and often enter buildings through windows and doors. This species invades houses and it is common in northern USA. It feeds on a variety of plants and trees, including alfalfa, and red, white, and pin oaks.

**Strawberry root weevil, *Otiorhynchus ovatus*** Adults are 5.0–6.5 mm long, dark brown to blackish brown, and have a short and wide snout. The adults do not fly. The larvae feed on plant roots. When populations peak during spring and fall, adults will enter houses and other buildings. They are attracted to indoor sites where water is available. Closely related species, *O. sulcatus*, and *O. rugostriatus*, also occur indoors. These short-nosed weevils are distributed in the UK, and they can be present in large numbers in gardens and around the perimeter of houses and other buildings, and sometimes move inside.

**European wood weevil, *Pentarthrum huttoni* (Fig. 5.7c)** Adults are 2.5–5.0 mm long, blackish brown to reddish brown. The snout length is about two-thirds the length of the pronotum; antennae are nine-segmented. Full-grown larvae are about 3.5 mm long and 1 mm wide; they are legless, strongly convex, and covered with fine spines and setae. Exit holes in the wood surface are irregular in shape and 1.5–2 mm diameter. Eggs are laid about 4 days after mating. Eggs are deposited singly in cracks and crevices, or in holes excavated by the female, which are sealed with a white substance from the ovipositor. Egg-laying continues for about 3 months; fecundity is 20–30 eggs. Hatching is in about 16 days, and the first-stage larvae use egg-bursters on the sides of the first three abdominal segments. Larval development includes five instars and lasts 6–8 months. Larvae tunnel parallel to the wood surface, but meandering tunnels are common. Larvae are capable of digesting cellulose and hemicellulose, but lignin is excreted in the frass. The full-grown larva constructs a pupal chamber, which may be lined with fungal hyphae; the pupal period is about 16 days. The adult beetle cuts an exit hole to the wood surface. Dead adults are frequently found at windows or around lights. This species is distributed in northern USA, Canada, and Europe.

**Pine weevils, pine borers, *Sipalinus* spp.** Adults are 14–20 mm long and brown to brownish black. The head and thorax are covered with tubercles and the femora are enlarged. These beetles infest pine (*Pinus* spp.) wood used in packing cases or structural wood. Larval development takes 1–3 months and adult beetles emerge from infested wood in use. These beetles are found throughout Europe and Asia.

**Granary weevil, *Sitophilus granarius* (Fig. 5.7d)** Adults are 2.5–4.5 mm long, dark brown to blackish brown, and without pale markings on the elytra. Pronotum has punctures that are distinctly elongate or oval and not round or irregular-shaped, as in *S. oryzae* and *S. zeamais*. Wings are greatly reduced or absent under the wing covers, and the adults do not fly. The male snout is shorter, wider, and has more punctures on the dorsal surface than is the case in the female. Full-grown larvae are 3.5–4.0 mm long, yellowish white, and with a dark brown head.

Eggs are deposited in a hole chewed into the grain kernel and a gelatinous material is deposited to seal the hole over the egg; fecundity is 36–254 eggs. Females deposit about 43 eggs at 17 °C, 100 eggs at 21 °C, and 268 at 25 °C; oviposition ceases at about 9.5 °C. Females do not oviposit on grain too small for complete development of a single larva. Larval development takes 57–71 days on wheat at 21 °C, and about 45 days at 25 °C and 75% RH. Larvae tunnel in the seed and pass through four instars; the last instar forms a pupal cell inside the grain kernel; the pupal period is 5–16 days. Females live for about 170 days. The adults and larvae of this weevil will feed on grain and cereal products, but it is primarily a pest of whole grain. It breeds in chickpeas, and all common grains such as maize, oats, barley, rye, wheat, kafir, buckwheat, and millet. *S. granarius* tolerates temperatures as low as 11 °C, and is well established in temperate regions of the world, and in cool upland regions of the tropics. This species is not usually a household pest because it infests whole grain.

**Rice weevil, *Sitophilus oryza* (Fig. 5.7f, g)** Adults are 3.0–4.6 mm long, reddish brown to blackish brown; elytra are marked with four reddish to yellowish brown spots. Wings are fully developed. The pronotum has punctures which are nearly circular or slightly elongate, less than twice as long as wide. It is closely related to *S. granarius*, but is distinguished by the circular punctures on the pronotum. The distribution of *S. oryza* and *S. zeamais* coincide, and they are difficult to separate. Full-grown larvae are about 4 mm long, yellowish white, and C-shaped. Eggs are laid singly in grain kernels after the female bites a small hole in the surface of the grain kernel; the hole is sealed with a waxy secretion; fecundity is 150–400 eggs. Hatching is in about 6 days at 25 °C and success is about 75%. Oviposition occurs between 15 and 30 °C. Highest oviposition is usually 1–2 weeks after emergence, then declines until the female dies in about 3 months. At a density of one weevil per 50 kernels of wheat, oviposition is maximal at 344–384 eggs per female at 25.5–29.1 °C and 14–20% moisture. No eggs are laid on grain with moisture content below 10%. Larval development is through four instars and takes about 26 days at 32 °C, 25 days at 29 °C, and 94 days at 18 °C and 70%. The pupal period lasts about 5 days, and adults remain in the pupal chamber for about 5 days before emerging. Temperature limits for development on wheat with moisture content of 14% are 15.2 °C and 34 °C. Adults live 4–5 months.

This species is nearly cosmopolitan in warm and cool temperate regions of the world. It feeds on beans, nuts, cereals, and cereal products such as macaroni and cake flour, rice, wheat products, and even fruits such as grape, apples, and pears. Direct damage is caused to cereal grain by one or more (maize) larvae feeding inside the grain. Adults can fly up to 1 km to infest stored grain or ripening fields of grain.

**Maize weevil, *Sitophilus zeamais* (Fig. 5.7h, i)** Adults are about 4 mm long and dark brown; the prothorax has round punctures. This species is closely related to *S. oryzae* and *S. granarius*. It is difficult to separate from *S. oryzae*, but is distinguished from *S. granarius* by the round punctures on the prothorax; these are oval in *S. granarius*. Eggs are deposited in the grain kernel; hatching is in about 6 days. Larval development on English wheat at 70% RH is 40 days at 25 °C, and 110 days at 18 °C; development is 35 days on hard red winter wheat at 27 °C and 69% RH. Natural mortality of first-stage larvae ranges from 3.5% at 80% RH to 30% at 50% RH. It is a common pest of grain crops in tropical regions around the world. Adults fly from granaries to fields to start infestations that continue in storage.

**Black wood weevil, *Tomolips quercicola*** Adults are about 3 mm long, cylindrical, and shiny black. The snout is slightly longer than the head, robust, and not curved; antennae are nine-segmented. This species occurs in eastern USA, west to Oklahoma and Texas, and also in Guatemala.

**Other Curculionidae** There may be local populations of weevils that peak at certain times and result in adults moving to other habitats. Some are attracted to lights at night, others simply move away from infested areas and encounter buildings. The acorn weevil, *Balanimus uniformis*, occasionally enters houses in California. *Trachyphloeus bifoveolatus* has been recorded in houses in eastern and western USA. Pea weevils, *Sitona lineata*, occur in large numbers in and around houses in August and September when the adult weevils leave fields and seek places for hibernation sites. Weevils reported infesting moisture-damaged and decayed structural wood include *Euophryum rufum*, *Pentarthrum huttoni*, *P. australe*, and *Caulotrupodes aeneopiceus*. Species of *Rhyncolus* are usually found under bark or associated with other weevils that feed under bark; *R. brunneus* occurs as infestations in structural softwood previously infested by *Hemicoelus gibbicollis*. Larvae and adults are found in the surface layers of infested wood; the frass is round, shiny, and yellowish brown.

## Dermestidae

Dermestids are 2–12-mm-long oval or elongate oval beetles with short, clubbed antennae. They usually have a distinct color pattern, and many are covered with fine setae or scales. Full-grown larvae are 4–12 mm long, brown and usually with long setae on the sclerites and posterior end; they are usually slow-moving. Their primitive habitat was probably feeding on decomposing animal matter, but, as the group evolved, feeding habits radiated to include other material. They are general scavengers on plant and animal material, including carrion, leather, furs, skins, museum specimens, wool and silk, and stored-food products. Larvae are capable of digesting keratin, which is a proteinaceous constituent of wool and other animal materials. Some dermestids visit flowers, and several species require pollen for successful egg production. Adults are capable flyers and can move indoors through doors and windows. About 55 species of dermestids are primary or secondary pests of stored foods. Many dermestid pests made the transition to household habitats from living in insect and bird nests and animal burrows. Dermestid beetles are among the commonest inhabitants of bird nests (especially sparrows), and *Anthrenus verbasci* is commonly found in abandoned wasp nests, including those of *Polistes* and *Dolichovespula*.

The family includes such well-known pests as hide and larder beetles, *Dermestes* spp., and carpet beetles, *Anthrenus* spp. and *Attagenus* spp., which feed on materials of animal origin. This diverse family also includes the kaphra beetle, *Trogoderma granarium*, which feeds exclusively on plant material and is a major pest of grain and cereal products. The larval stage of dermestids damages household materials; they can have an extended period of development and move to several locations while feeding. Full-grown larvae pupate in the last larval skin (*Anthrenus* spp.), or it sheds the skin (*Attagenus* spp.). *Dermestes* larvae bore into solid material to construct a pupal chamber, and then the last larval skin and larval debris is used to block the entrance to the chamber. Development of the adult is often followed by a short resting period within the last larval skin. After mating, the female lays eggs on a suitable substrate and then usually becomes positively phototactic and may be found at windows and indoor lights.

Some species of carpet beetles are named for the plants on which the adult has been found feeding. *Anthrenus scrophulariae* is named for the figwort, *Scrophularia* spp.; *A. verbasci* is named for the mullein, *Verbascum* spp., and a common name for this plant is mothmullein, because it apparently attracts moths. The bird nest carpet beetle *A. pimpinellae* is named for the pimpernel, *Anagallis* spp. (= *Pimpinella*), presumably because adults occur on the flowers of this plant. Larvae of *Attagneus* feed on a variety of dry proteinaceous materials. The most common habitats for dermestid larvae are bird nests and rodent nests, but they also feed in bee and wasp nests and spider webs. Adults are often found on flowers feeding on nectar and pollen. Many species have adapted to living indoors and successive generations are produced without access to outdoors. The species that are general feeders and tolerant of indoor conditions are household pests around the world.

Pest status of dermestid beetles is based primarily on the damage to commercial and household food materials, and the infestations that occur in the stored ingredients to the finished food products. Feeding on organic fabrics and other domestic materials results in economic loss and replacement costs. Infestations in museums result in losses to scientific collections of insects and vertebrates, and archived fabric and furs. Infestations may also cause medical problems. Setae on the larvae easily detach and produce allergic reactions, such as rhinitis and respiratory asthma. Infestations of *Dermestes* larvae indoors can expose sensitive skin and result in irritation. Similarly infested ship's cargo can release large quantities of larval skins in a confined space and cause irritated skin, conjunctivitis, and irritation of respiratory passages. The barbed larval setae of *Trogoderma* may occur in large numbers in infested grain, and cause allergic reactions if swallowed in food processed from this grain. *D. maculatus* and *D. lardarius* are pests in deep-pit poultry houses. The design of these commercial operations provides warm conditions and a readily available food supply for these beetles. Full-grown beetle larvae tunnel into woodwork and insulation in preparation for pupation, and these materials can become damaged and structurally

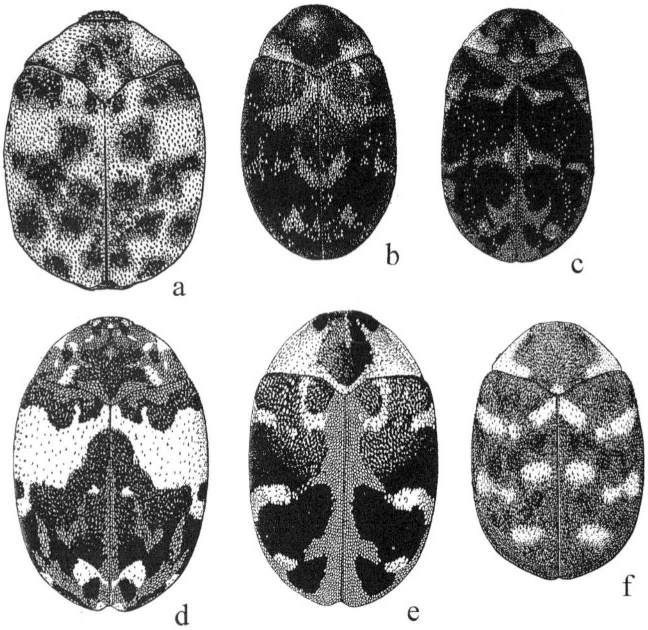


**Figure 5.8** Coleoptera: Dermestidae adults. (a) *Anthrenus flavipes*; (b) *A. fuscus*; (c) *A. museorum*; (d) *A. pimpinellae*; (e) *A. scrophulariae*; (f) *A. verbasci*.

weakened. Infested manure spread on agricultural fields results in the infestation of nearby houses.

**Asian carpet beetle, *Anthrenus coloratus*** Adults are 1.5–2.5 mm long and mottled white and black. Antennal segments 4–6 are elongate, and the visible abdominal sternite 1 has pale white scales, while sternites 2–5 have yellow anterior scales and pale white posterior scales. This species closely resembles the varied carpet beetle, but is distinguished by nine antennal segments, whereas *A. verbasci* has 11 antennal segments. It is originally from Asia, Africa, and Europe, but recently it has been recorded from western North America. It occurs in the nests of wasps and spiders, and infesting insect collections, seeds, and various plant products.

**Furniture carpet beetle, *Anthrenus flavipes* (= *A. vorax*) (Fig. 5.8a; 5.9i)** Adults are 2–3.5 mm long. The body is rounded oval and spotted yellow, white, and black on the dorsum, and white ventrally. Full-grown larvae are about 5 mm long; they are widest at the anterior end and become narrow toward the posterior. Eggs are laid in 1–3 batches containing up to 57 eggs; fecundity is 37–96 eggs. Hatching occurs in 9–16 days; no eggs hatch at 40 °C and development is slow at 20 °C. Adults overwinter, but no eggs are laid during cold weather. Larval development is 112–378 days depending on temperature; larvae that develop normally have six instars, but those developing slowly have 12 or more instars. Larvae feed in a limited radius and their cast larval skins can accumulate in one place, which gives the appearance of a severe infestation. The pupal period is 14–19 days, and the inactive period before emergence is 6–71 days. Development from egg to adult is 93–422 days, and adults live 30–60 days. Adults emerge in the spring and remain active during the warm season. Individuals that develop rapidly as larvae generally have a long adult life, and those that develop slowly have a short adult life.

This species is nearly cosmopolitan, but is probably native to the oriental region. It is named for its habit of attacking the animal-hair packing of stuffed furniture, which was a common practice in the early 1900s. It was first found in the USA in 1911 in upholstered furniture stuffed with horsehair imported from Russia. Larvae are sustained in modern households by feeding on a variety of organic substrates, including wool, silk, fur, feathers, and dry animal material. They need keratin in their diet and larvae will not fully develop if fed on pure wool unless it is impregnated with some other animal product. Larvae will feed on book bindings, chew holes in paper, and are attracted to and feed on dead animals.

***Anthrenus fuscus* (Fig. 5.8b)** Adults are 1.7–2.8 mm long and with a variable pattern of black, white, and orange-brown scales on the elytra. It is distinguished from the varied carpet beetle, *A. verbasci*, by having five antennal segments, whereas *A. verbasci* has 11 antennal segments. Eggs are laid on dead insects in late spring, and larval development usually extends to the following year. Indoors it is found damaging woolens, carpets, insect collections, and books. It is usually associated with spider webs where it feeds on insect debris that gathers in the web, or in corners where webs are built. Outdoors, larvae are found in peridomestic habitats, such as in outbuildings, sheds, barns, and stone walls. *A. fuscus* usually does not occur in bird nests, but is found associated with the sheet webs of the spider, *Tegenaria domestica*. This species occurs in North America, Europe, and Asia, including Japan.

**Museum beetle, *Anthrenus museorum* (Fig. 5.8c)** Adults are about 3 mm long and have yellow and white spots. It resembles *A. verbasci*, but the antennae of *A. museorum* are eight-segmented with a two-segmented club, whereas the antennae of *A. verbasci* are 11-segmented with a three-segmented club. Eggs are laid in late summer and larvae overwinter; pupation occurs in spring. Adults are active on flowers in sunlight, then become negatively phototactic and seek sheltered sites for oviposition.

There is usually one generation per year. This species is widely distributed, including North America, Europe, Australia, and New Zealand. In some regions, such as the UK, it lives primarily outdoors and is not a common indoor pest. Outdoors larvae are not usually found in bird nests, but have been found feeding on moth pupae in the nests of bees, and it occurs in the nests of spiders living in peridomestic habitats. In the sheet webs of *T. domestica* larvae are found in the egg sacs, feeding on eggs and spiders that failed to hatch. *A. museorum* larvae are capable of crawling on the sheet web of this spider, without being attacked. Pupation occurs among the remains of insects in the web or in old egg sacs. Indoors, this species attacks furs, woolens, carpets, silk, feathers, and skins. Larvae have also been recorded feeding on grain, wool, silk, and museum specimens, and dead cluster flies (*Pollenia rudis*) in attics.

***Anthrenus olgae*** Adults are 1.7–2.8 mm long and there are two distinct pale white bands on the elytra. Antennae are eight-segmented and have a distinct club. The abdominal sternites are uniformly covered with pale white scales. Distribution of this species was originally limited to Europe, but it has been introduced into the UK. It is not a common indoor pest, and has been recorded from insect and bird collections.

**Bird nest carpet beetle, panda carpet beetle, *Anthrenus pimpinellae* (= *A. pimpinellas lepidus*) (Fig. 5.8d)** Adults are 2–4.5 mm long and are mottled brown and white. The pronotum posterior margin has a patch of pale white scales at each side, which typically enclose a small oval, dark patch of scales. Full-grown larvae are about 5 mm long and blackish brown. Eggs are laid in the spring or early summer; fecundity is about 50 eggs; hatching occurs in 8 days at 26 °C and 15 days at 20–22 °C. Larval development takes 3–4 months, and the pupal stage lasts 8–10 days; larvae pupate in the last larval skin. Adults develop in the fall but overwinter and become active in the spring. They feed on pollen and nectar. This species is nearly cosmopolitan, and occurs indoors and outdoors in most regions of the world. In North America and Europe, it has been found damaging dried fish, woolens, and other animal materials. Outdoors, it develops in bird nests where the larvae feed on feathers, dead nestlings, or insect remains. Adults are common on flowers in the spring, then they move to bird nests or inside dwellings for oviposition.

**Guernsey carpet beetle, *Anthrenus sarnicus*** Adults are 2.6–3.2 mm long and have an indistinct mottled pattern. Antennae are 10-segmented, and segments 4–6 are rounded and beadlike. Abdominal sternites are uniformly covered with grayish-white scales. Eggs are laid in the spring, and fecundity is 50–80 eggs; hatching occurs in about 1.5 weeks. Larval development is completed in about 104 weeks at 15 °C, 40–50 weeks at 20 °C, and 10–18 weeks at 25 °C. The pupal period is 1.5–2.5 weeks. The adults live 3–14 weeks. This species was described from specimens collected in a house in Guernsey, Channel Isles, UK, and it has been found in domestic and commercial sites. It has been collected outdoors on flowers, and larvae have been recovered from house sparrow and pigeon nests, where it feeds on droppings, feathers, dead birds and insects. Larvae generally feed on animal materials, and indoors they infest fur, feather, skins, and dead insects.

**Common carpet beetle, marbled carpet beetle, buffalo carpet beetle, *Anthrenus scrophulariae* (Fig. 5.8e; 5.9h)** Adults are about 3 mm long. The body is oval, gray to black, and with a varied pattern of white and orange-red scales on the dorsal surface. The antennal club has segment 1 distinctly shorter than segment 2. Full-grown larvae are about 3 mm long, reddish brown, and with long setae. Eggs are laid singly or in batches of up to 36, and they are deposited directly on suitable food; fecundity is 40–60 eggs. Hatching occurs in 13–20 days, and the larval development period through about six instars is 60–80 days. First-stage larvae molt in about 13 days, second-, third-, and fourth-stage larvae molt in 8–10 days, while fifth- and sixth-stage larvae molt in 11–18 days. Pupae are formed in the last larval skin, and the pupal period is 7–12 days. The adult remains in the pupal chamber for 14–18 days, then it is active for about 30 days. Development from egg to adult takes about 95 days, and ranges from 89 to 108 days. Adults copulate and feed on the blossoms of white flowers; the females require nectar and pollen to stimulate oviposition. This requirement usually limits indoor infestations, and links long-term household infestations to adults moving in and out of dwellings. Once mated and fed, females fly into houses or they are carried in on flowers such as daisies (*Chrysanthemum*), wild asters, white roses, crepe myrtle (*Lagerstroemia indica*), and shrubs such as lilac (*Ceanothis*) and *Spiraea*. Once inside, or in an animal burrow or nest, the female lays eggs and dies. Larvae feed until the end of the warm season, when about 75% pupate, with the remainder overwintering.

This is a cosmopolitan household pest, but more common in north temperate regions, and less common in humid regions. In the UK it is only known from imported material and not known to occur outdoors; in other countries, such as Finland, it is only known to occur outdoors in bird and wasp nests and

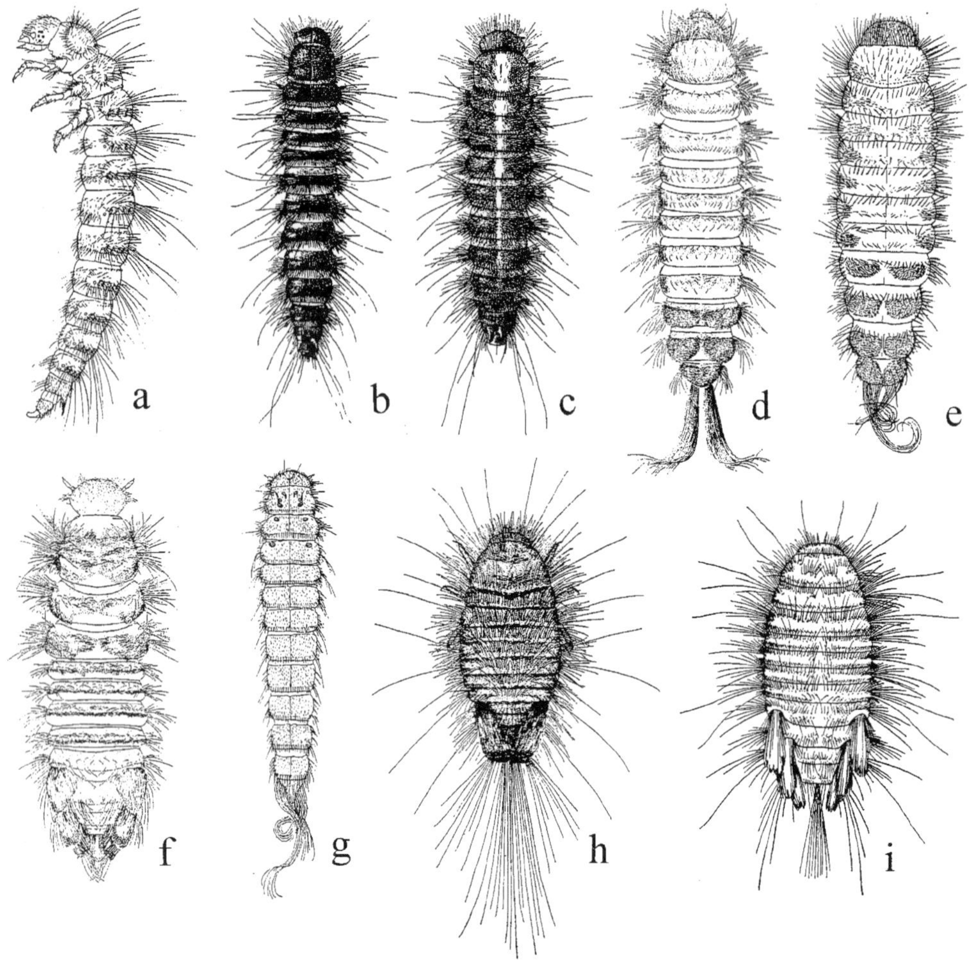


**Figure 5.9** Coleoptera: Dermestidae larvae. (a) *Dermestes ater*; (b) *D. lardarius*; (c) *D. maculatus*; (d) *Trogoderma variabile*; (e) *T. versicolor*; (f) *Attagenus unicolor*; (g) *Anthrenus verbasci*; (h) *A. scrophulariae*; (i) *A. flavipes*.

on dead animals. There seem to be two biological variants: one breeds indoors and does not visit flowers, and one breeds outdoors. Carpet beetle larvae feed on various animal materials, including wool, feathers, hair, and fur, museum specimens, and dried plants.

**Varied carpet beetle, varied cabinet beetle, small cabinet beetle, *Anthrenus verbasci* (Fig. 5.8f; 5.9g)** Adults are 2–3 mm long. The dorsal surface of the body has a pattern of white, black, and brownish yellow scales, and the ventral surface has grayish-yellow scales. Full-grown larvae are 4–5 mm long and have a series of light- and dark-brown transverse stripes. Late-stage larvae are broad toward the rear and narrow at the front, and at each side on the posterior end of the body there are tufts of long and short setae. When alarmed the larva erects these tufts of setae and rolls into a ball as a defensive response. Eggs are deposited singly or in batches; fecundity is about 40 eggs. Hatching is in about 18 days; it is in 30–35 days at 18 °C, 4–17 days at 24 °C, and 10–12 days at 29 °C. Larval development takes 222–323 days, and includes 5–16 instars; the development period is determined by temperature, humidity, and food quality. Successful larval development is between 15–25 °C. Pupation is in the last larval skin. Adults remain inactive in the pupal case for 1–8 days, then emerge. The pupal period is 10–13 days; it is 17–19 days at 18 °C, 10–12 days at 24 °C, and 7–8 days at 29 °C. Field strains of *A. verbasci* have an apparent diapause and must experience cold temperatures to produce adults in the spring; household strains produce adults in the fall. Diapause depends primarily on temperature. This species has one diapause at 25 °C and completes development in 1 year; there are two diapause stages at 15 °C, and the life cycle takes 2 years.

This species is cosmopolitan and occurs indoors and outdoors, but it is primarily a household pest on plant (dried fruits and nuts) and animal materials. *A. verbasci* is a pest of stored food materials and products, biscuits, cakes, seeds, wheat, maize, oats, rice, cayenne pepper, cacao, and dried cheese. It occurs in nests of birds, such as sparrows, martins, and swallows, and in wasp nests in attics, and under the siding of houses. It has also been found in bat roosts. In late spring and early summer, adults are found outdoors on as many as 30 species of flowering plants, but especially *Spiraea*, feeding on pollen. Adults from indoor populations have a negative attraction to light, but near the end of their oviposition period they become positively attracted to light. Adults from outdoor populations show a positive attraction to light. Adults are active flyers and often fly high above the ground. They enter houses through open windows, around eaves, soffits, and attic vents. This beetle is a common pest of insect collections, and the feeding larvae often leave a ring of frass around the hollowed-out insects. Adults have been reported to lay eggs in the dead insects collecting in light fixtures. Hymenoptera parasites recorded from *A. verbasci* include *Laelius trogodermis* and *L. anthrenivorus* (Bethylidae).

***Attagenus brunneus* (= *A. elongatus*)** Adults are 2.9–5 mm long and the body is yellowish brown to black. In the antennal club in the male, the apical segment is about three times the length of segments 1 and 2; the antennae are yellowish brown, and the apical segment may be dark brown. This species infests a variety of plant and animal materials, including woolens, felt, and processed foods such as dried buttermilk, dried milk, and peanuts.

***Attagenus cyphonoides* (= *A. alferii*)** Adults are 2.5–4.2 mm long. The body is uniformly yellowish brown to dark brown. Antennae and legs are pale brown. Females begin laying eggs 3 days after emerging at 22–25 °C and 45–50% RH; 80–90 eggs are deposited in 4–5 days and the female dies soon after. Larval development takes 6–7 months, and larvae hibernate and pupate in the spring. The pupal period lasts 8–10 days. This species is widely distributed and nearly cosmopolitan. In Central Asia it occurs in bird nests, especially sparrows, wasp and bee nests, and it damages furs, skins, woolen articles, and cereal products.

**Tobacco seed beetle, wardrobe beetle, *Attagenus fasiatus* (= *A. gloriosa*)** Adults are 3.6–5.8 mm long. The head and pronotum are uniformly covered with light-colored setae. Elytra have a transverse, subapical band of light-colored setae; the remainder of elytra is brownish black. This species is widely distributed in the warm regions of the world, and it occurs outdoors and indoors as a pest of stored-food products. It is known from a few locations in the USA, and it is an indoor pest in the UK. Larvae feed on wool, fur, skins, and other material of animal origin; it is also known to feed on dried milk, spices, and drugs. When it occurs with vegetable products, it is scavenging on other insects or the remains of grain insects.

**Two-spotted carpet beetle, fur beetle, *Attagenus pellio*** Adults are 3.6–6 mm long, reddish brown to black, and with the basal segments of the legs and antennae pale brown. Elytra have small patches of white to yellowish-white setae. Antennae have a three-segmented club; in males, the last antennal segment is enlarged. Full-grown larvae are about 6 mm long; they have a characteristic tuft of long setae on the last abdominal segment, and a banded appearance. Larvae avoid light and when disturbed often remain immobile in a slightly curved position. Eggs are laid directly on food material; hatching occurs in 22 days at 18 °C and in 6 days at 30 °C. Larval development is 65–184 days at 25–30 °C. The pupa period is 18 days at 18 °C and 6 days at 30 °C. Adults remain in the pupal case for up to 20 days before emerging and flying to flowers to feed on pollen and nectar. There is usually one generation per year, but development may be 6 months to 3 years. This species is cosmopolitan. It is the most common *Attagenus* in the UK, and occurs outdoors and indoors, and is found with the black carpet beetle, *A. unicolor*, infesting buildings. It is a common inhabitant of bird nests, and from these sites it enters structures. Larvae feed on animal material, such as wool, feathers, hair, and fur, and on cereal and vegetable products. Outdoors, larvae usually develop in bird nests, including pigeon, where they feed on feathers, droppings, and the remains of other insects. Household infestations may originate from outdoor sources.

***Attagenus smirnovi*** Adults are 2.3–4.0 mm long and elytra are brown (female) or reddish brown (male). The head, pronotum, scutellum, and ventral side are black or blackish brown; antennae and legs are light brown. Eggs are laid in April and May; fecundity is about 100 eggs. Hatching occurs in about 10 days and larval development is completed in 3 months. The pupal period is 8–13 days. The life cycle is completed in 113 days and the adults live for about 20 days at 24 °C and 70–80% RH. This species was described from specimens collected in Russia. It is probably indigenous to Kenya, Africa, where it is known to live in bird nests and bat roosts, and it occurs in farm stores and warehouses. It has been found in stored-food products in Ethiopia. It occurs indoors in western regions of

Europe, and the UK. Larvae feed on animal materials, including furs, hides, feathers, and dried insects. In Russia there is one generation a year in heated buildings.

**Black carpet beetle, *Attagenus unicolor* (= *A. megatoma*, *A. piceus*) (Fig. 5.9f)** Adults are 2.8–5.0 mm long and uniformly dark brown to black. The last segment of the male antennal club is somewhat pointed and as long as the entire remainder of the antenna; the last segment of the female antenna is not as long as the remainder of the antenna. Full-grown larvae are about 8 mm long and glossy brown; the head, thorax, and abdomen are covered with brown setae at the margins; a brush of hairs, 5–6 mm long, projects at the end of the abdomen.

Eggs are laid singly directly on food, usually in late fall and winter; there are distinct periods of egg-laying activity followed by 5–6 days of no egg-laying. Hatching is in 6–22 days in warm conditions; 20–24 days at 18 °C, 9–12 days at 24 °C, 5–8 days at 29 °C; at 24 °C and 20–93% RH hatching is 9–12 days. Fecundity is 61 eggs at 18 °C, 83 eggs at 23 °C, or 74 eggs at 29 °C. Larval development on rat fur and fishmeal is 269–639 days at about 26 °C for males, and 258–545 days for females. Males reared on feathers and oat flakes complete development in 455–577 days; females on feathers and oat flakes complete development in 509–580 days. At 25–27 °C the number of instars is 7–17, and at 21 °C the range is 8–15 instars. Larvae attain about 80% of their final weight during the first third of the last larval stage, and then enter a period of decreased development. Larvae are negatively phototropic, and will often crawl to the bottom of a food source or other substrate. The pupal period for males lasts 17 days at 18 °C, 9 days at 24 °C, or 5 days at 29 °C. For females it is 18 days at 18 °C, 9 days at 24 °C, or 6 days at 29 °C. Pupae that are 4.5–5.5 mm long produce males, and pupae 4.5–6.0 mm produce females. Mated males live about 17 days at 18 °C, and 40 days at 29 °C; mated females live 15–38 days. Adults feed on pollen of various plants outdoors, including *Spiraea vanhouttei*, *Daucus* spp., *Carola* spp., *Pastinaca sativa*, and *Viburnum dilatum*. Adults have been recorded on the fungus *Lenzites betulina*.

Larvae feed on animal materials, such as silk cloth, wool, feathers, hair, and fur, fishmeal, and cereal products. In Russia it has been reported to damage books, and in Japan it is a pest in the silk industry. Larvae often bore into stored-food containers, making these products vulnerable to further infestation by other pests. Outdoors it occurs in the nests of birds, wasps, and bees, and in rodent burrows and nests. When disturbed, larvae curl and remain motionless for a long period. Larvae move slowly and feed in secluded areas. This species is nearly cosmopolitan, and infests domestic and commercial locations.

**Australian carpet beetle, *Anthrenocerus australis*** Adults are 2–3 mm long, and blackish brown with pale markings. Full-grown larvae are about 4 mm long, and reddish brown; they are covered with setae, and have long setae posteriorly. It occurs in Australia and New Zealand where it attacks woolens, carpet, felt, and stored foods, including meat. In Belgium it has been recorded feeding on dried insects. In the UK it has been recorded from brushes, dried fruit, dried milk, oats, biscuits, hides, and fabrics. Outdoors it has been found in sparrow nests.

**Odd beetle, tissue paper beetle, *Thylodrias contractus*** This dimorphic species has few of the characteristics generally associated with the Dermestidae. Adult males are 2–3 mm long. The body is narrow, brown to light brown, and with scattered long setae. Elytra are yellowish brown and semitransparent; they are contiguous for about one-third of their length, then they separate to expose the abdomen. Hind wings range from fully developed to vestigial. Antennae appear 10-segmented and are longer than the head and pronotum combined. Females are 2–5 mm long. The body is broader than long and larviform. Antennae are nine-segmented and shorter than the head and prothorax combined. Elytra and wings are absent. Full-grown larvae have a transverse row of strong setae at the posterior edge of each segment, and those on the thorax are somewhat clubbed. When disturbed the larvae curl their body. It is known from the UK, North America, Europe, North Africa, Russia, and the Middle East.

The natural habitat for this species is the burrows or belowground nests of carnivorous animals in Central Asia. In the domestic habitat it infests dry animal matter, dead insects, feathers, wool, silk, and muslin cloth, and sometimes museum specimens. It has been reported eating and tunneling in tissue paper, and the common name is derived from this behavior. Larvae can live for long periods without food, and the typical life cycle requires 1 year. The long life cycle and wingless females may have facilitated the spread of this dermestid, first to domestic habitats in its native region and then to other regions of the world.

**Black larder beetle, *Dermestes ater* (Fig. 5.9a; 5.10a; 5.11d)** Adults are 7–9 mm long. Wings are uniform black or blackish brown, and the abdominal venter is brown and has medial and lateral brown spots on a dark-brown background. Full-grown larvae are 10–14 mm long and dark brown, the dorsal surface has numerous long setae; the urogomphi are broad at the base and directed posteriorly. This is one of the most

**Figure 5.10** Coleoptera: Dermestidae adults. (a) *Dermestes ater*; (b) *D. carnivorus*; (c) *D. frischii*; (d) *D. lardarius*; (e) *D. maculatus*; (f) *D. peruvianus*.

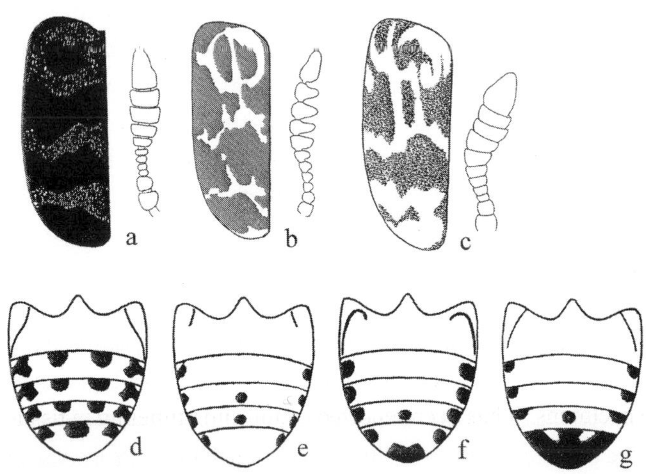


**Figure 5.11** Coleoptera: Dermestidae. (a) *Trogoderma glabrum*, left elytra and antenna; (b) *T. ornatum*, left elytra and antenna; (c) *T. inclusum*, left elytra and antenna; (d) *Dermestes ater*, abdomen venter (diagrammatic); (e) *D. carnivorus*, abdomen venter (diagrammatic); (f) *D. frischii*, abdomen venter (diagrammatic); (g) *D. maculatus*, abdomen venter (diagrammatic).

common *Dermestes* in the world, and is widely distributed in all zoogeographic regions. It feeds on a variety of animal and plant materials, but is most common in smoked meat, dried fish, bones, hides, animal skins, and cheese. In the USA, it is the most common pest of fishmeal-processing plants and warehouses. In tropical countries, it is found outdoors on dead birds, mammals, fish, and crabs. The life cycle and general habits are similar to *D. maculatus*.

***Dermestes carnivorus*** **(Fig. 5.10b; 5.11e)** Adults are 6.5–8.5 mm long. The head has white setae at the front; the apical abdominal sternites have white setae at the apex. Elytra have minute teeth at the apex, and elytra are separately rounded at the apex. This species has been recorded infesting hides, dried skins, dried fish, fishmeal, bones, Brazil nuts, grain residues, and oilcake. It is nearly cosmopolitan. It is an indoor pest in the UK.

**Fringed larder beetle, *Dermestes frischii* (Fig. 5.10c; 5.11f)** Adults are 6–10 mm long. The pronotum lateral edge has white setae; elytra are uniformly brownish black, and apical margins are not serrate. The abdomen venter has lateral dark spots on a grayish-white background; apical abdominal sternites have black setae at apex. Full-grown larvae are 10–14 mm long and dark brown; the dorsal surface has few long setae, and there is a median yellow stripe. Urogomphi are broad at the base and directed anteriorly. Eggs are laid in batches of 2–4 eggs, and hatching occurs in 2–3 days. Larval development is 4.4 days at 37 °C and 70% RH, 19.6 days at 35 °C and 90% RH, 23.4 days at 33 °C and 70% RH, and 170 days at 20 °C and 45% RH. The presence of salt in a fishmeal diet increases development time and reduces oviposition. Pupal period is 5–12 days.

It is nearly cosmopolitan in warm regions of the world. It commonly infests stored fishmeal; it may occur with *D. ater*, but the adults of *D. frischii* are more active. It also infests smoked meat, bone and bone meal, feathers, and skins. In the UK, this is the most common pest species of the genus, and it occurs in primarily commercial sites attacking animal products; outdoors it occurs on carrion and animal skins. Larvae will bore into material adjacent to their food source to form pupal chambers. Timber, cork, and mortar can be damaged.

***Dermestes haemorrhoidalis*** Adults are 6–9.5 mm long. The elytra have a dense fringe of setae that extend beyond the edge; elytra setae are dark reddish brown to black, and usually there are yellowish-white setae scattered singly at intervals. Full-grown larvae have urogomphi broad at the base and projecting posteriorly; the tip is distinctly sharp and bent. Eggs are laid directly on larval food; hatching is in 2.6 days at 32.5 °C and 65% RH, and in 11.6 days at 15 °C. Eggs do not hatch below 15 °C or over 32.5 °C. Larval development is optimal at 20 °C; there are 6–8 instars at 25 °C and 65% RH, and 7–9 instars at 30 °C; at 80% RH the number of instars is reduced to 6. Pupal period is 5.6 days at 30 °C and 17 days at 20 °C; it is not influenced by humidity. Larval and pupal development period is 38 days at 30 °C and 104 days at 20 °C. Adults live for 327 days at 17.5 °C, and oviposition lasts 7 months at 20 °C. This species is cosmopolitan, and it occurs in both outdoor and indoor populations. It feeds on skins, bone meal, dried fish, fishmeal, and some plant materials including grains. Outdoors it is known from carrion and wet bird nests.

**Larder beetle, *Dermestes lardarius* (Fig. 5.9b; 5.10d)** Adults are 7–9 mm long. The body is dark brown to black, with a pale yellow to yellowish white band across the elytra. The ventral surface and legs are covered with yellow setae. Full-grown larvae are 10–15 mm long and brown; the urogomphi are broad at the base and directed posteriorly. Eggs are laid singly on or near larval food; fecundity is 100–175, but 200–800 is recorded; hatching occurs in 3–9 days. Although they usually mate several times, only one mating is required for them to lay a maximum number of eggs. Few to no eggs are produced at 32.5 °C; the preoviposition period is 85–146 days at 20 °C. Larval development at 65% RH is through 6–8 instars at 20 °C and 5–10 instars at 25 °C. The number of larval molts decreases with increased humidity at 25 °C. The pupal period is about 8 days at 27.5 °C and 20 days at 17.5 °C. Duration of the pupal period is not influenced by humidity at 25 °C, but a reduction in temperature at 65% RH extends the period. Larvae complete development through the pupal stage in 45–50 days, at temperatures between 15 °C and 32.5 °C and 65% RH. Development increases to 145 days at 15 °C. Mated and fed adults live 23 days at 32.5 °C, and 250 days at 20 °C. At 25 °C, longevity ranges from 221 days at 40% RH to 290 days at 80% RH.

This species is cosmopolitan. In the UK it occurs indoors and outdoors. It is a minor household pest, but common in commercial warehouses. In deep-pit poultry houses it is a common pest, and damages structural wood. It occurs in pigeon nests and this habit provides access to indoor sites; the larvae are known to kill young pigeons, chicks, and ducklings. The larval stages feed on dry meat, fish, cheese, pet food, stored tobacco, and hides, but will also feed on plant material. Beetles indoors are found on the bodies of dead mice or birds. Outdoors, this species occurs on carrion, wasp nests, and in wet bird nests. Larvae typically leave the food source to pupate and often tunnel into wood, sometimes to a length of about 30 cm. This tunneling can result in structural damage. They can also penetrate sheet lead, and tin with some difficulty.

***Dermestes leechi*** Adults are 7–8.5 mm long. The head, pronotum, and elytra are uniformly reddish brown to black, and covered with grayish-yellow setae. Segment 1 of mid and hind tarsi is about twice as long as segment 2. This species occurs in southern Asia, including India and Pakistan; it is also known from Egypt and Russia, and has been introduced to the UK and other countries in infested materials. Outdoors in Russia it is known from animal burrows and in bird nests. Larvae and cast skins were found preserved in a 3000-year old Egyptian mummy.

**Hide beetle, bacon beetle, *Dermestes maculatus* (= *D. vulpinus*) (Fig. 5.9c; 5.10e; 5.11g)** Adults are 5.5–10 mm long. Elytra are uniformly dark brown to black and sometimes reddish brown, and the ventral surface is white. Elytra have an apical margin with small teeth, and the apex produced into a tooth at the suture edge. It resembles *D. lardarius*, but lacks the yellowish-white band on the elytra, and the ventral surface is white and not yellow. Full-grown larvae are 10–14 mm long, and dark brown. Eggs are laid singly or in batches of 17–25; fecundity is 200–800 eggs; hatching occurs in 2 days at 35 °C and 6 days at 21 °C. Oviposition is influenced by temperature and only slightly by RH between 55–75%. Females require a continuous supply of food and water to achieve maximum fecundity. Larval development is about 50 days and 7–9 instars at 21 °C and 75% RH on a diet of dried meat; it is 34 days and 6–9 instars at 55% RH and 27 °C, and about 19 days and six or seven instars at 35 °C. Pupal development is 5 days at 35 °C and 12 days at 21 °C; it is not influenced by RH. Adult life span is decreased by high temperature or low humidity. Males live about 169 days and females live 173 days at 21 °C and 75% RH. Pupal chambers are constructed after the larvae bore into available hard surfaces. The pupal chamber is enclosed within debris or the last larval skin.

This species is cosmopolitan and is common throughout the tropics; in temperate regions it is limited to heated facilities. It feeds primarily on animal protein, including feathers, fur, bone, cheese, and dried fish. Fishmeal and sun-dried fish are the most commonly infested material. Freshwater fish are more heavily infested, because marine fish have a higher salt content. Fish and fishmeal with a salt content above 10% are usually not infested. Adults are active flyers and are attracted to natural and ultraviolet light. In the UK, this is the most common pest species of the genus. It occurs in primarily commercial sites attacking animal products; outdoors it occurs on carrion and animal skins. It is a pest in deep-pit poultry houses in the

USA and the UK. It damages woodwork and insulation in these and other facilities. It has been used in museums for cleaning skeletons.

**Dermestes murinus** Adults are 7–9 mm long. The head, pronotum, and elytra are uniformly covered with black and white setae arranged in groups to form a mottled pattern. The antennal club is black. Full-grown larvae have urogomphi as long or nearly as long as segment 9, and slightly directed anteriorly. This species occurs nearly throughout Europe, including the UK, and Russia. In Europe, it lives outdoors on carrion of mammals, birds, and fish. It is primarily an outdoor pest in England, Wales, and Ireland, and is common and widely distributed. In Italy it has been recorded in houses, warehouses, and museums, and in Russia it is a pest of stored-food products.

**Peruvian larder beetle, *Dermestes peruvianus* (Fig. 5.10f)** Adults are 8–11 mm long. Elytra are without a dense fringe of setae that extend beyond the edge; setae on elytra are pale yellow. Setae on the head are inclined anteriorly towards the center between the eyes. Full-grown larvae have urogomphi narrow at the base and projecting posteriorly; the tip is distinctly sharp and bent. Eggs are laid singly directly on the food substrate; hatching occurs at 20–30 °C; there are few eggs laid below 20–25 °C and 80% RH. At 20 °C, 75 eggs are laid in 300 days. Larval development at 80% RH is completed in 60 days at 25 °C and in 300 days at 15 °C. This species is found in North, Central and South America, the UK, and continental Europe. It occurs in both outdoor and indoor populations, and it is the most common *Dermestes* indoors in the UK. It feeds on plant and animal material, including skins, bone meal, dried fish, fishmeal, and some plant materials, including grains. Outdoors, it occurs in carrion and wet bird nests, especially pigeon nests.

**Dermestes undulatus** Adults are 5–7 mm long. The head, pronotum, and the anterior portion of the elytra are covered with yellow setae, and some patches of black and white setae are interspersed. The antennal club is reddish brown. Full-grown larvae have urogomphi that are about half the length of segment 9. This species occurs nearly throughout North America and Europe, including the UK, and Russia. In most of its distribution this species lives outdoors on carrion of mammals, birds, and fish, but it is not common; it is primarily an outdoor pest in the UK. It is a stored-food product pest in some locations in continental Europe.

**Stored grain dermestid, wasp nest dermestid, *Reesa vespulae*** Adults are 2.5–4 mm long and elongate oval. Elytra are brown and have a yellowish-brown stripe. The antennae have a four-segmented club. Larvae have subequal pretarsal setae, and there is an absence of fine setae on an abdominal tergal band, behind a row of stout setae. Adults fly and readily move between infested sites. This species was introduced into Europe from North America, and it occurs in north-European countries, and the UK. It also occurs in New Zealand. Infestations occur in dead insects, museum material, and seeds of wheat, rye, tomato, capsicum, cock's foot (*Dactylus glomerata*), Timothy-grass (*Phleum pratense*), and dried plant material. Outdoors it has been found in bird and wasp nests. Reproduction is parthenogenetic. In the UK, adults of *R. vespulae* are found indoors at windows from April to October, with peak abundance in July.

**Colored cabinet beetle, *Trogoderma glabrum* (Fig. 5.11a)** The adult is 2–4 mm long, with dorsal setae that are brown, white, and golden; the wings are brown. Males have a five- to seven-segmented antennal club, and females have a four-segmented club. Eggs are deposited singly; hatching occurs in 4 days at 38 °C and in 12 days at 21 °C. Larval development is 24 days at 32 °C, 75–85 days at 25 °C, and 146 days at 22 °C. At 65–70% RH, development from egg to adult on wholewheat flour is 140 days at 21 °C, 50 days at 27 °C, 41 days at 32 °C, and 45 days at 38 °C. The pupal period is about 6 days, and the adults live 4–10 days. There are one or two generations per year. This species infests plant and animal material, but develops best on animal feeds, rolled barley, rice and rice flour, cottonseed meal, wheat, shelled corn, and poultry mash. Outdoors it occurs in sparrow's nests, and wasp and bee nests where the larvae feed on dead insects. It occurs in North America, Mexico, Europe, Caucasus, Kazakhstan, and Siberia; it is imported into the UK.

**Khapra beetle, *Trogoderma granarium*** Adults are 1.8–3.8 mm long. Elytra are yellowish brown to reddish brown, and have a brown and yellow color pattern. The pronotum is usually darker than the elytra, and the dorsal setae are pale yellow. Males have a dense fringe of setae on the apical margin of the last visible abdominal sternum; females lack this fringe of setae. Full-grown larvae are about 6 mm long, brown to yellowish brown. Eggs are laid singly on the surface or in crevices of the food. Hatching occurs in 4 days at 40 °C, 10 days at 25 °C; eggs hatch at all temperatures between 25–40 °C and 50–73% RH. Females, mated once, lay their full potential of eggs. Larval development is through four or five instars at 35 °C and 50% RH, 5–8 instars at lower temperatures; optimal conditions for larval development are 35 °C at 50–73% RH. Third-stage larvae can penetrate sound grain, and they feed on the germ and endosperm. Larvae are unable to penetrate undamaged pulses, but they will develop on bean flour. Full-grown larvae enter a form of diapause in which they continue to feed and molt, but do not pupate. They may remain in this state for as long as 6 years when food is present. Diapausing larvae often remain inactive in crevices away from food. This diapause is different from retrogressive molting, which is induced by lack of food. The incidence of diapause is influenced by the combined effects of crowding and exposure to temperatures of 30 °C or less. Pupal development requires about 3 days at 40 °C, and about 6 days at 25 °C. Adult beetles are short-lived, and apparently do not feed.

This species is a pest of stored grain, and is found in regions of the world with at least 4 months with temperatures above 20 °C and 50% RH. It prefers wholegrain and cereal products, but it will also feed on dried blood, dried milk, and fishmeal. The common name, khapra, describes its habit in India, which is occupying pores in bricks used to make grain storage buildings. This species is indigenous to India and Sri Lanka, but has spread throughout the world by grain commerce. It does not occur in the USA, Mexico, Ethiopia, South Africa, Russia, Malaysia, New Guinea, and Micronesia. In the UK it is only in heated conditions.

**Large cabinet beetle, mottled dermestid, *Trogoderma inclusum* (= *T. versicolor*) (Fig. 5.11c)** Adults are 2–5 mm long. Elytra are bicolored brown. Eyes have the medial margin distinctly notched. For other pest species of the genus the medial margin of the eye is entire or only slightly notched. Eggs are deposited directly on food source; hatching occurs in 4 days at 40 °C and 20 days at 20 °C. Larval development is optimal at 30–35 °C and 50–73% RH; there are four or five molts. Larvae do not complete development at 15 °C. Larvae can live 511 days at 50% RH, and they can survive for 6 days at 45–50 °C and 20–30% RH. Adult males at 73% RH live 10 days at 40 °C, and 38 days at 20 °C; mated females live 8–38 days. Adult life span at any temperature decreases with humidity. This species occurs in North America, the UK, continental Europe, the Mediterranean region, Egypt, India, and Russia. It is a pest of agricultural products, rice, rice bran, soybean meal, cottonseed, carobs, nuts, processed foods, dried milk, dried fruit, animal feeds, and animal detritus, such as dead insects. Outdoors it feeds on dead insects in beetle galleries and bee nests.

**Ornate cabinet beetle, *Trogoderma ornatum* (Fig. 5.11b)** Adults are 2–4 mm long; the body is elongate oval and black with red bands on the elytra. Full-grown larvae are about 10 mm long, reddish brown above, and pale white below and with short brown setae on the body. Eggs are laid singly or in batches directly on food material; fecundity is about 60 eggs; hatching occurs in about 12 days. Larvae usually remain at the oviposition site and move away only when food is not available. Larval development includes 8–12 stages and takes about 5 months; the pupal period is 11–17 days. Larvae and adults are present throughout the year in heated buildings, and there are two generations per year. This species is a common museum pest, and feeds on various dry animal and plant material. Outdoors it occurs in dead insects and debris in hollow trees, bird nests, and it occurs in the nests of wasps and bees that provision-nest their cells with spiders and insects.

**Warehouse beetle, *Trogoderma variabile* (Fig. 5.9d)** Adults are 3.2 mm long and reddish brown to blackish brown. Full-grown larvae are about 6 mm long, and yellowish white to dark brown; there are long setae on the posterior end of the abdomen. Eggs are laid singly in loose material, or in crevices of whole kernels of grain; at 70% RH hatching occurs in 6 days at 37.8 °C, and 8 days at 21.1 °C. The threshold for hatching is 18 °C, and fecundity is maximum between 27.5 and 30 °C. Females do not oviposit at temperatures below 15.6 °C or at 40 °C. Larval development includes an active diapause similar to that for *T. granarium*. There are normally six instars, but many full-grown larvae enter diapause. Nondiapausing larvae pupate within 7 weeks after hatching, but diapausing larvae may remain the same size for 2 years. The proportion of larvae entering diapause increases from 32 to 67% when they are exposed to 21–28 °C. Development of nondiapausing larvae requires 34 days, and most larvae pupate near the surface of the infested food. The pupal period is about 4 days, and newly formed adults spend 2–7 days within the last larval skin before emerging. Adult males live 9 days at 40 °C, and 50 days at 17.5 °C. Life span is maximal for both sexes at 12.8 °C.

This beetle occurs on a range of food materials, but prefers animal feed, whole kernels of barley and wheat, and grocery commodities such as noodles, oatmeal, wheat germ, and whole wheat flour. Other infested materials include dead animals, candy, cocoa, corn meal, fishmeal, flour, nut meats, and dried spices. Outdoors in Russia it occurs in bee nests. This species is probably indigenous to central Asia, but is now found in North

America, continental Europe, the UK, Russia, China, and parts of the Middle East.

**Other *Trogoderma*** There are 16 described species of *Trogoderma*, and many are similar in their biology and morphological features. *T. simplex*, *T. sternale*, and *T. grassmani* are known to infest stored food. *T. anthrenoides* has been reported from mud dauber wasp nests, and *T. primum* has been recorded from spider webs.

***Thorictodes heydeni* (= *Thaumaphrastus karanisensis*)** Adults are about 29 mm long. The body is reddish brown, shiny, and somewhat narrow and rounded. Eyes are very small or absent; the antennal club consists of three fused segments. Hind wings are absent. This unusual dermestid is a pest of stored cereals and cereal products. Adults and larvae feed on broken grain and grain flour. It has been recorded from cereals, rice, pulses, bones, and fishmeal. The life cycle takes 45–55 days at 27 °C and 65% RH. The earliest known specimen was found in a plant gall, which had been buried for 2000 years in Egypt. This species occurs in North and Central America, the UK, parts of continental Europe, the Middle East, and south Asia.

***Orphinus fulvipes*** Adults are 1.7–3.5 mm long and reddish brown; legs and antennae are light brown. Species in this genus are distinguished by a large, two-segmented antennal club, which is circular and flattened in the male, and oval and convex in the female. *O. fulvipes* is distributed along the sea coasts in the tropics, and occurs in Australia, Pacific Islands, Java, Madagascar, and it has been introduced into North America (Florida), Central and South America, and the UK, France, and Germany. Larvae feed on dried insects, book bindings, tobacco seeds, green cheese, nuts, cocoa, beans, ginger, cinnamon bark, sago flour, nutmeg, rice, and also in the galleries of termites.

***Megatoma undata*** Adults are 3.6–5 mm long and the body is black; there are two pale white bands on the elytra. Antennae are elongate and distinctly clubbed in the male, and with segment 1 longer than segments 2 and 3 combined. Female antennae are short and clubbed, but segments of club are nearly equal. This species occurs in the UK, continental Europe, and Siberia. It is occasionally found in houses and barns and infesting furs, and skins, and it is a scavenger in bees and bird nests and in spider webs, and on dead insects. Outdoors the larvae are found in the galleries of wood-infesting cerambycid beetles. Adults are active outdoors from April to June, and have been found indoors as early as February in Finland.

## Elateridae

Adults are 12–20 mm long, brown to dark brown, and usually with serrate antennae. Click beetles are named because of the spine on the prothorax that snaps into a groove on the mesosternum with an audible click. This mechanism enables an adult lying on its back to throw itself into the air and land on its feet. Most adults are strong flyers and can be found on flowers during the day, and many fly to lights at night. Larvae are known as wireworms because of their long, narrow, and well-sclerotized, brown bodies. Most species are phytophagous, but some are predators. *Pyrophorous* species have two light-producing spots on the posterior edge of the prothorax and one on the abdomen. Light produced from these organs is usually more intense than that of fireflies (Lampyridae). These beetles occur in southern USA and the tropics. *Agriotes* species are common in North America, the UK, Europe, and Asia, and adults often occur at lights at night. The larvae are pests of potato tubers; they excavate deep narrow tunnels in the tubers and can be carried into storage bins at harvest. The tunnels are secondarily infected by fungi and bacteria. Adults of the southern corn wireworm, *Monocrepidius vespertinus* (Fig. 5.2c), may be numerous during summer and often fly to lights at night.

**Eyed click beetles, *Alaus oculatus*, *A. myops*, *A. melanops*** Adults are 25–40 mm long, shiny black, and with two black eyelike spots on the prothorax. In *A. oculatus* the eye-spots are surrounded by a pale ring, but this ring is absent in *A. myops*. Full-grown larvae of *A. oculatus* are 50–65 mm long, smooth and yellowish brown; larvae of *A. myops* are about 35 mm long and dark brown. The adult of the western eyed click beetle, *A. melanops*, is about 35 mm long, dull black, and the eye-spots on the prothorax are small and oval. The distribution is northwestern USA. The eyed click beetles are primarily predators on wood-boring beetles in pines. They are often encountered, along with adults, in down logs, and in firewood piles.

**Click beetles, *Aeolus mellilus*, *A. dorsalis*** Adults are 8–15 mm long, brown to dark brown. Full-grown larvae are about 15 mm long. These are the wireworm species that typically infest turfgrass and are often attracted to outdoor and indoor lights at night. They fly to lights indoors and large numbers often die and collect in light fixtures. Adults of *A. mellilus* frequently enter buildings.

**Garden click beetle, *Athous niger*** Adults are about 12 mm long and dark brown. Larvae are common wireworms in gardens, where they feed on a variety of plant roots. The adults occur indoors. This species occurs in the UK and continental Europe.

**Lantern click beetle, *Pyrophorus schotti*** Adults are 20–22 mm long and brown. Pronotum has small reddish-yellow spots that are luminescent in living specimens. Adults are active at night. In flight, the thoracic light organs glow, and the abdominal light organs cast a light over the ground. The abdominal light organs are hidden when the wings are closed. Larvae are luminescent and carnivorous on beetle grubs in soil.

## Endomychidae

Adults are small, oval beetles that are 3–8 mm long; the body is smooth and shiny, and often brightly colored. The head is visible from above, the pronotum is excavated or grooved laterally, and the sides may extend forward. Some species have tarsi appearing four-segmented; for others tarsal segment 3 is small, and the tarsi appear three-segmented. Most species occur in natural habitats under bark and in fungi or decaying fruit.

**Hairy cellar beetle, *Mycetaea subterranea* (= *M. hirta*) (Fig. 5.4e)** Adults are 1.5–1.8 mm long; the body is bicolored brown to dark brown and shiny; the elytra is subshiny and with distinct pale setae. Antennae are about half as long as the body, and have a three-segmented club. The pronotum has lateral grooves and a raised edge. Full-grown larvae are 2.2–2.5 mm long and with scale-like setae on thoracic and abdominal segments; setae are slightly longer laterally. This species occurs in granaries and warehouses and it is associated with moldy grain. It is widely distributed, including North America, Europe, Asia, and Hawaiian Islands.

## Erotylidae

Pleasing fungus beetles are 3–20 mm long, elongate, and oval. Some have roof-shaped elytra. They are usually black with reddish yellow markings. Antennae are 11-segmented, and usually have a three-segmented club.

***Cypherotylus californica*** Adults are 16–18 mm long and black. Elytra are yellowish gray and with numerous small black spots. Adults are found in fungus-rotted wood, and in household stored-food products.

## Histeridae

Adults of this large family of predaceous beetles are 0.5–10 mm long, shiny black to blackish brown and sometimes metallic or with red markings, and with elbowed and clubbed antennae. Elytra are truncated, leaving the two apical segments of the abdomen exposed. When alarmed they retract the antennae and legs and remain motionless. Larvae are pale yellow and generally without sclerotization; the mandibles and palpi are prominent, and abdominal segment 9 is broad and bears two-segmented cerci. Species live in dung and decaying organic matter, some under bark or in the tunnels of wood-infesting beetles. Several genera are found in ant nests and some in termite colonies. *Carcinops pumilio* is the most common histerid found in stored products; other species include *Gnathonus nanus* and *G. nannetensis*.

## Hydrophilidae

Water scavenger beetles are 5–40 mm long, oval, and somewhat convex beetles that can be recognized by their clubbed antennae and long maxillary palps. Most species are aquatic. The aquatic species are black, and the metasternum is prolonged into a spine. These beetles are common in ponds and slow-moving streams. Adults are generally scavengers, but will attack live insects and other aquatic animals, including small fish; the larvae are predaceous on other insects. Several species are known to occur in outdoor swimming pools.

**Giant water scavenger beetle, *Hydrophilus triangularis* (= *Hydrous*)** Adults are 34–37 mm long, and with a prominent keel or metasternal spine on the middle of the venter. Eggs are enclosed in a silken case that is about 22 mm wide and 24 mm long, and has a horn-like mast; this case floats on the surface of the water or is attached to a floating leaf or other object. Larvae are predaceous on other insects and small animals. This species occurs generally throughout the USA. The adults fly to lights at night. Adults of a related species, *Tropisternus californicus*, are sometimes found in outdoor swimming pools, and may bite people.

## Lampyridae

Firefly beetles or lightning bugs are 5–20 mm long, brown with olive green or black markings, and sometimes marked with yellow or red. The head is somewhat concealed by the prothorax. Abdomen has ventral light organs in both sexes of most species; it is usually large in the male and small in the female. Larvae are found in leaf litter or on the soil surface; in some species the larvae are luminous. In spring, males usually fly above the ground, and the females of most species are on the ground or some other fixed site. Firefly light is produced by a mixture of chemicals similar to enzymes, resulting in the production of light and very little heat. For most species, the flashing is determined genetically and is a fixed pattern. The characteristic flashing pattern separates species that occur in the same habitat. Adults of some species are predaceous on other insects, and sometimes other fireflies. The importance of some species is their economic contribution to tourism, which is based on excursions to see large populations in synchronous flashing at night.

**Asian firefly, *Pteroptyx tener*** Several species of *Pteroptyx* are present throughout the year and have the behavior of aggregating in trees and synchronous flashing. This species occurs along the Selangor river, in the Kuala Selangor district in Malaysia. Related species in this region include *P. valida*, and *P. malaccae*.

**Blue Mountains firefly, *Atyphella lychnus*** Adults are about 9 mm long and the body is light brown with dull yellow stripes on the fore wings. This species occurs in the Blue Mountains of New South Wales, Australia. A related species, *Luciola costata*, is yellow with black tips on the front wings and occurs in northern Australia, including the Torres Strait Islands.

**Eastern firefly, *Photinus pyralis*** Adults are 10–14 mm long, brown, and the thorax has yellow margins. This species occurs in north- and southeastern USA.

**European firefly, *Luciola lusitanica*** This species occurs east of the Rhône river in southern Europe. It has light-producing organs in both sexes. The immature stages feed on land snails.

**Glow-worm, *Lampyris nocituca*** The adult male is about 28 mm long, light brown, and the pronotum covers the head. The female is about 23 mm long, brown, wingless and larviform, and the pronotum extends over the head. This species is distributed in the UK.

**Western firefly, *Ellychina californica*** Adults are 10–14 mm long, black or reddish black; the elytra are covered with fine, yellowish hairs. This species occurs in western USA.

## Languriidae

Adults are 2–10 mm long and the body is elongate and narrow. The pronotum is usually reddish brown and the elytra are black. Antennae are 11-segmented with a three- to six-segmented club. Adults are common on plants and the larvae of most species are stem borers. A few species are recorded as occurring in stored produce; the two species commonly reported are *Cryptophilus niger* and *Pharaxontha kirschii*.

## Lathridiidae

Lathridiids are 1.2–2.0 mm long; they usually have a rough surface, and the elytra have rows of pits and are broader than the thorax and head. Antennae are eight- to 11-segmented, and have a one- to three-segmented club. Most do not fly. These are referred to as fungus beetles or minute brown scavenger beetles. Natural habitats include plant debris, under bark, and in ant and termite nests. Larvae and adults feed exclusively on the hyphae and spores of molds, mildews, and fungi, and they can exist only in damp locations.

Pest status of these beetles is based on their presence in moldy material and their link with mold fungi. They occur on building materials, such as stacked lumber that is moist or exposed outdoors. Lathridiids often contaminate food and spread mold spores; there are records of infestations in brewer's yeast, and food quality problems caused by species of *Dienerella* (as *Cartodere*). Hygienic problems in a hospital have been caused by infestations of *Dienerella filum*. The fungi that are known to support infestations or are fed on by lathridiids include species of *Polysaccum*, *Ustilago*, *Arcticum*, *Trichothecium*, *Lycoperdon*, and *Tilletia*. Cultures of the molds *Mucor muccedo* and *Penicillium glaucum* have supported the common household lathridiid, *Dienerella filum*. Mixed cultures of the molds *Aspergillus*, *Mucor*, *Botrytis*, and *Penicillium* also support the feeding of larvae and adult beetles. However, *Penicillium* mold is preferred by some species.

**Household fungus beetle, *Adistemia watsoni* (Fig. 5.12b)** Adults are about 1.5 mm long and brown to dark brown. Coxae 1 and 2 are contiguous, and abdominal segment 1 is fused to the metasternum between the hind legs. Eggs are laid singly on the fungus substrate; however, oviposition does not occur under conditions of high humidity. Hatching occurs in about 6 days. Larvae remain in food source and migration is limited when food is plentiful. Larval development is about 13 days, and mortality increases with high humidity. Prior to pupation, last-stage larvae attach to the substrate; the pupal period is

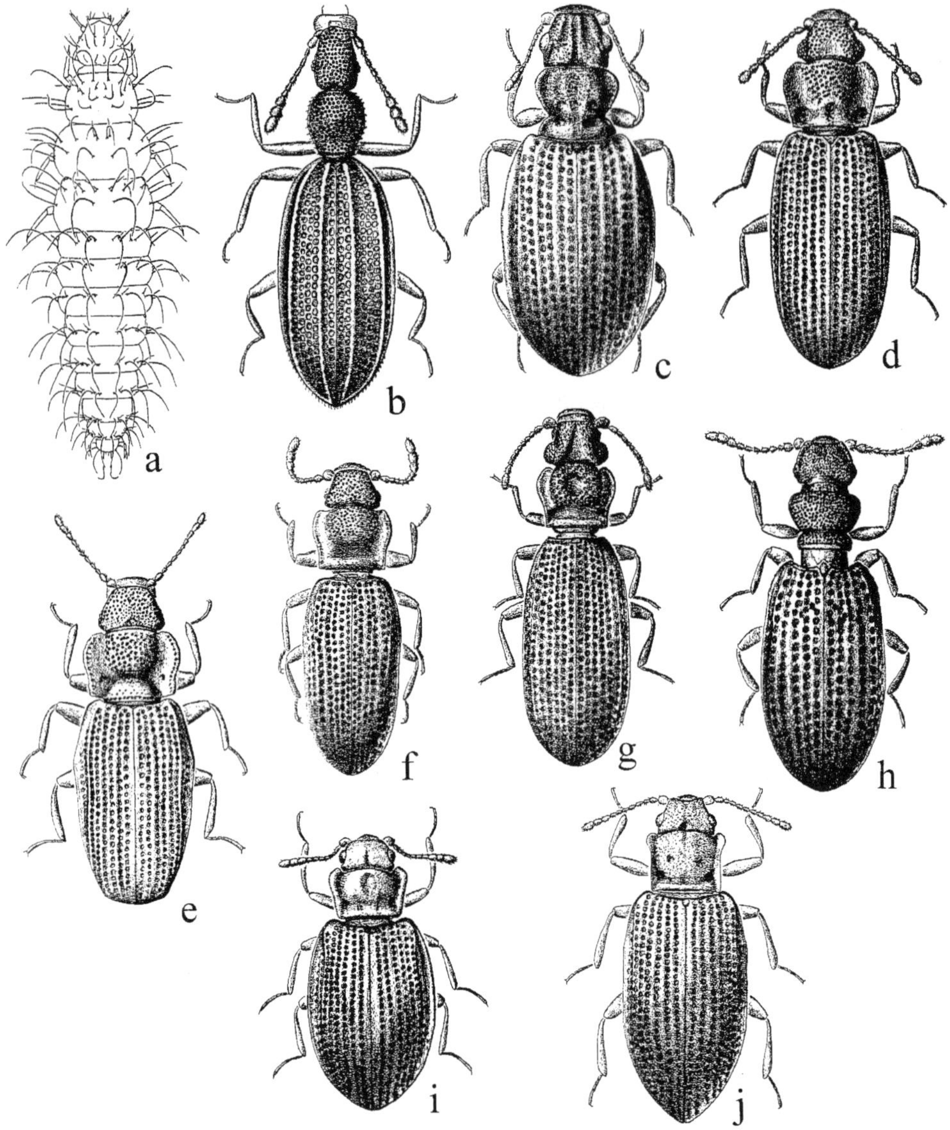


**Figure 5.12** Coleoptera: Lathridiidae. (a) *Aridus nodifer*; (b) *Adistemia watsoni*; (c) *Cartodere constricta*; (d) *Dienerella arga*; (e) *D. costulata*; (f) *D. filiformis*; (g) *D. filum*; (h) *D. ruficollis*; (i) *Lathridus minutus*; (j) *L. protensicollis*.

about 10 days. Development from egg to adult is 21–35 days at 25 °C. Adult mortality is 4% at 32 °C for 7 days; at 38 °C about 98% die within 7 days, and at 43 °C mortality is 100% within 24 h. Time for 100% mortality at 54 °C is 10 min, at 60 °C 5 min, and at 65 °C the time for 100% mortality is 4 min. Larvae and adults are negatively phototropic; adults have not been observed to fly. This species is distributed in North and South America, Europe, Canary Islands, Madeira, and Africa. It is a common household pest, and it is known to infest a variety of moldy materials, and to occur in outdoor populations (decaying potatoes).

**Swollen fungus beetle, *Aridus nodifer* (Fig. 5.12a)** Adults are 1.5–2.1 mm long and dark brown. Legs are pale brown to yellowish brown; the antennal club is three-segmented. Elytra have a very large and longitudinal swelling on the apical third; the male has a large tooth near the inner apex of hind tibia. Full-grown larvae are about 2 mm long, pale yellow, and with numerous long, recurved setae. Eggs hatch in 5–7 days; larval development is 20–28 days, and the pupal period is 3–4 days. This species occurs indoors and is cosmopolitan.

**Plaster beetle, *Cartodere constricta* (Fig. 5.12c)** Adults are 1.2–1.7 mm long and brown. The antennal club is two-segmented, and the pronotum is distinctly constricted. Full-grown larvae are about 2 mm long, pale yellow, and with a dark brown head. Eggs are laid singly, and hatching occurs in about 3 days. Larval development at 25 °C is 4–7 days, and the pupal period is 1–5 days. Larvae and adults are negatively phototropic, and when food is scarce or disturbed, the adults and larvae will disperse. Adults are capable of flight. This species is usually associated with infested grain and stored-food products; it has a cosmopolitan distribution.

**Dienerella arga (= Microgramme) (Fig. 5.12d)** Adults are about 1.4 mm long and the body is brown. The posterodorsal surface of the head has a triangular depressed area. The antennal club is three-segmented. Eggs are laid singly or in small batches; hatching occurs in about 5.5 days. Larval development is complete in about 13 days, and the pupal period lasts about 10 days. This species occurs in North America, Europe, and North Africa.

**Dienerella filiformis (= Microgramme) (Fig. 5.12f)** Adults are 1.2–1.4 mm long, and the antennal club is three-segmented and slightly enlarged. This species is distributed in the USA, Europe, Russia, and Japan. This species occurs indoors. A closely related species, *D. costulata* (Fig. 5.12e), has a broad pronotum and the three-segmented antennal club is slender. This species is distributed in North America, Europe, and Japan.

**Common plaster beetle, *Dienerella filum* (= *Microgramme*) (Fig. 5.12g)** Adults are 1.2–1.6 mm long and brown. The antennal club is two-segmented. The anterior half of the pronotal disk has a broad, moderately deep, oval depression; the head has a median suture. Other *Dienerella* (*Microgramme*) species have a three-segmented antennal club. Hind wings are lacking and adults do not fly. Full-grown larvae are 1.7–2 mm long and pale white. Pupae are about 1 mm long and yellowish white. Eggs are laid singly on the substrate; fecundity is about 20 eggs. Development is 36 days at 24 °C and 54 days at 18 °C; at lower temperatures, development takes as long as 5 months. The life cycle is completed in about 21 days at 24 °C. This species is the most common of the house-infesting lathridiids, and it is nearly cosmopolitan. They occur in damp, moldy warehouses and cellars, in areas near water pipes, around leaky windows, under moldy wallpaper, in moldy cereals, and moldy bread. They are known as plaster beetles because they occur in new or remodeled buildings that have recently been plastered. Adults cannot crawl on smooth or vertical surfaces, and they are often trapped in sinks and tubs. Adults appear 3–4 months or, rarely, 1 year after walls have been plastered. Moisture from plaster and warmth favors growth of molds, which is the primary food for these beetles. Plastered walls can support an infestation if wallpaper is applied soon after construction.

**Dienerella ruficollis (= Microgramme) (Fig. 5.12h)** Adults are about 1.5 mm long. The antennal club is three-segmented; the margin of the pronotum is constricted anteriorly and posteriorly. This species occurs indoors and outdoors, and it is distributed in North and Central America, Europe, North Africa, and New Zealand.

**Minute fungus beetle, square-nosed fungus beetle, *Lathridus minutus* (= *Enicmus*) (Fig. 5.12i)** Adults are 1.2–2.4 mm long and pale reddish brown to black; when black, the antennae and legs are reddish brown. The antennal club is three-segmented; elytra are rounded at the apex. Full-grown larvae are about 2.2 mm long, pale yellow, and have moderately sparse setae. The pupa is about 1.5 mm long. Eggs are laid singly, and at 18 °C hatching occurs in 5–6 days. The first two larval stages last 4–5 days at 18 °C; third-stage larvae feed for 3–4 days then wander for 2–3 days before pupating. The pupal period is 6–7 days, but may be 14–15 days. This species occurs indoors and is widely distributed.

**Other Lathridiidae** Numerous species of fungus beetles have been collected in buildings. In the UK and continental Europe, *Thes bergrothi* is a common species; it is 1.8–2.2 mm long and reddish brown. In northwestern USA, *Lathridus protensicollis* (Fig. 5.12j), is found indoors and outdoors.

## Lucanidae

Stag beetles are generally large, some reaching 60 mm long, and the males usually have large mandibles. Larvae live in decaying logs and stumps. They are distributed around the world, and are most common in the tropics. Adults are capable flyers and are often found at lights at night.

**American stag beetle, *Lucanus elaphus*** Adult males are 45–60 mm long; females are 30–35 mm. They are dark brown and shiny; legs and antennae are brownish black. Males have mandibles nearly as long as the body; mandibles of the female are not as long as the head. Larvae feed on decaying wood and other organic matter, and they are often found in old stumps, especially oak. Adults fly to lights at night.

**British stag beetle, *Lucanus cervus*** Adult males are 30–50 mm long and have large, antler-like mandibles. Females are about 30 mm long, and are without large mandibles. These beetles are distributed in the UK, and they feed in rotting tree stumps and decayed wood. A related species, the lesser stag beetle, *Dorchus parallelopipedus*, also occurs in the UK, and it is found in decayed wood. Adults fly to lights at night.

**Pinching beetle, *Pseudolucanus capreolus*** Adults are 22–35 mm long and brown to dark brown; the male does not have large mandibles. These beetles occur in eastern USA.

***Syndeus cornutus*** Adults are 8–10 mm long and uniformly brown; it has characteristically large and dentate mandibles. Larvae are C-shaped and feed on moist and decaying wood. In natural habitats, it lives in logs and decaying stumps of trees. In the urban environment, it infests decaying wood in moist (30% moisture content) subfloor timbers and joists.

## Lyctidae

Adult powderpost beetles are 2–7 mm long, reddish brown to black, and subcylindrical with a slightly prominent head which is constricted behind the eyes. Antennae are 11-segmented and claviform. The mandibles are well chitinized, broad at the base, and bidentate at the tip. Elytra have punctations and fine setae. Larvae limit their feeding to hardwoods, and feed on the sapwood of various hardwood species. Larvae utilize the starch, sugar, and proteins in wood, but they cannot digest the cellulose and hemicellulose in the cell walls.

Round emergence holes 2–3 mm in diameter and small amounts of powdery frass are characteristic of lyctid infestations. Frass usually falls from emergence holes and accumulates in small piles. Lyctid frass is fine and feels soft and not gritty when rubbed between the fingers; the frass of anobiids has a gritty texture. Lyctids attack oak, hickory, and ash, but other infested hardwoods include walnut, pecan, poplar, sweetgum, wild cherry, and several species of tropical hardwood. Bamboo, which is a grass, is often attacked because of its high starch content. Lyctids prefer the seasoned sapwood portion of hardwoods, and wood with moisture content between 8 and 32%, but the greatest activity occurs in wood with 10–20% moisture. They usually do not lay eggs in wood with a starch content less than 3%. In air-dried lumber much of the starch content is consumed in respiration by the ray cells, which remain alive for some time after cutting. Kiln drying kills the ray cells so that the starch content is fixed within the sapwood, and the wood is susceptible to lyctid attack for many years.

Oviposition takes place 2–3 days after mating, usually in the afternoon and early evening. Before oviposition adult females bite and perhaps chew a small amount of the wood to determine its starch and moisture content. The female extends her long, flexible ovipositor directly into vessels and pores of the wood. A preliminary examination of the vessel is made with the pygidial palps of the ovipositor, and further examination of the vessel is made by the vaginal palps. One or more eggs are deposited longitudinally in the vessel. Egg-laying usually lasts 1–2 weeks, with most of the eggs laid 7–8 days after mating. The diameter of the vessels in hardwood can be a limiting factor in lyctid infestations. They must be large enough to accept the ovipositor and the eggs. The average diameter of the ovipositor of *Lyctus brunneus* is 0.078 mm, and of *L. linearis* is 0.083 mm. Hardwoods with vessel diameters larger than this are subject to infestation. Softwoods are not infested since the exposed tracheids will not accept the *Lyctus* ovipositor. Examples of susceptible wood include a variety of temperate hardwoods, but beech and birch are not usually infested. Tropical wood infested includes obeche, agba, afara, mahogany, iroko, seraya, meranti, teak, and keruang.

Eggs are translucent white and cylindrical, with rounded ends. Eggs of *L. brunneus* are 0.8–1.25 mm long and 0.15–0.175 mm wide. Eggs typically have a threadlike process at the anterior end. Hatching occurs in 6–7 days at 26 °C, and 19–20 days at 15 °C. After leaving the egg chorion, the larva is generally facing the opening of the vessel. The larva feeds for a short time at the opening, then it tunnels into the wood. Larval food is the starch in the cells of the wood; the cell wall is not digested. Besides starch, certain sugars and protein are present in the larval food. First-stage larvae are white and straight-bodied, and with a pair of small spines at the posterior end. Larvae have a curved form and a slightly enlarged thoracic region after the first molt. Early-stage larvae usually tunnel with the grain of the wood; later stages have stronger mouthparts and tunnel at irregular angles. They are negatively phototropic and eat away from the wood surface. Full-grown larvae are usually less than 5 mm long. At this stage in development, they tunnel close to the surface of the wood and prepare a pupal chamber immediately below the surface. The pupal period lasts 12–30 days.

Adult beetles cut through the pupal chamber and surface of the wood to emerge, and a small amount of frass is expelled from the tunnel. During the day beetles remain in cracks and crevices in the wood and in the emergence holes, but become active at night; they fly and are attracted to light. The life span of the female is about 6 weeks, and males live 2–3 weeks. Development from egg to adult takes 9–12 months, but takes 7–8 months under favorable conditions of moisture and starch content of the infested wood. In warm regions, there are two

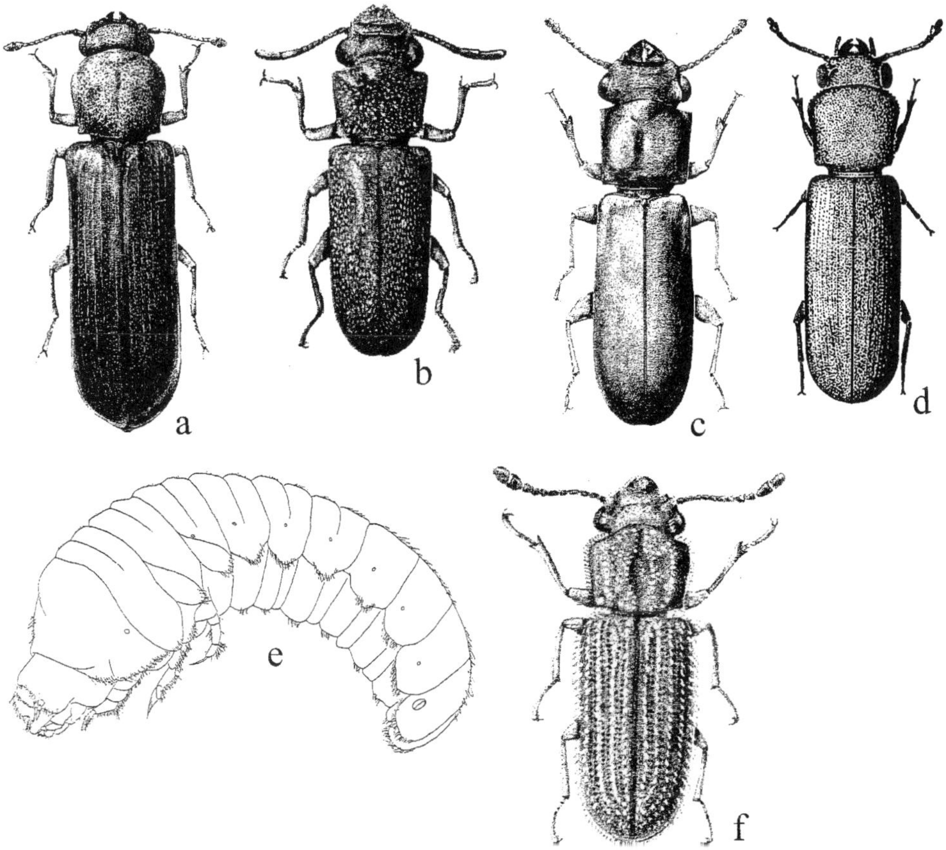


**Figure 5.13** Coleoptera: Lyctidae. (a) *Lyctus planicollis*; (b) *Lyctoxylon japonum*; (c) *Trogoxylon parallelopipedum*; (d) *Lyctus brunneus* (e) *L. brunneus* larva; (f) *Minthea rugicollis*.

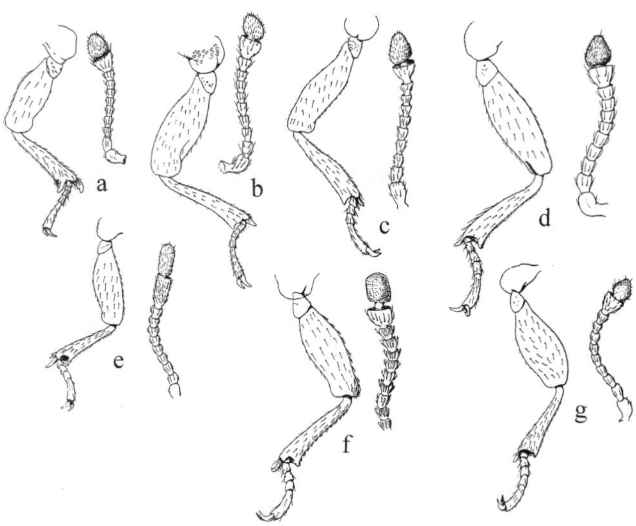


**Figure 5.14** Coleoptera: Lyctidae. (a) *Lyctus africanus* adult, prothoracic leg and antenna; (b) *L. brunneus*; (c) *L. cavicollis*; (d) *L. planicollis*; (e) *Minthea rugicollis*; (f) *Trogoxylon parallelopipedum*.

or three generations per year, but under adverse conditions the life cycle may extend to 4 years.

**Oriental powderpost beetle, *Lyctoxylon japonum* (Fig. 5.13b)** Adults are 1.5–2.0 mm long, and brown to reddish brown. The antennal club is elongate, with both segments longer than broad. This species is distributed in Asia, and may have become established in Central America. It infests bamboo in Asia.

**African powderpost beetle, *Lyctus africanus* (Fig. 5.14a)** Adults are 2.5–4 mm long and reddish brown. It resembles *L. brunneus*, but is distinguished by a distinct notch between the frontal and postclypeal lobe; this notch is absent in *L. brunneus*. Female *L. africanus* has a fringe of setae on the distal margin of sternite 4; in *L. brunneus* this fringe is absent. *L. africanus* may be endemic and restricted to the oriental and Ethiopian regions. It is often encountered at ports, and it infests food material, such as licorice, papaya seeds, and ginger roots. Other materials infested include hardwoods and mahogany plywood.

**Brown powderpost beetle, *Lyctus brunneus* (Fig. 5.13d, e; 5.14b)** Adults are 5–7 mm long and reddish brown to black; the body is somewhat flattened and with parallel sides. The head and mandibles are visible from above. Full-grown larvae are 5–7 mm long, yellowish white and with a distinct pair of enlarged and dark spiracles visible near the posterior end. The emergence hole is 1–1.8 mm diameter. Egg-laying takes place 2–3 days after copulation, and usually occurs at night. The female has a very long and slender ovipositor, and uses it to locate a vessel in the sapwood. During egg-laying the insect can bend the ovipositor 90°; typically 1–3 eggs are laid in each vessel selected. Fecundity is about 80 eggs. Hatching occurs in 19–20 days at 15 °C, 14–15 days at 20 °C, and 7–8 days at 26 °C. Larval development is 8–10 months, depending on environmental conditions and the wood infested. Larvae tolerate wood moisture content between 8 and 30%, and development is optimum at 16% moisture content and 25 °C. Full-grown larvae bore approximately 3 mm from the wood surface and form a pupal chamber; the pupal period lasts 12–21 days. Development from egg to adult is 1 year in unheated premises or outdoors, but extends to 2–4 years under adverse conditions. Emergence of adults takes place from May to September, with a peak in July in the UK. The adults are active flyers and are attracted to lights at night. Parasites of this species include *Eubadizon pallidipes* (Braconidae), *Sclerodermus domesticus*, *S. macrogaster* (Bethlyidae), *Tarsostenus univittatus* (Cleridae), and *Teretrius picipes* (Histeridae). This lyctid probably originated in North America, but it is now distributed around the world.

**Shiny powderpost beetle, *Lyctus cavicollis* (Fig. 5.14c)** Adults are 2.5–5 mm long and reddish to rust brown. The prothorax has the anterior portion narrower than the combined bases of the elytra. Adults closely resemble *L. planicollis*, but are distinguished by the slender antennal segments, and the oval and small antennal club. The antenna is large, and the club is prominent in *L. planicollis*. This species is widespread throughout North America.

**Small powderpost beetle, *Lyctus discedens*** Adults are 3–6 mm long, and brown to dark brown. Full-grown larvae are about 5 mm long, and yellowish white. Emergence holes are 1.5–2.0 mm diameter. This species is one of the major wood-infesting beetle pests in Australia, and it infests and reinfests sapwood of hardwoods.

**European powderpost beetle, *Lyctus linearis*** Adults are 3–7 mm long and dark brown, with glossy elytra. It resembles *L. cavicollis* in size and color, but can be distinguished by having single, rather than double rows of punctations of the elytral striae, and by its prominent frontal lobes. *L. linearis* is commonly encountered in commercial hardwood products, especially in seasoned hickory, oak, ash, and walnut. It occurs in eastern USA and in Europe. Parasites of this species include *Eubadizon pallidipes* (Braconidae), *Tillus unifasciatus*, and *Monophylla terminata* (Cleridae).

**Southern powderpost beetle, *Lyctus planicollis* (Fig. 5.13a; 5.14d)** Adults are 4–6 mm long and black to blackish brown. Prothorax has a broad, shallow median depression, and the elytral striae are separated by rows of long setae. Full-grown larvae are 3–5 mm long and yellowish white. This species is found throughout the USA and Mexico. The parasites reported for this species include *Hecabolus lycti*, *Monolexis lycti* (Braconidae), and *Tarsostenus univittatus* (Cleridae),

**Tropical powderpost beetles, *Minthea rugicollis*, *M. reticulata* (Fig. 5.13f; 5.14e)** Adults are 2.0–3.0 mm long and the body is uniformly brown. The antenna has the terminal segment slightly longer than the preceding segment. Abdominal sternite 5 of the female has a fringe of fine setae on the apical margin, but this fringe is absent on males. The emergence hole is 1–1.5 mm diameter. *Minthea rugicollis* and *M. reticulata* cause considerable damage to hardwoods in the tropics, and appear in imported hardwoods in some temperate-region countries. They attack seasoned and green timber of kempas, mernati, pulai, and seasoned rubberwood in Malaysia. Damage is distinguished from *Lyctus* species by the small size of the emergence holes. Parasites reported for this species include *Monolexis* spp., (Braconidae), *Cephalonomia* spp., *Sclerodermus* spp. (Bethylidae), *Cercocephala* spp. (Pteromalidae), and *Tarsostenus univittatus* (Cleridae).

**Parallel powderpost beetle, *Trogoxylon parallelopipedum* (Fig. 5.13c; 5.14f)** Adults are 2.5–4.2 mm long, reddish brown to brown, and with fine yellowish setae on the body that give it a velvety appearance. The prothorax lateral edge is straight and appears parallel. Adults are capable of rapid movement. It infests ash, hickory, and oak timbers, and bamboo. The winter is spent in the larval stage, and adult emergence and egg-laying occur in spring. Adults of the second generation emerge in late summer or fall. Development is completed in 4–5 months in recently sawed logs. Parasites reported from this species include *Monolexis lycti* (Braconidae), *Sclerodermus domesticus*, and *S. macrogaster* (Bethlyiidae). This species occurs in the USA and the UK.

## Lymexylidae

Adults are 9–15 mm long and narrow-bodied. The head is bent downward and narrowed behind the eyes to form a short neck; eyes are large and the legs are slender. The larva has an enlarged prothorax, well-developed legs, and abdominal tergite 9 is usually modified. Beetles in this family were serious pests at the time of wooden ships, especially those made of oak. They are often found under bark and in dead logs and stumps. These beetles cause much of the pinhole damage in commercial chestnut and oak.

**Chestnut timberworm, *Melittomma sericeum*** Adults are 10–15 mm long and the body is brown and covered with fine setae. Full-grown larvae are yellowish white and with large mouthparts. Tergite 9 of larva is enlarged and dark-pigmented, and has a scoop-like shape, with the margins bearing square teeth and many setae. These beetles infest chestnut and oaks, especially white oak. Eggs are laid in cracks on the surface of the wood. Larvae bore deep into the wood, enlarging their circular tunnels as they develop. Pupation occurs in enlarged cells near the surface of the wood, and the adults chew their way out.

**Sapwood timberworm, *Hylecoetus lugubris*** Adults are about 12 mm long and brown. Larvae are yellow with a light-brown head. Abdominal tergite 9 ends in an elongate, sclerotized spine that is bifurcate at the tip and has tooth-like projections along the edge. Eggs are laid on the surface and in crevices of the bark of dead trees or green logs. Larvae tunnel under the bark and across the sapwood of hardwoods, such as poplar, birch, basswood, and black walnut. The adults are active from April to July.

**Other Lymexylidae** *Hylecoetus dermestoides* attacks both hardwoods and softwoods in Europe and Siberia. *Lymexlon navale* attacks oak in Europe. In West Africa, *Atractocerus brevicornis* and *A. crassicornis* are pests of hardwoods.

## Meloidae

Blister beetles are 10–15 mm long, elongate, and soft-bodied. They are black, brown, or blue, but sometimes orange or marked with red or yellow. The head is bent downwards, and there is a narrow neck. Adults are phytophagous, and larvae are parasitic of wild bees and grasshopper eggs in soil. The body fluid of the adults contains a substance, cantharidin, which causes human skin to blister. When disturbed, blister beetles fill their trachea with air, close their spiracles, and build up body fluid pressure until breaks occur in the cuticle, usually at the joints. The defensive body fluid is released, but this type of reflex bleeding does not harm the insect. Handling live or dead adults can result in skin irritation and blisters.

**Striped blister beetle, *Epicauta vittata*** Adults are 9–15 mm long. They are black and marked with yellow; the head has two black spots on the top. The pronotum has two stripes, and the elytra have two or three black stripes. Adults feed on a variety of plants. Larvae feed on the egg masses of grasshoppers in the soil. This species is often associated with alfalfa. When it is harvested, adult blister beetles are crushed in the hay during bailing, causing the release of cantharidin into the hay. When horses eat the contaminated hay, they may develop severe colic and sometimes die.

**Spanish fly, *Lytta vulnerata*** Adults are about 20 mm long; the head and thorax are shiny, and mostly orange. Elytra, abdomen, and legs are black. Larvae often occur in wild bee nests. Adults are attracted to lights at night. The European species, *L. vesicatoria*, is the source of commercial cantharidin, which is used to treat urogenital diseases.

## Mycetophagidae

Hairy fungus beetles are 1.5–5.5 mm long, oval, and slightly flattened. The body is usually covered with fine setae, and is brown to black and sometimes marked with red or orange. The larvae and adults of nearly all species, including those associated with stored-food products, feed on fungi, hence the family name. They are not usually considered major grain pests, but they can be serious pests of rail-shipped grain. These beetles live in spilled grain along railroad tracks, and fly in large numbers in late summer and fall.

**Hairy fungus beetle, *Typhaea sterocorea* (= *T. fumata*)** Adults are 2.2–3.0 mm long, and brown to dark brown. The antennal club is three-segmented, the prothorax has basal pits, and the elytra have setae arranged in longitudinal rows. It is often confused with the drugstore beetle, *Stegobium paniceum*, but the antennae are distinctly clavate; in the drugstore beetle, the antennae are serrate. It occurs in houses, warehouses, retail stores, flourmills, and sometimes outdoors. It is frequently found in maize fields where it is attracted to decaying grain. It is a pest of stored grain and seeds, tobacco, peanuts, and cacao. It occurs as a pest in the litter in poultry houses in some areas of Europe. A closely related species, *T. decipiens*, has been recorded as a pest of stored grain in northern Europe (Germany, Denmark).

**Two-banded fungus beetle, *Alphitophagus bifasciatus*** Adults are about 3 mm long, elongate oval, and reddish brown and with two broad, black bands across the wing covers. Larvae feed on moist cornmeal and spoiled cereals. This species is cosmopolitan. It feeds on fungi and mold in refuse grain and grain products, including wet grain in the holds of ships.

**Four-spotted fungus beetle, *Mycetophagus quadriguttatus*** Adults are 3.5–4.0 mm long. The antennal club is four-segmented, and the prothorax has a distinct oval pit on each side near the base. The elytra have two large basal spots and two subapical spots. This species is cosmopolitan and infests various stored foods.

## Nitidulidae

Most nitidulids are 12 mm long, elongate or oval, and in many species the elytra are short and expose the terminal abdominal segments. They are found where plant fluids are fermenting; a few occur on or near the dried carcasses of dead animals. Twenty-two species, most of which belong to the genus *Carpophilus*, have been recorded from stored products around the world.

**Corn sap beetle, *Carpophilus dimidiatus* (Fig. 5.15a)** Adults are 2.0–3.5 mm long and reddish brown. It infests stored foods, especially dried fruits and cereals. In the field, it feeds on decaying fruit, and on the sap exuding from injured plants. Eggs are laid singly on the larval food, and hatching occurs in 2–4 days; fecundity is about 225 eggs. Oviposition is on food with 15–33% moisture; it does not survive in rice with less than 10% moisture. Larval development is 5–16 days, and the pupal period is 5–18 days. Development from egg to adult is 15 days at 32 °C and 49 days at 18 °C. Total development is 26 days at 32 °C and 90% RH. This species is distributed in the USA.

**Dried-fruit beetle, *Carpophilus hemipterus* (Fig. 5.15b, e)** Adults are about 3 mm long. The body is oval and black with two large brown spots on the elytra; the legs are reddish brown. It occurs on ripe and decomposing fruit in the field, and is a pest of dried fruit in storage. In date palm groves, *C. dimidiatus* and *C. hemipterus* infest fermenting dates on the ground. *C. hemipterus* also infests stored maize and maize meal, wheat, oats, rice, beans, nuts, copra, spices, and medical drugs. Eggs are laid singly; hatching occurs in 1–4 days; fecundity is 2134 eggs at 28 °C, and 461 at 14 °C. Larval development is 4–14 days, and the pupal period is 4–16 days. When reared on dates, development from egg to adult is about 12 days at 32 °C, and 42 days at 18 °C. Males live about 146 days, and females live 103 days. There are two or three generations per year.

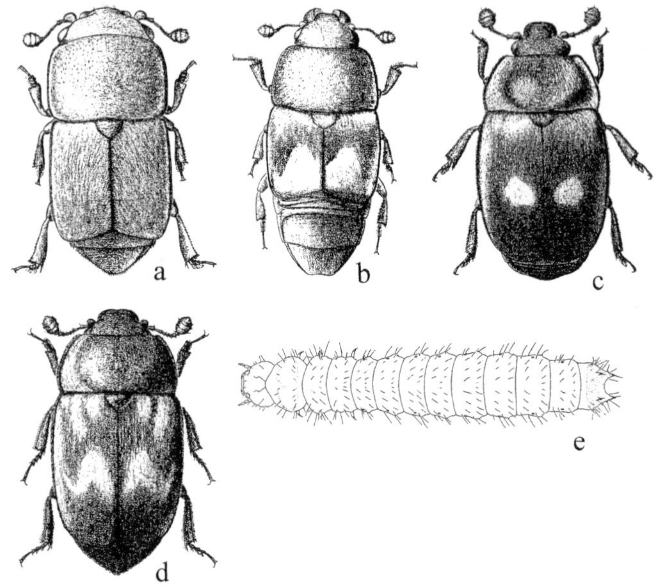


**Figure 5.15** Coleoptera: Nitidulidae. (a) *Carpophilus dimidiatus*; (b) *C. hemipterus*; (c) *Nitidula bipunctata*; (d) *N. ziczac*; (e) *C. hemipterus* larva.

**Red-spotted sap beetles, *Glischrochilus quadristignatus*, *G. hortensis*** Adults are 4–7 mm long, and the body is black and shiny. Each elytra has two red-orange spots; the spots often merge. Adults of *G. hortensis* are known to occur indoors in large numbers in the UK.

**Two-spotted sap beetle, *Nitidula bipunctata* (Fig. 5.15c)** Adults are 3–5 mm long; elytra are dark brown with a pair of pale spots at the middle. The male and female are externally similar. This species is often associated with carrion, but is occasionally found indoors feeding on meats, bread, and cake. It is distributed widely in the northern hemisphere. A closely related species, *N. ziczac* (Fig. 5.15d), has three longitudinal spots and a light median band on dark brown elytra. It is found on carrion, and occasionally reported indoors in North America.

**Other Nitidulidae** Dried fruits and other stored products around the world attract nitidulids that eventually become pests by establishing infestations. *Haptonchus luteolus* is one of the smallest species to infest dried fruit. It is about 3.4 mm long, and yellowish brown. The pineapple beetle, *Carpophilus humeralis*, is one of the largest nitidulids attacking dried fruit; it is 4–5 mm long and uniformly black.

## Oedemeridae

Adults range in length from 5 to 25 mm, and their color varies from pale brown to black, with red, yellow, or orange markings. Elytra are soft and sometimes covered with fine setae. Adults are attracted to lights at night and they are found on flowers or on old logs or stumps. Larvae are soft-bodied and they develop in wet, decaying wood, especially driftwood.

**Warf borer, *Nacerdes melanura* (Fig. 5.2f)** Adults are 7–15 mm long and reddish yellow with the ends of the elytra black; antennae are half the length of the body. Full-grown larvae are 12–16 mm long, yellowish white, and with dark brown mandibles. Thoracic legs of the larva are distinct, and there are conical protuberances (false legs) on abdominal segments 3 and 4. Eggs are laid in wet and decayed wood, including softwoods and hardwoods. Larval development is in irregular galleries in wood; pupation is in small chambers. The pupal period is 7–9 days, and adults live about 7 days. Adults emerge in large numbers about the same time every year, and are active from April to August outdoors, where they feed on flower pollen. Indoors they occur throughout the year. Larvae feed in wet wood and infest pilings under wharves and under buildings near the water. Boardwalks along the sea coast and wood floors in damp basements, boats, and greenhouse benches are infested. Thousands of adults may suddenly appear in or around a building. They fly to windows and other sources of light. This beetle occurs in North America along the Atlantic and Pacific coasts and around the Great Lakes, and it is known from some inland sites. It also occurs in the UK, and it is probably native to continental Europe.

**False blister beetle, *Oxycopis mcdonaldi*** Adults are about 8 mm long, and dark blue with an orange-red pronotum, which is rounded posteriorly and marked with a median longitudinal stripe. Full-grown larvae are about 7 mm long and yellowish brown, and with a dark brown head. Antennae are about as long as the body. Larvae are known to feed in moist and decayed wood, often along coastlines. Adults fly to lights at night, and if crushed on the skin they will cause blistering.

## Passalidae

These shiny black beetles are 30–40 mm long and are often called bessbugs or bess beetles. Adults prepare food, which is decayed wood, with their salivary secretions and feed it to the young. Passalids infest moisture-damaged and decayed logs in modern log houses and wood in contact with the ground in other buildings. Logs exposed to rain and roof runoff often have pockets of decay, and wood in the ground is subject to decay fungi. These sites can become infested with passalids. They do not increase the decay and damage to the wood, but contribute to the propagation of decay fungi. Three species occur in North America.

**Horned passalus, pegbug, *Odontotaenius disjunctus*** Adults are 30–40 mm long and shiny black, and with a short horn at the front of the head. Full-grown larvae are 37–40 mm long. The body is bluish white, and the head dark yellow. Metathoracic legs of the larvae are reduced to one segment, which gives the appearance of having only four legs. When disturbed the adults stridulate to produce a squeaking sound or the so-called bess notes. Larvae have stridulatory organs on the third pair of legs. These beetles are found in eastern North America in rotten logs and stumps of deciduous trees, and decaying logs in a modern log house.

## Phengodidae

Glowworm beetles are 10–30 mm long. Females are without elytra and wings, and luminous. Males have short elytra, and 12-segmented, doubly plumose antennae. Larvae are predators and live in leaf litter, under bark, and under boards in forested sites. Species occur in the USA and Canada.

**Glowworms, *Zarhipis piciventris*, *Z. riversi*, *Z. integripennis*** Adult female *Z. piciventris* is 30–50 mm long, larviform, and yellowish brown; males are about 10 mm long and black, but the mandibles, prothorax, bases of the antennae, and legs are reddish brown. This species occurs in the San Francisco Bay area of California. *Z. riversi* male is about 14 mm long and black; the thorax, scutellum, and abdomen except for the black tip are all yellowish red. *Z. integripennis* adults are 18–20 mm long; the male is pale yellowish brown, and the female yellow or tan. This species occurs in western USA.

**Pink glowworm, *Microphotus angustus*** The adult female is 10–15 mm long and larviform, somewhat flattened and pale pink. Elytra are vestigial and the antennae are nine-segmented.

The male is about 10 mm long, winged, and with a pink body; the prothorax and elytra are gray. It occurs in grassy areas of the foothills in the San Francisco Bay area of California. The adults are active on warm late summer and fall nights, and the light produced by these beetles is very bright.

## Platypodidae

Platypodids are known as pinhole borers or flatfooted ambrosia beetles. Adults are 2–8 mm long and brown to brownish black. The body is cylindrical and with the head slightly wider than the pronotum. Tarsi are very slender, with the first segment longer than the remaining segments combined; in scolytids the first tarsal segment is shorter than the remaining segments. Platypodids lack a spine or projection at the apex of the front tibia that is present in the scolytids. Tunnels made in wood by the adults are extensive, and they usually extend deep into the sapwood and hardwood. Females often die at the entrance to the system of galleries they have cut for the larvae, and their body protects developing larvae from predators and water loss (to maintain fungal growth). Weakened or recently felled trees are preferred for attack. Larval food and development are similar to scolytid beetles. Adult platypodids cut a new tunnel to the surface of the wood when emerging, and they leave behind a small amount of powdery frass. There are two or three generations per year for most species. They are common in tropical and subtropical regions; only species in the genus *Platypus* occur in North America.

**Asian pinhole borer, *Platypus parallelus*** Adults are 3.8–4.5 mm long and yellowish brown to brown. Elytra are dark brown at the apex in both males and females. Holes in the wood are about 1 mm diameter and they are usually stained black. The male makes a short tunnel in the bark of the tree or log, and then releases a pheromone to attract a female. Mating occurs on the surface, and then the female enters the tunnel and excavates the main gallery for the larvae. Frass is often extruded from the entry hole in long, compact strings; galleries are stained black by the fungal growth. Eggs are laid along the main gallery and larvae feed on the fungal mycelium that grows in the gallery system. Newly emerged adults exit the wood through the original entry hole. Peak emergence is in April in Malaysia, and there is apparently one generation per year. This species occurs in southern Asia, including Malaysia. The unseasoned hardwoods and softwood with a moisture content of about 40% that are susceptible to attack include *Cynometra* spp. (kekatong), *Gonystylus* spp. (ramin), *Hevea brasiliensis* (rubberwood), *Koompassia malaccensis* (kempas), *Myristica* spp. (penarahan), *Pinus* spp. (pine), *Shorea rugosa* (meranti bakau), and *Sindora* spp. (sepetir).

**Horizontal pinhole borer, *Austroplatypus incompterus*** Adults are about 4 mm long, and reddish brown to black. This Australian species attacks standing trees that have been injured by fire or stressed. The gallery system is constructed in one plane, and the fungus inhabiting the galleries breaks down cellulose. Large pieces of dimension lumber are subject to fracture when infested by these beetles.

**Oak pinhole borer, *Platypus quadrimaculatus*** Adults are about 4.5 mm long, and reddish brown. The elytra have parallel ridges, and the 3, 5, and 7 interspaces have tooth-like projections on the posterior of the male. This species attacks various species of hardwoods, especially oaks.

**Pinhole borer, *Platypus flavicornis* (Fig. 5.2e)** Adults are about 5 mm long and reddish brown. The head is flat anteriorly and has moderately long setae. Elytra parallel ridges and the 3, 5, 7, and 9 interspaces have tooth-like projections on the posterior of the male. This species attacks various species of pines and occasionally hardwoods. Dying trees and logs cut and remaining in the woods in the summer are commonly attacked. The adult bores a horizontal gallery in the sapwood, and it branches extensively. Often there are large amounts of frass present below the holes.

**Other Platypodidae** In Australia the pinhole borers encountered in forested areas and in structural wood include the polyphagous pinhole borer, *Platypus australis*; the mountain pinhole borer, *P. subgranosus*; and the omnivorous pinhole borer, *Crossotarus omnivorus*. In Europe, *Xyleborus saxeseni* (Fig. 5.2d) attacks a variety of hardwood and softwood timber. The larvae, pupae, and adults occur together in the brood tunnels. The island pinhole borer, *X. perforans*, is often found infesting wood from islands in the Pacific. Galleries are usually only in sapwood and they are not stained.

## Ptinidae

Spider beetles are 1.5–4.5 mm long, oval or cylindrical beetles that resemble large mites or small house spiders; the larvae are C-shaped. Spider beetles are cosmopolitan pests that are scavengers on dry animal and vegetable material. Larvae are in cracks and crevices in cabinets and in floorboards where they feed on organic debris, including wool dust, and rodent droppings. Full-grown larvae create shallow pupal

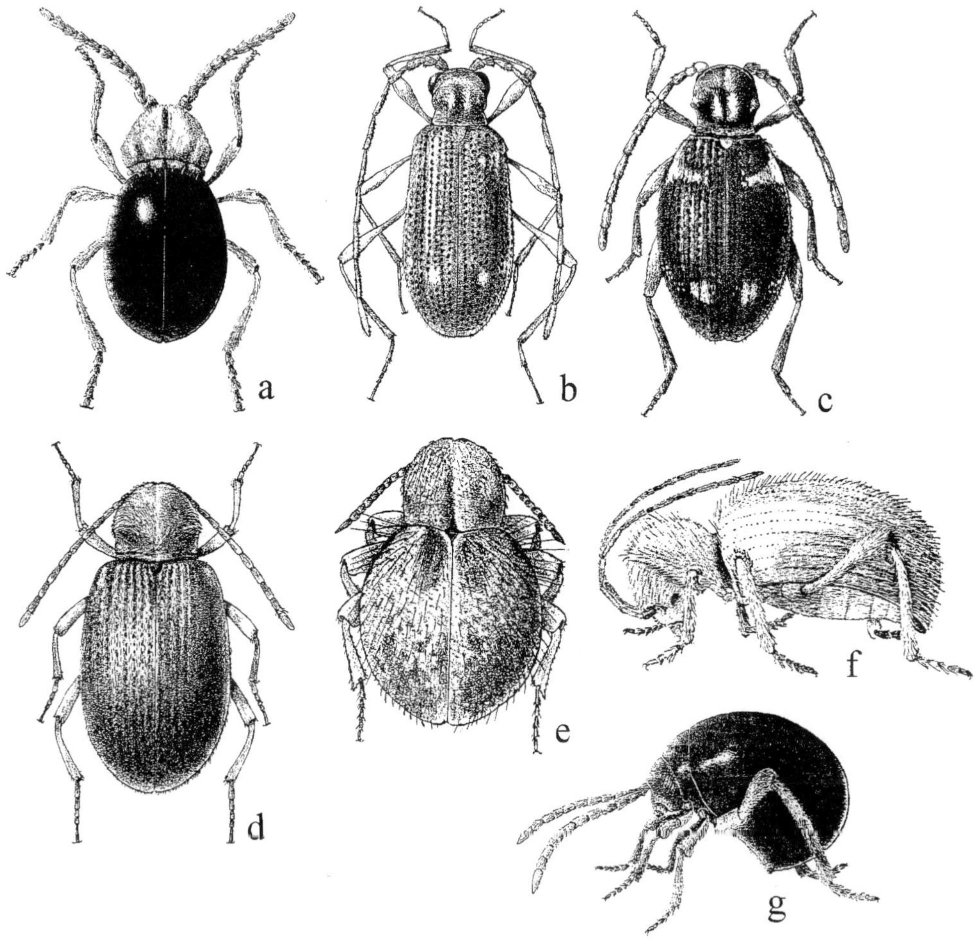


**Figure 5.16** Coleoptera: Ptinidae. (a) *Mezium americanum*; (b) *Ptinus fur* male; (c) *P. fur* female; (d) *P. ocellus*; (e) *Trigonogenius globulum*; (f) *Sphaericus gibboides*; (g) *Niptus hololeucus*; (h) *Gibbium psylloides*.

chambers in wood close to their food source. *Ptinus ocellus* larvae damage timbers in grain storage facilities, and *Sphaericus gibboides* (Fig. 5.16f) infests wooden cabinets. Ptinids are associated with premises rather than with a specific stored food. They find harborage in dark, moist crevices where spilled food accumulates. They lay eggs through cloth sacking or scatter eggs on the surface of bulked commodities. The larvae develop near the surface of the material where they construct feeding shelters and later thin-walled cocoons made from oral secretions. Adults have a daily rhythm of activity, but are usually most active at dusk.

Spider beetles are attracted to moisture and excrement, and are found in the nests of birds (house sparrows), and other animals. From these sites they move indoors. They are also found under bark, in rotting logs, and in caves. Larvae feed on wool, hair, feathers, fabric, and wood. Some species were probably first introduced into primitive food-storage sites by the wood used to form the storage racks. Mammals such as rats and mice and birds attracted to the stored food may have provided dung as additional food for these scavenging beetles. *Stethomezium*, *Mezium*, and *Gibbium* are probably subtropical in origin; they are capable of completing two or three generations per year in warm regions. The genera *Trigonogenius*, *Pseudeurostus*, *Tipnus*, and *Niptus* include species that survive cool temperatures; they lay more eggs at 20 °C than at 25 °C. *Eurostus hilleri* (= *Pseudeurostus*) occurs in Japan, the UK, and Canada, and is often found with rat and mouse droppings. *Ptinus* species males are usually strong flyers, and in some species both males and females fly. Nearly all of these species are associated with insect nests. *P. exulans* lives in spider nests in Tasmania. All species have a resting stage, usually the last instar, in their life cycle and this restricts them to one generation per year.

**Shiny spider beetle, *Gibbium psylloides* (Fig. 5.16h)** Adults are 1.7–2.0 mm long. The body has a distinct humped appearance, and the adult resembles a large mite. It is brownish red to black, with dense yellowish setae on the ventral surface. Larvae feed on woolens, leather, and tallow, and they occur in houses, hotels, warehouses, mills, and granaries. Larvae remain within the food during their development and do not leave to pupate. Pupal chambers are spherical and made within the infested food. The life cycle requires 22–42 weeks, and the adults usually live 30–40 weeks. This species is widely distributed in the tropics.

**American spider beetle, *Mezium americanum* (Fig. 5.16a)** Adults are 1.5–3.5 mm long, shiny reddish brown to black. Elytra are strongly convex to give this beetle a humped appearance. Larvae feed on dried animal material, seeds, mixed dry foods, tobacco, and woolen carpets.

**Golden spider beetle, *Niptus hololeucus* (Fig. 5.16g)** Adults are 3.0–4.5 mm long and pale yellow, with long, fine setae covering the fused elytra. Indoors, adults and larvae feed on woolens, linens, and silk; natural habitats include nests of bees, wasps, and birds. Eggs are laid in batches; fecundity is about 30 eggs. Hatching occurs in 11–20 days at 18–20 °C. Larval development is about 150 days, and adults live as long as 250 days. This species is probably originally from the region around the Black Sea, but is now distributed in Europe and North America.

**Brown spider beetle, *Ptinus clavipes*** Adults are 2.3–3.2 mm long. The male is uniformly dark brown and the female is blackish brown. Eggs are laid on the infested material, and hatching is in about 13 days at 22.5°C and 70% RH. Larval development is complete in 6–9 months; about 33% of the full-grown larvae enter diapause, which extends their life span to about 10 months. This beetle is a scavenger in cellars, attics, and storehouses, where it feeds on feathers, skins, rodent excrement, dried fruit, dry animal feed, grains, and sugar.

**Whitemarked spider beetle, *Ptnius fur* (Fig. 5.16b, c)** Adults are 2–4.3 mm long and covered with yellow setae. The female is larger and more rounded than the male, and has four white marks on the elytra; in some specimens the white marks on the female are joined to form stripes. The male pronotum has two rows of posteriorly directed, yellow setae that form a V-shape, and has two white marks posteriorly on the elytra. Development is completed in 30–90 days, depending on environmental factors. It commonly infests grain in warehouses. This species occurs in North America, but may be cosmopolitan.

**Australian spider beetle, *Ptinus ocellus* (Fig. 5.16d)** Adults are 2.5–4.0 mm long, and have fine, yellowish brown setae covering the elytra. Full-grown larvae are 3.5–4.0 mm long, fleshy, and covered with fine setae; they can roll into a ball when disturbed. Eggs are laid singly or in batches over a period of 3–4 weeks; fecundity is 100–1000. Oviposition usually occurs at 21 °C and 70% RH; eggs hatch in about 9 days. Larvae develop through three or four instars, and require about 60 days to complete development. Larvae are full-grown in about 93 days, and they pupate near the surface of infested material or crawl away to pupate. Adults are most active at night. Optimum conditions are 22–25 °C, and 80–90% RH. This species will not survive at temperatures above 28 °C, but develops at 10 °C and lays eggs at 5 °C. Infested food material includes nuts, beans, cacao, cayenne pepper, chocolate, corn, crabmeat, dried fruit, dried fish, poultry food, and hops; wood is sometimes invaded by full-grown larvae. The name *P. tectus* has been widely used for this species in Europe, but *P. ocellus* has been used in other regions. It is widespread and a common pest of household and commercial food commodities.

**Hairy spider beetle, *Ptinus villiger*** Adults are 2.2–4.0 mm long and reddish brown. Elytra have irregular white patches anteriorly and posteriorly. Full-grown larvae are about 3.8 mm long, yellowish white and with a dark-brown head. They are most often found in grain warehouses. Eggs are laid on the outside of flour bags or through the mesh of cloth sacs, and in flour debris; fecundity is about 40 eggs. Larval development is about 90 days, and full-grown larvae usually move away from the food source to pupate. Full-grown larvae burrow into wood close to the food source. Larvae often overwinter as pupae and continue development in the spring. This species is distributed throughout Europe, Asia, and North America.

**Globular spider beetle, *Trigonogenius globulum* (Fig. 5.16e)** Adults are about 2.5 mm long, pale brown to dark brown and covered with yellow and brown scales. Prothorax and abdomen are nearly globular.

## Scarabaeidae

Scarabs are generally robust as adults, and range in size from about 0.5 cm to more than 8 cm. They are distinguished by segmented antennae, which terminate in a large club of 3–7 segments that can be expanded like a fan or closed. Adults feed on living or decaying plants, and some species feed on dung. The C-shaped grubs usually have a sclerotized head and large mandibles. Grubs feed on plant material in soil, decaying vegetation, decaying logs, and dung. During the Egyptian Middle Kingdom scarabs were considered to have supernatural powers and were often featured in jewelry. Scarabs carved in stone were used to replace the hearts in Egyptian mummies.

*Phyllophaga* spp. are attracted to outdoor lights at night, and are often found at windows and door screens in spring and summer. Leaf chafers are often metallic-colored beetles, with head or wings gold- or chrome-colored. Flower scarabs are also brightly colored, and adults are often found on flowers and fruit around residential and commercial buildings. Elephant, Hercules, and rhinoceros beetle are names for scarabs that have large horns as adults. These beetles are among the largest insects; the males of some tropical species are 18 cm long. The atlas beetle, *Chalcosoma atlas*, is perhaps the largest insect and is found in southern Asia. Hercules beetle, *Dynastes tityus*, is the largest scarab in the USA; males are about 6 cm long. The elephant beetles, *Strategus* spp., are 35–50 mm long, and brown to dark brown. Males have three horns on the pronotum, and the females have two horns.

**May beetles, June beetles, *Phyllophaga* spp.** Adults are 12–23 mm long, and dark brown to brownish black. The body is usually slightly shiny, and the surface is sometimes covered with dense, pale setae. Eggs are laid in the soil in the spring; fecundity is 25–60 eggs. Larval development includes 3–5 instars, and a life cycle of 2 or 3 years is common for most species. Adults transform from the pupa in late summer or early fall; they emerge from the earthen pupal chamber the following spring. Adults are nocturnal and feed on the foliage of trees and shrubs. Adults of most species are attracted to lights at night. Larvae live in soil and feed on the roots of grasses, shrubs, and vegetable crops.

**Maybug, cockchafer, *Melolontha melolontha*** Adults are about 35 mm long. The body is blackish brown and elytra are brown; antennae are large and branched. Adults feed on foliage and flowers; larvae are called rookworms and found in soil feeding on plant roots. Adults are attracted to lights at night. This species occurs in the UK and continental Europe.

**Ten-lined June beetle, *Phyllophaga decemlineata*** Adults are 22–28 mm long. The body is brown and the elytra have broad white stripes; antennal segments are large. Larvae live in soil and feed on decaying organic matter. The adults are often found at lights at night.

**Eastern green June beetle, *Cotinis nitida*** Adults are 15–27 mm long, and the body is metallic green, and shiny. The pronotum and elytra are velvety green with the margins yellowish green. Full-grown larvae are 45–48 mm long and 11–12 mm in diameter; the body is yellowish white to greenish blue, and the head is reddish brown. Eggs are deposited in mid to late summer in soil rich in organic matter. Larvae continue to feed until soil temperature drops below about 12 °C; they overwinter deep in soil. In spring, they move close to the soil surface and feed. Larvae complete development in spring and form a subterranean pupal cell; adults emerge in June through August. This species occurs in eastern and midwestern USA. Adults of a related species, the green June beetle, *Cotinis mutabilis*, are metallic green, and the pronotum is deep green; elytra are pale, reddish brown. The ventral surface of the body and legs is bright metallic green. This species occurs in southwestern USA.

**Small June beetles, *Serica* spp.** Adults are 6–10 mm long, and the body is reddish brown to dark brown. Elytra have indistinct, but regularly spaced ridges. Adults resemble *Phyllophaga* except for their small size and the elytral ridges. Adults feed on the leaves of deciduous trees; they occur in large numbers, and damage ornamental and fruit trees in urban and suburban areas. During the day, they remain in undisturbed sites under leaf litter, and at night they feed on trees and fly to outdoor lights in large numbers.

**Hoplia beetle, *Hoplia laticollis*** Adults are 5.9–8.5 mm long; the head and thorax are reddish brown to blackish brown. Elytra are light brown and the surface is coarse; there are pale setae covering the body. The posterior tarsus has a single claw. Larval stages feed on grass roots in the soil. Large numbers of adults appear in spring and early summer. They are often seen flying low over turfgrass, and at lights at night. The swarming flights over turfgrass last for only a few hours.

**Japanese beetle, *Popillia japonica*** Adults are 10–11 mm long. The body is metallic green with reddish brown elytra; the abdomen has patches of white setae. Eggs are laid singly in the upper layer of soil of turfgrass in May, June, and July. Females are usually inseminated as they first emerge from the soil, and begin to lay eggs soon after mating. Fecundity is 40–60 eggs over a 4–6-week life span. Hatching occurs in 10–14 days. Development time for the first-stage larva is 2–3 weeks, and 3–4 weeks for the second-stage larva. Third-stage larvae continue to feed in the fall and are nearly full-grown before winter. They overwinter 5–15 cm below the soil surface, but some move deeper. Grubs move to the top layer of soil in the spring, and feed for 4–8 weeks on grass roots. They pupate in an earthen cell; the pupal stage lasts 7–17 days, and adults remain in the pupal cell for 2–14 days. Adults fly during the day, and aggregate on trees and shrubs. Plant species in the rose family are preferred. In the USA, adults emerge in May in southern Georgia, early to mid-June in North Carolina and Kentucky, and in July in Massachusetts. There is one generation per year throughout most of its range in the USA, but in some northern regions, a portion of the population takes 2 years to complete a generation.

This species is native to Japan, but it was introduced to the USA about 1911, possibly as grubs in soil around rhizomes of Japanese iris. It occurs naturally only on the main island of Japan. In Japan it is a minor agricultural pest, probably because there are limited amounts of turfgrass habitat and the presence of several natural enemies. Outbreaks of *P. japonica* occur in northern Japan where turfgrass has increased in urban areas. In the USA it is established in all states east of the Mississippi river, except for Florida.

Forty-nine species of natural enemies of *P. japonica* and related scarabs from Asia and Australia have been released into northeastern USA. Only a few became established: *Tiphia vernalis*, a tiphid wasp that parasitizes overwintered grubs in the spring; *T. popilliavora*, which attacks young grubs in late summer; and *Istocheta aldrichi*, a tachinid fly that parasitizes newly emerged adults. Females of *T. vernalis* use species-specific larval odor and frass in the turfgrass to locate Japanese beetle grubs.

**Eastern Hercules beetle, *Dynastes tityus*** Adults are 50–65 mm long, and the body is light green and mottled with large black markings. The male has three horns on the thorax; the middle horn, which is the largest, is curved down to meet an upward-curved horn on the head. The female has only a slight tubercle on the head. Larvae live in rotting logs and stumps, particularly in hardwoods. Adults feed on rotting fruit and have an offensive odor. This species occurs in eastern and midwestern USA.

**Western or Southwestern Hercules beetle, *Dynastes granti*** Adults are 55–70 mm long, and the body is dark green and with irregular spots. The pronotal horn of the male is large and extends over the head. Females lack a horn and are dark brown with scattered black on the elytra. This species occurs in western and southwestern USA.

**Rhinoceros beetle, *Xyloryctes jamaicensis*** Adults are about 25 mm long, and dark brown; males have a single upright horn on the head, and females have a small tubercle on the head. This species occurs in eastern and midwestern USA.

**Goliath beetle, *Goliathus goliathus*** Adults range in length from 80 to 115 mm, and this is the largest and most massive of all insects. It occurs in West Africa and surrounding regions. The larvae live in soil and feed on decaying organic matter. The adults are strong flyers, and their loud buzz when flying often attracts attention.

## Scolytidae

Scolytids are 6–8 mm long, cylindrical beetles, usually brown to blackish brown, and with a large antennal club. They are known as bark beetles, engravers, and ambrosia beetles (along with the Platypodidae). Adults and larvae feed on a mold-type fungus, known as ambrosia. They introduce this fungus into tunnels bored into the sapwood and sometimes heartwood of trees and logs. The fungus grows on the walls of the tunnels. A number of species breed in living trees, but dying or recently cut trees, logs, and pulpwood are usually preferred. Pest status of these beetles is based on sawn timber degraded by the presence of holes and staining. The dark-stained tunnels in the wood cause defects, which are called grease spots, steamboats, spot worms, and black holes. Damage is usually not sufficient to cause structural weakening of the wood.

Adults bore straight into the sapwood for several centimeters and push the wood boring out of the opening to the tunnel. Frass is white and like sawdust, which either falls loosely in piles or forms string-like masses, as if being squeezed from a tube. Once inside the sapwood, the larval tunnel often branches and follows the curvature of an annual ring or is unbranched and relatively straight. There may be short side tunnels of the same diameter in which the larvae feed and later pupate. Ambrosia beetle adults do not make an emergence hole in the infested wood; they generally emerge through the original entry hole made by the female. Several generations continue to extend the galleries if the wood retains high moisture content. There are two or three generations per year for most species.

Hardwoods and softwood trees are attacked. Maples (sugar, silver, and red maple), pecan, and cherry for use in furniture are commonly damaged. Other hardwoods infested include ash, beech, birch, chestnut, elm, hickory, oak, and walnut. Wood infested includes lumber for house construction, logs for modern log houses, hardwood and softwood for furniture, tools, picture frames, and ornamental items. All species prefer wood with moisture content of 30% and higher, because of the requirements of the ambrosia fungus. Seasoned wood is not infested.

**Eastern ambrosia beetle, *Xyloterinus politus*** Adults are 2.3–3.5 mm long, and dark brown to black. The pronotum is almost square, with spines in the front, and the anterior margin has 2–4 teeth. Elytra are reddish brown and covered with yellow setae. The entrance hole is about 1.6 mm diameter. The female excavates a gallery 2–45 mm into the sapwood of the host tree. Eggs are laid in secondary tunnels at right angles to the main gallery; secondary tunnels are about 14 mm long. Hatching occurs in 24 h, and larvae feed on fungi growing on the walls of the tunnel. As the larva grows it enlarges its tunnel by stripping off small fibers of wood with its mandibles. Frass containing wood fragments may be found in the main gallery and extruded or moved to the outside by the ovipositing female. Larval development is complete in about 29 days. Larval galleries differ from those of other ambrosia beetles in that they contain four rows of larval cradles, two above and two below the gallery. Adults emerge from the tree through the entrance hole made by the female. Adults overwinter and egg-laying begins in May and extends to August. Damage is usually done to recently cut trees, including hardwoods and softwoods, but hardwoods are preferred. Pine construction timbers often have stained galleries of this beetle. It is distributed in eastern and northeastern USA.

**Other ambrosia beetles: *Monarthrum fasciatum*, *M. mali*** Adults are 2.5–3 mm long; *M. fasciatum* have yellow bands across the elytra, *M. mali* are uniformly brown. These species develop in recently cut logs and lumber of practically all hardwoods throughout eastern USA.

**Coffee berry borer, *Hypothenemus hampei*** Adults are 1.6–2.5 mm long and dark brown to blackish brown; the body is setose. Eggs are laid in tunnels gnawed by the female in ripening coffee berries on the bush; females oviposit up to 20 eggs in one berry. Larvae complete development and pupate in the berry after drying and in the produce store. Adults are found in the product or at various locations in the store. This species also attacks beans. A related species, *H. liberiensis*, attacks maize in Nigeria, Africa.

**Ash bark beetles, *Lepersinus aculeatus*, *L. fraxini*** Adults are 2–3 mm long and have variegated markings. Adults differ from most other bark beetles by having a pattern of colors, which are produced by bands or spots of light-colored scales alternating with areas of dark scales. The antennal club is elongate and slightly compressed. Winter is spent in the adult stage in tunnels in the bark of living or felled trees or logs. Adults emerge in the spring to construct galleries between the bark and wood of host trees. Larvae feed in tunnels built away from the main gallery; pupation is in oval cells between the bark and wood. There are one or two generations per year. Adults of *L. fraxini* are common indoors in northern Europe (Denmark) in August and September, when ash firewood is brought indoors.

**Eastern juniper bark beetle, *Phloeosinus dentatus*** Adults are 2.2–2.8 mm long, brown to blackish brown, and covered with short, gray setae. Eggs are laid in short galleries that extend upward from the entrance hole. Larvae mine for a short distance across the grain, then mine upward with the wood grain. Infestations are usually found in saw logs or damage trees. This species attacks living red cedars infested with the root rot fungus, *Fomes annosus*. The most common host for this bark beetle is eastern red cedar (*Juniperus virginiana*), but it also attacks arborvitae and Atlantic white cedar. It is distributed in eastern USA, and west to Texas. Adults may emerge from structural logs in modern log houses in northeastern USA.

**Other *Phloeosinus*** Other juniper bark beetles include *P. taxodii*, the southern cypress beetle; *P. canadensis*, which also attacks eastern red cedar and arbovitae; *P. pini*, which attacks various pines; and *P. texanus*, which attacks Mexican juniper.

**Striped ambrosia beetle, *Trypodendron lineatum*** Adults are 3.0–3.5 mm long, and brown to black. The prothorax is reddish brown; the antennae are yellowish brown and with a large club. Elytra have several rows of punctures and range in color from completely black to yellowish brown. This species is widely distributed in North America, and it attacks recently felled logs of nearly all the coniferous trees. The galleries penetrate the sapwood and heartwood and have several branches.

**Fruit tree bark beetle, shot hole borer, *Scolytus rugulosus*** Adults are 2–3 mm long and blackish brown. Antennae, legs, and ends of the elytra are dark red to reddish brown. Full-grown larvae are about 3 mm long, yellowish white, and legless. Adults emerge in spring, and after mating, females bore round holes, about 1.3 mm diameter, through the bark of trees. They usually infest injured, dying, or dead wood. Eggs are deposited along main galleries in the wood below the bark. Development takes about 23 days and larvae excavate galleries at right angles to the main gallery in the cambium; pupation occurs at the end of the larval gallery. There are two or three generations per year, and larvae and pupae overwinter. Trees attacked include ash, elm, hawthorn, cherry, apple, peach, pear, and others. This species is native to Europe, but occurs throughout the USA. It commonly infests firewood logs, and adults are often found at windows during winter.

**Lesser shot hole borer, *Xyleborus saxeseni*** Adults are 2.3–3.5 mm long and yellowish brown to black. It lives in a wide variety of trees, including species of hardwoods and pine. Entry and exit holes are 1–3 mm diameter, and extend horizontally to the sapwood and end in a leaf-shaped cavity in which the ambrosia fungus grows. In fresh timber, the fibrous frass is compacted into short tubes, and these extend out of the entry hole. *Xyleborus* is represented by several species found in North America, Europe, and Asia. Many infest both coniferous and deciduous trees; dying, unhealthy, and felled trees are usually attacked.

**Western hemlock wood stainer, *Gnathotrichus sulcatus*** Adults are 2.6–3.0 mm long, and blackish brown with yellow appendages. The adults make a gallery about 1 mm diameter to a depth of 5–8 cm, and then the gallery branches in several directions to follow the annual rings. Fine white boring dust often collects in the crevices of the bark or at the base of infested trees or stacked logs. Females lay eggs in cavities on both sides of the main gallery. A blue stain fungus (*Ceratostomella* spp.) develops in the galleries. It is common in western North America, where it attacks spruce, fir, Douglas fir, hemlock, pine, cedar, and redwood.

**Red turpentine beetle, *Dendroctonus valens*** Adults are 5.5–9 mm long and reddish brown to dark brown. This species is similar in appearance to the black turpentine beetle, *D. terebrans*, but the head of *D. valens* is lighter in color. Full-grown larvae are 10–12 mm long and marked with brown tubercles on each side of the body. Eggs are laid in elongate masses in galleries excavated by females that have bored through the bark to the wood, usually in the basal part of the tree. The initial gallery extends horizontally or slightly upward, then the larva turns downward. Larvae feed gregariously away from the gallery, and a fully developed gallery is 30 cm or more wide. Pupation occurs in cells between the bark and the wood. Adults emerge through holes chewed through the bark. There are one to three generations per year in southern regions, but one or two in northern regions.

This species attacks trees of reduced vigor, but will also attack apparently healthy trees. Large populations develop in areas of logging or land clearing, and injured trees around construction sites or adjacent to piles of fresh lumber are frequently infested. Adults occur indoors in these areas. It is distributed in Canada and in all the coniferous forests of continental USA, except the Gulf Coast states.

**Other Scolytidae** The oriental bark beetles, *Hypothenemus* spp., attack a variety of trees. Adults are about 3 mm long, and uniformly dark brown to brownish black. Entry holes in the bark are about 0.5 mm diameter. Frass is a fine powder with some fragments of tree bark; larvae feed in cambium wood beneath the bark. The date stone borer, *Coccotrypes dactyliperda*, is a pest of unripe date palm fruits in the Mediterranean region. Adults occasionally emerge from ripe fruits in storage or in produce stores.

## Silphidae

Carrion beetle adults range in length from 3 to 35 mm, but the majority of the common species are about 10 mm long and colored orange, yellow, and black. The body is somewhat soft and flattened, antennae are clubbed, and the mandibles are prominent. Most species are associated with carrion or decaying organic matter. Some species are found on animal feces, including *Thanatophilus truncatus*, which has been collected on dog feces. Species in the genus *Nicrophorus* (= *Necrophorus*) are characteristic of the family. These are burying beetles; they excavate beneath the body of a dead animal, and after the body is buried, the beetles lay eggs on it. Silphids are common in undisturbed areas, but many fly to lights at night and come indoors.

**Carrion beetle, *Thanatophilus lapponicus* (= *Silpha lapponica*)** Adults are 9.4–14.0 mm long, black, and densely setose. Larvae are dark brown to black on the dorsal surface, and have urogomphi about twice the length of abdominal segment 10. Eggs are laid in the soil surrounding the carcass, usually in batches of 10; hatching occurs in 5–6 days. First-stage larvae molt in about 7 days, the second stage lasts 8–10 days, and the third stage 10–12 days. Adults of the first generation appear in June and July, adults of the second generation appear in late July through September, and these are the overwintering adults. It has been recorded injurious to furs, meats, and dried fish. This species is nearly cosmopolitan, and commonly occurs in northern Europe, Siberia, and nearly throughout North America.

**Garden carrion beetle, *Thanatophilus ramosa* (= *Silpha*)** Adults are 12–18 mm long, black, and with a velvet sheen. Full-grown larvae are about 16 mm long and shiny black. This species feeds primarily on decomposing vegetable matter, but also attacks garden and field crops, and grasses. It often occurs in lawns, and the adults fly to lights at night. This species is common in western North America.

## Staphylinidae

The rove beetles are 2–20 mm long, and are characterized by the short elytra. Usually there are 2–6 abdominal segments exposed behind the short wings. Most species are narrow, and parallel-sided, and with the head projecting forward. Most of the common species are found on carrion, but some are in fungi and a variety of other habitats. In Tomsk Territory, Southwest Siberia, 58 staphylinid species were collected in the suburbs and 35 species from town parks and gardens; the highest diversity of species occurred in the parks and green areas of housing estates. The most common species were *Philonthus decorus* and *P. rotundicollis*.

**Hairy rove beetle, *Creophilus maxillosus*** Adults are 12–21 mm long and the body is shiny black. Abdominal segments 2, 3, and sometimes 4 are densely covered with yellow setae, and there is a similar yellow band across the short elytra. This species occurs on dead animals, and is common on exposed corpses. It feeds on the maggots of carrion flies. It is widely distributed throughout Europe and North America.

**Pictured rove beetle, *Thinopinus pictus*** Adults are 15–18 mm long and wingless. The body is brownish yellow, with black curved and spotted markings on the thorax and the short elytra; mandibles are large and toothed. Larvae are marked similar to the adults, but thoracic segments are primarily black. Adults are abundant and very active on sandy ocean beaches at dusk just above the water line. This species is predaceous on insects and other arthropods. It occurs along the Pacific Coast of central and southern California.

**Paederus beetles, *Paederus ilsae*, *P. iliensis*, *P. fuscipes*, *P. kalalovae*, *P. balcanicus*, *P. lenkoranus*, and *P. cruenticollis*** Adults are 7–13 mm long, with a slender body and variously colored black, blue, and orange. Full-grown larvae are about 8 mm long, elongate, and the head and thorax are slightly bent. There are dark brown abdominal tergites, and long dorsal setae. Adults are often active during the day, but they seldom fly at this time. At night the adults of many species fly and are attracted to lights. The adults and immature stages of many species have toxins that produce an allergic skin reaction called dermatitis linearis. Most cases of dermatitis linearis are caused after sunset when people are at home and indoor lights are on. Adult beetles, attracted to lights, occasionally alight on people, and the toxin is released when the beetle is crushed on the skin. The toxin causes a delayed reaction, so that the dermal effect is not experienced in the first 24 h. Lesions usually appear in the form of a line on the skin; it is self-healing. These beetles are nearly cosmopolitan.

## Tenebrionidae

Darkling beetles or flour beetles are dark brown or black, not shiny, and they are usually slow-moving insects. These beetles are distinguished by the 5–5–4 tarsal formula, and notched eyes. They are primarily nocturnal, but many species are active during the day. Larvae are long, slender, and well sclerotized. They are scavengers on dead and decaying plant material. Several species occur in houses as accidental invaders, including *Metoponium abnorme*, *Coniontis subpubescens*, *C. parviceps*, and *Blapstinus* species. *Neatus tenebrioides* (Fig. 5.17c), is a large (10.5–14.5 mm long) species that occurs naturally under bark of dead trees and nests of squirrels and bumble bees, and also infests stored grain and cereals. The majority of pest species of tenebrionids are associated with stored foods.

Pest status is based on infestations of grains and cereals. Their presence in bird and mammal nests, in foods stored by small mammals, in caves, and in insect nests may predispose some of them to become pests in stored foods. Tenebrionids usually attack broken or damaged grains, because most of them cannot gnaw through the hard outer covering of whole grains. Food preferences for many species have not been determined, but the presence of mold or partially spoiled grain is often required for a sustained infestation. Many species live in wet and moldy habitats in nature. Pest species have been transported to all regions of the world.

**Lesser mealworm, *Alphitobius diaperinus* (Fig. 5.18h)** Adults are about 6 mm long and black or dark reddish brown. Full-grown larvae are about 20 mm long, brown to yellowish brown, and with pale bands on the margins of the sclerites. Eggs are deposited in batches of 4–12, and cemented to the surface with a clear secretion. At 32 °C, the daily oviposition rate is about six eggs per female, but the rate declines to four eggs per female at 38 °C, and one egg per female at 16 °C. Hatching is in 3–10 days; hatching success is 89% at 32 °C, 80% at 16 °C, and 77% at 38 °C. Larvae do not develop beyond the first stage at

10 °C; larval survival is 60% at 32 °C, 52% at 38 °C, and 27% at 16 °C. There is a positive correlation between larval development and presence of the fungus *Aspergillus*. Full-grown larvae form a pupal chamber in food material; pupal period is 6–10 days. When infestations are severe, full-grown larvae move away from the food substrate to pupate, and tunnel into wood or other material. Adult life span at 21–24 °C is over 400 days. Development from egg to adult takes 42 days at 38 °C and 97 days at 15.6 °C. This species is cosmopolitan and is found with damp and moldy grain and cereal products. Larvae and adults feed on dead animals. This species is nearly cosmopolitan.

**Black fungus beetle, *Alphitobius laevigatus* (Fig. 5.18i)** Adults are about 6 mm long, and shiny brown to blackish brown. It is similar in appearance to the lesser mealworm, *A. diaperinus*. It is distinguished by the curved sides of the thorax, which are coarsely and densely punctured. In the lesser mealworm, the sides of the thorax are nearly straight and sparsely punctured. *A. laevigatus* infests flour, bread, and maize in processing plants. This species is nearly cosmopolitan.

**Two-banded fungus beetle, *Alphitophagus bifasciatus*** Adults are about 3 mm long and elongate-oval. The body is reddish brown, and there are two black bands across the elytra. This species is cosmopolitan and a general feeder on fungi and molds. It often occurs as a scavenger in waste grain and milled products. Larvae occur in moist maize meal and in wet grain in ship holds and in storage.

**African fungus beetle, *Alphitobius viator* (Fig. 5.18j)** Adults are 5.5–6.6 mm long, brown, and the body is slender oval. Antennal segments 5–10 are asymmetrical, expanded laterally, and without setae at the margins. It is distributed in tropical Africa where it is associated with ginger roots, chili peppers, maize, and also reported from bone meal.

**Fig engraver beetle, *Apsena rufipes*** Adults are 4.2–7.0 mm long and the pronotum is broadly oval medially. The apical segment of the maxillary palp is broadly triangular. This species is known from California, and is associated with raisins and figs in the field and in storage.

**Fig darkling beetle, *Blapstinus discolor* (Fig. 5.17a)** Adults are 4.1–7.0 mm long and the body is blackish brown and distinctly oval. The head is prominent, eyes divided, and the elytra are pointed. Distribution is western Canada (British Columbia) and western USA, and it is associated with figs, grapes, raisins, peaches, persimmons, strawberries, dried fruits, tomatoes, sugar beets, peppers, and beans.

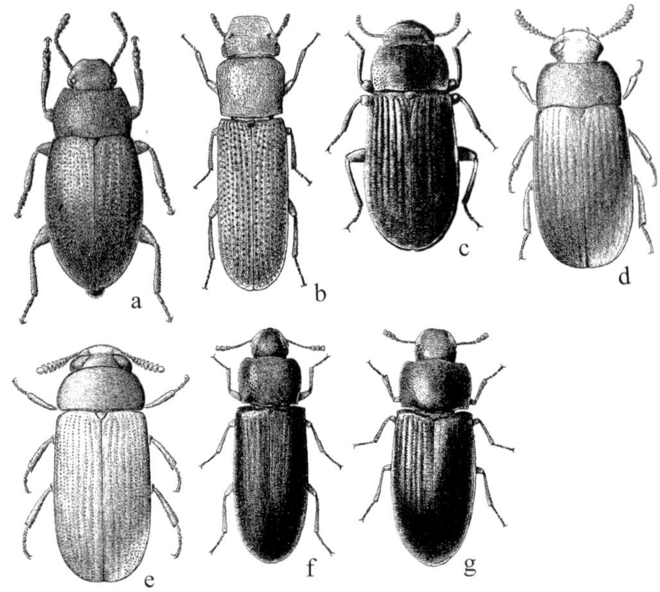


**Figure 5.17** Coleoptera: Tenebrionidae. (a) *Blapstinus discolor*; (b) *Latheticus oryzae*; (c) *Neatus tenebrioides*; (d) *Palembus dermestoides*; (e) *P. ocularis*; (f) *Tribolium audax*; (g) *T. madens*.

**Churchyard beetle, *Bleps mucronata*** Adults are 20–24 mm long, and black; the front and middle tarsi are five-segmented, and hind tarsi are four-segmented.

**Larger black flour beetle, *Cynaeus angustus* (Fig. 5.18g)** Adults are about 7 mm long. The body is reddish brown to black, and there are distinct longitudinal striations on the elytra. Full-grown larvae are about 15 mm long and brown. Eggs are laid singly and placed in crevices; hatching is in 3–4 days at 30 °C. Females lay four or five eggs per day and prefer food with moisture content of about 18%; fecundity is 360–450 eggs. Larval development takes 22–92 days at 30 °C, and there are 9–11 instars. Overcrowding slows larval development and sometimes prevents pupation. Full-grown larvae usually move away from the feeding substrate to pupate. The pupal period is 4–6 days at 30 °C. Development from egg to adult at 30 °C is 35 days on barley, 37 days on wheat, and 40 days on oats and maize; mortality ranges from 39 to 51% on these grains.

Prior to 1938, when it was discovered infesting stored grain in midwestern USA, this beetle was only known from debris at the base of yucca plants. Now this species is a common pest of farm-stored grain in northcentral USA. It feeds on damaged

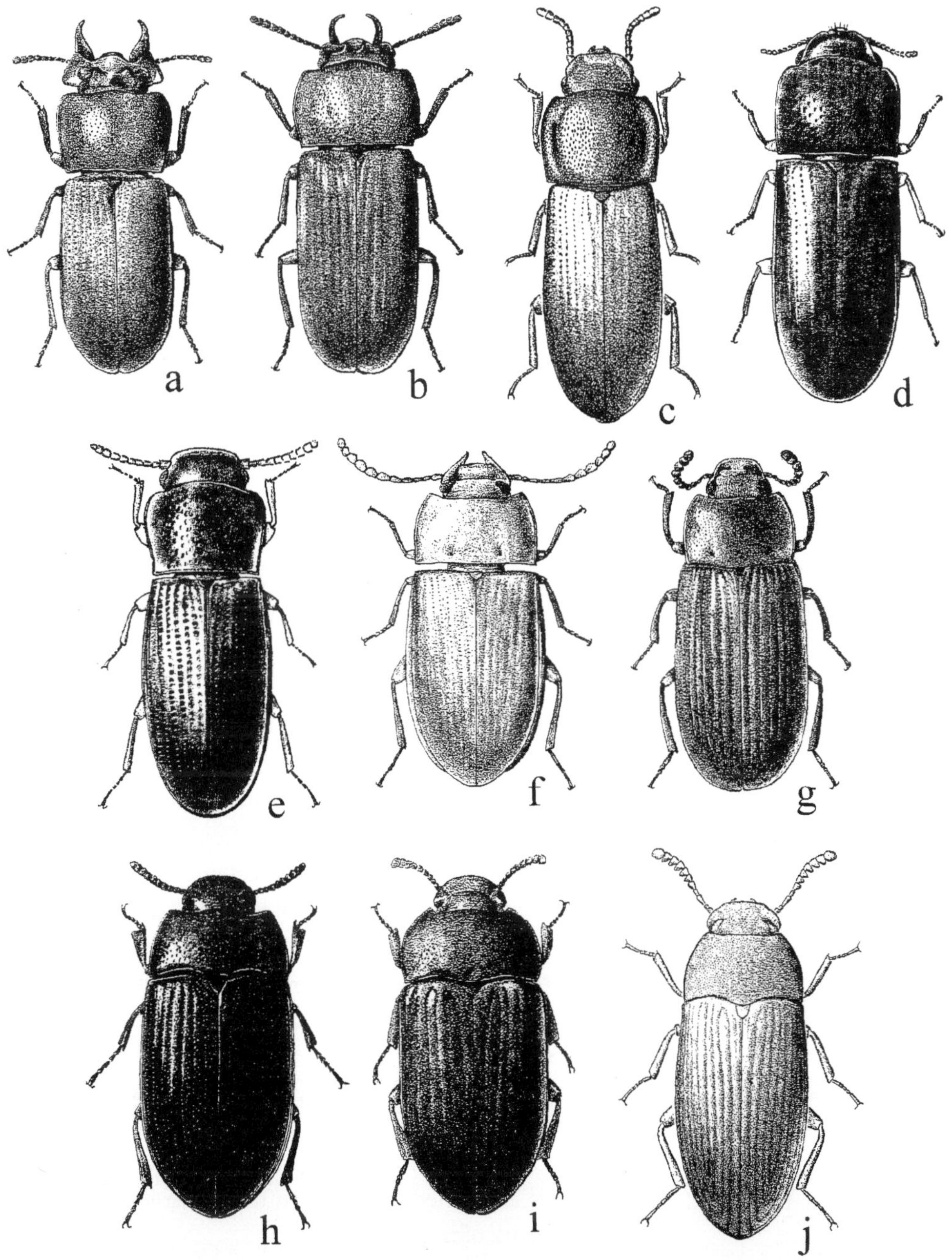

**Figure 5.18** Coleoptera: Tenebrionidae. (a) *Gnatocerus cornutus*; (b) *G. maxillosus*; (c) *Palorus laesicollis*; (d) *P. subdepressus*; (e) *P. ratzeburgii*; (f) *Sitophagus hololeptoides*; (g) *Cynaeus angustus*; (h) *Alphitobius diaperinus*; (i) *A. laevigatus*; (j) *A. viator*.

and undamaged grain, and it prefers maize and barley. It also feeds on fungi associated with insect-infested grain, and is able to reproduce on the fungus *Cladosporium cladosporioides*.

### Broad-horned flour beetle, *Gnatocerus cornutus* (Fig. 5.18a)

Adults are about 4 mm long and reddish brown; there are broad horns on the mandibles of the male. Eggs are laid singly on the food. Hatching occurs in 5 days at 30 °C and 8 days at 20 °C; fecundity is about 360 eggs in 5 months. Larval development takes about 39 days, and the pupal period is about 10 days at 27 °C. Development limits are 15–32 °C; optimum conditions are 24–30 °C. Larvae show a tolerance for fluctuations in temperature and humidity. It is a cosmopolitan pest of cereal and animal products, including flour, cornmeal, bran, and farina. It is usually associated with infestations of moths. *G. cornutus* larvae will feed on the eggs and injured adults, larvae, and pupae of the Mediterranean flour moth (*Anagasta kuehniella*).

**Slender-horned flour beetle, *Gnathocerus maxillosus* (Fig. 5.18b)** Adults are about 3 mm long and have a small horn on the mandible of the male. Eggs are laid singly on the food, and hatching occurs in about 4 days at 27.5–30 °C and 75% RH. Egg hatch is 92% at 27.5 °C and 64% at 30 °C. Larval development takes 34 days at 30 °C, and there are eight or nine instars. Larval mortality is 31% at 30 °C and only 1% at 27.5 °C. The pupal period is 6 days at 27.5 °C and 5 days at 30 °C. Development from egg to adult is 115–132 days, depending upon humidity. This species is distributed primarily in tropical and subtropical regions, and in warm temperate zones. It infests cereal grains, pumpkins, tamarind seeds, pulses, nutmeg, and peanuts.

**Longheaded flour beetle, *Latheticus oryzae* (Fig. 5.17b)** Adults are about 3 mm long, and the body is yellowish brown, slender, and somewhat flattened. Eggs are deposited individually on the food surface; hatching occurs in about 3 days at 40 °C, and 10 days at 25 °C. Humidity does not influence the incubation time, but high humidity reduces hatch. Larval development is 15 days at 35 °C and 85% RH; 17 days at 40 °C and 85% RH, and 95 days at 25 °C. The pupal period ranges from 3 days at 40 °C to 10 days at 25 °C. Minimum temperature for development is 25 °C, and a combination of 35 °C and 85% RH is optimum. This species is a cosmopolitan pest of wheat, rice, maize, barley, rye, and cereal products.

***Palembus dermestoides* (Fig. 5.17d)** Adults are 5.7–6.2 mm long and blackish brown. The head is prominent and antennae have broad segments, which are moderately asymmetrical. This species is considered pantropical. It is associated with maize, cornmeal, wheat flour, oats, rice bran, peanuts, nuts, yeasts, and bread. It has been reared and used as an oriental medicine.

***Palembus ocularis* (Fig. 5.17e)** Adults are 3.4–4.3 mm long and blackish brown. The pronotum is broad and head partly concealed. Distribution is pantropical, but it also occurs in the USA (Florida). It is associated with maize, oats, wheat, fishmeal, yeasts, various fruits (banana), and okra.

**Small-eyed flour beetle, *Palorus ratzeburgii* (Fig. 5.18e)** Adults are about 2.5 mm long and reddish brown. The sides of the head do not conceal the eyes from above. Eggs are laid singly or in small groups directly on the food substrate; hatching is in about 4 days at 30–40 °C, and 33 days at 17.5 °C. Larval development takes about 18 days at 32.5 °C and 55 days at 20 °C. Humidity threshold for development is about 50% RH at 20 °C, and about 20% RH at 25–30 °C. The pupal period is 4 days at 37.5 °C and 14 days at 20 °C. It is found in granaries, bakeries, and flourmills around the world, and feeds on flour, meal, maize, peanuts, and cassava meal.

**Depressed flour beetle, *Palorus subdepressus* (Fig. 5.18d)** Adults are about 3 mm long and uniformly brown. The sides of the front of the head extend backward to conceal the front portion of the eyes. Eggs are sticky and food particles adhere to them. Fecundity is about 303 eggs at 30 °C and 70% RH and 635 eggs at 30 °C and 80% RH. At 70% RH, eggs hatch in 3 days at 32.5 °C and 20 days at 17.5 °C. Larval development is through 7–9 instars; on wholewheat at 70% RH it is 28 days at 35 °C, and 96 days at 20 °C. Humidity threshold for development is 60–70% RH at 20 °C. Pupal chambers are constructed in the food material; the pupal period is 4 days at 35 °C and 14 days at 20 °C. This species infests wheat, rice, maize, sorghum, sago (*Metroxylon* spp.) flour, peanuts, illupe nuts (*Madhuca* spp.), pollards, and ginger.

**Other *Palorus*** These beetles are usually secondary invaders in the grain storage habitat. They are often found in old grain residues, and locations that have sustained infestations of *Sitophilus*. Larvae feed on yeast, molds, and all life stages, and the frass of primary pests. Adults of *P. genialis* are distinguished by the presence of pronotal punctures with setae; it is distributed in the oriental region where it infests rice and a variety of spices. *P. ficicola* and *P. cerylonoides* are found in Africa and the oriental region, and they infest rice, wheat, and peanuts. *P. cerylonoides* infests flour mills and in natural habitats it lives under the bark of trees. *P. laesicollis* (Fig. 5.18c) occurs in Africa where it is associated with broken maize kernels, oats, and under the bark of trees.

**Yellow mealworm, *Tenebrio molitor*, and dark mealworm, *Tenebrio obscurus* (5.19c–e)** Adults are about 14 mm long and the body is shiny, dark brown to black. The thorax is finely punctured and the forewings are longitudinally striated or grooved. Full-grown larvae are about 25 mm long, yellowish brown, and well sclerotized. Eggs are covered with a sticky secretion that causes flour and other particles to adhere to them. Egg-laying behavior is irregular, with infrequent periods between oviposition. *T. obscurus* females lay about 62 eggs per day, while *T. molitor* females lay about 40 eggs per day. Fecundity for *T. obscurus* is about 463 eggs, and for *T. molitor* about 276 eggs. Hatching is in 6 days at 30–35 °C and 17 days

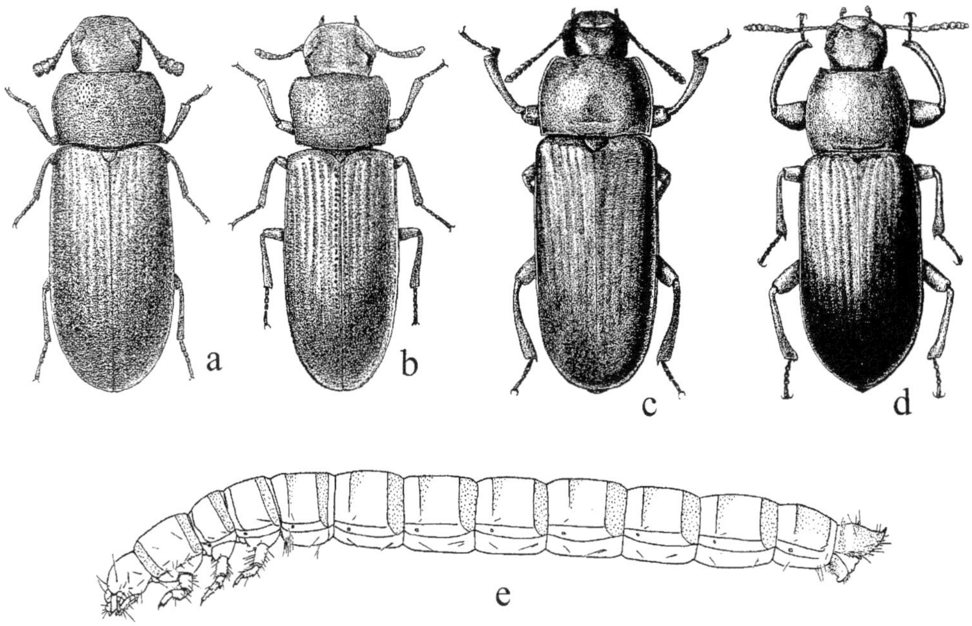


**Figure 5.19** Coleoptera: Tenebrionidae. (a) *Tribolium castaneum*; (b) *T. confusum*; (c) *T. molitor*; (d) *T. obscurus*; (e) *Tenebrio molitor* larva.

at 15 °C, and is most successful at 71% RH. Larval development is usually complete in 1 year, but some take 2 years. Larval molt for *T. obscurus* is 14 or 15 and 17–19 for *T. molitor*. Reared on graham flour and meat scraps, *T. obscurus* reaches the adult stage in 105–675 days, and *T. molitor* reaches the adult stage in 30–649 days. The pupal period is about 15 days; for *T. molitor* it is 7 days at 25–35 °C and 48 days at 15 °C. Full-grown larvae pupate or remain in the larval stage for many months with little change in size or appearance; both species overwinter as larvae. Mealworm larvae can survive for long periods without food or moisture. Adults have well-developed wings, and both species are attracted to lights.

The common names are derived from the general color of the immature stages. The larvae, pupae, and adults of these two beetles are similar in form, size and color. They are cosmopolitan, but *T. molitor* is common in north temperate regions, and *T. obscurus* is generally distributed in temperate and other regions. Both are nocturnal and prefer dark, moist habitats. Larvae feed on refuse grain, coarse cereal, and mill products. They also feed on animal material, such as meat scraps, dead insects, and feathers.

**Red flour beetle, *Tribolium castaneum*; confused flour beetle, *Tribolium confusum* (Fig. 5.19a, b)** Adults are about 3.5 mm long and reddish brown. *T. castaneum* and *T. confusum* are very similar in appearance and morphological separation is difficult. The confused flour beetle, *T. confusum*, was named because of the confusion pertaining to its identity. Antennae of *T. castaneum* have a three-segmented club; antennae of *T. confusum* have a four-segmented club. Full-grown larvae are 6–7 mm long, yellowish white, head and terminal segment brown, and with distinctly pointed urogomphi.

Eggs are laid directly on flour or other food material, and they are attached to the surface with a sticky substance; about 90% of the eggs laid are viable. The oviposition period of *T. castaneum* is 148 days at 27 °C and 174 days at 21 °C; the oviposition period of *T. confusum* is 235 days at 27 °C and 277 days at 21 °C. Fecundity for both species is about 950 eggs; it is about 84 eggs at 25 °C and 30% RH, 413 eggs at 23 °C and 30% RH, and 539 eggs at 32.5 °C and 70% RH. Larval development is influenced by food, temperature, and humidity. The number of larval instars ranges from five to 11, but usually there are seven or eight. Optimum temperature for *T. castaneum* is 35–37 °C at 70% RH, and development is completed in 19–22 days, and 30 days or less at 30 °C and 30% RH. Minimum temperature for development is 20–22.5 °C; the maximum is 37.5–40 °C. Optimum conditions for *T. confusum* are 32.5 °C and 70% RH, and development is completed in 25 days, and 30 days or less at 30–35 °C, and 30% RH. *T. castaneum* and *T. confusum* feed on certain species of seed-borne fungi. The red flour beetle is capable of completing development on eight species of fungi, and the confused flour beetle on seven species. Development is also enhanced by predation and feeding on other insects in grain. Population growth rates increase when food is supplemented with eggs or adults of the Indian meal moth, *Plodia interpunctella*. Pupae are formed in sheltered locations in the infested material; the pupal period is 4–6 days. Adult *Tribolium* live for about 3 years, and females lay eggs for more than 1 year. Males of *T. castaneum* at 20–27 °C live about 547 days; females live about 226 days. Males of *T. confusum* at 20–27 °C live about 634 days; females live 447 days.

These two species readily infest plant and animal products. Larvae have been recorded from grain, flour and other cereal products, peas, beans, cacao, nuts, dried fruits, medical drugs, spices, milk chocolate, dried milk, and hides. Wholewheat flour and milled products of grain are favored foods. Larvae or adults do not attack sound or undamaged grain. Adults of *T. castaneum* can fly short distances, but the adults of *T. confusum* have not been observed to fly. *Tribolium* species have been associated with grain and flour for thousands of years. A *Tribolium* species was recovered from a jar probably containing grain in an Egyptian pharaoh's tomb. Prior to becoming a household pest, *Tribolium* species are believed to have lived under tree bark. *T. confusum* is probably African in origin, and *T. castaneum* originated in the Indo-Australian region.

**American black flour beetle, *Tribolium audax* (Fig. 5.17f)** Adults are about 5 mm long. The body is black, slender, and the head has many confluent punctures between the eyes. It is distributed in North America, and associated with wheat flour, barley, grain and grain products, bee nests, and in logs.

**Black flour beetle, *Tribolium madens* (Fig. 5.17g)** Adults are about 5 mm long. The body is robust, shiny black to blackish brown. The head has no confluent punctures between the eyes. This species occurs primarily in Europe and North Africa, and also from Canada and western USA. It is associated with flour, cornmeal, grains, and seeds.

**False black flour beetle, *Tribolium destructor*** Adults are reddish brown and similar in appearance to the confused flour beetle. It is distributed in North America, Europe, and Africa. It has been found in animal-food pellets, flour, and other stored products, including flour, bran, rolled oats, semolina, alfalfa meal, sunflower seeds, grains, and poultry feed.

**Giant flour beetle, *Tribolium brevicorne*** This species is distributed in western Canada and western USA, and is associated with oats, chicken feed, and mixed feeds. It has been found in the nest of leaf-cutting bees.

## Trogistidae

Trogistids range in length from 2.6 to 20 mm. Most are brown to brownish black; a few are bluish green. Adults in the subfamily Trogossitinae are elongate beetles, with the head about as wide as the pronotum, and the pronotum widely separated from the base of the elytra (waist-like appearance). These beetles have large and powerful mandibles, and are primarily predators on insects under bark, and a few are stored-food pests.

**Cadelle, *Tenebroides mauritianicus*** Adults are 6–10 mm long. The body is elongate, somewhat flattened, and shiny black. The prothorax is distinctly separated from the body, giving a waist-like appearance. Larvae are yellowish white to gray, with a black head, and black prothoracic dorsal plate. The abdomen has two horny black projections (urogomphi) at the tip, they are surrounded by black plate. Eggs are deposited in batches of 10–60 in food materials or inserted in crevices; the interval between egg batches is 10–14 days; fecundity is 436–1319 eggs. Hatching is in about 7 days at 28 °C and 15–17 days at 21 °C. Females that emerge in summer lay eggs the same year, but cease during winter, and resume oviposition in the following spring. Females that emerge in spring complete their egg-laying and die before winter. Larvae molt three or four times, but as many as seven molts is possible when the development period is long. Larval development is completed in about 69 days on maize, wheat, or graham flour; development on barley flour takes about 90 days; development on rice is not completed until the following summer. Larvae do not complete development on refined white flour. Development interrupted by hibernation takes 271–410 days. The pupal period is 8 days at 29 °C and 25 days at 21 °C. Pupae are formed in a separate chamber in the infested material. Full-grown larvae burrow into softwood timber adjacent to the infestation site. The pupal period is 8 days at 29 °C and 25 days at 21 °C. In tropical climates there are three generations per year; in temperate regions there are two generations per year. Larvae and adults feed on nuts, seeds, dried fruits, and vegetables, as well as grains and their milled products. Adults also feed on other insects in the habitat, including larvae of the cheese skipper, *Tribolium* flour beetles, and larvae and adults of their own species. This insect is cosmopolitan and is a common pest in granaries, storehouses, and mills where it infests flour, meal, and grain. This is one of the longest-lived of the insects that attack stored grain, which increases the potential for damage by the adults and larvae, and the potential for dispersal to other locations.

**Siamese grain beetle, *Lophocateres pusillus*** Adults are 2.6–3.2 mm long, broadly oval, and blackish brown. Elytra have longitudinal ridges, and the anterior margin of the pronotum is straight and the edges have distinct points. Full-grown larvae are about 5.4 mm long and yellowish white; there is a distinct tubercle between the strongly curved urogomphi. Eggs are laid in fan-shaped clusters of 11–14 in cracks and crevices. Hatching fails at temperatures below 15 °C or above 37 °C. Larval development is completed in about 54 days at 23 °C and 75% RH, 49 days at 35 °C and 95% RH, and 180 days at 20 °C and 95% RH; there are four larval instars. Full-grown larvae construct a pupal chamber in the food. This species attacks a variety of legumes, rice and rice products, cereals, cocoa beans, coffee beans, nutmeg, cassava, sweet potatoes, and dried fruits. Sound rice kernels are resistant to attack, but are attacked if slightly damaged.

## Bibliography

### General

Armstrong, R. K. and J. L. Winfield. Staphylinidae dermatitis on Okinawa. *J. Med. Entomol.*, **5** (1968), 362.

Boving, A. G. and F. C. Craighead. An illustrated synopsis of the principal larval forms of the order Coleoptera. *Entomol. Am. (New Series)*, **11** (1931), 1–351.

Costa, C., S. A. Vanin, and S. A. Casari-Chen. *Larvae de Coleoptera do Brasil*. São Paulo: Museu de Zoologica, Universidade São Paulo, 1988.

Crowson, R. A. *The Biology of the Coleoptera*. London: Academic Press, 1981.

Evans, M. E. G. The jump of the click beetle (Coleoptera: Elateridae) – energetics and mechanics. *J. Zool., Lond.*, **169** (1973), 181–94.

Freud, H., K. W. Harde, and G. A. Lohse. *Die Käfer Mitteleuropas*, vol. 1. Krefeld, Germany: Goecke & Evers, 1965.

Hudson, G. V. *New Zealand Beetles and their Larvae*. Wellington: Ferguson and Osborne, 1934.

Nakane, T., K. Ohbayashi, S. Nomura, and Y. Kurosawa. *Iconographia Insectorum Japonicorum. Colore Naturali Edita*, vol. II. *Coleoptera*. Tokyo: Hokuryukan, 1963.

Svihla, A. Two-spotted lady beetles biting man. *J. Econ. Entomol.*, **45** (1952), 134.

### Wood-infesting

Baker, J. M. Wood boring weevils in buildings. *Timberlab. News*, **4** (1970), 6–7.

Becker, G. Beitrage zur Kenntis Hausbockkafers. *Z. Hyg. Zool.*, **34** (1942), 83–107.

Ein Kam bekannter Kafer im Bauholz breitet sich aus. *Holz-Zentralbl.*, **77** (1951), 1895–6.

Predators and parasites of wood-destroying insects in buildings. *Verhandl. Deutsch. Ges. Angew. Entomol.* (1954), 76–85.

Ecology and physiology of wood-destroying Coleoptera in structural timber. *Mater. Org.*, **12** (1977), 141–60.

Bois, P. J. Wood moisture content in homes. *J. Forest. Prod.*, **9** (1959), 427–30.

Cymorek, S. Beitrage zur Kenntis der Lebenweise und des Schadauftretens holzzerstorender Insecken. *Z. Angew. Entomol.*, **55** (1964), 84–93.

Dominik, J., J. Wazny, and M. Czajnik. *The More Important Wood-destroying Insects Found in Buildings in Poland*. International Research Group on Wood Preservation, document no. IRG/WP/128, 1974.

Eaton, R. A. and M. D. C. Hale. *Wood: Decay, Pests and Protection*. London: Chapman and Hall, 1993.

Fisher, R. C., F. R. Cann, and E. A. Parkin. A survey of the damage caused by insects to hardwood timbers in Great Britain. *Forest Prod. Res. Bull.*, **16** (1932), 9.

Hickin, N. E. *The Woodworm Problem*, 3rd edn. East Grinstead, UK: Rentokil Library, 1981.

Liese, W. (ed.) *Biological Transformation of Wood by Microorganisms*. Berlin: Springer, 1975.

Parkin, E. A. The digestive enzymes of some wood-boring beetle larvae. *J. Exp. Biol.*, **17** (1940), 364–77.

Silverborg, S. B. Fungi associated with the decay of wooden buildings in New York State. *Phytopathology*, **43** (1953), 20–2.

Spencer, G. J. The insects attacking structural timbers and furniture in coastal British Columbia. *Proc. Entomol. Soc. B. Columbia*, **55** (1958), 8–13

Tooke, F. G. C. Beetles injurious to timber in South Africa. *Sci. Bull.*, **293** (1949), 1–95.

### Food- and fabric-infesting

Alferi, A. Les insects de la tombe de Tutankhamun. *Bull. Soc. Entomol. Egypte, Fasc.*, **3–4** (1931), 188–9.

Bousquet, Y. *Beetles Associated with Stored Products in Canada: An Identification Guide*. Research Branch, Agriculture Canada, publication 1837. Ottawa: Agriculture Canada, 1990.

Cline, L. D. and H. A. Highland. Minimum size holes allowing passage of adults of stored-product Coleoptera. *J. Ga. Entomol. Soc.*, **16** (1981), 525–31.

Cotton, R. T. and N. E. Good. Annotated list of the insects and mites associated with stored grain and cereal products, and their arthropod parasites and predators. U.S.D.A. miscellaneous publication 258. Washington, DC: USDA, 1937.

Essig, E. O., W. M. Hoskins, E. G. Linsley, A. E. Michelbacher, and R. F. Smith. A report on the penetration of packing materials by insects. *J. Econ. Entomol.*, **36** (1943), 822–9.

Freeman, J. A. Problems of infestations by insects and mites of cereals stored in western Europe. *Annali Tecn. Agric.*, **8** (1973), 509–30.

Gerhardt, P. D. and D. L. Lingren. Penetration of various packaging films by common stored product insects. *J. Econ. Entomol.*, **47** (1954), 282–7.

Hinton, H. E. *A Monograph of the Beetles Associated with Stored Products*, vol. I. London: British Museum (Natural History), 1954.

Hinton, H. E. and A. S. Corbet. *Common Insect Pests of Stored Food Products*. British Museum (Natural History). Economic Series no. 15. London: British Museum, 1963.

Linsley, E. G. Natural sources, habitat and reservoirs of insects associated with stored food products. *Hilgardia*, **16** (1944), 187–224.

Protection of dried packaged foodstuffs from insect damage. *J. Econ. Entomol.*, **37** (1944), 337–79.

Mallis, A., A. C. Miller, and R. C. Hill. Feeding of four species of fabric pests on natural and synthetic textiles. *J. Econ. Entomol.*, **55** (1958), 351–5.

The attraction of stains to three species of fabric pests. *J. Econ. Entomol.*, **51** (1959), 248–9.

Mallis, A., B. T. Burton, and A. C. Miller. The attraction of salts and other nutrients to the larvae of fabric insects. *J. Econ. Entomol.*, **51** (1962), 382–4.

Mound, L. A. (ed.) *Common Insect Pests of Stored Food Products: A Guide to the Identification*, 7th edn. London: British Museum (Natural History), 1989.

O'Farrell, A. F. and P. M. Butler. Insects and mites associated with storage and manufacture of foodstuffs in Northern Ireland. *Econ. Proc. R. Dublin Soc.*, **3** (1948), 343–407.

### Anobiidae (stored food)

Armitage, F. D. and B. Verdcourt. The preference of *Stegobium paniceum* (L.) (Coleoptera, Anobiidae) for certain drugs. *Entomol. Mon. Mag.*, **83** (1947), 133.

Farag, F. A. and A. Y. Ismail. Biological studies on the developmental stages of the cigarette beetle *Lasioderma serricorne* Fab. *Iraqi J. Agric. Sci.*, **4** (1986), 63–8.

Griffith, C. F. Biometric studies on the drug-store beetle, *Stegobium paniceum* L. (Coleoptera, Anobiidae). *Entomol. Mon. Mag.*, **82** (1946), 186–91.

Howe, R. W. A laboratory study of the cigarette beetle, *Lasioderma serricorne* (F.) (Coleoptera, Anobiidae) with a critical review of the literature on its biology. *Bull. Entomol. Res.*, **48** (1957), 9–56.

Kuwahara, Y., H. Fukami, R. Howard, R. Ishii, F. Matsumura, and W. E. Burkholder. Studies on the isolation and bioassay of the sex pheromone of the drug store beetle, *Stegobium paniceum* (Coleoptera: Anobiidae). *J. Chem. Ecol.*, **1** (1975), 413–22.

LeCato, G. L. Infestation and development by the cigarette beetle in spices. *J. Ga. Entomol. Soc.*, **13** (1978), 100–5.

Lefkovitch, L. P. and J. E. Currie. Factors affecting adult survival and fecundity in *Lasioderma serricorne* (F.) (Coleoptera, Anobiidae). *J. Stored Prod. Res.*, **3** (1967), 199–212.

Milne, D. L. A study of the nutrition of the cigarette beetle, *Lasioderma serricorne* F. (Coleoptera: Anobiidae) and a suggested new method for its control. *J. Entomol. Soc. S. Afr.*, **26** (1963), 43–63.

Pant, N. C. and G. Fraenkel. Studies on the symbiotic yeasts of two insect species, *Lasioderma serricorne* F. and *Stegobium paniceum* L. *Biol. Bull.*, **107** (1954), 420–32.

Sivik, F. P., J. N. Tenhet, and C. D. Delmar. An ecological study of the cigarette beetle in tobacco warehouses. *J. Econ. Entomol.*, **50** (1957), 310–16.

Tobin, E. N. and L. W. Smith, Jr. Note on the mating behavior of the cigarette beetle (*Lasioderma serricorne* (F.): Anobiidae). *Ent. News*, **82** (1971), 23–5.

White, R. E. The Mexican book beetle, *Catorama herbarium*, established in the United States. *Ann. Entomol. Soc. Am.*, **56** (1963), 280–5.

### Anobiidae (wood-infesting)

Adlung, K. G. Das Auffinden der Brutplatze durch den Weichen Klopkfer (*Ernobius mollis* L.). *Z. Angew. Zool.*, **44** (1957), 173–86.

Allen, B. An Account of the Scarabaeus Galeatus Pulsator, or the Death-Watch; taken August 1695. *R. Soc. Lond. Phil. Trans.*, **20** (1695), 376–8.

Baker, J. M. Digestion of wood by *Anobium punctatum* DeG. and comparison with some other wood-boring beetles. *Proc. R. Entomol. Soc. Lond. C*, **8** (1969), 33.

Baker, J. M., R. A. Laidlaw, and G. A. Smith. Wood breakdown and nitrogen utilisation by *Anobium punctatum* Deg. feeding on Scots pine sapwood. *Holzforsch.*, **24** (1970), 45–54.

Baker, J. M., R. A. Laidlaw, and G. A. Smith. *Nicobium castaneum* (Coleoptera, Anobiidae), a pest in wood materials and works of art. In Walters, A. H. and E. H. Hueck-van der Plas (eds.) *Bio-deterioration of Materials*, vol. 2, pp. 408–15. London: Applied Science, 1972.

Becker, G. Okologische und physiologische Untersuchungen uber die holzzertorenden Larven von *Anobium punctatum* DeGeer. *Z. Morphol. Okol. Tiere*, **39** (1942), 98–152.

Behrenz, W. and G. Technau. Versuche zur bekampfung von *Anobium punctatum* mit Symbionticiden. *Z. Angew. Entomol.*, **44** (1959), 22–8.

Birch, M. C. and J. J. Keenlyside. Tapping behavior is a rhythmic communication in the death watch beetle (*Xestobium rufovillosum*). *J. Insect Behav.*, **4** (1991), 257–63.

Bletchly, J. D. Aspects of the habits and nutrition of the Anobiidae with special reference to *Anobium punctatum* de Geer. Beihefte zu Material und Organismen (International Symposium Berlin-Dahelm 1965). *Mater. Org.*, **2** (suppl.) (1966), 371–81.

Seasonal differences in nitrogen content of Scots pine (*Pinus sylvestris*) sapwood and their effects on development of the larvae of the common furniture beetle (*Anobium punctatum* de Geer). *J. Inst. Wood Sci.*, **4** (1969), 43–7.

Casimir, J. M. The common furniture beetle. Aust. Forest, **19** (1955), 26–34.

Cymoreck, S. Beitrage zur kenntis der Pochkaferarten *Anobium punctatum* Deg., *Anobium hederae* Ihss, *Anobium inexpectatum* Lohse (Coleoptera, Anobiidae). *Entomol. Blatter Biol. Syst. Kafer*, **53** (1957), 87–94.

*Nicobium castaneum* (Col., Anobiidae), a pest of wood materials and works of art. Proceedings of the 2nd Institute of Biodeterioration, Symposium Lunteren 1971. *Biodet. Mat.*, **2** (1972), 408–15.

Cymoreck, S. and D. Bauer. Prufung der Bekampfungwirkung von Holzschutzmitteln gegen *Anobium punctatum* DeGeer. *Holz Roh-Werkstoff*, **22** (1964), 304–8.

Español, F. Notas sobre Anóbidos. 11. Los anóbidos de las islas Canarias. *Pub. Inst. Biol. Apl.*, **37** (1964), 95–115.

Fisher, R. C. Studies of the biology of the death watch beetle (*Xestobium rufovillosum* de Geer). I. A summary of the past work and a brief account of the development stages. *Ann. Appl. Biol.*, **24** (1937), 600–13.

Studies of the biology of the death watch beetle (*Xestobium rufovillosum* de Geer). II. The habits of the adults with special reference to the factors affecting oviposition. *Ann. Appl. Biol.*, **25** (1938), 155–80.

Studies of the biology of the death watch beetle (*Xestobium rufovillosum* de Geer). III. Fungal decay in timber in relation to the occurrence and rate of development of the insect. *Ann. Appl. Biol.*, **27** (1940), 545–6.

Studies of the biology of the death watch beetle (*Xestobium rufovillosum* de Geer). IV. The effect of type and extent of fungal decay in timber on the rate of development of the insect. *Ann. Appl. Biol.*, **28** (1941), 244–60.

French, J. R. J. The effect of temperature and humidity on the life cycle of the common furniture beetle, *Anobium punctatum* (De Geer). *J. Inst. Wood Sci.*, **28** (1971), 31–5.

Gardiner, Ph. The morphology and biology of *Ernobius mollis* L. (Coleoptera, Anobiidae). *Trans. R. Entomol. Soc.*, **104** (1953), 1–24.

Hatch, M. H. *Hadrobregmus gibbicollis* infesting woodwork. *J. Econ. Entomol.*, **39** (1946), 274.

Hickin, N. E. The common furniture beetle, *Anobium punctatum* (DeG.) (Coleoptera, Anobiidae): some notes on its outdoor occurrence. *Entomologist*, **86** (1953), 216 17.

Hockey, M. J. Anobiid beetles in timber and buildings in Queensland. Queensland Dept. of Forestry, timber note no. 26, 1989.

Kimmins, D. E. Notes on the life-history of the death-watch beetle. *Proc. Soc. Lond. Entomol. Nat. Hist. Soc.*, (1933–34), 133–7.

Linscott, D. Susceptibility of imported and newly seasoned home-grown hardwoods to *Anobium punctatum* (Deg.). *J. Inst. Wood Sci.*, **5** (1971), 36–9.

Linsley, E. G. Woodboring habit of drugstore beetle. *J. Econ. Entomol.*, **35** (1942), 452.

Meyer, O. E. On adult weight, oviposition preference, and adult longevity in *Anobium punctatum* (Coleoptera, Anobiidae). *Z. Angew. Entomol.*, **66** (1970), 103–12.

Moore, H. B. Development and longevity of *Xyletinus peltatus* under constant temperatures and humidities. *Ann. Entomol. Soc. Am.*, **61** (1968), 1158–64.

Parkin, E. A. The larvae of some wood-boring Anobiidae (Coleoptera). *Bull. Entomol. Res.*, **24** (1933), 33–68.

Payne, N. M. Injury to lumber by *Hadrobregmus carinatus* Say. *J. Econ. Entomol.*, **29** (1936), 1027.

Serdjukova, I. R. Investigation of digestive enzymes of some wood-boring beetle larvae (Coleoptera, Anobiidae). *Zool. Z.*, **72** (1993), 43–51.

White, R. E. The Anobiidae of Ohio. *Ohio Biol. Sur. (N. S.)*, **1** (1962), 1–58.

Eight new North American species of Anobiidae with keys and notes (Coleoptera). *Proc. Entomol. Soc. Am.*, **65** (1976), 533–6.

North American *Euvrilletta* (Coleoptera: Anobiidae) – transfer of taxa from *Xyletinus*, two new species, and a key. *Coleopterists Bull.*, **39** (1985), 185–93.

White, P. R. and M. C. Birch. 1987. Female sex pheromone of the common furniture beetle *Anobium punctatum* (Coleoptera: Anobiidae). *J. Chem. Ecol.*, **13** (1987), 1695–706.

Williams, L. H. Responses of *Xyletinus peltatus* (Harris) (Coleoptera: Anobiidae) larvae to favorable and unfavorable temperatures. *Mater. Org.*, **12** (1977), 59–67.

Wood moisture levels affect *Xyletinus peltatus* infestations. *Environ. Entomol.*, **12** (1983), 135–40.

Williams, L. H. and H. M. Barnes. How *Xyletinus peltatus* beetles affect strength of southern pine floor joists. *Environ. Entomol.*, **8** (1979), 304–6.

Williams, L. H. and J. K. Mauldin. Survival and growth of the anobiid beetle, *Xyletinus peltatus* (Coleoptera, Anobiidae) on various woods. *Can. Entomol.*, **113** (1981), 651–7.

Williams, L. H. and J. D. Waldrop. *Xyletinus peltatus* seasonal flight, diel activity, and associated environmental influences. *Ann. Entomol. Soc. Am.*, **71** (1978), 567–74.

Williams, L. H., H. M. Barnes, and H. O. Yates. Beetle (*Xyletinus peltatus*) and parasite exit hole densities and beetle larval populations in southern pine floor joists. *Environ. Entomol.*, **8** (1979), 300–3.

### Anthribidae

Autuori, M. Dados biologicos sobre o *Araecerus fasciculatus* De Geer (Coleoptera, Anthribidae). *Rev. Entomol.*, **1** (1931), 52–62.

Childers, C. C. and R. E. Woodruff. A bibliography of the coffeebean weevil, *Araecerus fasciatus* (Coleoptera: Anthribidae). *Bull. Entomol. Soc. Am.*, **26** (1980), 384–94.

El-Sayed, M. T. On the biology of *Araecerus fasciculatus* De Geer (Coleoptera, Anthribidae), with special reference to the effects of variations in the nature and water content of the food. *Ann. Appl. Biol.*, **22** (1935), 557–77.

Mphuru, A. N. *Araecerus fasciatus* De Geer (Coleoptera: Anthribidae): a review. *Trop. Stored Prod. Inf.*, **26** (1974), 7–15.

### Bostrichidae, Bruchidae, Buprestidae

Arora, G. L. and T. Singh. The biology of *Callosobruchus chinensis* (L.) (Bruchidae: Coleoptera). *Res. Bull. Punjab Univ.*, **21** (1970), 55–66.

Beeson, D. F. C. and B. M. Bhatia. On the biology of the Bostrichidae. *Ind. Forest Rec. Entomol.*, **2** (1937), 223–323.

Birch, L. C. Two strains of *Calendra oryzae* L. (Coleoptera). *Aust. J. Exp. Biol. Med. Sci.*, **22** (1944), 271–5.

The influence of temperature on the development of the different stages of *Calandra oryzae* L. and *Rhizopertha dominica* Fab. (Coleoptera). *Aust. J. Exp. Biol. Med. Sci.*, **23** (1945), 29–35.

Ekbom, P. *Buprestis haemorrhoidalis* Herbst and *B. rustica* L. (Coleoptera, Buprestidae) as pests of buildings in Finland. *Finn. State Agric. Res. Board*, **178** (1959), 14–19.

Fisher, W. S. A revision of the North American species of beetles belonging to the family Bostrichidae. U.S. Department of

Agriculture miscellaneous publication 698. Washington, DC: USDA, 1950.

Gauss, R. Der Bambusbohrer, *Dinoderus minutus* Fabr., in Deutschland! *Anz. Schadlingskde.*, **31** (1958), 74–5.

Kapoor, S. Nutritional studies on *Rhizopertha dominica* F. (Bostrichidae: Coleoptera). I. Effects of various natural foods on larval development. *Ind. J. Entomol.*, **26** (1964), 289–95.

Larson, A. O. and P. Simmons. Notes on the biology of the four spotted bean weevil, *Bruchus quadrimaculatus* Fab. *J. Agric. Res.*, **27** (1923), 99–105.

Linsley, E. G. Delayed emergence of *Buprestis aurulenta* from structural timbers. *J. Econ. Entomol.*, **36** (1943), 348–9.

Middlekauff, W. W. Delayed emergence of *Polycaon stouti* LeC. from furniture and interior woodwork. *Pan-Pacific Entomol.*, **50** (1974), 416–17.

Plank, H. K. and R. H. Hageman. Starch and other carbohydrates in relation to powder-post beetle infestation in freshly harvested bamboo. *J. Econ. Entomol.*, **44** (1951), 73–5.

Potter, C. The biology and distribution of *Rhizopertha dominicia* (Fab.). *Trans. R. Entomol. Soc. Lond.*, **83** (1935), 449–82.

Raina, A. K. *Callosobruchus* spp. infesting stored pulses (grain legumes) in India and a comparative study of their biology. *Ind. J. Entomol.*, **32** (1970), 303–10.

Sandner, H. K. Investigation of the ecology of the lesser grain borer, *Rhizopertha dominica* (F.) (Coleoptera: Bostrichidae). *Ekol. Polska*, **B5** (1959), 181–5.

Shires, S. W. Ability of *Prostephanus truncatus* (Horn) (Coleoptera: Bostrichidae) to damage and breed on several stored food commodities. *J. Stored Prod. Res.*, **13** (1977), 205–8.

Smith, D. N. Prolonged larval development in *Buprestis aurulenta* L. (Coleoptera: Buprestidae). A review with new cases. *Can. Entomol.*, **94** (1962), 586–93.

Southgate, B. J. Biology of the Bruchidae. *Annu. Rev. Entomol.*, **24** (1979), 449–73.

Spencer, G. J. *Chrysophana placida* infesting buildings in British Columbia (Coleoptera: Buprestidae). *Proc. Entomol. Soc. Br. Columbia*, **61** (1964), 10.

Spilman, T. J. False powderpost beetles of the genus *Dinoderus* in North America (Coleoptera, Bostrichidae). *Coleopt. Bull.*, **36** (1982), 193–6.

Utida, S. Collective oviposition and larval aggregation in *Zambrotes subfasciatus* (Boh.) (Coleoptera, Bruchidae). *J. Stored Prod. Res.*, **8** (1967), 111–26.

### Cerambycidae

Baake, A. On the house longhorn beetle (*Hylotrupes bajulus* L.) in Norway. *Medd. Fra. Norske Skolgforsokdvesen*, **56** (1960), 263–79.

Becker, G. Sinnesphysiologische Untersuchungen uber die Eiablage des Hausbockkafer. *Z. Verlag. Physiol.*, **30** (1944), 253–99.

Der naturliche Schytz des Laubholzes gegen Hausbockkaferlarven und sein Ursache. *Z. Angew. Entomol.*, **30** (1944), 391–417.

Beitrage zur Okologie der Hausbockkaferlarven. *Z. Angew. Entomol.*, **31** (1949), 135–74.

Der Einfluss des Eiweiss-Gehalts von Holz auf das Hausbocklarven-Wachstum. *Z. Angew. Entomol.*, **51** (1963), 368–90.

Einfluss von Ascomyceten und Fungi imperfecti auf Larven von *Hylotrupes bajulus* (L.). *Mater. Org.*, **3** (1968), 229–40.

Uber die Verbreitung des Hausbockkafers *Hylotrupes bajulus* (L.) Serville (Col., Cerambycidae). *Z. Angew. Entomol. I. Mitt.*, **61** (1968), 253–81.

Uber die Verbreitung des Hausbockkafers *Hylotrupes bajulus* (L.) Serville (Col., Cerambycidae). *Z. Angew. Entomol. II. Mitt.*, **67** (1970), 99–102.

Die Verbreitung des Hausbockkafers in der Bundesrepublik Deutschland. *Holz-Zbl.*, **100** (1974), 1469–70, 1475.

Uber die Verbreitung des Hausbockkafers *Hylotrupes bajulus* (L.) Serville (Col., Cerambycidae). *Z. Angew. Entomol. II. Mitt.*, **80** (1976), 222–5.

Becker, W. B. *Prionus laticollis* (Drury) in a subterranean wooden duct for telephone cables. *J. Econ. Entomol.*, **35** (1942), 608.

Becker, G., T. Hof, and O. Walchli. Der Einfluss von Trocknung, Holzeigenschaften und Temperatur auf Schutzmittel-Giftwerte gegen Hausbock-Eilarven. *Holz Roh- Werkstoff*, **28** (1970), 186–93.

Brammanis, L. Der Hausbock (*Hylotrupes bajulus* L.) in Lettland. *Z. Angew. Entomol.*, **30** (1943), 372–80.

Breidbach, O. Zur Morphologie der Flugmuskulatur von *Hylotrupes bajulus* (L.) und *Strangalia maculata* (PODA) (Cerambycidae, Coleoptera, Insecta). *Beitr. Elektronenmikroskop. Direktabb. Oberfl.* **16** (1983), 497–504.

Zur Stridulation von *Hylotrupes bajulus* (L.) (Coleoptera, Cerambycidae). *Verh. Dtsch. Zool. Ges.*, **1983** (1983), 187.

Studies on the stridulation of *Hylotrupes bajulus* (L.) (Cerambycidae, Coleoptera): communication through support vibration–morphology and mechanics of the signal. *Behav. Proc.*, **12** (1986), 169–86.

Brimblecombe, A. R. The occurrence of the European house borer in Queensland. *Queensland Agric. J.*, **76** (1953), 303–9.

Cannon, K. F. and W. H Robinson. Distribution and biology of the old house borer in Virginia. *Entomol. News*, **93** (1982), 173–6.

The North American biotype of the old house borer, *Hylotrupes bajulus* (L.). *Proc. Entomol. Soc. Wash.*, **85** (1983), 104–9.

Wood consumption, growth, and respiration of the old house borer larvae, *Hylotrupes bajulus*. *Mater. Org.*, **20** (1985), 311–19.

Cymorek, S. *Hylotrupes bajulus* – Verpuppung und Flug, deren Klimaabhangigkeit und Beziehung zur Artverbreitung. *Z. Angew. Entomol.*, **62** (1968), 316–44.

Dominik, J. Badania nad rozprzestrzenieniem supszczela (*Hylotrupes bajulus* L., Cerambycidae, Col.) na ternie Polski wschodniej I nad niektorymi cznnikami sprzyjajacymi jego wystepowaniu. *Folia Forestalia Polonica, Ser. B.*, **4** (1962), 179–226.

Doppelreiter, H. Evidence for a female sex pheromone in the house longhorn beetle, *Hylotrupes bajulus*. *Z. Angew. Entomol.*, **88** (1979), 56–9.

Dubois, R. Le coleoptere cerambycide *Hylotrupes bajulus* L. dit le capricorne des maisons. *Bull. Soc. Agric. Sci. Arb. Sarthe*, **5** (1957), 187–202.

Duffy, E. A. J. *A Monograph of the Immature Stages of British and Imported Timber Beetles (Cerambycidae)*. London: British Museum (Natural History), 1953.

*A Monograph of the Immature Stages of African Timber Beetles (Cerambycidae)*. London: British Museum (Natural History), 1957.

*A Monograph of the Immature Stages of Neotropical Timber Beetles (Cerambycidae)*. London: British Museum (Natural History), 1960.

*A Monograph of the Immature Stages of the Australasian Timber Beetles (Cerambycidae)*. London: British Museum (Natural History), 1963.

*A Monograph of the Immature Stages of the Oriental Timber Beetles (Cerambycidae)*. London: British Museum (Natural History), 1968.

Durr, H. J. R. *The Morphology and Bionomics of the European Houseborer,* Hylotrupes bajulus *(Coleoptera, Cerambycidae)*. Union of South Africa Department of Agriculture Entomological Memoir vol. 4. Pretoria: The Government Printer, 1954.

*The European House Borer* Hylotrupes bajulus (L.) *(Coleoptera: Cerambycidae) and its Control in the Western Cape Province*. Department of Agriculture African Bulletin No. 337. Pretoria: The Government Printer, 1954.

*The Morphology and Bionomics of the European Houseborer,* Hylotrupes bajulus *(Coleoptera: Cerambycidae)*. Entomological Memoir vol. 4, no. 1, pp. 1–136. Pretoria: The Government Printer, 1957.

Ekbom, P. Husbockens forekomst och skadegorelse I Finland. *Flyvehullet*, **182** (1957), 80–4.

French, J. R. J. Occurrence and control of the European house borer in New South Wales. *Austral. Forestry*, **33** (1969), 11–18.

Higgs, M. D. and D. A. Evans. Chemical mediators in the oviposition behavior of the house longhorn beetle *Hylotrupes bajulus* (L.). *Experimentia*, **34** (1977), 46–7.

Jensen, K. Die Verbreitung des Hausbocks in Denmark. *Der Prakt. Schadlbek.*, **8** (1956), 59–60.

Jensen-Storch, Sv. Eine Ubersicht uber die Entwicklung der Hausbockfrage in Danemark und eine Methode zur Untersuchung des relativen Wertes der verschiedenen Konservierungsflussigkeiten als Bekampfungs- und Vorbeugungsmittel. *Anz. Schadlingsk.*, **7** (1932), 101–21.

Kaufman, R. E. U. Notes on the occurrence of *Hylotrupes bajulus* L. in England. *Entomol. Mon. Mag.*, **83** (1947), 37–40.

Kennedy, R. and M. G. Jeffries. The Two-Toothed Longhorn *Ambedontus tristis* (F.) (Col., Cerambycidae) Breeding in Leicestershire. British Wood Preserving Asso., News Sheet No. 138, 1975.

Klausnitzer, B. and F. Sander. *Die Bockäfer Mitteleuropas: Cerambycidae*. Wittenberg Lutherstadt: A. Ziemsen Verlag, 1978.

Knusson, P. and A. Baake. The present distribution of *Hylotrupes bajulus* L. (Coleoptera, Cerambycidae) in Norway and its abundance in some districts. *Nordsk Ent. Tidskr.*, **14** (1967), 94–102.

Korting, A. Zur Entwicklung und Schadtatigkeit des Hausbockkafers, *Hylotrupes bajulus* (L.) in Dachstuhlen verschiedenen Alters. *Anz. Schadlingsk.*, **39** (1962), 150–3.

Experimentelle Untersuchungen zur Entwicklung von *Hylotrupes bajulus* (L.) (Hausbockkafer) in Tannenholz. *Mater. Org.*, **5** (1970), 241–54.

Hausbockbefall und jahreszeitliche Schwankung der Holzfeuchtigkeit im Dachstuhl. *Prakt. Schadingsbekampfer*, **27** (1975), 137–8.

Kuhne, H. Lebenweise und Umweltabhangigkeit des Grubenhalsbocks (*Criocephalus rusticus* (L.), Cerambycidae, Col.). *Mater. Org.*, **10** (1975), 161–201.

Lea, R. G. *House Longhorn Beetle: Geographical Distribution and Pest Status in the UK*. Building Research Establishment information paper IP/8/94. UK: Stewkley Press, 1994.

Linsley, E. G. Ecology of Cerambycidae. *Annu. Rev. Entomol.*, **4** (1959), 99–138.

Mateus, T. J. E. Principalis aspectos do problema da proteccao das conservaaco madeiras em Portugal, em special contro o *Hylotrupes bajulus*, pp. 7–23. Publication no. 20. Lisbon: Ministeria das Obras Publicas, Lab Engenharia Civil, 1952.

Menon, K. D. Longhorn beetle attack on timber in buildings. *Malay. Forest*, **17** (1954), 143–6.

Milligan, R. H. The European house borer *Hylotrupes bajulus* L., a possible pest to New Zealand. *N.Z. Forest Res. Notes*, **25** (1961).

Susceptibility of New Zealand podocarb timbers to *Hylotrupes bajulus* L. (Coleoptera, Cerambycidae). *N. Z. Entomol.*, **3** (1967), 52–6.

Neu, W. *Funde von* Hylotrupes bajulus *(Hausbock) in Istanbul*. Festschrift 60, pp. 500–2. Riga: Geburtstag Strand, 1938.

Nielsen, H. Hausbocke (*Hylotrupes bajulus*). Vorkommen und Bekampfung in Norwegen. *Flyvehullet*, **181** (1957), 25–7.

Ozer, M. Ahsap mobily ada zazar yapen ev teke bocegi *Hylotrupes bajulus* L. uzerinde incelemeler. *Plant-Prot. Bull.*, **3** (1963), 16–20.

Pallaske, M. Autokologische Untersuchungen (Aktivitatun Orientierung) an Larven von *Hylotrupes bajulus* (L.). *Verh. Ges. Okol.*, **10** (1983), 619–23.

Palli, F. and A. Gambetta. Sue due Cerambicidi distributtori del legno (*Hylotrupes bajulus* L.; *Hesperophanes cinerus* Vill.). *Consiglio Ric. Centro Legno*, **5** (1962), 5–30.

Parkin, E. A. The occurrence of the house longhorn beetle, *Hylotrupes bajulus* (L.) in England. *Forestry*, **8** (1934), 150–4.

Rasmussen, S. Effects of the microclimate on the growth and metamorphosis of *Hylotrupes* in Denmark. *Oikos*, **12** (1961), 173–94.

Growth of larvae of the house longhorn beetle *Hylotrupes bajulus* in constant conditions. *Oikos*, **12** (1967), 173–94.

Saraiva, A. C. Combate ao *Hylotrupes bajulus* L. em bairos urbanos. *Tecnico*, **31** (1957), 270.

Schultze-Dewitz, G. *Beobachtungen uber die Bestandigkeit der Gewichtszunahme beim Wachstum der Larven des Hausbockes* (Hylotrupes bajulus L.), pp. 44–53. Berlin: Wanderversamml. Deutscher Entomologen, 1957.

Schwarz, L. U. and A. Rensch. Uber Borsaure als Vorbeugungsmittel gegen den Hausbockkafer. *Arch. Hyg. Bacteriol.*, **129** (1943), 293–311.

Vasic, K. Problem predohrane I isuzbijonova kucne strizibube (Fragen zum Schutz gegen Hausbock und zu seiner Bekampfung). *Zasista Bilja (Belgrad)*, (1962), 83–91.

White, M. G. The house longhorn beetle, *Hylotrupes bajulus* (L.) (Coleoptera: Cerambycidae) in Great Britian. *Forestry*, **27** (1954), 31–40.

Wichmand, H. *Hylotrupes bajulus* L. in Denmark. *Anz. Schadlingsk.*, **7** (1931), 13.

## Cleridae, Coccinellidae

Ashman, F. Factors affecting the abundance of the copra beetle, *Necrobia rufipes* (Deg.) (Col., Cleridae). *Bull. Entomol. Res.*, **3** (1962), 671–80.

Herman, J. and R. Cukerman. Contact dermatitis due to ladybirds. *Practitioner*, **226** (1982), 311.

Hodek, I. *Biology of the Coccinellidae*. The Hague: Junk, 1973.

Koch, R. L. and W. D. Hutchinson. Phenology and blacklight tapping of the multicolored Asian lady beetle (Coleoptera: Coccinellidae) in a Minnesota agricultural landscape. *J. Entomol. Sci.*, **38** (2003), 477–80.

Simmons, P. and G. W. Ellington. The ham beetle, *Necrobia rufipes* DeGeer. *J. Agric. Res.*, **30** (1925), 845–65.

## Cucujidae

Arborgast, R. T. Population parameters for *Oryzaephilus surinamensis* and *O. mercator*: effect of relative humidity. *Environ. Entomol.*, **5** (1976), 738–42.

Ashby, K. R. The life-history and reproductive potential of *Cryptolestes pusillus* at high temperatures and humidities. *Bull. Entomol. Res.*, **52** (1961), 353–61.

Bishop, G. W. The comparative bionomics of American *Cryptolestes* (Coleoptera: Cucujidae) that infest stored grain. *Ann. Entomol. Soc. Am.*, **52** (1959), 657–65.

Chang, S. S. and R. Loschiavo. The influence of some fungi in flour, and humidity on the survival and development of *Cryptolestes turcicus* (Coleoptera: Cucujidae). *Can. Entomol.*, **103** (1971), 261–6.

Currie, J. E. Some effects of temperature and humidity on the rates of development, mortality and oviposition of *Cryptolestes pusillus* (Schonherr) (Coleoptera, Cucujidae). *J. Stored Prod. Res.*, **3** (1967), 97–108.

David, M. H. and R. B. Mills. Development, oviposition and longevity of *Ahasverus advena*. *J. Econ. Entomol.*, **68** (1975), 341–5.

David, M. H., R. B. Mills, and D. B. Sauer. Development and oviposition of *Ahasverus advena* (Walt.) (Coleoptera, Silvanidae) on seven species of fungi. *J. Stored Prod. Res.*, **10** (1974), 17–22.

Davies, R. G. The biology of *Laemophloeus minutus* Oliv. (Coleoptera, Cucujidae). *Bull. Entomol. Res.*, **40** (1949), 63–82.

Finlayson, L. H. The biology of *Cephalonomia waterstoni* Ghan (Hymenoptera, Bethylidae), a parasite of *Laemophloeus* (Coleoptera, Cucujidae). *Bull. Entomol. Res.*, **41** (1950), 457–90.

Halstead, D. G. H. A revision of the genus *Orzaephilus* Ganglbauer, including descriptions of related genera (Coleoptera: Silvanidae). *Zool. J. Linn. Soc.*, **69** (1980), 271–374.

Howe, R. W. The biology of two common storage species of *Oryzaephilus* (Coleoptera, Cucujidae). *Ann. Appl. Biol.*, **44** (1956), 341–55.

LeCato, G. L. Sawtoothed grain beetle: population growth on peanuts stimulated by eating eggs or adults of the Indian meal moth. *Ann. Entomol. Soc. Am.*, **66** (1973), 13–65.

Lefkovitch, L. P. The biology of *Cryptolestes ugandae* Steel and Howe (Coleoptera, Cucujidae), a pest of stored products in Africa. *Proc. Zool. Soc. Lond.*, **128** (1957), 419–29.

The biology of *Cryptolestes capensis* (Walt.) (Coleoptera, Cucujidae), a pest of stored products in Africa. *Bull. Entomol. Res.*, **53** (1962), 529–35.

The biology of *Cryptolestes turcicus* (Grouvelle) (Coleoptera, Cucujidae), a pest of stored and processed cereals. *Proc. Zool. Soc. Lond.*, **138** (1962), 23–35.

Lefkovitch, L. P. and R. H. Milnes. Interaction of two species of *Cryptolestes* (Coleoptera, Cucujidae). *Bull. Entomol. Res.*, **54** (1963), 107–12.

Loschiavo, S. R. Food selection by *Oryzaephilus mercantor* (Coleoptera: Cucujidae). *Can. Entomol.*, **108** (1976), 827–31.

Loschiavo, S. R and D. Sabourin. The merchant grain beetle, *Oryzaephilus mercator* (Silvanidae: Coleoptera), as a household pest in Canada. *Can. Entomol.*, **114** (1982), 1163–9.

Payne, N. M. Life history and habits of the flat grain beetle (*Laemophloeus minutus* Oliv.). *J. N.Y. Entomol. Soc.*, **54** (1946), 9–12.

Thomas, E. L. and H. H. Shepard. The influence of temperature, moisture, and food upon the development and survival of the saw toothed grain beetle. *J. Agric. Res.*, **60** (1940), 605–15.

Williams, G. C. Observations on the life history of *Laemophloecus minutus* (Ol.) (Coleoptera, Cucujidae) when bred on various stored cereals and cereal products. *Bull. Entomol. Res.*, **45** (1954), 341–9.

Woodroffe, G. E. The status of the foreign grain beetle, *Ahasverus advena* (Walt.) (Coleoptera, Silvanidae), as a pest of stored products. *Bull. Entomol. Res.*, **53** (1962), 537–40.

## Cryptophagidae

Coombs, C. W. and G. E. Woodroffe. A revision of the British species of *Cryptophagus* (Herbst) (Coleoptera: Cryptophagidae). *Trans. R. Entomol. Soc. Lond.* **106** (1955), 237–82.

Woodroffe, G. E. and C. W. Coombs. A revision of the North American *Cryptophagus* Herbst (Coleoptera: Cryptophagidae). *Misc. Publ. Entomol. Soc. Am.*, **2** (1961), 179–211.

## Curculionidae (stored food)

Essig, E. O. Origin of the bean weevil, *Mylabris obtectus* Say. *J. Econ. Entomol.*, **22** (1929), 858–61.

Evans, D. E. The capacity for increase at low temperatures of several Australian populations of *Sitophilus oryzae* (L.). *Austr. J. Ecol.*, **2** (1977), 55–67.

The capacity for increase at low temperatures of several Australian populations of the granary weevil, *Sitophilus granarius* (L.). *Austr. J. Ecol.*, **2** (1977), 69–79.

Ewer, R. F. The effect of grain size on the oviposition of *Calandra granaria* Linn. (Coleoptera, Curculionidae). *Proc. R. Entomol. Soc. Lond. A*, **20** (1954), 57–63.

Howe, R. W. The biology of the rice weevil, *Calandra oryzae* (L.). *Ann. Appl. Biol.*, **39** (1952), 168–80.

Jay, E. G., H. L. Musen, and G. C. Pearman. Damage to stored soybeans by the cowpea weevil, *Callosobruchus maculatus* (F.). *Ga. Entomol. Soc.*, **8** (1973), 164–7.

Longstaff, B. C. Biology of the grain pest species of the genus *Sitophilus* (Coleoptera: Curculionidae): a critical review. *Prot. Ecol.*, **2** (1981), 83–130.

Maceljski, M. and Z. Korunic. Contribution to the morphology and ecology of *Sitophilus zeamais* Motsch. in Yugoslavia. *J. Stored Prod. Res.*, **9** (1973), 225–34.

Potter, C. The biology and distribution of *Rhizopertha dominica* (Fab.). *Trans. R. Entomol. Soc. Lond.*, **85** (1936), 449–82.

Reddy, D. B. Ecological studies of the rice weevil. *J. Econ. Entomol.*, **43** (1950), 203–6.

Richards, O. W. Observations on grain weevils, *Calandra* (Coleoptera, Curculionidae). I. General biology and oviposition. *Proc. Zool. Soc. Lond.* **117** (1947), 1–43.

### Curculionidae (wood infesting)

Buck, F. D. *Pentarthrum huttoni* Woll. (Coleoptera, Curculionidae) and some imported Cossoninae. *Entomol. Mon. Mag.*, **9** (1948), 152–4.

Hammad, S. M. 1955. The immature stages of *Pentarthrum huttoni* Woll. (Coleoptera: Curculionidae). *Proc. R. Entomol. Soc. Lond. (A)*, **30** (1955), 33–9.

Hum, M., A. E. Glaser, and R. Edwards. Wood boring weevils of economic importance in Britain. *J. Inst. Wood Sci.*, **8** (1980), 201–7.

Pitman, A. J., S. M. Cragg, and G. Sawyer. *An Investigation of the Nutritional Physiology of the Wood Boring Weevil* Euophryum confine *Brown*. The International Research Group on Wood Preservation, document no. IRG/WP 94–10082. 1994.

### Dermestidae

Amos, T. G. Some laboratory observations on the rates of development, mortality, and oviposition of *Dermestes frischii* (Kug.) (Coleoptera: Dermestidae). *J. Stored Prod. Res.*, **4** (1968), 103–17.

Archer, T. L. and R. G. Strong. Comparative studies on the biologies of six species of *Trogoderma*: *T. glabrum*. *Ann. Entomol. Soc. Am.*, **68** (1975), 105–14.

Armstrong, J. W. T. On Australian Dermestidae. Part II. The genus *Troderma* Berthold. *Linn. Soc. N. S. Wales Proc.*, **67** (1942), 321–30.

Ayappa, P. K., P. S. Cheema, and S. L. Perti. A life history of *Anthrenus flavipes* LeC. (Col., Dermestidae). *J. Econ. Entomol.*, **50** (1957), 469–71.

Azab, A. K., M. F. S. Tawfik, and N. A. Abouzeid. The biology of *Dermestes maculatus* De Geer (Coleoptera: Dermestidae). *Bull. Soc. Entomol. Egypt*, **56** (1972), 1–14.

Factors affecting development and adult longevity of *Dermestes maculatus* De Geer (Coleoptera: Dermestidae). *Bull. Soc. Entomol. Egypt*, **56** (1972), 21–32.

Back, E. A. and R. T. Cotton. The black carpet beetle, *Attagenus piceus* (Oliv.). *J. Econ. Entomol.*, **31** (1938), 280–6.

Badawi, A. The biology of two species of *Trogoderma* existing in Egypt (Coleoptera, Dermestidae). *Bull. Soc. Entomol. Egypt*, **57** (1973), 239–46.

Baker, J. E. Influence of nutrients on the utilization of woolen fabrics as a food for larvae of *Attagenus megatoma* (F.) (Coleoptera: Dermestidae). *J. Stored Prod. Res.*, **10** (1974), 155–60.

Banks, H. J. Distribution and establishment of *Trogoderma granarium* Everts (Coleoptera: Dermestidae): climate and other influences. *J. Stored Prod. Res.*, **13** (1977), 183–202.

Barber, H. S. On the odd, or tissue paper beetle supposed to be *Thylodrias contractus*. *Ann. Entomol. Soc. Am.*, **40** (1947), 344–9.

Beal, R. S. Biology and taxonomy of the Nearctic species of *Trogoderma* (Coleoptera: Dermestidae). *Calif. Univ. Pub. Entomol.*, **10** (1954), 35–102.

Synopsis of the economic species of *Trogoderma* occurring in the United States with description of a new species (Coleoptera: Dermestidae). *Ann. Entomol. Soc. Am.*, **49** (1956), 559–66.

*Coleoptera: Dermestidae. Insects of Micronesia*, vol. 16, pp. 109–35. Honolulu, Hawaii: Bernice P. Bishop Museum, 1960.

Bellemare, E. R. and L. Brunelle. Larval and pupal development of *Dermestes maculatus* (DeGeer) under controlled conditions of temperature and relative humidity. *Can. Entomol.*, **82** (1950), 22–4.

Bhattacharya, A. K. and N. C. Pant. Growth and development of khapra beetle, *Trogoderma granarium* Everts (Coleoptera, Dermestidae) on pulses. *Bull. Entomol. Res.*, **59** (1969), 383–8.

Bry, R. E. Feeding studies of larvae of the black carpet beetle (Coleoptera, Dermestidae) on wool/synthetic blend fabrics. *J. Ga. Entomol. Soc.*, **10** (1975), 284–6.

Burges, H. D. The effect of temperature, humidity and quality of food on the development and diapause of *Trogoderma parabile* Beal. *Bull Entomol. Res.*, **51** (1961), 685–96.

Studies on the dermestid beetle *Trogoderma granarium* Everts. *Bull. Entomol. Res.*, **53** (1962), 193–213.

Coombs, C. W. The effect of temperature and relative humidity upon the development and fecundity of *Dermestes lardarius* L. (Coleoptera, Dermestidae). *J. Stored Prod. Res.*, **14** (1978), 111–19.

The development, fecundity and longevity of *Dermestes ater* DeGeer (Coleoptera: Dermestidae). *J. Stored Prod. Res.*, **17** (1981), 31–7.

Greenwald, M. Studies on the biology of four common carpet beetles. Part II. The old-fashion carpet beetle (*Anthrenus scrophulariae* L.). Agricultural Experimental Station of the University of Cornell Mem., vol. 240, pp. 58–75. Ithaca, NY: Cornell University Press, 1941.

Griswold, G. H. Studies on the biology of four common carpet beetles. Part I. The black carpet beetle (*Attagenus piceus* Oliv.), the varied carpet beetle (*Anthrenus vorax* Waterh.). Agricultural Experimental Station of the University of Cornell Mem., vol. 240, pp. 1–57. Ithaca, NY: Cornell University Press, 1941.

Hadaway, A. B. The biology of the dermestid beetles, *Trogoderma granarium* Everts and *Trogoderma versicolor* (Cruetz.). *Bull. Entomol. Res.*, **46** (1956), 781–96.

Hall, E. and C. Russell. Dermestid beetles as an aid to cleaning bones. *J. Mammal.*, **14** (1933), 372–4.

Hamalainen, M. and I. Mannerkoski. Occurrence and distribution of dermestids (Coleoptera: Dermestidae) in Finland. *Notulae Entomol.*, **64** (1984), 167–84.

Hilali, M., H. K. Dahle, and K. Aurstad. Life history and food-spoiling enzymes of *Dermestes lardarius* (L.). *Norsk Entomol. Tidsskr.*, **19** (1972), 25–32.

Hinks, W. D. The odd beetle, *Thylodrias contractus* Motschulsky (Col., Dermestidae) in Britain. *Entomol. Mon. Mag.*, **86** (1950), 148.

Hinton, H. E. Natural reservoirs of beetles of the family Dermestidae known to infest stored products, with notes on those found in spiders' webs. *Proc. R. Entomol. Soc. Lond. Ser. A*, **18** (1943), 33–42.

Jefferies, M. G. The occurrence of *Dermestes* species (Coleoptera: Dermestidae) in deep pit poultry houses in Britain. *Entomol. Gaz.*, **30** (1979), 207–12.

Lingren, D. L., L. F. Vincent, and M. E. Krohne. The khapra beetle, *Trogoderma granarium*. *Hilgardia*, **24** (1955), 1–36.

Loschiavo, S. R. Life-history and behavior of *Trogoderma parabile* Beal (Coleoptera, Dermestidae). *Can. Entomol.*, **92** (1960), 611–18.

Adult longevity and oviposition of *Trogoderma parabile* Beal (Coleoptera, Dermestidae) at different temperatures. *J. Stored Prod. Res.*, **3** (1967), 273–82.

Mertins, J. W. Life history and morphology of the odd beetle, *Thylodrias contractus*. *Ann. Entomol. Soc. Am.*, **74** (1981), 576–81.

Metcalf, C. L. *Thylodrias contractus* Mots. *J. Econ. Entomol.*, **26** (1933), 509–10.

Minar, J. Synanthropisation and spreading of Dermestidae (Insecta: Coleoptera). In Robinson, W. H., F. Rettich, and G. W. Rambo (eds.) *Proceedings of Fourth International Conference on Urban Pests*, p. 675. Blacksburg, VA: Pocahontas Press, 2002.

Mroczkowski, M. Dermestidae (Coleoptera) from the Kazakh SSR. *Ann. Zool. Instytut Zool. Polska Akad. Nauk*, **20** (1962), 229–59.

Distribution of the Dermestidae (Coleoptera) of the world with a catalogue of all known species. *Ann. Zool. Instytut Zool. Polska Akad. Nauk*, **26** (1968), 16–191.

Dermestidae Skormikowate (Insecta: Coleoptera) [Dermestidae, Hide Beetles] (in Polish). *Fauna Polski*, **4** (1975), 1–163.

Nair, K. S. and A. K. Desai. Some new findings on factors inducing diapause in *Trogoderma granarium* Everts (Coleoptera, Dermestidae). *J. Stored Prod. Res.*, **8** (1972), 27–54.

Novák, I. and P. Verner. Faunistic records from Czechoslovakia. Coleoptera (Dermestidae): *Reese vespulae*. *Acta Entomol. Bohemoslov.*, **87** (1990), 479.

Partida, G. J. and R. G. Strong. Comparative studies on the biologies of six species of *Trogoderma*: *T. variable*. *Ann. Entomol. Soc. Am.*, **68** (1975), 115–25.

Rees, B. E. Taxonomy of the larvae of some North American species of the genus *Dermestes* (Coleoptera: Dermestidae). *Proc. Entomol. Soc. Wash.*, **49** (1947), 1–14.

Roth, L. M. and E. R. Willis. The oviposition of *Dermestes ater* DeGeer, with notes on bionomics under laboratory conditions. *Am. Midl. Nat.*, **44** (1950), 427–47.

Spencer, G. J. Dead *Pollenia rudis* (Fabr.) as hosts of dermestids. *Can. Entomol.*, **60** (1928), 283.

Strong, R. G. Comparative studies of the biologies of six species of *Trogoderma*: *T. inclusum*. *Ann. Entomol. Soc. Am.*, **68** (1975), 91–104.

Takio, M. Morphology and ecology of *Dermestes vorax* Motschulsky (Dermestidae). *Bull. Seric. Exp. Sta. Jpn.*, **9** (1937), 167–84.

White, G. D. and H. E. McGregor. Epidemic infestations of wheat by a dermestid, *Trogoderma glabrum* (Herbst). *J. Econ. Entomol.*, **50** (1957), 382–5.

Woodroffe, G. E. and B. J. Southgate. An investigation of the distribution and field habits of the varied carpet beetle, *Anthrenus verbasci* (L.) (Coleoptera, Dermestidae), in Britain with comparative notes on *A. fuscus* and *A. museorum* (L.). *Bull. Entomol. Res.*, **45** (1954), 575–83.

Zhantiev, R. D. Dermestids (fam. Dermestidae) of the fauna of U.S.S.R. Moscow: Moscow University, 1976.

### Histeridae, Lathridiidae

Hinton, H. E. The Lathridiidae of economic importance. *Bull. Entomol. Res.*, **32** (1941), 191–247.

The Histeridae associated with stored products. *Bull. Entomol. Res.*, **35** (1945), 309–40.

Morgan, P. B., R. S. Patterson, and D. E. Weidhaas. A life-history study of *Carcinops pumilio* Erichson (Coleoptera: Histeridae). *J. Ga. Entomol. Soc.*, **18** (1983), 353–9.

### Lyctidae

Alston, A. M. On the young larvae of *Lyctus brunneus* Steph. *Ann. Appl. Biol.*, **9** (1922), 187–96.

On the method of oviposition and the egg of *Lyctus brunneus* Steph. *J. Linn. Soc. Lond. Zool.*, **35** (1923), 581–97.

Christian, M. B. Biology of the powder-post beetles, *Lyctus planicollis* LeConte and *Lyctus parallelopipedus* (Melsh.). Part I. *La. Conserv. Rev.*, **9** (1940), 56–9.

Biology of the powder-post beetles, *Lyctus planicollis* LeConte and *Lyctus parallelopipedus* (Melsh.). Part II. *La. Conserv. Rev.*, **10** (1941), 40–2.

Cymorek, S. Experimente mit *Lyctus*. Holz. U. Organismen (Int. Sympos. 1965). *Beih. Mater. Org.*, **1** (1966), 391–413.

Ueber den Einfluss der Holzdichte auf die Entwicklung von Holzinsekten und Versuche daruber mit *Lyctus brunneus* (Steph.) in Pressholz. *Mater. Org.*, **2** (1967), 195–205.

Gerberg, E. J. A revision of the New World species of powder-post beetles belonging to the Family Lyctidae. *USDA Technical bulletin* no. 1157. Washington, DC: USDA, 1957.

Parkin, E. A. A study of the food relations of the *Lyctus* powder-post beetles. *Ann. Appl. Biol.*, **23** (1936), 369–400.

The moisture content in timber in relation to attack by *Lyctus* powder-post beetles. *Ann. Appl. Biol.*, **30** (1943), 130–42.

Rosel, A. Oviposition, egg development and other features of the biology of five species of Lyctidae (Coleoptera). *J. Aust. Entomol. Soc.*, **8** (1969), 145–52.

Schmidt, H. Holzwirtschaft und *Lyctus*-Splintholzkafer. *Holzforsch. Holzverwertung*, **15** (1963), 63–6.

Smith, R. H. The effect of wood moisture content on the emergence of the southern lyctus beetle. *J. Econ. Entomol.*, **48** (1955), 770–1.

### Mycetophagidae, Nitidulidae, Oedemeridae

Balch, R. E. Notes on the warfborer (*Nacerda melanura* L.). *Can. Entomol.*, **69** (1937), 1–5.

Balzer, A. I. The life history of the corn sap beetle in rice. *J. Econ. Entomol.*, **35** (1942), 606–7.

Dobson, R. M. The species of *Carpophilus* Stephens (Coleoptera, Nitidulidae) associated with stored products. *Bull. Entomol. Res.*, **45** (1954), 389–402.

Notes on the taxonomy and occurrence of *Carpophilus* Stephens (Coleoptera, Nitidulidae) associated with stored products. *Entomol. Mon. Mag.*, **95** (1960), 156–8.

Essig, E. O. The dried fruit beetle *Carpophilus hemipterus* (L.). *J. Econ. Entomol.*, **8** (1915), 396–400.

Hicken, N. E. *Nacerdes melanura* L. (Coleoptera, Oedomeridae) at a considerable height above ground. *Entomol. Mon. Mag.*, **88** (1953), 107.

Lefkovitch, L. P. Some observations on the life cycle of *Carpophilus dimidiatus* (F.) (Coleoptera, Nitidulidae) on wheat bran. *J. Stored Prod. Res.*, **2** (1966), 163–5.

Parsons, C. T. Revision of Nearctic Mycetophagidae (Coleoptera). *Coleopt. Bull.*, **29** (1975), 93–108.

Pitman, A., A. Jones, and E. Gareth Jones. The warf borer, *Nacerdes melanura* L: a threat to stored archeological timber. *Studi. Conserv.*, **38** (1993), 274–85.

Spencer, G. J. An unusual record of the wharf borer *Nacerda melanura*, in buried piling at Vancouver, British Columbia (Coleoptera: Oedemeridae). *Entomol. Soc. Br. Columbia, Proc.*, **43** (1947), 7–8.

The warf borer in a Vancouver branch library. *Proc. Entomol. Soc. Br. Columbia*, **54** (1957), 44–5.

Stejskal, V. Faunistic records from Czechoslovakia. Coleoptera (Latridiidae): *Latridus pseudominutus*. *Acta Entomol. Bohemoslov.*, **88** (1991), 79.

### Ptinidae

Belles, X. and D. G. Halstead. Identification and geographic distribution of *Gibbium aequinoctiale* Boieldieu and *Gibbium psylloides* (Czenpinski) (Coleoptera: Ptinidae). *J. Stored Prod. Res.*, **21** (1985), 151–5.

Brown, W. J. A key to the species of Ptinidae occurring in dwellings and warehouses in Canada (Coleoptera). *Can. Entomol.*, **72** (1940), 115–22.

Coombs, C. W. and G. E. Woodroffe. Some factors affecting the longevity and oviposition of *Ptinus tectus* Boieldieu (Coleoptera: Ptinidae) which have relevance to succession among grain beetles. *J. Stored Prod. Res.*, **1** (1965), 111–27.

Some factors affecting larval development in *Ptinus tectus* Boieldieu (Coleoptera: Ptinidae) relevant to succession among grain beetles. *J. Stored Prod. Res.*, **6** (1970), 199–216.

Ewer, D. W. and R. F. Ewer. The biology and behavior of *Ptinus tectus* Boiel. (Coleoptera: Ptinidae), a pest of stored products. III. The effect of temperature and humidity on oviposition, feeding, and duration of life cycle. *J. Exp. Biol.*, **18** (1942), 290–305.

Grace, J. K. A spider beetle, *Sphaericus gibboides* Boieldieu (Coleoptera: Ptinidae), tunneling in wood in service. *Pan-Pac. Entomol.*, **61** (1985), 288–90.

Gray, H. E. Some stored product pests in Canada with special reference to the hairy spider beetle, *Ptinus villiger* Reit. *Entomol. Soc. Ont. Rept.*, **65** (1934), 59–66.

Hall, D. W. and R. W. Howe. A revised key to the larvae of the Ptinidae associated with stored products. *Bull. Entomol. Res.*, **44** (1953), 85–96.

Hickman, V. V. Notes on the biology of *Ptinus exulans*. *J. Entomol. Soc. Aust. (N.S.W.)*, **8** (1974), 7–14.

Hinton, H. E. The Ptinidae of economic importance. *Bull. Entomol. Res.*, **31** (1940), 331–81.

The Ptinidae of economic importance. *Bull. Entomol. Res.*, **31** (1941), 331–81.

Hisamatsu, S. The Ptinidae of Japan (Coleoptera). *Aghea*, **11** (1970), 14–20.

Howe, R. W. Studies on beetles of the family Ptinidae. I. Notes on the biology of species in Britain. *Entomol. Mon. Mag.*, **85** (1949), 137–9.

Studies on *Ptinus tectus* in a warehouse. *Bull. Entomol. Res.*, **41** (1950), 371–94.

Howe, R. W. The biology of *Tipnus unicolor* Pill and Mitt. *Entomol. Mon. Mag.*, **91** (1955), 253–7.

The biology of *Ptinus pusillus* Strum. *Entomol. Mon. Mag.*, **92** (1956), 331–3.

The biology of *Ptinus hirtellus* Strum and some notes on *P. latro*. *Entomol. Mon. Mag.*, **92** (1956), 369–73.

The development period of *Ptinus clavipes* Panz. form *mobilis* (= *P. latro* auct.). *Entomol. Mon. Mag.*, **94** (1958), 236–7.

Conclusions and additional remarks. *Bull. Entomol. Res.*, **50** (1959), 287–326. [The biology of 14 species of spider beetles is summarized in this paper.]

Howe, R. W. and H. D. Burges. Studies on beetles of the family Ptinidae. VI. The biology of *Ptinus fur* (L.) and *P. sexpunctatus* Panzer. *Bull. Entomol. Res.*, **42** (1951), 499–513.

Studies on the beetles of the family Ptinidae VII. The biology of five ptinid species found in stored products. *Bull. Entomol. Res.*, **43** (1952), 153–86.

Studies on beetles in the family Ptinidae. IX. A laboratory study of the biology of *Ptinus tectus* Boield. *Bull. Entomol. Res.*, **44** (1953), 461–516.

The biology of five ptinid species found in stored products. *Bull. Entomol. Res.*, **43** (1952), 153–86.

The biology of *Mezium affine* Boieldieu. *Entomol. Mon. Mag.*, **91** (1953), 73–5.

Manton, S. M. The larvae of Ptinidae associated with stored products. *Bull. Entomol. Res.*, **35** (1945), 341–65.

Papp, C. S. An illustrated and descriptive catalogue of the Ptinidae of North America. *Deutsch. Ent. Z.*, **9** (1962), 367–423.

### Scarabaeidae, Scolytidae, Staphylinidae

Armstrong, R. K. and J. L. Winfield. Staphylinidae dermatitis on Okinawa. *J. Med. Entomol.*, **5** (1968), 362.

Balachowsky, A. *Coléoptères Scolytides. Faune de France*, vol. 50. Paris: Paul Lechevalier, 1949.

Frank, J. H. *Paedrus, sensu lato* (Coleoptera: Staphylinidae): an index and review of the taxa. *Insecta Mundi*, **2** (1988), 97–159.

Frank, J. H. and K. Kanamitsu. *Paederus, sensu lato* (Coleoptera: Staphylinidae); natural history and medical importance. *J. Med. Entomol.*, **24** (1987), 155–91.

Janbaksh, B. and A. Ardalan. Rove beetles (Coleoptera: Staphylinidae) and their medical importance. *Iran. J. Publ. Health*, **6** (1977), 70–7.

Maddock, D. R. and C. Fehn. Human ear invasions by adult scarabaeid beetles. *J. Econ. Entomol.*, **51** (1958), 546–7.

McCrae, A. W. R. and S. A. Visser. *Paedrus* (Coleoptera: Staphylinidae) in Uganda I. Outbreaks, clinical effects, extraction and bioassay of the vesicating toxin. *Ann. Trop. Med. Parasitol.*, **69** (1975), 109–20.

Nizbakhtzadeh, M. R. and S. Tirgari. Two medically important beetles of Fars province, Iran. *Iran. J. Publ. Health*, **28** (1999), 56–69.

Penchenier, L., J. Mouchet, B. Cros *et al.* Invasions de *Phaederus sabaeus* (Coleoptera: Staphylinidae) en Afrique Centrale 1. Aspects entomologiques et epidémiologiques. *Bull. Soc. Pathol. Exotique*, **87** (1994), 45–8.

Tirgari, S. and M. R. Nizbakhtzadeh. *Paedrus* beetles (Coleoptera: Staphylinidae), an urban problem in Iran. In Jones, S. C., J. Zhai, and W. H. Robinson (eds) *Proceedings of the 4th International Conference of Urban Pests*, pp. 401–7. Blacksburg, VA: Pocahontas Press, 2002.

Whelan, P. I. and T. Weir. Skin lesions caused by *Paedrus australis* Guerin-Meneville (Coleoptera: Staphylinidae). *J. Aust. Entomol. Soc.*, **26** (1987), 287–8.

### Tenebrionidae, Trogistidae

Applebaum, S. W. The suitability of groundnuts for the development of *Tribolium castaneum* (Herbst) (Coleoptera, Tenebrionidae). *J. Stored Prod. Res.*, **5** (1969), 305–10.

George, C. R. The effects of malnutrition on growth and mortality of the red rust flour beetle, *Tribolium castaneum* (Coleoptera: Tenebrionidae) parasitized by *Nosema whitei* (Microsporidia: Nosematidae). *J. Invert. Pathol.*, **18** (1971), 383–8.

Hafeez, M. A. and G. Chapman. Effects of temperature and high relative humidity on the rate of development and mortality of *Latheticus oryzae* Waterhouse (Coleoptera, Tenebrionidae). *J. Stored Prod. Res.*, **1** (1966), 235–42.

Halstead, D. G. H. Biological studies on species of *Palorus* and *Coelopalorus* with comparative notes on *Tribolium* and *Latheticus* (Coleoptera: Tenebrionidae). *J. Stored Prod. Res.*, **2** (1967), 273–313.

Some observations on the biology of *Lophocateres pusillus* (Klug) (Coleoptera: Trogistidae). *J. Stored Prod. Res.*, **4** (1968), 197–202.

Hinton, H. E. A synopsis of the genus *Tribolium* Macleay, with some remarks on the evolution of its species-groups (Coleoptera, Tenebrionidae). *Bull. Entomol. Res.*, **31** (1948), 331–81.

Howe, R. W. The effect of temperature and humidity on the rate of development and mortality of *Tribolium castaneum* (Herbst) (Coleoptera, Tenebrionidae). *Ann. Appl. Biol.*, **44** (1956), 356–68.

The effects of temperature and humidity on the rate of development and the mortality of *Tribolium confusum* Duval (Coleoptera, Tenebrionidae). *Ann. Appl. Biol.*, **48** (1960), 363–76.

Koura, A., M. A. El-Halfawy, and S. I. Bishara. On the biology of *Tenebrio molitor* L. in Egypt (Coleoptera: Tenebrionidae). *Bull. Soc. Entomol. Egypt*, **56** (1972), 297–300.

LeCato, G. L. Red flour beetle: population growth on diets of corn, wheat, rice or shelled peanuts supplemented with eggs and adults of the Indian meal moth. *J. Econ. Entomol.*, **68** (1975), 763–5.

LeCato, G. L. and B. R. Flaherty. *Tribolium castaneum* progeny production and development on diets supplemented with eggs or adults of *Plodia interpunctella*. *J. Stored Prod. Res.*, **9** (1973), 199–203.

Leech, H. B. Black flour beetle, *Tribolium madens* Charpi in British Columbia (Coleoptera, Tenebrionidae). *Can. Entomol.*, **75** (1953), 40.

Morison, G. D. Notes on the broad-horned flour beetle [*Gnatocerus (Echocerus) cornutus*, Fabr.]. *Proc. R. Phys. Soc. Edinb.*, **2** (1925), 14–18.

Preiss, F. J. and J. A. Davidson. Adult longevity, preoviposition period and fecundity of *Alphitobius diaperinus* in the laboratory (Coleoptera: Tenebrionidae). *J. Ga. Entomol. Soc.*, **6** (1971), 105–9.

Sinha, R. N. Development and mortality of *Tribolium castaneum* and *T. confusum* (Coleoptera: Tenebrionidae) on seed-borne fungi. *Ann. Entomol. Soc. Am.*, **59** (1966), 192–201.

Wilson, T. H. and F. D. Miner. Influence of temperature on development of the lesser mealworm, *Alphitobius diaperinus* (Coleoptera: Tenebrionidae). *J. Kansas Entomol. Soc.*, **42** (1969), 294–303.

# 6 Collembola, Dermaptera

## COLLEMBOLA
### Introduction

Springtails are 0.75–6 mm long, wingless and soft-bodied insects. The abdominal segments are fused and there are not more than six segments visible. Antennae have 4–8 segments. The common name is derived from a forked structure, the furcula, which helps to propel them through the air. The furcula is on the fourth ventral abdominal segment, and when at rest it is folded under the abdomen. Jumps are made by forcefully extending the furcula downward and backward to strike the substrate, which lifts the insect. The jumping ability is strongly developed in terrestrial springtails, which jump to escape from predators. *Entomobrya dorsalis* is 2 mm long, but it can travel a distance of more than 16 cm. When disturbed, most springtails can be airborne within 50 ms, and *Allacma fusca* takes only 12 ms to respond with a jump. On the ventral side of abdominal segment 1 there is an eversible vesicle, called the collophore. This hold-fast structure provides for water absorption from moist substrates. Eyes are reduced to 1–8 ommatidia or absent. Mouthparts are mandibulate or suctorial and styletlike. Collembola have simple metamorphosis and the immature stages resemble the adults.

Mating behavior ranges from males randomly placing stalked spermatophores in the environment for females to discover, to males and females involved in elaborate courtship and mating involving a stalked spermatophore, or direct transfer to the female. Females do not retain sperm during ecdysis, and must collect a spermatophore after each molt to continue to lay fertilized eggs. In *Sminthurus viridis*, the male deposits spermatophores and the female takes it up alone, or with a little encouragement from the male. Males in Sminthuridinae and Bourletiellinae have clasping organs on their antennae, and these are used in mating. The male *Sphaeridia pumilis*, after some preliminary courtship, clasps the antennae of the female and positions her ventral surface to his, then transfers a drop of sperm to the female. Copulation lasts for 15 min or more.

Eggs are smooth and spherical, usually yellowish white, and deposited singly or in small batches directly on a moist substrate. Eggs are sometimes laid on top of those deposited by conspecific females. Up to 10 000 eggs may accumulate in a single site in laboratory populations of *Proisotoma minuta*. Single eggs may be covered with liquid fecal material from the female. Hatching is in 20–26 days; fecundity is based on nutrition and access to a male, and may be 400 eggs. First-stage immatures are white with dark pigment in the region of the eyes if present. There are 6–8 molts before nymphs achieve maximum size; development is completed in about 48 days. Full-grown springtails live for about 15 days. There are multiple generations per year. There is a tendency for gregariousness and massing of large numbers of adults and nymphs for short periods. This behavior is usually associated with abundant food, favorable environmental conditions, or migration.

Pest status is due to their presence in small or large numbers indoors and outdoors, and to the ability of some species to cause dermatitis. Springtails inhabit moist locations and most feed on decaying plant material, fungi, pollen, algae, and arthropod feces. Several species occur in commercial sites, and cause damage to cultivated mushrooms and greenhouse plants. Those that occur in households are associated with moist or wet conditions, but some species have protective scales on their body and can persist in dry environments. Conditions favorable for springtails indoors include high humidity, mold, mildew, or other wet or moist organic matter. Species that occur outdoors may move when conditions are unsuitable. About 20 species have been reported indoors. *Entomobrya nivialis* is a small, cosmopolitan species and it has been reported to cause human dermatitis.

## Entomobryidae

These are slender and long springtails. In the large subfamily Entomobryinae, abdominal segment 4 is larger than segment 3, and the antennae are four-segmented. There are 1130 described species in this subfamily. In the Orchesellinae, the abdominal segments are nearly equal in size. Some species have long antennae and legs, and distinct body coloration. This is a cosmopolitan family, and the included species occur in numerous habitats, including on the soil surface, in leaf litter, caves, and indoors. Several species, including *Entomobrya multifasciata*, occur in greenhouses. More than 100 species of *Pseudosinella* (Fig. 6.1b) are found in caves, but some species occur indoors. Many of these species lack eyes.

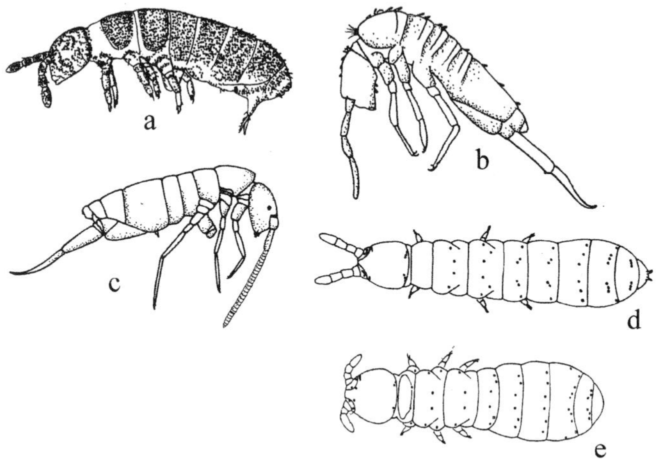


**Figure 6.1** Collembola. (a) *Proisotoma frisoni*; (b) *Pseudosinella* sp.; (c) *Tomocerus flavescens*; (d) *Onychiurus armatus*; (e) *O. fimetarius*.

***Entomobrya atrocincta* (Fig. 6.2a)** Adults have four-segmented antennae inserted on the anterior portion of the head, in front of the eyes. The pronotum is reduced, and the body has stripes or dark markings anteriorly; posteriorly the body is uniformly colored. This species is cosmopolitan and has been reported from indoor flowerpots and infesting powdered milk.

***Entomobrya griseoolivata* (Fig. 6.2b)** Adults have four-segmented antennae inserted on the anterior portion of the head, in front of the eyes. The pronotum is reduced, and the body lacks strips or other markings. This species is cosmopolitan and is a common domestic species.

**Household springtail, *Entomobrya kanaba* (= *E. unostrigata*)** Adults are about 1.5 mm long. The body is slender and brownish black to gray, and the antennae are long. Large numbers and extensive infestations have been reported in domestic habitats, such as in kitchens, bathrooms, and clothes closets. Large numbers may also occur around the outside of buildings. They may be dispersed from one location to another in household materials.

***Entomobrya nivalis* (Fig. 6.2c)** Adults have four-segmented antennae inserted on the anterior portion of the head, in front of the eyes. The pronotum is reduced. The body has stripes or dark markings on the head, thorax, and abdomen, and the head has a dorsal spot. This species is cosmopolitan and is a common domestic species. A closely related species, *E. purpurascens* (Fig. 6.2d), occurs in buildings in North America and Europe.

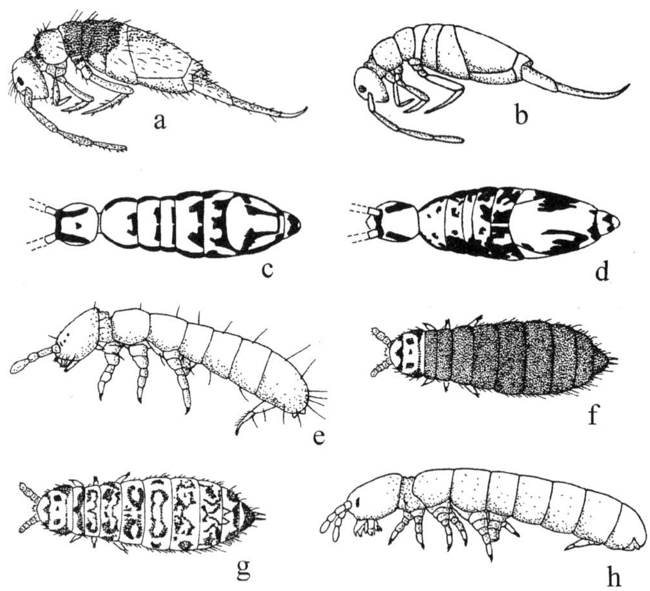


**Figure 6.2** Collembola. (a) *Entomobrya atrocincta*; (b) *E. griseoolivata*; (c) *E. nivalis*; (d) *E. purpurascens*; (e) *Folsomia quadrioculata*; (f) *Hypogastrura pseudarmata*; (g) *H. armata*; (h) *Isotomodes tenuis*.

***Lepidocyrtinus domesticus* (Fig. 6.3c)** The adult body is slender and with scales; antennae are four-segmented and segment 4 is annulate. This species is cosmopolitan and commonly found in houses.

***Lepidocyrtus curvicollis* (Fig. 6.3b)** The adult has segment 4 of antennae normal and not annulate, and the eye has eight ocelli. This species is cosmopolitan and occurs in houses and cellars. Other *Lepidocyrtus* in domestic habitats includes *L. cinereus*, which is reported from potting soil of indoor plants in California.

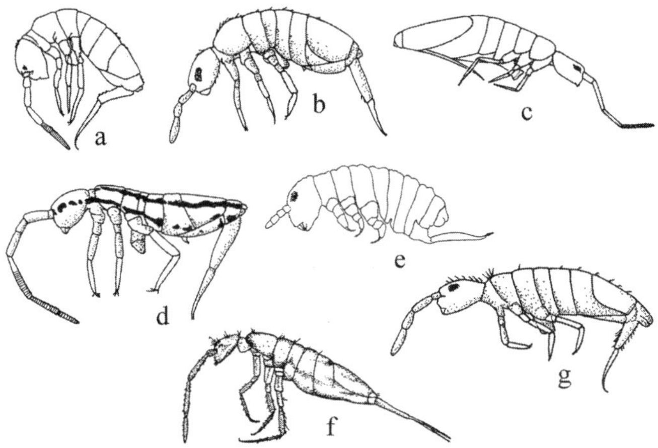


**Figure 6.3** Collembola. (a) *Heteromurus nitidus*; (b) *Lepidocyrtus curvicollis*; (c) *Lepidocyrtinus domesticus*; (d) *Orchesella albosa*; (e) *Podura aquatica*; (f) *Seira buski*; (g) *S. platani*.

***Orchesella albosa* (Fig. 6.3d)** Adults have six-segmented antennae inserted on the anterior portion of the head, in front of the eyes; the pronotum is reduced, and body scales are absent. This species occurs in North America and Europe; it has been reported infesting houses and on the human body.

**Domestic springtail, *Seira domestica*** This species is yellowish white, and with fine pale setae on the body. It is cosmopolitan and found in damp basements and cellars.

***Seira buski* (Fig. 6.3f)** The adult body has scales and the color is entirely dark blue. This species is distributed in North America and Europe and is associated with houses. A related species, *S. platani* (Fig. 6.3g) has the body marked with blue. This species is cosmopolitan and is also associated with houses and stored foods.

**Other Entomobryidae** *Willowsia buski* is often associated with the exterior of buildings. *Heteromurus nitidus* (Fig. 6.3a) is cosmopolitan and common in houses.

## Hypogastruidae

These springtails are 1–2 mm long and have short appendages; the furcula may be reduced or absent. Body color ranges from pale yellow to purple, blue, or green. This is a large family; there are more than 500 species in 39 genera and worldwide distribution. Species occur in numerous habitats, including under bark, the seashore, and commercial mushroom beds. They can be extremely abundant in the percolating filters of sewage treatment works. Members of this family are the springtails most often found in large aggregations.

***Hypogastrura armata* (Fig. 6.2g)** Adults are dark-gray to black, and the head and legs are reddish brown. The body is mottled with white and dark markings. This species is cosmopolitan and associated with houses; it sometimes occurs in swimming pools. Adults of a related species, *H. pseudarmata* (Fig. 6.2f), are uniformly dark and the head is mottled. This species has also been reported swarming on and in houses.

**Sewage springtail, *Hypogastrura purpurescens*** Adults have the antennae inserted anterior to the eyes, and the pronotum is well developed. This species is common in the gravel of percolating filters at sewage treatment works, where adults and immatures feed on the biological film, especially during winter. Eggs are laid on the substrate and hatch in 4–15 days; the first molt occurs in about 1 week and other molts occur at regular intervals. Females mature in 6–7 weeks, and the total life span is 7–8 months.

***Hypogastrura viatica*** Adults are about 2 mm long. The body is uniformly dark blue, and the antennae are shorter than the head. It feeds on organisms on the filtering sand in sewage treatment plants, and large populations can develop in treatment plants and move to surrounding buildings. This species is cosmopolitan, and is common in Europe, Asia, South America, and North America.

**Snow fleas, *Hypogastrura nivicola*, *Istoma saltans*** Adults of *H. nivicola* are 1–2 mm long. They are dark brown to black, and sometimes have dense setae on the body. *H. nivicola* occurs in North America and has been found in large numbers feeding on algae and fungal spores on the surface of snow. It is sometimes found in buckets of tree sap collected from maple trees in winter. *I. saltans* is black and it occurs on alpine snowfields and glaciers in northern Europe.

## Isotomidae

These collembolans have fine setae or hairs on their body, but no scales; the abdominal segments are all of equal length. There is a range of body colors and they are often striped or banded. This is a cosmopolitan family and many species live in soil or are associated with fresh water. *Isotomurus palustris* is widespread and one of the most frequently recorded springtails; it occurs along the margins of ponds and streams. *Cryptopygus antarctus* is abundant in the Antarctic, and populations of *Folsomia candida* are used to assess the effects of chemicals on nontarget soil invertebrates. One species, *Metisotoma grandiceps*, is carnivorous on other collembolans.

***Folsomia quadrioculata* (Fig. 6.2e)** Adults are about 3 mm long. Antennal segments 3 and 4 are not subsegmented; the anus is positioned ventrally. This species is distributed in North America and Europe; it has been reported indoors associated with potted flowers.

***Isotomodes tenuis* (Fig. 6.2h)** Adults are about 3 mm long. The body is slender; antennal segments 3 and 4 are not subsegmented; ocelli are absent, and the anus is positioned ventrally. This species is distributed in North America and it occurs indoors associated with potted plants.

***Proisotoma frisoni* (Fig. 6.1a)** Adults are about 2 mm long. The body is mottled; the anus is terminal. This species is distributed in North America and it occurs indoors associated with potted plants.

## Onychiuridae

These springtails are elongate and slender; most of them are 1–2 mm long and pale white, but *Tetrodontophora bielanensis* can be 9 mm long. Members of this family occur in leaf litter or soil. They lack eyes and the furcula is almost always absent, but they have defense pores called pseudocelli which are small areas of thin cuticle through which defensive fluid can be extruded. *Onychiurus* has 42 described species, and most are northern in distribution.

***Onychiurus armatus*, *O. fimetarius* (Fig. 6.1d, e)** Adults are about 2 mm long. The pronotum is well developed; the eyes are absent. The postantennal organ of *O. armatus* is simple and without branched vesicles; the postantennal organ of *O. fimetarius* is elaborate and has branching vesicles. These species are cosmopolitan and they are commonly found indoors, often associated with potted plants.

***Onychiurus cocklei*** Adults are about 1.8 mm long and golden yellow. It occurs in large numbers on snow and the surface of pools in British Columbia and Alaska.

## Poduridae

The single species in this family, *Podura aquatica* (Fig. 6.3e) is one of the most common springtails in the northern hemisphere. It often occurs in large numbers along streams, in freshwater ditches, and ponds. The abdomen is six-segmented, and the eyes are limited to 6–8 ocelli.

## Sminthuridae

This is a large and cosmopolitan family and the species are frequently encountered and often brightly colored. These springtails are 0.4–2.7 mm long and have a characteristic globular-shaped body. The abdominal segments are indistinct, and the antennae are longer than the head. Most species are exceptional jumpers. The subfamily Bourletiellinae contains the widespread species, *Bourletiella hortensis*, which is common in peridomestic gardens and some agricultural fields. Males in Sminthuridinae have their antennae modified for clasping females during mating. These small (about 1 mm long) springtails are often on the surface of fresh water. The Sminthuridinae are mainly distributed in the northern hemisphere and the tropics. The Lucerne flea, *Sminthurus viridis*, has been transported to many regions of the world where it is a pest of field clover.

**Seedling springtail, *Bourletiella hortensis*** The adult female is about 1.8 mm long; the male is about 1.2 mm long. The body is mottled blue to purple, and the area between the eyes is yellow to orange. The abdomen has scattered pale spots. This species is common in spring and early summer, and it is known to attack seedlings.

## Tomoceridae

Members of this family are widespread and common, primarily in the northern hemisphere. Most species have long antennae and the last two antennal segments in some species are ringed, making them appear to be composed of a large number of segments. These springtails have abdominal segment 3 only slightly longer than segment 4, and antennae in which segment 3 is longer or much longer than segment 4. This enables some species to roll their antennae into a coil. *Pogonognathellus longicornis* (= *Tomocerus*) is about 6 mm long and widely distributed in northern Europe.

**Silver springtail, *Tomocerus flavescens* (Fig. 6.1c)** Adults are 6–10 mm long, and with a silvery gray to grayish blue body. This large springtail is common on moist ground, in gardens, and in houses and greenhouses. It is cosmopolitan.

***Tomocerus minor*** This species is one of the several springtails that occur on sewerage filter beds. It molts at intervals of 1–4 weeks and lives about 1 year.

# DERMAPTERA
## Introduction

Earwigs are elongate, slightly flattened insects. They have long antennae, a shiny exoskeleton, and a characteristic pair of movable forceps at the posterior of the abdomen. They are wingless or winged, the latter with a front pair of short, wing covers; the hind wings project slightly beyond the front pair of wings. They have gradual metamorphosis, and chewing mouthparts. Earwigs are primarily nocturnal, and they feed on plant and animal material. Some winged species are good flyers, but others fly only rarely.

Eggs are shiny white, oval, and deposited in batches of 25–50 in burrows excavated in soil in spring. Females usually remain with their eggs until hatching, and then give additional care to the young. Females provide food for first-instar nymphs and nymphs usually remain in the nest site until the second instar develops. Family groups of adults and nymphs may be found. There are 4–6 nymph instars and most species become full-grown in late summer. Earwigs are generally gregarious, and often gather in large groups in narrow harborages. They mate in fall and overwinter as adults.

Pest status is based on their numbers around the outside of buildings. Some are attracted to lights at night, and small to large numbers of adults may enter buildings around doors and windows. The European earwig and the ring-legged earwig produce a secretion containing quinones, which can stain and sometimes irritate skin if they are handled. The name earwig, which is commonly used in many countries, arose from the superstition that they enter people's ear during the night. This may have been true when straw was used in bedding. The straw provided harborage for earwigs, and occasionally one of these insects would be found in the ear of a sleeper. The forceps are generally large in males and small in females. They are apparently used for defense, capturing and holding prey while eating in carnivorous species, probing narrow crevices, and to some extent for folding and unfolding the wings, and as a sexual display in mating. Some earwigs can inflict a painful pinch with their forceps.

## Carcinophoridae

These earwigs are characterized by being wingless, and with the combined length of antennal segments 4–6 longer than the basal segment. The male forceps are asymmetrical, with the right more strongly curved than the left.

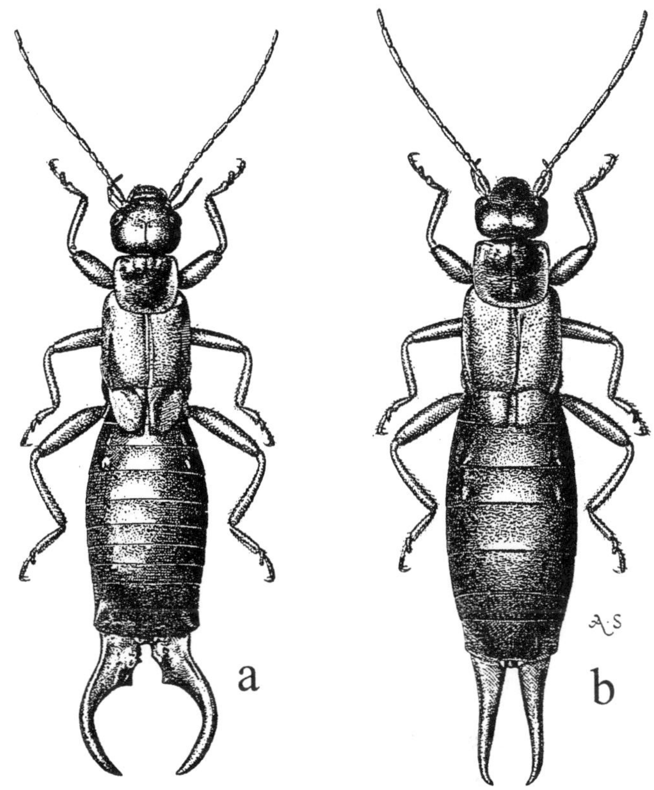

**Figure 6.4** Dermaptera (a) *Forficula auricularia* male; (b) *F. auricularia* female.

**Ring-legged earwig, *Euborellia annulipes*** Adults are 9–18 mm long. The body is dark brown to black, and the abdomen is yellowish brown; it is wingless. There are dark bands around the femora and tarsi of each leg; the antennae have 15–16 segments, and segments 3–5 are pale. The forceps of both sexes are short. Eggs are kidney-shaped and before hatching turn brown. Hatching occurs in about 14 days; fecundity is 47–52 eggs. Development from egg through five instars takes about 80 days. Some females live as long as 7 months; females in populations may outnumber males 4:1. This species occurs in southern USA, the UK, and continental Europe. It occasionally enters houses. This and a related species, *E. cincticollis*, fly to lights at night.

**Seaside earwig, *Anisolabis maritima*** Adults are 20–25 mm long. The body is brown to black; legs are pale yellow; it is wingless. Male forceps are short, and unequally developed; in the female, the forceps are symmetrical. Adults and nymphs are found in beach debris, and around seafood-processing plants in piles of shells and debris. Eggs are laid in the fall. This species is distributed along seashores in North America and Europe.

**Littoral earwig, *Anisolabis littorea*** Adults are about 22 mm long, dark brown, and wingless. Forceps of the female are symmetrical, and the males are asymmetrical, with one strongly curved. This species is found in seaside kelp, and along stones on sea beaches, above the high-tide level. It also occurs further inland and sometimes is found indoors. Adults overwinter and avoid light. They prey on other arthropods. This species occurs in Australia and New Zealand.

## Forficulidae

Adults in this family have an enlarged second tarsal segment, and a lack of dense setae on the underside of the tarsi. This is a large group of moderately sized earwigs.

**European earwig, *Forficula auricularia* (Fig. 6.4a, b)** Adults are about 16 mm long. The body is reddish brown, and the legs are pale yellow. Males have different-sized forms, and their forceps range in size from 3.5 to 9.5 mm. Eggs are about 1 mm long, elliptical, and pale yellow. Eggs may be deposited in cavities in the soil, and laid in fall or spring, or both to produce two generations in a year. Eggs laid in the fall hatch in about 73 days, and eggs laid in the spring hatch in about 20 days. Females remain in the earthen cavity and with the eggs until the nymphs hatch, and then provide food until the second-instar nymph. Development of the four instars is about 68 days in the field, and 31 days in the laboratory at 16–21 °C 60–70% relative humidity (RH). Adults appear in late summer. Abdominal glands secrete an odorous liquid. They feed on green plants and insects. This earwig is abundant in gardens and organic mulch around buildings. This species was introduced into the USA in 1912 and is widely distributed in urban habitats; it also occurs in Europe.

Both males and females produce an aggregation pheromone, which may explain the clumped distribution of this species, and why they are found in broods during winter. Their aggregation behavior also contributes to their pest status in the urban environment. Adults and nymphs gather in preferred harborages around the perimeter and inside buildings. Aggregation may be beneficial in terms of their role as predators of aphids, mites, insect eggs, and caterpillars, since they will aggregate in areas of high prey density. The parasitic tachinid (Tachinidae) flies, *Digonichaeta setipennis* and *Racodineura antiqua*, have been introduced to the USA, Australia, and other regions to control this earwig. *D. setipennis* deposits larvae into earwig harborages and it searches for a host; *R. antiqua* deposits seed-like eggs on potential food, and when eggs are eaten the fly larvae hatch.

**Spine-tailed earwig, *Doru aculeatum*** Adults are 10–12 mm long; the body is brown with pale markings. The male has a short median thorn-like spine on the terminal abdominal segment. It occurs at outdoor lights at night. This species is distributed in the eastern USA.

**Linear earwig, *Doru lineare*** Adults are 9–11 mm long, and shiny black to brown. The male has a short median spine on the terminal abdominal segment, between the bases of the forceps. This species will fly to lights at night. It is distributed in the eastern USA.

## Labiduridae

These earwigs are winged, and the combined length of antenna segments 4–6 is longer than the basal segment. The male forceps are symmetrical, not strongly curved, and the right and left forcep are nearly equal size.

**Striped earwig, *Labidura riparia*** Adults are 20–25 mm long. The body is pale brown to reddish brown with black markings, the abdomen is banded, and the forceps are yellowish brown. Male forceps are long and usually toothed, and the last abdominal segment has two spines. Female forceps have small basal teeth. Eggs are deposited in earthen burrows constructed by females. Females deposit three or four batches of eggs per year in favorable conditions. Females groom the eggs to remove fungi from their surface; hatching occurs in about 7 days, and the first-stage nymphs feed on the empty eggs. Nymphs remain in the egg chamber until the second molt. Development of the immature stages is 49–60 days at 26 °C. These earwigs are insect predators, but also feed on other material, although not usually plants. They are attracted to lights at night. This species occurs in the southeastern USA.

## Labiidae

Species in this family have antennae with 10–16 segments, the males do not have strongly curved forceps.

**Small earwig, *Labia minor*** Adults are 4–7 mm long, dull black, and covered with fine yellow setae; the antennae are pale. It occurs in North America, the UK, and Europe. Adults are commonly found at lights at night.

# Bibliography

## COLLEMBOLA

The published information on Collembola is extensive, and, like the pest species, it is representative for all regions of the world. The bibliography provided by Salmon (1964–65) is helpful.

Salmon, J. T. An index to the Collembola. *Bull. R. Soc. N. Z.* **1** (1964), 1–144 (1964); **2** (1964), 145–644; **3** (1965), 645–651.

### General

Arnaud, P. H. and T. W. Davies. *Entomobrya kanaba* (Wray) (Collembola: Entomobryidae): an indoor household pest in central California. *Pan-Pacific Entomol.*, **56** (1980), 155–6.

Bellinger, P. F. A new family of Collembola (Arthropoda, Tracheata). *Carib. J. Sci.*, **21** (1985), 117–23.

Betsch, J.-M. Elements pour une monographie des Collemboles Symphypleones (Hexapodes, Apterygotes). *Mem. Mus. Nat. Hist. Natur., Serie A*, **116** (1980), 1–227.

Bowden, J., I. H. Haines, and D. Mercer. Climbing Collembola. *Pedobiologia*, **16** ( 1976), 298–312.

Brackenberry, J. and H. Hunt. Jumping in springtails: mechanism and dynamics. *J. Zool.*, **229** (1993), 217 36.

Brummer-Korvenkontio, M. and L. Brummer-Korvenkontio. Springtails (Collembola) on and in snow. *Mem. Soc. Pr. Fauna Flora Fennica*, **56** (1980), 91–4.

Butcher, J. W., R. Snider, and R. J. Snider. Bioecology of edaphic Collembola and Acarina. *Annu. Rev. Entomol.*, **16** (1971), 249–88.

Christiansen, K. Bionomics of Collembola. *Annu. Rev. Entomol.*, **9** (1964), 147–78.

Folsom, J. W. The economic importance of Collembola. *J. Econ. Entomol.*, **26** (1933), 934–9.

Hopkin, S. E. *Biology of the Springtails (Insecta: Collembola)*. Oxford: Oxford University Press, 1997.

Lawrence, P. N. Collembola (springtails) of sewage filters. *Waste Water Treat.*, **13** (1970), 106–109.

Maria Mutt, J. A. M. Swarming of *Entomobrya urostrigata* (Insecta: Collembola) in South Holland, Cook County, Illinois. *Trans. Ill. State Acad. Sci.*, **71** (1978), 236–7.

Pichard, S. Contribution à l'étude de la biologie de *Podura aquatica* (Linné) Collembole. *Bull. Biol.*, **108** (1973), 191–9.

Richards, W. R. Genetic classification, evolution and biogeography of the Sminthuridae of the World (Collembola). *Mem. Ent. Soc. Can.*, **53** (1968), 1–54.

Scott, D. B. The economic biology of Collembola. *J. Econ. Entomol.*, **46** (1953), 1048–51.

Scott, H. G., J. S. Wiseman, and C. J. Stojanovich. Collembola infesting man. *Ann. Entomol. Soc. Am.*, **55** (1962), 428–30.

### Geographic distribution

Christiansen, K. and P. Bellinger. *The Collembola of North America North of the Rio Grande*. Grinnell, IA: Grinnell College, 1980–81.

Gisin, H. *Collembolenfauna Europas*. Geneva: Musée d'Histore Naturelle, 1960.

Gough, H. J. A key for the identification of the families of Collembola recorded from the British Isles. *Entomol. Mon. Mag.*, **113** (1977), 193–7.

Greenslade, P. and J. E. Ireson. Collembola of the southern Australian culture steppe and urban environments: a review of their pest status and key to identification. *J. Aust. Entomol. Soc.*, **25** (1986), 273–91.

Lippert, G. and H. Butler. Taxonomic study of Collembola of West Virginia. *West Virginia Univ., Agric. Exp. Sta. Bull.*, **643T** (1976), 46.

Maria Mutt, J. A. M. Genera of Collembola (Insecta) in Puerto Rico: keys, diagnosis and general comments. *J. Agric. Univ. Puerto Rico*, **60** (1976), 113–28.

Maria Mutt, J. A. M. and P. F. Bellinger. *Catalog of Neotropical Collembola*. Gainesville, FL: University of Florida, 1989.

Maynard, E. A. *A Monograph of the Collembola or Springtail Insects of New York State*. Ithaca, NY: Comstock Press, 1951.

Mills, H. B. *A Monograph of the Collembola of Iowa*. Iowa State College, monograph no. 3. Ames, IA: Iowa State College, 1934.

Paclt, J. On South and Central African Collembola. *J. Entomol. Soc. South Africa*, **29** ( 1967), 135–47.

Palcios, J. G. Diagnosis y clave para determinar las familias de los Collembola de la región Neotropical. *Man. Guias Est. Microartr. (UNAM, Mexico)*, **1** (1990), 1–15.

Rapport, E. H. The geographical distribution of Neotropical and Antarctic Collembola. *Pacific Inst. Monogr.*, **25** (1971), 99–118.

Scott, H. G. Collembola from Pennsylvania. *Ent. News*, **70** (1959), 81–3.

Collembola: pictorial keys to the Nearctic genera. *Ann. Entomol. Soc. Am.*, **54** (1961), 104–13.

Snider, R. J. An annotated list of the Collembola (springtails) of Michigan. *Michigan Entomol.*, **1** (1967), 179–234.

Sterzynska, M. Collembola from urban areas (Warsaw, Poland). In Dallai, R. (ed.) *Third International Seminar on Apterygota*. Siena: University of Siena, 1989.

Uchida, H. Tentative key to the Japanese genera of Collembola, in relation to the world genera of this order. *Sci. Rep. Hirosaki Univ.*, **18** (1971), 64–76.

Tentative key to the Japanese genera of Collembola, in relation to the world genera of this order. *Sci. Rep. Hirosaki Univ.*, **19** (1972), 19–42.

Tentative key to the Japanese genera of Collembola, in relation to the world genera of this order. *Sci. Rep. Hirosaki Univ.*, **19** (1972), 79–114.

## DERMAPTERA

### General

Ashford, R. W. 1970. Observations on the biology of *Hemimerus talpoides* (Insecta: Dermaptera). *J. Zool. Lond.*, **162** (1970), 413–18.

Behura, B. K. The biology of the European earwig, *Forficula auricularia* Linn. *Ann. Zool.*, **1** (1956), 117–42.

Bharadwaj, R. K. Observations on the bionomics of *Euroborellia annulipes*. *Ann. Entomol. Soc. Am.*, **59** (1966), 441–50.

Bishop, F. C. Injury to man by earwigs (Dermaptera). *Proc. Entomol. Soc. Wash.*, **63** (1961), 114.

Brindle, A. Earwigs attracted to light. *Entomol. Rec. J. Var.*, **83** (1971), 149–52.

Buxton, J. H. and D. S. Madge. Artificial incubation of eggs of the common earwig, *Forficula auricularia* L. *Entomol. Mon. Mag.* **110** (1974), 55–7.

Caussanel, C. Principales exigences cophysiologiques du forficule de sables, *Labidura riparia* (Derm. Labiduridae). *Ann. Soc. Entomol. France*, **6** (1970), 589–612.

Chopard, L. *La Biologie des Orthoptères*. Paris: Lechevalier, 1938.

Davies, R. G. The postembryonic development of *Hemimerus vicinus* Rehn and Rhen (Dermaptera: Hemimeridae). *Proc. R. Entomol. Soc. Lond. (A)*, **41** (1966), 67–77.

Eisner, T. Defense mechanisms of arthropods. II. The chemical and mechanical weapons of an earwig. *Psyche*, **67** (1960), 62–70.

Essig, E. O. A small insect that stings severely. *Science*, **75** (1942), 242–3.

Evans, K. A. and V. Longépé. The European earwig: getting the best of both worlds ? In Wildey, K. B. (ed.) *The 2nd International Conference on Insect Pests in the Urban Environment*, pp. 163–7. Edinburgh, Scotland: Heriot-Watt University, 1996.

Fulton, B. B. Some habits of earwigs. *Ann. Entomol. Soc. Am.*, **17** (1924), 357–67.

Gross, H. R. Jr. and W. T. Spink. Flight habits of the striped earwig, *Labidura riparia*. *Ann. Entomol. Soc. Am.*, **64** (1971), 746–8.

Guillet S., N. Josselin, and M. Vancassel. Multiple introductions of the *Forficula auricularia* species complex (Dermaptera: Forficulidae) in eastern North America. *Can. Entomol.*, **132** (2000), 49–58.

Klostermeyer, E. C. The life history and habits of the ring-legged earwig, *Euborellia annulipes* (Lucas) (order Dermaptera). *J. Kans. Entomol. Soc.*, **15** (1942), 13–18.

Knabke, J. J. and A. A. Grigarick. Biology of the African earwig, *Euborellia cincticollis* (Gerstaecker) in California and comparative notes on *Euborellia annulipes* (Lucas). *Hilgardia*, **41** (1971), 157–94.

Lamb, R. J. Effects of dispersion, travel, and environmental heterogeneity on populations of the earwig, *Forficula auricularia* L. *Can. J. Zool.*, **53** ( 1975), 1855–67.

Parental behavior in the Dermaptera with special reference to *Forficula auricularia* (Dermaptera: Forficulidae). *Can. Entomol.*, **108** (1976), 609–19.

Lamb, R. J. and W. G. Wellington. Life history and population characteristics of the European earwig, *Forficula auricularia* (Dermaptera: Forficulidae), at Vancouver, B. C. *Can. Entomol.*, **107** (1975), 819–24.

Morgan, W. P. Notes on the functions of the forceps of earwigs. *Proc. Indiana Acad. Sci.*, **33** (1923), 303–6.

Further observations on the function of the earwig forceps. *Proc. Indiana Acad. Sci.*, **34** (1924), 347–8.

Nutting, W. L. Notes on the occurrence of four adventive earwigs in Arizona. *Pan-Pac. Entomol.*, **36** (1960), 202–4.

Popham, E. G. The anatomy in relation to feeding habits of *Forficula auricularia* L. and other Dermaptera. *Proc. Zool. Soc. Lond.*, **133** (1959), 251–300.

Saupanor, B. Une phéromone d'agregation chez *Forficula auricularia*. *Entomol. Exp. Appl.*, **62** ( 1992), 285–91.

Schlinger, E. I., R. van den Bosch, and E. J. Dietrick. Biological notes on the predaceous earwig *Labidura riparia* (Pallas), a recent immigrant to California (Dermaptera: Labiduridae). *J. Econ. Entomol.*, **52** (1959), 247–9.

Shepard, M., V. Waddill, and W. Kloft. Biology of the predacious earwig, *Labidura riparia* (Dermaptera: Labiduridae). *Ann. Entomol. Soc. Am.*, **66** (1973), 837–41.


### Geographic distribution


Brindle, A. A revision of the subfamily Labidurinae (Dermaptera: Labiduridae). *Ann. Mag. Nat. Hist.*, **9** ( 1966), 239–69.

The Dermaptera of the Caribbean. *Stud. Fauna Curacao Caribb. Islands*, 1971.

Harz, K. and A. Kaltenbach. *The Orthoptera of Europe*, vol. III. The Hague: W. Junk, 1976.

Hebard, M. The Dermaptera and Orthoptera of Illinois. *Ill. Nat. Hist. Surv. Bull.*, **20** (1934), 125–279.

The Dermaptera and orthopterous families Blattidae, Mantidae, and Phasmidae of Texas. *Trans. Am. Entomol. Soc.*, **68** (1942), 239–319.

Hincks, W. D. Dermaptera. The earwigs of South Africa. *South Afr. Animal Life*, **4** (1957), 33–94.

Hoffman, K. M. Earwigs (Dermaptera) of South Carolina, with a key to the eastern North America species and a checklist of North American fauna. *Proc. Entomol. Soc. Wash.*, **89** (1987), 1–14.

Langston, R. L. and J. A. Powell. The earwigs of California. *Bull. Calif. Insect Surv.*, **20** (1975), 1–25.

Lucas, W. J. Notes on British Orthoptera (including Dermaptera) in 1924. *Entomologist*, **58** (1925), 81–6.

Marshall, J. A. and E. C. M. Haes. *Grasshoppers and Allied Insects of Great Britain and Ireland*. Essex: Harley Books, 1988.

Reichardt, H. Catalogue of New World Dermaptera (Insecta). Part I: introduction and Pygidicranoidea. *Papeis Avulsos Zool. S. Paulo*, **21** (1968), 183–93; Part II: Labioidea, Carciniphoridae. *Papeis Avulsos Zool. S. Paulo*, **22** (1970), 35–46; Part III: Labioidea, Labiidae. *Papeis Avulsos Zool. S. Paulo*, **23** (1971), 83–109.

Saki, S. *Dermapterorum Catalogus Preliminaris*. Parts I–VIII. Tokyo: Daito Bunka University, 1970–73.

Steinmann, H. Dermaptera: Catadermaptera I. *Tierreich*, **102** (1986), 1–345.

*World Catalog of Dermaptera*. Dordrecht: Kluwer Academic, 1989.

Vickery, V. R. and D. K. McE. Kevan. The grasshoppers, crickets, and related insects of Canada and adjacent regions. Ulonata: Dermaptera, Cheleutoptera, Notoptera, Dictuoptera, Grylloptera, and Orthoptera. *Insects Arachnida Can.*, **14** (1986), 1–918.

# 7 Diptera, Ephemeroptera

## DIPTERA
### Introduction

Flies are one of the most recognized pests in the urban environment. The fore wings of dipterans are developed, but the hind wings are reduced to small knobbed structures called halteres. Adult flies are active during the day, sometimes at dawn or dusk, and are usually attracted to the odors from sites suitable for larval feeding and development. The larvae are known as maggots, and although this is the primary feeding stage, it is often in a different habitat from the one visited by adults.

Domestic and peridomestic habitats attract fly species that feed as adults or larvae on decaying organic material. Many of these species came from populations in natural habitats separate from the urban environment. Food storage indoors often results in small amounts of ripe and decaying substrate that attract adult flies. Female fungus gnats and fruit flies can detect alcohols, acetic acid, and other volatile compounds from these materials, and follow the odor to an oviposition site. Dead and decaying organic matter is quickly identified as an oviposition site by adult scatopsides, sepsides, and phorids. Adults of several fly species are found indoors in the fall and winter. These include overwintering cluster flies and face flies that spend the winter in attics, wall voids, and other rooms. Chlorpoids and other small flies that occur in large numbers at windows in the fall do not overwinter, but are gathered in large numbers by prevailing winds and are carried in through open windows. Female mosquitoes use building basements and other protected sites to spend the winter, emerging sometimes to take a blood meal.

Peridomestic habitats and their organic substrates attract flies from populations in natural or undisturbed areas. Primitive families, such as the midges, chironomids, and bibionids, utilize wet organic matter as a breeding habitat, and sites with this substrate can be found scattered in urban environments. A large number of mosquitoes have adapted to breeding in standing water collected by domestic containers outside urban buildings. The original habitats for some species of flies have disappeared, or they have been abandoned in favor of artificial sites. Muscids, and blue and green bottle flies utilize the predictable presence of decaying garbage, carrion, and feces from domestic dogs and cats in urban environments.

Mouthparts of the adults are variable. Most species have sucking mouthparts; some blood-feeding groups have what can be considered as a piercing–sucking or piercing–lapping arrangement. In others the adults feed by sponging and lapping liquids from surfaces. In some of the primitive and advanced flies the mouthparts are nonfunctional, and the adults may live a short time. The immatures are the primary feeding stage for most species and usually have well-developed chewing mouthparts.

Eggs are usually deposited directly on the food source of the immature stages; the females of some species deposit live larvae. Hatching usually occurs in 1–3 days, and larval development is completed in 1–3 weeks. Larvae generally utilize a temporary food source, and complete development while conditions remain suitable. Larvae are generally legless and in most groups they are somewhat wormlike. The immature stages of some of the aquatic families, such as mosquitoes and blackflies, are unusual. In the primitive families of flies the larval head is well developed, usually sclerotized, and the mandibles move laterally. In the higher flies the head of the maggot is not sclerotized and indistinct, and the mandibles or mouthhooks are internal and move vertically. The mouthhooks in higher flies are accompanied by large muscles and they can be used effectively to scrape food into the mouth, and in some species they aid in locomotion, on a substrate or through the air.

There are usually four larval instars, but in higher flies the last instar is indistinct and completed within the puparium. The pupal stage of the primitive flies is exposed, but in the higher flies the pupal stage is passed in the last larval skin, the puparium. Full-grown larvae usually move a short distance away from their food substrate to molt and form the puparium, a process called pupariation (but inaccurately called pupation). The fourth instar is nonfeeding and within about 24 h molts to the pupal stage. The adult emerges from the puparium through the anterior end.

Pest status may be restricted to one stage or shared by both the adult and maggots. Mosquitoes, black flies, and deer flies are pests by their feeding habits, and their ability to transmit disease organisms. Other species are pests merely by their presence indoors or outdoors. Species in domestic habitats are associated with decaying plant or animal material, and the larvae are saprophagous. Pest status of a few species is based on their presence in large numbers, such as overwintering cluster flies, face flies, or lovebugs swarming over roads and around buildings. Species inhabiting peridomestic habitats also achieve pest status by large numbers, as a nuisance or because of the blood-feeding habits of females.

## Anisopodidae

These small flies are gray to brown and resemble small crane flies (Tipulidae). Adults are 6–10 mm long, the head has ocelli (except for *Anisopus*), and the antennae are nearly as long as the thorax. The thorax is without a V-shaped mesonotal suture. Legs are long, and the wing membrane is light brown and has scattered light and dark areas. Larvae are pale yellow or mottled orange, and elongate; each segment has a narrow constriction forming a circular swelling at the anterior end. Larvae of most species feed in decaying vegetation; larvae of *Anisopus* occur in dung and sewage. Adult anisopodids are most common in spring in areas where there is decaying organic matter.

### Window gnat, *Sylvicola fenestralis* (= *Anisops*) (Fig. 7.1a)

Adults are about 6 mm long, their wings are dark brown with scattered pale spots, and the antennae are longer than the head. Full-grown larvae are yellowish brown and 12–15 mm long yellowish white and nearly cylindrical; the head is small and with dorsal eye-spots. Each body segment has a narrow constriction at the anterior end, which forms an enlarged ring; abdominal segment 10 terminates in five short tubercles. Adults are often found indoors at windows; outdoors they occur in small swarms late in the day or at dusk. Larvae infest the biological film of the filter bed in sewage treatment plants. Indoors they feed on rotting fruit, vegetables, stored tubers, and fermenting material such as cider and wine. Eggs are laid in batches of about 150, which form a gray mass on the surface of the substrate. Hatching occurs in about 4 days. Development from egg to adult takes about 35 days at 20 °C; the egg stage lasts 4 days, the larval stage lasts 20 days, and the pupal period is 8 days. At 10.5 °C development is completed in about 88 days, and 50 days are spent in the larval stage. The full-grown larva usually moves away from the breeding substrate to a dry location to pupate. Adults live about 7 days. This species occurs widely in the USA and Europe.

## Asilidae

Robber flies are characterized by abundant and long setae on the head and legs; their body is stout and the abdomen is pointed posteriorly. Some species are small and slender flies. Adults are abundant in many habitats where they are predators of flying insects, including wasps, bees, dragonflies, grasshoppers, and other flies. The larvae of most species live in soil and prey on the larvae of insects. Adults occur in peridomestic habitats, and may be found capturing insects at lights at night.

**_Hyperechia_ spp.** Adults are about 18 mm long and the body is black with a distinct pale white to yellowish white band on the posterior margin of the thorax. The head has long, yellowish white setae on the face; legs are black and have long setae. These flies resemble carpenter bees (*Xylocopa*) with their body coloration and long setae. Adult *Hyperechia* prey on *Xylocopa* species, and they are often found resting near the entry holes of carpenter bees. These species are distributed in Africa.

## Athericeridae

These flies are separated from the Ragionidae by the lack of apical setae or spurs on the front tibiae, and a rounded antennal segment 3. The wings of most species are banded, such as in *Spaniopsis*, while the wings of rhagionids (*Symphoromyia* spp.) are not, but may have dark spots at the front margin. Three genera contain species that suck blood: *Atherix* and *Suragina* in North America, Europe, and Japan; *Suragina* in Asia and South America; and *Spaniopsis* in Mexico and Australia. The immature stages are aquatic and breeding sites are usually small streams in undisturbed areas. Adult *Atherix* are gregarious and occur in large numbers to lay eggs in masses on the branches and leaves of trees overhanging streams. *Atherix* species have been recorded biting people in Japan.

## Bibionidae

These flies are 6–15 mm long and characterized by an enlarged abdomen and short legs; the antennae are slightly longer than

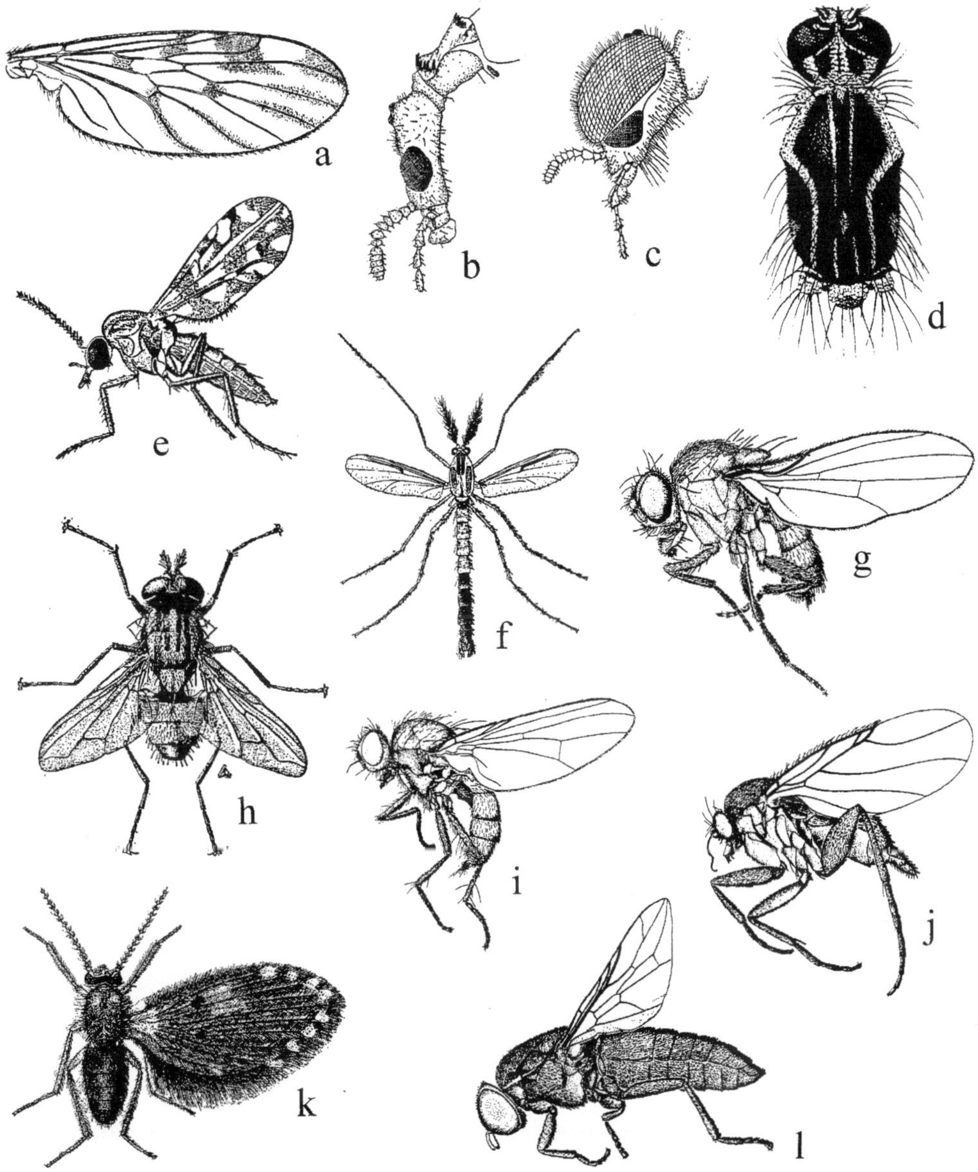


**Figure 7.1** Diptera. (a) *Sylvicola fenestralis* wing (Anisopodidae); (b) *Dilophus febrilis*, female head (Bibionidae); (c) *D. febrilis*, male head; (d) *Aedes aegypti*, female head and thorax (Culicidae); (e) *Culicoides* sp. (Ceratopogonidae); (f) *Chironomus* sp. (Chironomidae); (g) *Drosophila* sp. (Drosophilidae); (h) *Musca domestica* (Muscidae); (i) *Fannia canicularis* (Muscidae); (j) *Megaselia scalaris* (Phoridae); (k) *Clogmia albipunctatus* (Psychodidae); (l) *Scenopinus* sp. (Scenopinidae).

the head, and the eyes are usually large. The body is often covered with fine setae. Some species have color dimorphism: females are often reddish brown, while males are black. Full-grown larvae are 12–24 mm long and nearly cylindrical, and yellowish white with a dark brown head (Fig. 7.2e). There are rows of fleshy processes on the abdominal segments; the spiracles on segment 9 are large. Larvae often feed in large numbers at the roots of grasses, cereals, and in decaying vegetation. Several species are agricultural pests. The urban pest species are *Dilophus febrilis* and *Bibio marci* and *B. hortulanus*.

After mating, females dig a small cavity in the soil about 5 cm deep with their well-developed tibiae and with the tarsi folded back. A mass of 200–300 eggs is laid in the cavity and the female dies. Hatching occurs in about 30 days and the early-stage larvae often remain together in a mass while feeding. Late-stage larvae have a well-developed head and mandibulate mouthparts. They are 12-segmented and each of the nine abdominal segments has a number of conical processes. There are 10 pairs of spiracles; the spiracles on segment 12 have three openings in *Dilophus* and two openings in *Bibio*. Full-grown larvae form chambers in the soil in which to pupate. The pupal period for

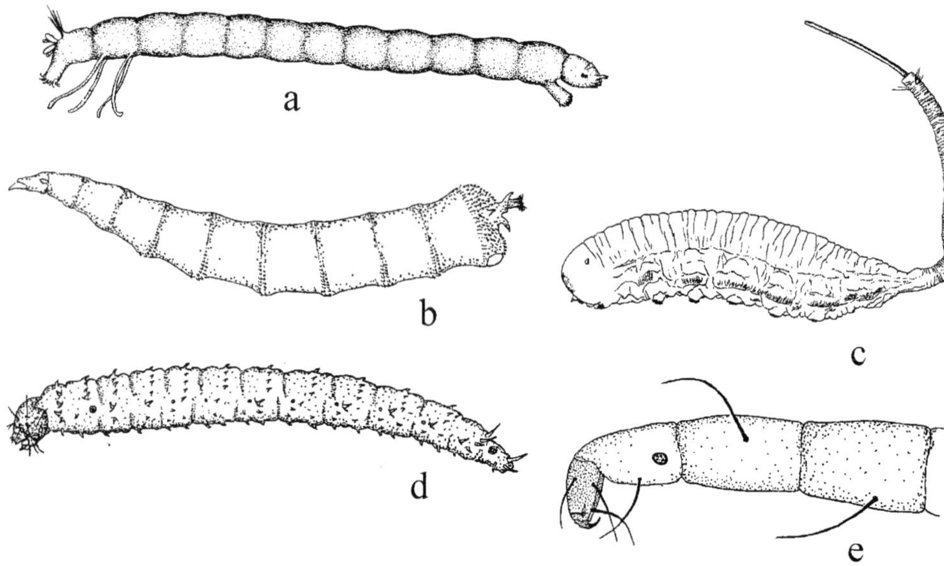


**Figure 7.2** Diptera larvae. (a) *Chironomus* sp. (Chironomidae); (b) *Sepsis* sp. (Sepsidae); (c) *Eristalis* sp. (Syrphidae); (d) *Scenopinus* sp., lateral view of the head and anterior segments (Scenopinidae); (e) *Bibio* sp. (Bibionidae).

most species lasts about 3 weeks and the adults dig their way out of the soil cavity to the surface.

The March fly is *Bibio albipennis* in North America, and the St. Mark's fly is *B. marci* in Europe. Both species emerge in large numbers for a few weeks during March and April (April 25 is St. Mark's day); they sometimes create a nuisance around buildings and along roadways. Other bibionids occur in spring, including *B. vestitus*, *Dilophus febrilis*, *D. occipitalis*, *D. ornatus*, and the lovebug, or telephone bug, *Plecia nearctica*. Several of these species have more than one generation per year, and emerge in large numbers in the fall.

**March fly, *Bibio albipennis*** Adults are 6–15 mm long; the body is reddish brown and covered with fine setae. The head and eyes are large, and the upper facets of the eyes are enlarged in both sexes. Antennae are six- to eight-segmented; the wings have a black spot near the middle of the anterior margin. Adults are weak flyers and are often found on flowers taking nectar. Larvae are cylindrical, legless, and with strong chewing mouthparts; they are often gregarious in moist soil. Eggs are laid on moist or wet vegetation, or directly on the soil surface. Females deposit 200–300 eggs in the few days they are alive; hatching occurs in 2–4 days. Development takes about 14 days; the pupal period is about 10 days. Soil temperature may regulate development time. Adults emerge in the morning and, with increased air temperature, they fly and mate. Swarms consist of males and females seeking mates; mating adults leave the swarm and remain in copula for several hours. This species occurs in the USA.

**St. Mark's fly, *Bibio marci*** Adults are 10–12 mm long, with a wing length 8–12 mm. The body is reddish brown to black and covered with fine setae; it is exceptionally long in the male. The thorax is shiny; wings of the male are translucent, while the wings of the female are opaque, both with a darkened costal vein. There is a single large spine on each front tibia. This is the largest bibionid in the UK; adults emerge and swarm in March and April.

**Blossom fly, *Bibio nervosus*** Adults are 6–12 mm long and black to reddish black. There are numerous hairs or fine setae on the body, and a black spot near the middle of the anterior edge of the wing. This fly is common on blossoms of fruit trees in the spring in western USA. Large numbers occur on trees around buildings and houses.

**Fever fly, *Dilophus febrilis* (Fig. 7.1b, c)** Adults are about 6 mm long and have a black body and legs, and dark wings. The thorax is shiny, and it is not densely setose in the male. The fore tibia has a single spine placed distant from the main group, and a circle of spines at the tip, and a double ring of spines at the front of the thorax. The wings of the male are translucent and with a distinct stigma; female wings are opaque with pale tips. The male pupa has three pointed processes anteriorly on the head, and a pair of processes at the anterior end; the female pupa has one process anteriorly. Adults have been recorded in large swarms on foliage in residential areas, and they often enter buildings. This species is common throughout the warm months, and there are several generations per year. This species occurs in the UK and perhaps regions of continental Europe. A closely related species, *D. femoratus*, has 11-segmented antennae and the wings of male and female are translucent.

**Lovebug, *Plecia nearctica*** Adults are 12–15 long; males are black with black wings, females are black to reddish black, and have dark-gray to black wings. Mating adults remain coupled in flight and this behavior is probably the basis of their common name, lovebug. They emerge in large numbers in April–May, August–September, and sometimes for a brief period in December. Their presence often creates problems with automobile travel and outdoor activities. Adults may be somewhat attracted to automobile exhaust fumes, and gather on or near busy highways. Eggs and body fluids from adult females crushed against car windows are a nuisance and can impair vision. In September 1969, an emergence of *P. nearctica* in Florida was estimated to extend over 25% of the land area of the state; some adult flies from this emergence were carried by the wind to altitudes of 300–400 m.

Eggs are laid on soil or in wet and decaying vegetation. Females deposit about 300 eggs during their lifetime; hatching is in 2–4 days. Development takes about 14 days, and the pupal period is about 7 days. The full-grown larva usually moves away from the breeding substrate to a dry location to pupate. First-generation adults emerge in April and May and the second-generation adults emerge in August and September. The flight period is about 4 weeks, depending on weather conditions. Mating occurs in flight and continues until the male dies or disengages. Males live 2–3 days and females live 7–10 days. *P. nearctica* was first collected in Florida in 1949. This species is common in all states bordering the Gulf of Mexico, including Georgia and South Carolina, and parts of Central America.

## Calliphoridae

Blow fly adults are 8–10 mm long, and generally have a metallic blue, green, coppery green, or greenish-black sheen to the thorax and abdomen. An exception is the cluster fly, *Pollenia rudis*, which is grayish black. In general, the flies that are metallic blue are *Calliphora*, and those that are metallic green are *Lucilia*. Full-grown larvae are 10–18 mm long and yellowish white; there are complete bands of small spines around the body on most segments. Larval anterior spiracles have 6–12 lobes; posterior spiracles are located dorsally, but in some species the spiracles are in a slight caudal depression. The posterior segment of the larva is usually surrounded with cone-shaped tubercles, and the posterior spiracles have a complete peritreme, and three slits. Larvae usually feed on dead animals or garbage containing meat; some are scavengers on excrement and plant material, while others are parasites. Female calliphorids can locate a suitable substrate for oviposition in almost any habitat.

Oviposition begins 3–7 days after emergence. At 24 °C the preoviposition period ranges from 4–5 days for *Calliphora* to 6–7 days for *Phormia*. Eggs are usually deposited in large batches with as many as 200 eggs. Females are very efficient in locating suitable breeding sites, and if many adults are present there may be an excessive number of eggs deposited in one location. Adults rarely enter buildings to oviposit, but they will enter through a door or window when attracted to a breeding substrate; they will oviposit at night. Hatching occurs in 12–16 h. Females retain the eggs for long periods when suitable breeding sites are not available; when laid, these eggs hatch in a short time. *C. vicina* and *C. vomitoria* will deposit living, newly hatched larvae. Maggots generally avoid light and burrow into the substrate to feed. Full-grown larvae cease feeding and migrate to find a site to pupariate; they prefer loose soil and will travel 25–30 m to find a suitable location. The pupal period is 5–9 days.

Pest status is based primarily on their presence around people and food. They are strong flyers, and will travel far from their breeding site; these flies are strongly attracted to organic odors. Adults are most active during warm and sunny weather. Some species freely enter buildings and become a nuisance by flying at lights and windows. Adults indoors are a threat to food sanitation. *Lucilia* species are common in and around houses, and adults can enter openings as small as 3.2 mm. Blow fly larvae will develop in carcasses of dead birds or rodents in attics and wall voids. Small animals can support a large number of larvae, and within a few weeks a large number of adults will emerge indoors. There are usually several generations per year.

Calliphorids are commonly used in maggot therapy, which is the treatment of wounds with live fly larvae. Blow fly species used for this type of medical treatment have some important characteristics. Their larvae feed only on necrotic or damaged tissue and not on sound, healthy tissue; larvae usually remain at the wound site and do not crawl to search for other sites; they develop rapidly; they are relatively easy to rear *in vitro*, and the eggs are easily sterilized. The most widely used calliphorid in wound therapy is *Lucilia* (= *Phaenicia*) *sericata*. This species feeds only on necrotic tissue and is ideally suited for use in removing dead tissue from wounds. Other calliphorids used by physicians for treating wounds include *Calliphora vicina*, *Chrysomya rufifacies*, *Lucilia cuprina*, *L. illustris*, *Phormia regina*, and *Protophormia terraenovae*. Early physicians learned that the larvae often found in wounds improved healing and reduced scarring; this type of therapy has been used for centuries. Placing maggots in wounds has been used in societies around the world, including aboriginal tribes in Australia, Hill Peoples of Northern Burma, and probably the Mayans of Central America. The founder of modern maggot therapy is William Baer (1872–1931), professor of orthopedic surgery at Johns Hopkins School of Medicine, Baltimore, Maryland, USA. Following his experience with maggots in the wounds of soldiers, he designed a therapy of using sterilized maggots to treat a variety of wounds. This method is successful in treating chronic or acutely infected wounds, including bone infections, abscesses, carbuncles, and ulcers.

**Congo floor-maggot, *Auchmeromyia senegalensis*** Adults are 9–13 mm long and the mesonotum has two longitudinal black lines. The abdomen is primarily black apically; abdominal segment 2 in the male is 1.5 times longer than segment 3, and segment 2 in the female is twice as long as segment 3. Full-grown larvae are about 14 mm long and yellowish white in the first stage, but red after feeding; the body has many folds in each of the segments. The posterior segment ends in a sharp angle, and dorsally on this segment are the posterior spiracles. Adult flies remain in shaded areas near human habitations, and they feed on discarded fruits and vegetables. Eggs are laid in batches of about 50 in dry soil in shade, in the nests of burrowing animals, or indoors in the soil floor of dwellings. Larvae remain in cracks and crevices in the soil, but move out at night to suck blood from people sleeping directly on the floor. The bite from the maggots is not painful and larvae are able to feed without being disturbed. Maggots are not capable of climbing to reach individuals sleeping in cots above the floor. Development takes about 14 days under suitable conditions, but can be extended for several months if the larvae do not find food. This species occurs in Africa.

***Calliphora uralensis*** Adults are 10–14 mm long. They are metallic blue with pale setae forming stripes on the thorax, and with yellowish-white spots on the abdomen. Eggs are usually laid on excrement, but also on meat and fish. More than 200 eggs may be deposited at one time; hatching occurs in about 20 h. Development is completed in 10–14 days, and the pupal period lasts 7–10 days. Adults live for about 3 months, and overwintering is in the pupal stage. This is a common fly in houses and food-handling establishments in the eastern regions of Europe and Russia during summer; they are uncommon in winter.

**Bluebottle fly, *Calliphora vicina* (= *C. erythrocephala*)** Adults are about 10 mm long and have a bluish-black thorax, and a metallic-blue abdomen. The region of the head below the eyes is orange; in the closely related species, *C. vomitoria*, it is black. Eggs are laid in batches of up to 180 directly on the larval substrate; fecundity is 500–700. Hatching is in about 88 h at 10 °C, 38 h at 12.5 °C, 19 h at 19 °C, and 14 h at 25 °C. First-instar larvae develop in about 49 h at 12.5 °C, 22 h at 19 °C, and 18 h at 25 °C; second-instar larvae develop in about 58 h at 12.5 °C, 23 h at 19 °C, and 19 h at 25 °C; third-instar larvae develop in about 65 h at 12.5–19 °C, and 26 h at 25 °C. There may be 5–14 days postfeeding in search of a pupariation site. The pupal period lasts about 41 days at 10 °C, 28 days at 12.5 °C, 14 days at 19 °C, and 11 days at 25 °C. Puparia are formed in the soil and adults can emerge from puparia buried 40 cm in sandy soil. Overwintering is by adults, larvae, and pupae. Adults live about 30 days, but 188 days has been reported. Females oviposit on fresh, decaying, or cooked meat, and on human excrement. Adults appear in early spring, but they are uncommon in summer; there may be a peak of adults in the fall. These are slow-flying and loud-buzzing flies, and they often enter houses. This species is widely distributed in North America, Mexico, Europe, and in Asia, and Australia.

**Blow fly, *Calliphora vomitoria*** Adults are about 10 mm long and they have a bluish-black thorax and a metallic-blue abdomen. The region of the head below the eyes is black, but in *C. vicina* it is orange. Eggs are laid on the larval food and they hatch in 6–48 h; larval development is complete in about 15 days. Third-stage larvae move to a dry substrate to pupate; the pupal period is about 11 days. Adults live about 35 days. This species is widely distributed in North America and Europe.

**Oriental latrine fly, *Chrysomya megacephyla*** Adults are about 11 mm long, and the thorax and abdomen are greenish blue with purple reflections. The first abdominal segment is black. Full-grown larvae are about 12 mm long, smooth, and yellowish white. Larvae feed on decomposing animal matter, and excrement, including human urine. Eggs are laid in a mass on or in a suitable substrate; fecundity is about 400 eggs; hatching occurs in about 10 h. Development takes about 4 days, and the pupal period lasts about 4 days. Adults live 54–90 days at 25–29 °C and 75% relative humidity (RH). This calliphorid occurs from the Middle East to southern and eastern Asia,

Australia, and New Zealand. It is a common pest in houses and outdoor markets, where adults come to meats and various household foods.

**Hairy maggot blow fly, *Chrysomya rufifacies*** Adults are 10–12 mm long and the thorax is uniformly green or violet blue; the parafacial and facial bristles are reddish brown. Full-grown larvae are about 14 mm long and brownish yellowish; the body has a median row of fleshy tubercles on each segment, which gives the maggot a somewhat hairy appearance. Small spines are present on some of these projections, especially those dorsally. In *C. albiceps*, a related species, the stalks of the projections have no spines. The peritreme of the posterior spiracle is very wide and the edges of the gap are forked; the slits are short and nearly fill the spiracular plate. This calliphorid produces unisexual progeny, which is an unusual feature among higher flies. *C. rufifacies* has a temperature tolerance that corresponds with its tropical origin and general distribution. Eggs are laid singly or in batches; fecundity is 210–368 eggs. Eggs fail to hatch at 9 °C, but at 15 °C eggs hatch and larvae develop but fail to pupate. Larvae grow and develop normally at 40 °C. Adults live 23–30 days. The temperature threshold for adult flight activity is 13 °C; this species has a flight range of 0.5–1.6 km/day (Australia). *C. rufifacies* is not a primary invader of tissue, and feeds as a scavenger or secondary carrion fly. It is native to the tropics but has spread around the world with commerce. It is known from India, Japan, Hawaii, Australia, Central and South America (Argentina), and North America (Arizona, Texas, Florida). It is common in some urban areas in India, where it occasionally enters houses; it also enters houses in Australia, but less so in Japan. In some regions, it causes primary and secondary myiasis in animals. It has a role in forensic entomology. It has been recovered from human cadavers in Costa Rica, California, Arizona, Texas, and Florida. This species is closely related to *C. albiceps*, which is distributed in Africa, and they may be conspecific.

**Tumbu fly, *Cordylobia anthropophaga*** Adults are 9–12 mm long and dull yellowish white to light brown, and with two dark gray dorsal longitudinal thoracic stripes. Wings are light brown. The four visible abdominal segments are about equal-sized. Full-grown larvae are 11–15 mm long, yellowish white, and covered with dark spines. Eggs are laid in batches on dry soil in shady sites, especially sites contaminated with urine or feces of humans and other animals (rodent, dog, monkey); oviposition sites are usually indoors, in the earthen floor of huts and other dwellings. Eggs are also laid on drying laundry hung out of direct sunlight; infestations develop if these eggs are not killed. Females lay eggs in batches of 200–300, and fecundity is about 500 eggs. Hatching occurs in 1–3 days, and first-stage larvae can remain alive for 9–15 days without food. Larvae usually remain concealed just below the soil surface during the day. Movement or body heat on the soil activates the larvae. They move at night and attack people or small animals sleeping on the earthen floor. They use their mouthhooks to attach directly to a host, or to clothing first then move to the host; larvae usually penetrate the skin to the depth of the posterior spiracles. Development is completed in 8–9 days, and the full-grown larva leaves the boil-like swelling on the skin and drops to the soil to form a puparium. Adult flies emerge in 8–15 days. Rats may be the normal host and probably serve as the natural reservoir for this fly. Dogs are also infested, and severe infestations cause death of the animal. This species occurs in Africa. The related species, *Cordylobia rohaini* (Lund's fly), closely resembles the Tumbu fly in appearance and life history, but attacks humans less frequently. It occurs in tropical Africa, especially in areas of rain forest.

***Cynomyopis cadaverina*** Adults are 9–14 mm long, the thorax is metallic bluish black, and the abdomen is shiny bluish green. The seasonal peaks in populations of this fly are in early spring and late fall, but adults occasionally appear during warm spells in winter. Adults enter houses readily. Larvae breed in carrion, excrement, and other decaying organic material. Eggs are deposited in batches of 25–50, and hatching is in about 24 h. Development is complete in about 6 days, and the pupal period lasts about 4 days. Overwintering is in the pupal or adult stage. It is found throughout North American, Mexico, and northern Europe.

***Lucilia illustris*** Adults are 5–10 mm long, and the thorax is bluish green with bronze and purple reflections. Full-grown larvae are 11–13 mm long and yellowish white; the body is somewhat peg-shaped and with dark bands of spines encircling segments 2–9, but the bands may be incomplete dorsally. Prothoracic spiracles have 6–8 lobes; the posterior spiracles have a complete peritreme. It generally occurs outdoors, but will come indoors under adverse weather conditions. Females oviposit on dead animals, and feces (animals and human). Larvae feed on these substrates, and on the wool and flesh of sheep. This species is often found with *L. sericata* on carcasses of animals. Larvae have been used in wound therapy. This species occurs in North America and Europe.

**Greenbottle fly, *Lucilia sericata* (= *Phaenicia*)** Adults are about 12 mm long and metallic green, with yellowish or coppery reflections. Full-grown larvae are about 14 mm long and may be colored slightly purple. Eggs are often deposited in large numbers in one location; fecundity is about 250 eggs. Hatching occurs in about 24 h; it is about 23 h at 22 °C and 18 h at 29 °C. Larval development is completed in 4–5 days at 24 °C. First instar develops in about 27 h at 22 °C and 16 h at 29 °C; second instar develops in about 22 h at 22 °C and 16 h at 29 °C; the third instar takes about 22 h at 22–29 °C. There may be 4.5–9 days postfeeding in search of a pupariation site. The pupal period is about 143 h at 22 °C and 130 h at 29 °C. The common breeding medium is carrion, but also includes decaying garbage and manure. Larvae generally confine their feeding to dead tissue; this species is commonly used in modern wound therapy. This species is widely distributed, and probably cosmopolitan.

***Phaenicia cuprina*** Adults are 6–8 mm long and the thorax is metallic green. Larvae are primarily scavengers and they are usually associated with wet garbage and decaying organic matter. Eggs hatch in about 24 h, and larvae feed for about 6 days before moving away and entering the ground to form the puparium. The pupal period lasts about 7 days. Adults often come indoors and they are strongly attracted to lights; they will fly to ultraviolet lights. This species is widely distributed.

**Black blow fly, *Phormia regina*** Adults are 6–11 mm long and the thorax is black with a bluish-green luster, and it has black, longitudinal stripes. The abdomen is shiny, and bluish green. Full-grown larvae are 12–17 mm long and yellowish white. Bands of small spines encircle segments 1–8, and segments 9 and 10 have a band of ventral spines. Prothoracic spiracles have 10 or 11 short lobes; the posterior spiracles are in a depression surrounded by six tubercles. The posterior spiracle peritreme appears incomplete. Eggs are often deposited in large numbers in one location; fecundity is about 250 eggs. Hatching occurs in about 24 h; it is about 20 h at 22 °C and 18 h at 29 °C. Larval development is completed in 4–5 days at 24 °C. First instar develops in about 25 h at 22 °C and 12 h at 29 °C; second instar develops in about 25 h at 22 °C and 15 h at 29 °C; the third instar takes about 25 h at 22–29 °C. There may be 5.2–9 days postfeeding in search of a pupariation site. The pupal period is about 116 h at 22 °C and 99 h at 29 °C. Eversion of the pupal respiratory horns occurs by 26.5 h at 22 °C and by 22 h at 29 °C. Third-stage larvae often pupariate on the surface of the breeding substrate, unless it is very wet or exposed to bright light. Adults often overwinter in protected locations. *P. regina* is common in early spring when temperatures are cool and is less abundant in the summer. Larvae feed on carrion and decaying plant material, and on the wool and flesh of sheep. Larvae have been used in wound therapy. This species is widely distributed in North America, Mexico, and Europe.

**Cluster fly, attic fly, *Pollenia rudis*** Adults are 4–8 mm long. They have a broad thorax covered with golden-yellowish setae; the wings overlap when at rest. This calliphorid lacks the shiny blue or green thorax, which characterizes most of the flies in this family. Large numbers of adults enter buildings in the fall, and remain there, relatively inactive, throughout the winter. In early spring adults leave overwintering harborages during warm and sunny days; mating occurs in spring. Eggs are laid singly in the soil, and hatching occurs in about 3 days. Larval stages are predaceous on earthworms. First-stage larvae seek out earthworms (*Allolobophora* spp.); larvae enter through the spermiducal opening and other pores. There is usually one fly larva per earthworm. Development is completed in 27–39 days, and the puparium is formed in the soil. There may be four generations per year in the USA. In late summer and fall, large numbers of adults gather on the sun-warmed sides of buildings, and then move through cracks and crevices to enter attic space and wall voids. Natural hibernation sites for this fly probably include animal burrows, under bark, and in cavities in trees and down logs. Earthworm populations, which may be as many as 420–500 per square meter in urban and suburban soils, ensure the larval stage of this fly an abundant food resource. Urban and suburban buildings provide the adult cluster fly suitable overwintering harborages throughout the distribution of the species. It occurs throughout North America, and from Ireland to Europe and North Africa, to Siberia and China.

**Other *Pollenia*** Adults of *P. atramentaria*, *P. dasypoda*, *P. intermedia*, *P. varia*, and *P. vespillo* are found at animal feces, but are not known to overwinter in large numbers. The Japanese cluster fly, *P. japonica*, is widely distributed in Japan. Adults of this species lay eggs in soil and larvae feed on earthworms.

***Protophormia terraenovae*** Adults are 8–12 mm long and the thorax is dark blue, the legs are black, and the abdomen is greenish blue. The front in the male is less than one-fifth the head width. Full-grown larvae are 10–16 mm long and yellowish white. This species has a northern distribution; it is usually confined to areas north of the Tropic of Cancer. It breeds in carrion and decaying organic matter, and it has been used in wound therapy. It is distributed in North America and Europe.

## Ceratopogonidae

These dark-colored and slender midges are 1–3 mm long and have dark or spotted wings (Fig. 7.1e). Females of most species suck blood of vertebrates or other insects, and their bite is painful. Males do not feed on blood. Females that attack humans are sometimes called no-see-ums or sand flies. They occur in coastal areas where they can rapidly colonize sandy areas in land development sites. During warm weather, these biting midges can occur in large numbers, and their biting behavior can affect tourism and commercial forestry operations. Larvae are aquatic or semiaquatic, and live in the mud or wet sand bordering stagnant water, brackish water, and fresh, flowing streams. Larvae are 3–5 mm long and pale yellow, or translucent. They feed on plants, and attack midge larvae and other small insects.

Most of the species that feed on warm-blooded animals belong to the genera *Culicoides*, which are generally distributed, and *Leptoconops*, which are small, black flies adapted to tropical and subtropical climates. These two genera can be distinguished on the basis of wing veins and female antennae. In *Culicoides*, the median vein is forked and connected to the anterior veins by a distinct cross-vein, the fine setae on the wing is interspersed with large setae, and the antennae are 14-segmented. In *Leptoconops*, the median vein is unforked and there is no cross-vein, there are only small setae on the wing, and female antennae are 11- to 13-segmented.

Pest status of *Culicoides* species is based on their biting habits and disease transmission. Worldwide, more than 50 arboviruses have been isolated from *Culicoides*, most within the families Bunyaviridae (20 viruses), Reoviridae (19 viruses), and Rhabdoviridae (11 viruses). Many of these viruses have been isolated from other arthropod groups, and their association with *Culicoides* may be incidental. The only significant viral pathogen of humans transmitted by *Culicoides* is Oropouche virus. *Leptoconops* are distributed in the tropics and subtropics. Although they do not spread disease, they occur in such large numbers that they are important biting pests, especially in coastal areas where they can become established in sandy areas near resorts. Their numbers and their habit of biting during the day can be a problem for tourism. *Leptoconops* bites are painful and reactions to them may last for several days.

In the UK and northern Europe there are several species that are pests, including the widely distributed species *Culicoides impunctatus*, *C. nubeculosus*, *C. obsoletus*, and *C. punctatus*. Other pest species include *C. vexans*, and *C. heliophilus*; coastal species include *C. fascipennis*, *C. reconditus*, *C. halophilus*, and *C. maritimus*. Species that have pest status in coastal districts of Italy include *Leptoconops irritans*, *L. kerteszi*, and *L. bezzii*; in the Camargue region of southern France, *C. circumscriptus* and *C. nubeculosus*, and *C. maritimus* are pests.

In the USA, the pest species include *C. furens* and *C. haematopotus* in eastern and Gulf Coast states; in midwestern and western states, *C. biguttatus*, *C. pilliferus*, *C. sanguisuga*, *C. obsoletus*, *C. tristriatulus*, and *C. alaskaensis* are pests. *L. kerteszi* and *L. torrens* are pests in California. *Ceratopogon stellifer* is an important pest in Arizona and New Mexico. *Austrosimulium pestilens* is a serious pest in Australia.

**No-see-ums, punkies, *Culicoides* spp. (Fig. 7.1e)** Adults are 1–3 mm long; they are gray to grayish black, and usually have spotted or patterned wings. Eggs are laid in batches on moist or wet substrates, and hatching occurs in 2–7 days. Larvae are wormlike and move in a serpentine motion. Development takes from 4–5 days to 3–4 weeks; the time varies with substrate and ambient temperature. In temperate regions most species overwinter as fourth-stage larvae in diapause. Larvae feed on vegetation, but some species are predaceous on nematodes and small arthropods in their habitat. Pupae are free-floating or attached to debris; the pupal stage lasts 2–3 days to 3–4 weeks, depending on species and temperature. Adults are crepuscular, but some species fly and feed during the day. Adult females live 10–20 days, and during this time take multiple blood meals. More than 1400 species have been identified, and they occur on all continents, but not the Hawaiian Islands, Antarctica, and New Zealand. The majority of *Culicoides* suck blood and attack mammals (including humans) and birds. About 50 species are associated with the spread of pathogens and parasites to humans and other animals. Larvae live in wet habitats, including pools, streams, marshes, bogs, beaches, swamps, tree holes, and animal dung.

***Culicoides crepuscularis*** Adults have a wing length of 1.45 mm, and the wings have a pattern of light and dark spots. Legs are brown with no color banding, and the hind tibia has four spurs. This species occurs in a variety of habitats across the USA, including fresh, salt, and alkaline water–soil habitats, and in sun or shade. There is one generation per year, and adults are active from June to September. They are common pests in many habitats.

**Salt marsh punkie, *Culicoides furens*** Adults are 2–2.5 mm long and dark gray, with white markings on the thorax. Wings are dark with scattered transparent spots. Larvae occur in salt marshes. Adults usually travel away from the breeding site and are severe biters. This species is known from salt marshes along the Atlantic and Gulf coasts from Massachusetts to the West Indies and south to Brazil, and along the Pacific coast from Mexico to Ecuador. It breeds year-round in tropical areas, and is a common pest in these regions.

**Scottish biting midge, *Culicoides impunctatus*** Adults are 1–2 mm long and grayish black. Adults are active from late April through September; there are two generations per year and each is about 6 weeks long. Females do not require a blood meal to oviposit their first batch of eggs (autogenous), but blood meals are necessary for subsequent egg-laying. Blood meals are usually derived from large livestock, but where these hosts are scarce, female *C. impunctatus* shift their feeding habits and attack humans. This species is the most common biting midge in the UK. It is a severe pest of humans and livestock, and large populations are common in the Scottish Highlands during summer. It has an adverse influence on tourism, and commercial forestry. In Scottish folklore the deciduous shrub *Myrica gale* (bog myrtle) has repellent properties against *C. impunctatus*. This plant occurs throughout the British Isles, and is especially common in bogs and wet heaths of Scotland and Ireland.

***Culicoides variipennis variipennis*** Adults have a wing length of 1.97 mm; the wings have a complex and irregular pattern of pale and dark spots. Legs are brown and the front and middle femora have preapical and subapical pale rings; the comb on the hind tibia has 6–8 spines. Palp segment 3 of the female is long and slightly swollen. There are two generations per year. This species is active from May to October. It is widespread across the USA.

**Other *Culicoides*** One of the most abundant and annoying biting midges in the woods of North America is *Culicoides sanguisuga*. Adults feed at dusk and during the night, and they are active during the day in shaded, humid wetlands. These flies are attracted to lights at night, and they enter houses through window screens to bite people.

**Biting midges, blood-sucking midges, *Forcipomyia (Lasiohelea)* spp.** Adults are about 3 mm long, and black to grayish black. Wings have long setae or hairs on the surface, but they are not mottled or patterned. The last tarsal segment has a large empodium (at least in the female) and there are strongly curved claws. Mouthparts of the female have sharp teeth. Larvae are terrestrial and live in sandy soil; larvae of many species have a well-developed proleg on thoracic segment 1 and abdominal segment 9, and there are branched abdominal setae dorsally on all segments. *Forcipomyia* contains the subgenus *Lasiohelea* with about 150 species; many of these species bite humans and livestock. *F. anabaenae* is a common pest species; it has been recorded biting humans in Singapore and Thailand, and it has been recorded from Queensland, Australia, and Japan.

**Bodega black gnat, *Leptoconops kerteszi*** Adults are 2.5–3 mm long and black, except for the legs, which are brown to pale brown. Larvae are found in the surface layers of porous and sandy soils. Adults emerge in spring and actively feed at that time. Males often form large mating swarms, usually near potential oviposition sites. Eggs are deposited singly or in small batches on wet soil or sand; fecundity is about 50 eggs. Hatching is in about 12 days. Development time is 8–10 months, and the pupal period is 8 days. Eggs that hatch from April to August produce larvae that pupate in March to August of the following year. Females live about 10 days and take as many as four blood meals; males live about 4 days. Adults emerge from early March until September, and the last swarms occur in October. Dispersal is limited to about 2 km from the breeding site. There is one generation per year. This day-biting midge is named for Bodega Bay, California, where it was first investigated in 1934. It was found in wet, organic sand along the banks of freshwater streams that entered Bodega and Tomales bays. It occurs in western North America and the Mediterranean region of North Africa, and its common name varies with the region.

**Valley black gnat, *Leptoconops torrens*** Adults are 1.5–2 mm long and black; they resemble *L. kerteszi*. Adults are active during the day in May and bite humans, domestic animals, poultry, and wild birds. They will get under clothing and often bite where it fits tightly. Females lay 60–70 eggs in cracks in soil. Development is about 2 years, but they may enter diapause and emerge in 3 years. Larvae are found in summer in soil with a moisture content of 17–20%, a salt concentration of 400 ppm, and pH 9.6. This species was originally recorded breeding in clay and adobe-clay soils found along the western side of the Sacramento Valley, and in isolated deposits in the San Joaquin Valley, California. This species occurs in western North America.

## Chaoboridae

These aquatic insects are very similar to mosquitoes, but differ in having short mouthparts. Males and females do not bite. They are called phantom midges because of the pale to almost transparent larvae of the common *Chaoborus* species. Full-grown larvae are 9–12 mm long and nearly transparent except for black eyes and the pigmented hydrostatic organs on the enlarged prothorax and posterior end. The head has large and raptorial antennae that terminate in spines. The caudal segment has a ventral anal fan of setae (fin), a pair of ventral hooks, and anal gills. Larvae are predaceous on aquatic arthropods, and may kill a large number of mosquito larvae while feeding. Adults live only a short time, but may occur in large numbers on houses and other buildings after emerging from lakes, ponds, and reservoirs.

**Clear Lake gnat, *Chaoborus astictopus* (= *C. lacustris*)** Adults are 4–5 mm long with a brown body. Males have plumose antennae and a slender abdomen; antennae of females are not plumose, and their abdomen is enlarged. Adults appear in early May and occur in small to large numbers until mid-September. Adults can be numerous and create health problems in residential and resort areas. Eggs are deposited on the water surface, but sink to the bottom; hatching occurs in about 24 h. Larvae feed on plankton, mosquito larvae, and other animals in the water. Development takes 11–25 days; the pupal stage is 2–3 days. Larval stage 3 and 4 and the pupa remain in mud at the bottom of the habitat. Larval development is usually completed the following year. Adult emergence begins at about 23:00 h and extends to about 6:00 h. Adults spend 36–48 h resting on the ground or vegetation around the shoreline before laying eggs. This species is known from large numbers that occur along the shores of Clear Lake, in northern California.

**Lake Nyasa gnat, *Chaoborus edulis*** Adults are about 4 mm long with a dark brown body. Swarms of adults of this species occur over Lake Victoria and Lake Nyasa in East Africa, and they can appear as a black cloud, which moves with the wind. These swarms are a seasonal nuisance, and sometimes alarming and dangerous for people who may be sensitive to arthropod allergens. Fisherman in small boats may be covered and the breathing air filled with these midges. In the past, bodies of these midges were compressed into kungu cake and eaten by local people.

## Chironomidae

These midges are 1–10 mm long; they resemble mosquitoes, but their mouthparts are not elongated and may be absent. Adults are gray to black, the thorax usually has a midline furrow, and the male antennae are plumose (Fig. 7.1f; Fig. 7.2a). Larvae of most species are aquatic and live in tubes or cases composed of fine particles of the substrate. Larvae of many *Chironomus* species are red because of hemoglobin in their blood, and they are known as bloodworms. Larvae are important food for other aquatic organisms. Eggs are laid in water, usually in masses or strings; the terrestrial species oviposit on wet soil or other surfaces. Young larvae establish their protective tubes within a few hours of hatching. Larvae feed on algae, plant cells, and miscellaneous debris. They remain in the water and obtain oxygen through small gills. Full-grown larvae pupate within their tube. Emergence of adults usually occurs early in the day, and the flies rest on vegetation until dusk; adults live 2–3 days. Males form large mating swarms in the evening when there is only slight wind. Females enter the swarm to find a mate, then the pair fly to nearby vegetation to copulate.

Chironomid adults usually form mating swarms 1–10 m above the ground and orient over an area of terrain that has a more color-conspicuous feature than the surround area. This may be an area of bare soil, a tree, a road or footpath, or shoreline. The swarming time is often species-specific, and is influenced by time of day, temperature, and wind velocity. In Japan, swarms of *Chironomus dissidens* are typically 1–3 m above the ground and begin forming around 18:30 h when the air temperature is about 21 °C; swarms of *C. plumosus* are 1–5 m above the ground and begin forming around 19:00 h when the air temperature is about 17.8 °C; and swarms of *Propsilocerus akamusi* are 1–10 m above the ground and begin forming around 15:30 h when the air temperature is about 16.7 °C. Swarms usually become more compact after forming and they last about 30 min, depending on the species.

Pest status is based on the presence of very large numbers of adults. Chironomid swarms often limit human activity outdoors because the adults can be inhaled or fly into the mouth, eyes, or ears. Chironomids are among the most abundant flies in aquatic habitats, and their emergence in large numbers from impounded water and wastewater stabilization lagoons can result in hypersensitivity in the human population in the surrounding areas. Midges attracted to lights at night cause problems in residential areas, and they may enter houses in large numbers during the day and night. Adult midges congregate in shaded sites and release egg masses, which stain surfaces.

Swarms of midges over highways and other roads can cover the windshield and headlights of automobiles. Swarming and the odor of the accumulation of dead adult midges can be a nuisance and cause considerable economic loss to hotels and tourism. The parthenogenetic midge, *Paratanytarsus grimmi*, has caused problems in urban water-supply pipes.

Chironomids associated with sewerage treatment facilities include *Spaniotoma minima*, *S. perennis*, *Metiocnemus hirticolis*, *M. obscuripes*, and *M. longitarsus*. Larvae of *Spaniotoma* spp. are usually distributed throughout the filter bed, but *Metiocnemus* larvae tend to occur in the upper layers. Several *Metiocnemus* species are predaceous on the eggs and pupae of other chironomids in the sewerage filter bed. The eggs are laid in gelatinous masses and full-grown larvae produce cocoon-like cases in which to pupate.

**Lake fly, *Chironomus plumosus*** Adults are about 12 mm long and light brown with green or yellow markings; the thorax may be reddish brown. Full-grown larvae are 20–25 mm long and slightly red with a light-brown head. There is a single ventral proleg on segment 1; segment 10 has one pair of short, gill-like lobes; segment 11 has two pairs of long gills, and segment 12 has two pairs of anal gills. Egg masses are usually placed on the water, where they swell then sink. Eggs laid on buildings cause staining. Hatching depends on the water temperature at the bottom, and varies from 3 days at 24 °C to 14 days at 9 °C. The tubes of the first- and second-stage larvae are close to the surface of the mud at the bottom. The tubes of the fourth-stage larva are U-shaped, and may extend deep in the mud. Development takes about 4 weeks, and the pupal period lasts 6–10 days. When pupation is complete the pupa leaves the larval burrow and swims to the surface and the adult emerges within 30 s. Empty pupal cases often accumulate in windrows on the water surface and along the shoreline. This species is common over much of the USA, and can be a serious problem in lakefront communities. In the area of Lake Suwa in central Japan, massive swarms of this species cause problems for local residents and tourists. This is one of the largest chironomid species. Adults may occur in large numbers and are attracted to lights at night. A related species, *C. (Lobochironomus) dissidens*, also occurs in large numbers in the summer along the shore of Lake Suwa, Japan.

**Nimitti, *Cladotanytarsus lewisi*** Swarms of adults occur in parts of the Nile valley in northern and central Sudan. Seasonal swarms of this species often result in outbreaks of allergic rhinitis and bronchial asthma in areas surrounding the breeding sites.

***Tokunagayusurika akamusi*** Swarms of this species occur around Lake Suwa, Japan. Densities of larvae at the lake bottom (profundal region) are about 700–4800 per square meter. The adults swarm from the end of October to early November. They form masses 1–10 m above terrain that has more conspicuous color than the surrounding area, such as along footpaths or lake shoreline. Swarms begin forming about around 15:30 h, and become more compact as daylight decreases. Swarms do not occur when wind velocity is above 5 m/s or the temperature is less than 12 °C.

**Other Chironomidae** There are numerous pest species of midges in urban environments around the world. In Japan, *Chironomus yoshimatsui*, *Rheotanytarsus kyotensis*, and *Tokunagayusurika akamusi* are pests, and *Polypedilum kyotoense* has been linked to bronchial asthma. Other pest species are: *Cricotopus bicinctus* (polluted rivers), *Tokunagayusurika akamusi* (lakes, ponds), and *Tanytarsus* spp. (saline water). In the USA, *Chironomus crassicaudatus*, *C. decorus*, *C. plumosus*, *C. riparius*, *C. utahensis*, and *Glyptotendipes paripes* are major pest species. Species that emerge in large numbers from water impounded in urban and rural areas include *Procladius freeborni*, *P. sublettei*, *Tanypus grodhausi* (sewage ponds), *T. neopunctipennis* (salt marshes), and *Apedilum subcinctum* and *Cricotopus sylvestris* (storm drains). In the UK, *Polypedilum nubeculosum*, *G. paripes*, *Tanypus punctipennis*, and several species of *Procladius* and *Chironomus* cause nuisance complaints. In Australia, *Polypedilum nubifer*, *Kiefferulus longilobus*, *Tanytarsus barbitarsis*, *T. semibarbitarsus*, *T. oyamai*, and several species of *Chironomus* emerge in large numbers and cause a nuisance by their presence. In Europe, *Chironomus balatonicus*, *C. plumosus*, *C. salinarius*, *Cryptochironomus redekei*, and *Polypedilum scalaenum* are nuisance pests.

## Chloropidae

Chloropid adults are about 5 mm long and are without bristles and fine setae. Some species are yellow and black, and often have green eyes, from which the family gets its name. Larvae of most species feed in grass stems, a few are scavengers in decaying vegetation and excrement, some are parasitic on other arthropods (spiders), and some are predaceous. *Liohippelates* species are called eye gnats because they frequently gather around the head and feed on secretions around eyes, and other sites. Eye gnats are commonly found in suburban areas bordering woods or agricultural fields. Adults are attracted to wounds, scabs, blood, sebaceous material, and body orifices, especially around the eyes. They do not bite, but feed with their sponging mouthparts on liquids at the surface. The labellum of the mouthparts has spines that may produce tiny scratches on the eyeball while the female fly is feeding. This assists the entry of pathogenic organisms, such as the causative agent for pinkeye, carried on their tarsi. These flies are very annoying, especially to small children, because the adults generally fly close to the ground.

**Eye gnat, *Liohippelates collusor*** Adults are about 3 mm long and black; the thorax is smooth and shiny. Eggs are deposited in batches of about 50 at first, followed by a second but small batch in about 7 days. They are laid on or below the surface of moist soil; hatching occurs in 2–3 days. Oviposition sites also include excrement mixed with soil, decaying meats, fruits, and other vegetable material. Development takes about 11 days, depending on substrate, moisture, and temperature; on human feces it is about 11 days, on canine excrement 8.7 days, and on decaying fruit 17 days. During cold weather, the larval and pupal period may last 3 weeks or longer. Adults are strong flyers; they can fly with and against the wind, and may disperse about 6 km from breeding sites. This species has a wide distribution in southern USA where winters are mild. Adults are present throughout the year in the desert and foothill region of California, and are annoying during April through November. During months of peak activity, they are noticeable in early morning and late in the afternoon, in the sun and shade.

**Other *Liohippelates*** The dominant species in southeastern USA include *L. pusio* and *L. bishopii*. Adult *L. impressus* are reddish orange. This pest is distributed from sea level to 1800 m elevation in southern California. It occurs in adjacent states east to Texas, and in Mexico and the Virgin Islands. *L. flavipes* is a pest species in Jamaica.

**Eye fly, *Siphunculina funicola*** This chloropid is distributed in India, primarily along the seacoasts and mountains. It occurs nearly year-round, with maximum abundance in summer and in short periods after the southwest monsoons. Adults come to feed on the liquid secretions from the eyes, lips, nose, and ears of humans and animals. Adults gather in large numbers indoors and are active on hot, sunny days. Larvae breed in soil and in fresh cow dung, in decomposing organic matter, and excrement. Eggs are deposited on the surface of soil; fecundity is 40–50 eggs. Hatching occurs in about 3 days. Development is completed in 4–7 days, and the pupal period lasts 2–7 days. Adults live about 30 days. There are two or three generations per year.

**Yellow swarming fly, *Taumatomyia notata*** Adults are about 4 mm long and yellow with black markings on the thorax and abdomen. Full-grown larvae are about 2.5 mm long and pale yellow. Eggs are deposited in the soil, around the base of grasses. Larvae feed on the root aphid, *Pemphigius bursarius*. There are two generations during the summer, the second of which normally spends the winter as pupae in the soil. Some adults emerge prematurely, especially during warm spells in fall. Cold temperatures at ground level force the flies into the air. The adults are easily swept up by strong winds and carried across the suburban countryside and urban areas. Windstreams carrying these flies pass over roofs, and eddies are formed under eaves on the lee side of buildings. Flies trapped in the downdrafts of air are carried into open windows on upper floors. This species is a household pest in the fall in the UK, an estimated cluster of 12–14 million was once collected indoors. A related species, *Thaumatomyia annulata* (as *Chlorpisca*), is known to occur in buildings in USA.

## Coelopidae

Seaweed flies are 5–12 mm long and dark brown to black. The dorsum of the thorax is flattened, and the body and legs have numerous bristles. They occur along the seashore where seaweed (kelp) has washed on shore in large amounts. The larvae breed in the seaweed, and the adults often swarm over it in large numbers, and they are sometimes carried inland where they can be nuisance pests around houses. Shore birds may be attracted to and feed on the flies gathering at the seaweed. Other flies are associated with accumulations of seaweed, including species of *Fucellia* (Muscidae), *Thoracochaeta* (Sphaeroceridae), and *Helcomyza* (Dryomyzidae).

**Seaweed fly, *Coelopa frigida*** Adults are 3–11 mm long, slightly flattened, and dull black or brown. The head and legs have strong bristles, and the thorax has long bristles at the sides only; the wings are clear to slightly brown. Full-grown larvae are about 16 mm long and yellowish white; there are fine spines on the anterior margins of segments 1–3. Anterior spiracles have 16–17 lobes; posterior spiracles are widely separated and surrounded by setae. There are star-like spines above the anal lobes. Larvae feed in aggregations (as many as 600 per liter) in piles of wet seaweed (*Laminaria*, *Fuscus*) on the shoreline in protected places. Seaweed along the tidemark dries quickly and is usually not infested. Eggs are laid in 3–5 batches directly on the surface of decaying seaweed; fecundity is about 80 eggs. Hatching occurs in about 3 days. Development takes about 28 days and the pupal period lasts about 14 days. Under laboratory conditions of 25 °C, development takes about 12 days. Adults either remain on wet seaweed during cold weather or hover 2–3 m above this substrate in large swarms. Sometimes they travel or are carried by wind inland away from the shore, and may be a nuisance in commercial and residential areas. Adults are attracted by the odor of the chemicals trichlorethylene and carbon tetrachloride, which are sometimes used in the process of commercial dry cleaning. This fly is found along most of the European coastline from Biscay to the Arctic, and also in eastern and western North America and Asia.

## Culicidae

Mosquitoes are important blood-feeding insects around the world. They are vectors of several human diseases, including malaria, yellow fever, encephalitis, dengue, and filariasis. Adults are 3–9 mm long with scales on their body and wings, and they have a long, slender proboscis. Females of most species take a blood meal, while males and the females of other species feed only on nectar and other plant juices. Larval stages are aquatic and occur in a variety of habitats, including standing or still water in natural sites and artificial containers. Larvae of most species feed on algae, but a few are predaceous.

Pest species in urban and rural areas belong to four genera: *Aedes*, *Anopheles*, *Culex*, and *Psorophora*. Adult males and females of *Anopheles* are distinguished by their long maxillary palps, and in the resting position, the body of adult *Anopheles* is at a 30–45° angle to the surface. Adults of the other genera have short palps, and in the resting position, their body is nearly parallel to the surface. The tip of the abdomen of female *Aedes* is usually pointed, and the thorax has white markings; in *Culex* the tip of the abdomen is usually blunt, and the thorax has no white marks. *Psorophora* species are 6–9 mm long and have long, erect scales on the hind tibia. Larvae of *Anopheles* lack a breathing tube at the end of the abdomen; while feeding they lie parallel to the water surface. *Anopheles* larvae are usually found in marshes and other places where there is considerable vegetation. Larvae of other genera have a breathing tube, and they hang down from the water surface. *Aedes* and *Psorophora* larvae are usually in woodland pools and salt marshes; *Culex* larvae are usually in artificial containers.

Mosquito-borne encephalitis viruses have become a major health concern in urban and suburban areas in many regions of the world. Encephalitis is caused by bacterial and viral agents. The resulting inflammation of the brain can result in high fever, seizures, prolonged illness, and sometimes death. The viruses carried by mosquitoes that cause encephalitis in people are essentially animal diseases. They are occasionally transmitted to people, but are usually circulated by blood-feeding mosquitoes in bird and small-animal populations. The most common strains of encephalitis affecting urban populations include: West Nile virus (WNV), which occurs throughout Europe, parts of Africa and Asia, and the USA; Japanese encephalitis, which occurs throughout Asia to India; Murray Valley and Kupiin virus, which occurs in Australia; and Rocio encephalitis and Venezuelan equine encephalitis, which occur in South America. There are several of these diseases in the USA, including eastern (EEE) and western (WEE) equine encephalitis, and St. Louis (SLE) and LaCrosse encephalitis.

Mosquitoes in several genera are vectors for encephalitides; many are bird-feeding species that occasionally bite humans. The most common vectors are species that have successfully adapted to urban habitats or secondary habitats near urban areas. EEE may be transmitted to humans by *Aedes sollicitans*, *Ae. vexans*, and *Coquillettidia perturbans*; WEE is primarily transmitted by *Culex tarsalis*. SLE is transmitted by several urban *Culex* species, including *Cx. tarsalis*, *Cx. quinquefasciatus*, and *Cx. nigripalpus*. LaCrosse encephalitis is transmitted by the tree hole mosquito, *Ae. triseritatus*. WNV has been isolated from more than 40 mosquito species: the most common are *Culex* species, including the common house mosquito, *Cx. pipiens*, and *Cx. univittatus*, *Cx. modestus*, *Cx. quinquefasciatus*, and *Cx. vishnui*.

Female mosquitoes may fly from their breeding site in search of a blood meal, but males usually remain close to the breeding site. Female flight range varies with species, time of year, wind direction, and other factors. Most adult mosquitoes disperse only 100–200 m from their emergence site. *Aedes aegypti* fly 25–100 m and *Anopheles* species fly about 2 km. Wind above 6.4 km/h permits only downwind movement, and velocities above 9.7 km/h inhibit flight of most adults. Dispersal records for *Aedes* mosquitoes include: *Ae. nigromaculatus*, 45 km downwind; *Ae. sollicitans*, 177 km away from the seacoast and at sea off the Atlantic Coast; *Ae. dorsalis*, 61 km downwind; *Ae. taeniorhynchus* 40 km downwind.

### *Aedes*

Most of the adults have distinct patterns on the thorax and abdomen formed by yellow, white, or silvery scales, and the legs often have white rings. Adults of most species bite during the day or early evening. Biting occurs outdoors and adults usually rest outdoors before and after feeding. Eggs are laid singly on a wet substrate, but they are resistant to drying and can withstand desiccation for months and, in some species, for several years. Winter is spent in the egg stage, and hatching occurs with spring flooding. Marshes and ground pools are breeding sites for many species, but some of the important disease vectors utilize household or domestic containers in peridomestic habitats. Several floodwater and salt marsh *Aedes* are serious pests in urban and rural areas around the world, including *Ae. taeniorhynchus*, *Ae. sollicitans*, and *Ae. dorsalis*. Larvae of many species live in clean drinking water. *Ae. aegypti* breeds in pots and water-storage jars located inside or outside houses. *Ae. albopictus*, which is a vector for dengue in southern Asia, breeds in natural and domestic containers, such as cooking pots and used vehicle tires. Spread of the Asian tiger mosquito, *Ae. albopictus*, to the USA and parts of Europe is linked to commercial distribution of used tires. In northern Europe (Lithuania), the most common *Aedes* biting humans in urban areas are: *Ae. communis*, *Ae. cantans*, *Ae. punctor*, and *Ae. cataphylla*. *Aedes* species such as *Ae. sierrensis* and *Ae. triseriatus* most often breed in water-filled tree hole cavities.

*Aedes* are vectors of yellow fever, dengue, and encephalitis viruses. The arbovirus causing yellow fever occurs in Africa and tropical areas of the Americas, and the major mosquito vector is *Ae. aegypti*. Several *Aedes* species vector filarial worms, including *Ae. togoi* in China, *Ae. polynesiensis* and *Ae. pseudoscutellaris* in the Polynesian region, and *Ae. niveus* in Thailand.

**Aedes aegypti (Fig. 7.1d)** Adults are about 6 mm long and the body is black with white markings. The thorax has several longitudinal silvery-white lines, resembling a lyre; the legs are black and with white bands. This is the common yellow fever mosquito, and it is widely distributed throughout tropical and subtropical regions. It does not usually occur in areas where night temperatures are 20 °C or less. This day-biting mosquito is adapted to living in the urban environment, and is often found breeding and feeding around and in buildings. Females lay eggs on wet surfaces adjacent to the water line. Eggs withstand desiccation, and can remain dry but viable for months. When flooded, some eggs hatch immediately while others delay a few days, thus hatching is spread over days or weeks. Larval development depends on temperature, and ranges from 7 days to about 7 months. There are two or three generations per year.

The nominal species *Aedes aegypti* exists in several subspecies forms, which are characterized by morphological and behavior differences. The African subspecies *Ae. aegypti formosus* has black abdominal tergites, and it rests and feeds outdoors in nonurban areas. The species with primarily peridomestic and domestic habits are *Ae. aegypti aegypti*, which has white scales on the abdomen and breeds in domestic water-storage pots outside of houses, and *Ae. aegypti queenslandensis*, which breeds in water-storage pots inside houses.

This species is believed to have migrated from West Africa to the New World in the fifteenth through seventeenth centuries aboard slave ships. It may have first invaded Portugal and Spain before reaching the western hemisphere on European ships. The evolution of domestic traits in this originally feral species was crucial for enabling *Ae. aegypti* to occupy and reproduce in water-storage jars in the holds of these sailing vessels. Yellow fever was absent from urban settlements in the western hemisphere until the arrival of *Ae. aegypti*, which is the only known vector of urban epidemics of this disease. The first documented epidemic of yellow fever in this region occurred in the Yucatan in 1648, although it may have appeared in Haiti in 1495. In the seventeenth through the nineteenth centuries, yellow fever occurred in seaports on the Atlantic coast, as far north as Philadelphia and New York. The yellow fever virus may have been reintroduced to the New World periodically by passengers, especially slaves, on ships coming from West Africa.

**Aedes albopictus** Adults are about 5.5 mm long and the body is black with white markings. The thorax has a broad, silvery white band in the center, and the legs are banded with white. Full-grown larvae are about 8 mm long and yellowish white to gray; the head is yellowish white and slightly longer than broad, and has few markings. Eggs are deposited in batches containing 42–88 eggs per blood meal; in natural habitats females produce 0.2–2.1 batches; the highest number of eggs produced after a single blood meal is 147. Fecundity is 300–345 eggs, but as many as 950 have been recorded. Hatching is in 6–7 days at 30 °C and about 10 days at 21 °C. Mortality is nearly 100% when eggs less than 12 h old are dried, but only 60% for eggs 16 h old. After 3 months at 25 °C and 70–75% RH egg survival is about 50%; 78–99% of *Ae. albopictus* eggs from temperate Asia and USA survive exposure to −10 °C for 24 h. Maximum longevity of an egg is 243 days. Larval development is about 19 days at 14 °C and the pupal period is about 4 days at 15 °C; at 25 °C larval development is 8 days and the pupal period is 1 day. In the laboratory at 25 °C and adequate food, larval development is 5–10 days. Development is prolonged when food amount is suboptimal: 58 days for males and 24 days for females. Females live 4–8 weeks in the laboratory at 25 °C and 30% of RH females may live 3–6 months. Time from

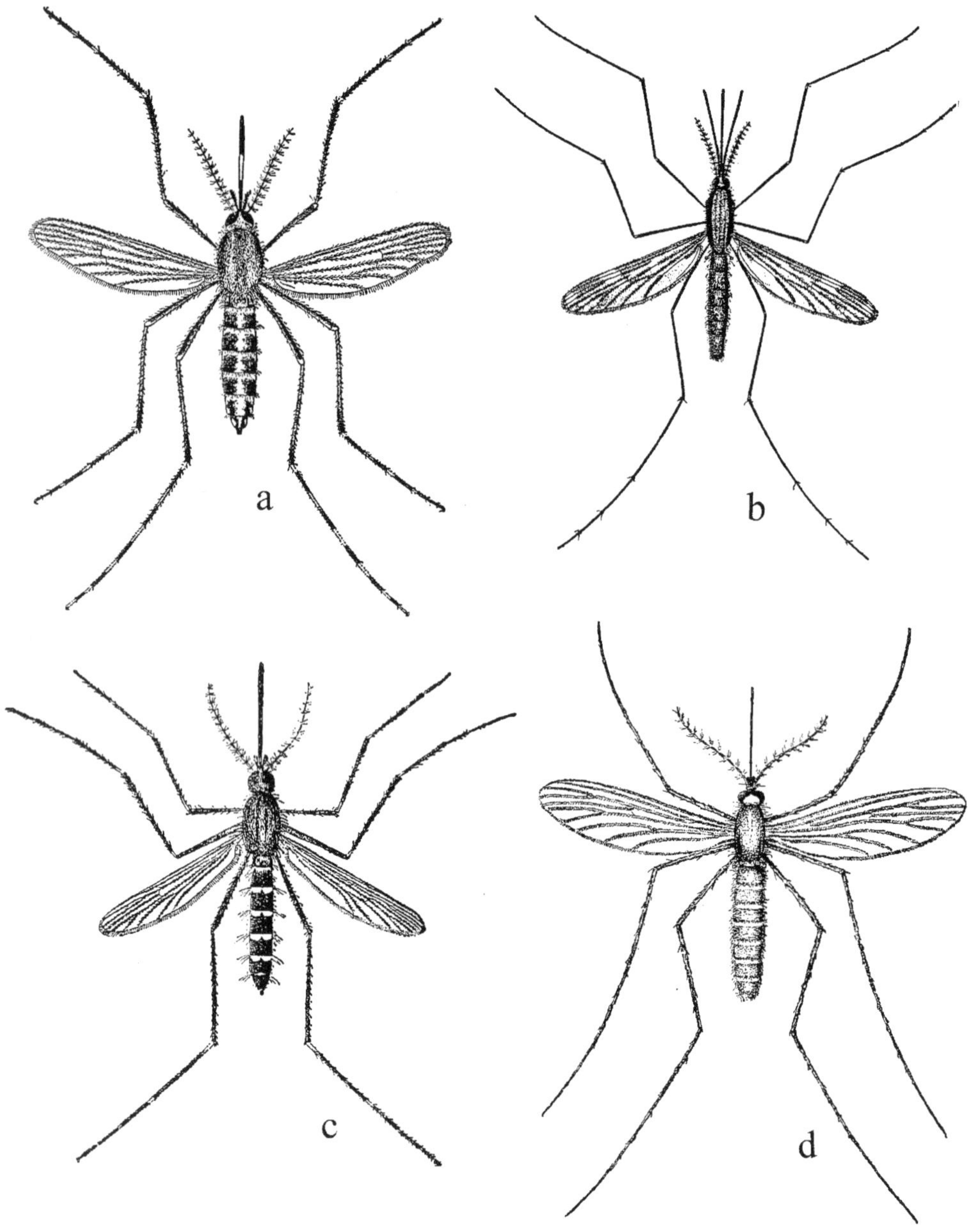


**Figure 7.3** Diptera: Culicidae. (a) *Aedes sollicitans*; (b) *Anopheles punctipennis*; (c) *Aedes vexans*; (d) *Culex pipiens*.

emergence to first blood meal at 25 °C is about 2.5 days for a population from Japan, time is 3.5–4.5 days when reared at high larval densities. Females become receptive to mating 48–72 h after emergence.

Adults fly close to the ground and do not fly in strong winds, and wind-aided dispersal is limited. Dispersal is 90–183 m from breeding sites, and maximum dispersal is about 434 m. The potential for dispersal is greatest for females from high-density larval rearing sites. It is a day-biting mosquito, and there may be an early-morning and late-afternoon peak in biting. Females usually bite at ground level, and prefer to feed around the ankles and knees. The native range of this species is the oriental region and India, but also extends to Mauritius, Seychelles, and Madagascar. Based on presence of diapause in established populations, the North American *Ae. albopictus* apparently originated in temperate Japan. The *Ae. albopictus* biotype occurring in southeastern Brazil has no diapause, and it probably originated from a tropical site. In Guatemala, it breeds in small water containers in urban and suburban sites of the Caribbean harbor cities of Puerto Barrios and Puerto de Santo Tomás de Castilla. *Ae. albopictus* is a vector of dengue fever in epidemics in Southeast Asia, southern China, and Japan. It is capable of transmitting Japanese B encephalitis, and is susceptible to infection with Getah virus, a virus affecting horses.

*Ae. japonicus* is a related species, and is known primarily from Japan and Korea. It has also spread in used tires to other parts of the world, including eastern USA. In the last 20 years, widespread distribution of the Asian tiger mosquito has been facilitated primarily by shipments of used tires and casings.

**Aedes sollicitans (Fig. 7.3a)** Adults are 5.5–6 mm long and brown to dark brown; the thorax is golden brown, and the pleura silvery white. The proboscis and tarsi are white-banded; abdominal segments have yellowish-white bands at the base, and a yellowish-white longitudinal stripe medially. Full-grown larvae are 8–9 mm long and yellowish white to gray; the head is yellowish white, 1.5 times as broad as long; the thorax is wider than long. Eggs are deposited in three batches containing 14–115 eggs; there is a 4–7-day interval between batches, and a blood meal is required for each batch. Eggs hatch in a range of salinity, including concentration of salt up to 25% greater than seawater. Larval development is in water temperatures 10–39 °C. Females live 25–35 days in a laboratory. Breeding is in coastal marshes and salt pools where marsh grasses grow, but infrequent in marshes that are submerged for 25 days or more each month. Inland sites for this mosquito include habitats retaining or inundated with salt water, such as from natural salt deposits, oil well, mines, and swimming pools. A percentage of emerging females disperse from the breeding site, and mass dispersal flights occur. Long-distance dispersal often brings females 13–17 km and as far as 166 km (at sea) from breeding sites. Urban areas that are a long distance from *A. sollicitans* breeding sites can be affected by this mosquito. However, this species usually remains outdoors, and does not enter a building to bite. It is a blood-feeder on any animal; females take a blood meal on the second day after emergence, and feed twice before oviposition. This species feeds during the day and night, and records of severe and intense attack by large numbers are well known. This species is distributed in the coastal marshes of eastern and southeastern USA, and in salt pools of all eastern states.

**Aedes vexans (Fig. 7.3c; Fig. 7.4c)** Adults are 4–5.5 mm long, brown, with golden-brown scales scattered on the thorax. The proboscis is dark brown to blackish brown toward the tip. The abdomen is dark brown with yellowish white bands at the base of segments; the band is constricted medially and laterally. Wings are dark brown. Full-grown larvae are 6–8.5 mm long, yellowish brown to gray throughout; the head has distinct, symmetrical ventral markings. Eggs are deposited singly on the surface of water and they sink to the bottom, or eggs may be laid in moist soil at the edge of pools (subject to inundation in the spring). Overwintering is in the egg stage, and prolonged freezing or drying has little detrimental effect on egg viability; eggs may survive 1–2 years. Hatching is erratic or uneven. It may be induced by submergence in the spring or summer, depending on water temperature, an interval of drying prior to submergence, and embryo development time. Larval habitats are fresh water over a layer of decaying vegetation, and free of filamentous algae; flooded meadows or urban grassy areas are suitable. Development is 7–25 days at 22 °C, 5–21 days at 27 °C, and 4–6 days at 32 °C; development is prevented at 0–10° C. The pupal stage lasts 3–9 days at 15–22 °C, and 1–3 days at 27–37 °C; there is 50% pupal mortality at 40 °C. Larvae are frequently found in small pools and in large concentrations, such as 500+ in 0.6 liters, but this does not adversely affect development. This is a migratory species that disperses individually or in mass movements in which both males and females fly from the breeding site. Mass dispersals have created clouds of mosquitoes moving towards the glowing skyline of suburban lights. Females are capable of moving 24–48 km from the breeding site, and wind seems to have little or no influence on the direction of flight. This species breeds in woodland pools and in temporary pools in urban areas. It breeds throughout the year, but is usually more abundant in late summer, when it may be the dominant species in some

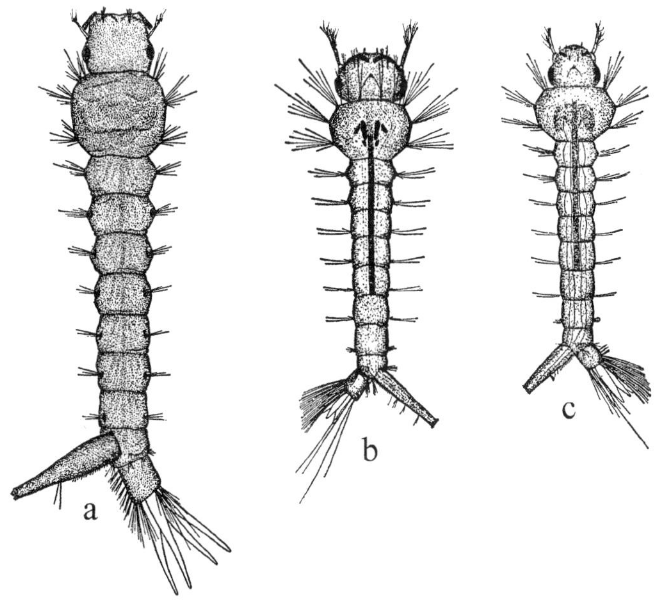


**Figure 7.4** Diptera. Culicidae larvae. (a) *Psorophora ciliata*; (b) *Culex pipiens*; (c) *Aedes vexans*.

areas. It readily moves indoors to bite. This is one of the most abundant and widely distributed mosquitoes: it occurs in the Nearctic and Palearctic regions south of 55 °N latitude, African west coast, and the oriental region, south and east to Samoa (170 °W longitude). It is a common pest in urban and suburban areas.

### *Anopheles*

Adults have spotted wings and the arrangement of blocks of dark and pale scales on the wings provides characters for species identification. The dorsal and ventral surface of the abdomen is nearly or entirely without scales. In males and females, the palps are about as long as the proboscis, and in males the palps are enlarged apically. *Anopheles* species are major vectors of malaria around the world. The genus is distributed nearly worldwide, except for Micronesia (but they are on Guam) and Polynesia. Adults of most species are crepuscular or nocturnal, but the time and location for seeking a blood meal vary. *An. albimanus*, which is a malaria vector in Central and South America, bites mainly outdoors (exophagic) from sunset to about 21:00 h. The *An. gambiae* species, which are vectors in Africa, bite mainly after 23:00 h and mostly indoors (endophagic). Before and after feeding some *Anopheles* will rest indoors (endophilic); others will rest outdoors (exophilic) on vegetation or in natural cavities.

Anopheline mosquitoes overwinter as fertilized females, and in spring they oviposit and die. Breeding is generally continuous throughout the warm months. Eggs are laid singly but in batches of 100 or more; fecundity is 400–500 eggs. Eggs float freely on the water surface and accumulate around floating objects or the shoreline. Hatching occurs in 1–3 days. Larvae do not have a long breathing siphon, and remain parallel with the water surface. They feed by moving their mouthparts back and forth to sweep the undersurface of the water. The respiratory siphons of the pupae are also short. The adults are not strong flyers and dispersal is usually accomplished by short flights in low vegetation. The flight of these mosquitoes produces a low-pitched hum that is almost inaudible unless they are close to the ear. The adults usually bite at night, inside or outside houses.

Pest status of these mosquitoes is based on their nuisance biting and their ability to vector disease. There are more than 400 described species of *Anopheles* worldwide, and about 40 are known to vector malaria. There are 12 epidemiological zones of malaria – North, Central, and South America, Mediterranean, North Eurasian, Afro-Arabian, Afrotropical, Indo-Iranian, Indo-Chinese, Malaysian, Chinese, Australasian – and each has one or more *Anopheles* species as vectors of the disease. *An. quadrimaculatus* is a common vector of malaria in North America. Certain species of *Anopheles* transmit filarial worms of *Wuchereria bancrofti*, *Brugia malayi*, and *B. timori*, all of which cause filariasis in humans. The mosquitoes involved in disease transmission differ according to region, but many of the principal malaria vectors are also involved. O'nyong nyong, an arbovirus that occurs in Africa, is spread by species of the *An. gambiae* complex and *An. funestrus*.

Urban habitats are utilized as breeding sites by several *Anopheles* species. *An. plumbeus* is widely distributed in northern Europe, south to Sicily and east to the Caucasus. It breeds in tree holes in natural habitats; in the urban environment it breeds in flowerpots and household containers. *An. maculipennis* is the dominant species in permanent water bodies in towns in Lithuania. The adults feed during the day and night, and are commonly found in urban greenspace and undisturbed areas. *An. claviger* is distributed from Europe to northern Africa and east through southern Russia. In temperate regions it breeds in woodland ditches and pools, but in southern regions it commonly breeds in underground cisterns and wells. Adults usually feed outdoors, but in the UK it enters houses to bite in spring. *An. algeriensis* is widely distributed in Europe and south to northern Africa; it typically breeds in marshy areas but will invade houses to bite humans.

There are several *Anopheles* species complexes, composed of various forms with differing morphological, physiological, and behavioral characteristics. In addition to the complexes presented here, there is the *An. punctulatus* complex in tropical Australasia (New Guinea, Solomon Islands), *An. culicifacies* complex (four species), *An. dirus* complex (eight species), *An. leucophyrus* complex (22 species), *An. maculatus* (four species) of the oriental region, the *An. marshalli* complex (four African species), and the *An. gambiae* complex (six species).

***Anopheles maculipennis* complex** This species complex is distributed across Europe and North America, in some regions member species are vectors of malaria, and it is called the European malaria mosquito (complex). This group of mosquitoes was the first species complex to be delimited; now there are at least 15 species considered within the complex. At first they were distinguished using egg morphology and wing characteristics; larval salivary gland chromosomes are currently used. *An. maculipennis* is distributed in Europe, except the UK, Northern Scandinavia, Greece, and Italy; *An. messeae* has nearly the same distribution, but includes the UK and excludes Spain. *An. labranchiae atroparvus* occurs in Europe, from Spain east to the Caspian; *An. labranchia labranchiae* occurs along the shore of the western Mediterranean; *An. sacharovi* occurs in the eastern Mediterranean, and the Middle East to the Caspian; *An. subalpinus* occurs in the Pyrenees, Alps, and Balkan mountains, and the Elburz mountains of Iran; *An. melanoon* occurs in Corsica, Albania, and southern Italy.

Adults are about 6 mm long and brown. They have black-spotted wings, and a brown patch at the apex of the wing fringe. The palp is entirely black. Eggs for most species are laid in fresh water in ditches, pools, and the margins of rivers. Eggs of *An. atroparvus*, *An. labranchiae*, and *An. sacharovi* are laid in brackish water, each with its optimum degree of salinity and temperature. Eggs are dropped randomly over the water surface. They are deposited in batches, and fecundity is about nine batches per female. A single batch may contain 86–197 eggs, and the largest number of eggs in one batch may be 312. Hatching occurs in 2–3 days, but is temperature-dependent and can extend to 7 days (*An. sacharovi*). Some of the species in the complex are resistant to drying and cold temperatures. Development varies in duration according to the time of year and abundance of food; it is complete in 27–31 days at 13–15 °C, 15–20 days at 19–21 °C, and 10–14 days at 22–24 °C. The pupal period is variable; 2 days is the shortest time, and the minimum critical temperature is about 10 °C. Larvae do not tolerate freezing and will die at 0 °C for 6 h, and last-instar larvae dies when submerged for 30 min (*An. sacharovi*) or 4 h (*An. labranchiae*). Adults overwinter and become active and begin laying eggs in spring. There are two or three generations per year. Females may be active and feed during the day or night and enter quarters occupied by humans or domestic animals. Hibernation sites may be selected on the basis of temperature, humidity, and proximity to animals; in Siberia, *An. messeae* selects dark cold cellars where temperature varies from 3 to 7 °C. Dispersal flights are for mating, feeding, ovipositing, and hibernation. In Netherlands, premating flights of *An. atroparvus* may be 14 km to leeward, and feeding flights may be 2.5–2.9 km.

**Anopheles punctipennis (Fig. 7.3b)** Adults are 4–5.5 mm long, but there is considerable variability in size. The head is dark brown with yellowish-white scales at the margins of the eyes. Wings have black scales on the veins, and a yellowish-white spot and three small spots on the outer third of the wing. Legs have white spots on the tips of the femora and tibia. Full-grown larvae are 6–6.5 mm long, gray to brownish gray; there is a longitudinal white stripe on the dorsum. Eggs are laid singly on the water surface and usually occur together in groups of 3–20 eggs; this species may produce morphologically different eggs in the winter and summer. Eggs mature in batches of about 200, and fecundity is about 2000 eggs. Hatching is irregular, but occurs in 2–6 days at 20–25 °C. Adults feed outdoors at night. Larvae are in weedy ponds, springs, streams, and shallow pools; they may be in flowing water. Development takes 12–14 days during warm months; in winter it is 18–25 days. The pupal period is 2–4 days, but is extended to 6 days at 8–13° C. Adults feed during the early evening and usually outdoors, but will feed in the shade during the day. Adults overwinter in protected locations, and are active during winter. This species is generally distributed in North America, from southern Canada to Mexico and from coast to coast; it has the greatest range of any *Anopheles* in the Nearctic region. Adults attack humans and animals outdoors, but do not usually enter buildings. Breeding may be continuous in southern regions; in the north adults overwinter in protected locations, such as in culverts and outbuildings in urban areas, and in hollow trees and tree holes in natural habitats.

**Anopheles pseudopunctipennis** Adults have distinct white patches on the wings, and a dark-gray abdomen. There are yellowish-white spots on the legs, and white rings at the base of the palpi. This species breeds in sunlight habitats with algae, including pools, puddles, and along the edges of streams. It feeds indoors and outdoors on humans and domestic animals; it typically rests outdoors. It is distributed in Central and South America, from Argentina to southern USA. *An. albimanus* is a related species and an important malaria vector in this region.

**Anopheles quadrimaculatus complex** In North America the *An. quadrimaculatus* complex is represented by *An. quadrimaculatus*, which is widely distributed east of a line from southern Ontario to eastern Texas; *An. freeborni* occurs west of the Rocky Mountains, *An. earlei* occurs in southern Canada and northern USA, *An. occidentalis* is restricted to the coast of California and British Columbia, and *A. aztecus* occurs in the interior of Mexico.

Adults are about 5.5 mm long and light brown. The median thoracic stripe is yellowish brown and somewhat indistinct. Wings have brown scales, and four distinct patches of brown scales, which distinguish it from other *Anopheles*. Full-grown larvae are about 9 mm long; the head is rounded and longer than wide; the thorax is about as long as wide. The abdomen has six pairs of fan-lift tufts of setae on segments 2–7. Eggs are laid singly around emergent vegetation; hatching depends on temperature and the optimum is 33 °C, at which hatching is in about 24 h. Larvae do not generally occur in large bodies of water, but prefer small pools with emergent vegetation. In lakes or other large bodies of water they are usually limited to the vegetated margins. Emergent vegetation, leaves, sticks, and other material on the water surface provide suitable conditions for the larvae. Development is affected by temperature, depth of the water, and food. The minimum temperature for development is about 7 °C and the lowest lethal temperature is about 35 °C. Larvae complete development in 8–10 days at 27 °C; the pupal period is 37–40 h at 28 °C. Larvae may survive 1–2 days in moist soil above the water line. Adults often rest in dark or shaded locations, including outbuildings and cellars. This species overwinters as an adult, and is active and flies during winter warm spells. It bites humans and domestic animals indoors and outdoors, and rests in either. Adults generally remain close to their breeding site during summer. In the southern part of its range, brood peaks may be 20–30 days apart, and there are 8–10 annual broods with the first appearing about 30 days after the last frost, and the last brood late in the year. In spring and fall females disperse long distances, as much as 20 km, to locate breeding sites (spring) or suitable overwintering sites (fall).

### *Culex*

Adult thorax, legs, and wings are uniformly covered with brown scales, but some white scales may be present on the abdomen. These mosquitoes are generally distributed in urban environments around the world, but they are not common or absent in extreme northern parts of temperate zones. Most species breed in ground pools, puddles, and agricultural fields; some utilize domestic containers in urban areas. *Cx. pipiens fatigans* (= *Cx. quinquefasciatus*) is a vector of filariasis, and breeds in waters polluted with organic debris, such as household refuse and excreta. This species is associated with urbanization, and towns with poor or inadequate drainage and sanitation support large populations. *Culex* mosquitoes are night biters, and they commonly rest indoors before and after biting.

***Culex pervigilans*** This species breeds throughout the year, and hibernates in the larval or adult stage. Eggs are laid in both fresh and partly saline water. The anal papillae are large in fresh water and small in saline water. This is the most prevalent species in New Zealand.

***Culex pipiens* complex (Fig. 7.3d; Fig. 7.4b)** This is the common house mosquito and it is a large species-group consisting of several urban pests: *Cx. p. pallens*, *Cx. p. pipiens*, *Cx. p. fatigans* (= *Cx. p. quinquefasciatus*), and *Cx. p. molestus*. Each species has adapted to a particular habitat or ecological zone. *Cx. p. molestus* and *Cx. p. fatigans* only occur in urban environments, while *Cx. p. pipiens* can live in both agricultural and urban environments and is the world's third most commonly distributed mosquito (*Aedes aegypti* and *A. vexans* are more cosmopolitan, but less abundant). *Cx. p. fatigans* evolved in tropical zones; and *Cx. p. pallens* evolved in temperate zones; larvae of both live in highly polluted urban habitats. *Cx. p. pipiens* extends across Europe and Asia between 30 and 60 °N latitude, and extends up the coast of Norway as far as the Arctic Circle. In the Nearctic region this form occurs between 30 and 55 °N latitude. It ranges into eastern Africa, South Africa, and Dakar. *Cx. p. molestus* is distributed nearly worldwide (North America, Europe, northern Asia, Mediterranean, Africa, and east coastal Australia). *Cx. p. pallens* occurs in the islands of the Pacific and Japan, and coastal California. *Cx. p. fatigens* is nearly circumtropical.

Adults are 4–5.5 mm long and brown, with yellow scales around the eyes; the thorax and legs are brown. The abdomen is blackish brown and with yellowish bands, and the apical segment may be yellowish brown. Full-grown larvae are 7–8 mm long; the anal siphon is pale brown and about four times as long as broad, and tapered on the terminal half. Eggs are laid in masses or rafts on the water surface; they are a single layer of erect eggs placed side by side. Oviposition is often in domestic and discarded containers and ground pools. For *Cx. p. molestus*, oviposition is in dark sites where the water is foul, as in cesspits and sumps. Hatching varies according to temperature, but over an optimum range of 21–35 °C hatching occurs in 36–26 h. Water temperatures near 0 °C are usually lethal, but *Cx. p. pipiens* may survive a few days at 2–5 °C. Larvae live in a variety of urban habitats, including street drains, gutters, domestic containers, cess pits, and drains. Temperature affects larval development, and generally no development occurs at 5 °C and 34 °C. Development takes 60 days at 10 °C, 45 days at 15 °C, and 10 days at 25 °C. The pupal period is 2–3 days; in Egypt the pupal period is 76 h at 20 °C, 53 h at 24.7 °C, and 26–36 h at 30–33 °C.

Larval habitats are variable according to locality, season of the year, and population. In the northern part of its range, larvae are in ground water pools and artificial containers. In southern and warmer parts of its range larvae are primarily in containers. *Cx. p. molestus* seem to prefer dark and confined locations. *Cx. p. pipiens* along the seaboard of northeastern USA are in polluted ground water; along the southeastern seaboard the preferred sites are protected places, such as street drains, catch basins, and other pools of foul water. In western USA (California) they are found in foul water in street drains, gutters, and domestic containers; water with algae is preferred. In Europe and Africa foul water in containers in shaded and unshaded sites are infested. Clear water of ditches, pools, ponds, and recently flooded depressions contain larvae in Norway and Denmark. Larvae ingest almost anything organic, including diatoms, filamentous algae, algal cysts, flagellates, bacteria, and other organic matter. Growth does not occur when the diet is dead organic matter.

### *Culiseta*

Adults in this genus are large and the tarsi lack pulvilli. Wings of several species have one or two dark spots. Eggs are generally produced in rafts on the water or leaf surface. *Culiseta* eggs do not withstand desiccation, but those of *Cs. morsitans* may remain viable for 6 months on damp surfaces. Larval habitats are usually pools with submerged vegetation, but in Europe, *Cs. longiareolata* occurs in rock pools and in domestic habitats, such as wells. These are often considered as winter mosquitoes, with many species found from September through May or June. In coastal areas, some species develop throughout the year. Most of these species are distributed in temperate regions, and some, including *Cs. alaskaensis*, *Cs. incidens*, and *Cs. silvestris*, occur in northern or subarctic areas. Species in temperate regions usually overwinter as larvae or hibernating adults. Overwintering adults of the European species *Cs. annulata* periodically take blood meals.

### *Mansonia*

Adults are recognized by the speckled pattern of pale and dark scales on their wings, and their pale-banded legs. The species here are primarily found in wet tropical regions, but some range to temperate latitudes, such as Tasmania to the south and Sweden to the north. The larval stage is long and the number of generations per year is usually limited to two or three in the tropics and one in temperate regions. Larvae penetrate the roots of submerged plants with their siphon to tap air cells. A variety of plants is used, but water lettuce, *Pista stratoites*, is the most common and preferred by some *Mansonia* species.

Urban environments may provide limited breeding sites for *Mansonia* species, but the adults are strong flyers and may travel several kilometers from the breeding site in search of a blood meal. *M. perturbans* is distributed in North America south to the Gulf Coast, and in Europe from the UK and Sweden south to the Middle East. Females readily enter houses at night, and it is a consistent pest wherever it occurs. *M. uniformis* is distributed from West Africa eastward through the Indian subcontinent to Japan and the Australian region. Females will typically travel far for a blood meal, and enter houses soon after dark. *M. indubitans* is distributed in northern South America, from Brazil to Peru; while it is a forest mosquito in the Amazon Valley, it readily invades houses in Peru. *M. africana* commonly occurs indoors after dark in Central Africa.

***Mansonia pertubans*** Adults are 5.5–6.5 mm long and dark brown; the thorax is dark brown, and with numerous yellowish scales. The proboscis has a median white band; legs have white bands on base of tarsi; the abdomen has indistinct, narrow white bands at the base of segments. Full-grown larvae are 6–7.5 mm long, yellowish white with a tinge of green. Larvae remain below the surface in the debris at the bottom, or attached to plant roots, and are rarely at the water surface. In the UK, larvae attach to the roots of *Acorus*, *Ranunculus*, *Glyceria*, and *Typha*. The Norwegian form is found exclusively with *Sparganium*. In the USA, a variety of plants provide suitable attachment, including species of *Typha*, *Limnobium*, *Pistia*, and *Saggitaria*. Attachment is often for a long time, but larvae often relocate. Air may be obtained at the surface, from dissolved oxygen in the water, or from air cells in submerged plants; satisfactory development requires submerged plants. Pupae generally remain submerged and attached to plant roots; they are at the surface prior to adult emergence. Eggs are laid in rafts on the water surface; the number of eggs per raft is about 195, but may be as many as 308 eggs. Hatching occurs in 6–7 days in the field, and 6 days in the laboratory at 20 °C. Development takes 83–117 days, while the pupal period takes about 5 days. There is one generation per year, and the majority of larvae complete development during the first season. Other larvae overwinter and complete development in the spring of the second year. Adults are active and feed during the day and night, and readily enter houses to feed; the greatest activity is at dusk. Both sexes fly long distances, and have been recorded 17 km from breeding sites. This species is distributed in North America, from southern Canada to the Gulf of Mexico. In Europe it occurs from Sweden to the UK, and extends to the Middle East.

### *Psorophora*

These are primarily Neotropical mosquitoes, but several species, including *Psorophora columbiae*, *P. discolor*, *P. ferox*, and *P. ciliata*, occur in North America. The adults are large, ranging from 6 to 9 mm long. Larvae of some species are predaceous, feeding on other mosquito larvae and other animals in temporary pools. Eggs are deposited singly on wet substrates, and they can withstand months of desiccation, then hatch when flooded (thus their name floodwater mosquitoes), which often gives rise to enormous numbers of adults. Breeding sites are flooded pastures and sometimes rice fields. Larvae of some species are predators. Adults bite during the day and night.

***Psorophora ciliata* (Fig. 7.4a)** Adult wing span is 8–9 mm and the body is dark brown to black. The head and thorax have patches of white and yellow scales. The abdomen has patches of purplish blue, iridescent scales. The proboscis and tarsi are banded white. Larvae breed in rain pools and other temporary standing water. Full-grown larvae are 10–12 mm long, but occasionally 15 mm long. The body is pale gray and the thorax is nearly as long as broad; the abdominal segments 1–7 are large and with small tufts of setae set in small depressions. The anal siphon is large and broad at the base. Larvae are predators on aquatic invertebrates and other mosquitoes, including other *P. ciliata*, in the habitat. First-stage larvae do not feed. Eggs are deposited on moist soil in woodlands and meadows; eggs are laid in small batches of 6–8. Hatching is in about 5 days, but in general hatching occurs when dried sites are flooded. Development takes 4–10 days; the pupal period is 1–3 days. Females feed outdoors and generally do not enter structures or barns. This species occurs in transient pools, ditches, and agricultural fields in eastern and southeastern USA, and its range extends to Canada and South America. It is a severe biter and capable of penetrating several layers of clothing with its proboscis. Gallinipper is a common name for this mosquito.

***Psorophora confinis*** Adults have the proboscis, tarsi, and abdomen with white bands; abdominal bands are triangular. Eggs are laid on soil or vegetation above the existing water line and they are able to withstand a long period of desiccation. Hatching occurs in 4–5 days when submerged soon after deposition, after being on the surface for weeks; hatching occurs within minutes after submergence. Development in summer may last 4 days, and the pupal period lasts 1–2 days. Adults live for 1–2 months, and they have a flight range of about 16 km. Temporary pools can produce large numbers of adults in a short period, and breeding occurs in agricultural fields. The common names rice-field mosquito and glades mosquito have been used for this species. The adults have a painful bite, and feed during the day and night; adults are attracted to lights at night. This species is widely distributed in the USA, and its range extends to Central and South America.

## Drosophilidae

Fruit flies or vinegar flies are small, yellowish-brown flies (Fig. 7.1g) that gather around ripe and decaying fruit, fungi, and decaying vegetation indoors and outdoors. There are about 3000 described species in 60 genera. There are 1595 species in the genus *Drosophila*. The majority of species occupy outdoor habitats in tropical regions. There are about 10 species closely associated with urban habitats, indoors and outdoors. Adults are 3–4 mm long and yellowish brown. These flies are distinguished by wing venation, and plumose antennae. Larvae are yellowish white, and with the posterior spiracles close together on a stalk. The small size of the adult gives them access to food sources unavailable to some other flies, but their weak flight limits their activity to protected locations. There are numerous species, and indoor populations are usually represented by natural populations outdoors.

Adults and larvae feed on yeasts and bacteria, which occur in fermenting substances rich in carbohydrates. There are distinct kinds of feeding and breeding habits in this family of flies: fruit-feeders and general scavengers. Although some species are restricted to one of these categories, many are not. *Drosophila immigrans* and *D. funebris*, both common as indoor pests, are regarded as fruit-feeders and general scavengers, respectively, but are also found in fungi. *D. repleta* and *D. immigrans* can occur in large urban populations, infesting restaurants, kitchens, and hospitals where they breed in various forms of waste, including feces. Adults are attracted to a variety of organic compounds that are found in fermenting organic material. These include amyl and ethyl alcohol, acetic acid (thus the name vinegar flies), lactic acid, and ethyl acetate. Mixtures of some of these are more attractive to adult flies than pure substances.

Eggs are about 0.5 mm long, and on the surface there are one or two pairs of long filaments; these serve as respiratory filaments and to maintain the eggs on the surface of soft or fluid substrates. Eggs hatch in about 24 h at 25 °C, and first-stage larvae begin feeding on the substrate. The first stage lasts about 1 day, the second stage about 2 days, and the third stage about 4 days. Third-stage larvae are about 4.5 mm long and have distinct black mouthparts. The mouthparts have one tooth on the first-stage larva, two or three teeth on the second-stage larva, and 9–12 teeth on the third-stage larva. Larvae are nearly translucent and the internal organs are visible. Large tracheal tubes extend the length of the body, fat bodies appear as pale-white sheets, the coiled intestine is yellowish brown, and Malpighian tubules are dark yellow. Embedded in the fat bodies are dark gonads; these are large in larvae that will produce males, but small in larvae that will produce females.

Third-stage larvae usually move from the feeding substrate to a dry surface to form the puparium. Puparia may be tightly attached to substrates with casein-like glue that can often resist commercial cleaning operations. Within the puparium a fourth-stage larva completes development in about 12 h. The pupal period is about 1 day, and the adult fly emerges in 4–6 days. Under ideal conditions and with adequate food, adults live about 40 days; however in some species they may live for only 6 or 7 days. When there is sufficient food for adult females, they may begin laying eggs 2 days after emergence. Egg production increases to a maximum on day 4 and 5, and is maintained for 3–10 days. The total number of eggs laid varies, but as many as 3168 have been recorded during a 70-day female life span.

Predators and parasites of fruit flies include several species of wasps. The cynipids *Pseudeucolia bochei* and *Phaenocarpa tabida* are known to use *Drosophila* as hosts. The female *P. bochei* lays an egg inside fruit fly larva. The egg hatches in about 48 h and the wasp larva completes development in the fly larva and pupa, and emerges from the fly puparium in about 3 weeks. The life history of *Phaenocarpa tabida* is similar, except for a larval diapause that extends the development of this cynipid by several weeks. Parasitism by *Pseudeucolia bochei* in the wild is mainly of *D. melanogaster*, *D. subobscura*, *D. hydei*, *D. funebris*, and *D. littoralis*. Adult wasps produced from the large *Drosophila* species, such as *D. funebris* and *D. littoralis*, are often large.

***Drosophila ambigua*** Adults are about 3 mm long and the body is dark brown; the wings are translucent. There are eight rows of acrostichal bristles, and the palps have one strong terminal bristle and one subterminal bristle. Abdominal tergites are dark brown without anterior bands. Development from egg to adult takes 19–25 days at 18 °C. This species occurs in urban habitats in the UK and northern Europe. It has been collected in greenhouses, fruit stores, and houses.

***Drosophila busckii*** Adults are 2–2.5 mm long and yellowish brown. The mesonotum is yellowish brown with five dark brown to black longitudinal stripes; the middle stripe splits posteriorly. Development from egg to adult takes about 21 days at 18 °C, and 12 days at 25 °C. It is cosmopolitan and commonly found indoors. It has been collected in chicken coops, greenhouses, around decaying plant material outdoors, and from *Russula* mushrooms.

**Dark fruit fly, *Drosophila funebris* (Fig. 7.5e, f)** Adults are 3–4 mm long and reddish brown. The thorax has eight rows of acrostichal bristles, and three sternopleural bristles. The abdomen is dark brown with a yellow band along anterior margin of at least the first four tergites. Males are lacking sex comb setae on the forelegs. Full-grown larvae are about 7 mm long; spinous areas completely surround the anterior portion of each body segment, but may be more pronounced as a ventral patch. The brownish-black posterior spiracles are on an extension equal to the length of the last segment. The last body segment and bases of the posterior spiracles are covered with fine setae. Eggs are laid singly directly on the substrate, and they have two threadlike filaments at one end. Hatching occurs in 12–24 h. Development takes 23–28 days at 18 °C, and 13–15 days at 25 °C. Numerous generations occur each year, and indoor populations may persist year-round. This is a cosmopolitan species associated with decaying and fermenting organic material, including human feces indoors and outdoors. There is a record of this species causing intestinal myiasis in humans.

***Drosophila hydei*** Adults are 3–4 mm long and dark brown. The mesonotum is brown or gray with a green tinge; nearly all bristles on the mesonotum are inserted on a dark spot; there are 6–8 rows of acrostichal bristles. Development from egg to adult is completed 23–28 days at 18 °C. It is a cosmopolitan species and usually found associated with urban structures, including houses, and the perimeter of vegetable gardens.

***Drosophila immigrans*** Adults are 3–3.5 mm long and pale brown, and with 6–8 rows of acrostichal bristles. The mesonotum is yellowish brown, and there are 6–8 rows of acrostichal bristles. The foreleg femur has a row of about 10 small black spines on the inner side. The wing is clouded at the apex; the abdomen is yellow and with two large, dark-brown triangles on each tergite. Development from egg to adult is completed in 18–23 days at 18 °C. It is a cosmopolitan species, found in food stores and warehouses.

**Common fruit fly, laboratory fruit fly, *Drosophila melanogaster*** Adults are about 3.5 mm long; the thorax and abdomen are yellowish brown, and with bright red eyes. Full-grown larvae are 5–6 mm long, pale yellow to white, and the last 10 segments of the body have patches of spines. Posterior spiracles are yellowish brown, and on a short projection. Adults indoors are found close to the breeding site, and rarely at windows and lights. Larvae remain buried in the food substrate, with only their posterior spiracles exposed to the air. Eggs are laid singly or in small groups directly on the substrate; fecundity is about 600 eggs. Hatching occurs in 12–24 h. Development from egg

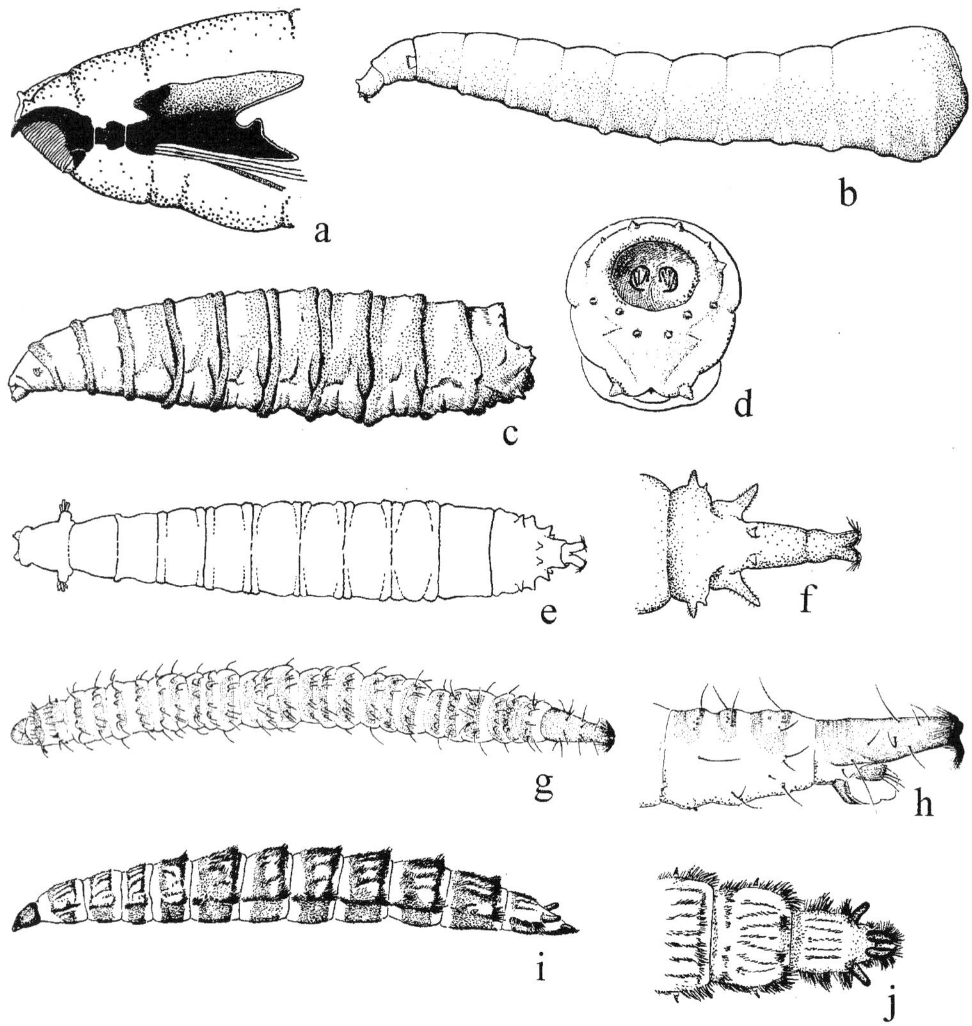


**Figure 7.5** Diptera larvae. (a) *Musca domestica*, cephalopharyngeal skeleton (Muscidae); (b) *M. domestica*; (c) *Sarcophaga crassipalpis* (Sarcophagidae); (d) *S. crassipalpis*, posterior view of terminal segment and posterior spiracles; (e) *Drosophila funebris* (Drosophilidae); (f) *D. funebris*, dorsal view of posterior segments; (g) *Psychoda* sp. (Psychodidae); (h) *Psychoda* sp., lateral view of posterior segments; (i) *Scatopse notata* (Scatopsidae); (j) *S. notata*, dorsal view of posterior segments.

to adult is 18–20 days at 18 °C, and 8–10 days at 25 °C. Third-stage larvae move from the substrate to a slightly drier location to form the puparium. Adults emerge in about 4 days, and they mate and begin egg-laying within 2 days. Adults live about 30 days in humid conditions. Outdoor populations are in garbage, and decaying vegetable material in natural areas. This species is nearly cosmopolitan.

***Drosophila obscura*** Adults are 2–3 mm long and brown. The mesonotum is brown with dark-brown stripes; there are two sternopleural bristles. The abdominal tergites are dark brown without anterior bands, and males have two sex combs on each foreleg. This species is widespread in outdoors and indoor habitats and it is common in urban areas.

**Brown fruit fly, *Drosophila repleta*** Adults are 3–4 mm long and reddish brown. Abdominal tergites have a yellow band along the anterior margin of at least four tergites. The males are without sex comb setae on forelegs. This species is associated with decaying and fermenting organic material, indoors and outdoors. It is widely distributed in urban environments.

***Drosophila simulans*** Adults are 2–3 mm long and the body is brown. The mesonotum lacks distinct stripes; the maxillary palps have two strong bristles. The yellow abdominal tergites have a dark-brown band along the posterior margins; these bands are not interrupted in the median line. This species is widespread and occurs indoors with decaying and fermenting plant material.

**Drosophila subobscura** Adults are 2–3 mm long, the thorax is blackish brown, and the abdomen is pale brown. Maxillary palps have only one conspicuous bristle; the upper tarsal comb has 10–15 bristles and the lower tarsal comb has 9–13 bristles. Adults of this species are known from a wide variety of habitats, including urban and rural gardens, and the interior of buildings. It is very common in late summer and fall. The daily pattern of activity of this species includes peaks in the evening. It is distributed throughout the UK and continental Europe.

**Black fruit fly, *Drosophila virilis*** Adults are 2.5–3 mm long and the thorax and abdomen are dark brown to black. Wings are dusky, and the eyes are dark. Full-grown larvae are about 4 mm long and pale yellow, the posterior spiracles are brown, and the mouthparts are black. The egg has four long filaments. The puparium is cylindrical, dark brown to black, and the pupal respiratory horns are short. This species occurs in Asia and is often associated with decaying organic material indoors.

## Ephydridae

Shore fly adults are 3–5 mm long and gray to black. Some have patterned wings and have sparse setae on the thorax and abdomen. Adults are found in moist sites, such as marshes, along the shore of ponds and streams, and seashore. Sometimes they occur in very large numbers, and there may be swarms of adults in one place for a short period of time. Larvae are aquatic or semiaquatic and some species occur in brackish or strong saline water. Ephydrid larvae generally have their posterior spiracles at the tip of a forked process, which adapts them to remain submerged in liquid habitats.

Large numbers of larvae and puparia may occur along the shore of freshwater pools and saltwater lakes. The leaf and stem mines made by the larvae of *Hydrellia* species become extensive and damage crops of watercress, rice, barley, and other irrigated cereals. *Helaeomyia petrolei* breeds in pools of crude petroleum. Algae and other organic matter in the oil is the larval food source; the oil is not digested. The brine fly, *Ephydra gracilis*, breeds along the shore of Great Salt Lake in Utah, and similar habitats in western USA. The larvae of *Ochthera* are predators on immature Chironomidae. In the urban environment, several species are associated with excrement and corpses, including *Hecamede persimilis* and *Chlorichaeta tuberculosa*. Larvae of *Discomyza maculipennis* penetrate deep into decaying organic matter, including body tissue. *D. incurva* is a parasite of *Helix* spp. land snails. *Tichomyza fusca* occurs around cesspits and areas where urine accumulates in soil, and it is known from cadavers in the early stages of decomposition. Adult *Chlorichaeta tuberculosa* is attracted to perspiration and eyes, and is a frequent pest.

**Cesspool ephydrid, urine fly, *Tichomyza fusca* (= *Teichomyza*)** Adults are about 4 mm long and brownish black; the wings are slightly dusky. Full-grown larvae are 10–13 mm long, cylindrical, and somewhat translucent. The body is covered with small black setae, and each of the thoracic segments except segment 1 and the abdominal segments anterior to the seventh bears an ambulatory process (proleg) on each side, near the middle. Each anterior spiracle ends in finger-like processes. The posterior spiracles are on long stalks projecting backwards. Eggs are laid singly or in batches directly on the infested substrate; egg-laying peaks when females are 17–27 days old; fecundity is about 614 eggs over the maximum life span of about 93 days. Hatching is in about 4 days. Larval development is completed in about 15 days, and the pupal period is about 15 days at 20 °C and 85% RH. The average generation time under laboratory conditions of 20 °C and 85% RH is about 58 days. This species is associated with decaying organic matter, and is common in feces. Larvae are frequently found in outdoor urinals and associated with human excrement. Larvae are gregarious, and when numerous they may block the drain pipes of septic tanks. A closely related species, *Scatella stagnalis*, has an egg to adult development time of about 16 days at 20 °C and 11 days at 25 °C. Fecundity for *S. stagnalis* is about 315; females live for about 15 days and males for 22 days. These species occur in Europe, including western Russia, the UK, and Ireland.

## Hipposboscidae

These unusual flies are blood-sucking ectoparasites on birds and mammals. Both males and females suck blood, but their bite is not painful. They are called louse flies when they occur on birds, and keds when they occur on mammals. They are 4–7 mm long, somewhat flattened, and winged or wingless. Most of the winged forms are dark brown; the legs are widely separated, and the abdomen of the female is membranous. The palpi are elongate and slender, and form a sheath around the proboscis. All species are larviparous, females rear one larva at a time internally; development is completed in 3–8 days. The anterior end of the larva is enclosed in the anterior part of the uterus and receives nourishment from a specialized gland (milk gland). When the larva is full-grown the abdomen of the female is distinctly swollen; when the larva is released from the abdomen it immediately forms a puparium. Formation of the puparium takes about 1 h and development to adult takes 20–30 days; the pupa may overwinter. Puparia may be in the nest material or attached to the hair or feathers of the host.

Pest status is based on the movement of lice flies indoors and biting people. This usually occurs in the spring before the host birds return to the nest sites, and the adult flies have emerged and are seeking a blood meal. Deer keds infrequently bite people when they occur in barns. Humans are not a preferred host for these insects, but there are records of some louse fly species biting people, including *Crataerina pallida*, *Hippobosca equina*, *H. camelina*, *H. variegata*, *H. rufipes*, *Melophagus ovinus*, *Lipoptena cervi*, *L. depressa*, and the pigeon fly, *Pseudolynchia canariensis*.

**Swift louse fly, *Crataerina pallida*** Adults are about 6 mm long and light to dark brown, and the body is somewhat flattened, leathery, and with strong setae. This species is found on swifts (*Apus apus*), but it also occurs on swallows and martins. Feeding occurs every 3–6 days; the adult louse imbibes nearly its own weight in blood from the host. Emergence begins at the end of April or early May when swifts return to breed, and peaks by the time the young birds are sufficiently feathered to serve as adequate hosts. The louse population declines before the birds migrate. The seasonal infestation lasts about 3.5 months. Pupae overwinter in the deserted nest, and the next generation of adult flies emerges after the birds return in the following summer. If the nest is not reoccupied, the flies will crawl away in search of food; they can live for long periods without food. They may enter rooms of houses that have swifts nesting under the eaves. They sometime find harborage in bed frames and in the sheets and blankets.

**Martin louse fly, *Crataerina hirundinis*** Adult females are about 6.2 mm long, pale to dark brown. The body is flattened and the abdomen leathery and without distinct segments. This species is parasitic on house martins (*Delichon urbica*), and sometimes swifts, and sand martins (*Riparia riparia*). It feeds daily on host birds, and adult flies will move between adjacent nests to feed. Emergence of the adult louse flies is in May when the host birds return to nest sites. The population peaks in August and September, but may extend into September or later. A few lice may be carried away with the birds in the feathers, but louse flies are not known from the winter nesting sites. This species may have two generations per year, perhaps due to the extended breeding season of the house martin. These louse flies may enter rooms indoors and bite people in the spring, before the host birds have returned.

**Deer keds, *Lipopterna depressa*, *L. cervi*, *L. mazamae*, *Neolipoptena ferrisi*** Adults are 4–6 mm long, winged or wingless, and the body is reddish brown. They have well-developed wings when they emerge from the puparium, and may fly in search of a host. The puparia are shiny brown to black, and rounded. Emergence of adults begins towards the end of summer, and continues in fall. Infestation is high through the winter. Wings are shed when they become established on a host. Larvae deposited by the female are yellowish white, and there is a black cap at the posterior end (which includes the spiracles). Some larvae are deposited while the female is still on the host animal, and they may be trapped in the fur for a short time, but most larvae are deposited while the host is resting. These species are found on European deer, mule deer, Virginia white-tailed deer, and wapiti. *Lipopterna mazamae* occurs on deer in South and Central America, and in southwestern USA. In fall, winged individuals may fly to barns, and there they sometimes bite humans.

## Muscidae

Adult muscids are 4–8 mm long and have a wing span of 13–15 mm; they are strong flyers. The body is usually blackish gray to black and their sponging mouthparts are well developed. The eyes are widely separated in the female and usually contiguous in the male. Females deposit eggs, but some species deposit first-stage larvae on suitable substrates. Larvae are smooth, cylindrical, and anteriorly tapered; larvae of *Fannia* have lateral processes. Larvae are 6–8 mm long, and feed as scavengers, but some are carnivorous on other insects in their food substrate. Development is rapid and overwintering may be as an adult, full-grown larva, or pupa. There are many domestic and peridomestic pests in this family. Most of the common species are cosmopolitan.

Pest status is based on their association with organic waste material and the transfer of pathogenic organisms to humans and food, and on the ability of some species to bite and suck blood. Many of these flies, especially the *Musca* species, do not bite, but are simply a nuisance around animals and humans. Muscids travel 2–3 km from their breeding site to find food or habitats for oviposition. *Hydrotaea* larvae are primarily saprophagous (except for the final instar), and the adults are mostly sweat flies with a preference for mammalian blood. Muscids in the urban environment often utilize garbage and excrement as a food source, and they have been associated with humans for centuries. Several species are active throughout the year, but develop slowly in cold months.

**Durn fly, *Dasyphora cyanella*** Adults are about 9 mm long and greenish blue; the thorax has two longitudinal dark stripes. This metallic muscid looks like a bluebottle fly (Calliphoridae). Eggs are laid in batches of 25–30 eggs, and deposited below the surface of cow dung in fields. Hatching occurs in 1–3 days. Larval development is completed in about 28 days, and the pupal period lasts 21–28 days. Adults are active from May to November, after which they hibernate indoors in buildings. They seem to prefer large and open structures, such as barns. This species is distributed in the UK and northern Europe.

**Little house fly, *Fannia canicularis* (Fig. 7.1i)** Adults are about 6 mm long and the body is blackish gray with brownish-yellow markings on the abdomen; the thorax has three brown longitudinal stripes. Full-grown larvae are 7–8 mm long and yellowish white to reddish brown, and slightly flattened; there are projections on the lateral and dorsal aspect of most segments. The anterior end is tapered and the posterior end is rounded. The anterior spiracles have six or seven lobes; the posterior spiracles are on small tubercles. Eggs are laid directly on the substrate, and are morphologically adapted for wet conditions. Hatching is in 30–40 h during warm temperatures of summer, and 3–4 days at other times. Development is complete in 8–10 days, and the pupal period lasts 9–10 days. Full-grown larvae leave the substrate and move to dry locations to form the puparium. Adults overwinter in protected locations outdoors and indoors, but they usually overwinter as pupae in the soil. In temperate regions, overwintering adults become active in late February, and emergence from overwintering pupae usually occurs in March. Natural populations of *F. canicularis* and other closely related species occur in nests of bees (*Bombus* spp.), decaying snails, and vegetation, and sites that have nitrogenous droppings and rich, decaying organic material. Males and females are attracted to honeydew and are often found at aphid-infested trees and shrubs. This fly is cosmopolitan, and it is one of the most abundant of all the flies associated with humans. They are found indoors and out, associated with excreta, vegetables, fruits, and beverages. It is usually more abundant in spring and fall; high temperatures and dry breeding media suppress populations.

**Latrine fly, *Fannia scalaris*** Adults are about 7 mm long. The thorax and abdomen are bluish black, and the abdomen has a dark median stripe. The middle tibia has a tubercle near the middle, and the coxae on the midlegs have two apically bent setae. Full-grown larvae are 6–8 mm long and may be brown to yellowish brown; the body is flattened. There are long lateral projections on all segments, and the projections on segment 8 are longer than the length of segment 7 and 8. Eggs are deposited directly on the substrate; hatching occurs in about 24 h. Development is completed in about 6 days, and the pupal stage lasts about 9 days. This species is cosmopolitan and it is associated with animal and human excrement. It resembles *F. canicularis* and has similar larval morphology and habits. The life cycle is similar to *F. canicularis*; and the larvae occur in excrement, cadavers, bird nests, and decaying fungi. When it occurs indoors it is usually associated with human or animal feces.

**Other *Fannia*** Several species are found in the urban environment and associated with human or undisturbed habitats. *F. benjamini* is a pest in suburban and recreational areas of California; here they annoy people by their attraction to perspiration and mucous secretions. *F. femoralis* is a small fly and it occurs throughout the USA and northern South America. *F. incisurata* occurs in North America, Canada, and from Iceland to Japan and North Africa, and from Mexico and neotropical regions. It is common in excrement and wet substrates. *F. leucostica* is found in North America and Europe, and it prefers dry excreta.

**Sweat fly, *Hydrotaea metorica*** Adults are 4–6 mm long; the males are shining black, females are dull gray. Wings are light gray and the halteres are black. The abdomen is dull gray with a median black band formed from connected triangles. Eyes of the male are narrowly separated. The full-grown larva has transverse rows of setae on abdominal segments 2–8; there are three prospiracular lobes, and the posterior spiracles are weakly sclerotized. Larval development takes about 39 days in summer. The puparium is pale orange, and slightly shining; pupal respiratory horns are very long, and bright orange red. Overwintering is in the larval stage. In Europe there are two generations per year, with adults appearing in May and June, and in August and September. Adults are attracted to perspiring cattle and humans. This species is distributed from Spain and Majorca to northern Scotland, to Sweden and Finland eastwards to Siberia and south to Israel. In North America it occurs from Arizona and New Mexico to British Columbia and Maine. It is a pest in Siberia, where it enters houses to feed on perspiration and blood. A related species, *H. pelluscens*, breeds in cow dung and the adults are a persistent sweat-fly around humans.

***Hydrotaea dentipes*** Adults are about 6 mm long and shining black to dull gray. Full-grown larvae are about 13.6 mm long and yellowish white; the posterior spiracles are reddish brown and elevated. Eggs are laid in batches beginning in late March; fecundity is 170–200 eggs. Development to the third-stage larvae requires 1–2 days, and this stage is predatory on other insects in the substrate. Third-stage larvae complete development in about 7 days, and the pupal period lasts 2–3 weeks. Development is completed in 23–25 days at 16–28 °C. There are usually two generations per year. In the agricultural environment, larvae breed in organic material, including dung of pig, horse, cow, and humans. In the urban environment larvae are in rubbish dumps, cesspools, carrion, and manure heaps. Third-stage larvae are predators of *Musca domestica*, *Stomoxys calcitrans*, and the larvae of other pest species of flies in animal dung. This species is nearly cosmopolitan, except for subarctic tundra and desert regions.

**Face fly, *Musca autumnalis*** Adults are 4–6 mm long and resemble the house fly. Eyes of the male face fly are nearly contiguous at the top; in the house fly they are not. Eggs are laid singly or in batches of 7–36, and females produce four or five batches in their lifetime; fecundity is about 230 eggs. Hatching occurs within 24 h. Larvae complete development in 5 days at 20 °C, and 2.5 days at 35–40 °C. Full-grown larvae move to drier sites to form the puparium; the pupal period is 7–10 days. Adults live about 10 days, but those of the last generation live for several months. Adults are attracted to undisturbed cattle excrement within 2 h after deposit; they usually do not visit manure that is in piles, or mixed with straw, hay, or urine. Face fly fecundity is affected by the entomoparasitic nematode *Paraiotonchium autumnalis*. This nematode is host-specific and infests face fly larvae in dung. Infected adult flies have one or more nematodes in their hemocoels, and eventually nematode larvae invade and disrupt the ovaries of female flies. *M. autumnalis* is native to Europe, western Asia, and eastern Africa, but now occurs in Canada and northern USA.

Face flies hibernate in large numbers as unmated adults in buildings, mammal burrows, and in other protected places, whether heated or not. Both sexes exhibit adult facultative diapause, which is characterized by fat hypertrophy and cessation of ovarian development. Overwintering groups have a sex ratio of 1:1. Mating occurs after they emerge from the overwintering site; females then disperse for egg-laying. Face flies tend to aggregate and overwinter in the same sites year after year; perhaps they are stimulated by or attracted to volatile chemicals remaining in the habitat. However, houses with continued problems with overwintering face flies may not be in areas where there are herds of cattle. The adults of *M. autumnalis* can fly long distances to find or return to suitable hibernation sites. Mild winter temperatures may limit their overwintering ability, especially along the Atlantic and Gulf Coast of the USA. Warm temperatures in winter may exhaust the metabolic reserves of post-diapausing flies, and reduce the number of reproductive adults the following spring. Fluctuations in populations may be linked to a variety of environmental factors.

**House fly, *Musca domestica* (Fig. 7.1h; Fig. 7.5a, b)** Adults are 4–8 mm long and there are four stripes lengthwise on the thorax; the wings are translucent. Full-grown larvae are 12–13 mm long and yellowish white; the body is smooth and slightly shiny. There is a patch of small spines ventrally between abdominal segments 1 and 7, but absent on the thoracic segments. Anterior spiracles are yellowish white and have six or seven orange-yellow openings, but not on distinct lobes. The posterior spiracles have a complete peritreme and the 3 openings are sinous. This is probably the most widely distributed insect pest, and it is associated with humans around the world. It is most abundant during the warm season, but it may overwinter as adults in temperate regions and remain a pest throughout the year. In North America and Europe it is common from July through September; in South America and Australia it is common from October through February and March.

Eggs are deposited in batches of 75–150; they are usually piled into masses and there are several deposits at intervals of 3–4 days. Females may deposit as many as 21 batches of eggs for 31 days after emergence. Females need mate only once to fertilize all the eggs laid in their lifetime. Under warm conditions the eggs hatch in 8–12 h. Larvae complete development in about 5 days, and the pupal period lasts about 4 days. Larvae spend 3–4 days in a migratory stage prior to forming the puparium. During this time, they usually move to a substrate that is drier than their feeding place. Adult flies live about 30 days during warm months, but this may extend to 60 days. Overwintering occurs as a larva, pupa, or an adult; in protected and moist locations, adults may live from October to April. The potential distance for adult dispersal is 27–1080 m in urban habitats, and 270–1530 m in rural habitats, but a distance of 9 km has been recorded. In urban localities, the usual distance traveled by *M. domestica* is about 400 m. Adults are inactive at temperatures below 7.2 °C, and temperatures below 0 °C are lethal. They remain alive for long periods at temperatures in the range of 10–26.6 °C.

Adult activity is diurnal, and reaches a peak between 14:00 and 16:00 h, which usually corresponds to the hottest and driest portion of the day. Adults are inactive and at rest during night, but they will move to artificial light during the day or night. Flight occurs at an air temperature of 11.6 °C, and reaches maximum intensity at 32.2 °C. Above 32.2 °C, flight declines rapidly and ceases at the thermal death point of 44.4 °C. High and low temperatures are lethal to *M. domestica* adults when humidity is high. Adults live longest at 15.5 °C and 42–55% RH, and they require two or three feedings of liquid food each day. Their sponging mouthparts restrict them to utilizing liquids or foods soluble in salivary secretions. Adult food sources include milk, sugar, blood, and other substances, such as feces and decaying organic matter; a source of water is also important. Adult flies do not emerge from puparia when exposed to temperatures below 11 °C for 20 days, or 8.8 °C for 24 h.

Larvae of *M. domestica* survive best in compost-like mixtures of decaying vegetable material enriched with dung or animal material, which is the basic formula for household garbage. Feeding larvae prefer a temperature range of 30–35 °C. Larvae prefer dung of pigs, horses, and humans to that of cows, which is preferred by the closely related species *M. autumnalis*. The long-term association between *M. domestica* and the production of garbage in the human household gives this species a secure future in the urban environment.

Morphological differences exist between *M. domestica* populations throughout the world. *M. d. nebulo* and *M. d. vicina*, sometimes called the Egyptian house fly, are two notable subspecies. *M. d. nebulo* is a common house fly throughout the Ethiopian ecological region, and is found in outdoor markets and houses, and is the most common household fly in southern India. *M. d. vicina* breeds primarily in horse dung in urban areas, and in donkey dung in rural areas.

In the development of insecticide resistance, the house fly is one of the most mutable pest species. It has the ability to develop resistance to representatives of all the major classes of chemical insecticides. All of the resistance-related mechanisms, which include enhanced metabolic degradation, diminished target site sensitivity, reduced rates of cuticular penetration, sequestration of toxicants, and behavioral changes enabling avoidance of toxic residues, have been demonstrated in this species. In many cases, several of these resistance factors exist concurrently.

**Bazaar fly, *Musca sorbens*** Adults are about 6 mm long and the thorax is grayish black. The female abdomen is grayish black; the terminal abdominal segments of the male are yellowish brown. Full-grown larvae are about 12 mm long and yellowish white; the puparium is pale orange. Eggs are laid closely packed and cemented together in rows, usually in batches of about 30. Preferred oviposition sites are human excrement, garbage, carrion, and other decaying organic material. Substrates may attract numerous females. As many as 42 000 larvae have been collected from 1 kg of human feces, which indicates that several females deposited eggs at this site. Females usually feed for a short time on the excrement substrate after ovipositing. Development takes 2.5 days at 23 °C, and about 15 days at 17–20 °C. Puparia are usually formed in dry locations away from the larval feeding site. Adults live 14–20 days, but the life span may be longer at lower temperatures. This species is widely distributed in the western hemisphere tropics and subtropics, and in Africa, the Middle East, India, and from China to Australia. It is a common household fly pest in India. The adults occur at outdoor markets and bazaars, food stores, and houses year-round; there is a population peak in the spring and one at the end of summer.

**Other *Musca*** This large genus has over 3900 species worldwide, and many are associated with rural and urban areas. *M. fasciata*, *M. vetustissima*, *M. vitripennis*, and *M. patoni* are attracted to sores or any body secretion. The bush fly, *M. vetustissima*, is a major pest over much of Australia, *M. biseta* is an equal pest in Africa, and *M. conducens* enters houses in India. These species commonly settle in large numbers on people living in unsanitary conditions, and eventually individuals habituate to the presence of flies crawling on the face and head. *M. osiris* and *M. tempestiva* breed in cow dung and the adults are troublesome sweat flies outdoors in deserts and villages; *M. osiris* extends across north Africa to Egypt, and north to Suffolk, UK; *M. tempestiva* extends from Africa to China and Japan, and north to the Channel Islands.

**False stable fly, *Muscina stabulans*** Adults are 8–10 mm long and dark gray; the abdomen is reddish brown. This species has sponging mouthparts, and the adults do not bite. The larval stages breed in a variety of material, including human excrement, manure, and garbage. Eggs are deposited on tainted food, and 140–200 eggs may be deposited in 2–3 days. The first- and second-stage larvae are primarily saprophagous, but the third stage is a predator on larvae of other flies in the substrate. Development is completed in 15–25 days, and there are two or three generations each year. Overwintering is in the pupal stage. Natural populations occur in bird and animal nests in natural areas. Other related species, such as *Muscina assimilis* and *M. pabulorum*, have similar habits, but do not enter houses. This species is nearly cosmopolitan. The pest status of *M. stabulans* is based on the large numbers that can breed in the urban environment, and the abundance of adults indoors.

**Bronze dump fly, *Ophyra aenescens*** Adults are about 5 mm long and yellowish green. Mating occurs on the second day after emergence. Eggs are deposited in batches of about 74 and hatching occurs in about 12 h; fecundity is 276–438 eggs. Development is completed in about 5 days, and the pupal period is 4 days. The life cycle is completed in about 14 days at 27 °C. Males live 15–18 days and females live 20–35 days. The adults are capable of dispersing about 6 km from their breeding site. Females are attracted to carrion, animal manure, and human feces. Larvae are saprophagous during the first stage, but second- and third-stage larvae are predators and attack other fly larvae in the substrate. This species has been used as a biological control agent for *Musca domestica* in commercial poultry houses, and urban refuse disposal sites. It first appeared in Spain in 1966, and in Germany in 1971, in Denmark in 1972, and in the 1970s it spread through central Europe. This species is nearly cosmopolitan in refuse disposal sites. Adults are commonly found indoors in the southern part of its range.

**Black dump fly, *Ophyra leucostomata*** Adults are 4–5 mm long, black and glossy; they have few setae on the body and the wings are slightly opaque. Adults prefer to rest on vegetation around breeding sites, and in northern regions they remain in the sunlight. Larvae develop in decaying organic matter, including human and livestock feces, and household kitchen garbage. First-stage larvae are saprophagous, but second- and third-stage larvae are predators and attack other fly larvae in the substrate. They are common predators of *Musca domestica* larvae, and may kill 2–20 maggots per day. *O. leucostomata* is distributed in North America and Europe.

**Other *Ophyra*** Dump flies are distributed around the world; they are often found around garbage disposal sites and are often abundant in urban areas. The adults of some species readily move indoors, while others remain outdoors. *O. anthrax* is widely distributed in the Caucasus and Central Asia, and occurs around toilets, animal carcasses, and in houses. *O. capensis* is found in southern Europe and has habits similar to *O. leucostoma*. In the Australian region, *O. nigra* is closely associated with humans. *O. ignava* larvae have been recorded from corpses of animals and humans (at the phase of ammoniacal fermentation).

**Stable fly, *Stomoxys calcitrans*** Adults are 4–6 mm long and they have a hardened proboscis projecting from the lower part of the head. Both males and females suck blood. Full-grown larvae are 11–12 mm long and the body is yellowish white and smooth. The ventral bands between abdominal segments 1 and 7 have ridges and scattered setae. The anterior spiracle is light yellow and has five or six lobes; posterior spiracles have a reduced periterme and irregular-shaped openings. Adult feeding habits are linked to domesticated animals and occasionally humans. Larval feeding habits are varied, and they may not always include feeding on dung. In warm climates adults remain in the open, but in temperate regions they are usually in animal stables or other protected sites. *S. calcitrans* is cosmopolitan in temperate and tropical regions, and a common pest of livestock and humans. This species probably originated in a warm climate, and it became associated with humans when animals were domesticated and stables were built. It is a biting fly pest in the urban habitats, and along coastal areas where its breeds along beaches (sometimes it is called the beach fly).

Eggs are laid in batches of 40–80; fecundity is about 600 eggs. Hatching occurs in 12–24 h during summer; in fall it is in 1–4 days. Development time varies with temperature, and may be 12 days to over 30 days. Full-grown larvae move from the substrate to dry locations to form the puparium. The pupal period is 2–20 days, depending on temperature. Adults live about 30 days, and males usually live longer than females. Adults are active from May to October, and they are pests in late summer and fall. They are strong flyers and may be found indoors as far as 1 km from cattle. They seek shelter at night, but do not bite after dark. Feeding on animals is preferred; attacks on humans are usually accidental except along beaches.

**Other *Stomoxys*** The other *Stomoxys* species common in peridomestic habitats in Africa are: *S. niger niger*, *S. n. bilineatus*, *S. inornatus*, *S. taeniatus*, and *S. varipes*. The diurnal activity of *S. n. niger* includes a peak of activity in the morning between 08:00 and 11:00 h, and a peak in the evening between 17:00 and 18:00 h; there is little activity during the middle of the day.

## Mycetophilidae

Fungus gnats are 3–6 mm long and slender, delicate flies. Adults are black, yellowish brown, and sometimes reddish brown. The head is small, the antennae are long and slender, the thorax is dorsally enlarged, and the coxae are elongated. Larvae are slender, smooth, and pale yellow to white, but with a brown to black sclerotized head. These flies are associated with moist and decaying organic material and fungi, although some are phytophagous and some predaceous. They occur indoors infesting moist or wet potting soil and decaying vegetables. Adults usually remain close to the breeding site and prefer habitats with high humidity and little air movement. They are sometimes at windows and may be attracted to artificial lights. Eggs are deposited singly in crevices in the substrate; hatching occurs in 2 days and larvae feed in the immediate area of the eggs. Development is completed in about 8 days, and pupae are formed in dry locations in the substrate, usually near the surface; the pupal period is about 3 days. The adults live about 2 weeks, but sometimes less when conditions are dry.

***Leia bimaculata*** Adults about 6 mm long, with wing length 4–6 mm; the body is usually black or grayish black, but the abdomen may be pale. Bands on abdominal tergites are usually broad in the middle, especially the tergite on segment 2. This species is widely distributed in the UK and perhaps continental Europe. It has been bred from decaying *Russula* mushrooms, and may occur at windows indoors.

***Mycetophila fungorum*** Adults are 4–6 mm long and the body is light to dark brown; the wings are translucent to pale brown. This species is common in woodlands, but also occurs in peridomestic habitats. It is widely distributed in North America. A closely related species, *M. ichneumonea*, is about 2.7 mm long and is reddish brown. It also occurs in peridomestic habitats in North America.

***Mycomya mendax*** Adults are about 6 mm long, and yellow and black. Larvae feed on decayed wood and fungi; adults may occur indoors. This species occurs from British Columbia to California.

## Phoridae

Phorids have a characteristic hump-backed appearance and wing venation that is limited to strong veins at the front margin. Adults are 3–5 mm long; the thorax is generally brownish black, the abdomen brown to yellowish brown. Adults are strong flyers, but have an erratic movement behavior when they land on surfaces. They are sometimes attracted to decaying organic matter and chemical solvents, and can be found at windows and lights. Full-grown larvae are 4–6 mm long, yellowish white, and they have a typical maggot shape. However, some are flattened and have projections dorsally and laterally on the body segments. Mouthparts of the larvae are similar to those of higher flies. The puparium has elongate and sometimes dark pigmented respiratory horns at the anterior end. These horns appear several days after pupariation. Phorids are usually associated with decaying animal matter, but some are parasites, predators, and many species live in bee, ant, and termite nests. The majority of species common in the urban environment are cosmopolitan.

Pest status is based primarily on their presence in small or large numbers indoors. The erratic walking and flying behavior of adults and their presence around food are a nuisance. Phorids associated with household habitats include species from several genera: *Diplonevra*, *Dohrniphora*, *Megaselia*, and *Spinophora*. Ovipositing females may be attracted to different substrates, but the basic life cycle and larval development are similar for most phorids. Species of the large genus *Megaselia* are common indoors, and *M. scalaris* is a cosmopolitan species that is known to infest a great variety of habitats, from decaying plant and animal material to causing intestinal myiasis in humans.

Phorids that regularly come to cadavers and carrion have had only a limited role in providing forensic evidence. Phorids associated with carrion may be feeding on organic liquids; some species may deposit eggs and larvae feed on the substrate. Phorids recorded at dead animals include *Diplonevra perefrinia* from dead bats in Malaysia; *Megaselia rufipes* and *M. scalaris* from dead rodents; and *Metopina oligoneura* and *Triphleba nudipalpis* from meat buried 15–20 cm in soil. Meat buried 0.5 m in soil attracts phorids as early as 4 days after placement, and the species include *Conicera tibialis*, *Megaselia rufipes*, *Metopina heselhausi*, *Triphleba autumnalis*, *T. dudai*, and *T. nudipalpis*. Phorids reported to occur at corpses include *Conicera tibialis* from a corpse buried 10 months, and adults were reported swarming over a buried animal; *Megaselia abdita* from a corpse exposed 90–98 days at 10 °C; *M. rufipes* from a 4-month-old corpse, and from a corpse exposed 52 days; *M. scalaris* from a corpse exposed 4–6 months and treated with a bleaching agent, and in a 90–98-day-old corpse; and *T. hylinata* from a corpse exposed 25 days, and in buried corpses. Larvae of *T. lugubris* are common scavengers in wasp colonies that are beginning to decline.

### *Conicera*

The adults are generally small, 0.8–1.6 mm long, black flies with distinct wings and antennae. The life history of some of the 20 species in the genus is known, and includes association with decaying organic matter, dead animals, and ants. Distinguishing characters include: the third wing vein is unforked, mid tibiae without dorsal bristles besides the basal pair, hind tibia without a hair-seam, and antennal segment 3 of male is conical.

**Coffin fly, *Conicera tibialis*** Adults are about 3 mm long, and the body is dull black; wings are translucent. Larvae are yellowish white, the mouthparts are distinctly black, and the posterior spiracles are brownish black. This species is widely distributed and may be cosmopolitan. It is commonly found associated with dead and decaying organic matter, and is known from corpses. Adults may occur in large numbers indoors.

### *Dohrniphora*

These flies are associated with fungi and decaying organic matter. Adults are attracted to organic matter, including dead animals and insects; larvae feed in a variety of organic substrates, and may cause intestinal myiasis in humans (*Dohrniphora cornuta*). Several species are associated with ants and termites. The adults are characterized by having isolated bristles on the tibiae, the third wing vein is forked, the hind tibia has a hair-seam, and the hind femur of the male has sensory organs near the base.

***Dohrniphora cornuta*** The adult male is 1.9–2.3 mm long, and the female is 1.5–2.4 mm long. Frons are brownish black (male) to pale brown (female); palpi are reddish yellow in the female, brown in the male; the proboscis is short in the male, and long and geniculate in the female. The thorax is blackish brown, the pleura bicolored – dorsally brown, ventrally yellow. As regards the male abdominal tergites: 1 is yellowish orange with a black spot, 2 is black basally with a yellow stripe, 3–5 are black, and 6 is yellowish orange and black posteriorly. As regards the female abdominal tergites: 1 is yellowish brown, 2 is brown, 3 and 4 are dull black and trapezoid or rectangular, and 7 is black. Full-grown larvae are 2.5–5.4 mm long and yellowish white; segments 2–9 have lateral projections and dorsal spines; the terminal segment has six long projections, with posterior spiracles medially. Eggs are laid singly or in batches of 3–5; at 22–24 °C hatching occurs in 24 h. Development takes 5–8 days, and the pupal period is 13–15 days at 22–24 °C. Full-grown larvae usually leave the feeding substrate to a dry location to form the puparium. The puparium is 3.6–5.2 mm long and is boat-shaped; it tapers to a point at the anterior end, and is broadly rounded at the posterior. Pupal respiratory horns are 0.36–0.41 mm long, slender, and flattened with a row of small setae along both edges from the base to the tip. Larvae are known to feed on a variety of decaying organic material, and there has been one case of human myiasis. This species is widely distributed and probably cosmopolitan.

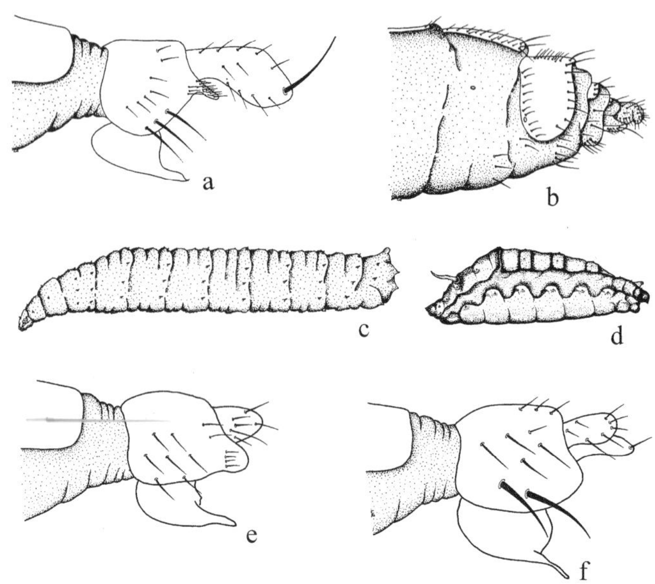


**Figure 7.6** Diptera: Phoridae. (a) *Megaselia scalaris*, lateral view of male genitalia; (b) *M. scalaris*, lateral view of terminal segments of female abdomen; (c) *M. scalaris*, third-instar larva; (d) *M. scalaris* puparia; (e) *M. agarici*, lateral view of male genitalia; (f) *M. halterata*, lateral view of male genitalia.

### *Megaselia*

This genus has nearly 1400 described species and about 45% of all phorid species are in *Megaselia*. The larvae feeding habits are extremely varied, and range from saprophagous to phytophagous. Larvae of a large number of species feed on decaying organic matter and occur in domestic and peridomestic habitats. Adults are characterized by having the third antennal vein forked, tibiae without isolated bristles, and the hind tibiae have a single hair-seam.

***Megaselia agarici* (Fig. 7.6e)** Adults are about 2.5 mm long and uniformly brown to yellowish brown; the halteres are yellow. Costa is 38% of wing length and the end of the R4+5 vein has four sensilla arranged in a line along the edge of the vein. Male genitalia (epandrium) is without strong bristles. Female sternum 6 is elongate and with four bristles. Full-grown larvae are about 4 mm long and yellowish white, posterior spiracles are pale, and they have black mouthparts. This species is often found associated with wild, uncultivated mushrooms in the USA and Europe. It is not typically found in cultivated, commercial mushrooms.

**Mushroom phorid, *Megaselia halterata* (Fig. 7.6f)** Adults are about 2.5 mm long and uniformly brown to yellowish brown. Costa is 37% of the wing length and the end of the R 4+5 vein has four sensilla arranged in two groups and not along the edge of the vein. Male genitalia (epandrium) has 2–3 strong bristles laterally on right. Female sternum 6 is absent. Full-grown larvae are about 4 mm long and yellowish white, posterior spiracles are pale, and they have black mouthparts. Eggs are laid singly or in batches of 6–10; hatching occurs in about 2 days. Development at 20–24 °C takes 10–14 days, and the pupal period is about 12 days. Adults are seasonally abundant and large numbers can occur around commercial mushroom production facilities. Outbreaks of this fly may occur and adults often enter commercial and residential buildings near the breeding site. This species is not found in wild, uncultivated mushrooms. It occurs in the USA and Europe. Other species associated with cultivated mushrooms include *M. bovista* and *M. nigra*.

***Megaselia rufipes*** Adults are about 3.5 mm long. The body is dark brown, and with the dorsal abdominal sclerites dark brown; the wing is clear to dusky. Full-grown larvae are 3.5–4.5 mm long and yellowish brown; the body is somewhat flattened and broadly oval at the posterior end. There are plumose tubercles on segments 2–10; on each segment there are four small tubercles dorsally and a single large tubercle laterally. This species is cosmopolitan, and often found at decaying organic matter indoors and outdoors. The larval mouthparts include the presence of ridges in the floor of the pharynx, indicating that this species is saprophagous.

**Megaselia scalaris (Fig. 7.1j; Fig. 7.6a–d)** Adults are 1.7–2.3 mm long and the body is yellowish brown; dorsal abdominal sclerites are yellow to yellowish brown. Wings are clear to dusky, and the halteres are yellow. Full-grown larvae are about 4 mm long and pale yellow; the posterior spiracles are dark brown. Eggs are usually laid at the edge of the substrate, and females oviposit regularly for about 30 days. Fecundity is about 600 eggs, but can be as many as 1000. Hatching occurs in about 24 h; it is about 31 h at 22 °C and 18 h at 29 °C. Larvae immediately enter the substrate to feed. Larval development is complete in about 73 h at 22 °C and 50 h at 29 °C. First-stage larvae can live for 5–6 days without feeding. Third-stage larvae search for a dry pupation site for about 60 h at 22 °C and 33 h at 29 °C. Pupal respiratory horns appear in 1–2 days. The pupal period is about 15 days at 22 °C and 7 days at 29 °C. Adults live about 30 days. Total larval development time under a photoperiod of light–dark (LD) 12–12 at 32 °C is about 13 days, 17 days at 27 °C, and 37 days at 21 °C.

Adults indoors are found close to the breeding site and often at windows and lights near the site. When there are large populations, adults may occur far from the breeding site. Larvae remain buried in the food substrate, and puparia are usually found in a dry location adjacent to infested material. Outdoor infestations are common in dead animals located in moist and humid environments. Indoors, this species feeds in decaying plant and animal matter, including sewage and household organic waste. This phorid is common in the urban environment, it occurs in a variety of decaying organic substrates, and can persist throughout the year. It is often found in hospital and health care facilities, where it infests organic substrates, such as human waste. Larval feeding habits of *M. scalaris* are extremely varied. In addition to dead and decaying animal matter, larvae have been reported feeding on onion bulbs, culture media, animal feces, and various species of laboratory insects.

This phorid has been incorrectly considered parasitic on at least 10 different insects, including the gypsy moth, cotton leafworm, and boll weevil. Females of *M. scalaris* are capable of detecting injured individuals (field or laboratory), and laying eggs on that tissue. The feeding habits of larvae of this phorid are saprophagous. Close examination of the larval mouthparts shows the presence of pharyngeal ridges, and these ridges in the floor of the pharynx indicate saprophagy. The food of a saprophage is particulate matter in a liquid, and the pharyngeal ridges act as a filter to concentrate the food material and eliminate excess liquid.

### *Phalacrotophora*

Species in this genus are 1.2–2.7 mm long, and the body is yellow to black; the head (frons) is generally shiny and with coarse punctures. Antennal segment 3 is globular or oval, and often with a dorsal arista. Several species are parasites or predators of coccinellid beetles (Coccinellidae). *Phalacrotophora longifrons* has been found hovering over the burrow of the pigeon tremex, *Tremex columba*, and it has been reared from the nests of *Crabro* spp. (Sphecidae). *P. philaxyridis* is a pupal parasite of *Harmonia axyridis*, the Asian ladybird beetle. It has been recorded from Japan, and is probably distributed wherever the Asian ladybird beetle is found. *P. berolinensis* and *P. fasciata* are parasites or predators of several species of coccinellid beetles, including species in *Adalia*, *Anatis*, *Aphidecta*, *Calvia*, *Chilocorus*, *Coccinella*, *Exochomus*, *Harmonia*, *Myzia*, *Psyllobora*, and *Vibidia*.

***Phalacrotophora nedae*** Adults are 1.4–1.8 mm long. The thorax, abdomen and legs are yellow to dull yellow; antennal segment 3 is oval, orange, and the arista is short. The wing is 1.7–2.0 mm long, and pale yellow; the costa is 41% of wing length. This species has been bred from several beetle species, including larvae of *Neda marginalis* and the pupae of *Neocalvia anastomosans*. It is distributed in North and South America. It is closely related to the European species, *P. fasciata*, which is also known to attack coccinellids.

***Phalacrotophora halictorum*** Adults are 1.4–2.5 mm long. The thorax and abdomen are blackish brown; the legs are yellowish brown. The wing is 1.9 mm long, yellowish brown and narrow; the costa is 40% of the wing length. This species has been reared from the larvae of *Nomia melanderi* (Halictidae), and found in the the nest of *Halictus ligatus* and *Lasioglossum zephyrum* (Halictidae).

***Phalacrotophora epeirae*** Adults are 1.2–3.0 mm long. The thorax and abdomen are yellowish brown to reddish brown; the legs are yellowish brown. Abdominal tergites 2 and 3 have a black spot laterally. The mesopleura has numerous small setae and one long spine. The wing costa is 52% of the wing length. This species has been reared from the eggs of *Epeira* spp. spiders. This species is distributed widely in North America, but may occur in other regions.

### *Spiniphora*

Species in this genus are distributed in North America, Europe, and Asia. The larvae of several species are saprophagous, and are known from dead and decaying animals. Distinguishing features include: mesopleura without bristles, mid-tibiae with strong anterior bristle, third wing vein is forked, and hind tibia lacks a hair-seam.

***Spiniphora bergenstammi*** The adult male is 1.7–2.3 mm long, and the female is 2–2.7 mm long. The thorax is black to brownish black, sometimes reddish brown; abdominal tergites are black with yellow seams, and the venter is yellowish brown. Antennal segment 3 is oval, and orange; the palpi are yellow. Legs are yellow to yellowish brown. The wing is opaque gray, and the costa of the male is about 48% of wing length, while the costa of the female is 50–54% of wing length. Larvae are known from a variety of decaying animal matter, including mollusks and vertebrates (rodents), both indoors and outdoors. This species is distributed in North and South America, and Europe, and perhaps Asia.

## Piophilidae

This family includes slender flies that are about 5 mm long and glossy to metallic black or blue, and with few bristles and setae. Antennae are short and the head is round. The larval stages are general scavengers on animal material. They are distributed worldwide, and the most economically and medically important species is the cheese skipper, *Piophila casei*. Other species of *Piophila* live in carrion.

**Cheese skipper, *Piophila casei* (Fig. 7.7a)** Adults are 3.5–4 mm long and the body is black, but the lower part of the head, the antennae, and parts of the legs are yellow. The wings lie flat over the abdomen when at rest. Full-grown larvae are 6–9 mm long, and slender, cylindrical, and somewhat translucent. There are transverse rows of fine setae on the anterior ventral margins of segments 5–11. The anterior spiracles have 8–10 lobes; the posterior spiracles are light brown and located on a rounded tubercle.

This fly is known primarily as a pest of stored cheese (*casei*, Latin for cheese) and preserved meats. Larvae prefer protein-rich but generally dry animal material, particularly preserved meats, fish, and cheese. They have been recorded at dead animals and corpses. Natural populations occur outdoors, and they are frequently found in bird nests. Eggs are laid on the surface of the food, and females deposit about 150 eggs during a period of 3–4 days; fecundity is about 500 eggs. Hatching is in 24 h at 27–32 °C, or may be delayed for 2–3 days. Larvae crawl into and feed in small cracks in the cheese substrate, and into connective tissue between muscle bundles of preserved meat. They are often found deeply within preserved food. Larval development is completed in 5 days at 27–32 °C, or it can be extended to 12–14 days. Full-grown larvae move to a dry portion of the substrate to pupate. The pupal period is about 8 days at 25 °C and 6 days at 30 °C. Adults live for 3–4 days at 24 °C and about 21 days at 15 °C. Larvae can survive for 1–3 h at 51 °C, but immersion in water at 54.5 °C is lethal. Full-grown larvae can live for 6 months at 9–10 °C; partially grown larvae can survive at −15 °C for 64 h. Adults can live for 30 days, without food, at 9–10 °C.

The larva of *P. casei* is known as the cheese skipper, because when larvae are in an exposed situation they are capable of suddenly leaping into the air. They can rise 15–20 cm in the air above the surface, and travel a horizontal distance of 25 cm. This leap is achieved by bending the body into an arc until the anterior mouthhooks can grip the posterior tip of the abdomen. At the posterior end of *Piophila* there are two small papillae

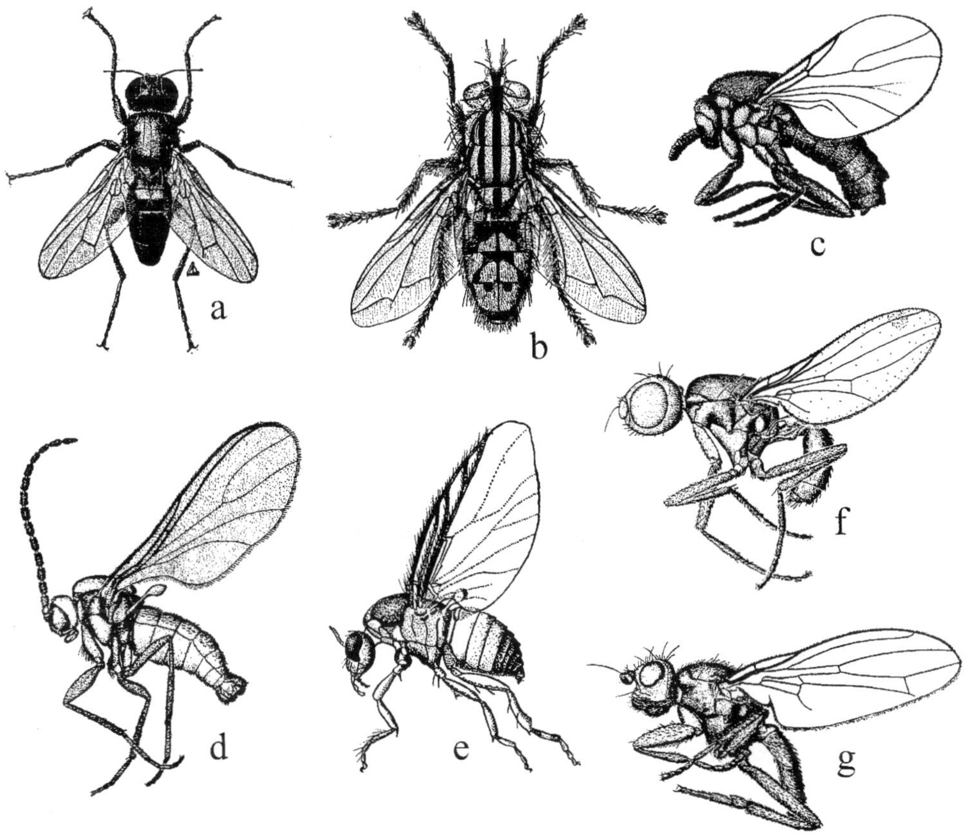


**Figure 7.7** Diptera. (a) *Piophila casei* (Piophilidae); (b) *Sarcophaga cruentata* (Sarcophagidae); (c) *Scatopse* sp. (Scatopsidae); (d) *Sciara* sp. (Sciaridae); (e) *Simulium* sp. (Simulidae); (f) *Sepsis* sp. (Sepsidae); (g) *Leptocera* sp. (Sphaeroceridae).

that provide a place for the mouthhooks to grasp. Once the mouthhooks are attached, the longitudinal muscles contract and the grip is suddenly released, and the larva is flung into the air, perhaps out of danger from a predator or to a more favorable feeding site.

**Other Piophilidae** Several species are associated with the human environment and are considered synanthropic. *Protopiophila latipes* is distributed worldwide, and it has been recorded from slaughterhouses, and animal carcasses (cat, wild deer). *P. contecta* is distributed in the oriental region, and the adults are attracted to decayed meat. *Stearibia nigriceps* has been reported from slaughterhouses, meat factories, poultry farms, and a human corpse. In Japan, *S. nigriceps* was reared from the bones of a whale, and the carcass of a deer. *Liopiophila varipes* has been found in food-processing plants and poultry farms, and adults are attracted to decayed meat.

## Platypezidae

Smoke flies are 2–5 mm long, usually black or brown, and with the tarsi on the hind legs distinctly enlarged. Most species are found in damp woods associated with decaying organic matter and fungi. Some species are called smoke flies because they are attracted to the smoke from open wood fires. The common species in the USA associated with smoke are *Microsania imperfecta*, the eastern smoke fly, and *M. occidentalis*, the western smoke fly. Although they are not common, adults and larvae may be found in wet and decaying organic matter indoors.

## Psychodidae

Psychodids are about 5 mm long and usually the legs and wings are covered with fine setae. Some species may resemble moths by their small size, and the densely setose wings. The family is divided into two groups: moth flies and sand flies. Moth fly females are not blood-feeding, and have wings that are held roof-like over the body; the larvae are aquatic (Fig. 7.5g, h). Sand fly females are blood-feeders, and do not hold their wings roof-like. The larvae of sand flies are not aquatic. Larvae of both groups feed on decaying plant and animal material. Sand flies are found mainly in the tropics and subtropics, with a few species occurring in temperate regions in both the northern and southern hemisphere. The females have long mouthparts capable of puncturing skin and sucking vertebrate blood. The human-biting genera are *Phlebotomus* in the eastern hemisphere, and *Lutzomyia* in the western hemisphere. They are vectors of the protozoan parasites, *Leishmania*, which are the cause of leishmaniasis (kala-azar), and they vector a bacterium, *Bartonella bacilliformis*, which is the cause of bartonellosis (Oroya fever) and sand fly fever virus.

Moth flies have a cosmopolitan distribution. They are known as filter flies and sewer flies because they often breed in sewage treatment plants, and drain flies because they frequently occur in clogged drains indoors and sometimes outdoors. The domiciliary genera are *Psychoda* and *Telmatoscopus*. *Psychoda* species have the terminal antennal segments reduced in size, but *Telmatoscopus* species do not. Species in these two genera occur as natural populations in urban and suburban environments. Moth flies are found in shady and undisturbed sites along stream banks, in wet decaying vegetation, bird nests, and tree holes.

Psychodids are commonly associated with the material in the filter beds of sewage treatment plants. The larvae feed upon the organic film that covers the surface of the filters. This film contains algae (*Ulothrix*), fungi (*Phormidium*), bacteria, and protozoans. The fly larvae are beneficial because they break down the gelatinous-like film that forms on the filter bed, and produce small fecal pellets that are easily carried away in the water that drains the filter beds. Adult moth flies can be serious pests when large numbers of them move to adjacent neighborhoods. They may be present year-round in buildings close to a sewage treatment facility. Live and dead adult moths may cause asthmatic reactions in people at the site or in other areas, such as around outdoor lights and doors where the adults aggregate, and their dead bodies collect.

**Sewer fly, *Clogmia albipunctatus* (= *Telmatoscopus*) (Fig. 7.1k)** Adults are 4–5 mm long, and the thorax and abdomen are uniformly brown to grayish brown. Wings have two dark spots close to the base, and they have 8–9 white spots along the margin. Full-grown larvae are 8–10 mm long. The dorsal sclerites are dark brown to black; the posterior breathing tube is long and well sclerotized. This is a common insect in the filter beds of municipal sewage treatment plants, and in clogged drainpipes in residential buildings. In natural habitats, this species occurs along stream banks, in the mud of tree holes and rain barrels, and in wet and decaying leaves. It is widely distributed in the UK and northern Europe.

**Filter fly, *Psychoda albipennis*** Adults are about 1.2 mm long and the wings are uniformly gray and unmarked. The body is gray and distinctly hairy, and the antennae have 15 segments. Larvae are common in sewage treatment plants, in the filter beds, and on the bacterial film. They occur in decaying vegetation, and in septic drains. The adults feed on nectar and polluted water. This species is generally distributed throughout the UK and continental Europe.

**Drain fly, *Psychoda alternata*** Adults are about 2.5 mm long and the body is uniformly gray and covered with fine setae or hairs; antennae are 15-segmented. Full-grown larvae are 4.5–6 mm long and yellowish white to light brown. There are sclerotized areas on the dorsum of all the larval segments. The head is dark brown and the terminal segment is sclerotized. Eggs are laid in batches of 20–100 directly on decaying substrates, such as the gelatinous-like surface that is in clogged drains or around fixtures in bathrooms and kitchens, and the surface of sewage disposal beds. Hatching occurs in about 48 h and larval development is complete in 9–15 days. The pupal period lasts 20–40 h. During warm weather and indoor conditions the life cycle takes 21–27 days to complete. There may be successive generations throughout the year, depending on the availability of a suitable substrate. The adults are weak flyers, and indoors they remain close to the breeding site, usually resting on walls in dark locations. They are attracted to lights at night. Outdoors, adults are found on foliage, usually in moist and shaded sites; they will feed on nectar and polluted water. It is common year-round, but may be more abundant when there is abundant food for the larvae. This species is generally distributed throughout North America, and it may occur in other regions.

**Pacific drain fly, *Psychoda pacifica*** Adults are 2–2.3 mm long and they have brownish-gray wings. This species is common in domestic habitats in the spring in western USA, and occurs from southern California to Alaska.

**Other *Psychoda*** The species that commonly occurs indoors along the Pacific coast, ranging from Alaska to southern California, is *P. phalenoides*. The adult is about 2 mm long, and has 13-segmented antennae and brownish-gray wings. There is a dense covering of setae on the wings and body. Adults of *P. satchelli* and *P. cinerea* are pale yellow and commonly found associated with the filter beds of sewage treatment plants. *P. satchelli* has 14-segmented antennae, and *P. cinerea* has 16-segmented antennae. Both species are generally distributed throughout the USA. *P. severini* has also been recorded from sewage treatment plants. It is a parthenogenetic species and unmated females lay eggs directly on the filter beds.

## Rhagionidae

Snipe flies are 6–9 mm long and brown to dull gray; the wings are opaque and usually with dark spots or bands. The head is somewhat rounded; the antennae are distinctly three-segmented and with segment 3 somewhat kidney-shaped and with a subterminal, slender arista. The abdomen is relatively long and tapering, and the legs are long. These flies are common in wet and wooded areas; some species are aquatic, and some live in decaying vegetation where they are predaceous on other insects. Adult males and females have mouthparts similar to tabanids and can inflict a painful bite. *Symphoromyia* species are blood-suckers and will bite humans; they are generally associated with wooded and undisturbed areas, and urban populations are not common. This large genus occurs in North America and Europe.

***Symphoromyia atripes*** Adults are 5–8 mm long and with a gray-black body and reddish-brown legs. Adults occur in sunny areas where they readily attack humans; they have a painful bite. This species has a widespread distribution in western North America, from Alaska and British Columbia to Colorado and California.

**Other *Symphoromyia*** Several species are common in wooded areas, and may occur in suburban habitats. S. *kincaidi* females are about 6 mm long, dull black, and with a reddish brown head. This species bites humans and has a distribution similar to S. *atripes*.

## Sarcophagidae

Sarcophagids are also called flesh flies. They are 10–14 mm long and the body is robust and dull grayish black; they have three dark stripes on the dorsum of the thorax. Most are parasitic on arthropods or mollusks, some on vertebrates, and several are associated with carrion and excrement. The common species are characterized by a black and gray checkerboard pattern on their abdomen. Larvae are yellowish white and are characterized by having the posterior spiracles somewhat concealed in a pronounced depression or cleft at the posterior end. Eggs develop within the female's body and she is able to deposit first-stage larvae directly on a potential larval food source. This can decrease the development time and permit the larvae to utilize a temporary food source. There are several species that regularly occur in the urban environment.

**Banded flesh fly, *Eumacronychia decens*** Adults are 4–9 mm long and the body is gray, the thorax has dark longitudinal lines. The abdomen is a white and black checkerboard, and the tip of the abdomen is orange brown. Females enter and lay eggs in the cells of wasp nests, after first destroying the wasp eggs. The developing fly larvae consume the food provided by the wasps, and complete development in the wasp nest. There are several generations per year.

**Red-eyed flesh fly, *Sarcophaga aldrichi*** Adults are about 12 mm long and distinguished by a grayish-black body and red eyes. This species occurs throughout southern Canada, and eastern USA. Adults are often numerous outdoors and may be a nuisance. Larvae feed on insects, and they are common predators of pupae of the forest tent caterpillar (*Malacosoma disstria*). S. *aldrichi* may be abundant when there is a large population of this caterpillar in the area. A related species, S. *houghi*, is found in southern USA where it is a frequent predator of the forest tent caterpillar and elm spanworm, *Ennomos subsignarius*.

***Sarcophaga crassipalpis* (Fig. 7.5c, d)** Adults about 14 mm long and a with black stripes on the thorax, and a black and white checkered pattern on the abdomen. The tip of the abdomen is orange brown. Full-grown larvae are 20–22 mm long and yellowish white. The abdomen is covered with small papillae, and there are bands of setae around each segment. The anterior spiracles have 12–15 lobes; the posterior spiracles are in a deep cavity, and the peritreme is incomplete. Larvae infest ham and other prepared meats, and it has been found in wounds and bedsores. This species occurs in North America, Europe, South America, South Africa, and Australia.

**Redtailed flesh fly, *Sarcophaga cruentata* (= S. *haemorrhoidalis*) (Fig. 7.7b)** Adults are 10–14 mm long and the body is blackish gray. Male genitalia is reddish brown, and the eyes are reddish brown. Larvae are primarily scavengers, and feed on dead insects, carrion, and excrement. On carrion, the larval development is about 6 days, and the pupal period is 8–10 days. Adults live about 30 days and there are several generations per year. This species is widely distributed in North and South America, Europe, Africa, and Asia.

**Asian flesh fly, *Sarcophaga peregrina*** Adults are 10–14 mm long and the body is black to grayish black. The male genitalia is usually brown. Females are attracted to a variety of substrates, including fresh meat and carrion, and human excrement. It is closely associated with humans, indoors and outdoors throughout its range. This sarcophagid is distributed in southern to eastern Asia and Australia, and also in the Pacific islands.

## Scatopsidae

These flies are 1.5–3 mm long, and with a black or brownish black body (Fig. 7.7c). The antennae are slightly longer than the head, and the femora of the front legs are enlarged. Wing veins are thick near the costal margin, and the remaining veins are weak. Wing venation and general appearance are similar to the Phoridae. Full-grown larvae are 3–5 mm long and yellowish brown to brown; they are somewhat flattened and tapered anteriorly and posteriorly. The anterior and posterior spiracles are on short stalks. Some species have lateral tubercles. Larvae are known to feed in human excrement and the adult flies often occur in large numbers around this type of substrate. Other species, such as *Coboldia formicarium* and *Scatopse transversalis*, are associated with ant colonies, and have vestigial wings.

**Minute black scavenger fly, *Coboldia fuscipes*** Adults are about 3 mm long, and the body is smooth and black. The wings have a dark spot at the tip, and the front legs are distorted in shape. This species has been found in large numbers indoors. Larvae feed in decaying organic matter, including the waste from canneries, wineries, rotting vegetables, human and livestock manure.

**Scavenger fly, *Scatopse notata* (Fig. 7.5i, j)** Adults are about 3.5 mm long; the wing length is 3.5–4 mm. The body is shiny black, the pleura of the thorax and abdominal segment 1 is yellow, and there is a yellow spot above the wing base. Full-grown larvae are 4–5 mm long and slightly depressed; the body is yellowish brown and the head brown. Short setae are arranged in a pattern on the abdominal segments; the dorsum of segment 8 has a dense fringe of setae. Larvae feed in wet and decaying organic matter. Development is completed in about 14 days, and full-grown larvae move to the edge of the substrate to pupate. The pupa has two anterior spiracles. Several generations can be completed in a year. This species is widely distributed, and commonly aggregates indoors. It is nearly cosmopolitan.

## Scenopinidae

Window flies are about 6 mm long and usually black and with only a few bristles and setae on their thorax and abdomen; the antennae lack a style or an arista (Fig. 7.1l). The thorax is enlarged and the abdomen is flattened and somewhat bent downward. When viewed from the side the fly has a humped-back appearance. At rest the wings lie parallel over the body. Full-grown larvae are about 20 mm long and yellowish white, and very narrow (Fig. 7.2d). They are found indoors feeding on live insects in various microhabitats, and outdoors they prey on a variety of insects, and some feed on fungi in decaying wood.

**Window fly, *Scenopinus fenestralis*** Adults are 4–5.5 mm long, the thorax is grayish yellow, and the abdomen is blackish brown and somewhat flattened. Legs are gray, and the femora are brown. Full-grown larvae are 20–22 mm long, narrow, and yellowish white; the head is brown. Abdominal segments are subdivided to give the appearance that there are 17 segments. Anterior and posterior spiracles are round and surrounded by a pigmented boarder or peritreme. The larvae of these flies feed on the larvae and pupae of beetles and moths in stored grain, and they prey on larvae of carpet beetles and clothes moths in household habitats. The long, smooth worm-like larvae are active in sites that have accumulations of dead insects, such as the sills of windows and doors, and along the edges of household carpeting. Natural populations occur in bird nests where similar food sources are found. Linneaus noticed this species at windows in urban houses. This species is nearly cosmopolitan.

***Scenopinus glabrifrons*** Adults are 4–6 mm long and the body is grayish yellow. The front of the head is polished, and the halteres are pale white. The legs are brownish orange. This species is nearly cosmopolitan.

## Sciaridae

Adults are 1–4 mm long, and they are generally black or brownish black and their wings are dark (Fig. 7.7d). The antennae are long and slender, and their eyes are joined above the antennae. Adults are short-lived and are usually found in dark, humid locations. Larvae feed on decaying organic material, including damaged plant roots, rotting leaves, and fungi. They often occur indoors in the moist or wet soil for potted plants, and outdoors in a variety of microhabitats, including rotten wood under bark, commercial and wild fungi, and in bird nests. The rich organic soil of houseplants, greenhouse plants, and in agricultural fields is a suitable food for the larvae and large populations can develop in these sites. *Plastosciara perniciosa* is the most common greenhouse pest, and it occurs indoors. *Lycoriella solani* is a pest of commercial mushrooms, and also occurs indoors, along with *L. auripila* and *Bradysia brunnipes*.

Adults come to these organic substrates from populations in natural or undisturbed areas. Their small size gives them access to domestic habitats through narrow openings around doors and windows.

**Fungus gnats, *Sciara* spp., *Lycoriella* spp. (Fig. 7.7d)** Adults are 2–4 mm long and have a black body; the wings are dusky to nearly black. Wing length is 2.7–6 mm in *Sciara* spp. Larvae are white and slightly transparent; the body is 12-segmented and the head is black and well sclerotized. Fungus gnats occur indoors in the moist or wet soil used for potted houseplants. Mating is on the surface of moist, organic substrates; there are no mating swarms. Eggs are deposited in crevices in batches of up to 30 eggs; fecundity is about 175 eggs. Hatching occurs in 4–7 days. Larval development through four instars is completed in 2–3 weeks. A pupal chamber is formed from debris in the substrate; the pupal stage lasts about 7 days. Adults live about 10 days and feed on moisture in soil. These species are widely distributed around the world.

**Other Sciaridae** The larvae of *Sciara militaris* and *S. thomae* have the habit of aggregating and moving together in large numbers, probably a result of females depositing numerous eggs at one time in one location. When larvae are full-grown and ready to pupate, the aggregation may move together in a long column (snakeworm, armyworm) over the ground. Mass movements have been described that were 8–10 cm wide and 2–3 m long. The larvae crawl over each other so that the column advances about 25 cm/min. Species that are recorded as pests of commercial mushrooms, and may also occur in buildings close to these sites, include *Lycoriella solani*, *L. auripila*, *L. agarici*, and *Bradysia brunnipes*. Several species of *Bradysia*, including *B. amoena*, and *Lycoriella* infest greenhouses, and *Plastosciara perniciosa* larvae damage greenhouse crops, and they have been reported from decaying peat used as the lining of an old safe. *B. tritici* is known as the moss fly by orchid growers. Numerous species have been recorded from bird nests.

## Sepsidae

Sepsids are about 4 mm long, shining black or reddish-brown flies; they look somewhat like winged ants or small wasps (Fig. 7.2b; Fig. 7.7f). The head is spherical and the abdomen narrow at the base. Many species have a dark spot along the costal margin of the wing. Larvae live in decaying organic materials, and are commonly found in excrement. Most species of *Themira* develop in the mud alongside ponds and streams, particularly where this is enriched by dung or sewage effluent. *T. putris* and *T. nigricornis* will feed on human excrement, sludge, and sewer seepage. Other sepsids associated with human feces include *T. leachi*, *T. nigricornis*, *Meroplius minutus*, and *Sepsis fulgens*, *S. pectoralis*, *S. punctum*, and *S. violacea*.

Eggs have long respiratory filaments and are laid directly on the larval food; hatching occurs in about 2 days. Larvae are able to jump on the surface of infested material by grasping and releasing their posterior, similar to the hop made by cheese skipper (*Piophila*) larva. Larval development is completed in 7–32 days, and generally depends on temperature and humidity. Pupariation occurs in the larval food or in the soil beneath the infested material. The pupal stage lasts 8–13 days in warm weather, but may be 19 weeks in winter.

**Scavenger fly, *Sepsis pectoralis*** Adults are about 4.5 mm long and black. Full-grown larvae are 5–6 mm long, slender and pale white except for yellowish-white spiracles. Segments 5–10 have ventral, transverse rows of setae. The anterior spiracles are inconspicuous and with 5–7 lobes. The posterior spiracles are distinctly separated on sclerotized projections. The caudal segment has four tubercles surrounded by a broad band of spines. This species occurs in rotting garbage and household waste.

***Sepsis violacea*** Adults are about 4 mm long shiny black. Wings are gray and have a dark spot at the anterior margin near the tip. Larvae breed in animal and human excrement, they occur in decaying food, and they are found at windows indoors.

**Black scavenger fly, *Themira putris*** Adults are about 5 mm long, shining black, and with a distinctly round head. The abdomen of the adult is constricted at the base, and it appears somewhat wasplike. This species typically breeds in rich organic matter, such as carrion, manure, and the sludge of sewage treatment plants. Occasionally, the mass emergence of adults can infest neighborhoods and adults can occur indoors and outdoors.

## Sphaeroceridae

Small dung fly adults are 1.5–5 mm long and the body is usually dull black or brown; the basal segment of the antenna is nearly round (Fig. 7.7g). Hind tarsal segment 1 and sometimes 2 is enlarged in both males and females. Adults often occur in large numbers around organic waste and manure; *Copromyza equina* is commonly found on manure. Large numbers of sphaerocerids can occur indoors when sewer or septic systems are infested.

Larvae live as scavengers in excrement and decaying vegetable matter, seaweed, or fungi. Common species found associated with septic or sewer systems are *Leptocera caenosa* and *L. empirica* (= *L. pectinifera*). These, and a *Leptocera* species taxonomically near *L. gracilipennis*, are limited to peridomestic habitats, and rarely live in natural habitats. Other species, such as *L. limosa*, *L. forceps*, *L. lutosa*, *L. urodela*, and *L. vagans*, are abundant outdoors, but do not infest sewer and septic systems and rarely occur indoors.

**Small dung fly, *Leptocera caenosa*** Adults are about 3.5 mm long and dull black. Wings are translucent and the body setae are distinct. Full-grown larvae are about 5 mm long, yellowish white, and the body is nearly devoid of spines. Mouthparts are well-sclerotized. Anterior spiracles have about six lobes, and the posterior spiracles are contiguous on short tubercles, and surrounded by a circle of fine setae. Adults are generally unresponsive to light and their movement consists of short runs and occasionally short flight or hops. Mating occurs on the ground in daylight or darkness. Eggs are laid singly on the substrate, and females lay eggs for about 45 days; fecundity is 982–1328 eggs. Hatching occurs in about 48 h. Larval development is 13 days at 22 °C; in adverse conditions, development can extend to 63 days. The pupal period lasts 7–11 days at 22 °C. Adult males live about 32 days and females live about 79 days. This species occurs in North America and Canada, and is widespread in Europe and New Zealand.

This species is considered a domestic pest in the region it occurs. It does not generally occur outdoors, and is known only to breed in sewer or septic systems in the urban environment. It is native to Europe, and there it occurs in buildings and the holds of ships. This fly has been carried around the world, perhaps on infested ships, and in new regions it may displace native species. The adults are capable of entering openings as small as 0.05 mm diameter, and can locate and enter sites that contain animal waste.

***Leptocera fuscipennis*** Adults are 1.5–2 mm long and brownish black to dull black. Wings are light gray. Full-grown larvae are about 4 mm long and pale white; the puparium is about 3 mm long. Development from egg to adult takes about 14 days at 25 °C. This species occurs in Eurasia, North America, and North Africa. In Japan, adults become numerous following the rainy season in July and September. Soil in an area near Tokyo (close to Narita airport), an industrial waste treatment facility, becomes a breeding site, and large numbers of adults are carried by wind to outdoor lights around houses and commercial buildings in the area.

## Simuliidae

These flies are known as buffalo gnats, black gnats, and black flies (blackflies). They are 2–5 mm long, with a short body and variable in color from gray to dark yellow (Fig. 7.7e). The thorax is strongly developed and has a humped appearance. Larvae develop in relatively nonpolluted, fast-flowing streams. Both sexes possess elongated mouthparts, but the mandibles of the females are broad, and blade-like, and capable of piercing animal skin. The mouthparts of the male are weak and not adapted for piercing skin. Females of many species are blood-feeders, but some (*Simulium*) suck the blood of insects. Adults occur in large numbers during late spring and early summer, and they are some of the most persistent human-biting insect pests. These flies are cosmopolitan and abundant in all zoogeographic regions from arctic to equatorial latitudes. Urban habitats for these flies are limited, but populations often occur in rivers and streams in cities, especially those with reduced pollution. Artificial habitats for these flies in urban areas are provided by ornamental fountains and landscape ponds, which have flowing water and sufficient organic material.

Pest status is based on large numbers of these flies annoying people outdoors. Black flies are active and bite during the day and usually when there is little surface wind. They do not bite at night outdoors, or in the daytime indoors, and they enter buildings only accidentally. The bite from a simuliid results in a reddened, swollen area; the site may itch and remain irritated for several days. The only known disease transmitted to humans by Simuliidae is onchocerciasis, river blindness, which is caused by the filarial worm *Onchocera volvulus*. However, around the world simuliids are best known for their biting and nuisance swarming behavior. They are sometimes called buffalo gnats in the USA, because when viewed from the side they resemble a buffalo, 'potu' fly (*Simulium indicum*) in the Himalayas, and 'no-no' (*S. buissoni*) in the Marquesas Islands.

Not all black flies suck blood, even among those with fully formed biting mouthparts. Human-biting behavior is generally restricted to three genera: *Prosimulium*, *Austrosimulium*, and *Simulium*. The identification of species involved in pest outbreaks is sometimes difficult. The taxonomy of simuliids includes the use of a species complex for an assemblage of sibling species, which can only be distinguished using non-morpohological (chromosomal) criteria. Chromosome data often provide the only reliable means of identifying individual members of a complex.

Eggs are deposited in large numbers directly into streams or rivers, or on rocks or other objects projecting out of the water. Females lay 200–500 eggs singly in the water or in masses on vegetation or other objects near the water. Hatching occurs in 3–7 days, but the eggs laid in the fall may not hatch until the following spring. Eggs of disease vectors and pest species will not develop unless the female has had a blood meal before ovipositing. Larvae are 10–15 mm long and brown to gray, cylindrical and 12-segmented. They move on rocks and debris under water using a looping motion and use an anterior proleg that is modified into a prehensile, toothed disk, and a posterior sucker. Larvae cover their place of attachment under water with silken threads produced by the salivary glands. They feed on diatoms, algal filaments, animal plankton, and other organic matter that is collected on brush-like structures at their mouth opening. There are six or seven instars, and development takes about 10 days. Cocoons and pupae are attached to rocks and debris under water. Adults emerge under water, but they are carried to the surface in air bubbles; they fly to vegetation and mate soon after. Adults live about 3 weeks. Most *Simulium* species are capable of dispersing 15–35 km from their breeding site, and some species make wind-assisted migrations of 150–225 km. *S. arcticum* has occurred in pest numbers at a distance of 225 km from its breeding site.

**Buffalo gnat, *Cnephia pecuarum*** This is the simuliid originally called the buffalo gnat. It is a serious pest to humans and livestock throughout the region. The common name, buffalo gnat, may be linked to the general shape of this fly, or to its association with livestock. This species is distributed in southern USA, primarily in the Mississippi river basin.

**Western black fly, *Prosimulium exigens*** Adults are 2–4.5 mm long and dull black to grayish black. They crawl over the face, neck, and hands, and often bite. This species occurs along streams in May and June in the Pacific northwest of the USA.

**Golubatz fly, *Simulium colombaschense*** Adults are about 2.5 mm long and grayish black. Larval development is 20 days at 20–25 °C; the pupal stage is about 5 days. This species occurs in middle and southern Europe, in the Danube river basin, and is a periodic pest of livestock and humans. In 1923 two large swarms of this species appeared in southern Romania and caused the death of 16 474 domestic animals, including cattle, horses, and pigs.

**Turkey gnat, *Simulium meridionale*** Adults are about 2 mm long and grayish to brownish black. It is distributed from eastern to southeastern USA, and in spring it is a common pest of humans, domestic animals, and poultry. Adults bite the combs and wattles of poultry, and cause symptoms similar to the disease cholera. This species is also known as the cholera gnat. These flies can occur in large numbers and may be a serious pest.

**White-stockinged black fly, *Simulium venustum* complex** Adults are distinguished from other common species by having silvery-white tibiae. Adults occur in large numbers during June and July and persist throughout the summer. In the Adirondack mountains of New York, it appears later in the year than *Prosimulium hirtipes*, and is not considered a serious pest of humans. People in the region say that, when the black flies put on their white stockings in June, the trouble with black flies is about over. This means that the most common species then is *S. venustum*, which is not a serious biting pest. This species is distributed in Canada and eastern USA.

**Common black fly, *Simulium vittatum* complex** Adults are 2–3 mm long and dark gray to velvety black. It does not ordinarily bite humans, but attacks livestock and other domesticated animals. However, this species will fly around the heads of people in agricultural areas or working around large animals. This species is widely distributed in North America and Iceland.

**Other *Simuliidae*** *Simulium quadrivittatum* is an important biting species in various habitats in Belize and Panama. *S. arakawae* usually has a localized distribution, but is often a serious biting pest in Japan. *S. tuberosum* (species complex) occurs in Scotland and northern Europe and is a pest throughout this region. In the UK, *S. bredign* may become established in landscape ponds and ornamental fountains and waterfalls. The Blanford fly *S. posticatum* occurs primarily in southern UK, and can occur in large numbers as a local pest. In southeastern Lithuania, the most common blood-sucking species is *Byssodon maculatus*. This species overwinters in the egg stage, there is one generation per year, and the adults are active until the end of June. *Cnephia pecuarum* is the southern buffalo gnat, and it is a severe pest of people and livestock in the Mississippi valley and other regions in southern USA. Other species that attack humans include *S. erythrocephalum* in central Europe, and *S. aokii* in Japan.

## Stratiomyidae

Soldier flies are 10–18 mm long and some have brightly colored stripes on the abdomen. Generally, the abdomen is flattened and partially covered by the wings at rest. Larvae occur in a variety of habitats. Some are aquatic and feed on algae; others are found in decaying organic matter, including dung and feces. Larvae are flattened and the surface of their integument is rough and slightly granular. The posterior spiracles are together and usually concealed in a terminal cleft.

**Soldier fly, *Hermetia illucens*** Adults are 16–20 mm long and black with translucent spots on the second abdominal segment. The apex of the abdomen is sometimes reddish brown; tarsi are white to pale yellow, and the wings are uniformly dusky. Full-grown larvae are 23–27 mm long and reddish brown; the head is elongated and yellowish brown to dark brown. The cuticle is leathery and covered with fine setae. The body is distinctly depressed and tapering toward the anterior end. Abdominal segments 1–7 have lateral spiracles, and there are wartlike organs posterior of spiracles on segments 2–5. The caudal segment is flattened with a transverse spiracular cleft at the anal end, and a longitudinal, slit-like anal opening. Larvae breed in decaying organic matter, including vertebrate corpses and feces. This species occurs in North, Central and South America, and Pacific islands.

## Syrphidae

These flies are 6–12 mm long and brightly colored yellow and black. They are often seen hovering around flowers. Others are 12–20 mm long, brown to dark brown flies associated with decaying organic matter. Many species resemble bees and wasps; some look like honey bees and some bumble bees. The drone fly, *Eristalis tenax*, resembles a honey bee in appearance and sound. Larvae of this species are found in decaying organic matter. Aristotle theorized that honey bees could be generated from the dead carcass of an ox. Indeed, this substrate may have provided a breeding site for the saphrophagous larvae of *E. tenax*, and the honey bees he saw may have been the adult flies.

Syrphid larvae feed on a variety of materials, including primary invaders of living plants, predators of aphids and other homopterans, secondary feeders on decaying vegetation and animal feces, and living in the nests of ants, termites, and bees. Larvae of species associated with wet decaying organic matter and animal feces have a long extension of the body and the posterior tracheal trunk (Fig. 7.2c). These are called rattail maggots. The end of the rattail is the respiratory tube. This tube consists of three segments; when contracted, the second and third are telescoped into the first. When fully distended the respiratory tube is usually several times the length of the body. This larval form is found in species with saprophagous larvae, including *Eristalis tenax*, *E. dimidata*, *E. aeneus*, *E. arbustorum*, and *Helophilus pendulus*.

***Eristalis arbustorum*** Adults are about 11 mm long and dark brown to blackish brown. The head is produced anteriorly, the face is entirely covered with fine, pale white setae, and the eye lacks a zone of dense pile in the median third. Abdominal segments 2–4 have narrow apical, yellow bands; there are prominent yellow spots on abdominal segment 2 and sometimes segment 3. Abdominal segment 3 has a prebasal and a preapical black band. Middle tarsi are reddish brown at the base. Full-grown larvae are 15–18 mm long, cylindrical (excluding caudal portion); the cephalic end is blunt and the caudal end terminates in an elongated breathing tube. The cuticle is yellowish white, translucent, and covered with fine setae. Abdominal segments 1–6 have short, ventral prolegs bearing numerous hooks. Prothoracic spiracles are pigmented and with 10–12 openings. The caudal breathing tube is about 50 mm long, and terminates in a circle of plumose setae. Larvae of this species have been found in organic-rich water from sewer systems and animal waste.

**Rattailed maggot, drone fly, *Eristalis tenax* (Fig. 7.2c)** Adults are about 15 mm long and dark brown to yellowish brown. The head is produced anteriorly, the thorax is covered with fine setae, the abdomen is shiny and has black and yellow markings. Full-grown larvae are 17–20 mm long, pale yellow to yellowish white, with distinct prolegs on the mesothorax and abdominal segments 1–6. The larvae have anal gills and a long respiratory tube, which provides the common name of rattail. The puparium retains the long rattail, and the pupal respiratory horns are long. The long tail functions as a type of snorkel, allowing the larva to take air from the surface while feeding on the bottom. The mouthparts consist of coarse and fine filters to remove particles of food that enter the mouth. Water is sucked in and then forced out to clean the filters and prevent clogging. This species is distributed worldwide, except for some tropical regions. In about 1870 it spread through Asia to western North America, and by 1884 it was common throughout the USA. It reached New Zealand in 1888. Larvae occur in wet compost, decaying animal carcasses, and feces; they are common in urban and farm sewerage water and septic tanks. Larvae are extremely resistant to adverse environmental conditions.

## Tabanidae

Tabanids are known as horse flies, breeze flies, deer flies (*Chrysops*), gadflies (livestock-associated species of *Tabanus*), clegs (*Haematopota*), and greenheads (marshland species of *Tabanus*). They are generally large flies, ranging in length from 6 to 30 mm, and have a brown, black, or gray body, but can be yellow, green, or metallic blue. In many species, particularly of *Chrysops* and *Haematopota*, the eyes are colored and marked with patterns. Tabanids are pests of animals and livestock, but many species bite humans. The females are blood-sucking, but the males have no mandibles, and can only feed at flowers. Females of *Philoliche longirostris* have mouthparts that are about 40 cm long and take nectar from flowers, but also bite humans and cattle. Tabanids are strong flyers, and can travel at speeds of 21 km/h. The Afrotropical species *Tabanus taeniola* has been recorded on boats 1.5 km from shore and on the islands of Madagascar and Aldabra, and this suggests that it is capable of flying long distances. *T. atratus* has been found on oil drilling rigs 8 km offshore. Tabanids are found in all zoogeographic regions.

Pest status is usually limited to the species that attack humans. Residential areas located close to large marshland and streams may experience problems with these flies during late spring and summer. Tabanids often occur in large numbers and they may cause economic losses by affecting laborers and tourists. Some people are very sensitive to the bites of these flies, and require antihistamines to treat the swelling medically. Bites of certain species, such as *Diachlorus ferrugatus* of southern North America, and the widespread neotropical species *Lepiselaga crassipes*, can be particularly painful. *Haematopota* species attack people and large animals at about waist-height, and often bite the wrist and hands of a standing person. The head and neck are more often the target of the brightly colored female *Chrysops*. In North America, they are called deer flies because they are typically active in open woodland, in which deer are found. They typically bite humans on the head or at the back of the neck. Hosts are located by sight, odor, carbon dioxide emission, and possibly body heat. Adult tabanids are attracted to large objects, and traps can be used to exploit this tendency.

Eggs are long and spindle-shaped, and they are deposited on foliage at the edges of streams, lakes, or marshes. Some species oviposit on rocks or other objects extending over flowing streams. Females deposit masses of 100–800 eggs, which may be protected by a whitish secretion; a few species lay their eggs singly or in scattered groups. Eggs hatch in 5–7 days, and the larvae drop to the water or tunnel in wet soil and begin feeding. Larvae are primarily carnivorous and cannibalistic, preying on insect larvae, snails, earthworms, and other soft-bodied animals in the habitat. The first-instar larva usually does not feed. Some *Chrysops* species are considered saprophagous, and they are found feeding on a variety of decaying organic material. Larvae have 6–13 instars, and most species have a development period of several months to several years. Pupation takes place close to or within the dry parts of the substrate inhabited by the larva. The pupal stage lasts 7–21 days, and the adults live for about 30 days. Most temperate species have one generation per year, but in tropical zones two or three generations per year are typical.

**Green-headed horse fly, *Hybomitra lasiophthalma*** Adults are about 14 mm long. The head is brown, and it has purple eyes with green bands; the apical half of the abdomen is brown. Adults fly from May to September, and usually near water; males may be found on flowers. These flies commonly bite wet skin. In spite of the species name, the eyes do not appear hairy. This species occurs in eastern USA.

**Clegs, dunflies, *Haematopota* spp.** Adults are 6–12 mm long, gray to brown, and they have speckled wings, which are held roof-like over the abdomen when the fly is at rest. Eyes are iridescent and with zig-zag bands of gold, red, and green. Eyes of males meet at the top of the head, and the upper region facets are larger than those in the bottom. Larvae are carnivorous and eat other insects, small crustaceans, and snails in the habitat. The strong mouthparts of the larvae are sharp, and can pierce human skin. They are known to rasp the skin of frogs and suck blood. While hunting their prey, larvae sometimes bite people working in paddyfields and other farm locations. Clegs are noted for their silent approach when landing to bite. They are associated with antelopes, and cattle, and they are common in Europe. *Haematopota pulvialis* is a common species in the UK and continental Europe, and is a serious pest of cattle and often attacks humans.

### *Chrysops*

Adult deer flies are 6.5–10 mm long, black or brownish black. They usually have dark spots on the wings, antennal segment 3 lacks a basal tooth-like process, and there are apical spurs on the hind tibiae. These tabanids are distinguished from clegs, which have banded eyes and spotted wings. The marshland breeding sites of many deer fly species bring them close to human habitation. Their habit of hovering around the head and shoulders and biting people is the basis of pest status to people and small animals.

**Common deer flies, *Chrysops* spp.** Adults are 6–12 mm long and the body and wings have bands or stripes. There may be regional breeding sites and deer fly species that occur regularly, and cause problems by feeding on humans. Larvae are generally found in the wet soil along the margins of ponds and marshy areas. They are subcylindrical and yellowish white; there are fine, longitudinal ridges along the body. There are rounded swellings on segments 4–10, and the posterior spiracles are on a extended spine. Species that occur in North America include: *C. callidus*, *C. carbonarius*, *C. flavidus*, and *C. vittatus*. Their abundance is dependent on seasonal conditions and suitable habitats. *C. callidus* adults are 7–9 mm long, black, and have mid dorsal yellow, triangular patches on the abdomen, and there are large yellow spots on the sides of the abdomen near the base. The wings have brown markings. It is active from spring to fall. *C. carbonarius* adults are 7–8 mm long, black, and have yellowish-gray and greenish-gray markings. The wings have black markings. Adults are active during spring and summer. *C. flavidus* adults are 8–12 mm long, brown to blackish brown, and with light-brown markings. *C. vittatus* adults are 6.5–10 mm long, yellowish brown, and the abdomen has four dorsal black stripes. The wings have black markings.

**Western deer fly, *Chrysops discalis*** Adults are 8–10.5 mm long. The male is black with yellowish-gray spots on the abdomen; the female is yellowish gray with black spots on the abdomen. This large deer fly occurs in western USA and western Canada, east to North Dakota and Manitoba.

**Black deer fly, *Chrysops niger*** Adults are 6.5–8.5 mm long. They are black with some white hairs covering the body, and the wings have black markings. Adults are active from May to September in areas near marshland. This species is generally distributed in USA.

**African horse (deer) fly, *Chrysops silaceus*** This species lives in the tree canopy of the African rain forest. It descends to ground level and enters houses to feed and is a common pest to people indoors. The adults are apparently attracted to the smoke from wood-fueled cooking fires. It bites humans and other vertebrates, including monkeys, rodents, and some reptiles, in the tropical forest. This species is responsible for transmitting the filarial nematode worm *Loa loa*, the cause of human loiasis in forested areas of Central and West Africa.

### *Tabanus*

Adult horse flies are 10–25 mm long, black or brownish to grayish black, and usually without dark spots on the wings. Antennal segment 3 has a basal tooth-like process, and there are no apical spurs on the hind tibiae. Full-grown larvae are 23–35 mm long and yellowish white, but with light-brown bands on the anterior margins of most or all segments. Some species have six prolegs with apical hooks on segments 5–9, and there are similar but smaller prolegs on segments 4 and 10. Posterior spiracles are usually on a short tube. Some species of *Tabanus* are commonly called greenheads and they have bright green eyes when alive.

**Black horse fly, *Tabanus atratus*** Adults are 16–28 mm long, and uniformly black with slight gray or bluish-black markings. The abdomen sometimes shows a distinct bluish-black color, and this may be the "blue-tailed fly" of the well-known American folk ballad. Full-grown larvae are 55–60 mm long and pointed at both ends; they are yellowish white, but with extensive brown areas consisting of a dense layer of fine, short setae. There are rounded tubercles on abdominal segments 1–7. The posterior spiracles are located within a vertical slit at the tip of a cone-shaped siphon. This species occurs in the USA east of the Rocky Mountains, and in Mexico.

**Striped horse fly, *Tabanus lineola*** Adults are 12–15 mm long with a gray body and a brown head. The thorax is striped brown and gray; the abdomen has a pale and narrow, median dorsal stripe. Eggs are laid in batches of 50 or more; hatching is in about 4 days. Larval development is completed in about 49 days and the pupal period is about 8 days. Females have a preoviposition period of about 9 days. It occurs in eastern and southeastern USA.

**Saltmarsh greenhead, *Tabanus nigrovittatus*** Adult females are about 10 mm long. This tabanid is found in eastern North America, and is often a pest by biting and annoying people in seashore recreational areas. Adults emerge from June to August, and in eastern states the population peaks in late June and early August. Outbreaks of greenheads in eastern USA may be mixed populations of *T. nigrovittatus* and *T. conterminus*. These two species can usually be separated by adult body size, but precise differentiation requires electrophoretic analysis.

## Tipulidae

Crane fly adults are 10–25 mm long and have a 6–75 mm wing span. Crane flies are usually brown and have long legs and long wings. The giant tropical species of *Ctenacroscelis* has a wing span of about 10 cm. *Holorusia rubiginosa* is the largest crane fly species in the USA; it has a wing span of about 7.5 cm. The head of tipulids may be pointed anteriorly, and the mouthparts are usually reduced to only the maxillae and maxillary palps. The adults do not feed and live for only a few days. These flies are common in urban and rural areas, and they are sometimes considered large mosquitoes. In spite of their long wings, they are not strong flyers, and their wings and legs are easily broken off. Crane flies are usually found in damp habitats with abundant decaying vegetation. Some species breed in grassland and pastures, and their larvae have a thick integument and are called leatherjackets.

Many of the small-sized species fly at dusk, and a number are attracted to lights at night. They often collect in large numbers outside windows and around lights. Species of locally abundant tipulids may occur indoors, depending on the season. These flies are usually active in the spring, and many fly after sunset. Some of the small-sized species form mating swarms, but males of large species simply search for females. Prevailing winds can carry weak-flying adults from breeding sites and deposit them in commercial and residential areas.

**Giant western crane fly, *Holorusia rubiginosa*** Adults are 25–35 mm long, and the wing span is about 75 mm. The body is reddish brown to dark green with some white markings on the sides of the thorax; the wings are clear. Full-grown larvae are about 55 mm long, dull brown, and the cuticle is somewhat leathery. It occurs from southern California to British Columbia, and may be found at lights at night.

**Wood-boring tipulid, *Ctenophora vittata angustipennis*** Adults are 17–25 mm long. The body is black and yellowish red, the head and antennae are black, and wings are pale brown. The abdomen is reddish brown with a black stripe. Full-grown larvae are about 39 mm long and 5.2 mm diameter, and pale brown. The eggs are black. This species inhabits wet, decaying wood. It occurs in northwestern USA.

**European crane fly, daddy-long-legs, *Tipula paludosa*** Adults are 14–25 mm long, and brown to dark brown. Larvae feed on organic matter in pastures, grain fields, and residential lawns, and they often cause economic damage. Adults are attracted to lights at night. A closely related species, *T. oleracea*, is abundant and shares pest status and the common name with *T. paludosa*. They are both common in the UK.

**Common crane fly, daddy-long-legs, leatherjacket, *Tipula trivittata*** Adults are 18–26 mm long and brown to grayish brown. Wings are mottled brown, black, and white. Full-grown larvae are about 30 mm, grayish brown to dark brown, and the integument appears leathery. This species develops in leaf litter and wet organic material, especially along streams and woodland pools. Adults emerge in spring and begin laying eggs soon after they fly. There are two or three generations per year, but adults are most often found in spring or early summer.

**Other wood-boring tipulids** Species of *Tipula*, *Ctenophora*, and *Dictenidia* deposit eggs on the surface of moist wood or insert the eggs below the surface with the end of their abdomen. Females of *Tanyptera frontalis* have a long and slender ovipositor, and are capable of depositing eggs in wood. Larvae of this species bore in soft or decayed wood that has not disintegrated, especially willow and beech. Larvae of *Macromastix holochlora* occur in decaying forest trees and in rotting wood of old buildings.

# EPHEMEROPTERA
## Introduction

Mayfly adults are 5–35 mm long, soft-bodied insects with a cylindrical, elongate shape. They have two or three long abdominal tails. Wings are membranous: the front pair is triangular, and the hind pair is small and rounded. At rest, the wings are usually held upright over the body. Antennae are small, and the mouthparts are vestigial and adults do not feed. Immature stages are aquatic, with plumose or leaflike gills on the sides of the abdomen, and usually with three long tails. Nymphs feed primarily on algae and diatoms in a variety of aquatic habitats. Full-grown nymphs rise to the surface of the water and molt to a winged form, called a subimago, which flies only a short distance. The subimago molts within 24 h to the adult stage. Adults of many species in a region may emerge at the same time from lakes, rivers, and streams in a region. They gather in large numbers in which mating takes place. Mating occurs in flight and egg-laying begins soon after the male and female separate. Eggs are laid on the water surface or on vegetation. Nymph development requires 1–2 years.

Pest status of mayflies is linked to the mass emergence of subimagos and adults from lakes and rivers near urban and suburban areas. Large numbers of dying mayflies accumulate along roads and streets; sometimes piles of dead insects may be 1.2 m deep. This concentration of insects in one place often causes problems with automobile and train traffic. Polarized light may be the attraction that brings mayfly adults to water surfaces, and certain kinds of asphalt can reflect polarized light, which may be the reason why mayfly adults gather on roadways. The shed skins and dried bodies of dead mayflies break into small pieces and they are carried by the wind. When this material is inhaled, it can cause allergic reactions in people sensitive to insects. The overenrichment of streams and rivers from farm and urban runoff can result in large numbers of mayfly nymphs in streams. The nymphs of an Asian species of *Povilla* burrow into submerged wood pilings and may cause structural damage and collapse.

## Ephemeridae

This family is distinguished by the large, 10–30 mm, size of the adults and larvae that live in streams and lakes. The family is widespread on all continents except Australia; it is present in New Zealand and Madagascar.

**Giant mayfly, fishfly, Green Bay fly, *Hexagenia limbata*, *H. bilineata*** Adults are 18–30 mm long, excluding tails. The body is yellowish brown, and the head and thorax have brown markings. The wings are pale yellow and the margin of the hind wing is brown. Nymphs are 17–33 mm long, excluding tails; they have brown markings on a yellowish-brown body. The larval gills are large and feathered. *Hexagenia* larvae burrow in silt deposits of lakes, ponds, streams, and rivers throughout North America.

Eggs are laid as females rest on the surface of the water with their wings spread; rarely females fly above the water and oviposit as their abdomens touch the surface; fecundity is about 4000. Hatching depends on temperature, but usually occurs in 14 days, but eggs may remain dormant for long periods. Nymphs are in the soft bottom of lakes and streams; they live in U-shaped burrows at a depth of about 12 cm. In the northern part of their distribution, development is completed in 3–6 weeks. Emergence of adults is in late afternoon or at dusk. The subimago stage lasts 24–48 h. Larvae of these species are often sold as fish bait. Massive emergence along waterways occurs during warm months. *Hexagenia* species are known as the Green Bay fly because of the large number of adults that once emerged from Green Bay of Lake Michigan. At one time, *Hexagenia* adults would annually invade the city of Green Bay, Wisconsin. However, pollution of the lake has greatly reduced their populations.

**Coffinfly, *Ephemera guttulata*** Adults are 18–23 mm long, excluding tails, and are pale white, with brown forelegs. Wings are clear to amber with numerous dark veins. This species lives in shallow lakes and along rivers with sandy or silty bottoms. Subimagos emerge from May to August. Sometimes large numbers emerge at one time and collect around street lights. It occurs in eastern USA and eastern Canada.

## Palingeniidae

These are the spiny-headed burrowing mayflies, and their habits include living in the bottom of large streams and rivers. These are relatively large mayflies. Their wings have more than four longitudinal cross-veins, and the legs are well developed in both sexes. The male pronotum is short and about three times as wide as long; the caudal filaments of the female are shorter than the body.

**Long-tailed mayfly, Tisza mayfly, *Palingenia longicauda*** Adults are about 10 cm including the long caudal filaments. The body is yellowish brown and the wings of the female are nearly clear; the membranes of the front wings of the male are light blue. Eggs are laid on the water surface and they sink to the bottom; hatching occurs in about 45 days. Larvae live in tunnels on the bottom, and may occur in dense groups of 400 larval tunnels per square meter. Development may take as long as 3 years. Females usually fly 1–3 km before laying eggs. This is one of the largest mayflies in Europe, and it is primarily distributed in the Tisza river, Hungary. Adult emergences can be very large and adults can become numerous in urban areas along the Tisza river.

**Other mayflies** In the USA, the mayflies reported as occurring in large numbers include *Ephoron album* and *Ephemera simulans*. In Latin America, *Tortopus* spp. have been reported in large numbers.

# Bibliography

**DIPTERA**

**General**

Aradi, M. P. and F. Mihalyi. Seasonal investigations of flies visiting food markets in Budapest. *Acta Zool. Acad. Sci. Hung.*, **17** (1971), 1–10.

Blakitana, L. P. On wintering and some other aspects of the biology and ecology of synanthropic flies in northern regions of the Kirghiz Republic. *Med. Parazit. Moscow*, **31** (1962), 424–9.

Bohart, G. E. and J. L. Gressitt. Filth-inhabiting flies of Guam. *Bernice P. Bishop Museum, Honolulu, Bull.*, **104** (1951), 52.

Bryan, E. H. A review of the Hawaiian Diptera, with descriptions of new species. *Proc. Hawaiian Entomol. Soc.*, **8** (1934), 399–468.

Cole, F. R and E. I. Schlinger. The flies of western North America. Berkeley, CA: University of California Press, 1969.

Colyer, C. N. and C. O. Hammond. *Flies of the British Isles*. London: Warne, 1951.

De Souza, A. M. and Arício Xavier Linhares. Diptera and Coleoptera of potential forensic importance in southeastern Brazil: relative abundance and seasonality. *Med. Vet. Entomol*, **11** (1997), 8–12.

Dethier, V. G. *To Know A Fly*. New York: Holden-Day, 1963.

Downes, J. A. The swarming and mating flight of Diptera. *Annu. Rev. Entomol.*, **14** (1969), 271–98.

Greenberg, B. *Flies and Disease*, vol. 1. *Ecology, Classification and Biotic Associations*. Vol. 2. *Biology and Disease Transmission*. Princeton, NJ: Princeton University Press, 1971, 1973.

Flies as forensic indicators. *J. Med. Entomol.*, **28** (1991), 565–77.

Haines, T. W. Breeding media of some common flies. I. Urban areas. *Am. J. Trop. Med. Hyg.*, **54** (1953), 261–2.

Havlik, B. and B. Batova. A study of the most abundant synanthropic flies occurring in Prague. *Acta. Soc. Entomol. Czech.*, **48** (1961), 1–11.

Hayahsi, A. and S. Shinonaga. *Flies: Ecology and Control* (in Japanese). Tokyo: Bun-Eido, 1979.

Illingworth, J. F. Insects attracted to carrion in Hawaii. *Proc. Hawaiian Entoml. Soc.*, **5** (1923), 280–1.

Insects attracted to carrion in southern California. *Proc. Hawaiian Entomol. Soc.*, **6** (1927), 397–401.

Ishijima, H. Revision of the third stage larvae of synanthropic flies of Japan (Diptera: Anthomyidae, Muscidae, Calliphoridae and Sarcophagidae). *Jpn. J. Sanit. Zool.*, **18** (1967), 47–100.

Kawai, T. (ed.) *An Illustrated Book of Aquatic Insects of Japan* (in Japanese). Tokyo: Tokai University Press, 1985.

Keh, B. Scope and applications of forensic entomology. *Annu. Rev. Entomol.*, **30** (1985), 137–54.

Kubo, K. Common species of flies in houses in Manchuria. *Tokyo Med. News*, **2180** (1920).

Liu, D. and B. Greenberg. Immature stages of some flies of forensic importance. *Ann. Entomol. Soc. Am.*, **82** (1989), 80–93.

Liu, S., H. Chen, and J. Lien. A brief study of the bionomics of fly breeding in Keelung City, Taiwan. *J. Formosan Med. Assoc.*, **56** (1957), 417–24.

Lu, B. S. (ed.). *Identification Handbook for Medically Important Animals in China* (in Chinese). Beijing: Science Press, 1982.

Lysenko, O. and D. Povolny. The microfauna of synanthropic flies in Czechoslovakia. *Folia Microb.*, **6** (1961), 27–32.

Mackerras, I. M. The zoogeography of the Diptera. *Aust. J. Sci.*, **12** (1950), 157–61.

Manson-Bahr, P. E. C. and D. R. Bell (eds). *Manson's Tropical Diseases*, 19th edn. London: Baillière Tindale, 1987.

McAlpine, J. F. (ed.) *Manual of Nearctic Diptera*, vol. 2. Research Branch, Agriculture Canada monograph no. 28. Ottawa: Research Branch, Agriculture Canada, 1987.

McAlpine, J. F., B. V. Peterson, G. E Shewell *et al. Manual of Nearctic Diptera*, vol. 1. Research Branch, Agriculture Canada monograph no. 27. Ottawa: Research Branch, Agriculture Canada, 1981.

Meng, C. H. and G. F. Winfield. Studies on the control of fecal-borne diseases in North China. V. A preliminary study of the density, species make up, and breeding habits of the house frequenting population in Tsinan, Shantung, China. *Chinese Med. J.*, **2** (suppl.) (1938), 463–86.

Studies on the control of fecal-borne diseases in North China. XVI. An approach to the quantitative study of the house frequenting population. D. The breeding habits of the common North China flies. *Phil. J. Sci.*, **79** (1950), 175–200.

Mihalyi, F. Flies visiting fruit and meat in an open-air market in Budapest. *Acta Zool. Hung.*, **9** (1966), 153–64.

Seasonal distribution of the synanthropic flies in Hungary. *Ann. Hist. Nat. Mus. Nat. Hung.*, **59** (1967), 327–44.

Oldroyd, H. *The Natural History of Flies*. London: Weidenfeld and Nicolson, 1964.

Oosterbroek, P. *The Families of Diptera of the Malay Archipelago*, Leiden: Brill Academic Publications, 1988.

Quarterman, K. D., J. W. Kilpatrick, and W. Mathis. Fly dispersal in a rural area near Savannah, Georgia. *J. Econ. Entomol.*, **47** (1954), 413 19.

Quarterman, K. D., W. Mathis, and J. W. Kirkpatrick. Urban fly dispersal in the area of Savannah, Georgia. *J. Econ. Entomol.*, **47** (1954), 405–12.

Roberts, D. W. and J. M. Castillo. Bibliography on pathogens of medically important arthropods: 1980. *Bull. World Health Org.*, **58** (suppl.) (1980), 1–197.

Roberts, D. W. and M. A. Strand. Pathogens of medically important arthropods. *Bull. World Health Org.*, **55** (suppl.) (1977), 1–419.

Roberts, D. W., R. A. Daoust, and S. P. Wraight. *Bibliography of Pathogens of Medically Important Arthropods*: 1981. VBC/83.1. Geneva: World Health Organization, 1983.

Roberts, M. J. The structure of the mouth parts of some calyptrate dipteran larvae in relation to their feeding habits. *Acta Zool.*, **52** (1971), 171–88.

Robinson, W. and V. H. Norwood. The role of surgical maggots in the disinfection of osteomyelitis and other infected wounds. *J. Bone Joint Surg.*, **15** (1933), 409–12.

Roy, D. N. On the number of eggs of the common house-frequenting flies of Calcutta. *Ind. J. Med. Res.*, **26** (1928), 531–3.

Sehgal, B. S. and P. Kumar. A study of the seasonal fluctuations in fly populations in two villages near Lucknow. *Ind. J. Med. Res.*, **54** (1966), 1175.

Service, M. W. (ed.). *Demography and Vector-Borne Diseases*. Boca Raton, FL: CRC Press, 1989.

Sherman, R. A., M. J. R. Hall, S. Thomas. Medicinal maggots: an ancient remedy for some contemporary afflictions. *Annu. Rev. Entomol.*, **45** (2000), 55–81.

Shura-Bura, B. L., A. D. Shaikov, E. V. Ivanova *et al.* Migration of synanthropic flies to the cities from open fields (in Russian). *Med. Parasit. Parasitarn. Bolezn*, **25** (1956), 368–72.

Shura-Bura, B. L., E. V. Ivanov, A. N. Onutshin, A. Ia Glazunov, and A. D. Shaikov. Migration of flies of medical importance in Leningrad district (in Russian). *Entom. Obozzren.*, **35** (1956), 334–46.

Smith, K. G. V. *A Manual of Forensic Entomology*. London: British Museum (Natural History) and Ithaca, NY: Cornell University Press, 1986.

Sychevskaia, V. I. On changes in the daily dynamics of the specific composition of flies associated with man in the course of the season. *Entom. Obozren.*, **41** (1962), 545–53.

Trofimov, G. K. A brief review of synanthropic flies (Muscidae, Calliphoridae and Sarcophagidae) of the Talysh region of the Caucasus. *Entom. Obozren.*, **44** (1965), 357–61.

Uemoto, K. Studies on the dispersion of flies at winter season. *Jpn. J. Sanit. Zool.*, **11** (1960), 95–101.

Usinger, R. L. and V. R. Kellen. The role of insects in sewage disposal beds. *Hilgardia*, **23** (1955), 263–321.

Vincent, C., D. K. McE. Kevan, M. Leclercq, and C. L. Meek. A bibliography of forensic entomology. *J. Med. Entomol.*, **22** (1985), 212–19.

Williams, R. W. A study of the filth flies in New York City, 1953. *J. Econ. Entomol.*, **47** (1954), 556–63.

World Health Organization. *Geographical Distribution of Arthropod-borne Diseases and Their Principal Vectors*. WHO/VBC/89.967. Geneva: World Health Organization, 1989.

Zumpt, F. and P. M. Patterson. Flies visiting human faeces and carcasses in Johannesburg, Transvaal. *South Afr. J. Clin. Sci.*, **3** (1952), 92–106.


### Bibionidae


Buschman, L. L. 1976. Invasion of Florida by the 'lovebug' *Plecia nearctica* (Diptera: Bibionidae). *Florida Entomol.*, **59** (1976), 191–4.

Cherry, R. and R. Raid. Seasonal flight of *Plecia nearctica* (Diptera: Bibionidae) in southern Florida. *Florida Entomol.*, **83** (2000), 94–6.

Edwards, E. E. The fever fly *Dilophus febrilis* L., and methods for control of its larvae in cultivated lawns. *Ann. Appl. Biol.*, **28** (1941), 34–8.

Hardy, D. E. Studies in New World *Plecia* (Bibionidae: Diptera). Part 1. *Kansas Entomol. Soc.*, **13** (1940), 15–27.

Hetrick, L. A. Biology of the 'love bug,' *Plecia nearctica* (Diptera: Bibionidae). *Florida Entomol.*, **53** (1970), 23–6.

Lovibond, B. The fever fly, *Dilophus febrilis*. *J. Board Greenkeeping Res.*, **5** (1938), 271–3.


### Caliphoridae


Barnes, H. F. Some facts about *Pollenia rudis* Fabr. *Vasculum*, **10** (1924), 34–58.

Baumgartner, D. L. Spring season survey of the urban blowflies, (Diptera, Calliphoridae) of Chicago, Illinois. *Great Lakes Entomol.*, **21** (1988), 119–21.

Davies, L. Seasonal and spatial changes in blowfly production from small and large carcasses at Durham in lowland northeast England. *Med. Vet. Entomol.*, **13** (1999), 245–51.

Decoursey, R. M. A bionomical study of the cluster fly, *Pollenia rudis* (Fab.). *Ann. Entomol. Soc. Am.*, **20** (1927), 368–84.

Gagne, R. J. *Chrysomya* spp., Old World blow flies (Diptera: Calliphoridae), recently established in the Americas. *Bull. Entomol. Soc. Am.*, **27** (1981), 21–2.

Garrison, G. L. Rearing records of *Pollenia rudis* Fab. (Dipt., Muscidae). *Entom. News*, **35** (1924), 135–8.

Hagmann, L. E. and G. W. Barber. Overwintering habits of *Phaenicia sericata* (Mg.). *J. Econ. Entomol.*, **41** (1948), 510.

Hall, D. G. *The Blowflies of North America*. College Park, MD: Thomas Say Foundation, 1948.

Keilin, D. On the parasitism of the larvae of *Pollenia rudis* (Fabr.) in *Allolobophora chlorotica* Savingny. *Proc. Entomol. Soc. Wash.*, **13** (1911), 182–4.

Lane, R. P. An investigation into blowfly (Diptera: Calliphoridae) succession on corpses. *J. Nat. Hist.*, **9** (1975), 581–8.

Mariluis, J. C. and J. A. Schnack. Ecología de una taxocenosis de Calliphoridae del area Platense (provincia de Buenos Aires) (Insecta, Diptera). *Ecosur*, **12–13** (1985–6), 81–91.

Mihalyi, F. Contribution to the knowledge of the genus *Pollenia* R.-D. (Diptera: Calliphoridae). *Acta Zool. Acad. Scient. Hungaricae*, **22** (1975), 327–33.

A new key for Hungarian *Lucillia* species (Diptera: Calliphoridae). *Ann. Hist.-Nat. Mus. Natl. Hungarici*, **69** (1977), 181–4.

Norris, K. R. The bionomics of blow flies. *Annu. Rev. Entomol.*, **10** (1965), 47–68.

Daily patterns of flight activity of blowflies (Calliphoridae: Diptera) in the Canberra district as indicated by trap catches. *Aust. J. Zool.*, **14** (1966), 835–53.

Nuorteva, P. 1963. Synanthropy of blowflies in Finland. *Ann. Entomol. Fenn.*, **29** (1963), 1–49.

The flying activity of blowflies (Diptera: Calliphoridae) in subarctic conditions. *Ann. Entomol. Fenn.*, **31** (1965), 242–5.

Local distribution of blowflies in relation to human settlement in an area around the town of Forssa in South Finland. *Ann. Entomol. Fenn.*, **32** (1966), 128–37.

Nuorteva, P. and T. Vesikari. The synanthropy of blowflies (Diptera, Calliphoridae) on the coast of the Arctic Ocean. *Ann. Med. Exp. Fenn.*, **44** (1966), 544–8.

O'Flynn, M. A. The succession and rate of development of blowflies in carrion in southern Queensland and the application of these data to forensic entomology. *J. Aust. Entomol. Soc.*, **22** (1983), 137–48.

Ogata, K., Nagai, N., Koshimizu, N., Kato, M., and Wada, A. Release studies on the dispersion of the house flies and the blow flies in the suburban area of Kawasaki City, Japan. *Jpn. J. Sanit. Zool.*, **11** (1960), 181–8.

Pimentel, D. and B. Epstein. The cluster fly, *Pollenia rudis* (Diptera: Calliphoridae). *Ann. Entomol. Soc. Am.*, **53** (1960), 553–4.

Reid, J. A. Notes on house-flies and blow-flies in Malay. *Inst. Med. Res. Malaya Bull.*, **7** (1953), 26.

Reiter C. Zum Wachstumsverhalten der Maden der blauen Schmeisfliege *Calliphora vicina*. *Z. Rechtsmed*, **91** (1984), 295–308.

Schoof, H. F. and G. A. Mail. Dispersal habits of *Phormia regina* in Charleston, West Virginia. *J. Econ. Entomol.*, **46** (1953), 258–62.

Sychevskaia, V. I. Biology and ecology of *Calliphora vicina* R.-D. in central Asia. *Zool. Zhurn.*, **44** (1965), 552–60.

Wallman, J. F. A key to the adults of species of blowflies in southern Australia known or suspected to breed in carrion. *Med. Vet. Entomol.*, **15** (2001), 433–7.

Webb, J. L. and R. H. Hutchison. A preliminary note on the bionomics of *Pollenia rudis* (Fabr.) in America. *Proc. Entomol. Soc. Wash.*, **18** (1916), 197–9.

Wijesundra, D. P. The life history and bionomics of *Chrysomyia megacephala* (Fab.). *Ceylon J. Sci.*, **25** (1957), 169–85.

Williams, H. and A. M. M. Richardson. Growth energetics in relation to temperature for larvae of four species of necrophagous flies (Diptera: Calliphoridae). *Aust. J. Ecol.*, **9** (1984), 141–52.

### Ceratopogonidae, Chironimidae

Aldrich, J. M. The dipterous genus *Symphoromyia* in North America. *US Natl Museum*, **49** (1916), 113–42.

Ali, A. Nuisance chironomids and their control. *Bull. Entomol. Soc. Am.*, **26** (1980), 3–16.

Perspectives on management of pestiferous Chironomidae (Diptera), an emerging global problem. *J. Am. Mosq. Control Assoc.*, **7** (1991), 260–81.

Anderson, L. D., E. C. Bay, and M. S. Mulla. Aquatic midge investigations in southern California. *Proc. California Mosquito Control Assoc.*, **33** (1965), 31–3.

Armitage, P. D., P. S. Cranston, and L. C. V. Pinder (eds) *The Chironomidae: Biology and Ecology of Non-biting Midges*. London: Chapman and Hall, 1995.

Austin, M. D. The insect and allied fauna of cultivated mushrooms. *Entomol. Mon. Mag.*, **69** (1933), 16–19.

Bay, E. C., L. D. Anderson, and J. Sugerman. The abatement of a chironomid nuisance on the highways of Lancaster, California. *Calif. Vector Views*, **12** (1965), 29–32.

Blanton, F. S. and W. W. Wirth. *Arthropods of Florida and Neighboring Land Areas. 10. The Sandflies* (Culicoides) *of Florida (Diptera: Ceratopogonidae)*. Gainesville, Florida: Florida Department of Agriculture and Consumer Services, 1979.

Boorman, J. British *Culicoides* (Diptera: Ceratopogonidae): notes on distribution and biology. *Entomol. Gaz.*, **37** (1986), 253–66.

*Culicoides* (Diptera: Ceratopogonidae) of the Arabian peninsula, with notes on their medical and veterinary importance. *Fauna Saudi Arabia*, **10** (1989), 160–224.

Brydon, H. W. The Clear Lake gnat and its control in Clear Lake, California during 1954. *J. Econ. Entomol.*, **49** (1956), 206–9.

Chu, F.-L. On the blood-sucking midges (Diptera, Ceratopogonidae) from the coastal regions of south-eastern China (in Chinese). *Acta Entomol. Sinica*, **24** (1981), 307–13.

Cranston, P. S., M. O. G. El Rab, and A. B. Kay. Chironomid midges as a cause of allergy in the Sudan. *Trans. R. Soc. Trop. Med. Hyg.*, **75** (1981), 1–4.

Darby, R. E. Midges associated with California rice fields, with special reference to their ecology (Diptera: Chironomidae). *Hilgardia*, **32** (1962), 1–206.

Debenham, M. L. *An Annotated Checklist and Bibliography of Australasian Region Ceratopogonidae (Diptera, Nematocera)*. University of Sydney School of Public Health and Tropical Medicine monograph series (Ent.) vol. 1, pp. 1–671.

Deonier, C. C. Biology of the immature stages of the Clear Lake gnat (*Chaoborus astictopus*) (Diptera: Culicidae). *Ann. Entomol. Soc. Am.*, **36** (1943), 383–8.

Fontain, R. E., D. H. Green, and L. M. Smith. Ecological observations of the valley black gnat, *Leptoconops torrens* Townsend. *J. Econ. Entomol.*, **50** (1957), 764–7.

Foulk, J. D. Ecology and control of *Leptoconops kerteszi* biting gnats in southern California. *Proc. Calif. Mosquito Control Assoc.*, **35** (1967), 88.

Grodhaus, G. Chironomid midges as a nuisance. II. The nature of the nuisance and remarks on its control. *Calif. Vector Views*, **10** (1963), 27–37.

Identification of chironomid midges commonly associated with waste-stabilization lagoons in California. *Calif. Vector Views*, **14** (1967), 1–12.

Hirabayashi, K. Studies on massive flights of chironomid midges (Diptera, Chironomidae) as nuisance insects and plans for their control in the Lake Suwa area, Central Japan. 1. Occurrence of massive flights of *Tokunagayusurika akamusi* (in Japanese, English summary). *Jpn. J. Hyg.*, **46** (1991), 652–61.

Studies on massive flights of chironomid midges (Diptera, Chironomidae) as nuisance insects and plans for their control in the Lake Suwa area, Central Japan. 2. Quantative evaluations of the nuisance of chironomid midges (in Japanese, English summary). *Jpn. J. Hyg.*, **46** (1991), 662–75.

Hirabayashi, K. and T. Okino. Massive flights of chironomid midges (Diptera) as nuisance insects around a hyper-eutrophic lake in Japan. A questionnaire survey to tourists. *J. Kansas Entomol. Soc.*, **71** (1998), 439–46.

Itoua, A., M. Cornet, G. Vattier-Bernard, and J. Trouillet. Les *Culicoides* (Diptera, Ceratopogonidae) d'Afrique Centrale. *Cahiers ORSTOM (Entomol. Méd. Parasitol.)*, **25** (1987), 127–34.

Kitaoka, S. Japanese *Culicoides* (Diptera: Ceratopogonidae) and keys for the species (in Japanese). I. *Bull. Natl Inst. Animal Health*, **87** (1984), 73–89.

Japanese *Culicoides* (Diptera: Ceratopogonidae) and keys for the species (in Japanese). *II. Bull. Natl Inst. Animal Health*, **87** (1984), 91–108.

Langton, P. H., P. S. Cranston, and P. D. Armitage. The parthenogenetic midge of water supply systems, *Paratanytarsus grimmi* (Schneider) (Diptera: Chironomidae). *Bull. Entomol. Res.*, **778** (1988), 317–28.

Lewis, D. J. Chironomidae as a pest in the Northern Sudan. *Acta Trop.*, **13** (1956), 142–58.

Observations on Chironomidae at Khartoum. *Bull. Entomol. Res.*, **48** (1957), 155–84.

Lien, J.-C. and C.-S. Chen. Seasonal succession of some common species of the genus *Culicoides* (Diptera, Ceratopogonidae) in eastern Taiwan. *J. Formosan Med. Assoc.*, **82** (1983), 399–409.

Linley, J. R. and J. B. Davies. Sandflies and tourism in Florida and the Bahamas and Caribbean area. *J. Econ. Entomol.*, **64** (1971), 264–78.

Linley, J. R., A. L. Hoch, F. P. Pinheiro. Biting midges (Diptera: Ceratopogonidae) and human health. *J. Med. Entomol.*, **20** (1983), 347–64.

Smith, L. M. and H. Lowe. The black gnats of California. *Hilgardia*, **18** (1948), 157–83.

Tabaru, Y., K. Moriya, and A. Ali. Nuisance midges (Diptera: Chironomidae) and their control in Japan. *J. Am. Mosq. Control Assoc.*, **3** (1987), 45–8.

Yamagishi, H. and H. Fukuhara. Ecological studies on chironomids in Lake Suwa. I. Population dynamics of two large chironomids, *Chironomus plumosus* L. and *Spaniotoma akamusi* Tokunaga. *Oecologia*, **7** (1971), 309–27.

## Chlorpodidae, Coelopidae

Bassett, D. C. J. *Hippelates* flies and streptococcal skin infection in Trinidad. *Trans. R. Soc. Trop. Med. Hyg.*, **69** (1970), 138–47.

Burgess, R. W. The life history and breeding habits of the eye gnat, *Hippelates pusio* Loew, in the Coachella Valley, Riverside County, California. *Am. J. Hyg.*, **53** (1951), 164–77.

Egglishaw, H. J. Studies on the family Coelopidae (Diptera). *Trans. R. Entomol. Soc. Lond.*, **112** (1960), 109–40.

Gayden, D. M. and R. T. Adkins, Jr. Effect of cultivation on emergence of eye gnats (*Hippelates* spp.) in South Carolina. *J. Econ. Entomol.*, **62** (1969), 312–14.

Hall, D. G. Some studies on the breeding media, development, and stages of the eye gnat, *Hippelates pusio* Loew (Diptera: Chloropidae). *Am. J. Hyg.*, **16** (1932), 854–64.

Legner, E. F., L. Moore, and R. A. Medved. Observations on prediction of *Hippelates collusor* and distribution in southern California of associated fauna. *J. Econ. Entomol.*, **64** (1971), 161–8.

Mulla, M. S. The breeding niches of *Hippelates* gnats. *Ann. Entomol. Soc. Am.*, **55** (1962), 389–93.

Oviposition and emergence of the eye gnat *Hippelates collusor*. *J. Econ. Entomol.*, **56** (1966), 768–70.

Mulla, M. S. and R. B. March. Flight range, dispersal patterns and population density of the eye gnat *Hippelates collusor* (Townsend). *Ann. Entomol. Soc. Am.*, **52** (1959), 641–6.

Sabrosky, C. W. Chloropids swarming in houses. *J. Econ. Entomol.*, **33** (1940), 946–7.

Schwartz, P. H. Some observations of the reproduction of the eye gnat, *Hippelates pusio*. *J. Med. Entomol.*, **2** (1965), 141–4.

Womeldorf, D. J. and E. W. Mortenson. Occurrence of eye gnats (*Hippelates* spp.) in the Central San Joaquin Valley, California. *J. Econ. Entomol.*, **55** (1962), 457–9.

## Culicidae

Apiwathnasorn, C. *A List of Mosquito Species in Southeast Asia*. Bangkok: Faculty of Tropical Medicine, Mahidol University, 1986.

Bates, M. *The Natural History of Mosquitoes*. New York: Macmillan, 1949.

Berner, L. Notes on the breeding habits of *Aedes (Stegomyia) aegypti* (Linneaus). *Entomol. Soc. Am.*, **40** (1947), 528–9.

Bohart, R. M. and R. K. Washino. *Mosquitoes of California*. Berkeley, CA: University California Press, 1978.

Bram, R. A. 1967. Contributions to the mosquito fauna of southeast Asia II. The genus *Culex* in Thailand (Diptera: Culicidae). *Contrib. Am. Entomol. Inst.*, **2** (1967), 296.

Brown, A. W. A. Studies on the responses of the female *Aedes* mosquito, Part VI: The attractiveness of coloured cloths to Canadian species. *Bull. Entomol. Res.*, **45** (1954), 67–78.

Bruce-Chwatt, L. J. *Essential Malariology*, 2nd edn. London: Heinemann Medical Books, 1985.

Carpenter, S. J. and W. J. LaCasse. *Mosquitoes of North America*. Berkeley, CA: University of California Press, 1955.

Christopher, R. Aedes aegypti, *The Yellow Fever Mosquito: Its Life History, Bionomics and Structure*. Cambridge: Cambridge University Press, 1960.

Clarke, J. L. Studies of the flight range of mosquitoes. *J. Econ. Entomol.*, **36** (1943), 121–2.

Clements, A. N. *The Physiology of Mosquitoes*. Oxford: Pergamon Press, 1963.

*The Biology of Mosquitoes*, vol. 1. Oxford: Pergamon Press, 1992.

Conway, G. R., M. Trpis, and G. A. H. McCleland. Population parameters of the mosquito *Aedes aegypti* (L.) estimated by mark–release–recapture in a suburban habitat in Tanzania. *J. Anim. Ecol.*, **43** (1974), 289–304.

Cornel, A. J. and R. H. Hunt. *Aedes albopictus* in Africa? First records of live specimens in imported tyres in Cape Town. *J. Am. Mosq. Cont. Assn.*, **7** (1991), 107–8.

Craven, R. B., D. A. Eliason, B. B. Francy *et al.* Importation of *Aedes albopictus* and other exotic mosquito species into the United States in used tires from Asia. *J. Am. Mosquito Control Assoc.*, **4** (1988), 138–42.

Crovello, T. J. and C. S. Hacker. Evolutionary strategies in life table characteristics among feral and rural strains of *Aedes aegypti* (L.). *Evolution*, **26** (1972), 185–96.

Eades, R. B. Recovery of *Aedes atropalpus* from used tires to United States ports. *Mosq. News*, **32** (1972), 113–14.

Gillett, J. D. *Mosquitoes*. London: Weidenfeld and Nicholson, 1971.

Harrison, G. *Mosquitoes, Malaria and Man: History of Hostilities since 1880*. London: John Murray, 1978.

Hawley, W. A., P. Reiter, R. S. Copeland, D. B. Pumpini, and G. B. Craig. *Aedes albopictus* in North America: probable introduction in tires from northern Asia. *Science*, **236** (1987), 1114–16.

Horsfall, W. R. *Mosquitoes: Their Bionomics and Relation to Disease*. New York: Ronald Press, 1955.

Kon, Y. S., D. I. Dobrosmaislor, and Z. L. Ginzburg. Mosquito larvae in the tunnels of the Moscow underground railway. *Rev. Appl. Entomol. Ser. B.*, **32** (1944), 121.

Laird, M. *The Natural History of Larval Mosquito Habitats*. London: Academic Press, 1988.

Lindsay, S. W., J. H Adiamah, J. E. Miller, R. J. Pleass, and J. R. M. Armstrong. Variation in attractiveness of human subjects to malaria mosquitoes (Diptera: Culicidae) in Gambia. *J. Med. Entomol.*, **30** (1993), 368–73.

Lu, B. L. and B. S. Li. Identification of Chinese mosquitoes. In Lu, B. S. (ed.) *Identification Handbook for Medically Important Animals in China* (in Chinese). Beijing: People's Health Publication, 1982.

Maibach, H. I., W. A. Skinner, W. G. Strauss, and A. A. Kahn. Factors that attract and repel mosquitoes to human skin. *J. A. M. A.*, **196** (1966), 263–6.

Marshall, J. F. *The British Mosquitoes*. London: British Museum (Natural History), 1938.

Mattingly, P. F. Contributions to the mosquito fauna of southeast Asia. XII. Illustrated keys to the genera of mosquitoes (Diptera, Culicidae). *Contrib. Am. Entomol. Inst.*, **7** (1971), 1–84.

Mattingly, P. F. and K. L. Knight. The mosquitoes of Arabia I. *Bull. Br. Mus. (Nat. Hist.) Entomol.*, **4** (1956), 91–141.

Morlan, H. B. and R. O. Hayes. Urban dispersal and activity of *Aedes aegypti*. *Mosq. News*, **18** (1958), 137–44.

Reeves, W. C. Ecology of mosquitoes in relation to arboviruses. *Annu. Rev. Entomol.*, **10** (1965), 25–46.

Reisen, W. K., R. P. Meyer, C. H. Tempelis, and J. J. Spoehel. Mosquito abundance and bionomics in residential communities in Orange and Los Angeles counties, California. *J. Med. Entomol.*, **27** (1990), 356–67.

Reiter, P. and Sprenger, D. The used tire trade: a mechanism for the world-wide dispersal of container breeding mosquitoes. *J. Am. Mosq. Cont. Assoc.*, **3** (1987), 494–501.

Travis, B. V. Studies of mosquito and other biting insect problems in Alaska. *J. Econ. Entomol.*, **42** (1949), 451–7.

Trpis, M. and W. Hausermann. Demonstration of differential domesticity of *Aedes aegypti* (L.) (Diptera, Culicidae) in Africa by mark–release–recapture. *Bull. Entomol. Res.*, **65** (1975), 199–208.

Wesley, C., Jr. and A. W. Morrill, Jr. Air and insect penetration of insect screens. *Mosq. News*, **16** (1956), 204–6.

Wood, D. M., P. T. Dang, and R. A. Ellis. The insects and arachnids of Canada. Part 6. The mosquitoes of Canada: Diptera: Culicidae. *Res. Branch Agricult. Can. Pub.*, **1686** (1979), 1–390.

Xue, R. D. and D. R. Barnard. Human host activity in *Aedes albopictus*: influence of mosquito body size, age, parity and time of day. *J. Am. Mosq. Control Assoc.*, **12** (1996), 58–63.

### Drosophilidae, Ephydridae

Basden, E. B. The distribution and biology of Drosophilidae (Diptera) in Scotland, including a new species of *Drosophila*. *Trans. R. Soc. Edinb.*, **62** (1954), 602–54.

Bastock, M. and A. Manning. The courtship of *Drosophila melanogaster*. *Behavior*, **8** (1955), 85–111.

Burla, H. Systematik, Verbreitung und Oekologie der *Drosophila* Arten der Schweiz. *Rev. Suisse Zool.*, **58** (1951), 23–175.

Carson, H. L. and H. D. Stalker. Natural breeding sites for some wild species of *Drosophila* in eastern United States. *Ecology*, **32** (1951), 317–30.

Dayson-Hudson, V. R. D. The daily activity rhythm of *Drosophila subobscura* and *D. obscura*. *Ecology*, **37** (1968), 562–7.

Dekker van Klinken, R., G. H. Walter, and M. K. Ross. Drosophilidae (Diptera) of Australia's Northern Territory: ecology and biogeography. *Aust. J. Entomol.*, **41** (2002), 236–42.

Demerec, M. (ed.) *Biology of* Drosophila. New York: Wiley, 1950.

Dobzhansky, Th. and C. Paven. Local and seasonal variations in relative frequencies of species of *Drosophila* in Brazil. *J. Anim. Ecol.*, **19** (1950), 1–14.

Frydenberg, O. The Danish species of *Drosophila* (Dipt.). *Entomol. Meddelelser*, **27** (1956), 249–94.

Hadron, E., H. Burla, H. Gloor, and F. Ernst. Beitrag zur Kenntnis der *Drosophila*-Fauna von Sudwest-Europe. Zeitschrift fur indukt *Abstammungs Verebungslehre*, **84** (1952), 758–61.

Harrison, F. P., L. P. Ditman, and W. E. Bickley. Habits of *Drosophila* with reference to animal excrement. *J. Econ. Entomol.*, **47** (1954), 935.

McEvey, S. F., Potts, A., Rogers, G., and Walls, S. J. A key to Drosophilidae (Insecta: Diptera) collected in areas of human settlement in southern Africa. *J. Entomol. Soc. S. Africa.*, **51** (1988), 171–81.

Speith, H. T. The breeding sites of *Drosophila lacicola* Patterson. *Science*, **113** (1951), 232.

Stevenson, R. and H. St. C. James. Temperature as a factor in size and activity of wild populations of *Drosophila*. *J. Tenn. Acad. Sci.*, **28** (1953), 43–8.

Sturtevant, A. H. The classification of the genus *Drosophila*, with descriptions of nine new species. *Univ. Texas Publ.*, **4213** (1942), 5–51.

Vanninen, I. Biology of the shore fly *Scatella stagnalis* in rockwool under greenhouse conditions. *Ent. Exp. Appl.*, **98** (2001), 317–28.

Vibe-Peterson, S. Development, survival and fecundity of the urine fly, *Scatella (Teichomyza) fusca* and predation by the black dumpfly, *Hydrotaea aenescens*. *Ent. Exp. Appl.*, **87** (1998), 157–69.

Wheeler, M. R. Geographical survey of Drosophilidae: Nearctic species, pp. 99–121. In Ashburner, M. (ed.) *The Genetics and Biology of* Drosophila. New York: Academic Press, 1981.

### Muscidae

Awati, P. R. Bionomics of houseflies IV. Some notes on the life-history of *Musca*. *Ind. J. Med. Res.*, **8** (1920), 80–8.

Barber, G. W. A note on migration of larvae of the house fly. *J. Econ. Entomol.*, **12** (1919), 466.

Brookes, V. J. and G. Fraenkel. The nutrition of the larva of the housefly, *Musca domestica* (L.). *Physiol. Zool.*, **31** (1958), 208–23.

Charlwood, J. D. and S. Sama. The age structure, biting cycle and dispersal of *Stomoxys niger* Macquart (Diptera: Muscidae) from Ifakara, Tanzania. *Afr Entomol.*, **4** (1996), 274–7.

Chillcott, J. G. A revision of the Nearctic species of Fanniinae (Diptera: Muscidae). *Can. Entomol.*, **92** (suppl. 14), (1960).

Curran, C. H. The parasitic habits of *Muscina stabulans* Fabricius. *J. N.Y. Entomol. Soc.*, **50** (1942), 355–6.

Dakshinamurty, S. The common housefly, *Musca domestica* L., and its behavior to temperature and humidity. *Bull. Entomol. Res.*, **39** (1948), 339–57.

Hafez, M. A study of the biology of the Egyptian common house-fly; *Musca vicina* Macq. (Diptera: Muscidae). *Bull. Soc. Fouad Entomol.*, **25** (1941), 163–89.

Hafez, M. and M. A. Attia. Studies on the ecology of *Musca sorbens* Weid Egypt. *Bull. Soc. Entomol. Egypte*, **42** (1958), 83–121.

Harris, R. L., P. D. Grossman, and O. H. Graham. Mating habits of the stable fly. *J. Econ. Entomol.*, **59** (1966), 634–6.

Ilse, D. and L. Nulherkar. Mating reactions in the common Indian house fly, *Musca domestica nebulo.*, *Curr. Sci.*, **23** (1954), 227–8.

Il'Yasenko, M. and L. Ya. Migration of flies (*Musca domestica vicina*) in rural areas. *Med. Parazitol. Parazitarn. Bolezni*, **33** (1964), 9–13.

Keiding, J. *VII. The House Fly. Biology and Control.* Report no. WHO/BC/86.937. Geneva, Switzerland: World Health Organization, 1986.

Killough, R. A. and D. M. Mc Kinstry. Mating and oviposition studies of the stable fly. *J. Econ. Entomol.*, **58** (1965), 489–591.

Levinson, Z. H. Food of housefly larvae. *Nature*, **188** (1960), 427–8.

Morgan, C. E. and G. D. Thomas. Annotated bibliography of the horn fly, *Haematobia irritans* (L.), including references on the buffalo fly, *H. exigua* (de Meijere), and other species belonging to the genus *Haemotobia*. *US Dept. Agric. Miscell. Publ.*, **1278** (1974), 1–134.

Supplement I: annotated bibliography of the horn fly, *Haematobia irritans irritans* (L.), including references on the buffalo fly, *H. irritans exigua* (de Meijere), and other species belonging to the genus *Haemotobia*. *US Dept. Agric. Misc. Publ.*, **1278** (suppl. I), (1977), 1–138.

Morgan, C. E., G. D. Thomas, and R. D. Hall. Annotated bibliography of the stable fly, *Stomoxys calcitrans* (L.), including references on the other species belonging to the genus *Stomoxys*. *Univ. Missouri-Colombia Agric. Exp. Station Res. Bull.*, **1049** (1983), 1–190.

Ogata, K. and T. Suzuki. Release studies on the dispersion of the lesser house fly, *Fannia canicularis*, in the residential area of Babai (in Japanese). *Boytu-Kagabu*, **25** (1960), 51–7.

Oo, K. N. Carriage of enteric bacterial pathogens by houseflies in Yangon, Myanmar. *J. Diarr. Dis. Res.*, **7** (1989), 81–4.

Parker, R. R. Dispersion of *Musca domestica* Linnaeus under city conditions in Montana. *J. Econ. Entomol.*, **9** (1916), 325–54.

Paterson, H. E. The *Musca domestica* complex in Sri Lanka. *J. Entomol.*, (B), **43** (1974), 247–59.

Paterson, H. E. and K. R. Norris. The *Musca sorbens* complex: the relative status of the Australian and two African populations. *Aust. J. Zool.*, **18** (1970), 231–45.

Patton, W. S. Studies on the higher Diptera of medical and veterinary importance. A revision on the genera of the tribe Muscini, subfamily Muscinae, based on a comparative study of the male terminalia. I. The genus *Musca* Linneaus. II. A practical guide to the palaearctic species. *Ann. Trop. Med. Parasitol.*, **27** (1933), 135–56, 327–45, 397–430.

Pickens, L. G., N. O. Morgan, J. G. Harstock, and J. W. Smith. Dispersal patterns and populations of the house fly affected by sanitation and weather in rural Maryland. *J. Econ. Entomol.*, **60** (1967), 1250–5.

Pickens, L. G., N. O. Morgan, and R. W. Thimijan. House fly response to fluorescent lamps: influenced by fly age and nutrition, air temperature and position of lamps. *J. Econ. Entomol.*, **62** (1969), 536–9.

Pont, A. C. A review of the Fanniidae and Muscidae of the Arabian peninsula. *Fauna Arabia*, **12** (1991), 312–65.

Ranade, D. R. Some observations on the range of flight of the common Indian housefly, *Musca domestica nebulo* Fabr., *J. Anim. Morph. Physiol.*, **3** (1956), 104–8.

Sabrosky, C. W. House-flies in Egypt. *Am. J. Trop. Med. Hyg.*, **1** (1952), 333–6.

Savage, E. P. and H. F. Schoof. The species composition of fly populations at several types of problem sites in urban areas. *Ann. Entom. Soc. Am.*, **48** (1955), 251–7.

Schoof, H. F. and E. P. Savage. Comparative studies of urban fly populations in Arizona, Kansas, New York, and West Virginia. *Ann. Entomol. Soc. Am.*, **48** (1955), 1–12.

Schoof, H. F. and R. E. Shiverly. Multiple release studies on the dispersion of *Musca domestica* at Phoenix, Arizona. *J. Econ. Entomol.*, **47** (1954), 830–8.

Schoof, H. F., Siverley, R. E., and Jensen, J. A. Housefly dispersion studies in Metropolitan Areas. *J. Econ. Entomol.*, **45** (1952), 675–83.

Schoof, H. F., G. A. Mail, and E. P. Savage. Fly production sources in urban communities. *J. Econ. Entomol.*, **47** (1954), 245–53.

Shinonaga, S. and R. Kano. *Fauna Japonica. Muscidae (Insecta: Diptera)*, vol. 1. Tokyo: Academic Press of Japan, 1971.

Siverley, R. E. and Schoof, H. F. Utilization of various production media by muscoid flies in a metropolitan area. I. Adaptability of different flies for infestation of prevalent media. *Ann. Entomol. Soc. Am.*, **48** (1955), 258–62.

Simmons, S. W. and W. E. Dove. Breeding places of the stablefly or 'dog fly' *Stomoxyes calcitrans* (L.) in northwestern Florida. *J. Econ. Entomol.*, **34** (1941), 457–62.

Skidmore, P. *The Biology of the Muscidae of the World*. Dordrecht: W. Junk 1985.

Somme, L. On the overwintering of house-flies (*Musca domestica*) and stable flies (*Stomoxys calcitrans*) in Norway. *Norsk. Ent. Tidsskrift*, **11** (1961), 191–223.

Teskey, H. J. A review of the life history and habits of *Musca autumnalis* DeGeer (Diptera: Muscidae). *Can. Entomol.*, **92** (1960), 360–7.

On the behavior and ecology of the face fly, *Musca autumnalis* (Diptera: Muscidae). *Can. Entomol.*, **101** (1969), 561–76.

Thomson, R. C. M. Observations on the biology and larvae of the Anthomyidae. *Parasitology*, **29** (1937), 273–358.

Trofimov, G. K. The bazaar fly *Musca sorbens* Wd. (Diptera, Muscidae) in Azerbajan. *Entomol. Obozren.*, **42** (1963), 757–64.

Weismann, R. Untersuchungen uber das physiologische. Verhalten von *Musca domestica* L. versichiedener Provenienzen. *Schweiz. Entomol. Gesell. Mitt.*, **20** (1947), 484–504.

West, L. S. *The Housefly: Its Natural History, Importance, and Control*. Ithaca, NY: Comstock, 1954.

West, L. S. and O. B. Peters. An annotated bibliography of *Musca domestica* Linneaus. London: Dawsons of Pall Mall, 1973.

### Mycetophilidae, Phoridae, Piophilidae, Psychodidae, Rhagionidae

Abonnenc, E. Les phlébotomes de la région thiopienne (Diptera, Psychodidae). *Mém l'ORSTOM*, **55** (1972), 1–289.

Abonnenc, E. and D. M. Mintner. Bilingual keys for the identification of the sandflies of the Ethiopian region. *Cahiers ORSTOM (Entomol. Méd.)* **5** (1965), 1–63.

Abul-hab, J. and S. A. Ahmed. Revision of the family Phlebotomidae (Diptera) in Iraq. *J. Biol. Sci. Res. (Baghdad)*, **7** (1984), 1–64.

Artemiev, M. M. Sandflies (Diptera, Phlebotominae) of Afghanistan. Part 2. Distribution according to landscapes (in Russian, English summary). *Med. Paraz. Parazitol.*, **52** (1983), 25–33.

Ashford, R. W. Sandflies (Diptera: Phlebotomidae) from Ethiopia: taxonomic and biological notes. *J. Med. Entomol.*, **11** (1974), 605–16.

Berland, L. Sur le parasitisime des Phorides (Dipteres). *Bull. Soc. Zoo. Fr.*, **57** (1933), 529–30.

Borgmeier, T. Revision of the North American phorid flies. Part I. (Dipt. Phroidae). *Studia Entomol.*, **6** (1963), 1–256.

Revision of the North American phorid flies. Part II. The species of the genus *Megaselia*, subgenus *Aphiochaeta* (Diptera, Phoridae). *Stud. Entomol.*, **7** (1964), 257–416.

Revision of the North American phorid flies. Part III. The species of the genus *Megaselia*, subgenus *Megaselia* (Diptera, Phoridae). *Stud. Entomol.*, **8** (1965), 1–160.

Brown, B. V. and L. W. Morrison. New *Pseudacteon* (Diptera: Phoridae) from North America that parasitizes the native fire ant *Solenopsis geminata* (Hymenoptera: Formicidae). *Ann. Entomol. Soc. Am.*, **92** (1999), 308–11.

Buxton, P. A. British Diptera associated with fungi. 2. Diptera bred from Myxomycetes. *Proc. R. Entomol. Soc., Lond. (A)*, **29** (1954), 163–71.

British Diptera associated with fungi. III. Flies of all families reared from about 150 species of fungi. *Entomol. Mon. Mag.*, **96** (1961), 61–94.

Chandler, P. J. Notes on British Fungus Gnats of the subfamily Ditomyiinae, Bolitophilinae, Diadocidiinae, Manotinae and Keroplatinae (Dipt., Mycetophilidae). *Entomol. Mon. Mag.*, **113** (1978), 31–44.

Colyer, C. N. 'Swarming' of Phoridae (Diptera). *J. Soc. Br. Ent.*, **5** (1954), 22–7.

Davidson, I. H. *Sandflies of Africa South of the Sahara: Taxonomy and Systematics of the Genus* Sergentomyia. Johannesburg: South African Institute for Medical Research, 1990.

Disney, R. H. L. *Scuttle flies: Diptera, Phoridae, except* Megaselia. *Handbook for the Identification of British Insects*, vol. 10, part 6. London: Royal Entomological Society, 1983.

*Scuttle Flies: Diptera, Phoridae, genus* Megaselia. *Handbook for the Identification of British Insects*, vol. 10, part 8. London: Royal Entomological Society, 1989.

*Scuttle Flies: Phoridae*. London: Chapman and Hall, 1994.

A new species of Phoridae (Diptera) that parasitises a widespread Asian ladybird beetle (Coleoptera: Coccinellidae). *Entomologist*, **116** (1997), 163–7.

Duckhouse, D. A. Psychodidae (Diptera, Nematocera) of southern Australia: subfamily Psychodinae. *Trans. R. Entomol. Soc. Lond.*, **118** (1966), 153–220.

Edwards, F. W. British Fungus-Gnats (Diptera, Mycetophilidae). With a revised generic classification of the family. *Trans. Entomol. Soc. Lond.*, **1924** (1925), 505–670.

Eisfelder, I. Beitrage zu Kenntnis der Fauna in hoheren Pilzen. *Z. Pulzk.*, **16** (1954), 1–12.

Hackman, W. and M. Meinander. Diptera feeding as larvae on macrofungi in Finland. *Ann. Zool. Fenn.*, **16** (1979), 50–83.

Hanson, W. J. The breeding places of *Phlebotomus* in Panama (Diptera, Psychodidae). *Ann. Entomol. Soc. Am.*, **54** (1961), 317–22.

Hoy, J. B. and J. R. Anderson. Behavior and reproductive physiology of blood-sucking snipe flies (Diptera: Rhagionidae: *Symphoromyia*) attacking deer in northern California. *Hilgardia*, **46** (1978), 113–68.

Iwasa, M. The Piophilidae (Diptera) of Japan. *Med. Entomol. Zool.*, **49** (1998), 33–9.

Kern, R. Asthma due to sensitization to a mushroom fly (*Aphiochaeta agrici*). *J. Allergy Clin. Immunol.*, **61** (1938), 10–16.

Lane, R. P. The sandflies of Egypt (Diptera: Phlebotominae). *Bull. Br. Museum (Nat. Hist.) Entomology*, **52** (1986), 1–35.

Laštovka, P. A study on the last instar larvae of some Czechoslovak *Mycetophila* (Diptera, Mycetophilidae). *Acta Univ. Carol. Biol.*, **1970** (1971), 137–76.

Lewis, D. J. and W. Buttiker. Some ecological aspects of Saudi Arabian phlebotomine sandflies (Diptera: Psychodidae). *J. Aust. Entomol. Soc.*, **21** (1983), 37–54.

Madwar, S. Biology and Morphology of the immature stages of Mycetophilidae (Diptera, Nematocera). *Phil. Trans. R. Soc. B*, **227** (1937), 1–110.

Maeta, Y. Some biological studies on the natural enemies of some coccinellid beetles. (II). *Phalacrotophora* sp. *Tohoku Konchu Kenkyu*, **4** (1969), 1–6.

Mallis, A. and R. J. Pence. The Pacific drain fly in homes. *J. Econ. Entomol.*, **34** (1941), 586–7.

Mills, H. B. An outbreak of the snipe fly *Symphoromyia hirta*. *J. Econ. Entomol.*, **36** (1943), 806.

Nagatomi, A. Studies in the aquatic snipe flies of Japan. Part V. Biological notes (Diptera, Rhagionidae). *Mushi*, **36** (1962), 103–49.

Ordman, D. Sewage filter flies (*Psychoda*) as a cause of bronchial asthma. *South Af. Med. J.*, **20** (1946), 32–5.

Quate, H. W. A revision of the Psychodidae (Diptera) in America north of Mexico. *Univ. Calif. Publ. Entomol.*, **10** (1955), 103–273.

Robinson, W. H. Old and new biologies of *Megaselia* species (Diptera, Phoridae). *Stud. Entomol. Petropolis*, **14** (1971), 321–48.

*Megaselia* (*M.*) *scalaris* (Diptera: Phoridae) associated with laboratory cockroach colonies. *Proc. Entomol. Soc. Wash.*, **77** (1975), 384–90.

Identification of *Megaselia agarici* (Lintner) and *Megaselia halterata* (Wood) (Diptera: Phoridae). *US Dept. Agric. Coop. Plant Pest Rep.*, **3** (1978), 14–16.

Ryan, L. *Flebótomos do estado do Pará, Brasil. (Diptera: Psychodidae: Phlebotominae.) Documento Técnico Inst. E. Chagas, Belem*, vol. 1. Belém: Minist. De Saude.

Satchell, G. H. The ecology of the British species of *Psychoda* (Diptera: Psychodidae). *Ann. Appl. Biol.*, **34** (1947), 611–21.

Shemanchuk, J. A. and J. Weintraub. Observations on the biting and swarming of snipe flies (Diptera: *Symphoromyia*) in the foothills of southern Alberta. *Mosq. News*, **21** (1961), 238–43.

Truitt, G. W. The mushroom fly as a cause of bronchial asthma. *Ann. Allergy*, **9** (1951), 513–16.

Trumble, J. T. and R. L. Pienkowski. Development and survival of *Megaselia scalaris* (Diptera: Phoridae) at selected temperatures and photoperiods. *Proc. Entomol. Soc. Wash.*, **81** (1979), 207–10.

Turner, W. J. A case of severe human allergic reaction to bites of *Symphoromyia* (Diptera: Rhagionidae). *J. Med. Entomol.*, **15** (1979), 138–9.

Webb, D. W. The Nearctic Athericidae (Insecta: Diptera). *J. Kansas Entomol. Soc.*, **50** (1977), 473–95.


#### Scatopsidae, Scenopenidae, Sciaridae


Bovien, P. The larval stages of Scatopse (Diptera Nematocera). Vidensk. *Meddr Dansk Naturh. Foren.*, **99** (1935), 33–43.

Edwards, F. W. A synopsis of British Bibionidae and Scatopsidae. *Ann. Appl. Biol.*, **12** (1925), 263–75.

Enderlein, G. Die phyletischen Beziehungen der Lycoriiden (Sciaridae) zu den Fungivoriden (Mycetophiliden) und Itonididen (Cecidohyiiden) und ihr systematische Gliederung. *Arch. Naturgesch.*, **77** (suppl. 3) (1911), 116–201.

Kelsey, L. P. A revision of the Scenopinidae (Diptera) of the world. US National Museum Bulletin 277. Washington, DC: Smithsonian Institute, 1969.

Meade, A. B. and E. F. Cooke. Notes on the biology of *Scatopse fuscipes* (Meigen) (Diptera: Scatopsidae). *Ent. News*, **72** (1961), 13–18.

Mihalyi, F. Rearing flies from feces and meat, infected under natural conditions. *Acta Zool. Hung.*, **11** (1965), 153–64.

Stephan, W. A. A generic revision of the family Sciaridae (Diptera) of America north of Mexico. *Univ. Calif. Publs. Entomol.*, **44** (1966), 1–77.


#### Sepsidae, Simulidae, Sphaeroceridae, Syrphidae


Adler, P. H. and K. C. Kim. The black flies of Pennsylvania (Simuliidae, Diptera). Bionomics, taxonomy, and distribution. *Penn. State Univ. Coll. Agricult. Bull*, **856** (1986), 1–88.

Buckton, G. B. The Natural History of *Eristalis tenax* or the drone fly. London: Macmillan, 1895.

Crosskey, R. W. *The Natural History of Blackflies*. Chichester: John Wiley, 1990.

Crosskey, R. W. and W. Buttiker. Insects of Saudi Arabia. Diptera: Fam. Simuliidae. *Fauna Saudi Arabia*, **4** (1982), 398–446.

Davies, D. M. Seasonal variation of Tabanids (Diptera) in Algonquin Park, Ontario. *Can. Entomol.*, **91** (1959), 548–53.

Davies, L. Key to the British species of Simuliidae (Diptera) in the larval, pupal, and adult stages. *Freshwater Biol. Assoc. Sci. Pub.*, **24** (1968), 1–126.

Dawe, G. and S. McGlashan. The ecology of urban hoverflies in relation to spontaneous and managed vegetation. *Br. Ecol. Soc. Bull.*, **8** (1987), 168–71.

De Villiers, P. C. *Simulium* dermatitis in man – clinical and biological features in South Africa. *South Afr. Med. J.*, **71** (1987), 523–5.

Dixon, T. J. Key to and descriptions of the third instar larvae of some species of Syrphidae (Diptera) occurring in Britain. *Trans. R. Entomol. Soc. Lond.*, **112** (1960), 345–79.

Fredeen, F. J. H. Outbreaks of the black fly *Simulium arcticum* Malloch in Alberta. *Quaest. Entomol.*, **5** (1969), 341–72.

Fredeen, F. J. H. and M. E. Taylor. Borborids (Diptera: Sphaeroceridae) infesting sewage disposal tanks, with notes on the life cycle, behavior and control of *Leptocera* (*Leptocera*) *caenosa* (Rhondani). *Can. Entomol.*, **96** (1964), 801–8.

Gidgel, E. F. and F. H. Grauer. Acute and chronic reactions to black fly bites (*Simulium* fly). *Arch. Dermatol. Syphilol.*, **70** (1954), 609–15.

Gilbert, F. S. 1986. *Hoverflies, Naturalist Handbook no.* 5. Cambridge: Cambridge University Press.

Jamnback, H. Recent developments in control of blackflies. *Annu. Rev. Entomol.*, **18** (1973), 281–304.

Kim, K. C. and R. W. Merritt (eds) *Black Flies: Ecology. Population Mangement, and Annotated World List*. University Park, PA: Pennsylvania State University, University Park, 1987.

Mangan, F. A key and selected notes for the identification of larvae of Sepsidae (Diptera) from the temperate regions of North America. *Proc. Entomol. Soc. Wash.*, **79** (1977), 338–42.

Palmer, R. W. and F. C. De Moor. Annotated records of blackfly (Diptera: Simuliidae) distribution in southern Africa. *Afr. Entomol.*, **6** (1998), 223–51.

Schneider, F. Bionomics and physiology of aphidophagous syrphid larvae. *Annu. Rev. Entomol.*, **14** (1969), 103–24.

Schofield, S. W. Black fly host location: a review. *Can. J. Zool.*, **64** (1986), 1041–53.

Schofield, S. W. and J. F. Sutcliffe. Human individuals vary in attractiveness for host-seeking black flies (Diptera: Simuliidae) based on exhaled carbon dioxide. *J. Med. Entomol.*, **33** (1996), 102–8.

Humans vary in their ability to elicit biting responses from *Simulium venustum* (Diptera: Simuliidae). *J. Med. Entomol*, **34** (1997), 64–7.

Stubbs, A. E. and S. J. Falk. *British Hoverflies: An Illustrated Identification Guide*. London: British Entomological and Natural History Society, 1983.

**Tabanidae, Tipulidae**

Chvala, M., L. Lyneborg, and J. Moucha. *The Horse Flies of Europe (Diptera, Tabanidae)*. Copenhagen: Entomological Society of Copenhagen.

Hayakawa, H. A key to the families of Japanese tabanid flies with a checklist of all species and subspecies (Diptera, Tabanidae). *Jpn. J. Sanit. Zool.*, **36** (1985), 15–23.

Krinsky, W. L. Animal disease agents transmitted by horse flies and deer flies (Diptera: Tabanidae). *J. Med. Entomol.*, **13** (1976), 225–75.

Lewis, D. J. The Tabanidae of the Anglo-Egyptian Sudan. *Bull. Entomol. Res.*, **44** (1953), 175–216.

Mackerras, I. M. The classification and distribution of Tabanidae (Diptera). *Aust. J. Zool.*, **2** (1954), 431–54.

The Tabanidae (Diptera) of Australia. I. General review. *Aust. J. Zool.*, **4** (1956), 376–407.

McCrae, A. W. R. Unique record of crane flies biting man. *Uganda J.*, **31** (1967), 128.

Mitzmain, M. B. The biology of *Tabanus striatus* Fabr., the horsefly of the Philippines. *Philippine J. Sci.*, **8** (ser. B, no. 3) (1913), 197–218.

Murdoch, W. P. and H. Takahasi. The female Tabanidae of Japan, Korea and Manchuria. The life history of the family, morphology, classification, systematics, distribution, evolution and geologic history of the family Tabanidae (Diptera). *Mem. Entomol. Soc. Wash.*, **6** (1969), 1–230.

Stone, A. The bionomics of some Tabanidae (Diptera). *Ann. Entomol. Soc. Am.*, **23** (1930), 261–304.

Tashiro, H. and H. H. Schwardt. Biological studies of horse flies in New York. *J. Econ. Entomol.*, **46** (1953), 813–22.

Theodor, O. Tabanidae of Israel. *Israel J. Zool.*, **14** (1965), 241–57.

Wang, Z.-M. *Economic Insect Fauna of China 26. Diptera Tabanidae* (in Chinese). Beijing: Science Press, 1983.

Wilkerson, R. C., J. F. Butler, and L. L. Pechuman. Swarming, hovering, and mating behavior of male horse flies and deer flies (Diptera: Tabanidae). *Myia*, **3** (1985), 515–46.

**EPHEMEROPTERA**

Alba-Tercedor, J. and A. Sanchez-Ortega. *Overview and Strategies of Ephemeroptera and Plecoptera*. Gainesville, FL: Sandhill Crane Press, 1991.

Berner, L. *The Mayflies of Florida*. Gainesville, FL: University of Florida Press, 1950.

Berner, L. and M. L. Pescador. The mayfly family Baetiscidae (Ephemeroptera). Part I. In Flannagan, J. F. and K. E. Marshall (eds) *Advances in Ephemeroptera Biology*. New York: Plenum Press, 1980.

Britt, N. W. Biology of two species of Lake Erie mayflies, *Ephron album* (Say) and *Ephemera simulans* Walker. *Bull. Ohio Biol. Surv. (New ser.)*, **1** (1962), 1–70.

Brittain, J. E. Biology of mayflies. *Annu. Rev. Entomol.*, **27** (1982), 119–47.

Burks, B. D. The mayflies or Ephemeroptera of Illinois. *Ill. Nat. Hist. Surv. Bull.*, **26** (1953), 1–216.

Campbell, I. C. (ed.) *Mayflies and Stoneflies: Life Histories and Biology. Series Entomologica* 44. Dordrecht: Kluwer Academic, 1990.

Edmunds, G. E., Jr and W. P. McCafferty. The mayfly subimago. *Annu. Rev. Entomol.*, **33** (1988), 509–29.

Edmunds, G. E., Jr, S. L. Jensen, and L. Berner. *The Mayflies of North and Central America*. Minneapolis: University of Minnesota Press, 1976.

Flannagan, J. F. and K. E. Marshall (eds) *Advances in Ephemeroptera Biology*. New York: Plenum Press, 1980.

Fremling, C. R. Biology of a large mayfly, *Hexagenia bilineata* (Say) of the Upper Mississippi river. *Iowa Agr. Home Econ. Exp. Sta. Res. Bull.*, **482** (1960), 842–52.

Rhythmic *Hexagenia* mayfly emergences and the environmental factors which influence them. *Verh. Internat. Verein. Limnol.*, **15** (1964), 912–16.

Harker, J. *Mayflies. Naturalists' Handbooks 13*. Slough, UK: Richmond, 1989.

Swarm behavior and mate competition in mayflies (Ephemeroptera). *J. Zool.*, **228** (1992), 571–87.

Henson, E. B. Aquatic insects as inhalent allergens: a review of American literature. *Ohio J. Sci.*, **66** (1966), 529–32.

Hubbard, M. D. *Mayflies of the World: A Catalog of the Family and Genus Group Taxa*. Gainesville, FL: Sandhill Crane Press, 1990.

Hubbard, M. D. and W. L. Peters. Ephemeroptera. In Hurlbert, S. H. (ed.) *Biota Acuática de Sudamérica Austral*. San Diego, CA: San Diego State University, 1977.

Landholt, P. and M. Sartor (eds) Ephemeroptera and Plecoptera biology, ecology, systematics. Fribourg: Mauron, Tingley & Lachat.

Macan, T. T. The study of stoneflies, mayflies and caddisflies. London: Amateur Entomologist's Society, 1982.

Nedham, J. G. Burrowing mayflies of our larger lakes and streams. *Bull. US Bur. Fish.*, **36** (1920), 269–92.

Needham, J. G., J. R. Traver, and Y.-C. Hsu. *The Biology of Mayflies*. Ithaca, NY: Comstock, 1935.

Peters, W. L. and P. G. Peters (eds) *Proceedings of the First International Conference of Ephemeroptera*. Leiden: E. J. Brill, 1973.

Savolainen, E. Swarming in Ephemeroptera: the mechanism of swarming and the effects of illumination and weather. *Ann. Zool. Fenn.*, **15** (1978), 17–52.

Simpson, S. J. and G. C. McGavin. *The Right Fly: An Angler's Guide to Identifying and Matching Natural Insects*. London: Aurum Press, 1996.

# 8 Hemiptera, Homoptera

## HEMIPTERA
### Introduction

Insects in the orders Hemiptera and Homoptera differ principally in the structure of the wings, and in the location of the mouthparts on the head. The front wings of Homoptera have a uniform texture, either leathery or membranous, and the mouthparts arise at the posterior of the head. The front wings of Hemiptera are divided into a thickened basal portion and a membranous apical portion, and the mouthparts arise from the front of the head. Some classifications include these insects in one order, Hemiptera, with two suborders: Heteroptera (previously Hemiptera) and Homoptera. Here the orders are considered separate.

Hemiptera may be wingless, brachypterous, or a combination of long-winged and brachcypterous forms in one species. These bugs have a piercing and suctorial proboscis that is attached anteriorly and flexed posteriorly when not in use. The primary food for the terrestrial species is plant fluids; some utilize insect or animal blood. Most aquatic species are predators on insects and other small animals. Several species are important plant and animal disease vectors, and are economically and medically important, especially in the tropics. Eggs are usually laid on or inserted in plants, or they are placed in crevices or they are attached to features of the substrate, which are near potential food for the immature stages. There are usually five nymph stages. Overwintering often occurs in the adult stage.

Pest status of this group is based on various behavior patterns. Some species, such as boxelder bugs, overwinter in large numbers around and sometimes in structures, and are pests by their presence or the allergic reactions they cause. Occasional or accidental biting by reduviids and other bug species and blood-feeding by bed bugs can be a serious problem, and difficult to control.

### Acanthosomatidae

This small family is closely related to the Pentatomidae, but separated from them by having two instead of three tarsal segments. There are several species that overwinter as adults in large numbers in peridomestic and domestic habitats. In Japan, numerous species occur around buildings or indoors, including *Acanthosoma denticaudum*, *A. forficula*, *A. haemorrhoidale angulatum*, *A. labiduroides*, and *A. spinicolle*; *Dichobothrium nubium*, *Elasmostethus putoni*, *E. amurensis*, *E. signoreti*, and *E. humeralis*, and *Sastragala easkii*.

***Elasmostethus humeralis*** Adults are about 11 mm long and the body is light green. The front wing margin and scutellum are reddish brown anteriorly; the wing membrane is opaque white. Legs are light green and the antennae are lightly banded. This species commonly overwinters around the perimeter of buildings in Japan, and occasionally occurs indoors.

### Alydidae

These bugs are similar in appearance to coreids, but the head is broad and nearly as long as the pronotum. They have large metathoracic scent glands and the adults and nymphs produce a strong and offensive odor when handled.

***Alydus* spp.** *Alydus* spp. have been reported entering houses in large numbers from nearby fields of groundsel (*Senecio*) in southeastern USA. Nymphs of these broad-headed bugs are slender and resemble ants; adults in the field resemble spider wasps (Pompilidae) in their shape and behavior. *Alydus pilosulus* feeds on several legumes, and soybean is a preferred host. This species occurs throughout the USA.

## Anthocoridae

These bugs are predaceous on small arthropods, such as mites, thrips, and aphids. They occur on foliage, flowers, under bark, and in leaf litter. Some species live in bird nests. They are elongate oval, 2–5 mm long, and most are black with white markings. Anthocorids are distributed nearly worldwide. Adults of several species, including *Lyctocoris campestris* and *Anthocoris kingi*, have been reported to bite people indoors. *Xylocoris* species are associated with stored products in warehouses and bakeries, and they are known to bite people.

**Pirate bug, *Lyctocoris campestris*** Adults are about 3.5 mm long. The head, thorax, and abdomen are black, and wings are yellowish white. They often occur in houses and granaries, and in the nests of birds and mammals. They are known to bite people while sleeping, particularly those on bedding stuffed with straw or grass.

**Flower bug, *Orius insidious*** Adults are about 2 mm long, and have pale-yellow wings and a black head, pronotum, and scutellum. This species is widely distributed and usually found on flowers and other vegetation. It is important as a predator on the eggs and larvae of some pest insects. Adults are encountered in gardens, or when people handle ornamental flowers brought indoors. The bite from this bug, and *O. tristicolor*, another common anthocorid, is very painful.

***Xylocoris curistans*** Adults are 2.2–2.5 mm long and shiny dark brown. Wings are light yellow to dark brown. It usually occurs in the brachypterous form in which the wing covers extend only to abdominal segment 3. Natural habitats for this species include under the bark of fallen and rotting logs where it probably feeds on springtails and mites. It occurs in warehouses and food-processing sites, and feeds on stored-product pests. Infestations can become very large, and this usually results in people being bitten. This species is probably native to Europe, but it has been introduced into other regions of the world, including the USA.

## Belostomatidae

Giant water bugs are aquatic and prey on snails, small fish, tadpoles, young frogs, and insects in the water. They occur in North and Central America, South Africa, India, and parts of Asia. These are large insects and can reach a length of 80 mm; two Central American species, *Lethocerus grandis* and *L. maximus*, are more than 100 mm long. Adults are elongate oval and flattened, with raptorial front legs; the other legs are ciliated and shaped for swimming. Adults and nymphs crawl on the bottom substrate or swim. They obtain air by extending the tip of their abdomen to the water surface, and bringing air to a cavity beneath the wings. The adults are capable flyers and can move between small and large ponds.

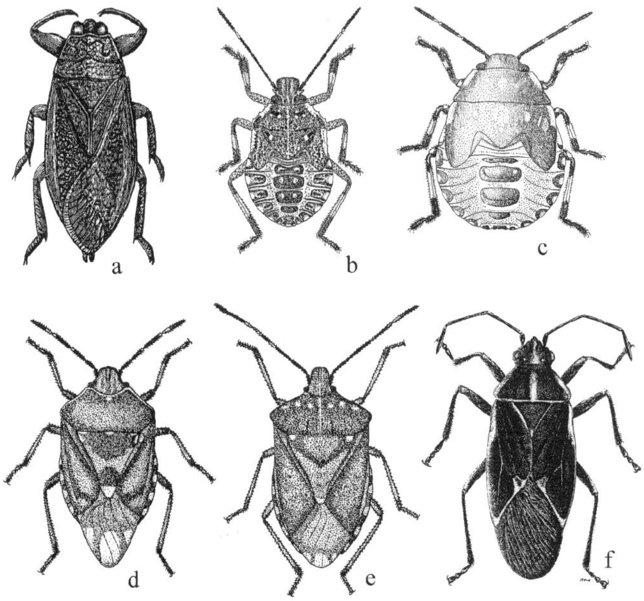


**Figure 8.1** Hemiptera: Belostomatidae, Pentatomidae, Rhopalidae. (a) *Lethocerus americanus* adult (Belostomatidae); (b) *Halyomorpha mista* nymph (Pentatomidae); (c) *Menida scotti* nymph (Pentatomidae); (d) *H. mista* adult; (e) *M. scotti* adult; (f) *Boisea trivittata* adult (Rhopalidae).

**Giant water bug, electric light bug, *Lethocerus americanus* (Fig. 8.1a)** Adults are 45–60 mm long and 20–25 mm wide; the body is dark brown to slightly greenish brown. The hind legs are narrow and the eyes are strongly divergent. Eggs are laid in spring; they are deposited in masses on stems of emergent vegetation in ponds and temporary pools. Occurrence in streams is infrequent. Development from nymph to adult takes about 40 days. There may be more than one generation per year. Adults overwinter in water, usually buried in the bottom substrate. During summer, adults feed on other insects, fish, frogs, salamanders, and freshwater snakes. This species occurs in central and northern North America and Canada. Related species with similar habits include *L. uhleri*, common in southern USA and coastal northeastern Mexico, *L. cordofanus* in Africa, and *L. indicus*, which occurs from Australia to India.

Adult *Lethocerus* species fly readily during summer, and often travel long distances between aquatic sites. In rural and residential areas they are attracted to lights at night. They are called electric light bugs because this may be the only place people ever see them. The most common species gathering at outdoor lights are *L. griseus* and *L. uhleri*. These insects can inflict a painful bite if handled or accidentally encountered under lights at night. Australia's largest species is *L. insulanus*, and it is about 70 mm long.

**Other Belostomidae** Adults of *Belostoma flumineum* are 20–25 mm long, light brown, and the legs are mottled with dark-brown markings. This species occurs throughout North America, and the adults are in flight from May to October.

## Cimicidae

Cimicids are characterized by a short head that is broadly attached to the prothorax, an oval body, and well-developed eyes. Adults are 3–13 mm long and 3–5 mm broad, yellowish brown to reddish brown, and their body is somewhat flattened. They are wingless, but have small wing pads. Antennae are four-segmented, and the three-segmented proboscis rests in a groove beneath the head and thorax. Nymphs and adult males and females are obligate feeders on vertebrate blood. They are temporary ectoparasites on bats, birds that roost together in small to large numbers, and people. Included in this family are the bed bugs, swallow bug, pigeon bug, and poultry bug. These insects live in cracks and crevices separately from their hosts, and move to the host for periodic blood meals. Large populations of cimicids develop when conditions and suitable hosts are available. These bugs are distributed worldwide, but mostly in the tropics and subtropics. Bed bugs have been closely associated with humans for centuries, and are nearly dependent on humans to survive. Prehistoric human communities encountered these insects in caves, and here their adaptation to feeding on humans probably occurred.

Cimicidae includes 91 described species. Most are associated with birds or bats, but the two bed bug species, *Cimex lectularius* and *C. hemipterus*, are associated with humans. Although one population of *C. lectularius* has been found with cave bats in Afghanistan, this species is considered one of the few insects associated only with humans. It is distributed in North America, Europe, and Russia. *C. hemipterus*, the tropical bed bug, is common in tropical regions of Asia, Africa, and Central America where it attacks humans and chickens, and rarely bats. In general, *C. lectularius* occurs in temperate regions and *C. hemipterus* in tropical regions, but there is some hybridization, and in some regions, such as Brazil, populations of *C. lectularius* are increasing.

*Haematosiphon inodorus* occurs in Central America and *Ornithocoris toledoi* in Brazil. *C. pipistrelli* is associated with birds in Europe and *C. columbarius* occurs in poultry houses and pigeon cotes. *Oeciacus hirundinis* infests martin nests; it sometimes bites humans when adults and nymphs move inside houses when nests are abandoned. *Hesperocimex sonorensis* is an ectoparasite of purple martins (*Proge subis*) in southwestern USA and northern Mexico. *Leptocimex boueti* is associated with bats in West Africa, but has also been reported biting people in native huts. Other bat bugs include *Stricticimex parvus*, which also feeds on humans in bat-caves in Thailand.

*Cimex* spp. takes a large blood meal in a short time, and this behavior permits them to feed infrequently. Bed bug adults can engorge in about 5 min and take up to 1.8–24 μg of blood per second. The blood meal weighs from 3.7 (first-stage nymph) to 4.9 (fifth-stage nymph) times the body weight of the unfed insect. Molting and egg production require a blood meal, so feeding is directly linked to development. In spite of their blood-feeding habits, cimicids are not vectors of any disease pathogens, although there is a report of their vectoring hepatitis B virus in Senegal. Investigations showed that the human immunodeficiency virus (HIV) was retained in the tropical bed bug, *Cimex hemipterus*, for a week. However, the virus did not replicate, was not detected in excreta, and was not transmitted to other animals after feeding.

Mating behavior is not elaborate in these bugs and the process of insemination is considered as traumatic. The female bed bug has no functional genital opening, but there is a site between abdominal segments 5 and 6 that is modified to accept the reproductive organ of the male. The male has a C-shaped intromittant organ that curves around the tip of the abdomen, and directed to the left. This organ punctures the abdominal cuticle of the female at the modified site on the abdomen. In copulation, the male mounts the female, and curls his abdomen over the right side of hers, and inserts his intromittant organ into the copulatory pore in the female's abdomen. This pore leads to a bag-like organ called the mesospermalege or the organ of Berlese or Ribaga, which serves to collect and store sperm. The sperms bore their way through the bag and migrate through the body cavity to the oviducts. They enter and fertilize the eggs, and the eggs pass normally down the oviduct. The wound created by the male during mating quickly heals and leaves a scar.

**Eastern bat bug, *Cimex adjunctus*** Adults are about 6 mm long and oval; the body is brown to dark brown. This species is morphologically similar to other *Cimex* spp. It occurs throughout eastern and midwestern regions of the USA. Large infestations may occur with bat populations in the attics of buildings. Adults are known to move from household harborages near roosting bats to occupied rooms and bite people during the night.

**Pigeon bug, *Cimex columbarius*** This species is morphologically similar and closely related to *C. lectularius*. It is primarily distinguished by the ratio between head width and antennal segment 3: in *C. lectularius* the ratio is 1.45, and in *C. columbarius* it is 1.78. Hybridization occurs between these two species, but the offspring have low fertility. *C. columbarius* occurs in Europe and feeds primarily on domestic pigeons, but also feeds on the pied flycatcher (*Muscicapa atricapilla*) in birdhouses.

**Tropical bed bug, *Cimex hemipterus* (Fig. 8.2b)** Adults are about 5 mm long, oval, and somewhat flattened. The apical margins of the wingpads are distinctly rounded. These bugs are reddish brown to light red depending on recent blood-feeding. The pronotum has the anterior margin moderately concave (it is deeply curved in *C. lectularius*); the sides of the pronotum are not dilated, and are fringed with sparse and nearly straight setae. Females lay 2–3 eggs per day; fecundity is 78–125 eggs. Hatching occurs in about 3 days. Development from egg to adult takes about 20 days: first instar lasts about 4 days, second and third instar about 3 days, fourth about 4 days, and the fifth instar lasts about 6 days. Adult females live about 41 days and males about 198 days.

*C. hemipterus* and *C. lectularius* are closely related and hybridization between the two occurs. However, female *C. lectularius* mated with a male *C. hemipterus* die, perhaps as a result of the traumatic insemination by the male genital clasper puncturing the female abdomen. When the male *C. lectularius* mates with a *C. hemipterus* female the results are not harmful.

**Bed bug, *Cimex lectularius* (Fig. 8.2a, c, d)** Adults are 4–6 mm long, oval, and somewhat flattened; the wingpads are small. These bugs are dark brown to reddish brown depending on whether they have taken a blood meal recently. The pronotum has the anterior margin deeply concave (it is moderately curved in *C. hemipterus*). The body is covered with microscopic setae, which give the abdomen a slightly banded appearance. Nymphs are morphologically similar to the adults; the small wingpads appear at the last larval molt. The abdomen cuticle of nymphs is relatively thin and displays the color of the partly digested blood inside. The cuticle of the adult is thick and dark brown.

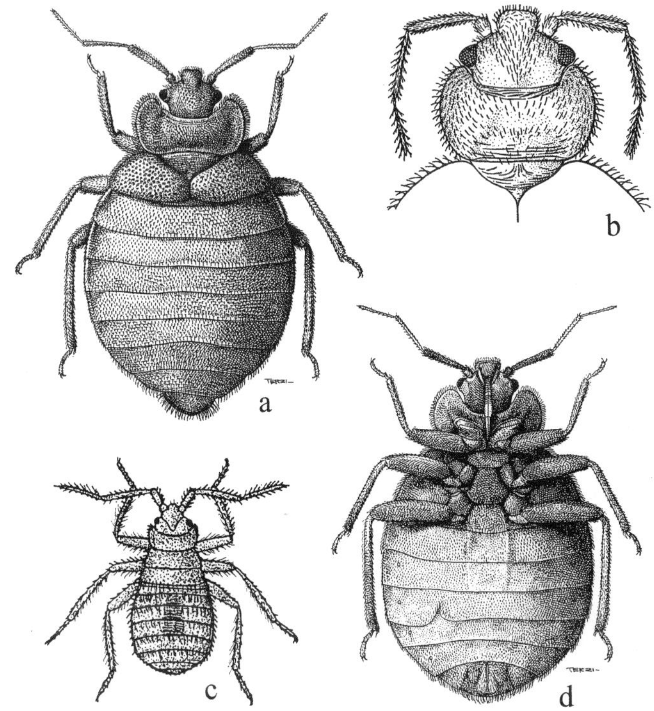


**Figure 8.2** Hemiptera: Cimicidae. (a) *Cimex lectularius* male; (b) *C. hemipterus*, head and prothorax; (c) *C. lectularius* nymph; (d) *C. lectularius* female.

Eggs are about 1 mm long, yellowish white, and with a distinct cap at one end. Females deposit eggs in batches of 10–50; fecundity is about 350, but ranges from 200 to 500. Maximum egg-laying occurs between 21 and 28 °C; eggs do not hatch above 37 °C or below 13 °C. Hatching is in 6–17 days at 21 °C, and in 30–40 days at 14 °C. A transparent substance attaches the eggs to the substrate, and the empty chorion remains on the substrate after hatching. Development from egg to adult is through five nymph instars, and takes 14 days at 28 °C, and 118 days at 15 °C; development may be further extended in unfavorable conditions. The lowest temperature at which bed bugs complete development is 13 °C. Nymphs can survive long periods without feeding: the first instar can live 114 days without a blood meal; second instar, 171 days; third instar, 214 days; fourth instar, 234 days; and, the fifth instar can survive 161 days. With frequent opportunity to feed, adults live about 360 days at 18–20 °C, 105 days at 27 °C, and about 70 days at 34 °C. Adult males live for about 176 days and females about 277 days without feeding. At 7 °C and 90% RH, unmated and unfed females live for 550 days, but at 23 °C and 90% RH they survive for 181 days. Under favorable conditions, there are three or four generations per year, but in unheated rooms in north temperate climates, there is one generation per year.

A blood meal is required between each molt, and before egg development. Adults do not seek food at temperatures below 9 °C; between molts they will feed in about 6 days at 15 °C to 24 h at 25 °C. Bed bugs prefer harborages with rough surfaces, and if not available in the bedstead or mattress they will find suitable sites along baseboards close to the bed, and move from these to the host. Feeding is nocturnal and usually peaks soon after sunset and before dawn, but they will feed during the day if conditions are favorable. In cool weather, the adults and nymphs may remain active in the harborage for 2–4 weeks. Carbon dioxide and warmth are the environmental features used in their host-finding behavior. These bugs will respond to temperature gradients only 2° above normal. Adults produce an aggregation pheromone and an alarm pheromone. When alarmed, bed bugs can move away at 2 cm/s, and typically emit an odor. Heavily infested sites are often characterized by a distinct odor attributed to the scent glands of adults, and accumulations of nymph and adult bug excrement in harborages. Bed bug odor is a complex mixture of chemicals, but mainly hexanol and octenol. They frequently defecate brownish-black deposits of partly digested blood in the harborage. The odor of a bed bug infestation is similar to the aroma of the spice coriander, which is a name derived from *koris*, the Greek word for bug. *C. lectularius* (*cimex*, Latin for bug; *lectul*, Latin for bed) is a common household pest in most parts of the world. Populations and pest status frequently fluctuate, perhaps depending on living conditions and the increased movement of people around the world.

Heat and cold influence bed bug development and survival. Adults are killed by exposure to −17 °C for 2 h or −18 °C for 1 h. Recently fed bugs are more susceptible to temperature extremes than partially starved ones. Exposure for 2 h at −15 °C is lethal to about 75% of eggs. At −9 °C eggs die in 30–60 days. The thermal death point for adults is 1 h at 44 °C and 24 h at 40 °C; eggs are killed at about the same temperatures.

There is little evidence that bed bugs are vectors of any human pathogen, but experimentally they can be infected with hepatitis B, HIV, and may be the agent for Chagas disease, *Trypanosoma cruzi*. Daily feeding of large numbers of bed bugs can contribute to anemia in infants. Domestic infestations that remain for long periods can be distressing, and bites can become infected after scratching. Allergic reactions to the bites are common. In some situations, *C. lectularius* has been found as an ectoparasite and remaining in clothing of vagrants.

**Western bat bug, *Cimex pilosellus*** This species is distinguished from *C. lectularius* by long setae on the body of the adult and large nymphs, and the thorax, which extends forward on both sides of the head. It does not feed on humans, but bugs from infestations associated with bats may spread to occupied rooms in houses. This species occurs in western Canada and in western USA.

**Poultry bug, *Haematosiphon inodorus*** This species has long legs, and a long beak that extends to the hind coxae. It is sometimes called the Mexican chicken bug. It is a parasite in the roosts of birds, including owls, eagles, condors, and chickens. However, it has been reported to spread from roosts to dwelling spaces where it may bite people. The species is distributed in western and southwestern USA and Central America.

**Swallow bugs, *Oeciacus hirundinis*, *O. vicarius*** Adults are about 5 mm long and brown to reddish brown; the body has numerous long setae. The middle coxae are widely separated and the beak does not extend to coxa 2. Antennal segments 3 and 4 are equal or nearly equal in length. These two species feed on swallows and occasionally attack humans. *O. hirundinis* occurs throughout western and central Europe, and south into North Africa. *O. vicarius* occurs in western North America and occasionally in some eastern states and eastern Canadian provinces. It is not associated with swallows in their South American wintering quarters. *O. vicarius* will also feed on the house martin, and some other bird species.

*O. vicarius* assemble in the empty nests of swallows in the spring (April). When migrating swallows return to the nests from their wintering quarters, the bugs take blood meals. Sometimes while feeding they are carried to other (new) nests on the body of birds. Eggs are deposited within 24 h of feeding. Eggs are laid singly or in small batches of about 16 eggs; hatching occurs in 3–5 days. Development from egg to adult takes about 60 days. They overwinter as nymphs or adults. These bugs are most active when the host birds are nesting, but show a peak of activity after the birds leave the nest site. For about 9 months of the year, swallows are absent from the nest site and the bugs have limited access to food. Some find alternative hosts, such as other birds or bats, but most species survive on food reserves in their body. However, many do not survive the winter without food. *O. vicarius* adults and nymphs enter houses from nests built near or on houses; in the absence of other hosts, they bite humans. Large numbers of these bugs enter houses, schools, and commercial buildings in response to the removal of bird nests.

*O. hirundinis* usually feeds on other bird species, as well as mammalian hosts, than swallows. In Europe, there are two generations per year. In spring, when martins return to nesting sites, there is a peak in the bug population, primarily from those that overwintered at the nest site. In winter, there is a peak in the nymph population. Although the martins have left the nest site by then, there are other birds, such as house sparrows, that serve as hosts.

## Coreidae

Leaf-footed bugs are large, elongate insects with the head narrower than the pronotum. Some species have the hind tibiae expanded and with spines in the males. They have well-developed scent glands and give off a distinct odor when disturbed or handled. Coreids are primarily plant-feeders but some are predaceous on other insects. There is usually one generation per year, and adults overwinter in garden debris and leaf litter. Pest status of the few peridomestic species is based on their overwintering habits. The number of individuals is usually small, but their presence indoors is unacceptable.

**Squash bug, *Anasa tristis*** Adults are 13–18 mm long, grayish brown, with the edges of the abdomen orange, or striped orange and brown. Antennae are uniformly dark brown or yellow marked with dark spots; at the base of the antenna is a small tubercle. Venter and legs are mottled yellow and brown. Immatures are slightly green and sometimes covered with a white, powdery material. Eggs are laid on plant leaves singly or in batches of 17–19. Females lay about 13 eggs per day at 30 °C, and about four eggs per day at 20 °C. Development from egg to adult takes about 24 days at 30 °C. There are 1–3 generations per year, depending on location: 1 in the east, 1.5 in midwest, and 3 in western USA. Adults overwinter in natural habitats under loose bark and leaves. They are known to invade houses, especially basements, in the fall. They do not bite people, but are a nuisance by their presence. This bug feeds on a variety of cultivated crops in the genus *Cucurbita*, particularly summer squash and pumpkin. It is distributed throughout most of the USA and in Mexico.

***Thasus acutangulus*** Adults are 35–40 mm long, and the body is orange and black. The nymphs are orange, yellow, and black. Adults and nymphs are commonly found on mesquite and other trees in southeastern North America and Central America. Nymphs form aggregations on trees and shrubs, and as a group they give off noxious secretions when disturbed. Aggregations on ornamental trees can be a nuisance around houses and commercial buildings.

**Western conifer-seed bug, *Leptoglossus occidentalis*** Adults are about 18 mm long; the body is reddish brown to black, and has pale zigzag stripes across the forewings. Hind legs are long and hind tibiae are flattened. Abdominal segments are orange and black striped. Eggs are laid in rows on the needles of coniferous trees. Nymphs feed on needles, ripening seeds of pine trees, and become adults in late summer. Adults overwinter in protected locations, including around and in buildings in urban and rural areas. They usually gather on south- and west-facing sides of structures, then move to cracks and crevices, or indoors. They do not usually occur in large numbers. This species is distributed in western USA and feeds on the seeds of Douglas fir trees, and the needles and green cones of other pine trees. It has expanded its range to eastern USA, and frequently occurs indoors in midwestern and eastern states.

**Other Coreidae** In Japan, there are several species that overwinter around the perimeter of buildings, including *Acanthocoris sordidus*, *A. striicornis*, *Cletus trigonus*, *Homocerus unipunctatus*, *Hygia opaca*, and *Plinachtus bicoloripes*.

## Corixidae

Adults are 5–18 mm long, with an elongate oval and usually dark-gray body. These bugs are called water boatmen. They swim on their ventral surface and use the long posterior legs to move through the water. These insects inhabit freshwater ponds and lakes; one species occurs in brackish water. Most species feed on algae; a few are predators of aquatic invertebrates. Water boatmen obtain air at the water surface and carry a supply under the wings or around the body. The name of the typical genus, *Corixa*, is evidently from the Greek word *coris*, meaning a bug. It was probably used because these insects produce an odor similar to the bed bug.

Eggs are attached to the stems of aquatic plants, sometimes by means of a glutinous substance. *Corixa* is a large genus and distributed nearly worldwide, and has species in diverse habitats. Adults are strong flyers and travel between favorable habitats. They can leap into the air from the water, take flight, and move between aquatic habitats at night. They often collect in large numbers at outdoor lights in summer. Masses of eggs from *Arctocorixa abdominalis* and *A. mercenaria* are collected and used as food by some Mexican people. Adult corixids are collected and used as food in Mexico and Egypt.

## Cydnidae

These bugs resemble stink bugs (Pentatomidae) in general appearance, except that they are more oval and have spiny tibiae. It is a large family with species that live in leaf litter and in soil at the base of plants. Some species are associated with ants. Several species are commonly found at lights at night. In the Amami Islands of Japan, large numbers of *Aethus indicus* in houses have been reported. There were no reports of this bug biting people indoors.

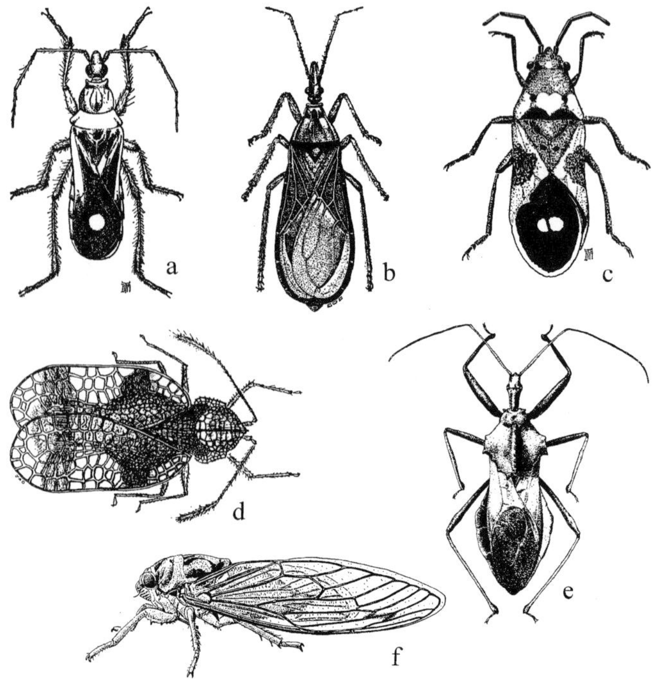


**Figure 8.3** Hemiptera: Lygaeidae, Reduviidae; Homoptera: Cicadidae. (a) *Rasahus thoracicus* (Reduviidae); (b) *Triatoma protracta* (Reduviidae); (c) *Lygaeus reclivatus* (Lygaeidae); (d) *Stephanitis pyriodes* (Tingidae); (e) *Arilus cristatus* (Reduviidae); (f) *Tibicen* sp. (Cicadidae).

## Lygaeidae

Adults range in size from 2 to 18 mm, and are characterized by the veins on the wing. They have four-segmented antennae, and have four or five distinct veins on the membranous portion of the wing. Many species are marked with spots or bands of red, white, or black. These bugs feed primarily on seeds and plant sap, but a small number are predaceous. Various species of lygaeids are known to bite humans.

**Birch catkin bug, *Kleidocerys resedae geminatus*** Adults are 5–11 mm long and brownish red; nymphs are dark red and black. This species feeds on the catkins of yellow and gray birch, and seed capsules of azaleas, rhododendron, and Japanese andromeda (*Piers japonica*). Adults and nymphs may occur in large numbers, and they can be carried indoors. These bugs have a strong odor when handled or crushed. This species occurs in eastern USA.

**Milkweed bug, *Lygaeus equestris*** Adults are 10–13 mm long, and distinctly marked black, white, and red, and with a pale white spot on the wing membrane. This species is a common pest in Europe and Asia (Japan) where, in the fall, large numbers of adults fly to buildings seeking an overwintering site. The adults do not bite, but are a pest by their numbers and disagreeable odor.

**Common milkweed bug, *Lygaeus reclivatus* (Fig. 8.3c)** Adults are 9–11 mm long; the head, thorax, appendages, and spots on the abdomen are grayish black; the abdomen and marks on the dorsum are red. Wing membranes are black with white markings. Nymphs are red and black. This insect feeds on various species of milkweed plants (*Asclepias* spp.). Hibernating adults are often observed in large numbers, and sometimes around buildings.

**False chinch bug, *Nysius ericae*** Adults are 3–4 mm long, light to dark gray; nymphs are pale gray with a reddish-brown abdomen. This species occurs on grasses, and it is widely distributed in North America. Large populations occur in the fall. Overwintering may be as nymphs or adults, but the adults have the best survival rate. In the fall, large numbers of adults and nymphs seek additional food when the host grasses decline, and they move to overwintering sites. They may gather on ornamental plants and around buildings. Some reports of *N. ericae* as a pest around buildings may actually be *N. niger*, since these two are closely related.

**Grass bug, *Nysius raphanus*** Adults are 3–5 mm long, black, and with pale white wings. Nymphs and adults feed primarily on grasses in pastures. In urban areas they occur in vacant lots and weedy areas. Large numbers are known to move to the perimeter of structures and enter when their preferred habitat is removed or disturbed. They fly to lights at night.

**Spotted milkweed bug, *Oncopeltus fasciatus*** Adults are 10–15 mm long and often brightly colored red or orange, and black, with three large black areas on the dorsum, and with pale white spots on the wing membrane. Nymphs are usually red with black appendages. These bugs feed on the seeds of milkweed (*Asclepias* spp.) and other plants. Large populations may develop in undisturbed areas. Adults overwinter in protected sites and often fly in large numbers to the perimeter of buildings. The distribution ranges from Brazil to Canada, and it is reported in western USA.

**Other Lygaeidae** Lygaeid bugs found indoors include *Ischnodemus falicus*, which occurs around houses in southern USA (North Carolina), and the chinch bug, *Blissus leucopterus*. Although the chinch bug occurs in agricultural areas, large numbers sometimes enter neighboring houses. *Neacoryphus lateralis*, a large red and black lygaeid, is known to infest buildings. *Eremocoris borealis* adults and nymphs have been reported in large numbers around and in houses. A related species, *E. ferus*, has also been recorded indoors. Species of the genus *Leptodemus* in the Middle East and eastern Mediterranean may occur in swarms and bite humans. In desert areas of Egypt, *Leptodemus bicolor ventralis* bites, causing irritation and fever. In Kuwait and Sudan, *Nysius* spp. and *Pamera* spp. land on exposed skin and bite humans; in Japan *N. plebius* occurs indoors. In Brazil, *Clerada* spp. bite people indoors.

## Miridae

This is the largest family of bugs and it contains species that feed primarily on plants and are destructive pests. Adults are elongate oval, 4–14 mm long, and with two distinct cells on the wing membrane. They are active flyers. Some species are predaceous on other insects and have the ability to deliver a painful bite to humans. In Japan, *Phytocoris ohataensis* and *Lygocoris idoneus* often occur around the perimeter of buildings.

Mirids will pierce human skin when it is either moist with perspiration or dry. A large number of species are known to bite and suck blood. The reasons for this behavior are unclear, and it is likely that they are seeking nourishment. The majority of records of mirids biting are from North American species, but species from other regions have been reported biting; these are indicated in the list below.

*Atractotomus miniatus*
*Blepharidopterus angulatus* (UK)
*B. provancheri*
*Brachynotocoris puncticornis* (Algeria)
*Camptotylus yersini* (Sudan)
*Campylomma verbasci*
*Campyloneura virgula*
*Chlamydatus albidus* (North Africa)
*Creontiades pacificus* (Malaysia)
*C. pallidus* (Sudan)
*Cyrtorhinus lividipennis* (Japan)
*Deraeocoris manitou*
*D. nebulosa*
*D. quercicola*
*D. ruber* (UK)
*Haematocapsus bipunctatus* (Nigeria)
*Hyaliodes harti*
*H. vitripennis*
*Irbisia brachycera*
*Lopidea chandleri*
*L. marginata*
*L. robiniae*
*Lygocoris communis*
*L. omnivagus*
*Lygus lineolaris*
*L. pratensis* (UK)
*Maurodactylus albidus* (North Africa)
*Microphylellus tsugae*
*Orthotylus flavoparsus*
*O. ramus*
*Pilophorus crassipes*
*Plagiognathus obscurus*
*P. politus*
*P. repetitus*
*Pseudatomoscelis seriata*
*Reuteria irrorata*
*Rhinocapsus vanduzeei*
*Sejanus albsignatus* (New Zealand)
*Taedia scrupea*
*Trigonotylus brevipes* (East Africa)

**Azalea plant bug, *Rhinocapsus vanduzeei*** Adults are 8–16 mm long, reddish orange and black. Nymphs and adults are often found on or around ornamental azalea plants and other shrubs. They may be attracted to perspiration and carried indoors on clothing or on skin. Adults have been reported to bite people; the bite is painful and produces a small welt.

## Naucoridae

Creeping water bugs are 9–13 mm long and light brown to brown; the body is broadly oval and somewhat flattened. The head is rounded and the front femora are enlarged. They are common in still water where they occur in emergent vegetation. They feed on small aquatic animals, and deliver a painful bite to humans.

**Western water creeper, *Ambrysus signoreti*** Adults are about 8 mm long and yellowish brown with dark markings. It ordinarily lives in pools of still water in streams and small bodies of water. They sometimes occur in outdoor swimming pools in southern California. The adults will bite if handled.

## Nepidae

Waterscorpions can be distinguished by the presence of a long respiratory tube at the end of the abdomen, and the raptorial front legs. The body shape is flat and broad (*Nepa*) or rounded and linear (*Laccotrephes*, *Ranata*). These bugs live in aquatic environments where they are carnivorous on other insects and small animals. They usually remain in shallow water among vegetation, with only the breathing tube piercing the water surface. They have well-developed wings but seldom fly. These bugs occur in ponds or other small bodies of water in peridomestic habitats. Some *Laccotrephes* and *Ranata* species may be useful in controlling mosquito larvae in the urban environment. They will bite when handled. *Ranata* stridulates under water by rasping areas on the front leg coxae and coxal cavities.

***Laccotrephes grossus* (= *L. kohlii*)** Adults are 30–35 mm long and 7–9 mm wide. Body is brown to dark brown. Eggs are laid in spring and again in summer; females deposit about 1.6 eggs per day, and fecundity is about 100 eggs. Hatching is in 12–13 days. Development of the five instars takes about 40 days. Adults and nymphs overwinter. This species occurs throughout China, southeastern Asia, and India. A closely related species, *L. maculatus*, occurs in Java, India, Taiwan, and Japan. Both *L. grossus* and *L. maculatus* are efficient predators of mosquito larvae.

***Nepa apicauda*** Adult body is about 16 mm long, and the breathing tube is about 6 mm long. The body is brownish gray, broad, and somewhat flattened and thin. The front legs are distinctly raptorial. Adults and nymphs live in shallow bodies of water with emergent vegetation. Eggs are inserted into dead or decaying plants; they are distinctly oval and have seven filaments at the top. This species occurs in eastern USA.

***Ranata fusca*** The adult body is about 24 mm long and nearly uniformly brown to dark brown. Eggs are laid in spring through summer, and hatching occurs in about 11 days. Development in the laboratory from egg to adult takes about 36 days at about 27 °C and a light/dark ratio of 16:8. Predation of mosquito larvae by nymphs may reach 33 per hour. Adults overwinter. There are two generations per year. This species is distributed across northern USA and southern Canada.

## Notonectidae

These bugs are called back swimmers, and they prefer to live in still water with emergent vegetation, but some inhabit slow-moving streams. They have a thick, boat-shaped body and swim on their back, with the ventral surface contacting the water surface. Adults are usually found floating at the surface with the long oar-like legs outstretched. They can quickly submerge or leap into the air and fly. They are predators of insects and small animals. Eggs are attached to emergent plants or debris with a glutinous substance, or inserted in leaves or stem by a piercing ovipositor (some *Notonecta* spp.). *Notonecta unifasciata* and *N. indica* eggs are collected and used as food by people in towns near Mexico City. Large numbers of adults are collected and ground into flour for cooking.

Adult notonectids are often brightly colored with areas of the body blue, violet, red, orange, and white. They have sharp mouthparts and can inflict a painful bite if handled. These insects can be serious pests in fish hatcheries where they prey on small fish, and are occasionally found in swimming pools where they bite people. In some regions in Europe (Germany) they are known as waterbees.

**Back swimmer, *Notonecta undulata*** Adults are 10–13 mm long, and vary from brown to blackish brown. Eggs are elongate, about 1 mm long, and laid on the surface of aquatic plants. Development is through five instars; there are at least two generations per year. Adults fly at night and large numbers can be found around lights. They are often found in swimming pools, sometimes a long distance from an aquatic habitat. This is a widely distributed species in North and South America.

## Pentatomidae

These bugs are called stink bugs because of the disagreeable odor they emit when handled or disturbed. The adult is usually broad, oval, with the head and thorax forming a triangle; the antennae are five-segmented. Many are brightly colored or conspicuously marked. Most species are plant-feeders, but some feed on other insects, and some on both. Pest status is based on overwintering of adults of several species that occur in urban and suburban areas. Individuals move toward sheltered sites in late summer and fall; adults are strong flyers and can move long distances. Some species are consistent pests in the fall and winter because they return to the same sites yearly. More than 30 000 individual stink bugs have been collected inside buildings, and numbers nearly equal to that collected outside. Several species that overwinter in large numbers are found in Asia.

**African cluster bug, *Agonoscelis puberula*** Adults are 8–10 mm long. The body is pale yellow with a red tinge and with black punctures arranged in a pattern of irregular dark stripes. The head is somewhat elongate. The hemelytra membrane has dark radiating stripes. This species is native to southern and eastern Africa, and extends northward to the Arabian peninsula. It has been introduced into the USA (Arizona, Florida, New Mexico, Texas), Mexico, Dominican Republic, and the islands of Jamaica and Hispaniola. In South Africa this species overwinters in buildings and on fruit trees.

**Black and red stink bug, *Cosmopepla bimaculata*** Adults are 5–7 mm long, shining black with bright red or yellow stripes on the pronotum, and two red spots on the scutellum near the tip. It is found on flowers of various plants, including ragweed, goldenrod, and mints. Adults fly from May to October, and are found throughout the USA. Adults often occur indoors.

**Marmorated stink bug, Asian stink bug, *Halyomorpha halys*** Adults are 16–18 mm long; the body is variegated (marmorated) dark and pale brown. Nymphs are grayish black and with a white spot on the apical third of the antenna. Adults overwinter in large numbers inside and outside buildings. The flight of adults to new or previously infested sites begins in the fall (October) and lasts for about 3 weeks. Movement to overwintering sites around and in buildings begins during the day when the temperature reaches about 25 °C after an overnight temperature of about 15 °C. This species occurs in China, Korea, and northern Japan (excluding Hokkaido), and it has been introduced into eastern USA (Pennsylvania).

**Brown marmorated stink bug, *Halyomorpha mista* (Fig. 8.1b,d)** Adults are 14–18 mm long and the body is dark purple-green, with a spotted yellow and brown pattern (marmorated). Antennae are black, and have yellowish-brown bands on segment 4, and at the base of segment 5. Adults and nymphs feed on a variety of fruit trees and grasses. It is often a serious pest of fruit trees, and it may be a pest of agricultural crops such as peas, cucumber, and kidney beans. There are two generations per year. Large populations occur and adults of the second generation seek overwintering sites around and in buildings. This species is distributed in northern Japan (excluding Hokkaido), Korea, and China.

**Spotted stink bug, *Menida scotti* (Fig. 8.1c, e)** Adults are 9–11 mm long; the body is brown to dark brown, with a copper-like luster; there is a median pale spot behind the pronotum. Antennae and legs are banded with white or light brown, and the lateral edges of the abdomen have distinct pale white marks. Adults and nymphs feed on a variety of trees, but they are not considered a pest. Adults overwinter and often occur in large aggregations in the fall. This species is distributed in northern Japan, Korea, and northern China. Large populations occur yearly, and masses of adults gather around the perimeter and inside buildings in the fall. A closely related species, *M. violacea*, has a transverse pale band on the posterior of the thorax. These two species occur in Japan (excluding Hokkaido), and in some northern regions they overwinter in large numbers inside and outside buildings.

**Predaceous stink bug, *Perillus bioculatus*** Adults are about 9 mm long and the body is black with red markings on the dorsum. The color-based variety, *P. clanda*, is black or brown with white markings on the dorsum. This species is predaceous on caterpillars, including armyworms and cutworms, and it also feeds on beetle larvae. Adults often occur around buildings and indoors in fall and winter. It is distributed throughout the USA.

**Black paddy bug, *Scotinophara coarctata*** Adults are 8–9 mm long and brownish black, and with a few yellow spots on the thorax. The tibiae and tarsi are pink. Nymphs are brown and with a green abdomen. Eggs are laid in batches of 40–50; fecundity is 200–300 eggs. Hatching occurs in 4–7 days. Development from egg to adult is 25–30 days and there are five nymph stages. Nymphs and adults feed primarily on rice plants. Adults live 1–3 months, and they are strongly attracted to lights at night. This species often appears in large numbers in urban and suburban areas. A related species, *S. punctatonevosus*, occurs around the perimeter of buildings in fall.

**Other Pentatomidae** Large populations in fall and the overwintering habits of many stink bug species bring them to the perimeter or inside buildings. For some species, this behavior is repeated yearly, but for others it is infrequent. In Japan, *Menida musiva*, *Pentatoma japonica*, and *P. rufipes* often occur in large numbers around buildings in the fall. Other species that commonly enter houses in Japan include *Picromerus lewisi*, *Pinthaeus sanguinipes*, and *Dinorhynchus dybowski*. Large numbers of *Brochymena affinis*, *B. quadripustulata*, and *B. sulcata* have been reported overwintering around buildings in western USA. Large aggregations of *Carbula pedalis* have been reported at buildings in Africa. Some species gather at lights at night and sometimes enter houses; these include *Chlorochroa sayi*, and *Thyanta accera*. *Banasa dimidiata* is bright green with dark reddish markings across the posterior half of the pronotum and over the wing. It is often found at outdoor lights at night.

## Reduviidae

Assassin bugs are primarily predators of other insects, but some feed on humans. These bugs are widely distributed around the world, but most are tropical. They are usually brown to blackish brown, but some are brightly colored red, orange, or yellow; nymphs are often brightly colored and spiny. These bugs have an elongate head that is narrow behind the eyes, and they have a three-segmented proboscis. They are primarily diurnal, but some species hunt and feed at night. The subfamily Triatominae has 118 species, and 105 occur in the western hemisphere from northern USA to southern Argentina. *Linshcosteus* is confined to the Indian subcontinent, and seven species of *Triatoma* occur only in southeastern Asia. These bugs are generally distinguished by their mouthparts, insertion of the antennae on the side of the head, and ocelli behind the eyes in the adults. The body, legs, and antennae of the nymphs are sometimes covered with particles of fine material (up to 10 μm diameter), including house dust, which adhere to a viscous substance on the body surface. Nymphs apply material to their body using their hind legs, and it is replaced after each molt. Apparently this conceals the nymphs in their natural habitats.

Pest status of this family is based on the habits of species that occasionally bite people, and some that are obligate feeders on vertebrate blood. *Triatoma* and *Panstrongylus* require blood for development and occur indoors. They are important disease vectors. The life cycle requires 300 days from egg to adult; but for some it is short and there are 2–3 generations per year. Many species are only known from poultry and livestock barns, and mud or adobe houses. Adults of species not known to complete their life cycle indoors are attracted to lights at night, and they occasionally bite people. Reduviids are sometimes called kissing bugs. This term originated in 1899, when *Reduvius personatus* was reported to feed on the faces and especially the lips of sleeping people.

The subfamily Triatominae contains about 110 species. Many of these species occur in natural (sylvatic) habitats and are in the burrows and nests of vertebrates or in vegetation. Several species occur in peridomestic and domestic habitats and some are vectors of Chagas disease (American trypanosomiasis). This disease occurs primarily in Central and South America, but cases are known from the USA. The Brazilian, Carlos Chagas, discovered and named the flagellate protozoan, *Trypanosoma cruzi* (originally described as *Schizotrypanum cruzi*), which causes the disease. He determined the epidemiological elements of the disease, and identified the probable hemipteran vector.

The domestic and peridomestic habitats frequented by some Triatomines are an important link to their role as disease vectors for trypanosomiasis. A range of animal reservoirs provides *T. cruzi* access to both the natural and domestic environment. The sylvatic form of *T. cruzi* circulates in natural areas among species of small animals and wild species of triatomine bugs. The disease does not harm wild animals and the triatomine carriers do not occur in domestic or peridomestic habitats. The domestic form of *T. cruzi* involves humans, animals, and triatomines common in domestic habitats. The Chagas disease protozoan is capable of developing in most triatomine bugs, but only 10–12 species are synanthropic and serve as vectors for humans or domestic animals. Dogs, cats, squirrels (*Sciurus*), house mice, wood rat (*Neotoma fiscipes macrotis*), and the black rat (*Rattus rattus*) are important domiciliary reservoirs. *T. cruzi* does not infect birds. The important reservoir in urban, suburban, and rural areas is the opossum, *Didelphis marsupialis*. This prolific animal has great adaptive capacity, and it is distributed throughout the Americas.

Chagas disease exists in a domestic form and is most common in rural areas, although it has spread to suburban and urban communities where there is poor housing. The development and spread of this disease are linked with the settlement and urbanization of Latin America. As early settlers moved inland they cleared land and constructed simple houses with adobe walls and palm-thatched roofs. Stored food in the houses attracted small animals, which may have been infected with *T. cruzi*, and adult or nymphs of triatomine bugs may have come into the houses on small animals or flying at night. Eggs of *Rhodinus* may be on the palm leaves used for roof thatching. The abundance of harborage in the earthen walls and roof thatch and the availability of vertebrate hosts provided suitable conditions for the survival of domestic bug populations, and the spread of the disease organisms to humans. Domestic bug populations can be 200–300 individuals in different stages, and each bug feeding every 4–9 days. Biting rates can be as high as 20 bites per night, and a loss of 1–3 ml of blood per person. The disease is transmitted in the feces of bugs, which defecate on the skin of the host while feeding. Trypanosomes in the bug feces are transferred by fingers from the surface of the skin to eyes, mouth, or nose where entry of the infectious agent takes place.

Control programs reduce or eliminate populations of the domestic vectors, such as *Triatoma infestans*, of Chagas disease. However, there are sylvatic species that occupy the vacant niche, and serve as vectors for *Trypanosoma cruzi*. In regions of Brazil where *T. infestans* has been controlled, *T. sordida*, *Panstrongylus megistus*, and *Rhodinus neglectus* have become vectors, and in Uruguay *Triatoma rubrovaria* has filled the niche of *T. infestans*. In Argentina, the sylvatic species *T. guasayana* can vector Chagas disease, and it can become established in domestic habitats.

**Wheel bug, *Arilus cristatus* (Fig. 8.3e)** Adults are 28–36 mm long and with a formation on the thorax similar to a semicircular wheel with 8–12 cogs. Nymphs have red and reddish-yellow markings. Eggs are laid in the fall in masses of 100 or more; oviposition sites include tree bark, twigs, fences, and the eaves of buildings. Development begins in the spring and there are five nymph instars; there is one generation per year. This bug often occurs in home vegetable and flower gardens in the fall, and flies to lights at night to feed. When handled they can inflict a painful bite, and a reaction to the bite can last for hours or days. It occurs throughout the USA and Central America.

**Thread-legged bug, *Emesaya brevicoxa*** Adults are 30–35 mm long, and the head is slender, and the front legs long and raptorial. The body is brown and the wings are about one-third the length of the abdomen. The coxae of the front legs are twice as long as the head. This species occurs in California. *E. brevipennis* is a closely related species; the adults are 33–37 mm long, and it is widely distributed in the USA. Thread-legged bugs resemble walking sticks (Phasmida, Phasmatidae). The adults and nymphs have very slender bodies and long legs and wings, and appear fragile. They are often found hanging by their hind legs from rafters in barns and sheds, and feed on insects trapped in spider webs, or the spider.

**Black corsair, *Melanolestes picipes*** Adults are 15–20 mm long and black. The wings of the female are often reduced to small pads. There are two color forms: entirely black, and black with red markings on the abdomen (which is sometimes considered as *Melanolestes abdominalis*). The nymphs are colored red and black. There is one generation per year and adults and nymphs overwinter. Adults (long-winged forms) are attracted to lights at night. Adults can deliver a painful bite. This species occurs in the USA.

***Panstrongylus megistus*** Adults are 21–34 mm long and the body is blackish brown to black. The pronotum has four reddish-brown spots, and there is a series of reddish-brown spots on the abdomen. This species is common in domestic habitats and is a vector of Chagas disease in Brazil. Carlos Chagas reported successful transmission of *T. cruzi* by *P. megistus*, but believed it was through the bite; later, transmission was demonstrated to be through the feces of the vector.

**Delicate assassin bug, *Ploiaria domestica*** Adults are 7–8 mm long, wingless, and with raptorial front legs. This species occurs in houses in southern Europe. Adults capture small insects at night.

**Two-spotted corsairs, *Rasahus biguttatus*, *R. thoracicus* (Fig. 8.3a)** Adults are about 20 mm long and have a slender, yellowish-tan body. Wings are black with a pale yellow, circular spot on the membrane of each front wing. There is one generation per year, and overwintering is in the nymph stage. Adults are attracted to lights at night and occasionally enter buildings. *R. biguttatus* occurs nearly throughout North and Central America; *R. thoracicus* occurs in California, Arizona, and Central America.

**Masked hunter, dust bug, kissing bug, *Reduvius personatus*** Adults are 17–22 mm long and brownish black. The covering of debris on the nymphs, which may be more than one-quarter of the insect body weight, is the origin of the specific name *personatus*, which means masked. Eggs are deposited singly or in small clusters in the harborage. Development is generally slow and depends on available food and environmental conditions. There are one or two generations per year. This bug occurs indoors and it is considered a predator of bed bugs and other household insects. It is known to bite humans on the lips, which swell – hence the origin of the name kissing bugs. They also occur outdoors, such as in the nests of the wood rat (*Neotoma fiscipes macrotis*). They have been found in nests of cliff swallows where they prey on the cimicid, *Oeciacus vicarius*, which feeds on the blood of these birds. *R. personatus* occurs in buildings associated with bat roosts, and it may prey on the bat bug, *Cimex adjunctus*. This species was introduced to North America, and it occurs in Central America and Europe.

***Rhodnius prolixus*** Adults are about 20 mm long and reddish brown. The head is very long and slender, and the antennae are inserted near the apex. Clypeus is widened apically. Eggs are bright red and laid in batches of about 14; fecundity is about 280 eggs. Microenvironmental conditions for development are 27–30 °C and 60–85% relative humidity (RH). Development from egg to adult in the laboratory is 80–100 days. There are five instars and each requires a blood meal at each molt. The adult needs blood for reproduction, and males and females live 5–6 months. Exposure to temperatures above 35 °C or below 22 °C results in abnormal growth or death. This species is the principal vector of Chagas disease in Venezuela and Columbia. It is highly adapted to the domestic environment, and primarily occurs in rural houses with roofs of palm leaves.

**Spider predator, *Stenolemus lanipes*** Adults are 9–11 mm long and pale yellow. The head is wide and the eyes are large; there is a slender "neck" and the lobes of the prothorax are formed to resemble a dumbbell. This reduviid is associated with the webs of the common house spider, *Achaeranea tepidariorum*. The adult bugs are in or near the spider webs, and nymphs are in harborages adjacent to webs. It preys on spiders and insects that are trapped in the web. This species is distributed in southeastern USA.

**Conenose bugs, chinches, bareiros, benchuca, chirimacha, vinchua, *Triatoma* spp.** Adults are usually less than 30 mm long and the body is brown to brownish black. Head, body, and legs have only short setae. Eggs are laid singly or in small batches, hatching occurs in 10–30 days, and fecundity is 100–300 eggs. A blood meal is usually required for oviposition. Development from egg to adult is through five instars and takes 3–4 months. The nymph stages feed on the same range of host animals as do the adults. Fifth-instar nymphs take 300–1000 mg of blood when feeding to repletion. During or soon after feeding, the bugs begin to defecate. The feces of nymphs contain a pheromone that attracts unfed nymphs. Most species of *Triatoma* are nocturnal, and in houses feed on sleeping inhabitants. They are most abundant in bedrooms, which may contain 50% of the bugs in a house. The feces of infected adults and nymphs contain the flagellate protozoan, *Trypanosoma cruzi*. The walls of infested houses may have black and white streaks of bug feces. Adults live for 3–6 months. The *Triatoma* species that are important vectors of Chagas disease in Latin America are: *T. infestans*, *T. dimidiata*, *T. sordida*, and *T. brasiliensis*; *T. barberi* is an important pest in shantytowns around Mexico City. Other species involved in transmitting the disease in the region are *Rhodnius prolixus*, *R. pallescens*, and *Panstrongylus megistus*.

***Triatoma infestans*** Adults are 18–28 mm long, dark brown to black, and with red markings on the abdomen. Eggs are laid singly or in small numbers in cracks and crevices indoors; hatching occurs in about 21 days at 27 °C and 60% RH. Fecundity is 150–250 eggs, and peak production is about 11 weeks after the first oviposition. Development from egg to adult is about 20 weeks; nymph stage 5 can last about 7 weeks. Females live about 16 weeks, and males live about 26 weeks. This species occurs in domestic and peridomestic populations in South America, as far south as Argentina. This is the most important vector of *Trypanosoma cruzi* in South America. Domestication of this species is presumed to be pre-Columbian. Then dispersal followed pre-Inca tribal migrations from Bolivia to southern Peru and northern Chile. In the early 1900s, *Triatoma infestans* probably accompanied migrant workers across Paraguay and Argentina to coffee farms near São Paulo, Brazil. Once this bug has been passively transported into an area, probably among household material, it may actively disperse among houses by its ability to fly about 200 m.

***Triatoma protracta* (Fig. 8.3b)** Adults are 16–19 mm long and the body is blackish brown to black, with brown markings on the lateral edges of the abdomen. The wing membrane is opaque. It frequently enters houses or the tents of campers and is responsible for painful bites to humans. This species is distributed in southwestern USA and northern Mexico.

***Triatoma sanguisuga*** Adults are 18–24 mm long and the body is reddish brown to black. This species occurs throughout the southern part of the USA and as far north as Illinois and along the western seaboard of Mexico. It is a nocturnal hunter, and is sometimes called the big bed bug. It often occurs in houses where it feeds on bed bugs and other insects, as well as mammals. It often bites people.

**Other Reduviidae** Many species of reduviids bite people if handled or accidentally encountered. This behavior has been reported for *Zelus renardii*, *Rhynocoris ventralis*, and *Rasahus* spp.

## Rhopalidae

These insects are usually less than 10 mm long, and distinguished by many veins on the membranous portion of the forewings. They differ from coreids and other hemipterans in lacking functional scent glands. They are usually found on weeds and grasses, but one species, the boxelder bug, feeds on trees. All are plant-feeders, usually on green ripe seeds. In Japan, several species overwinter in small to large numbers around the perimeter of buildings, including *Coriomeri scabriocornis*, *Rhophalus maculatus*, and *Stictoplerus punctatonervosus*.

**Brown grass bug, *Arhyssus crassus*** Adults are about 9 mm long, and uniformly brown. The brown grass bug is known to be abundant on plants growing in uncultivated land adjacent to residential housing, and to move from these sites and invade structures in summer. This bug produces an objectionable odor, and when it occurs indoors it can stain fabric with excrement. It occurs in western USA. A closely related species, *A. scultatus*, has also been reported to occur in houses in California.

**Boxelder bug, *Boisea trivittata* (= *Leptochoris*) (Fig. 8.1f)** Adults are 10–14 mm long, and somewhat flat and elongate. The body is gray-brown to black with three red lines on the pronotum and forewing; the abdomen is usually red. The young bugs are bright red and become marked with black when about half-grown. Early-stage nymphs are red, and become marked with black in instar 4 and 5. Eggs are laid singly or in batches of about 11 eggs on the bark and leaves of the host tree; eggs may be laid on the ground in grass and leaf litter. Hatching occurs in 11–14 days and the eggs turn from yellow to dark red; fecundity is 200–300 eggs. Development from egg to adult takes about 60 days, and there are 1–3 generations per year. Adults and large nymphs of the last generation seek an overwintering site in late fall. They are frequently found in large numbers around building foundations and ground-floor windows. They gather on the south and west sides of buildings where the sun heats exposed surfaces; they are sensitive to small temperature differences and select the warmest. Adults are capable of flying 3–4 km in search of suitable overwintering sites. Usually only adults survive winter. Adults of the first generation feed on fallen boxelder seeds on the ground or on low vegetation. They feed on female trees once the seeds begin to form.

*B. trivittata* is native to western and southwestern USA, where it develops on boxelder trees, *Acer negundo*. In 1880, it was known from Mexico and the western USA; by the 1930s it spread to midwestern states, and now the range includes southern Canada and eastern USA. This species spread eastward as its primary host the boxelder was planted as an urban ornamental and shade tree. Adults and nymphs on *B. trivittata* feed on seeds, leaves, and twigs of boxelder and other maples, including silver maple (*Acer saccharium*), sycamore maple (*A. pseudoplatanus*), and ash (*Fraxinus*). It will also feed on young fruits such as apple, pear, peach, plum, and grape. These bugs will feed on each other during molting, and they will bite people.

**Western boxelder bug, *Boisea rubrolineata* (= *Leptochoris*)** Adults are 10–14 mm long, and somewhat flat and elongate. The body is gray-brown to black with red lines on the pronotum and forewing; the abdomen is usually red. Adults have three longitudinal red stripes. This boxelder bug has been reported from bigleaf maple (*Acer macrophyllum*) and is known to damage almonds and pistachios. The biology of this species is similar to *B. trivittata*.

**Grass bug, *Corizus validus*** Adults are about 8 mm long, and yellow to light brown. It is common in grassy areas and gardens in summer. Adults overwinter under the bark of trees, but they are known to enter houses in fall. It has been reported to bite people when indoors.

**Soapberry bug, *Jadera haematoloma*** This Neotropical species feeds primarily on the seeds of sapindaceous plants, including the soapberry tree and the goldenrain tree. The adults may damage trees and, like *Boisea trivittata*, they often cluster in large numbers on houses during cool weather.

## Tingidae

Lace bugs are usually less than 5 mm long, flattened, and they have a reticulated surface, which may be lace-like in appearance. The head is hidden beneath the pronotum, which projects laterally, and the abdomen is completely hidden by the reticulated wings. All species are plant-feeders, and many are associated with ornamental trees, frequently sycamore, and shrubs in the urban environment. Several species of the large genus *Corythucha*, including *C. cydoniae* and *C. ciliata*, and *C. morrilli*, have been reported to bite people. The azalea lace bug, *Stephanitis pyriodes* (Fig. 8.3d), is common wherever azaleas are grown. The adults may be numerous on ornamental azaleas around buildings.

**Sycamore lace bug, *Corythucha ciliata* (= *Corythuca*)** Adults are about 3.7 mm long and pale white; the ventral side is light brown. Nymphs are brownish white and the body has many long spines. Adults and nymphs occur on and under loose bark of sycamore trees (*Plantus occidentalis*), and on ash and hickory. Eggs are deposited singly or in groups adjacent to veins on the underside of the leaf; hatching is in 7–28 days. Development from egg to adult takes about 5 weeks. Adults overwinter under the loose bark. It is distributed throughout North America, chiefly east of the Rocky Mountains. Adults have been reported to bite people.

# HOMOPTERA

## Introduction

Homoptera are characterized by their piercing–sucking mouthparts and with the mouthparts arising from the back of the head. In some species, the adult mouthparts are vestigial or lacking. As the name suggests, the wings are usually of the same size and texture, but in some the forewings are thickened and often pigmented. At rest, the wings are folded roof-like over the abdomen. All homopterans feed on plant fluids and produce quantities of honeydew as adults and nymphs. Many species are pests of outdoor ornamental trees, shrubs, and indoor plants. Pest status is limited to species that infest plants in the domestic and peridomestic environment, or species that occur in large numbers in peridomestic habitats and have adults or nymphs that enter buildings. Except for some aphids (Aphididae), mealybugs (Pseudococcidae), and whiteflies (Aleyrodidae) that habitually infest houseplants, homopteran species do not usually complete development and persist indoors.

## Aleyrodidae

Whiteflies are 2–3 mm long and resemble tiny moths. Adult wings are covered with a white wax that gives a dirty-white appearance. Nymphs and adults suck plant sap and can cause economic damage to plants. They are active on the undersurface of leaves, primarily citrus and greenhouse plants. The life cycle includes eggs, a first stage that is active, successive stages that are sessile, and a thick-bodied last instar with eye-spots that is referred to as the pupa. Adults can fly and, when plants are disturbed, they move from the leaves and fly around the plant. They are common on potted indoor houseplants.

**Silver-leaf whitefly, *Bemisia argentifolii*** Adults are 0.9–1.2 mm long, with opaque white wings and yellowish-white body. Immature stages are yellowish white. Females lay 50–400 eggs on the underside of leaves. Development from egg to adult takes 18–60 days, depending on conditions. There are many generations per year. This species occurs in southern USA and may be a serious pest of greenhouse plants and ornamentals outdoors.

**Greenhouse whitefly, *Trialeurodes vaporariorum*** Adults are about 1.5 mm long and with pale yellow bodies and opaque white wings. Females lay eggs singly on the underside of leaves. Eggs are elongate and yellowish green; they are attached to the leaf by a short, slender pedicel. Immatures are flattened and semitransparent pale green. Development from egg to adult takes about 1 month. Adults live for about 20 days, and feed less than the immature stages. There are multiple generations per year. This is a common pest in greenhouses and on a variety of houseplants in houses and commercial buildings. This species is nearly worldwide in distribution.

**Other Aleyrodidae** Ornamental plants and shrubs may be fed on by a variety of whitefly species. The adult azalea whitefly, *Pealius azaleae*, is about 1.4 mm long and pale yellow with opaque white wings. It is a worldwide pest of flowering azaleas.

## Aphididae

Aphids are soft-bodied insects, 1–3 mm long, with long antennae. The body is oval and has a short or long pair of projections at the posterior end of the abdomen. There are winged and wingless forms, depending on the species and time of year. The life cycle includes periods when young are deposited by females, and periods when eggs are produced. Overwintering is usually in the egg stage. Infestations can increase rapidly on succulent houseplants and plant vigor can decline. Damage to plants includes curling, discoloring, and wilting of leaves and flowers. Some species are vectors of plant diseases and are economically important agricultural pests.

**Cotton aphid, *Aphis gossypi*** Immature and winged forms are 2–3 mm long, usually dark green, occasionally yellow or dark brown to black. In some populations, the nymphs and adults may be covered with a white wax. Adults overwinter in plant debris and begin laying eggs early in spring. Sexual forms are not developed in some populations. There are multiple generations per year under household or greenhouse conditions. All forms feed on leaves and buds of ornamental shrubs and houseplants. This aphid has wide distribution in North America.

**Green peach aphid, *Myzus persicae*** Winged forms are 3–4 mm long, pale to light green and with black marking on the abdomen. Immature and wingless forms are 2–3 mm long, yellow, pink, or green. Development throughout the year is parthenogenetic in warm climate regions; sexual reproduction occurs in north temperate regions. Eggs are laid in fall and they overwinter; hatching occurs in spring. Nymphs and mature forms feed on nearly all above-ground parts of plants, and usually occur in aggregations of all life stages. There may be 30 generations per year outdoors in favorable conditions, but fewer occur indoors on houseplants. This aphid is distributed throughout North America, and is a common pest of outdoor ornamental and houseplants.

## Cicadellidae

Adult leafhoppers are 4–9 mm long, slender, and with front wings often brightly colored. They are active as nymphs and adults, and characterized by running sideways; the adults can fly long distances. They are all plant-feeders and many are economically important pests of agricultural crops. Some species occur in large numbers, especially in late summer. Pest status of leafhoppers in the urban environment is limited to a few species that are attracted to lights at night, and species that occur in small numbers indoors. There are limited reports of leafhoppers biting people.

**Green flat-headed leafhopper, *Xerophloea viridis*** Adults are 6–7 mm long, wedge-shaped, and bright green. The head is flat and the tips of the wings at rest are perpendicular. They are common on grasses in western USA. They can occur in large numbers, and there is one report of a massive invasion of a town in Nebraska by this leafhopper.

**Sharp-headed grain leafhopper, *Draculacephala mollipes*** Adults are 6–10 mm long and slender. The head is sharply pointed, green, and the pronotum and scutellum are yellow. This species overwinters as adults and the overwintering form is often dark green or brown. It has been reported attracted in large numbers to outdoor lights at night.

## Cicadidae

Cicadas are medium to large insects characterized by having a proboscis arising from the head, three-segmented tarsi, and membranous wings with numerous, distinct veins. This group has representatives in many regions of the world and it contains the largest homopterans: some are 50 mm long. A characteristic of cicadas is the ability of males to produce loud sounds, primarily for courtship. Periodical cicadas, *Magicicada* spp., occur primarily in eastern USA. They have red eyes, red wing veins, and a 13- or 17-year life cycle. Pest status of periodical cicadas is based on the sound produced by large numbers of adult males, and broken twigs, which were weakened by oviposition slits made by female cicadas. They rarely enter houses and do not bite people.

**Annual cicada, dog-day cicada, *Tibicen* spp. (Fig. 8.3f)** Adults are 50–60 mm long and have dark-green, brown, and black markings. Development takes 5–7 years, but the broods of one or more species overlap, so adults are present each year. Annual cicadas appear in trees each year in mid to late summer. Females lay eggs in the twigs and small branches of trees and shrubs; hatching occurs in about 30 days. First-stage nymphs drop to the ground and burrow to reach the roots of trees. They remain in the ground feeding on the sap of tree and shrub roots until they are ready for the last molt. The fifth-stage nymph leaves the soil, and crawls on to a tree trunk or other surface and sheds its skin. Adults live about 30 days. They rarely occur in large numbers, and their presence and singing are usually not considered objectionable.

**Other annual cicadas** In the USA there are several species of annual cicadas, and some of these are common in urban environments. *T. cinctifer* is about 25 mm long and has a brown head and thorax, and a black body. It occurs in Arizona, California, and New Mexico, and frequently damages ornamental trees. *T. pruinosa* and *T. chloromera* are common species in northcentral and eastern USA. In Europe, *Brachytrypes megacephalus* and *Tibicen (Lyristes) plebejus* are the largest and noisiest species. In Australia, *Macrotristria* contains several common species, including the fiddler cicada, *M. angularis*, which is dark brown and spotted with yellow. The genus *Henicopsaltria* includes the mottled brown cicada, *H. eydouxi*, and the white-banded cicada, *H. fullo*.

**Periodical cicadas, *Magicicada* spp.** Adults have red eyes and the wing veins are black or slight reddish black. Females have a long, sword-like ovipositor. Eggs are inserted into twigs of trees and shrubs in batches of about 24; fecundity is 100–450 eggs. Hatching occurs in 3–5 days and first-stage nymphs drop to the ground and burrow in the soil to a depth of 45–60 cm. Nymphs feed by sucking plant sap from the roots of shrubs and trees. Development requires 13 or 17 years, but there is variability in broods and adults may appear a year early or late. The fifth nymph instar leaves the soil and adults emerge within 24 h. Mating begins during the first and second weeks of emergence. A few days after emergence males begin singing. The sound is loud and constant during the day, it begins at dawn, and the volume increases with temperature. Singing of large numbers of males can be annoying to people living in wooded areas and around urban and suburban buildings with shade trees. Adults appear in late May and early June and live 4–6 weeks. The 17-year form is distributed in northern USA and the 13-year form in southern USA, but there is overlap. Periodical cicadas are native to USA east of the Great Plains.

There are three morphologically distinct species of 17-year cicada: *Magicicada septendecim*, *M. cassini*, and *M. septendecula*. They differ in size, coloration, song, mating behavior, and habitat preference; however, the ranges of many of these characters overlap. The same three species occur in the 13-year cicada, but they have been named separately: *M. tredecim*, *M. tredecassini*, and *M. tredecula*. Differences in these species are based on life cycle and morphology.

***Magicicada septendecim*** Adult body is 27–33 mm long, and the propleura and later extensions of the pronotum between the eyes and the wing bases are reddish brown. Abdominal sterna are primarily reddish brown or brownish yellow.

***Magicicada cassini*** The adult body is 20–28 mm long, and the propleura and the lateral extensions of the pronotum between the eyes and wing bases are black. Abdominal sterna are all black, or a few have a narrow band of reddish brown or yellowish brown on the apical third; this band may be narrow or broken medially. The last tarsal segment has the apical half or more black.

***Magicicada septendecula*** The adult body is 19–27 mm long, and the propleura and lateral extensions of pronotum between the eyes and the wing bases are black. Abdominal sterna are black at the base with a broad apical band of reddish yellow or brown on the posterior half of each sternum; this band is not broken medially. The last tarsal segment is entirely brown black or yellowish brown, or the apical third is black.

Periodical cicadas are divided into year classes, called broods, which are potential incipient species. They are isolated in time because the adults never meet, but their geographical distributions overlap. Some broods occupy only a small area, others occur in as many as 15 states; some are entirely southern, some entirely northern. Broods are loosely geographically contiguous. Membership in a brood is determined by the year of adult emergence. The 17-year broods are assigned the numerals I–XVII, and the 13-year broods are XVIII–XXX. Not all the numbered broods of 13- and 17-year cicadas are known to exist. Broods XI and XXII have become extinct, and broods XII, XV, XVI, XVII, XVIII, XX, and XIV–XXX have not been verified or are represented by limited data. There are three well-documented existing broods of 13-year cicadas and 12 broods of 17-year cicadas. The consistency of the life cycle in each brood is variable. For example, adults are commonly sighted 1 year before, or 1 year after, and sometimes 4 years ahead of large emergences of 17-year cicadas. The appearance of adults before or after the main emergence is probably linked to local environmental conditions, and the varying rates of growth of nymphs. The 17-year cicada nymphs develop more slowly than the 13-year nymphs.

**Cicadette montana** Adults are about 2 cm long and the body is blackish brown; the abdomen has brown bands at the segments. This is the only species in the UK and it occurs in the New Forest, where it is associated with hazel trees.

**Large brown cicadas, *Cryptotympana* spp.** Adults are about 34 mm long and the body is uniformly brown; the bases of the wings are light brown, while the apical two-thirds of the wing is clear. These cicadas occur in southern Asia and feed on a variety of small trees and shrubs.

## Coccidae

Scale insects (Superfamily, Coccoidea) are divided into groups, primarily depending on morphological features. Females are wingless and usually legless, and males have only a single pair of wings. Females and nymphs feed by sucking plant fluids; males have no mouthparts. Development of scale insects includes the production of eggs or live young, an active first stage, but the remaining stages are sessile. Soft-scale (family Coccidae) females are elongate oval, usually convex, and with a smooth exoskeleton or a wax covering. Legs are usually present, and antennae are absent or much reduced. Males are sometimes wingless.

**Brown soft scale, *Coccus hesperidum*** Females are oval, reddish brown to pale yellow, often with a brown marbled appearance dorsally. Eggs hatch when they are laid, and first-stage nymphs disperse in a few days. Development requires about 30 days under suitable conditions; there are several generations per year. Large infestations result in excessive honeydew on leaves. A fungus, *Capnodium* spp., grows on the honeydew and causes the leaf surface to appear black and sooty. This insect is a pest of ornamentals and houseplants around the world. It may occur on bay leaves in the UK and parts of Europe. When the bay leaves are harvested, the scale may be included on some of the leaves, especially if young leaves are taken.

## Pseudococcidae

Mealybugs have flattened, oval bodies that are covered with white, powdery wax, which may extend from the body. Adult females are 2–7 mm long; nymphs are small and indistinct. Adults and immatures suck plant sap and weaken plants when they occur in large numbers. Mealybugs are serious pests of houseplants. They occur as white, cottony, or waxy forms on the underside of leaves, on stems, and particularly in the angle where leaves are attached to stems. Cottony masses that contain eggs may also be present, but some species bear live young.

**Longtailed mealybug, *Pseudococcus longispinus*** Female and nymph forms are about 3 mm long, oval, and with a white, waxy coating. Adult females have long, white anal filaments, which are often as long as or longer than the body. Females produce live young. Populations of this insect can increase rapidly when conditions are suitable, and plants decline from the feeding of adults and nymphs. This species is cosmopolitan. It is common in greenhouses, and it is a pest on a variety of houseplants.

## Psyllidae

Psyllids are 2–5 mm long and resemble very small cicadas or aphids; the adults have stout legs and the hind pair are large, and for jumping. Adults are very active and jump or fly when disturbed. Nymphs of some species are covered with waxy material; others are concealed in galls on plant leaves. All species feed on plant sap. The pest status of these insects is short-lived and limited to the presence of large numbers of adults in fall or spring around the outside of houses. They do not infest houseplants.

Hackberry trees (*Celtis*) grow in a variety of soil types and are often planted as ornamental shade trees; the most common is *C. occidentalis*. These trees are attacked by many gall-forming psyllids, all of which belong to the genus *Pachypsylla*. Although the hackberry tree is cosmopolitan, *Pachypsylla* only occurs in North America. These insects are divided into species groups that form galls on leaves, and those that form galls on woody portions of the plant. Adults of the forms that cause wood-gall emerge in the spring. Adults of the leaf-gall forms emerge in the fall and overwinter; large numbers collect on windows, window screens, and doors in fall. Species in the chalcid genera *Psyllaephalgus* and *Torymus* are predaceous on *Pachypsylla* larvae in leaf galls. Predation rates range from 30 to 50%.

**Blister-gall psyllid, *Pachypsylla celtidisvesicula*** Adults are about 4 mm long and brownish green or black; wings are translucent. Adults emerge from overwintering sites and become active when the leaves of the host tree develop. Eggs are laid singly on leaves; hatching occurs in 7–10 days. Feeding by the immature stages stimulates the development of a blister-like gall on the top surface of the leaf; there is one psyllid nymph per gall. Development is completed and adults emerge in September and October. Prior to finding an overwintering harborage, large numbers of adults can occur on buildings close to hackberry trees. This species occurs primarily in eastern and midwestern USA.

**Hackberry nipple-gall psyllid, *Pachypsylla celtidismamma*** Adults are about 4 mm long and brownish green or black; wings are translucent. Adults emerge from overwintering sites and become active when the leaves of the host tree begin to develop. Eggs are laid singly on leaves; hatching occurs in 7–10 days. Feeding by the immature stages stimulates the development of a cylindrical gall about 6 mm high on the underside of the leaf; there is one psyllid nymph per gall. Development is completed and adults emerge in September and October. Large numbers of adults can occur on buildings close to hackberry trees. This species occurs primarily in eastern and midwestern USA.

**Other *Pachypsylla*** The hackberry twig gall-maker, *P. venusta*, deforms small twigs. The adults of this species may occur in large numbers outside buildings.

## Bibliography

### HEMIPTERA

Bernath, R. F. and T. H. Kunz. Structure and dynamics of arthropod communities in bat guano deposits in buildings. *Can. J. Zool.*, **59** (1981), 260–70.

Britton, W. E. The Hemiptera or sucking lice of Connecticut. *Conn. State Geol. Nat. Hist. Surv. Bull.*, **34** (1923), 807.

Brooks, A. R. and L. A. Kelton. Aquatic and semiaquatic Heteroptera of Alberta, Saskatchewan, and Manitoba (Hemiptera). *Mem. Entomol. Soc. Can.*, **51** (1967), 1–92.

China, W. E. and N. C. E. Miller. Check-list and keys to the families and subfamilies of the Hemiptera-Heteroptera. *Bull. Br. Mus. (Nat. Hist.) Entomol.*, **8** (1959), 1–45.

Dolling, W. R. *The Hemiptera*. Oxford: Oxford University Press, 1991.

Groves, E. W. Hemiptera-Heteroptera of the London area. Part IV *London Nat.*, **46** (1967), 82–104.

Henry, T. J. and R. C. Froeschner. *Catalog of the Heteroptera, or True Bugs, of Canada and the Continental United States*. Leiden: E. J. Brill, 1988.

Herring, J. L. and P. D. Ashlock. A key to the nymphs of the families of Hemiptera (Heteroptera) of America north of Mexico. *Fla. Entomol.*, **54** (1971), 207–13.

Hocking, B. Blood-sucking behavior of terrestrial arthropods. *Annu. Rev. Entomol.*, **16** (1971), 1–26.

Hsiao, T., S. Ren, L. Zheng *et al.* *A Handbook for the Determination of the Chinese Hemiptera-Heteroptera* (in Chinese). Beijing: Academia Sinica, 1977.

Hurlburt, S. H., G. Rodreguez, and N. Dios dos Santos. *Aquatic Biota of Tropical South America*. Part I. *Arthropoda*. San Diego, CA: San Diego State University, 1981.

Lawson, F. A. Identification of the nymphs of the common families of Hemiptera. *J. Kansas Entomol. Soc.*, **32** (1959), 88–92.

Miller, N. C. E. *The Biology of the Heteroptera*. London: Leonard Hill, 1956.

Parshley, H. M. On some Hemiptera from Western Canada. *Occ. Pap. Mus. Zool., Univ. Michigan*, **71** (1919), 1–35.

Péricart, J. *Hémiptères, Anthocoridae, Cimicidae, Microphysidae de l'Ouest-Paléarctique. Fauna de l'Europe et du Bassin Mediterranéean*, vol. 7. Paris, France: Masson, 1972.

Ryckman, R. E. 1979. Host reactions to bug bites (Hemiptera, Homoptera): a literature review and annotated bibliography Part I. *Calif. Vector Views*, **26** (1979), 1–24.

Ryckman, R. E. and D. G. Bently. Host reactions to bug bites (Hemiptera, Homoptera); a literature review and annotated bibliography. Part II. *Calif. Vector Views*, **26** (1979), 25–49.

Slater, J. A. and R. M. Baranowski. *How to Know the True Bugs (Hemiptera-Homoptera)*. Dubuque, IA: Wm. C. Brown, 1978.

Southwood, T. R. E. and D. Leston. *Land and Water Bugs of the British Isles*. London: Frederick Warne, 1959.

Usinger, R. L. Blood sucking among phytophagous Hemiptera. *Can. Entomol.*, **66** (1934), 97–100.

*Aquatic Insects of California, With Keys to North American Genera and California Species*. Berkeley, CA: University of California Press, 1956.

Weber, H. *Biologie der Hemipteren*. Amsterdam: Asher, 1968.

Zimmerman, E. C. *Insects of Hawaii*, vol. 3, *Heteroptera*. Honolulu: University of Hawaii Press, 1948.

#### Anthochoridae, Belostomatidae

Atkinson, M. D. Reports of human biting by the flower bug, *Anthocoris nemorum* (L.) (Hem., Anthocoridae). *Entomol. Mon. Mag.*, **126** (1990), 96.

Cullen, M. J. The biology of giant water bugs (Hemiptera: Belostomatidae) in Trinidad. *R. Entomol. Soc. Lond. Proc.*, **A44** (1969), 123–6.

DeCarlo, J. A. Consideraciones sobre la biologia de *Lethocerus mazzani*. DeCarlo. *Psyche*, **23** (1962), 143–51.

Dolling, W. R. *Dufouriellus ater* (Dufour) (Hemiptera: Anthocoridae) biting industrial workers in Britain. *Trans. R. Soc. Trop. Med. Hyg.*, **71** (1977), 355.

Duviard, D. Flight activity of Belostomaticae in central Ivory Coast. *Oecologica*, **15** (1974), 321–8.

Herring, J. L. Keys to genera of Anthocoridae of America north of Mexico, with description of a new genus (Hemiptera: Heteroptera). *Fla. Entomol.*, **59** (1976), 143–50.

Hoffman, W. E. Studies on the bionomics of the water bug *Lethocerus indicus* (Hemiptera: Belostomatidae) in China. *Verh. Int. Verein. Theor. Angew. Limnol.*, **5** (1931), 661–7.

Kelton, L. A. *The Anthocoridae of Canada and Alaska (Heteroptera: Anthocoridae). The Insects and Arachnids of Canada*, Part 4. Ottawa: Canadian Government Publishing Centre, 1978.

Ledger, J. A., J. B. Rossiter, and J. M. Ossthuizen. Anthocorid bug bites in a Transval goldmine. A case report. *South Afr. Med. J.*, **62** (1982), 69–70.

Melanby, K. The physiology and activity of the bed-bug (*Cimex lectularius* L.) in a natural infestation. *Parasitology*, **31** (1939), 200–11.

Menke, A. S. A review of the genus *Lethocerus* in North and Central America, including the West Indies (Hemiptera: Belostomatidae). *Ann. Entomol. Soc. Am.*, **56** (1963), 261–7.

Smith, K. G. V. Hemiptera (Anthocoridae and Miridae) biting man. *Entomol. Mon. Mag.*, **126** (1990), 96.

Stys, P. Cases of facultative parasitism of Lyctocorinae (Heteroptera: Anthocoridae) on man in Czechlosvakia. *Folia Parasitol. (Praha)*, **20** (1973), 103–4.

### Cimicidae, Corixidae, Cydindae

Aboul-Nasr, A. E. and M. A. S. Erakey. Behaviour and sensory physiology of the bedbug, *Cimex lectularius* L. to some environmental factors. Chemoreception (Hemiptera-Heteroptera: Cimicidae). *Bull. Soc. Entomol. Egypt*, **52** (1958), 353–62.

Back, E. A. Bat bug (*Cimix pilosellus* Horv.). *US Dept. Agric. Insect Pest. Serv. Bull.*, **20** (1940), 359.

Bell, W. and C. W. Schaefer. Longevity and egg production of female bed bugs, *Cimex lectularius*, fed various blood fractions and other substances. *Ann. Entomol. Soc. Am.*, **59** (1966), 53–6.

Bhat, H. R. A review of Indian Cimicidae (Hemiptera-Heteroptera). *Orient. Insects*, **8** (1974), 545–50.

Burton, G. J. Bedbugs in relation to transmission of human disease. *Public Health Rep.*, **78** (1963), 513–24.

Cragg, F. W. Observations on the bionomics of the bedbug (*Cimex lectularius* L.), with special reference to the relation of the sexes. *Ind. J. Med. Res.*, **11** (1923), 449–73.

Crissey, J. T. Bedbugs: an old problem with a new dimension. *Int. J. Dermatol.*, **20** (1981), 411–14.

Decoursey, R. M. Keys to the families and subfamilies of the nymphs of North American Hemiptera-Heteroptera. *Proc. Entomol. Soc. Wash.*, **73** (1971), 413–28.

Dickerson, G. and M. M. J. Lavoipierre. Studies on the methods of feeding of blood-sucking arthropods: II. The methods of feeding adopted by the bed-bug (*Cimex lectularius*) when obtaining a blood-meal from the mammalian host. *Ann. Trop. Med. Parasitol.*, **53** (1959), 347–57.

Eads, R. B., D. B. Fancy, and G. C. Smith. 1980. The swallow bug, *Oeciacus vicarius* Horvath (Hemiptera: Cimicidae), a human household pest. *Proc. Entomol. Soc. Wash.*, **85** (1980), 853–4.

Girault, A. A. The bedbug, *Clinocoris* (= *Cimex* = *Acanthia* = *Klinopilos*) *lectularia* Linnaeus. Part I. Life-history at Paris, Texas, with biological notes, and some considerations on the present state of our knowledge concerning it. *Psyche*, **12** (1905), 61–74.

The bedbug, *Cimex lectularius* Linneaus. Part II. Critical remarks on its literature, with a history and bibliography of pathogenic relations. *Psyche*, **13** (1906), 42–58.

Preliminary studies on the biology of the bedbug, *Cimex lectularius*. II. Facts obtained concerning the duration of its different stages. *J. Econ. Biol.*, **7** (1912), 163–88.

1914. Preliminary studies on the biology of the bedbug, *Cimex lectularius*, Linn. *J. Econ. Biol.*, **9** (1914), 25–45.

Hungerford, H. B. The Corixidae of the western hemisphere. *Univ. Kan. Sci. Bull.*, **32** (1948), 1–827.

Johnson, C. G. The ecology of the bed-bug, *Cimex lectularius* L., in Britain. *J. Hygiene*, **41** (1941), 345–461.

Jupp, P. G. and S. F. Lyons. Experimental assessment of bed-bugs (*Cimex lectularius* and *Cimex hemipterus*) and mosquitoes (*Aedes aegypti formosus*) as vectors of human immunodeficiency virus. *AIDS*, **1** (1987), 171–4.

Jupp, P. G. and S. E. McElligott. Transmission experiments with hepatitis B surface antigen and the common bedbug (*Cimex lectularius* L.). *South Afr. Med. J.*, **56** (1979), 54–7.

Lee, R. D. The biology of the Mexican chicken bug, *Haematosiphon inodorus* (Dugas) (Hemiptera: Cimicidae). *Pan-Pacific Entomol.*, **31** (1955), 47–61.

Levinson, H. Z. and A. R. Bar Ilan. Assembling and alerting scents produced by the bedbug, *Cimex lectularius* L. *Experientia*, **27** (1971), 102–3.

Mail, G. A. Infestation of a high school by *Oeciacus vicarius* Horv. *J. Econ. Entomol.*, **33** (1940), 949.

Marinkelle, C. J. *Cimex hemipterus* (Fabr.) from bats in Columbia, South America (Hemiptera: Cimicidae). *Proc. Entomol. Soc. Wash.*, **69** (1967), 179–80.

Myers, L. E. The American swallow bug, *Oeciacus vicarius* Hovarth (Hemiptera, Cimicidae). *Parasitology*, **20** (1928), 159–72.

Nagem, R. L. Ocorrência de *Cimex lectularius* L., 1758 (Hemiptera: Cimicidae) em algumas habitações humanes de Belo Horizonte e municipios vizinhos. *Rev. Bras. Entomol.*, **29** (1985), 217–20.

Negromonte, M. R. S., P. M. Linardi, and R. L. Nagem. Prevalência, intensidade e fluxo da infestação pro *Cimex lectularius* L., 1758 (Hemiptera, Cimicidae) em uma comunidade de Belo Horizonte, MG. *Rev. Bras. Entomol.*, **35** (1991), 715–20.

Newberry, K. The effects on domestic infestations of *Cimex lectularius* bedbugs of interspecific mating with *C. hemipterus*. *Med. Vet. Entomol.*, **3** (1989), 407–14.

Newberry, K. and E. J. Jansen. The common bedbug *Cimex lectularius* in African huts. *Trans. R. Soc. Trop. Med. Hyg.*, **80** (1986), 653–8.

Newberry, K. and Z. M. Mchumu. Changes in the relative frequency of infestations of two sympatric species of bedbugs in northern Natal and KwaZulu, South Africa. *Trans. R. Soc. Trop. Med. Hyg.*, **83** (1989), 262–4.

Newberry, K., Z. M. Mchumu, and S. Q. Cebekhulu. Bedbug reinfestation rates in rural Africa. *Med. Vet. Entomol.*, **5** (1991), 503–5.

Ogston, C. W., F. S. Wittenstein, W. T. London, and I. Millman. Persistence of hepatitis B surface antigen in the bedbug, *Cimex hemipterus* Fabr. *J. Infect. Dis.*, **140** (1979), 411–14.

Omori, N. Experimental studies on the cohabitation and crossing of two species of bedbugs, *Cimex lectularius* L. and *Cimex hemipterus* (F.) and on the effects of interchanging males of one species for the other, every alternate day, upon the fecundity and longevity of females of each species. *Acta Jpn. Med. Trop.*, **1** (1939), 127–54.

Overal, W. L. and L. R. Wingate. The biology of the bat bug *Stricticimex antennatus* (Hemiptera: Cimicidae) in South Africa. *Ann. Natal Mus.*, **22** (1976), 821–8.

Rafatjah, H. The problem of resurgent bed-bug infestation in malaria eradication programmes. *J. Trop. Med. Hyg.*, **74** (1971), 53–6.

Ryckman, R. E., D. G. Bently, and E. F. Archbold. The Cimicidae of the Americas and oceanic islands: a checklist and bibliography. *Soc. Vector Ecol. Bull.*, **6** (1981), 93–142.

Spencer, G. J. The status of the barn swallow bug, *Oeciacus vicarius* Horvath. *Can. Entomol.*, **62** (1930), 20–1.

The bedbugs of British Columbia. *Proc. Entomol. Br. Columb.*, **39** (1935), 23–9.

Taki, R., T. Yamaguchi and T. Kurihara. Mass occurrence of *Aethus indicus* (Hemiptera: Cydnidae) as a house-frequenting pest in the Amami islands. *Jpn. J. Sanit. Zool.*, **26** (1975), 61–3.

Usinger, R. L. *Monograph of Cimicidae (Hemiptera-Heteroptera)*, vol. 7. Entomological Society of America, College Park, MD: Thomas Say Foundation, 1966.

Usinger, R. L. and D. Povolný. The discovery of a possibly aboriginal population of the bed bug (*Cimex lectularius* Linnaeus, 1758). *Acta Mus. Moravia*, **51** (1966), 237–41.

## Lygaeidae, Miridae

Al-Houty, W. *Nysius* (Hem., Lygaeidae) sucking human blood in Kuwait. *Entomol. Mon. Mag.*, **126** (1990), 95–6.

Lewis, D. J. Hemiptera of medical interest in the Sudan Republic. *Proc. R. Entomol. Soc. A*, **33** (1958), 43–7.

Wheeler, A. G. Jr. *Biology of the Plant Bugs*. Ithaca: Cornell University Press, 2001.

## Naucoridae, Nepidae, Notonectidae

Bisht, R. S. and S. M. Das. Observation on aquatic insects as food of fishes and the predatory action of some aquatic insects on fish and fish food. *J. Inland Fish. Soc. India*, **13** (1981), 80–6.

Constantz, G. D. The mating behavior of a creeping water bug, *Ambrysus occidentalis* (Hemiptera: Naucoridae). *Am. Mid. Nat.*, **92** (1974), 234–9.

Gittleman, S. H. The ecology of some Costa Rican backswimmers (Hemiptera: Notonectidae). *Entomol. Soc. Am. Ann.*, **68** (1975), 511–18.

Lansbury, I. A review of the oriental species of *Ranata* Fabricius (Hemiptera-Heteroptera: Nepidae). *Tijdschr. Entomol.*, **116** (1972), 84–106.

LaRivers, I. A revision of the genus *Ambrysus* in the United States (Hemiptera: Naucoridae). *Univ. Calif. Publ. Entomol.*, **8** (1950), 277–338.

Radinovsky, S. Cannibal of the pond. *Natural Hist.*, **73** (1964), 16–25.

Rao, T. K. R. On the biology of *Ranata elongata* Fabr. (Heteroptera: Nepidae) and *Sphaerodema annulatum* Fabr. (Heteroptera: Belostomatidae). *Proc. R. Entomol. Soc. Lond. A*, **37** (1962), 61–4.

Sites, R. W. and B. J. Nichols. Life history and descriptions of immature stages of *Ambrysus lunatus lunatus* (Hemiptera: Naucoridae). *Ann. Entomol. Soc. Am.*, **83** (1990), 800–8.

Sites, R. W. and J. T. Polhemus. Nepidae of the United States and Canada. *Ann. Entomol. Soc. Am.*, **87** (1994), 27–42.

Tawfik, M. F. S., M. M. El-Hussieni, and H. A. Bakr. Ecological observations on aquatic insects attacking mosquitoes in Egypt. *Bull. Soc. Entomol. Egypt*, **66** (1986), 117–26.

## Pentatomidae, Psyllidae

Azeez, S. A. Mass invasion of *Carbula pedalis* Bergr. (Pentatomidae) and its control in Bulassa N. W. State. *Nigerian Entomol. Mag.*, **2** (1972), 125–6.

Barrera, M. Fauna del Nordoeste argentino. Observaciones biologicas sobre *Antiteuchus variolosus* Westwood (Hemiptera-Pentatomidae). *Acta Zool. Lillioana*, **30** (1973), 141–62.

Beckwith, R. D. Notes on the hackberry blister gall, *Pachypsylla celtidis-vesicula*. *Sci. Tree Topics*, **2** (1959), 21–2.

Linnavuori, R. Hemiptera of the Sudan, with remarks on species of the adjacent countries, 5. Pentatomidae. *Boil. Soc. Port. Cienc. Nat.*, **15** (1975), 5–128.

Thomas, D. B., J. E. Eger, W. Jones, and G. Ortega-Leon. The African cluster bug, *Agonoscelis puberula* (Heteroptera: Pentatomidae), established in the New World. *Fla. Entomol.*, **86** (2003), 151–3.

Watanabe, M. House-invading stink bugs and its anti-invading methods. *Household Insect Pests*, **17** (1995), 119–30.

Yang, W. I. *Economic Insect Fauna of China: Fascicle. 2. Hemiptera, Pentatomidae*. Beijing: Academia Sinica Science Press.

## Reduviidae

Ambrose, D. P. *Assassin Bugs*. New Hampshire: Science Publishers, Enfield, 1999.

Balahov, Yu. S. Interaction between blood-sucking arthropods and their hosts, and its influence on vector potential. *Annu. Rev. Entomol.*, **29** (1984), 137–56.

Barber, G. W. On the bite of *Arilus cristatus*. *J. Econ. Entomol.*, **12** (1919), 466.

Beard, R. L. The biology of *Anasa tristis* DeGeer with particular reference to the tachinid parasite, *Trichopoda pennipes* Fabr. *Conn. Agric. Exp. Stn. Bull.*, **440** (1940), 595–679.

Brener, Z. Biology of *Trypanosoma cruzi*. *Annu Rev. Microbiol.*, **27** (1973), 347–82.

Buxton, P. A. The biology of a blood-sucking bug, *Rhodnius prolixus*. *Trans. Entomol. Soc. Lond.*, **78** (1930), 227–36.

Canale, D. M., M. C. Cecere, R. Chuit, and R. E. Gürtler. Peridomestic distribution of *Triatoma garciabesi* and *Triatoma guasayana* in northwest Argentina. *Med. Vet. Entomol.*, **14** (2000), 383–90.

Caracavallo, R. U. and A. Martinez. Life cycles of some species of *Triatoma* (Hemiptera: Reduviidae). *Can. Entomol.*, **104** (1972), 699–704.

Chagas, C. Nova tripanomíase humana. Estudos sobre a morfologia e o ciclo evolutio do *Schizotrypanum cruzi*, n. gen., n. sp., agente etiológico de nova entidade mórbida do homen. *Mem. Inst. Oswaldo Cruz*, **1** (1909), 159–218.

Dujardin, J. P., M. Muñoz, T. Chávez *et al*. The origin of *Rhoninus prolixus* in Central America. *Med. Vet. Entomol.*, **12** (1998), 113–15.

Dye, C. The analysis of parasite transmission by bloodsucking insects. *Annu. Rev. Entomol.*, **16** (1992), 1–26.

Eberhard, W. G., N. I. Platnick, and R. T. Schuh. Natural history and systematics of arthropod symbionts (Araneae; Hemiptera; Diptera) inhabiting webs of the spider *Tengella radiata* (Araneae, Tengellidae). *Am. Mus. Novit.*, **3065** (1993), 1–17.

Ekkens, B. G. Nocturnal flights of *Triatoma* (Hemiptera: Reduviidae) in Sabino Canyon, Arizona I. Light collections. *J. Med. Entomol.*, **18** (1981), 211–27.

Ferreira, L. de C. and L. Deane. Encontro de um nová hematophago do homen com habitos domiciliares. *Brasil Med.*, **52** (1938), 1137–41.

Forattini, O. P. Biogeografia origem e distribuição de domiciliação de triatomíneos no Brasil. *Rev. Saúde Públ.*, **14** (1980), 265–99.

Gardiner, B. O. C. and S. H. P. Maddrell. Techniques for routine and large-scale rearing of *Rhodinius prolixus* (Hem., Reduviidae). *Bull. Entomol. Res.*, **61** (1972), 505–16.

Griffith, M. E. The blood-sucking conenose, or big bedbug, *Triatoma sanguisuga* (LeConte), in an Oklahoma City household. *Proc. Oak. Acad. Sci.*, **28** (1947), 24–7.

Grundemann, A. W. Studies on the biology of *Triatoma sanguisuga* (LeConte) in Kansas, (Reduviidae, Hemiptera). *J. Kansas Entomol. Soc.*, **20** (1947), 77–85.

Gómez-Núñez, J. C. Resting places, dispersal and survival of $CO^{60}$-tagged adult *Rhodnius prolixus*. *J. Med. Entomol.*, **6** (1969), 83–6.

Lent, H. and P. Wygodzinsky. Revision of the Triatominae (Hemiptera, Reduviidae), and their significance as vectors of Chagas' disease. *Bull. Am. Mus. Nat. Hist.*, **163** (1979), 127–520.

Rabinovich, J. E., J. A. Leal, and D. Feliciangeli de Piñero. Domiciliary biting frequency and blood ingestion of the Chagas' disease vector *Rhodnius prolixus* Stahl (Hemptera: Reduviidae), in Venezuela. *Trans. R. Soc. Trop. Med. Hyg.*, **73** (1979), 272–83.

Ramsey, J. M., R. Ordoñez, A. Cruz-Celis *et al*. Distribution of domestic Triatominae and stratification of Chagas disease transmission in Oaxaca, Mexico. *Med. Vet. Entomol.*, **14** (2000), 19–30.

Readio, P. A. Studies on the biology of the Reduviidae of American north of Mexico. *Univ. Kans. Sci. Bull.*, **17** (1927), 5–291.

Ryckman, R. E. The kissing bug problem in western North America. Bull. Soc. Vector Ecol., **6** (1981), 167–9.

Ryckman, R. E. and M. A. Casdin. The Triatomine of western North America, a checklist and bibliography. *Calif. Vector Views*, **23** (1976), 35–52.

Schofield, C. J. The behavior of Triatominae (Hemiptera: Reduviidae): a review. *Bull. Entomol. Res.*, **69** (1979), 363–79.

Schofield, C. J. *Triatominae: Biology and Control*. Bognor Regis West Sussex: Eurocommunica, 1994.

Schofield, C. J. and J. W. Patterson. Assembly pheromone of *Triatoma infestans* and *Rhodinus prolixis* nymphs (Hemiptera: Reduviidae). *J. Med. Entomol.*, **13** (1977), 727–34.

Schofield, C. J., M. J. Lehane, P. McEwan, S. S. Catala, and D. E. Gorla. Dispersive flight by *Triatoma sordida*. *Trans. R. Soc. Trop. Med. Hyg.*, **85** (1991), 676–8.

Schofield, C. J., N. G. Williams, M. L. Kirk, M. T. Garcia Zapata, and P. D. Marsden. A key for identifying faecal smears to detect domestic infestations of triatomine bugs. *Rev. Soc. Bras. Med. Trop.*, **19** (1986), 5–8.

Smith, F. D., N. G. Miller, S. J. Carnazzo, and W. B. Eaton. Insect bite by *Arilus cristatus*, a North American reduviid. *A. Med. Assoc. Arch. Dermatol.*, **77** (1958), 324–30.

Snoddy, E. L., W. J. Humphreys, and M. S. Blum. Observations on the behavior and morphology of the spider predator, *Stenolemus lanipes* (Hemiptera: Reduviidae). *J. Ga. Entomol. Soc.*, **11** (1976), 55–8.

Teo, S. K. and J. S. Cheah. Severe reaction to the bite of the Triatomid bug (*Triatoma rubrofasciata*) in Singapore. *J. Trop. Med. Hyg.*, **76** (1973), 161–2.

Usinger, R. L., P. Wygodzinsky, and R. E. Ryckman. Biosystematics of Triatominae. *Annu. Rev. Entomol.*, **11** (1966), 309–30.

Zeledón, R. and J. E. Rabinovich. Chagas' disease: an ecological appraisal with special emphasis on its insect vectors. *Annu. Rev. Entomol.*, **26** (1981), 101–33.

Zeledón, R., C. E. Valerio, and J. E. Valerio. The camouflage phenomena in several species of Triatominae (Hemiptera: Reduviidae). *J. Med. Entomol.*, **10** (1973), 209–11.

### Rhopalidae

Abbot, C. E. Cannibalism of *Leptocoris trivittatus* Say. *Bull. Brooklyn Entomol. Soc.*, **43** (1948), 112–13.

Barber, H. G. A new species of *Leptocoris* (Coreidae: Leptocorini). *Pan-Pacific Entomol.*, **32** (1956), 9–11.

Brannon, D. H. Boxelder bug (*Leptocoris trivittatus*). *US Dept. Agric. Coop. Econ. Insect Rpt.*, **11** (1961), 1100.

Carroll, S. P. and C. Boyd. Host race radiation in the soapberry bug: natural history. *Evolution*, **46** (1992), 1053–69.

Gollner-Scheiding, U. Revision der Afrikanischen Arten swie Bermerkungen zu weiteren Arten der Gattungen *Leptocoris* Hahn, 1833, und *Boisea* Kirkaldy, 1910. (Het., Rhopalidae). *Dtsch. Entomol. Z.*, **27** (1980), 103–48.

Hoebeke, E. R. and A. G. Wheeler, Jr. *Rhopalus* (*Brachycarenus*) *tigrinus*, recently established in North America, with a key to genera and species of Rhopalidae in eastern North America (Hemiptera-Heteroptera). *Proc. Entomol. Soc. Wash.*, **84** (1982), 213–24.

Knowlton, G. F. Boxelder bug observations. *J. Econ. Entomol.*, **37** (1944), 443.

Boxelder bug 'bites' man. *Bull. Brooklyn Entomol. Soc.*, **42** (1947), 169.

Schaefer, C. W. A re-assessment of North American *Leptocoris* (Hemiptera-Heteroptera: Rhopalidae). *Ann. Entomol. Soc. Am.*, **68** (1975), 537–41.

Schowalter, T. D. Overwintering aggregation of *Boisea rubrolineatus* (Heteroptera: Rhopalidae) in western Oregon. *Environ. Entomol.*, **15** (1986), 1055–6.

Slater, J. A. and C. W. Schaefer. *Leptocoris trivittatus* (Say) and *Coriomeris humilis* Uhl. in New England (Hemiptera: Coreidae). *Bull. Brooklyn Entomol. Soc.*, **58** (1963), 114–17.

Smith, R. C. and B. L. Shepherd. The life history and control of the boxelder bug in Kansas. *Trans. Kansas Acad. Sci.*, **40** (1937), 143–59.

Swenk, M. H. Boxelder bug (*Leptocoris trivittatus* Say). *US Dept. Agric. Insect Surv. Bull.*, **9** (1929), 407–14.

Tinker, M. E. The seasonal behavior and ecology of the boxelder bug *Leptochoris trivittatus* in Minnesota. *Ecology*, **33** (1952), 407–14.

Yoder, K. M. and W. H Robinson. Seasonal abundance and habitats of the boxelder bug, *Boisea trivittata*, (Say), in an urban environment. *Proc. Entomol. Soc. Wash.*, **92** (1990), 802–7.

### Tingidae

Baily, N. S. The Tingoidea of New England and their biology. *Entomol. Am.*, **31** (1951), 1–140.

Drake, C. J. and F. A. Ruhoff. Lacebugs of the world. a catalog (Hemiptera: Tingidae). *US Nat. Mus. Bull.*, **213** (1965), 634.

Heiss, E. Die amerikanische Platennetzwanze *Corythucha ciliata* – eine Advvenivart im Vormarsch auf Europa (Heteroptera, Tingidae). *Cat. Landesmus (n. F.)*, **84** (1995), 143–8.

Hoffman, R. L. A second case of lacebug bite (Hemiptera: Tingidae). *Entomol. News*, **64** (1953), 176.

Knowlton, G. F. Tingidae are biters. *Bull. Brooklyn Entomol. Soc.*, **53** (1958), 73.

### HOMOPTERA

Alexander, R. D. and T. E. Moore. The evolutionary relationships of 17-year and 13-year cicadas, and three new species (Homoptera, Cicadidae, Magicicada). *Misc. Publ. Mus. Zool. Univ. Mich.*, **121** (1962), 59.

Beirne, B. P. Leafhoppers (Homoptera: Cicadellidae) of Canada and Alaska. *Can. Entomol.*, (suppl. 2) (1956), 1–180.

Caldwell, J. S. The jumping plant lice of Ohio (Homoptera, Chermidae). *Ohio Biol. Surv. Bull.*, **6** (1938), 229–81.

Crawford, D. L. Monograph of the jumping plant lice or Psyllidae of the New World. *US Natl. Mus. Bull.*, **85** (1914), 186.

DeLong, D. M. The bionomics of leafhoppers. *Annu. Rev. Entomol.*, **16** (1971), 179–210.

Dybas, H. S. and M. Lloyd. The habitats of the 17-year periodical cicadas (Homoptera: Cicadidae: *Magicicada* spp.). *Ecol. Monogr.*, **44** (1974), 279–324.

Eastop, V. E. and D. H. R. Lambers. *Survey of the World's Aphids*. The Hague: Junk, 1976.

Ferris, G. F. *Atlas of the Scale Insects of North America*. Stanford, CA: Stanford University Press, 1937–55.

Forsythe, H. Y. Distribution and species of 17-year cicadas in broods V and VIII in Ohio. *Ohio J. Sci.*, **76** (1976), 254–8.

Hottes, F. C. and T. H. Frison. The plant lice, or Aphididae, of Illinois. *Ill. Nat. Hist. Surv. Bull.*, **19** (1931), 121–447.

Kennedy, J. S. and H. G. Stroyan. Biology of aphids. *Annu. Rev. Entomol.*, **4** (1959), 139–60.

Lambers, D. H. R. Polymorphism in Aphididae. *Annu. Rev. Entomol.*, **11** (1966), 47–78.

McKenzie, H. L. *The Mealybugs of California*. Berkeley, CA: University of California Press, 1967.

Moore, T. E. The cicadas of Michigan (Homoptera: Cicadidae). *Pap. Mich. Acad. Sci.*, **51** (1966), 75–96.

Morrison, H. and A. V. Renk. A selected bibliography of the Coccoidea. *USDA Agr. Res. Serv. Misc. Publ.*, **734** (1957), 222.

Ossiannilsson, E. Insect drummers: a study of the morphology and function of the sound-producing organs of Swedish Homoptera Auchenorrhyncha. *Opusc. Entomol.* (suppl. 10) (1949), 1–146.

Simon, C. Evolution of 13- and 17-year periodical Cicadas (Homoptera: Cicidadie: *Magicicada*). *Bull. Entomol. Soc. Am.*, **34** (1988), 163–76.

Smith, C. E. Bibliography of the Aphididae of the world. *N. C. Agr. Expr. Sta. Tech. Bull.*, **216** (1972), 717.

Smith, R. C. and R. S. Taylor. The biology and control of the hackberry psyllids in Kansas. *J. Kansas Entomol. Soc.*, **26** (1953), 103–5.

Way, J. T. Mutualism between ants and honeydew producing Homoptera. *Annu. Rev. Entomol.*, **8** (1953), 307–44.

Williams, M. L. and M. Kosztarab. Morphology and systematics of the Coccoidea of Virginia, with notes on their biology (Homoptera: Coccoidea). *Va. Polytech. Inst. State Univ. Res. Div. Bull.*, **74** (1972), 215.

# 9 Hymenoptera

## Introduction

Hymenoptera is a large order; the members occupy a large number of habitats. Ants, wasps, and bees have habits that range from solitary to species with complex social systems living in large colonies, and from plant-feeders to predators and parasites. Adults range from 0.2-mm-long parasites of insect eggs to 50-mm-long parasitic and predatory wasps. Antennae are long in social and parasitic species, with 13 segments in the male and 12 segments in the female, but may be as few as three segments. The mouthparts are mandibulate, but in many species there is also a tongue for lapping nectar from flowers. There are usually two pairs of wings, but many are wingless in one or both sexes or certain castes. They have complete metamorphosis, with egg, larva, pupa, and adult stages. The most advanced types of behavior and social structure are found in this order. The success of ants and the others that have evolved social behavior is based on nest building, colony housing, and a caste system. Some species control the sex of offspring by permitting or withholding fertilization; males are usually produced from unfertilized eggs, and females from fertilized eggs. Regulating the sex and size of individuals in large colonies is linked to a caste system and temporal and spatial partitioning of labor. For many species there are morphological adaptations and complex reproductive behaviors associated with socialization.

Pest status for hymenopterans is based on the medical threat and outcome of stings and bites of ants, bees, and wasps; the economic costs of repair and replacing structural timber damaged by carpenter ants and carpenter bees; and further economic damage to food caused by foraging ants. Aesthetic damage to ornamental turfgrass by ground-nesting bees, and large ant-nest mounds, and the general foraging behavior of solitary and social bees and wasps may also be extensive. The structural damage and economic costs of the wood-infesting *Camponotus* species equal or exceed that of subterranean termites in some north temperate regions, including northeastern USA and southern Canada. Carpenter bees have adapted to wood substrates available in urban environments, and have expanded their range and increased their economic and aesthetic pest status. In many regions of the world, urban populations of *Vespula* and *Paravespula* yellowjackets disrupt sanitation services, and diminish activities and general safety in recreational areas. The medical treatment for individuals attacked and the recurring costs of controlling stinging and biting hymenopterans further increases their pest status. A few hymenopteran species infest trees, and they may occur as larvae in structural wood. Parasites of household food moths, and cockroaches, can place them in direct contact with people indoors.

A hymenopteran sting can cause the death of humans and other animals, such as pets and livestock. This is the result of an extreme allergic reaction to the components of the venom, or to a large amount of venom entering the body at one time. In typical reactions, a sting produces a slight swelling and itching for 1–2 days at the sting site. Further stings may produce no reaction to persons who have become desensitized to the venom. However, in other cases a person may become more allergic or hypersensitive with successive stings and may reach a sensitivity level in which a single sting will result in death through acute anaphylaxis. The venoms of vespid wasps, some ants, and honey bees are chemically similar. They contain hyaluronidase, which is an enzyme that degrades the substances between cells, and phospholipase, which destroys cell membranes. The action of these two chemicals produces the immediate reaction of local swelling and redness. Pain-producing peptides, which are kinins, are known only from vespids such as yellowjackets and hornets, while ant venom often contains alkaloids, and honey bee venom contains a cell-destroying (cytolytic) protein called melittin.

## Aulacidae

These wasps are parasites of wood-nesting and wood-infesting insects. Adult aulacids are 12–15 mm long, excluding the ovipositor, and the thorax is black and the abdomen usually reddish brown. The abdomen is attached high on the thorax (propodeum), well above the bases of the hind coxae. The ovipositor is usually slightly longer than the abdomen. Adults are associated with infested trees and logs, and may emerge from firewood infested with beetles or xiphydriid wasps. *Aulacus* parasitizes larvae of the *Xiphydria* wood-wasps, and *Odontaulacus* and *Pristaulacus* parasitize wood-infesting beetle larvae. *Gasteruption* and *Rhydinofoenus* are parasitic in bee and wasp nests in twigs or wood. The common species in North America include *Pristaulacus rufitarsis*, which is a parasite of the hemlock borer and poplar borer and *P. bilobatus*, which also parasitizes the hemlock borer. *Aulacus burquei*, *A. digitalis*, *A. lovei*, and *A. pallipes* parasitize several species of wood-wasps (*Xiphydria*). In the UK, *A. striatus* attacks the larvae of *Xiphydria camelus*.

Female aulacids have a groove on the inner side of each hind coxa, and this forms a channel (*aula*, Greek for furrow) for the long ovipositor. During the process of oviposition, the female inserts her long and delicate ovipositor into infested wood; the hind coxae brought together around the ovipositor probably serve as support and guide. This feature of supporting the ovipositor between the hind coxae when it is in use is found in the braconid genus *Capitonius* and the ichneumonid genera *Labena* and *Apechoneura*. These genera are also parasites of wood-infesting beetle larvae.

## Andrenidae

These small bees are 6–8 mm long, mostly metallic black, but some are brown or reddish brown. The head and thorax are usually covered with dense setae. They have solitary or colonial habits, and large groups may occur together to utilize a favorable habitat. Individual females construct their own entrance burrows and cells in the ground, but there may be some common-use entry burrows. The below-ground brood cells are provisioned with nectar and pollen for developing larvae. Both sexes remain in the nest through late summer and winter and emerge in the spring.

***Andrena vicina*** Adults are 11–13 mm long, and the body is black to brownish black; there are yellow markings on the face. This species is commonly found visiting flowers of apple, wild cherry, and many other trees. Adults are active from April to October. Large numbers may establish nests in turfgrass and patches of bare soil. They damage or disrupt normal growth of turfgrass, and their presence is a health threat to some people and pets and a nuisance to others. Large colonies may occur year after year in the same location, especially if the nest site remains. This species is distributed in North America.

## Anthophoridae

This is the family of cuckoo bees, digger bees, and carpenter bees. It is a diverse group of species with a great range of habits and habitats. Cuckoo bees are considered cleptoparasites in the nests of many families of bees. They lack the hairlike setae and pollen baskets of other bees, and they resemble small wasps. Digger bees store a mixture of honey and pollen for larval food in burrows in the ground; a few are cleptoparasites on other members of the Anthophoridae. There are two primary genera of carpenter bees, *Ceratina* and *Xylocopa*. Most species collect pollen and make their nests in wood; some nest in the stems and branches of plants, and some in stumps, live trees, and structural wood in use. Beneficial aspects of the anthophorids include the pollination of a variety of trees and plants visited for pollen and nectar.

### Carpenter bees

*Xylocopa* (Fig. 9.1a) is a very large genus consisting of several hundred species and distributed throughout much of the world. Most species are in Africa and Neotropical regions, where this genus probably originates. Some *Proxylocopa* excavate nests in the ground, in bricks, or in similar substrates. The majority of carpenter bees gnaw nesting tunnels in dead wood, including seasoned hardwoods, softwoods, decaying wood in various locations, and sometimes in bamboo and other hollow-stem plants. Several species excavate nesting tunnels in structural wood used in building, bridges, water storage tanks, fences, and house siding materials. Some *Xylocopa* have been able to extend their geographic range to suburban and urban sites by using structural wood for nests. Some have been transported to areas where natural nesting sites are limited, but urban sites are readily available. In Europe, the range of *X. violacea* extends as far north as Paris, perhaps carried there in infested wood and sustained by available nest sites. In Asia, *X. phalothorax* and *X. latipes* are common species, and sometimes pests of structural timbers.

Pest status of carpenter bees is based on the threat of being stung – these are large bees – and the cosmetic and sometimes structural damage that results from their nests in structural wood. Males utilize aggressive and threatening behavior as they guard nest sites, but they are not able to sting. Females come

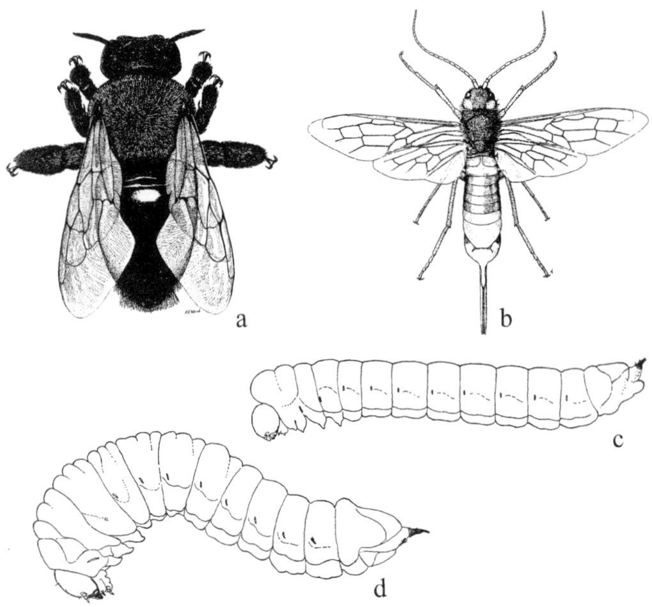


**Figure 9.1** Hymenoptera: Anthophoridae, Siricidae, Xiphydriidae. (a) *Xylocopa* sp. (Anthophoridae); (b) *Urocerus gigas* (Siricidae); (c) *Tremex columba* larva (Siricidae); (d) *Xiphydria maculata* larva (Xiphydriidae).

and go from the nest site during the day, but remain there at night; they are not usually aggressive, but are capable of stinging. In general, yearly nesting of carpenter bees causes annoyance from the sound of their gnawing and nest building in house siding. Unless there is an ongoing nest for several years, there is only minimal damage to infested wood. In southern Asia, *X. rufescens* is nocturnal and, where it is nesting, the loud buzzing and nest-building activity of the females can be heard throughout the night. Structural timbers in old or uninhabited buildings can be continuously infested by *Xylocopa*, and in some cases the wood becomes structurally unsound. Many species prefer natural sites, such as the wood of dead trees, stumps, and fallen logs. As natural or undisturbed areas for nesting become scarce, the abundance of some *Xylocopa* species may decrease, and those capable of nesting in structural wood may increase.

Frass or the pieces of wood shavings resulting from carpenter bee tunneling in wood are usually scattered below the opening to the nest site. These irregular pieces of wood removed by the queens are 1–3 mm long, fibrous, and irregularly shaped. The composition is only wood shavings, unlike the frass of the pavement ant (*Tetramorium caespitum*) and carpenter ants (*Camponotus* spp.), in which there may be fragments of dead ants, soil, and other nest refuse.

Natural food includes nectar and pollen from flowering plants. A few species in the western hemisphere are considered necessary pollinators of some agricultural and horticultural crops. In tropical regions, *Xylocopa* species greatly increase fruit setting in passionfruit (*Passiflora edulis*). There are several *Xylocopa* species in Central America, including *X. fimbrata* and *X. gualanensis*. On the Galapagos Islands, *X. darwini* is the only bee species, and it may play a role in establishing immigrant plant species. Among flowers visited and often pollinated by carpenter bees are many indigenous and introduced plants in the urban and rural landscape. Their nesting tunnels and pollen residues in unused nests can provide harborage and food for other arthropods, including spiders, mites, cockroaches, moths, and some stored-food beetles, such as dermestids, and tenebrionids.

Carpenter bees have relatively few predators and parasites. As regards Coleoptera, the eastern hemisphere (Old World) genus *Hornia* and the western hemisphere (New World) genus *Cissites* (Meloidae) are nest parasites of *Xylocopa*. They can destroy the contents of a few cells or an entire nesting tunnel. Referring to Diptera, *Anthrax* (Bombyliidae) and *Hyperechia* (Asilidae) contain species whose larvae are parasites of the larvae of *Xylocopa*. Various species of Ichneumonidae, Chalcididae, Leucospididae, Encyrtidae, and Evaniidae have been reared from carpenter bee larvae. Woodpeckers, Baltimore oriole, and house wren attack the nest sites or otherwise feed on these bees. Humans are major predators of carpenter bees in certain places of the world, particularly in Asia. Adults are sometimes collected for food, or used to prepare a medication for the treatment of sore throats in children.

**Little carpenter bee, *Ceratina dupla*** Adults are 6–8 mm long, and the body and legs are bluish green. Antennae are brown at the base and dark blue apically. Adults are often found at flowers. Nests are in the stems of pithy reeds and stems, such as sumac and brambles. However, they may use the exposed emergence hole to access the tunnels of wood-infesting beetles as a nest site, and they nest in decaying wood. This species usually does not nest in structural wood. There are about 20 species in the USA, and they are generally distributed.

**African carpenter bee, *Xylocopa caffra*** Adults are about 20 mm long and the body is black. The female has a yellow to yellowish white band posteriorly on the thorax, and another band anteriorly on the abdomen; the wings are brown. The male is nearly covered with a dense coat of yellowish-white setae; the wings are light brown. In the interior regions of South Africa, the markings are white instead of yellow or yellowish white. Nesting occurs in spring (October and November), and nest sites include tree branches or dry structural timber. The entrance hole to the nest is about 10 mm diameter and about 15 cm deep. Females provision the nest with pollen collected from pea, bean, and sage flowers. Nest construction is completed in about 30 days and development of the immature stages is complete in 3–4 weeks. Adults emerge in January and February, but they remain in the nest until the following spring. In eastern and coastal areas of southern Africa there are two or three generations of *X. caffra* per year. In the winter rainfall area at the southern tip of Africa there is one generation per year. Large mites in the genus *Dinogamasus* are found on the adults and in the nests of *X. caffra* and allied species. The ensign wasp (Evaniidae), *Gasteruption robustrum*, is a parasite of *X. caffra*, and is commonly found near the nests.

***Xylocopa californica*** Adults are 20–24 mm long, and the body of both sexes is metallic blue or metallic green. The pronotum of the male has white, yellow, or orange setae, and the entire first abdominal segment is covered with white setae. This species is known to nest in redwood and incense cedar in California and Oregon.

***Xylocopa iridipennis*** Adults are about 26 mm long, the body of both sexes is metallic blue, and the wings are blackish blue. The pronotum is blackish blue and without yellow marking. This species occurs in southern Asia, where it commonly nests in bamboo stems, and locally is known as the bamboo carpenter bee.

**Tropical carpenter bee, *Xylocopa latipes*** Adults are 23–26 mm long; the body is metallic, bluish black, and with green and purple reflections. Wings are bluish black at the base and dark brown at the apical third. The head, antennae, and legs are bluish black. This species occurs in tropical regions of southern Asia. In Malaysia, the structural wood susceptible to attack includes *Dyera costulata* (jelutong), *Agathis alba* (damar minyak), *Alstonia* spp. (pulai), and *Shorea* spp. (light red meranti). Tropical woods that are relatively safe from attack by this species include nyatoh, kapur, kempas, and mengkulang.

**Southern carpenter bee, *Xylocopa micans*** Adults are 14–27 mm long, and the head and thorax are black and shiny; there is a band of pale white setae at the base of the abdominal segments. Males and females overwinter in nests constructed in fence posts, wood rails, and the eaves of buildings. This species occurs in southeastern USA.

***Xylocopa tabaniformis orpifex*** Adults are 19–23 mm long, and the body is somewhat metallic, and the fine setae (pubescence) on the thorax are black in females and yellow in males; the males have yellow spots on the face. This species occur throughout eastern USA and west to Texas, Oklahoma, Kansas, and Missouri.

**Common carpenter bee, *Xylocopa virginica virginica*** Adults are 19–23 mm long. The body is somewhat metallic, fine setae (pubescence) on the thorax are black in females and yellow in males; the males have yellow spots on the face. This species occurs throughout eastern USA and west to Texas, Oklahoma, Kansas, and Missouri.

Adult males and females overwinter in a nonreproductive phase in nest tunnels or in protected locations outdoors. In the northern hemisphere, overwintering quiescence usually lasts from October through March. Most quiescent females are unmated. In early spring, mating occurs following territorial flights by the males; this is followed by nest construction. Although males mate mainly during spring, they may also defend territories and mate in fall. For nesting, females may use an old gallery without further burrowing, or create an entirely new gallery, or make a new one from an entrance used by several females. Adults that have overwintered die during the summer of the second year, but usually after they have mated and produced male and queen galleries and eggs for the next generation. In temperate regions, there is usually one generation per year.

Selection of the nest site may be elaborate, and usually includes dry softwoods in exposed locations, but may also include hardwoods, even hardwood tool handles; white-painted wood may be avoided. In construction of new nest sites, female bees gnaw into the wood surface to make a hole about 15 mm diameter, and in most kinds of wood females can tunnel 10–15 mm per day. Boring proceeds slowly when galleries extend across the wood grain, and is much faster with the grain. Direction of the gallery depends on the orientation of the wood; if the grain is oriented vertically, the galleries are vertical, and if the grain is oriented horizontally, the galleries are horizontal with respect to the ground. Galleries extend 30–45 cm in newly completed nests. New galleries are smooth and uniform throughout, but old ones are irregular and often have depressions or excavations along their sides. Galleries may be used by several generations of bees, and the system of tunnels in one piece of wood may become very elaborate. Sawdust is pushed out of the nest by the head, abdomen, or legs of the female, and it may pile up at the nest entrance. A distinctive flight behavior, called the bobbing dance, has been reported for *X. virginica*. The bees hover in flight in front of the opening to the nest, and they position their head to strike the wood around the nest opening. This flight and behavior may persist for several minutes. It is demonstrated primarily by males, and is probably linked to mating.

After excavating the nest site, provisioning individual cells in the gallery is begun. A mass of pollen and regurgitated nectar is placed at the end of the gallery, and the female deposits an egg on it. She then seals this portion of the gallery with a partition of chewed wood pulp. This procedure of provisioning, oviposition, and sealing is repeated until a series of about six cells is formed in the gallery. Eggs hatch in 2–3 days, and larval development is complete in about 15 days. The prepupal period is about 4 days and the pupal period about 15 days. Total development time is about 36 days and it is all spent within the cell. Following emergence from its cocoon, the adult bee spends 1–2 days in the cell grooming and feeding on the remaining food. The first bee to become an adult is usually in the first cell to be provisioned, which is at the end of the gallery. The adult cuts through the cell partitions of all the cells to emerge at the exterior of the nest. For the remainder of the season the adult bees gather pollen and nectar and store caches of food in the galleries for use during inclement weather. During this period, there is no mating and no nest construction.

### Digger bees

These small bees store a mixture of honey and pollen for larval food in burrows in the ground. Their long burrows are lined with a waxlike substance produced by the female during construction. Nests of North American *Anthophora* are usually built in steeply inclined or perpendicular soil banks exposed to the sun, but they are also excavated in other locations, including in outside walls of adobe houses in western USA.

**Mortar bees, *Amegilla* spp.** These bees normally make their nests in banks of hard soil or sand. In coastal areas they may nest in rock fissures. However, brick houses and buildings with brick foundation walls often serve as nests sites. Females take advantage of old or poor-quality mortar to construct burrows and brood cells. *Amegilla* females make their burrow deep into the mortar; other females dig lateral tunnels from the main burrow. Large numbers of females may use the same site. Damage to brick and mortar walls can be extensive, because they may return to the same site for several years. These species occur in Australia.

**Digger bees, *Centris* spp., *Ancyloscelis* spp.** Adults are 14–18 mm long. They are usually black with the abdomen (metasoma) red, yellow, or metallic green and covered with dense, yellow setae (*Centris*), or black with narrow white bands of setae on abdominal tergites (*Ancyloscelis*). Adults of *Centris* spp. either dig a burrow in the soil or use pre-existing holes in wood. Nests usually contain several brood cells that are provisioned with pollen and plant oil, which is collected from oil-bearing plants. Species that nest in holes in wood often line their cells with wood chips scraped from the existing gallery. This genus occurs in the deserts of North America, south to neotropical regions of Central America. *Ancyloscelis* spp. often nest in large aggregations. Adults usually dig nests in sloping ground or vertical embankments exposed to the sun; some species tunnel in the bricks of adobe buildings in southwestern USA.

**Eastern digger bee, *Ptilothrix bombiformis*** Adults are 12–18 mm long. The body is black and without markings, except for thick, pale setae on the head, thorax, and apex of each abdominal segment. They resemble small bumble bees (thus the species name), and build nests in bare soil. This species is common is suburban habitats and distributed in eastern and southeastern USA.

## Apidae

This family comprises medium-sized to large bees with long tongues, and setae that are branched or feathery and give a dense covering on the body. They usually live in large, well-organized communities consisting of sexually functioning females or queens, nonreproductive workers, and sexual males or drones. Queens and workers originate from fertilized eggs, and males from infertile eggs. These bees pollinate a variety of economically important crops and tree fruits. In this family are the well-known bumble bees and honey bees.

### Bumble bees

These bees commonly occur in peridomestic habitats and undisturbed areas. Bumble bees are 15–32 mm long, and the body is black, but it is often distinctly marked with black, white, and yellow setae that are branched or feathery. In *Psithyrus* bumble bees, which are social parasites of *Bombus*, the setae are less dense on the abdomen, and it appears shiny like the abdomen of *Xylocopa*, carpenter bees. Bumble bees are abundant in the temperate parts of North America, Europe, and Asia; they occur within the Arctic Circle, along the northern fringe of Africa, and in South America. They are generally absent from tropical regions, although a few species are found in Brazil. There are no indigenous bumble bees in Australia, and a few species have been introduced into New Zealand. There are about 50 species in North America, while 18 species have been recorded in the UK. Most species restrict their activity to undisturbed areas and remain unnoticed.

Hibernation (diapause) occupies a large part of the life of the bumble bee queen, and is an important aspect of the biology of this insect. The fertilized queens produced during the summer enter hibernation and reappear the following spring to form new colonies. In north temperate regions, bumble bees hibernate for 6–8 months, depending on the species and spring temperatures. They normally hibernate individually in small spherical or oval cavities excavated in the soil. Soil banks and lightly wooded slopes with a north or northwestern exposure are preferred sites, but they also hibernate in soil beneath trees, in rotten tree stumps, under leaf litter, and under piles of garbage. Preference is for protected sites with soil moisture above 50%. Bumble bees hibernating in banks or slopes burrow directly into the soil. They use their mandibles and legs to loosen the soil, and occasionally back out of the cavity to deposit soil out and away from the opening. This soil forms a mound at the entrance, and the soil from the lower parts of the burrow is used to block the entrance. During hibernation, the adults remain lying on their backs or in a crouched position. If disturbed in late summer or fall, they immediately become active. In mid-winter they become active only when warmed.

Nest sites are selected soon after the fertilized queens emerge in spring. Queens search for suitable places in clumps of vegetation or holes in the ground. Surface-nesting species will repeatedly drop to the ground and force their way into the ground cover to find a suitable place. Underground-nesting species frequently hover over holes in the ground, before landing and investigating more thoroughly. The site must provide protection from rain, and have a supply of grass, animal hair, leaves, and other fine material the queen can use to form into a nest. Most colonies are established in abandoned burrows of mice, voles, and shrews. In general, queens do not excavate a new nest chamber in soil, and do not forage for nesting material. Surface-nesting species usually establish nests in grassy areas, or in vegetation at the base of walls. When a nest site is selected, the chamber is lined with nest material and a small entrance is formed. The first few flights from the nest are usually in increasing circles and serve to orient the queen to the site before foraging begins.

Once the nest is prepared, the queen forages for pollen and deposits it on the floor of the nest chamber. Pollen collected by the queen is molded into a mass and small cavities or pockets are formed in the upper surface. One egg is deposited in each cavity or around the perimeter of the mass; the total number of eggs is 6–8. When egg-laying is complete, more pollen is added to cover the eggs; then the entire mass of eggs and pollen is covered with a protective layer of wax. In addition to the brood mass, at the entrance to the chamber the queen constructs a waxen receptacle (honey pot) for storing nectar. This is a food reserve for when she cannot leave the nest to forage. When not foraging or maintaining the nest site, the queen lies on the mass of pollen and eggs, usually facing the nest entrance and the nectar pot. An incubating queen presses her body against the mass and has rapid respiratory movements. Heat is produced in her thorax and transferred to the abdomen, then conducted to the mass of pollen and eggs. The mass and developing larvae are maintained at 30–32 °C. Eggs hatch in about 5 days, larvae complete development in about 14 days, and the pupal period is about 14 days. In large colonies, several hundred workers may be produced during the season, but they are not all alive at the same time. Workers can live for about 2 months; their longevity depends on their duties in the colony. Those that remain in the colony live longer than those that forage. The primary role of workers is foraging: they do not assist the queen in constructing egg cells for developing larvae.

The nest is a large community comprised of an egg-laying queen, workers, and developing larvae. The presence of more workers increases the food available for larvae and the body size of the workers increases, but they usually remain smaller than the queen. Males and females are produced in late summer and fall. Males are produced in advance of queens, although mixed batches of males, workers, and queens sometimes occur together. Males and workers die at the end of the season, after mating queens seek hibernation sites.

The pest status of bumble bees is based on their nesting habits and the stinging hazard of encountering a mature nest with a large number of workers. The large size of these bees and their relatively slow foraging flights, and their presence on ornamental flowers (in spring) may be threatening to people. Nests are usually located in sites away from human activity; however, nests are sometimes located in peridomestic habitats, such as in the turfgrass around structures or among the landscape plantings. Hibernating queens may be found in or around structures. Most bumble bee species are not aggressive, but workers will protect the nest from intruders. Temperament could be classified as mild if few bumble bee workers flew around intruders but did not sting when a full-sized colony with brood was disturbed during the day. Moderate-temperament species dispatched several workers to attack and sting the intruder, while aggressive species consistently attacked, stung, and followed intruders.

*Bombus* species superficially resemble one another, except for small differences in the color of the males and queens, and the body size of overwintering queens. Specimens of the male are necessary for accurate identification of most species. In North America, the native species can be divided into categories based on nesting time and colony productivity (including new queens). Late nesting species, which include *Bombus californicus*, *B. rufocinctus*, and *B. borealis*, form small colonies of 60–190 individuals. Large colonies are considered to have over 680 individuals, and in this category are *B. impatiens*, *B. affinis*, *B. occidentalis*, and *B. vosnesenskii*. Medium-sized colonies have a total of 150–430 individuals, and in this category are *B. bimaculatus*, *B. fervidus*, *B. pennsylvanicus*, *B. terricola*, and *B. caliginosus*.

**Common bumble bee, *Bombus bimaculatus*** Workers are 11–16 mm long. Setae on the pronotum, scutum, and abdominal segment 1 are bright yellow; wing veins are black, and the membrane is yellowish brown. Nests are on the surface of the ground, and rarely below ground or aerial; nest sites in peridomestic habitats include in sheds and outbuildings, and within stone walls. Colonies are small, and young queens are produced in July and August in the Pacific northwest. Their temperament is mild. This species occurs nearly throughout the USA and southern Canada.

***Bombus lucorum*** Workers are 9–16 mm long, and the body is black; it has a yellow band on the thorax and on abdominal tergum 2, and pale white setae at the tip of the abdomen. It is active from as early as February. Nests are built below ground in abandoned mouse burrows, often under the floors of garden sheds. This species is abundant throughout the UK, including Ireland.

***Bombus pratorum*** Workers are 9–14 mm long, with a yellow band on the thorax and on abdominal tergum 2, although the band on the abdomen may be absent. The setae on the body are long and somewhat uneven, particularly on the abdomen. Nests are established in early spring and end by the middle or end of July. Nests occur above and below ground, and frequently in inactive bird nests. This species is generally abundant throughout the UK.

**Eastern bumble bee, *Bombus ternarius*** Workers are 8–13 mm long. Setae on the head are black, there are pale yellow setae on the thorax with a fringe of black setae, and abdomen terga 1 and 2 have yellow setae and a reddish-brown band; the remainder is black. Nests are established primarily on the surface or below ground, rarely above ground. Nest sites in peridomestic habitats include in old mammal burrows, and sheds. Colonies are large and established in late spring. This species occurs in northeastern USA and British Columbia, south to South Carolina and Georgia.

***Bombus terrestris*** Workers are 10–14 mm long, with a yellow band on the anterior margin of the thorax, and a yellow band at the posterior margin. The abdominal terga 4 and 5 are white to yellowish white, and there is a narrow band on tergum 3. This bumble bee often uses its mandibles to cut a hole through the nectary wall of flowers, such as clover and broad beans. In New Zealand, nest development of *B. terrestris* continues through the winter, and nests can be established during winter and nearly throughout the year. Queens and workers are on the wing almost throughout the year, even during warm days of winter. Males are produced and mating occurs in summer. This species is widely distributed in the UK, and was introduced to New Zealand.

**Western bumble bee, *Bombus terricola occidentalis*** Workers are 17–27 mm long. The thorax has pale yellow setae on the anterior portion; the abdomen has two terga with white setae, and one with orange-yellow setae. Wing veins are black and the membrane is yellowish brown. Nests are established primarily on the surface or below ground, and rarely are aerial. Nest sites in peridomestic habitats include in old mammal burrows, and among railroad ties; nests are lined with grass, insulation, or bird feathers. Colonies are large and have about 300–1000 workers and males, and 32–146 queens; young queens are produced in August and September in the Pacific northwest. Their temperament is mild. This species occurs in western USA, including California, Arizona, New Mexico, and the Pacific northwest.

**Other *Bombus*** In North America, several species regularly nest in peridomestic habitats. *B. affinis* has below-ground or aerial nests with about 1000 workers, males, and queens; queens emerge in August and September. *B. borealis* has nests on the soil surface with about 200 individuals; queens emerge in August and September. *B. fervidus* is aggressive and establishes aerial or surface nests in woodpiles and in household materials.

They have about 200 individuals, and produce queens in August and September. *B. melanopygus* has aerial or surface nests in insulated house walls, household materials, lofts, and bird nests. Colonies have about 500 individuals, and produce queens in May and June. *B. pennsylvanicus* is aggressive and has aerial nests in lofts, sheds, and bird boxes. Colonies have about 200 individuals, and produce queens in August and September. *B. perplexus* establishes either aerial or surface nests in insulated house walls, woodpiles, and in sheds. Colonies are established in May and June; mature colonies have about 300 individuals and produce queens in August and September.

### Stingless bees

These bees are widespread and common in the urban environment. Their biology and habits are similar to honey bees; within the nests there are pollen and honey pots and brood cells. These bees often rob pollen and honey from other bee species. In some habitats, they are very common, with densities of 600 colonies per square kilometer. The common name is derived from the fact that their sting is reduced. Despite their inability to sting, they may be aggressive, and some species swarm out of their nest to attack intruders by crawling into hair, ears, nose, and eyes, and then biting. The most common nest sites are cavities in tree trunks, holes, and cavities in the ground. Among those nesting in the ground, *Plebia mirandula* and *Trigona buchwaldi* sometimes occupy abandoned mammal burrows. *P. latitarsis* and *T. ferricauda* often nest in termite nests, and *Paratrigona peltata* invades aerial nests of *Camponotus senex*.

### Honey bees

These bees are well known and recognized by their light brown color and characterized shape. This group contains one genus, *Apis*, which has traditionally been regarded as containing four or five species. The honey bee, *A. mellifera*, is endemic to the eastern hemisphere, but it has been domesticated and carried throughout the world. In general, *A. mellifera* colonies are in managed hives, and there are few successful or long-lasting feral colonies. However, escaped swarms may establish new colonies in tree cavities in undisturbed areas and occasionally in wall voids of houses and other domestic structures.

In 1956, an African strain of the honey bee was introduced into Brazil to obtain a strain of honey bee better adapted to tropical conditions. Mated queens and workers escaped and established feral colonies. This subspecies, *A. mellifera scutellata*, spread across South America at the rate of 300–500 km/year. By 1979, it had reached Columbia, by 1986 Mexico and in 1990 it entered the USA in southern Texas. The frequency of reproductive swarms and the ability to establish feral colonies is part of the successful spread of this honey bee. The best-known characteristic of the African honey bee is its more rapid arousal and the intensity with which it defends its nest. Although it stings more readily, its venom is no different from that of the European honey bee. These characteristics have utility in tropical climates, where nectar resources are unpredictable, and nest predators are common.

The pest status of honey bees is primarily based on the infrequent occurrence of nests established in the walls of structures. These bees are relatively docile, but large colonies within structures present a stinging hazard and large nests can cause damage by their presence. If large honey bee colonies in structures are controlled and the nests are left unattended, the stored honey and nest structure will attract insects such as wax moths, carpet beetles, meal moths, and other scavenger insects. The pest status of the African honey bee, known to the public as the killer bee, is based on the potential human and livestock fatalities due to bee stings. The fatalities are usually due to exposure to a large amount of bee venom, because of the aggressiveness of the African bee to attack and sting intruders.

## Bethylidae

These wasps are 2–5 mm long, and have a black or brown body and a narrow thorax; the antennae are moderately long, and have 12 or 13 segments. The front femora are thickened in the middle, and the abdomen has seven or eight visible segments. Females of many species are wingless and resemble ants; the males are winged. Some species are parthenogenetic, and the unfertilized eggs produce males. Many species are parasites of larvae of household pests, such as wood-infesting beetles and ants, and stored-grain beetles and moths. A number of species in the genus *Sclerodermus* are parasitic on the larvae of Cerambycidae. Species in the large genus *Pseudobrachium* are parasites of ants, and *P. mandibulare* is associated with carpenter ants (*Camponotus*). Species in *Cephalonomia* and *Neoscleroderma* are parasites of stored-grain beetles; and *Laelius trogodermatis* is a parasite of dermestid beetle larvae. Three genera, *Cephalonomia*, *Sclerodermus*, and *Epyris*, have species that are reported to sting people and sometimes cause a severe reaction, including respiratory problems, nausea, and headaches. Many species are cosmopolitan.

**Parasitic grain wasps, *Cephalonomia gallicola*, *C. watersoni*, *C. benoiti*** Adults are about 1.6 mm long, and the body is brown to brownish black. The tarsi are yellowish brown to brown, and the wings are clear. *C. watersoni* is a cosmopolitan parasite of the rusty grain beetle (*Cryptolestes ferrugineus*). *C. gallicola* is probably cosmopolitan and parasitizes larvae of several species of stored-food beetles. *C. benoiti* and *C. gallicola* occur and sting people indoors.

**Bethylid wasps, *Epyris californicus*, *Epyris* spp.** Adults are about 5 mm long and the body is black. It is believed to parasitize the larvae of tenebrionid (Tenebrionidae) beetles. There are many species of *Epyris*, and some are known to inflict a severe sting when brushed or crushed by clothing or sheets against the skin. This species is reported from northern California, but it may occur in neighboring states.

**Dermestid wasp, *Laelius trogodermatis*** Adults are about 4 mm long, with a black body and dusky wings. Full-grown larvae are about 4 mm long and yellowish white. The adult female searches for dermestid larvae and oviposits 1–6 eggs on a larva after first stinging it several times. Eggs hatch in about 2 days and larval development is complete in 10–14 days. Silken cocoons are spun outside the skin of the dermestid larva. The prepupal stage is about 10 days, and the pupal period is 8–10 days. Females may lay unfertilized eggs, which produce males.

**Scleroderma wasps, *Sclerodermus domesticus*, *S. abdominalis*** Adults are 3.5–4 mm long and the body is brown. The antennae have 13 segments, and the mandibles have three teeth. Females of these species are known to sting people. Besides the acute pain associated with the sting, there are small and local patches of irritated skin, and these patches are associated with intense itching.

## Colletidae

These are the plasterer and yellow-faced bees. The adults are 3.5–12 mm long and the body is black and may be marked with white or yellow bands. Plasterer bees in the genus *Colletes* nest in the ground or in crevices in bricks and stones. They plaster the sides of the nest with a secretion, which dries to form a cellophane-like lining that is unique among the bees. Yellow-faced bees make simple nests in plant stalks and insect burrows in wood.

**Common plasterer bee, *Colletes compactus compactus*** Adults are about 12 mm long and the head is marked with yellow; the abdomen has white markings on the apical portion of each segment. These bees commonly nest in the ground, but they will form nests in crevices in bricks and stone walls. The female typically digs a burrow in the soil, which extends 45–70 cm. From this main burrow, lateral branches 5–15 cm long are made; at the end of each is a brood cell. The brood cells are made, provisioned, and receive an egg, starting from the bottom to the surface. There may be 2–4 lateral branches from each main burrow. These bees make numerous nests in the mortar of the brick veneer of houses. The old or weak mortar may be excavated to accommodate a burrow and brood cells. The presence of these bees dislodging mortar on the sides of houses causes concern. This species is generally distributed in the USA.

**Other Colletidae** Species of *Colletes* occur in the UK and continental Europe and are a nuisance when they nest in the mortar and crevices between bricks of houses. *Colletes daviesanus* is a common pest of external brickwork in the UK. The nest is usually 5–10 cm deep and contains 2–8 separate cells. Species of *Hylaeus* also nest in mortar and in the burrows of wood-infesting beetles.

## Chrysididae

The cuckoo wasps are 6–12 mm long, and their body is usually brilliantly metallic blue or green with some red and gold colors. The integument is punctured and well sclerotized; when disturbed they can curl the body into a ball for protection. Most of the species in this family are parasites (cleptoparasites) of solitary bees and wasps, including ground-nesting species and mud daubers. *Chrysis fuscipennis* is a parasite of the mud dauber, *Sceliphron caementarium*. Adults of the mud wasp parasite, *Trichrysis tridens*, are 7–11 mm long, and the body is metallic blue-green or reddish brown.

## Encyrtidae

These small parasitic wasps are 0.5–3 mm long, and have a robust body and are usually fully winged. The body is metallic yellow to orange, red or brown. The four- to nine-segmented antenna of the female is cylindrical to slightly flat; in the male it is three- to eight-segmented. Wing venation is limited and the wings are clear. This family is employed in applied biological control programs of insect pests, and many encyrtid species have been used successfully against a variety of agricultural pests. Species in the genus *Cerchysius* are parasites of anobiid larvae, and *Blatticida pulchra* and *Blatticidella ashmeadi* are reported as parasites of cockroach oothecae.

**Cockroach eggcase parasite, *Comperia merceti*** Adults are about 3 mm long. The body is blackish brown with pale white bands on the legs and antennae. It is an ootheca parasite of *Supella longipalpa*, but does not attack *Blattella germanica* ootheca.

Eggs are deposited singly into the oothecae; 5–25 eggs may be deposited, and parasitized oothecae may have 1–50 punctures. Females prefer to lay eggs in oothecae that are about 2 weeks old, but eggs are laid in oothecae less than 1 week old, and in those in the green-band stage of development. Development is 30–41 days at about 20 °C, and usually all the cockroach eggs may be eaten, but when not, cockroach nymphs may successfully develop. Adults are very active and, when populations are large, they can be found running and hopping on surfaces; they continually touch the surface with their antennae. They are attracted to light and are often found at windows. This species occurs nearly throughout the USA, Hawaii, and Central and South America.

## Eulophidae

These small to extremely small wasps are 0.4–6 mm long, and the body is yellow to brown with some dark markings, or sometimes metallic. The marginal vein on the forewing is usually long. The majority of eulophids are parasites of concealed larvae, usually Lepidoptera, but also Diptera and Coleoptera.

**Eulophid cockroach eggcase parasite, *Aprostocetus hagenowii* (= *Tetrastichus*)** This wasp is an egg parasite of several species of domiciliary cockroaches, including *Periplaneta americana*, *P. brunnea*, *P. fuliginosa*, *P. australasiae*, *Blatta orientalis*, and *Eurycotis floridana*. Eggs are deposited into the ootheca after the female wasp explores the surface with her antennae and ovipositor. They oviposit into freshly deposited or old oothecae. In *P. americana*, development is completed in 23–56 days at 29 °C, but the time depends on the number of wasps in the ootheca. The larger the number of wasps, up to about 70 wasps per ootheca, the shorter the time required for development. All eggs are consumed when there is a large number of parasites, but when there are few, some cockroach nymphs complete development and hatch. Adult parasites emerge from the ootheca through 1–3 holes. In *B. orientalis*, an average of 66 offspring may develop from the eggs of one female wasp; in *P. americana*, there is an average of 103 offspring per female. The average number of parasites per ootheca in *E. floridana* is 648; in *P. americana* 204 parasites; and in *P. australasiae* it is about 50 parasites. Adults of this small parasite may be seen indoors when cockroach populations are large. It is probably cosmopolitan.

## Eupelmidae

These wasps are 1.5–20 mm long and they are usually metallic. The midlegs are attached distinctly behind the midline of the mesopleuron, adjacent to the hind legs and far from the front legs. Dead individuals are often contorted into a U-shape, with head and abdomen (metasoma) reflexed upwards. Species of *Anastatus* are endoparasitic on the eggs of Orthoptera, Blattaria, Hemiptera, and Lepidoptera. *Eupelmus atriflagellum* is an ootheca parasite of *Blattella germanica* in Australia.

**Eupelmid eggcase parasite, *Anastatus floridanus*** This wasp is an ootheca parasite of *Eurycotis floridana*, *Periplaneta americana*, and *Blatta orientalis*. Wasps may lay eggs in an ootheca carried by the female cockroach, or after the ootheca is deposited and is as many as 36 days old. Several *A. floridanus* females may lay eggs in the same ootheca. In *E. floridana*, development is completed in about 36 days at about 29 °C. Development time is regulated by the number of parasites in the ootheca. As many as 306 adult *A. floridanus* may develop from one cockroach ootheca, and there are 1–6 emergence holes in parasitized *E. floridana* oothecae. Parthenogenesis occurs in this species, and unfertilized eggs produce males. This species is widely distributed in the USA, Europe, and Asia.

**Cockroach eggcase parasite, *Anastatus tenuipes*** This wasp is an ootheca parasite of *Supella longipalpa*, and perhaps *Periplaneta americana*. Adults may be seen running rapidly on walls and other surfaces in buildings infested with the cockroach host. The adults usually hop across surfaces; they are rarely seen flying. The female wasp begins the egg-laying process by touching the ootheca with her antennae many times, then she inserts her eggs. Development is completed in about 32.6 days at 27.8 °C. Several females may attack the same ootheca, and 10–15 adult wasps may develop from one ootheca. Parthenogenesis occurs in this species, and unfertilized eggs produce males. It is widely distributed in the USA, Europe, and Asia.

## Evaniidae

The ensign wasps are 1.7–17 mm long. The body color is predominantly black or dark brown to reddish brown, sometimes with white markings. The abdomen (metasoma) is small and inserted high on the thorax (mesosoma). This unusual body shape, and the habit of moving the abdomen up and down when the wasp walks, gives the impression of a waving flag, hence the name ensign wasp. Biological information from Europe and North America indicates that most evaniids are parasites in the oothecae of cockroaches. Female evaniids generally run about quickly in locations where hosts deposit oothecae, and they may prefer those freshly deposited. Evaniid eggs are deposited singly in the ootheca. There are 3–5 larval instars; the first-stage larvae may feed on one cockroach egg, but later instars consume several eggs within the ootheca. The prepupal stage may diapause during an unfavorable season.

Some evaniid genera with species that parasitize domiciliary cockroach species include: *Brachygaster* with *Ectobius panzeri*; *Evania* with *Periplaneta americana*, *P. australasiae*, and *Blatta orientalis*; *Hyptia* with *Parcoblatta pennsylvanica*; *Prosevania* with *Periplaneta americana*, *P. australasiae*, *Blatta orientalis*, and *Blattella germanica*; *Szepligetella sericea* with *P. americana* and *P. australasiae*; and *Zeuxenia* with *Loboptera* spp.

**Evaniid eggcase parasite, *Evania appendigaster*** The adult is 5.5–7 mm long and the body is black; the wings are clear. It is the largest species in the family, the other species are usually less than 5 mm long. It is a parasite of several domiciliary cockroaches, including *Periplaneta americana*, *P. australasiae*, and *Blatta orientalis*; but some records may be in error, including those for *Blattella germanica* and *Leucophaea maderae*. A related evaniid species, *E. dimidiata*, is reported as a parasite of *B. orientalis* in Egypt.

Eggs are deposited singly, and usually there is only one per ootheca. Oviposition occurs after the female crawls over the ootheca surface, and rests with the axis of her body parallel to the axis of the ootheca as it lays on its side. As the ovipositor punctures the ootheca, the female may remain in place for about 15 min. The evaniid larva feeds on the cockroach eggs and development through five larval instars is completed within 60 days. The larva consumes the entire contents of the ootheca. The pupa is formed within the ootheca, and the adult wasp cuts an exit hole to emerge. Adults live 2–3 weeks with food and water. There are three or four generations per year, depending on environmental conditions. This species is widely distributed, and may be cosmopolitan.

**Ensign wasp parasite, *Prosevania punctata*** The adult is about 4 mm long, and the body is black. Eggs are deposited singly into an ootheca, and only one per ootheca. The female *Periplaneta americana* may show instinctive hostility towards evaniid females approaching a freshly deposited ootheca. Oviposition usually occurs after the female is oriented so that she is facing the keel of the ootheca, but egg-laying may occur in other positions. Females apparently do not distinguish between parasitized and unparasitized ootheca. There is no apparent preference for freshly deposited oothecae or those several weeks old, but the oothecae carried by cockroaches are not attacked. The development period in *Blatta orientalis* is 40–57 days; development in *P. americana* is about 127 days. The adult emerges from the ootheca through a 4-mm diameter hole. There are two or three generations per year. Adults live about 20 days. This species is widely distributed and occurs in eastern USA, Europe, and the Middle East, but may be cosmopolitan. It can occur in large numbers indoors, and achieve pest status.

**Other evaniids** Several species of *Hyptia* are parasites of *Parcoblatta* cockroaches, including *H. dorsalis*, *H. harpyoides*, *H. reticulata*, and *H. thoracica*. These wasps may be encountered indoors when fireplace logs with *Parcoblatta* under the bark are brought into houses.

## Formicidae

Ants are 1–20 mm long and the body color ranges from yellowish brown to black. They are characterized by a large head and elbowed antennae. The antennae of females and workers are 10–12-segmented, and 10–13-segmented in males. Abdominal segment 2, or segments 2 and 3, are separated as distinct nodes. Colonies are comprised of individuals in three castes: workers, which are apterous neuter females; queens, which are reproductive females that may be winged, dealate, or apterous; and males, which are usually winged. Ants are successful social insects that occur in all zoogeographic zones, and their predominance in some ecosystems is linked to their strongly developed social organization and complex behavior, and on their ability to exploit a wide range of food sources. Most species forage for small arthropods or other invertebrates, but they also feed on plant products by directly feeding at flowers, or indirectly by tending homopterans and ingesting the honeydew they produce.

All species form long-lasting colonies, which are constructed in soil, living or dead wood, or plant crevices. The nest contains one or more reproductive females and a large number of workers and immatures. Males and virgin females are produced at specific times during the year. New colonies are usually produced by means of a mating flight, where recently produced males and females fly from the parent nest and mate in the air or on the ground. After mating, the males die and one or more fertilized females shed their wings and form a brood chamber. Egg-laying begins soon after the brood chamber is built. Females do not feed during the development of the first brood. They nourish themselves and the first brood on the degeneration products of their wing muscles, fat body, and other energy reserves. Workers of the first brood leave the nest, forage, and feed the queen; they rapidly expand the nest and care for the next brood. The founding queen continues to lay eggs and remains in the nest. When the colony reaches a certain size the queen lays fertilized (diploid) eggs, which receive a special diet and treatment, and will develop into reproductive females. The queen lays unfertilized (haploid) eggs, which develop into males. The cycle of colony expansion and the production of reproductives may be repeated for many years.

The pest status of ants is based on their nesting and foraging habits. The urban environment provides a variety of soil types and conditions, such as open sun or shaded sites, wooded, field, or turfgrass, and these will be suitable for a large number of ant species. Food resources are not limited, since ornamental trees and shrubs usually sustain populations of aphids and other homopterans. These insects provide honeydew, which is the basic food for most ants. Nesting along building foundations can result in damage to structural members of buildings, and to exterior faces. The proximity of nests to buildings usually results in ants entering the building and finding food, then recruiting nestmates to the site. The persistence of ants foraging inside or the establishment of primary or satellite nests indoors is the basis of the pest status for most synanthropic species. Ant species that are associated with trees or other plants in natural areas often accompany them when they are planted as ornamentals in urban areas. *Cercopia obtusifolia* trees in urban areas of the subtropics and neotropics may have associated with them colonies of leaf-cutting ants, *Acromyrmex octospinosus*, *Cercopia*, and *Azteca* spp. In the USA, the rover ant *Brachymyrmex defilis* swarms in the spring and fall. Large numbers of winged females may collect in outdoor swimming pools.

The habit of foraging ants to visit a variety of food sources increases their potential to acquire pathogenic and food decay organisms. A large number of pathogenic organisms have been recorded from ant species, including bacteria:

*Acinetobacter baumannii*
*Aeromonas hydrophila*
*Agrobacterium radiobacter*
*Alcaligenes faecalis*
*Burkholderia cepacia*
*B. mallei*
*Chryseobacterium meningosepticum*
*Citrobacter frundii*
*Clostridium* spp.
*Enterobacter aerogenes*
*E. cloacae*
*E. durans*
*Enterococcus casseliflavus*
*Escherichia coli* O2
*Klebsiella pneumoniae*
*Micrococcus pyogenes aureus*
*M. albus*
*Morganella morganii*
*Proteus vulgaris*
*Pseudomonas aeruginosa*
*P. putida*
*P. stutzeri*
*P. vesicularis*
*Salmonella* spp.
*Serratia liquefaciens*
*S. marcescens*
*Shewanella putrefaciens*
*Sphingomonas paucimoblis*
*Staphylococcus aureus*
*Stenotrophomonas maltophila*
*Vibrio parahaemolyticus*
*V. fluvalis*

The fungi reported from ants include: *Aspergillus flavus*, *A. fumigatus*, *A. niger*, *Cladosporium wernickii*, *Fusarium oxysporum*, *Penicillium* spp., *Scopulariopsis* spp., and *Syncephalastrum* spp. The yeast reported from ants includes *Gerotrichum candidum* and *Trichosporon cutaneum*.

### *Acanthomyops*

The workers are 3–4.5 mm long and usually light brown to yellowish brown; the body is shiny, and with few setae. Antennae are 12-segmented and without a club; maxillary palps are three- or four-segmented; eyes are small. The epinotum is without spines and the pedicel is one-segmented. Workers are capable of emitting a pleasant odor similar to citronella when alarmed or crushed. Citronellal and citral are produced in the mandibular glands, and these chemicals probably function as an alarm pheromone. A formic acid odor may be detected when the gaster is crushed. They are nocturnal and nest below ground; they usually collect honeydew from aphids (*Prociphilus*, *Anuraphis*) that suck sap from plant roots. During winter, entire colonies, including workers and winged forms, may move next to house or building foundations. Nests are established in cracks and crevices in the foundation, but they rarely forage indoors. When reproductives swarm in large numbers in basements and around the outside of buildings, they may be mistaken for termites. Workers of *A. claviger* have been used in anting by starlings (*Sturnus vulgaris*), and winged males and females have been used in anting by bluejays (*Cyanocita cristata*). Anting is the behavior of some birds which place live ants within their feathers; the ants may remove parasites from the bird's body and feathers.

**Smaller yellow ant, *Acanthomyops claviger* (Fig. 9.2a)** Workers are 3–4 mm long and the head and thorax are brown; the gaster is dark brown. The scape does not extend beyond the posterior margin of the head, and the apical segments of the antenna are distinctly clavate. The pedicle is narrow and sharp at its peak. Nests are in open woodlands, pastures, and grassy fields. In the urban environment, nests may be in soil surrounding buildings. Colonies are of moderate size and probably with a single queen. Natural food is primarily honeydew from subterranean aphids and other homopterans. Winged reproductives emerge from August to November. Large

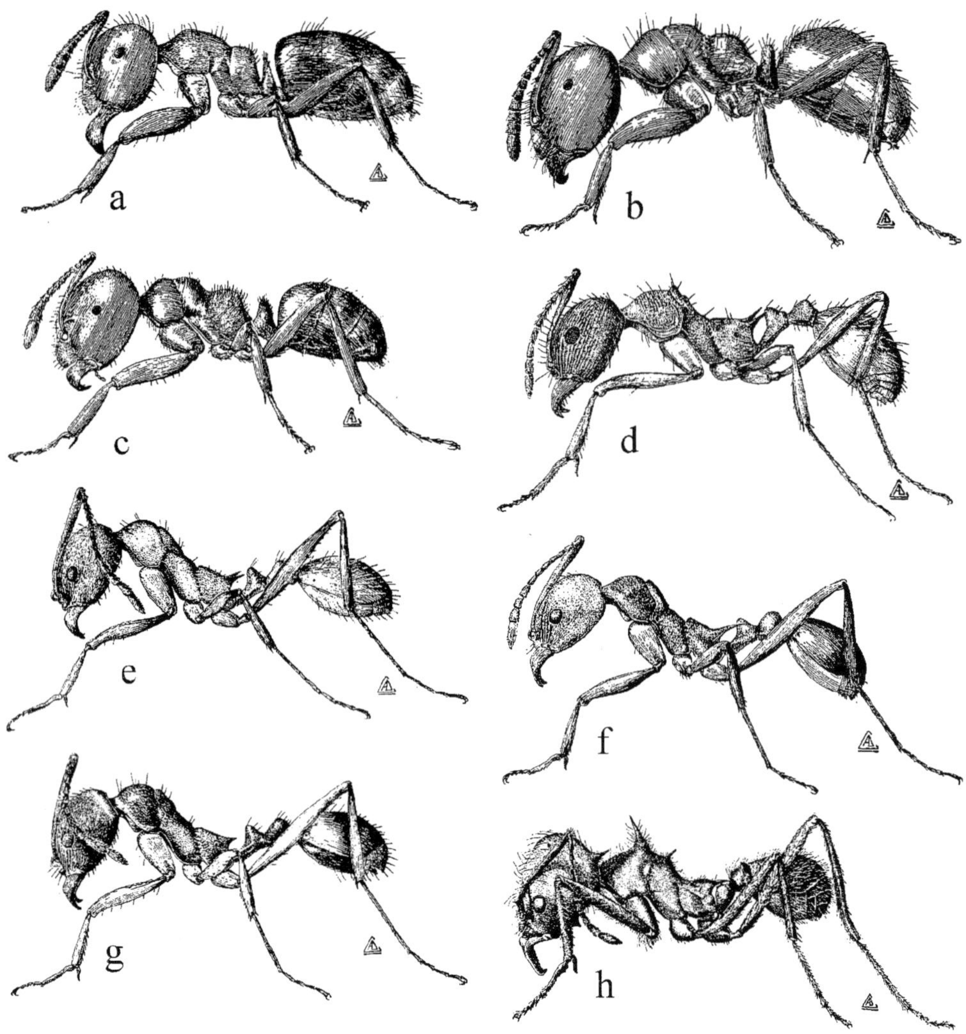


**Figure 9.2** Hymenoptera: Formicidae. (a) *Acanthomyops claviger*; (b) *A. latipes*; (c) *A. murphyi*; (d) *Aphaenogaster fulva*; (e) *A. lamellidens*; (f) *A. tennesseensis*; (g) *A. rudis*; (h) *Atta texana*.

numbers of males and females may fly from the nest at one time, and several of these flights may occur and involve 40 000–60 000 individuals. Winged forms may overwinter in nests and they are often released during warm periods in winter. This species is native to North America, and it occurs generally throughout the USA and southern Canada.

### Larger yellow ant, *Acanthomyops interjectus* (Fig. 9.3a, b)

Workers are 4–4.5 mm long and yellowish brown to brown. The scape extends beyond the posterior margin of the head. The dorsum of the gaster has long setae, which are largely confined to the first gastric segment, and to the posterior border of the succeeding segments. Nests are usually in exposed soil or under the cover of stones or logs. In open areas, the nests are in mounds. In the urban environment, nests may be around the perimeter and along the foundation of buildings, and under concrete slabs. Colony activity, such as nest construction and foraging, is primarily at night. Winged forms emerge from nests from March to July, and these individuals may have been produced in the colony the previous year. Indoors, swarming may occur from late fall to early spring. Natural food is honeydew produced by root-feeding homopterans and other subterranean insects. Indoors they forage for household foods. This species is native to North America, and it occurs generally throughout the USA and southern Canada.

**_Acanthomyops latipes_ (Fig. 9.2b)** Workers are 3.5–3.7 mm long and brown to dark brown. The scape extends to or slightly beyond the posterior border of the head; antennal segments 6–11 are enlarged. The pedicel is thick and blunt at the apex.

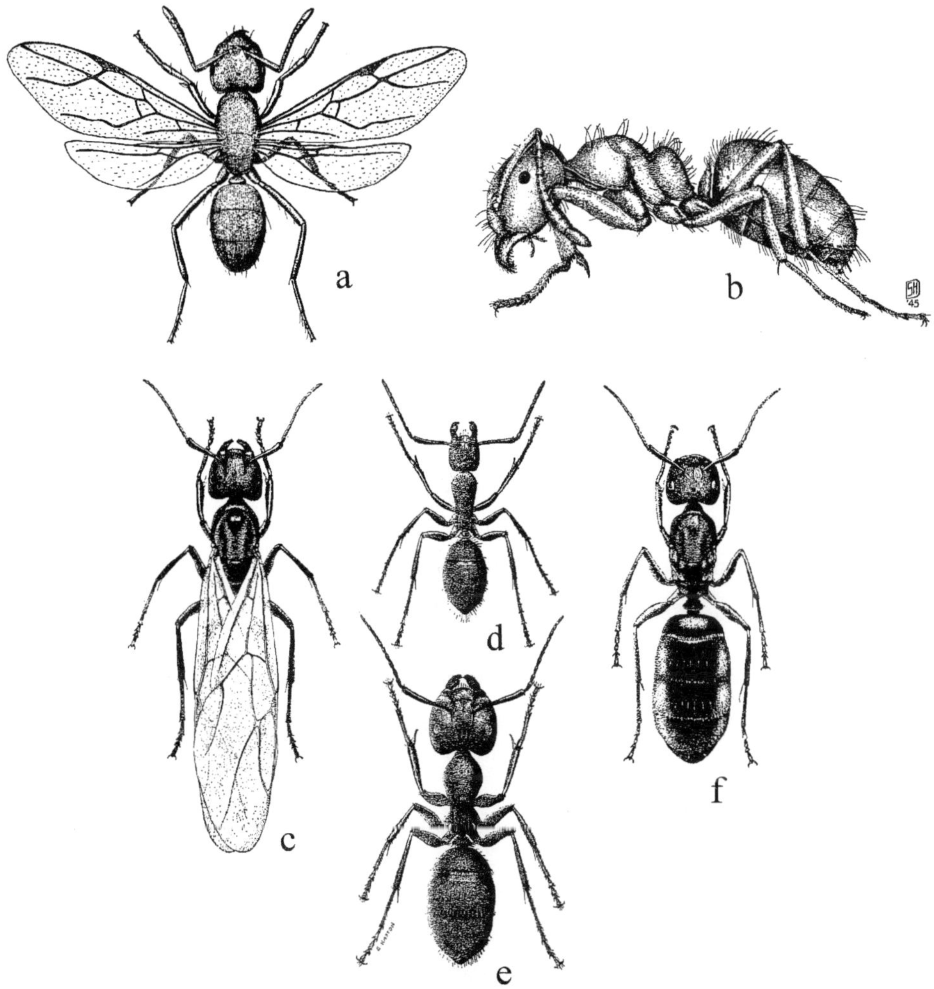


**Figure 9.3** Hymenoptera: Formicidae. (a) *Acanthomyops interjectus*, winged queen; (b) *A. interjectus*, worker; (c) *Camponotus pennsylvanicus*, winged queen; (d) *C. pennsylvanicus*, minor worker; (e) *C. pennsylvanicus*, major worker; (f) *C. pennsylvanicus*, dealate queen.

Setae on the body are long, especially on the gaster; setae are present on the cheeks, gula, and the pedicel. Nests are in open woodlands, grassy fields, and meadows, and usually in dry soils. In the urban environment, nests are often around building foundations, but it is infrequently found indoors. In open areas, nests are in earthen mounds that are 30–45 cm diameter and 2–8 cm high. Mounds have vegetation growing from them and a large number of entry and exit holes. Natural food is honeydew of subterranean aphids and other homopterans that are feeding on the roots of wild and domesticated plants. Winged forms are produced in the colony in summer, and they emerge in August and September. Some of the males and winged females may remain in the nest over the winter and emerge the following spring. New colonies are often formed by a fertilized queen becoming a temporary parasite in the nests of another species of ant, and most likely one of the *Lasius* species. This species is native to North America, and it ranges from Alaska and Quebec, Canada, south to New Mexico and South Carolina.

***Acanthomyops murphyi* (Fig. 9.2c)** Workers are 3–3.7 mm long and yellowish brown to brown. The eye has about eight ommatidia at its greatest diameter, which is about 0.10 mm. Fine setae are abundant on the epinotum and pedicel. The scape extends to the posterior margin of the head or slightly beyond. The apex of the petiole is blunt. Nests are in open woods and below stones and other objects on the ground. In urban areas, the nest may occur infrequently around the perimeter of buildings. Natural food is honeydew from subterranean aphids and mealybugs that are feeding on plant roots. Although this species may nest around buildings, there are no records of them foraging indoors. Winged forms emerge from the nest in June and July; these individuals may be produced in the colony the previous year and overwintered in the nest. *A. murphyi* is native to North America, and in the USA it ranges from Montana to Ontario, Canada, south to Colorado and Georgia. It is rare in all parts of its range.

*Aphaenogaster*

This large genus has several species that are common in urban habitats. Workers are 4–6.5 mm long and slender; it has long legs and 12-segmented antennae with an indistinct four-segmented club. The epinotum has prominent spines and the pedicel is two-segmented. These species usually have solitary foragers in leaf litter above ground and take a variety of foods. They nest in decayed wood, under stones, and in soil, where they often make an elevated area where they have deposited soil from the nest excavation. Nests have about 3500 workers and several queens.

**Aphaenogaster fulva (Fig. 9.2d)** Workers are 3.5–5.8 mm long and the body is pale brown to dark brown. The gaster and appendages are dark brown. The mesonotum has the anterior border strongly produced to form a transverse swelling, which may be depressed in the center. Epinotal spines are long and pointed upwards. Nests are in rotting wood, such as logs or stumps, and in the soil beneath stones and other objects. They may nest around the foundation of buildings and houses. Natural food is live and dead insects; they are not known to eat honeydew. This species is distributed in eastern USA.

**Aphaenogaster lamellidens (Fig. 9.2e)** Workers are 4.1–6.5 mm long; the head, thorax, and petiole are reddish brown. The gaster is yellowish brown, and antennal segment 1, femora, and tibiae are dark brown to blackish brown. The outer face of the frontal lobe has a flange, which projects posteriorly in the form of a tooth. Colonies have 500–3000 workers and they are established in wood and in structural sites. Natural food includes live and dead insects; they are not known to eat honeydew. This species is native to North America and occurs from Illinois to New York and south to Louisiana and Florida.

**Aphaenogaster tennesseenis (Fig. 9.2f)** Workers are 4.1–5.3 mm long, and the head, thorax, and petiole are reddish brown. The gaster is yellowish brown, and the appendages are dark brown. The mesonotum is strongly pronounced anteriorly, and the thorax is strongly sculptured. Epinotal spines are long and pointed, and there are few erect setae on the dorsal surface of the body. The female body appears polished. Nest sites include decaying logs and stumps, or in decayed sites in living or dead trees. This species is found in wooded areas. Natural food includes live and dead insects, and honeydew. This species is native to North America, and it occurs from South Dakota and Ontario, Canada, south to Oklahoma and Georgia.

**Aphaenogaster rudis (Fig. 9.2g)** Workers are 4.5–5 mm long and the body color is pale brown to dark brown, while the antennae and legs are light brown. The body is covered with erect setae. Sculpturing on the head varies from longitudinal ridges to punctate, and the pronotum may be punctate or with ridges. Epinotal spines are short, and seldom as long as the base of the epinotum. Nests are in exposed soil, leaf litter, under stones, or in logs or decaying wood. Colonies have 500–3500 individuals. There is one functional queen, but there may be 2–15 other females present, perhaps the result of the fusion of several colonies. Immature stages of this species pass the winter as eggs and larvae. Natural food includes seeds, pollen of ground-nesting bees, and other insects, which are killed and eaten. This species is native to North America, and it occurs from Illinois to Massachusetts, south to Colorado and Florida.

*Atta*

These are leaf-cutting ants. The reddish-brown workers are 1.5–12 mm long; the legs are long. Their body is usually densely punctate, and covered with fine setae. The head has distinct spines, and the antennae are 11-segmented and without a distinct club. The thorax has three prominent spines, the anterior pair being the largest; the pedicel is two-segmented, and the gaster is rounded. Nests are below ground and usually in well-drained sandy or sandy-loam soils. The workers cut pieces of leaves and usually carry them above their head back to the nest in long columns. In the nest, the leaves are made into a substrate for growing fungi; these ants feed on fungi growing in the nest. A mature colony contains thousands of individuals, including a large-headed soldier caste.

**Texas leaf-cutting ant, *Atta texana* (Fig. 9.2h)** Workers are 4–14 mm long and they are extremely polymorphic, with some workers being very large. The body color is dull brown to reddish or orange brown, and the body is densely punctate. The head is strongly bilobed and the mandibles are large and flattened with numerous teeth. Each occipital lobe has two spines: a large one at the posterior corner of the head, and a small one anterior to it. The thorax has three pairs of prominent spines, with the anterior ones the largest. The legs are very long. Nests are in the ground and usually in well-drained sandy or loamy soils. Colonies may have several functional queens. Winged reproductives emerge from April to June. Flights from the nest usually occur before dawn on still, moonless nights; the exit of males and females from the colony may last for several weeks.

The interior of the nest may extend 6 m, and it may occupy 420 $m^2$, and have 1000 entrance holes. Colonies may occupy the same nest site for more than 50 years. The interior of the nest contains chambers for growing a fungus on macerated plant material, primarily leaves. Adults and larvae eat the fungus. Minor workers cultivate the fungus and care for the brood larvae, medium-sized workers cut and transport the leaves and other plant material, and large workers guard the nest. Workers forage 90–180 m from the nest site and make conspicuous paths. During summer, they forage at night, but in other seasons they forage during the day. Workers also collect floral parts, seeds, caterpillar droppings, and other plant material. Indoors they forage for rice, meat, sugar, and bread. This species is native to North America, and it occurs in eastern Texas and western Louisiana, and south into Mexico.

### *Camponotus*

This is the largest ant genus, with about 1000 species worldwide, distributed in tropical and temperate regions. They are called carpenter ants because of the habit of some species to excavate nest galleries in wood. Carpenter ants do not eat wood. Many species in this genus do not nest in wood, but live in soil or plants. In North America, there are 23 species of *Camponotus* that are considered structural or nuisance pests, but only seven of these species cause severe damage to structural timbers. The others invade houses and other buildings and forage for food; they excavate wood, but the damage is limited. Workers are generalized scavengers on a variety of animal materials in natural and domestic habitats. They are typically established in small or large colonies, and sometimes colonies are divided into a parent and several satellite nests.

Pest status of carpenter ants is based on the damage by some species to live in standing trees in managed forests in the natural environment, and the structural damage and household nuisance of the several species in the urban environment. In north temperate regions, such as northern USA and southern Canada, carpenter ants replace subterranean termites as the most important wood-infesting insect in the urban environment. *Camponotus* infestations cause millions of dollars in repair, replacement, and control costs in the USA and Canada. In Europe, *C. herculeanus* causes the most damage to structures, and the value may be comparable to the amounts in the USA.

Adult *Camponotus* are 3–14 mm long, and may be all black, black and red, or reddish brown. Antennae are 12-segmented and without a club. The antennal scape is inserted some distance from the clypeal border, and this feature distinguishes the *Camponotus* from *Lasius* and *Formica*. Maxillary palps are six-segmented, labial palps are four-segmented, and ocelli are absent. The epinotum lacks spines and the abdominal pedicel is one-segmented. Large workers have heads disproportionately larger than those of small workers; these large, big-headed individuals are called major workers, and the small ones are called minor workers. The proportion of large workers (majors) in the colony is an indication of its age. The percentage of major workers is high in old colonies and low in young colonies.

Nests are in live or dead trees, in rotting logs or stumps, or in sound or moisture-damaged structural wood. Some species nest in soil beneath logs and leaf litter, such as *C. floridanus*. Workers excavate wood with their large mandibles; the fibrous shavings are removed and pushed out of the galleries through the opening, or packed into unused tunnels. Galleries are irregular in shape and are generally excavated in the direction of the wood grain, and in the sapwood portion of the wood. Walls of the galleries are smooth and have a sandpapered appearance. Part of nests may occur in soil, but these often serve as incubation sites for brood, and the actual nest may be adjacent to these soil chambers in roots or wood. Trees utilized for nest sites in Europe and Canada include the softwood species of fir *Abies* and spruce (*Picea*); in the Pacific northwest of the USA, western red cedar and Douglas fir are most often attacked by *C. modoc*. The main colony is usually in the base of the tree, and galleries extend into the roots and up the trunk.

Colonies are usually large and divided into small satellites and a main nest site with the parent colony. The parent colony is usually in a humid environment and contains the functional queens and developing brood, including the eggs and small larvae. In the satellites there are only workers, full-grown larvae, pupae, and sometimes winged reproductives; satellite colonies may be located in dry and warm locations. Colonies maintain contact with one another by movement of brood and workers along interconnecting pathways, which may be on the surface or in the soil. During winter, contact between the colony segments usually stops, but is re-established in spring. The number of satellite nests per colony is variable, and probably linked to species, environmental conditions, and colony age. Colonies of *C. herculeanus* and *C. noveboracensis* may have 8–13 satellites, 10 satellites for *C. floridanus*, while five satellites have been reported for *C. modoc* and *C. vicinus*.

The size and age of the colony when winged reproductives are produced are variable, and these features are difficult to determine. Estimates of the total number of workers in a colony when reproductives are produced are: 12 000–50 000 for *C. herculeanus*, 8900 for *C. noveboracensis*, 9000–12 000 for *C. modoc*, 3018–10 243 for *C. pennsylvanicus*, 2631 for *C. chromaiodes* (as *C. ferrugineus*), 150–8100 for *C. floridanus*, and 50 000 for *C. vicinus*. Estimates of colony age when reproductives are produced depend on the development rate of the colony, and whether there are multiple queens. In multiple-queen or polygynous colonies, overall growth is accelerated and reproductives may be produced sooner than in single-queen colonies. Reproductives are produced in *C. modoc* and *C. vicinus* colonies after a minimum of 10 years, and for *C. pennsylvanicus* it may take 3–5 years. Winged forms produced the preceding summer and overwintered in the nest usually leave to mate in the spring. There may be several mating flights, some with a large number of individuals and some with only a small number. They usually occur in the late afternoon of sunny, warm, and humid days.

Following mating, males die and queens seek a sheltered site to deposit a batch of 9–16 eggs over a period of 2–3 days; hatching is in 2–3 weeks. Small larvae are covered in fine setae with hooked tips. When larvae are stacked, the hooked setae interlock and clusters of larvae can be moved together by the queen or workers. The queen and the first batch of larvae are nourished by the metabolism of her fat bodies and flight muscles. Workers produced from the first brood are small (minor workers) and show little size variation. Colony growth with a founding queen is slow. When the first brood enters the prepupal stage in late summer, eggs for the second brood are laid. In fall, colony activities decrease and the parent and satellite segments of the colony enter a dormant phase to overwinter. After the first year, two periods of oviposition and larval development occur in colonies. *Camponotus* workers are polymorphic, and the food supply determines worker size. Major workers are not produced until the third season in colonies of *C. pennsylvanicus*, *C. chromaiodes*, and *C. modoc*.

Natural food includes honeydew, plant exudates, and live and dead insects and other arthropods. They also scavenge on the carcasses of dead vertebrates. Indoors they forage for sweets and high-protein foods. During spring and early summer, when brood production is at a peak, the workers prefer proteins, which are fed to developing larvae. In late summer and fall, workers prefer carbohydrates, which are used as energy sources by adults. In nonforested areas, carpenter ants follow chemical and physical trails on the ground. In forested areas, they often construct extensive tunnel systems through the soil litter. The majority of foraging is done at night. The peak foraging activity period for *C. modoc* is 20:00–4:00 h in the Pacific northwest. In California, foragers are active all day from May to October, but switch to nocturnal foraging in July and August. Foraging by *C. pennsylvanicus* may be influenced by ambient humidity and temperature may be important in June and August, whereas time of day is a determining factor in July.

Most species of carpenter ants are nocturnal foragers and rely on chemical trails for orientation to and from the nest to a food source. A trail pheromone, produced in the hindgut, is deposited by touching the tip of the abdomen to the ground as the workers move along the established trail. They also use a light compass and orient to the sun, the moon, or street lights at night. They also orient to landmarks such as trees, bushes, and rocks; the silhouette of a tree against the night sky may be sufficient to determine the way to the nest. Indoors, carpenter ants orient to artificial pathways formed by electrical wires and water pipes. On the outside of structures, they often travel along the branches of trees that contact the roof or siding.

Infestations indoors are usually satellite nests and they are connected to a main colony which is located outdoors in a living or partially dead tree, woodpile, or some other wood structure. Primary or satellite nests may be in roof timbers and remain unnoticed for many years because workers forage outdoors and primarily in the evening. However, nests may be established in wall voids, and the ants observed or heard during the day or evening, or winged swarmers observed at windows. The most frequent locations of nest sites in structures include exterior and interior wall voids, in the attic or other roof supports, in subflooring, in ceiling joists, and the sill plate or foundation piers. Chewing sounds made by large-headed workers tunneling in the wood and the rustling sounds of numerous workers moving in galleries can be heard 2–3 m away, especially at night when houses may be quiet and the ants are active. When disturbed by a knock on the wall near the nest, workers of *C. modoc* will often respond by striking walls of their tunnels with their mandibles and gasters, apparently to alert the colony.

The fibrous pieces of wood (frass) resulting from carpenter ant tunneling in wood are concentrated in unused galleries or mixed with other colony debris and pushed out of the nest through small openings (windows) to form small piles. Some colonies deposit frass several meters from the nest site.

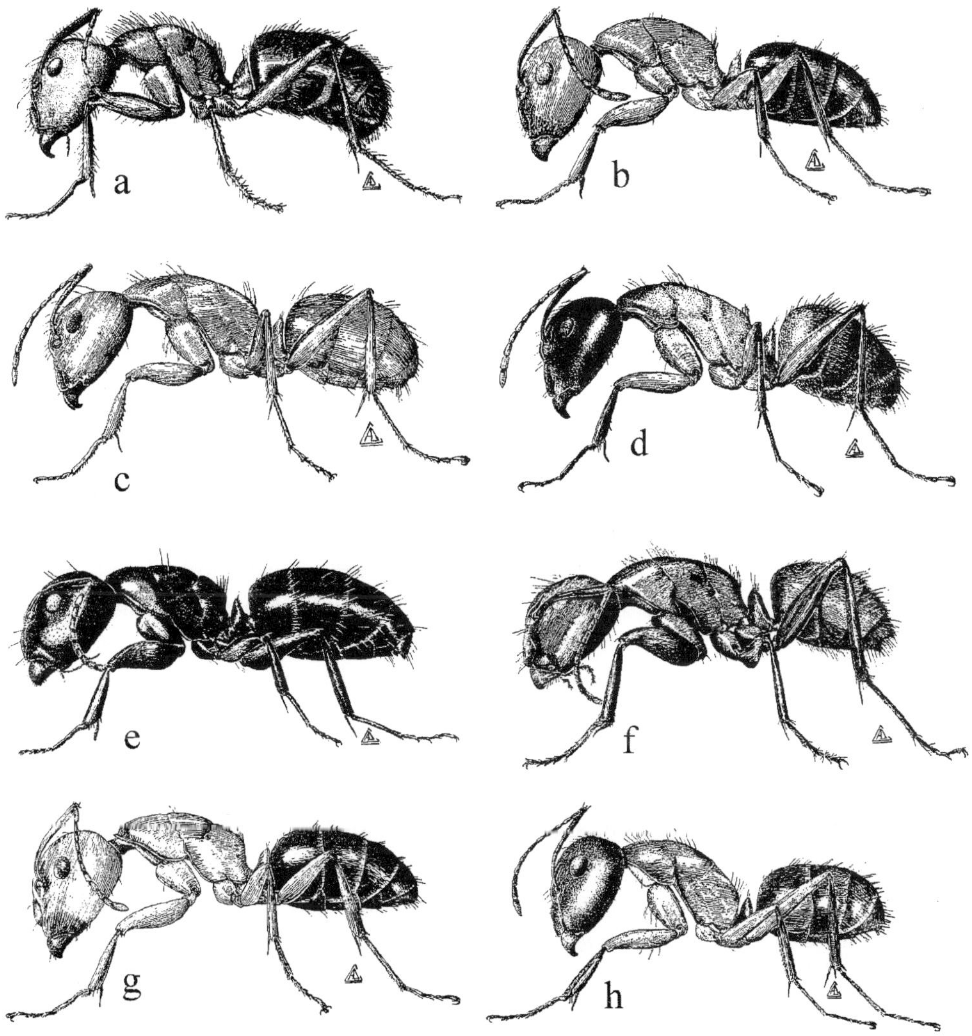


**Figure 9.4** Hymenoptera: Formicidae. (a) *Camponotus floridanus*; (b) *C. caryae*; (c) *C. castaneus*; (d) *C. chromaoides*; (e) *C. nearcticus*; (f) *C. pennsylvanicus*, (g) *C. marginatus decipiens*; (h) *C. tortuganus*.

The fibrous pieces of wood removed by the workers are about 3–5 mm long and the color depends on the condition of the wood. Tunneling in sound wood produces frass uniform in color; frass from moisture-damaged wood may be a mix of dark and light pieces. Carpenter ant frass also contains ant feces, small pieces of soil and gravel, and leftover pieces of food, such as seed coats and uneaten parts of insects, and parts of dead carpenter ants.

***Camponotus castaneus* (Fig. 9.4c)** Workers are 7–10 mm long, the body is shiny, and the color ranges from yellowish brown to yellowish red. The head and gaster are usually darker than the thorax. The head is as long as or longer than wide. Cheeks are without erect setae. The scape extends beyond the posterior margin of the head. The middle and hind tibia each have a row of graduated bristles. Colonies have 200–1000 individuals, and probably one functional queen. Workers forage at night, and occur indoors. This species is native to North America, and occurs from Iowa to New York, south to Texas and Florida.

**Red carpenter ant, *Camponotus chromaiodes* (= *C. ferrugineus*) (Fig. 9.4d)** Workers are 6–13 mm long; the head and most of the gaster are blackish brown to black. The thorax, pedicel, and base of the gaster are reddish brown. Setae on the gaster are golden or pale yellow. Near the apex of each middle and hind tibia there is a short row of graduated bristles. Nests are usually in wooded areas and established in dead standing trees, rotted logs, and stumps, and the galleries often extend into the soil. Indoor nests are excavated in moisture-damaged wood. Colonies have one queen and about 3000 workers when mature, producing winged females and males. Natural food is live and dead insects, and honeydew; they are not common indoors. This species is native to North America, and occurs from Nebraska to New York, south to Georgia. In 1798, Fabricius described this ant as *Formica ferruginea*. However, that name is a homonym of *F. ferruginea*, which was described earlier (1791). This species was moved to *Camponotus* by Roger in 1863, and the replacement name, *C. chromaiodes*, was provided by Bolton in 1995.

***Camponotus marginatus decipiens* (= *C. rasilis*) (Fig. 9.4g)** Workers are 4–9 mm long, the head, thorax, and pedicel are yellowish red, and the gaster is black or blackish brown. Gaster segment 1 is lighter in color than the remainder; the gaster is smooth and shiny. The clypeus and cheeks have shallow grooves, and the grooves on the cheeks lack short, erect setae. This species frequently nests in structural wood in houses. It is native to North America, and occurs from Nebraska south to Texas, South Carolina, and Florida. It is common in the Gulf Coast states.

***Camponotus discolor*** Workers are 3.5–7.5 mm long, and the head, thorax, and pedicel are yellowish red or reddish brown; the gaster is blackish brown to black. The anterior portion of the head has distinct punctures. Colonies are established in structural wood, but they are small and may not cause significant damage. There are no records of this species foraging indoors. It is a native species, and occurs from Kansas and Iowa to Ohio, south to Texas and Florida.

**Florida carpenter ant, bulldog ant, *Camponotus floridanus* (= *C. abdominalis floridanus*) (Fig. 9.4a)** Workers are 5.5–10 mm long and the head is opaque, but the thorax and gaster are shiny. The head is reddish brown, the thorax and pedicel are yellowish red, while the scape and gaster are blackish brown to black. Gaster segment 1 may be yellowish red. Legs have numerous erect or suberect setae, and the body has abundant, long setae. Nests are in rotting logs and stumps, and sometimes partly in the ground. Indoors the nests may be in kitchens, roofs, and porches. Colonies are moderate to large, and there is one queen. Winged forms emerge from June to August. Natural food includes live and dead insects, honey bee hives, and honeydew; workers forage during the day and night. Indoors they feed on a variety of household foods. This species is native to North America, and occurs from Alabama east to North Carolina and Florida; it apparently does not occur far inland. The small wasp, *Obeza floridana* (Eucharitidae) is a parasite of this carpenter ant.

**Northern carpenter ant, *Camponotus herculeanus*** Workers are 5–15 mm long, the head and gaster are dull black; the thorax is reddish brown to reddish black. The scape does not have erect setae, except for a small cluster at the tip; the apex of the scape reaches or slightly surpasses the posterior border of the head. Nests in natural areas are in standing trees, fallen logs, and tree stumps. In the urban environment nests occur in moisture-damaged and sound structural wood. Colonies are in damp or decaying wood; mature colonies have 12 000–50 000 workers. Egg-laying begins soon after the emergence of winged females and males, which is in May and early June. Eggs are present in the colony between July and September, and are probably laid continuously throughout summer. Larval development is during the summer; the main period of larval growth is June and July. Pupae are present in the colony from the end of June to the end of August, and the pupal period lasts 25–40 days. There may be little larval growth after July, although workers actively forage until November. Foraging during this time may be to build body food reserves, such as carbohydrates and lipids, for overwintering. The larval life cycle in some regions may require 2 years, with larvae spending two winters in the nest. Winged forms emerge during late summer, at the time larval growth stops. The winged forms overwinter in the nest and leave to mate in spring. Natural food is primarily honeydew, plant sap, and live and dead insects. Indoors they feed on sweets, raw and cooked meats, fruit, and other materials. This species occurs in North America, Europe, and Asia.

***Camponotus ligniperda*** Workers are 6–14 mm long; the thorax is bright yellowish red to brownish red. The gaster is shiny black with short setae; the base of the gaster is sometimes reddish brown. Nests are usually along borders of woodlands, under stones, or in dry stumps. In the urban environment nests may occur in structures. Natural food includes live and dead insects, and honeydew. Workers are aggressive and will freely bite and attack other *Camponotus* or *Formica* species. The large *C. ligniperda* workers can cut through the thorax or crush the head of their opponent with their strong mandibles. This species occurs from central Spain to western Russia and from Sweden south to Sicily.

***Camponotus maccooki*** Workers are 6–12 mm long, and the body is yellowish brown or amber-colored. The head is blackish brown and the gaster is partially or wholly blackish brown. This species is native to Mexico, but occurs in New Mexico, Arizona, California, and Colorado. It is a household pest in New Mexico.

**Spotted sugar ant, *Camponotus maculatus*** Major workers are about 9–12 mm long; the head and gaster are black; the abdomen has pale brown to brownish white markings. Minor workers are about 8 mm long and pale brown, and with pale brown markings on the abdomen. Nests are in the soil under stones or logs, and the nest entrance is often circled by loose soil. This species is nocturnal, and it is a household pest in Africa. In Saudi Arabia, it is the dominant ant species in natural, vegetable, and date palm habitats.

***Camponotus modoc* (= *C. herculeanus* subsp. *modoc*)** Workers are 8–14 mm long and black. The gaster is subobaque, except for a narrow band at the posterior edge of each segment; the anterior margin of the clypeus lacks a medial depression; the scape does not have erect hairs, and there are coarse and dense setae on the gaster. One queen usually founds a colony. They are usually established in damp or decaying wood; mature colonies contain 9000–12 000 workers. Egg laying begins soon after the major emergence of winged females and males, which is in April and extends to June. Eggs hatch in 2–5 weeks. Larval development is during the summer, pupae are present in the colony in July, and most of the workers have emerged by August and October. Egg production stops in August and September, and the larvae present in the colony overwinter with the queen. No food is consumed by the colony during the colony dormant phase, which is October through January. Egg production resumes and larvae develop into pupae in January and February, during the second year of a new colony. Cannibalism occurs as eggs hatch. Both eggs and first-stage larvae are consumed by the queen and workers, especially at the end of the dormant phase. Some of this food is fed to developing larvae. This species is native to North America, and occurs in the Pacific northwest. Nests in natural areas are in trees and tree stumps; in the urban environment nests occur in structural wood.

Foraging activity begins in May and ends in September, with the peak in July and August. Foraging is primarily nocturnal and the greatest activity is at 20:00–24:00 h and 24:00–4:00 h. Most of the food returned to the nest is liquid and carried by workers in their crop. Workers are unable to ingest solid particles in excess of 100 μm, and the majority of water-soluble proteins are imbibed in the field and brought back to the nest as liquids. The most common prey are crickets, grasshoppers, aphids, spiders, fly larvae, and caterpillars. In the domestic and peridomestic habitats, the most common food carried to the nest includes candy, syrup, apples, raisins, and pet food. Minor workers comprise the largest percentage of foragers. The major workers are involved in foraging early and late in the active phase of the colony, which is May through September.

***Camponotus nearcticus* (Fig. 9.4e)** Workers are 4.5–7.5 mm long and the body is shiny and black. Legs do not have setae, and the pedicel is narrow in profile. Colonies have 100–300 individuals, and nests are often in trees. Nests in houses are in roofing woodwork, and usually in wood with moisture damage. They seldom forage indoors. This species is native to North America and ranges from North Dakota to Ontario, Canada south to Colorado and Florida.

***Camponotus noveboracensis*** Workers are 6–13 mm long. The head and gaster are black; the thorax is reddish brown to yellowish red. Nests are usually in wooded areas, in deciduous and pine forests. Moist to wet dead wood is used as a nest site, and sometimes nests are located in soil adjacent to wood. Nests are sometimes made in structures, and workers forage indoors. Natural food includes live and dead insects, and honeydew. This species is native to North America, mainly in northern USA and southern Canada.

**Black carpenter ant, *Camponotus pennsylvanicus* (Fig. 9.3c–f; 9.4f)** Workers are 6–13 mm long and the body color is typically black; some individuals may have reddish-black thoracic pleuron, pedicel, and legs. Body setae are yellowish brown and moderately abundant, especially on the dorsum of the gaster. The scape extends past the posterior border of the head, except in the largest workers. The pronotum is flattened, and noticeably so in large workers. Nests are usually in live and dead standing trees, and in rotting logs and stumps. Indoors they frequently nest in moisture-damaged wood, but they also tunnel into sound wood. Colonies typically have one queen, and when the colony has 2000 or more workers it produces winged reproductives. It takes 3–5 years to achieve the size necessary to produce winged forms. Indoors where colonies remain active and forage for longer periods in the fall, the time needed to achieve this size may be reduced. Mature colonies produce males and winged females yearly. The number of ants in the colony engaged in foraging depends on colony size, and may range from 364 to 1329 ants. Outdoor and indoor colonies are active during the warm season, usually beginning in April and extending to November in some regions; they are inactive from about December to April. In winter, nests may contain up to 500 winged females and up to 8000 larvae.

Natural food is primarily live and dead insects, honeydew, the juices of fruit, and plant sap. Indoors they feed on sweets, raw and cooked meats, fruit, and various other materials. Depending on the food available, *C. pennsylvanicus* forms ground trails 2–3 cm wide, from the nest to relatively stable food sources, such as honeydew producers or indoors. Foraging is primarily at night, and usually peaks soon after sunset and extends into the night, and peaks again before sunrise; daytime foraging may occur. Foraging activity is generally influenced by temperature. This species is native to North America, and occurs from North Dakota to Quebec, Canada, and south to Texas and Florida.

***Camponotus tortuganus* (= *Camponotus maculatus* subsp. *tortuganus*) (Fig. 9.4h)** Workers are 6–11 mm long, and the body color is reddish brown. The head is darker than the thorax, and the gaster is dark brown to blackish brown, but there is much variation. The anterior margin of the clypeus extends forward as a prominent lobe, and the clypeus has a well-defined ridge. Tibiae of all legs lack erect setae, and the middle and hind tibia lack a row of bristles. Body setae are long and yellowish brown. Nests in natural habitats are usually in rotting wood and in the soil beneath stones. Indoors, nests are in moisture-damaged wood and persist for many years. This species is native to North America, and occurs in the southern half of Florida, including the Florida Keys, and Dry Tortugas.

***Camponotus vagus*** Workers are 6–12 mm long. The body is uniformly black, and with numerous setae; there are numerous setae on the gaster. Nests in natural habitats are in dry rotten wood among roots, and under stones in dry and sunny locations. Indoors, nests are in moisture-damaged wood. Natural food includes live and dead insects, and honeydew. Winged forms usually occur in July in central Europe. It is aggressive and workers will bite freely when nests are disturbed. This species is distributed in southern Europe, and is abundant in the Mediterranean area; it occurs from Portugal to southern Russia, and from Poland to the mountains of North Africa.

***Camponotus vicinus*** Workers are 6–14 mm long. The head is dull black to reddish black, the thorax is brownish red with black markings, while the gaster is shiny black. Nests in natural habitats are usually in dry and sunny areas, under stones and sometimes in dry decayed logs. Colonies are composed of as many as 100 000 workers, with multiple queens. This species is an occasional pest indoors. Natural food includes other insects and honeydew. It is native to North America and occurs from North and South Dakota west to the Pacific coast.

**Other *Camponotus*** The workers of the giant carpenter ant, *C. laevigatus*, are 8–10 mm long and shiny black, and this species is a nuisance in the Pacific coast and Rocky Mountain states. In Europe, the species causing structural damage include *C. herculeanus*, *C. ligniperda*, and *C. vagus*. The Hawaiian carpenter ant, *C. variegatus*, is pale brown and winged queens are about 1.2 cm long. Winged adults may be confused with swarming termites.

### *Crematogaster*

Workers are 2.5–4.0 mm long, and distinguished by their broad head and heart-shaped gaster. They usually have a reddish-brown head and dark brown to black abdomen. The upper surface of the gaster is flat, the ventral surface is curved, and it is distinctly pointed posteriorly. Antennae are 11-segmented with a three-segmented club. The epinotum has a pair of spines, and there is a two-segmented pedicel. Nests in natural habitats are above ground in posts and stumps; they do not nest in soil. Indoors, nests are in moisture-damaged structural wood. Colonies contain 2000–3000 workers. These ants are general feeders and are abundant in peridomestic habitats. The common name, acrobat ant and cocktail ant, is derived from their habit of raising or cocking their gaster over the head and thorax when disturbed.

***Crematogaster ashmeadi* (Fig. 9.5a)** Workers are 2.6–3.2 mm long. The body color is variable, ranging from reddish brown to blackish brown for the head and thorax. The head is shiny, and the dorsal surface of the thorax is finely punctate, but appears shiny; the side of the thorax is distinctly punctate. Epinotal spines are normally shorter than the distance between the base of the spines. Erect setae are sparse on the body, especially the thorax. Nests are usually above ground, in cavities in trees, logs, and stumps. Indoors they nest in structural wood, such as rafters, and posts. Natural food includes live and dead insects and honeydew. Indoors they are omnivorous, but prefer sweets and meat. This species is native to North America and occurs from Texas east to Virginia and south to Florida.

***Crematogaster bicolor* (= *C. laeviuscula*, *C. clara*) (Fig. 9.5c)** Workers are 3.3–4 mm long and yellowish brown to brown.

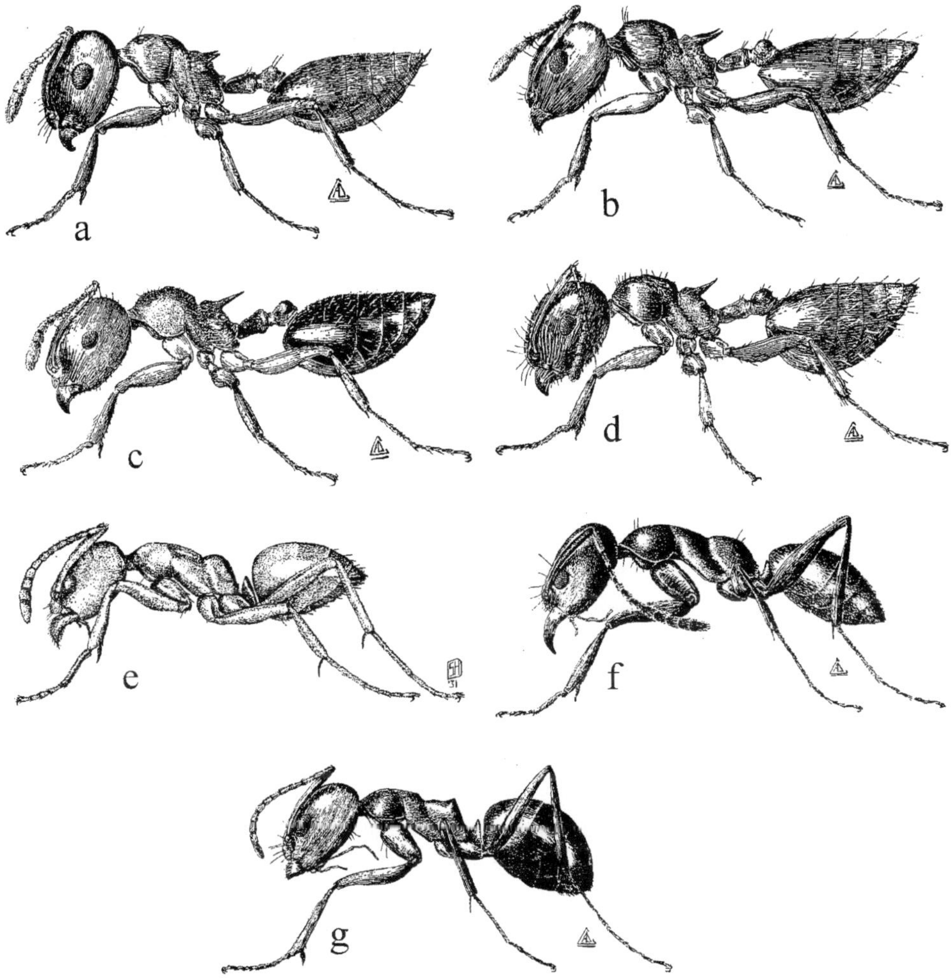


**Figure 9.5** Hymenoptera: Formicidae. (a) *Crematogaster ashmeadi*; (b) *C. cerasi*; (c) *C. laeviuscula*; (d) *C. lineolata*; (e) *Linepithema humile*; (f) *Forelius pruinosus*; (g) *Dorymyrmex insanus*.

The head is shiny except for sculpturing on the cheeks and clypeus. The scape extends beyond the posterior border of the head. The mesoepinotal impression is strong, and the sides and dorsum of the thorax have fine sculpturing. Epinotal spines are long and divergent, straight, or occasionally curved. There are long, erect setae on the head and anterior part of the thorax. Nests in natural habitats are tree branches, rotten stumps, and logs. Indoors they nest in wood in houses and feed on sweets and meat. Natural food is honeydew, and live and dead insects. This species is native to North America and occurs from Indiana to New Jersey and south to Texas and Florida.

***Crematogaster cerasi* (Fig. 9.5b)** Workers are 2.6–4 mm long and the head and thorax are reddish brown to almost black. The scape extends beyond the posterior border of the head. The mesoepinotal impression is strongly defined, constricting the thorax. Epinotal spines are divergent, long and straight, or occasionally curved. The sides of the thorax are sculptured, and slightly opaque. Erect setae on thorax are normally confined to a small cluster on each pronotal shoulder. Nests in natural habitats occur in the ground beneath objects, and in rotting stumps and logs; habitats include open fields, pastures, marshes, and woods. Indoor nests are in wood in roof, siding, ceiling, and porch, and commonly in door and window frames. Nests may be in cavities in wood made by other insects. Winged forms emerge in June to November. Workers in large colonies may be aggressive and when disturbed emit an objectionable odor. They tend honeydew-producing insects, and feed on live and dead insects. This species is native to North America and occurs from southern Canada through the eastern Rocky Mountains to New Mexico and east to Georgia.

**Acrobat ant, *Crematogaster lineolata* (Fig. 9.5d)** Workers are 2.5–3.5 mm long and yellowish brown to brown to blackish brown. The vertex of the head and around the eyes may be punctate. The scape extends beyond the posterior border of the head. The promesonotum with longitudinal grooves and with punctures, and sides of thorax are sculptured. Erect setae on the dorsum of the thorax are coarse or bristlelike. Epinotal spines are long and divergent. The pubescence on the body is short and close to the body. Nests are in soil or in logs and stumps of dead trees; they may nest in burrows of wood-infesting wasps. Workers bite aggressively, and give off an odor when alarmed. Winged forms emerge from June to September. Indoors they nest in woodwork. Natural food is honeydew, and live and dead insects; indoors they feed on sweets and meat. This species is native to North America, and occurs from southern Canada south along the Rocky Mountains and east to Florida.

**Crematogaster scutellaris** Workers are 4–5 mm long and dark brown. Antennae are 11-segmented, and the scape does not extend to the posterior border of the head. Nests in natural habitats are above ground and usually in trees and tree stumps. It is frequently introduced to northern Europe with cork from southern Europe and North Africa; it occurs in and around warehouses and cork factories in the UK.

*Dorymyrmex*

Workers are about 3 mm long and the body is uniform light brown to uniform dark brown to black. Antennae are 12-segmented and without a club. The maxillary palpus is unusually long and six-segmented. The epinotum has a posterior conical projection, but no spines. The pedicel is one-segmented, and narrow or thin in profile. Nests are usually in the open and crater-like. Anal glands produce an odor of rotten coconuts.

**Pyramid ant, *Dorymyrmex insanus* (= *D. pyramicus, Conomyrma insana*) (Fig. 9.5g)** Workers are about 3 mm long and the body color ranges from uniform light brown to uniform dark brown or blackish brown. The gaster is darker than the head and thorax. The scape extends slightly beyond the posterior border of the head; the pedicel is one-segmented. Nests are constructed in a variety of soil types in sunny open areas, rarely under an object. Nests are often located close to the nest of other ant species, including the harvesting ants *Pogomyrmex occidentalis* and *P. barbatus*. The 5–10-cm-wide nest opening is surrounded by excavated soil. Mounds may be numerous in lawns and commercial turfgrass. Colonies contain 2000–3000 individuals. Workers forage in trails and are generally fast-moving. Winged forms emerge in June, and reproductives overwinter in nests before emerging the next spring. Natural food includes honeydew, live insects, and other arthropods; indoors they forage for a variety of foods. This species is native to North America, and it is generally distributed over most of the USA, and in Mexico, South America, and the West Indies.

*Forelius*

Workers are 1.6–2.5 mm long, and the body is brown, reddish brown to black and brown. Antennae lack a club and the scape extends beyond the posterior margin of the head. This is a small genus with species closely related to *Linepithema*. Colonies in soil have 1–3 queens and are usually located in open sites. They are distributed in arid habitats in the new world. Workers are active during the day and may be considered thermophilic. Colonies are characterized by the high level of movement of the workers.

*Formica*

Workers are 2.5–8 mm long, and the body is brownish black to black, sometimes with reddish-brown head. Antennae are 12-segmented and without a club; the epinotum is without spines, and the pedicel is one-segmented. They forage during the day, and are general scavengers and predators on other insects and nectar feeders, and they collect honeydew from aphids. Some species are social parasites and young queens enter the nest of other ants and kill the resident queen. Many species build mounds over the nest, and colonies may contain 100 000 workers. In the urban environment, they nest in turfgrass and around the foundation of buildings.

**Allegheny mound ant, *Formica exectoides*** Workers are 3–6 mm long and the head is reddish brown and about as wide as long. The thorax and legs are reddish brown, and the gaster may be blackish brown. Nests are in the ground and are usually as mounds 1 m high and 2 m wide. Nests in natural habitats are usually along the edge of forested areas or in clearings. In the urban environment, nests may be along railroad tracks or in undisturbed sites in suburban areas. Worker ants remove vegetation in the area surrounding the mound. Workers are active from April to September and feed primarily on live and dead insects, and honeydew. Workers are aggressive and, if the nest or the mound is disturbed, workers will attack the intruder; they will actively bite. This species is native to North America, and occurs in northeastern states, south to Georgia and west to Iowa and Kansas.

**Formica francoeuri (= F. pilicornis)** Workers are 3–7 mm long and have a blackish-red head and gaster, and usually a red thorax. There are many erect setae on the legs, scapes, and gaster. Nests in natural areas are in soil, often under cover. In the urban environment, nests are in cracks in sidewalks and along the sides of buildings. They forage for honeydew from aphids and mealybugs on plants and trees. This species occurs in western USA.

**Common black ant, *Formica fusca*** Workers are 4.5–7.0 mm long and black; the legs are blackish brown. The gula, occiput, and mid-femora have erect setae, and there are occasionally one or two setae on the pronotum. Nests in natural habitats are in the ground, usually in banks, under stones, and in tree stumps. In the urban environment, nests may be along hedgerows and along the edge of undisturbed areas. Workers forage singly and are predators of other insects, but also feed on nectar from flowers, and aphid honeydew. Colonies are usually small, with up to about 500 workers and 1–3 queens. Winged forms are developed in June and July, and they emerge in July and early August. This species may be native to Europe, but it is nearly cosmopolitan. It occurs nearly throughout North America. It is common in southern England, Wales, and southwest Ireland, and it is local in north England and Scotland. It extends from Fennoscandia to Italy and from Portugal to Japan.

**California red-and-black field ant, *Formica moki* (= *F. occidua*)** Workers are 4–7.5 mm long, and the apex of the head is dusky brown. The thorax is red, and the gaster is silvery black. These ants move with a jerky running and stopping behavior. Nests in natural areas are in soil beneath rocks in open, unshaded areas. In the urban environment, nests are located in cracks in sidewalks, along the foundation of buildings, and at the base of trees. Natural food is honeydew from homopterans on shrubs and trees; they may forage indoors for household foods. This species is native to North America.

***Formica neoclara*** Workers are 3–6 mm long; the head is yellowish red ventrally and dark reddish brown dorsally; the thorax is reddish yellow to reddish brown, and the gaster is dark brown. Nests are not formed into a mound, but they may extend 1 m in circumference and have several nest openings. This species occurs along streets and parks, and along the foundations of buildings in western USA (Colorado). Natural food includes the honeydew of aphids and membracids, and live and dead insects. This species is native to North America and occurs from North and South Dakota to Washington, and south to New Mexico.

**Red ant, red wood ant, *Formica rufa*** Workers are 4.5–9 mm long and bicolored red and brownish black, but with a dorsal dark patch on the head and promesonotum; eyes have a few microscopic hairs. Long erect setae are abundant on the gula, clypeus, dorsum of the head, pedicel, and gaster, but never on the scape or posterior border of occiput. Nests are common in lowland woods and undisturbed areas. This species builds large mound nests with leaves and twigs; nests may be isolated or in small groups. Large colonies have 100 000–400 000 individuals and 100 or more queens. Foraging trails from a large nest may extend 100 m, and they are usually directed to aphid-infested trees. The primary food is aphid honeydew, but they also prey on other arthropods and small animals. Workers are aggressive and they will eject formic acid when disturbed. Winged forms emerge from the colony in May until early June. New nests are formed from colony splitting in the spring, but single queens may be adopted in *F. rufa* nests. This species may be native to Europe, but it is nearly cosmopolitan. It occurs in North America and continental Europe; it is common in the UK. A form of *F. rufa*, which has colonies with a single queen, occurs in continental Europe. It is sometimes referred to as *F. rufa rufa*. The workers are large and conspicuously hairy or setose; males may have large setae on the gena below the eye.

***Formica rufibaris*** Workers are 4–6 mm long and slender; they have a reddish brown head and thorax, and a blackish brown abdomen. Nests may be in forested areas, orchards, and gardens. This species occurs in western USA.

***Formica subpolita*** Workers are 3–4 mm long and shining brownish red or dark brown, and the thorax, pedicel, and legs are reddish brown. Nests are under stones in bare or grassy areas. This species is common in the San Francisco Bay area of California, but occurs throughout western USA.

*Hypoponeura*

These ants are characterized by their one-segmented maxillary palp, and the lack of ornamentation on the petiole. This genus includes a large number of species that nest below ground throughout the tropics. *Hypoponeura* species occur in southern and northern Europe.

***Hypoponeura punctatissima*** Workers are 3.5–3.8 mm long and reddish brown to dark brown. Antennae have 12 segments, which gradually increase in size to an indefinite club. The scape does not extend to the posterior border of the head. Eyes are minute and set forward, close to the mandibles. The pedicel is one-segmented. Colonies are often large, and winged forms emerge during August and September; wingless males remain in the nest. Natural food includes live and dead insects. Nest sites in Europe include rubbish dumps and sawdust heaps. Indoors they occur in heated buildings. This species is cosmopolitan, and it is widely distributed in Europe, the tropics, and subtropics.

### *Labidus*

These are legionary ants and they live in temporary nests in decayed logs and stumps, or in soil beneath stones and other objects. However, they may also nest around building foundations and under concrete floors. Workers are 3–10 mm long and brown to reddish brown. Antennae are very short, 12-segmented, and without a club; eyes are absent or very small. The body is usually smooth and shiny; the epinotum lacks spines. The pedicel is two-segmented, and usually there is a short spine on the ventral surface of gaster segment 1.

***Labidus coecus* (Fig. 9.6a)** Workers are 2.9–9.7 mm long and brown to reddish brown. The scape is short and extends to about the midpoint of the head. The head and abdomen are smooth and shiny, and the posterior half of the thorax is opaque. Nests in natural habitats are in decayed logs and stumps, in the ground under stones. In the urban environment, they nest around building foundations and under concrete slabs. Colonies are large and workers forage for arthropods, small mammals, and birds, which the ants kill and eat as carrion or refuse. Food with a high protein content is preferred, and they may enter houses when foraging. This species is native to North America, and occurs from the midwestern states south to Argentina.

### *Lasius*

These ants are common in agricultural fields and urban environments. Workers are 2–4 mm long and the body is brown, dark-brown, to almost black. Antennae are 12-segmented and without a club; the maxillary palps are six-segmented, and the terminal segments are approximately the same length. Mesoepinotal depression is distinct, with a pair of spiracles laterally. The pedicel is one-segmented and thin or narrow in profile, and it is almost hidden. Workers are capable of emitting a strong formic acid odor. These ants form large colonies in soil or under objects; each nest may contain 10 000 or more individuals. They are general scavengers and often enter buildings in search of sweets or meat.

**Cornfield ant, *Lasius alienus* (Fig. 9.6b)** Workers are 2–2.5 mm long and dark brown to blackish brown. The two terminal teeth on the mandible are not equal-sized. The scape extends by one-fifth its length beyond the posterior border of the head. The eye has about 16 ommatidia in its greatest diameter. Nests in natural habitats are under stones and in rotting logs in well-shaded woodlands, sometimes at the edges of wooded areas. Indoor nets may be in woodwork or sites within building foundations. Colonies are large and winged reproductives emerge during August and September. Natural food includes live and dead insects, nectar from flowers, and honeydew; they may tend honeydew-producing homopterans. Indoors they forage for meat and sweets. Individuals of this species are often infected with the parasitic fungus, *Laboulbenia formicarum*. This species is probably native to Europe, but occurs in North America from British Columbia to northern parts of California, and North Dakota to Nova Scotia south to Arkansas and Florida. The common name, cornfield ant, does not accurately describe its habitat, because it is not commonly found in agricultural fields or in turfgrass in the urban environment.

***Lasius brunneus*** Workers are 3.5–4.5 mm long; the head and thorax are pale reddish brown and the gaster is dark brown. The scape and tibiae do not have erect setae; ocelli are small but clearly visible. Nests occur in the interior of dead trees, but also in soil of hedgerows; indoors it may establish nests in the timbers of old houses and farm buildings. Colonies are initiated by single queens in cavities in old trees, but mated queens may return to the parent colony after the mating flight, which occurs in June and July. Natural food includes honeydew of tree aphids, including the large bark-feeding species of *Stomaphis*. This species occurs from Spain to Crimea and West Himalayas, Italy to Sweden. It also occurs in northern Europe and the UK.

***Lasius niger* var. *neoniger* (Fig. 9.6c)** Workers are 2–2.5 mm long and yellowish brown, rarely dark brown. The scape and tibiae have numerous long setae. The penultimate basal tooth of the mandible is smaller than the two flanking teeth. Nests in natural habitats are in open areas, under stones, or in exposed soil. In urban habitats, nests are most common in turfgrass around houses and buildings, and along roadsides. Nests have a small mound of soil surrounding a central crater and

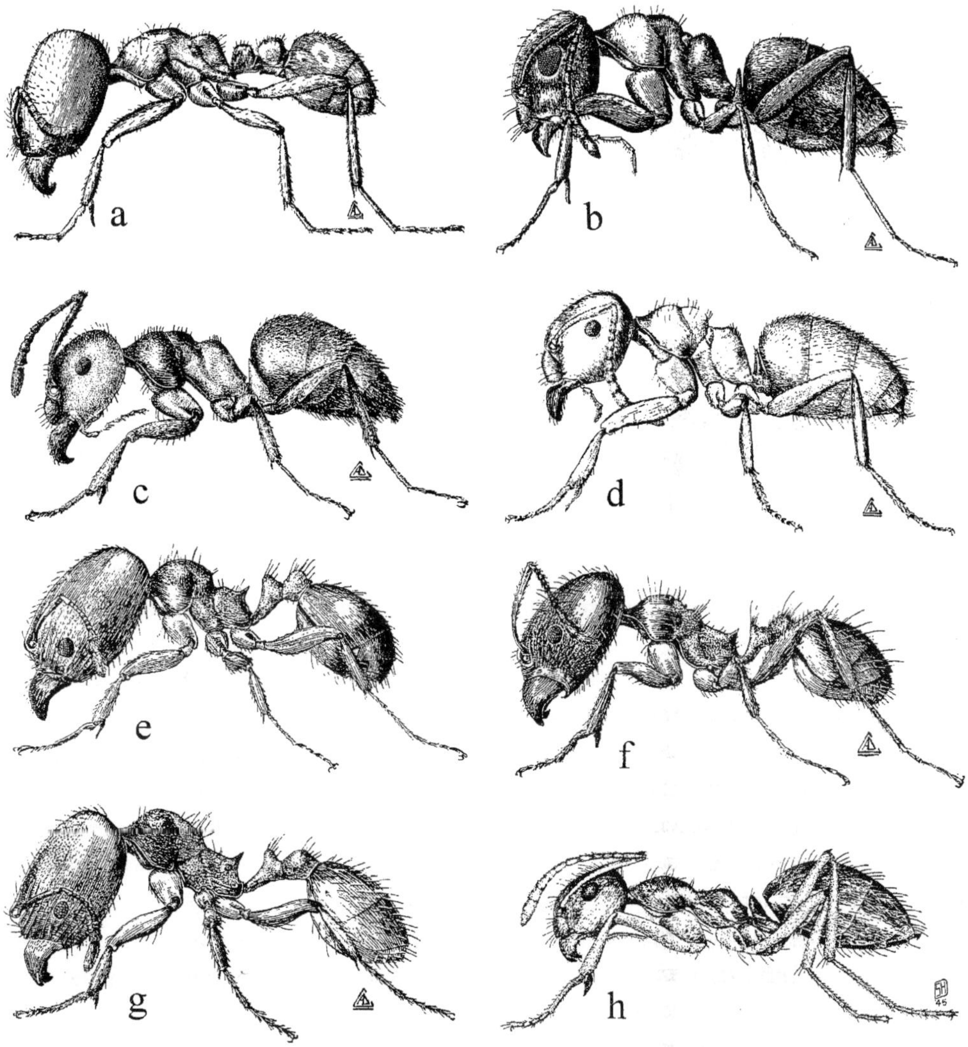


**Figure 9.6** Hymenoptera: Formicidae. (a) *Labidus coecus*; (b) *Lasius alienus*, (c) *L. niger* var. *neoniger*; (d) *L. umbratus*; (e) *Pheidole bicarinata*; (f) *P. dentata*; (g) *P. floridana*; (h) *Prenolepis imparis*.

opening; the mounds may be abundant in turfgrass. The nest is composed of a number of shallow, interconnected cambers. Colonies contain several thousand individuals and apparently one functional queen. Winged forms emerge in August and September; flights are usually after rain, and may be extensive. Flights from some colonies consist of only males, while others are a mixture of males and females.

**Black garden ant, *Lasius niger*** Workers are about 5 mm long and brownish black. Nests in natural habitats are in exposed soil or under objects; in peridomestic habitats nest sites include under concrete, along and under driveways, and around foundations. Indoor nests are found in hospitals, warehouses, and factories. Winged forms emerge in large numbers from nests in August and September. Colonies have a single queen. This species is native to Europe and the UK, but occurs through this region to Asia.

***Lasius umbratus* (Fig. 9.6d)** Workers are 3.3–3.7 mm long and light brown to brown. The scape and tibiae are normally without erect setae; the dorsal surface of the gaster has numerous fine setae. Nests in natural areas are in exposed soil, or under the cover of objects, and in or around rotting stumps or logs. Nests in peridomestic habitats occur around the foundations of buildings. Exposed nests are large earthen mounds in woodlands, and with grass growing from the mounds. In eastern North America this species occurs in woodlands in decaying wood or under stones. Colonies are large and have a single queen. Natural food is honeydew from aphids, and live and dead insects. Workers rarely forage indoors. Workers forage at night. Winged forms emerge from July to October, and males and winged females may remain in some parent nests during the winter. This species is native to Eurasia and North America; in the USA it occurs from Nova Scotia to the Gulf coast.

### *Linepithema*

Some species in this genus were originally placed in *Iridomyrmex* which, as now defined, is absent from the western hemisphere. Workers are small, ranging from 2 to 3 mm long, and the body is yellowish brown to brown. Antennae are 12-segmented and without a club. The eye is placed toward the median line of the head. The thorax is slender and with a constriction; the pedicel is one-segmented. Workers forage in narrow columns and eat a variety of plant and animal foods, including live and dead insects. Nests are in soil and colonies consist of 2000–3000 individuals.

**Argentine ant, sugar ant, *Linepithema humile* (Fig. 9.5e)** Workers are 2.2–2.8 mm long and uniformly light brown or brown. Winged males are 2.8–3.0 mm and dark brown; queens are 4–6 mm long. The body is slender and the head is distinctly oval or subtriangular. The clypeus is distinctly broader than long, convex in the middle, and with a broad notch on the anterior border. The eye has 12–14 ommatidia in its greatest diameter. The scape extends beyond the posterior border of the head. Nests in natural areas are in exposed soil and soil under cover, in rotten wood, and in cavities in trees. In the urban environment nest sites include refuse piles, bird nests, in wall voids, masonry voids, and in cracks in concrete slabs around the perimeter of buildings. Nests may be located 20–32 cm deep in the soil during dry or cold weather. During summer nests may be 2.5–5 cm below the surface. Nests can be numerous in urban areas, particularly in disturbed areas with permanent water supplies. Excessively dry or wet conditions often cause workers to invade houses.

Eggs are deposited throughout the year, but mostly during summer months; queens deposit 1–60 eggs per day. Hatching occurs in 27 days at about 21 °C and 22 days at about 23 °C. Oviposition does not occur below about 18 °C. There are four larval stages, and development lasts for about 15 days at about 24 °C and 27 days at 19.4 °C. The prepupal stage lasts about 4 days at 28 °C, while the pupal stage lasts for 16 days at about 24 °C and 14 days at about 20 °C. In late winter, large numbers of eggs are laid that will produce mainly sexual forms. Male eggs are produced throughout the year by mated queens, but are only reared every 3–4 months; at other times these eggs are eaten by workers. Males develop in early spring when there is little brood in the colony.

Colonies are usually large and contain hundreds of queens. This species does not have nuptial flights; newly emerged, virgin queens mate inside the parent nest, and join the colony as egg-laying queens. The production of males precedes that of the queens by about 2 weeks. If young queens are not present in the parent nest, winged males leave and seek other colonies; males are accepted into colonies if there are unmated virgin queens or queen pupae available. Young queens mate with males from their own nest and from other nests. New colonies are formed by budding, in which a queen and a small or large number of workers migrate to a new site. In winter, several colonies may combine to form a large colony in a favorable nesting site. *L. humile* exhibits little or no intraspecific or intercolonial hostility. During the warm season, foraging trails from many colonies often intersect, merge, and coexist without antagonism among workers. This often results in supercolonies or unicolonial populations where the entire local population, or the dispersed colonies (6000 km apart) in a large geographic region, is made up of one vast colony. Supercolonies of Argentine ants very often have numerous nests within a foraging territory. A local population may interact as one large colony; it rarely shares one large common nest during the warm season. A network of interconnected nests usually exists, and workers from these nests regularly interchange among them. Worker Argentine ants are not associated with one nest, but move randomly between nests, and may travel 40–60 m in 6 days, passing through four or five nests during this time.

Natural food is primarily honeydew from aphids and mealybugs; indoors it feeds on sweets, meats, fruit, eggs, dairy products, animal fats, and vegetable oils. They forage along regular paths and, when temperatures are between 5 and 35 °C, *L. humile* produces a trail pheromone, *cis*-9-hexadecenal, that orients workers to the food source; foraging trails may extend 75 m from a parent or satellite nest site. When searching for a new food source, *L. humile* foragers randomly explore a chemically unmarked area. Unlike most species of ants that produce a trail pheromone only when returning to the nest from a food source, the Argentine ant produces a trail continuously. Workers follow chemical rather than visual cues in the environment, which allows them to forage day and night.

Colonies will readily move to new locations when conditions or food sources change. Satellite nests can be quickly established near a food source, and forming temporary or satellite nests is an ongoing feature of Argentine colonies during summer months. Some of the satellite colonies may develop into large colony sites, some may merge back to the parent colony, and some may be abandoned. The foraging and aggressive behavior of *L. humile* often lead to the displacement of the native ant species in the region. This species will attack the colonies of ant species in the foraging territory, including species that are much larger than itself, such as the red harvester ant, *Pogomyrmex californicus*.

This species was first described as *Iridomyrmex humilis* by the Austrian myrmecologist, Gustav Mayr, in 1868 from foragers collected in Buenos Aires, Argentina. It may be native to Brazil and Argentina, but it is nearly cosmopolitan. *L. humile* is successful because of its large colony size, aggressive foraging behavior, multiple queens, and minimal intercolonial aggression. It has also been successful because it is attracted to disturbed habitats that are closely associated with human activity. It first appeared in the USA in 1891. It now occurs in the southern half of continental USA, and in Hawaii, and it has extended as far north as Washington, Oregon, Illinois, and Indiana.

**Forelius pruinosus (= Linepithema) (Fig. 9.5f)** Workers are 1.8–2.5 mm long and the body color is dark brown to blackish brown, while the gaster is light brown. The head is subrectangular and the anterior border of the clypeus is rounded; the scape extends beyond the posterior margin of the head. Nests in natural habitats are usually located in open habitats such as fields, meadows, and in entirely bare areas. Nests sometimes have a mound of soil surrounding the opening. Colonies are small to moderate-sized, and winged forms emerge from May to July. Natural food consists of honeydew, and live and dead insects; workers form distinct foraging trails, which are marked with a methyl-n-amyl ketone substance, which acts as a trial pheromone. Indoors, workers forage indoors for a variety of household foods, including sweets. This species is native to North America, and occurs from Wisconsin to New York, south to New Mexico and Florida; it also occurs in the West Indies.

### *Liometopum*

This is a small genus of ants, which occurs in Europe and Asia. The antennae are 12-segmented and the pedicel has one segment. The workers of most species can emit a disagreeable odor, which resembles the secretions of the odorous house ant, *Tapinoma sessile*.

**Velvety tree ant, California velvety tree ant, *Liometopum microcephalum* var. *occidentale*** Workers are 2.5–5.5 mm long and polymorphic. The two-color pattern is characteristic for this ant: the head and gaster are black and somewhat shiny, while the thorax is reddish brown. The abdomen is covered with fine pubescence that produces the velvety appearance. Nests in natural habitats are in hollow trees, tree stumps, and in the ground under stones; large nests are sometimes lined with a paperlike material. Indoors, nests are in cavities that have been excavated by workers. Sometimes they will tunnel into wood and produce a fibrous frass similar to carpenter ants (*Camponotus*). Nests may have cavities used as temporary resting sites for foraging workers. Colonies are large and probably have several functional queens. Natural food includes honeydew from aphids and scale insects, and also live and dead insects, and animals. Indoors they feed on a variety of household foods. Foraging trails may be 2–3 cm wide. These ants will bite, and may inject a fluid into the wound, which causes a severe stinging sensation. It is native to North America and occurs primarily in western and southwestern states from Colorado through Arizona, California, and the Pacific northwest.

**Other *Liometopum*** The silky carpenter ant, *L. luctuosum*, is a household pest in western states. Workers are 3–5 mm long and uniformly dark brown to blackish brown.

### *Monomorium*

These ants are relatively small and are distributed in temperate, tropical, and neotropical regions. Workers are 1.2–2 mm long, and the body is dark brown to black, but they are typically black. The body is almost entirely smooth and shiny, with few setae. Antennae are 11- or 12-segmented and with a distinct three-segmented club. The mesoepinotal region has a well-defined constriction, without spines; the pedicel is two-segmented. Nests are in soil and workers are predaceous and carnivorous on other insects, but also feed on honeydew, nectar, and pollen of certain plants. Indoors they feed on a variety of domestic food, and some tropical species may be restricted to domestic habitats.

**Singapore ant, *Monomorium destructor* (Fig. 9.7a)** Workers are 1.8–3 mm long and yellowish brown, and the gaster is dark brown. The body setae are long and slender, and nearly erect. The dorsal surface of the posterior border of the head has irregular sculpturing; the mesopleuron and side of epinotum are sculptured, but the remainder of the body is smooth and shiny. Nests are outdoors in tropical and neotropical regions, and indoors in temperate regions. Colonies are large and contain many functional queens. Workers are predaceous on other insects, but also tend honeydew-producing homopterans, and

**Figure 9.7** Hymenoptera: Formicidae. (a) *Monomorium destructor*; (b) *M. floricola*; (c) *M. minimum*; (d) *M. pharaonis*; (e) *Neivamyrmex nigrescens*; (f) *N. opacithorax*; (g) *Wasmannia auropunctatus*; (h) *Paratrichina longicornis*.

feed on seeds; they forage in trails. Indoors they feed on sweets, meats, oils, grease, and animal material. They may bite people indoors. This species is native to the oriental region, but it is nearly cosmopolitan in urban areas.

***Monomorium floricola* (Fig. 9.7b)** Workers are 1.4–1.8 mm long. The body is distinctly slender and bicolored: the head and gaster are dark brown or blackish brown, and the thorax, pedicel, and legs are pale brown, but the femora may be dark brown. The body is smooth, shiny, and setae and pubescence are sparse; the gaster is narrow at the base. Nests in natural habitats are above-ground in twigs and branches of trees, and under bark; it nests in live and dead wood. They may nest indoors. Colonies are usually large and have numerous functional queens, which lack wings. Natural food includes live and dead insects, honeydew, and nectar from flowers; indoors they feed on sweets. This species is native to the African or oriental region, and it occurs in southeastern USA, in Alabama and Florida, and in southern Asia, including Malaysia.

**Little black ant, *Monomorium minimum* (Fig. 9.7c)** Workers are 1.5–2 mm long and the body is almost entirely smooth and shiny, and dark brown to black, but typically black. Antennae are 11- or 12-segmented with a three-segmented club. The clypeus has a pair of longitudinal ridges, which often extend beyond the anterior margin of the clypeus and appear as teeth. Nests in natural habitats are usually outdoors in exposed soil or under objects, and in decaying wood. Indoor nests may be in structural wood or within the masonry of the foundation. Colonies usually contain several fertile queens, and winged reproductives emerge from June to August. This ant successfully competes with fire ants for food and territory. Natural foods include live and dead insects, honeydew, pollen, and nectar. Indoors they feed on meats, sweets, bread, grease, oils, cereals, fruit, and fruit juices. This species is native to North America and occurs in southeastern Canada and the northern and eastern regions of the USA.

**Pharaoh's ant, *Monomorium pharaonis* (Fig. 9.7d)** Workers are about 2 mm long and yellowish brown to reddish brown; erect hairs and pubescence on the body are sparse. Segments of antennal club gradually increase in size towards the apex of the club. The eye is relatively small, with 6–8 ommatidia at the largest diameter. The thorax has a well-defined mesoepinotal impression. The head, thorax, and pedicel are densely but weakly punctate; the clypeus, mandibles, and gaster are shiny. Nests are normally restricted to indoor sites in temperate regions, and location is in inaccessible sites, such as in light switches, behind baseboards, and in cabinets. Colonies are active all year and produce a large number of individuals. Development from egg to adult worker takes about 36 days, and from egg to reproductive takes about 41 days. Colonies have a large number of functional queens, but flights of reproductives from the nest are not known. New colonies are formed by the process of budding. This occurs when a young queen and a small number of workers split from the parent colony. Pharaoh ants make use of chemical trails when foraging. Through their stinger, workers deposit faranal from the Dufour's gland on the ground. This chemical orients other workers to the trail and the food source. Workers forage 24 h/day, in nearly all weather. Natural food includes live and dead insects; indoors they feed on sweets, meat, grease, and a variety of other materials. This species is native to the North African region, but it is nearly cosmopolitan in buildings and houses. Linnaeus described this ant in 1758 as *Formica pharaonis* from specimens collected in Egypt.

Pest status of *M. pharaonis* is based on infestation of buildings, especially urban apartment buildings, and health care facilities. The main problems caused by this species are skin irritation, skin lesions, contamination of instruments and machines, and germ transmission. Skin irritation results primarily when ants crawl under bandages; they cause itching by feeding around wounds. A large number of ants may be recruited to wound sites, causing extensive problems; they may be especially attracted to burn victims. Skin lesions occur when worker ants bite and feed directly on undamaged skin, such as hospital patients under sedation and newborn babies. Because of their small size and their ability to pass through tiny cracks, workers can infiltrate sterile hospital supplies, intravenous fluid delivery systems, and oxygen supply tubes.

A large number of pathogenic organisms have been recorded from this species, including bacteria: streptococci, enterococci, *Micrococcus pyogenes aureus*, *M. albus*, *Proteus vulgaris*, *Escherichia coli*, *Alcaligenes faecalis*, *Salmonella* spp., *Pseudomonas aeruginosa*, *Clostridium* spp., and *Bordetella bronchi-septica*.

**Other *Monomorium*** Household infestations and damage to rubber material by *M. latinode* occur in Java, and *M. barbatulum* occurs in granaries in Egypt. *M. intrudens* workers are 2–2.5 mm long and yellowish brown with a dark-brown gaster. This species is a pest in Japan.

### *Neivamyrmex*

These ants are distributed in North and South America. They often form temporary nest sites, and may occur around building foundations. Workers are 2.5–5.8 mm long and the body is brown to blackish brown; setae are abundant on thorax and abdomen. Antennae are short, and the 12 segments are slightly enlarged; the eyes are very small. The epinotum is without spines, and the pedicel is two-segmented. The legs are long. These ants may occur in large colonies. They are predaceous on other insects.

***Neivamyrmex nigrescens* (Fig. 9.7e)** Workers are 2.8–5.8 mm long and the body is pale brown to reddish brown to blackish brown; the body is densely covered with punctures. The scape extends to the posterior margin of the eye. Colonies contain 150 000–250 000 workers, and there is apparently one functional queen. Males emerge from September to November. New queens are not winged and mating may occur within the parent nest. New colonies are formed by budding, which occurs when a new queen leaves the parent nest with a small number of workers. Foraging occurs during the day, and the natural food is other insects, including the brood in nests of other ant species. It is native to North America and occurs from Virginia to California south into Mexico. Related species, *N. fallax* and *N. nigrescens mexicanus*, occur in the Gulf Coast states; they have been reported from domestic and peridomestic habitats, including in uncovered wells used for household water.

***Neivamyrmex opacithorax* (Fig. 9.7f)** Workers are 2.2–4.6 mm long and yellowish brown to reddish brown; the legs and pedicel are lighter in color than the head and thorax. The scape does not extend to the posterior margin of the eye. The thorax and pedicel are opaque and densely punctured, while the remainder of the body is smooth. Characteristics of the colony and feeding habits are similar to *N. nigrescens*, including the emergence of males from colonies from September to November. Workers may enter houses to forage from nests established outdoors.

*Paratrechina*

These ants are cosmopolitan, and species occur in urban environments from temperate to tropical regions. Workers are 2.4–4.5 mm long and the body is yellowish brown to blackish brown. The body has a dense covering of setae, which are long and may be especially dense on the gaster. The scape exceeds to the border of the head by at least one-fifth or more of the length of the scape. Legs are long and have long setae. Native species usually nest outdoors, while tropical species in temperate regions nest indoors. They forage during the day and night, and seek live and dead insects. Some species forage for sweets and they are sometimes called sugar ants.

**Hormiga loca, formiga cuiabana, crazy ant, *Paratrechina fulva*** Workers are 2.1–2.4 mm long and reddish brown. The mandibles are larger than half of the head with six teeth; the antennae are 12-segmented. Abdominal segments have rows of brown setae. Nests in natural habitats are usually in soil; various natural cavities in humid sites are used to raise the brood. They do not construct galleries or tunnels but often use nests constructed by termites, and soil galleries made by spiders, grubs, and small earthworms. Transitory nests are constructed during the rainy season, and are located in vegetation and often in unprotected sites and are moved daily. The composition of these nests includes about 50 workers and nearly the same number of developmental stages. Permanent nests in dry locations extend about 1 $m^2$ to a depth of about 40 cm. These nests contain multiple queens. Natural food includes liquid honeydew, and solid pieces of animal protein from birds, calves, snakes, and lizards. *P. fulva*-infested areas are often marked by a reduction in white grubs, leaf-feeding Lepidoptera, and termite colonies, when compared to non-infested areas. Indoors they feed on household foods with high protein content. The common name is derived from the erratic and random movements of the foraging workers. It is native to Brazil, but has been introduced into Columbia and perhaps other countries in South and Central America.

**Crazy ant, *Paratrechina longicornis* (Fig. 9.7h)** Workers are 2.2–3 mm long. The head, thorax, pedicel, and gaster are dark brown to blackish brown. The body is very slender, with extraordinarily long legs. The body is covered with long, coarse setae, which are grayish white. These ants are capable of moving rapidly and jumping, and the common name is derived from their habit of darting in and around objects, and their long legs, which emphasize these motions. Nests in natural habitats are in wet and dry locations, including in soil under objects, in trees and plants, and in rotten wood. In the urban environment, they nest indoors in wall voids, and outdoors around buildings, and sites such as trashcans and refuse dumpsters. They occur on the ground- and upper floors of commercial buildings and hospitals. Natural foods include honeydew, and a variety of live and dead insects; indoors they forage for various household foods. This species is native to Africa, but it has been spread by commerce to nearly all parts of the world. It occurs widely in disturbed urban areas in the tropics and subtropics around the world. It may be a pest species on ships, and in artificial ecosystems, such as greenhouses and Biosphere 2, near Tuscon, Arizona.

***Paratrechina vividula*** Workers are 2–3 mm long and yellowish brown to blackish brown. The setae are short and dense on all legs and the antennae. The scape extends to the mesonotum. This species occurs in heated glasshouses in botanic gardens in Denmark and other countries in northern Europe; it also occurs in the UK.

**Other *Paratrechina*** *Paratrechina pubens* are 2.5–3 mm long and brown to reddish brown. It is native to the Antilles Islands, and occurs as a peridomestic pest in Florida (West Palm Beach).

*Pheidole*

This is a cosmopolitan genus with species in tropical and temperate regions. They are characterized by large, broad-headed workers, some with a relatively small body and large head. Workers are 2.2–4.5 mm long and the body is yellowish brown to dark brown. Antennae terminate in a distinct three-segmented club. The thorax is punctate and the epinotum has distinct spines; the pedicel is two-segmented. Nests usually are under stones or in soil, sometimes around building foundations. Nests may contain several hundred workers; small workers have normal-sized heads, while larger ones have greatly enlarged heads with very large mandibles. Natural food includes seeds, which are brought back to the nest by small workers, and then crushed into fragments by large-headed workers.

***Pheidole bicarinata* (Fig. 9.6e)** Workers are 1.6–3 mm long and the body is yellowish brown to dark brown. The head is longer than broad and has a median furrow; the anterior portion of the head has longitudinal grooves, while the posterior portion is without grooves, and shiny. The scape extends to the midlength of the head. This species occurs in a range of habitats, including deserts, mountains, and beaches. Nests in natural habitats are in rotting wood, exposed soil, or under stones and other objects. Natural foods are honeydew, seeds, and small insects. Indoors, workers forage for meat, bread, and grease. This species is native to North America and occurs from Nebraska to New York and south to Arizona and Florida.

***Pheidole dentata* (Fig. 9.6f)** Workers are 2.4–3.8 mm long and light brown to dark brown. The head is bilobed with a median furrow; the anterior margin of the clypeus has a median emargination or tooth. The anterior portion of the head is sculptured and opaque, while the posterior half is smooth and shiny; the sides of the thorax, excluding the prothorax, are sculptured. The scape extends beyond the midlength of the head. The dorsal surface of the gaster has long, erect setae. Nests in natural habitats are in exposed soil, under objects, in logs and rotting wood. Natural food is live and dead insects, seeds, and honeydew. Indoors they forage for high-protein food, including meat, and grease; they also feed on sugary material such as fruit juices. This species is native to North America and occurs from Kansas toVirginia south to Texas and Florida.

**Coastal brown ant, big-headed ant, *Pheidole edax* (= *P. megacephala*)** Workers are 1.5–2.5 mm long and yellowish brown to dark brown. The head is bilobed and has a median furrow; colonies contain distinct large- and small-headed workers. Epinotal spines are short. Nests in natural habitats are in exposed soil; their large size and extensive tunnels dominate an area and often eliminate native ant species. In peridomestic habitats, the nests may undermine paving stones, small structures, and the root systems of some plants; indoors they nest in wall voids. Colonies have several functional queens, and colonies spread by groups of workers accompanying fertilized queens separating from the main colony. This process is referred to as budding. Natural foods include live and dead insects, honeydew, and sap-sucking insects. Indoors they feed on meat, grease, pet food, and other high-protein material; workers forage in kitchens and bathrooms during warm months. This species is native to Africa, but it is a pest in domestic habitats in coastal and inland sites in Australia and Hawaii.

***Pheidole floridana* (Fig. 9.6g)** Workers are 1.5–2.6 mm long and yellowish brown to very light brown. The head is bilobed with a median furrow; the scape extends to about midlength of the head, and a groove along the side of the head accepts the scape. The head is punctate and nearly opaque, except for a smooth and shiny area that may be the posterior third or fourth of the head, or only a narrow area around the eyes. Nests in natural habitats are in exposed soil or under objects; they also nest in logs and stumps, and in litter at the base of trees. In the urban environment, they nest around the perimeter of buildings and houses. This species may be confined to coastal areas. This species is native to North America and occurs from North Carolina to Florida and south through the Gulf Coast states to Texas.

**Other *Pheidole*** The small-sized species, *P. hyatti*, sometimes occurs indoors feeding on grease and meats; it occurs in the San Joaquin and Sacramento valleys of California and east to Texas.

*Pogonomyrmex*

These are the harvester ants, so called because they gather seeds and grasses. The workers are reddish brown and large, ranging from 4.5 to 12 mm long. They nest in the ground, and usually construct a mound that is cleared of vegetation around it to a distance of about 1 m. When disturbed, the workers will attack and sting. Mammalian toxicity is greater than any other insect venom; it is eight to 10 times as toxic as honey bee venom. The reaction to a sting is not localized, but spreads along the lymph channels to other regions, and lingers for a long period. It is not toxic to other insects, and appears to have evolved as a deterrent to vertebrate predators. This genus is distributed throughout North and South America, from Canada to Patagonia.

**California harvester ant, *Pogonomyrmex californicus*** Workers are 5.5–6 mm long and pale reddish brown; the thorax lacks spines. Nests in natural habitats are in exposed soil, usually in sunlight; in peridomestic habitats the nests may be in lawns and recreational turfgrass. The nests (mounds) damage highway shoulders. These are usually low and flat mounds surrounded by a vegetation-free area. Natural food consists of a variety of seeds; these ants do not usually forage indoors. Much of the foraging activity is in mid-summer and there is little or no activity in the winter. The maximum foraging activity occurs when ground surface temperature is 32–46 °C. Colonies are large and winged forms emerge during hot, clear days in June and July. Winged males from neighboring colonies will often display male aggregation syndrome. For this, large numbers of males will gather at prominent features of the landscape, such as the tops of trees and bushes, hilltops, tops of buildings, and the tops of chimneys of houses. Volatile pheromones secreted from male mandibular glands attract young queens to these sites, and 4–5-cm-diameter clusters of mating males and females are formed. When these mating clusters occur on top of chimneys of houses, ants often fall down the chimney and into the fireplace. They then crawl on to the floor and reform the mating clusters, often resulting in many of these clusters in the house. This species is native to North America, and occurs primarily in southwestern states.

### *Prenolepis*

Workers are 2–4 mm long and usually dark brown. Antennae are 12-segmented, and lack a club. The scape extends beyond the posterior border of the head by approximately one-half its length. The maxillary palp is six-segmented. The thorax is slender and divided into two parts by a constriction; the pedicel is one-segmented. These ants tend aphids and are general scavengers in narrow columns. Nests are built below ground, and colonies contain 2000–3000 workers. Workers usually forage at night, and individual workers may become so engorged with honeydew that their gasters are distended like a balloon.

**False honey ant, *Prenolepis imparis* (Fig. 9.6h)** Workers are 2–4 mm long. The body color is variable, ranging from yellowish brown to dark brown, and sometimes blackish brown; the head and thorax are often lighter in color than the gaster. The body appears shiny and smooth, and the setae are pale yellow and abundant on the head and gaster. Nests in natural habitats are located in moist clay or loamy soils in well-shaded sites. Indoor nests may be established in small cavities and wall voids. The nest has a single opening surrounded by a mound of earthen pellets. Colonies contain 2000–3000 individuals. Brood production is completed in late summer and no brood overwinters in the nest. Winged reproductives emerge in April and May; this species is one of the earliest to produce reproductive swarms in spring. New colonies are formed by fertilized queens, who establish a primitive nest in a soil cavity. Natural food includes dead insects, fruit juices, and plant sap, but is primarily honeydew. The abdomens of workers often become greatly distended and balloon-like; they may become so enlarged that they have difficulty walking. This engorgement behavior is similar to true honey ants (*Myrmecocystus*). Indoors they feed on meat, bread, sugar, and fruits. This species is native to North America, and it occurs from Nebraska to Ontario, Canada and south to Texas and Florida.

### *Solenopsis*

This is a large and cosmopolitan genus with species in tropical and temperate regions. They are often dominant ants in leaf litter in forest clearings and in disturbed habitats. Workers are 1.3–1.8 mm long, dark brown to black, and sometimes shiny. Antennae are 10-segmented with a two-segmented club. The thorax is slightly constricted with epinotal spines; the pedicel is two-segmented. Nests are usually deep in soil, and most species are mound builders; nests may occur indoors. Colonies are usually large and contain several thousand workers, and some species are very aggressive and will sting. This aggressive behavior and painful sting are the bases of their common name, fire ant.

Three fire ant species are native to North America. The desert fire ant, *Solenopsis aurea*, occurs in southwestern desert areas. The southern fire ant, *S. xyloni*, occurs in southern states, and *S. geminata*, the tropical fire ant, occurs from Texas to South Carolina. *S. geminata* has spread pantropical and now occurs in south Asia. Two fire ant species have been introduced into the USA from South America: *S. richteri* and *S. invicta*. The hybrid imported fire ant, *Solenopsis invicta* + *richteri*, was discovered in the mid-1980s and it occurs in portions of Mississippi, Georgia, and Tennessee.

Pest status of the red imported fire ant, *S. invicta*, is based primarily on its aggressive behavior and painful bite and sting. The importance of this ant extends across the agricultural and urban environments. This species typically nests in the ground in sunny areas, but it also occurs indoors in walls, in attics, and in various household materials. When nesting indoors or in spaces above ground, the ants usually bring in soil to the nest site, and this causes additional problems. This species seems to be attracted to electrical fields, and nests in electrical equipment can result in short circuits and interfere with key mechanical devices. The ants remove insulation from wires, causing short circuits between lines; they also bridge contacts on terminal blocks, causing direct electrical shorts between their bodies. However, the principal problem is interference when they enter relay switches. They may accumulate in such large numbers that the switch cannot function. This results in problems in equipment such as pump motors, air conditioners, and traffic signals.

Pets, domestic animals, and sometimes wildlife, can be attacked and injured by fire ants. Newborn animals are vulnerable to attack and may be killed. Worker ants are attracted to mucous areas and wounds, and young animals are often stung around their eyes, which can lead to blindness. Major workers in the colony rarely sting, but they will bite. Minor workers are very aggressive and will sting and bite. When nests are disturbed, the workers will leave the nest and attack the intruder. They will hold on to the skin with their mandibles, and pull it forward so that they can extend the gaster between their legs and sting repeatedly. The venom of the red imported fire ant is unusual in that it contains a necrotizing toxin, *trans*-2-methyl-6-*n*-undecylpiperidine (solenopsin A). Potential severe allergic reactions can occur when susceptible individuals are stung.

**Tropical fire ant, *Solenopsis geminata*** Workers are 2.4–6 mm long and the color is variable; some individuals are yellowish brown to reddish brown, others are blackish brown, and others are a combination of light and dark colors. The head of the major worker is enlarged, and more than twice as broad as the pronotum, and it is strongly bilobed posteriorly. The thorax is usually sculptured and opaque; segment 1 of the pedicel is narrow in profile. Nests in natural habitats are in the ground in open areas with dry or moist soil. Nests are surrounded by irregular piles of excavated soil. In undisturbed or agricultural sites, colony densities are usually 15–25 per hectare. Colonies are large, and they usually have only one functional queen. Winged reproductives emerge from May to June. This species is an important predator and natural food includes other insects. It is native to North America, but it has been displaced from its former range by the imported fire ant. It occurs from Texas to South Carolina, and Florida, and south to Costa Rica, and the West Indies. This species is pantropical and occurs in southern Asia, including from India to Malaysia.

**Red imported fire ant, *Solenopsis invicta* (Fig. 9.8b)** Workers are 3–6 mm long and yellowish brown to blackish brown. The head is less than twice as broad as the pronotum; the mandible has four teeth along the biting edge. The apex of the scape extends more than half the distance between the upper border of the eye and the posterior border of the head. A large mature colony may have 200 000–300 000 workers. Nests in natural habitats are built in all types of soil except forest and marshland. The nest is typically a 30–40-cm-high mound above ground. The mound has no entrance holes, but workers enter and exit through holes in tunnels that radiate 2–3 m from the mound. The number of nest (mounds) per hectare depends on whether the colony is a single- or multiple-queen form; the range is 50–75 mounds per hectare. Mating flights from mature colonies 6–8 months old occur in the spring and fall. The winged reproductives accumulate in the nest until environmental conditions are suitable for their release. From April to November a colony may produce 4000–6000 winged forms per year, and the number of queens produced can range from 40 000 to 300 000 per hectare. Natural food includes invertebrates, which they sting and kill; they also feed on dead animals, plant material, seeds, and honeydew. Indoors they feed on meat, grease, and other foods that have a high protein content. Workers are aggressive and sting repeatedly.

Multiple-queen colonies consist of a series of mounds containing numerous queens, with workers moving freely from one mound to another. One result of this phenomenon is a much greater density of mounds, ranging from three to 10 times the single-queen form, per hectare. Multiple-queen forms of *S. invicta* have been found in South America. This species was introduced into the USA from central Brazil between 1933 and 1945. Originally it was believed to be *S. saevissima richteri* with two color forms, one red and one black. The black form was described as a distinct species, *S. richteri*, and distinguished from the red form, *S. invicta*. The red imported fire ant occurs in 11 southern states from Virginia west to Texas. The northern movement of *S. invicta* may be restricted by prolonged cold weather. This species may be unable to reproduce in areas where the minimum yearly temperatures are less than −12.3 °C. Colonies transported to northern locations may survive in buildings or sewer systems, but they may not be able to reproduce and spread successfully. However, the multiple-queen form (polygyne) of this species may be able to survive above this temperature limit. A minimum yearly temperature of −17.8 °C may be the limit for this species.

**Thief ant, grease ant, piss ant, *Solenopsis molesta* (Fig. 9.8c)** Workers are 1.3–1.8 mm long and the body is smooth and shiny. The color ranges from yellowish brown to dark brown. The eyes are small, with 4–6 ommatidia or fewer. The scape extends more than half the distance between the eye and the posterior border of the head. The antennal club is large and elongate; it is approximately 1.3 times the combined lengths of the remainder of the segments after the scape. The mandible usually has four teeth. The body setae are moderately abundant and well distributed, long, and usually erect. Nests in natural habitats are in exposed soil or under the cover of stones and other objects, including rotting wood. Indoor nests are in various cavities in wood, masonry, and household materials. Colonies contain several hundred to several thousand individuals. In the laboratory, queens laid 27–387 eggs per day, and workers developed from egg to adult in about 52 days. Winged reproductives emerge from July to October. Natural food includes live and dead insects, seeds, and honeydew; indoors they feed on meat, sweets, ripened fruit, oils, and dairy products. They may prefer food with high protein content. This species is native to North America, and it occurs in eastern and central USA from

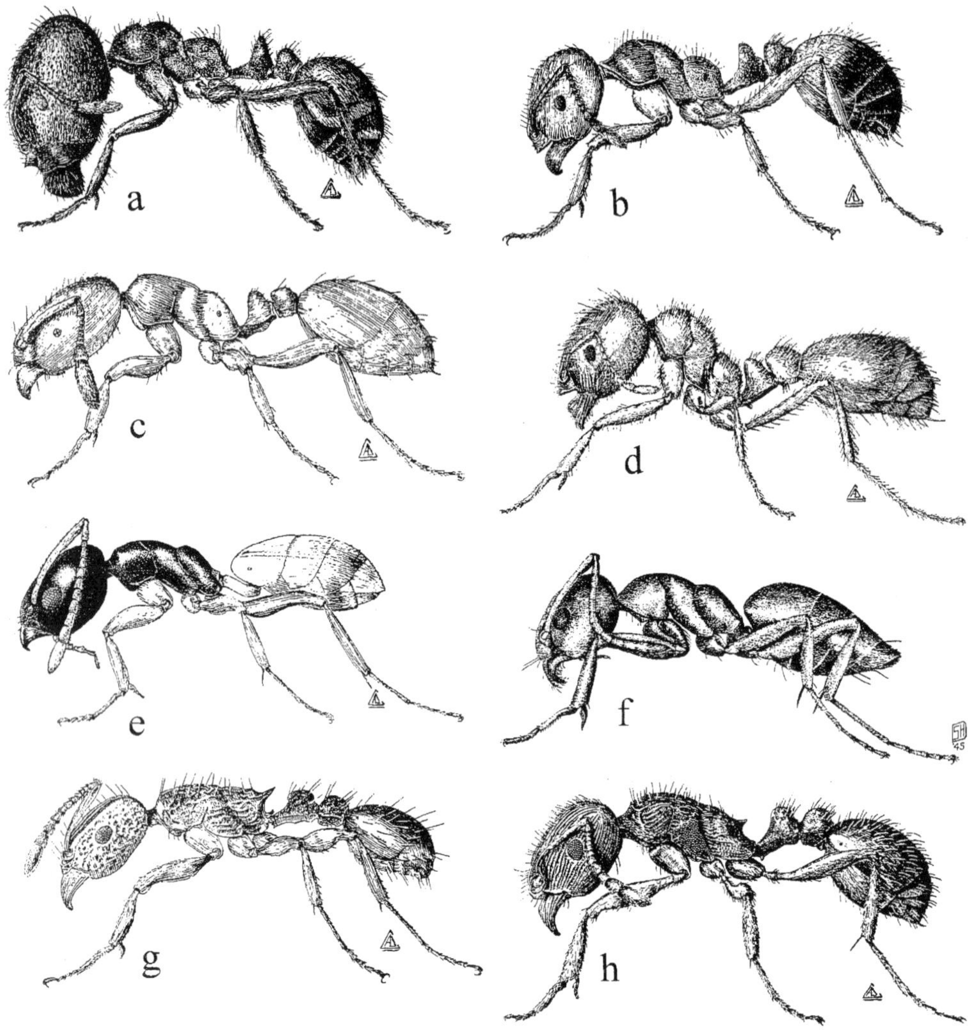


**Figure 9.8** Hymenoptera: Formicidae. (a) *Solenopsis geminata*; (b) *S. invicta*; (c) *S. molesta*; (d) *S. xyloni*; (e) *Tapinoma melanocephalum*; (f) *T. sessile*; (g) *Tetramorium guineense*; (h) *T. caespitum*.

southern Canada to the Gulf Coast. The name thief ant refers to the habit of nesting in or near the nests of other ants, which they rob of food and brood.

**Black imported fire ant, *Solenopsis richteri*** Workers are 2.8–6 mm long and the body is dark brown to blackish brown, except for a broad band of yellowish red at the base of the gaster. The head is less than twice as broad as the pronotum. The apex of the scape extends more than half the distance between the upper border of the eye and the posterior border of the head. Nests in natural habitats are in various types of soil. In open areas the excavated soil next to the nest opening may be 1 m high, and large mounds of soil are usually dome- or conical-shaped. The mounds of soil contain galleries and chambers above and below the soil level. Nests in peridomestic habitats occur in turfgrass and among ornamental plants and shrubs. Colonies are large and contain more than one functional queen. Winged reproductives emerge from December to February, but reproductives may be produced at other times. Natural food includes flesh of insects, mammals, birds, honeydew, fruits, and seeds. Indoors they feed on meat, grease, and other foods that have a high protein content. Workers are aggressive and sting repeatedly. This species is native to South America and naturally occurs in Brazil, Uruguay, and Argentina. It was introduced into North America and occurs in southern USA, from northeastern Mississippi and northwestern Alabama.

**Southern fire ant, *Solenopsis xyloni* (Fig. 9.8d)** Workers are 1.6–5.8 mm long and yellowish brown to reddish brown. The gaster is usually dark brown, and the body is densely covered with fine setae. The scape extends about halfway between the eye and the posterior border of the head. The surface of the mesopleuron is finely sculptured. This species is similar to

*S. invicta*, except it lacks mesopleuron striations, three teeth on the mandible, and the short antennal scape. Nests in natural habitats are in the ground in an exposed site or under stones or other objects; nests are surrounded with craters of loose soil. Indoors they nest in wood, masonry, and wall voids. Nests may be in turfgrass around buildings. In agricultural sites, the number of colonies rarely exceeds 15–25 per hectare. Colonies are large, and there may be several functional queens. They are very sensitive to ground vibrations, and when disturbed the workers will leave the nest and attack intruders. New colonies may be formed by budding, which is accomplished when a young queen and a small number of workers split from the parent colony. Natural food includes seeds, honeydew, live and dead insects, juices or sap of fruits and plants. Indoors it eats nuts, meat, grease, and fruits. They bite holes in fabrics, such as silk, linen, and nylon. This species is native to North America, but it has been displaced from its former range by the imported fire ant. It now occurs from South Carolina and Florida west to California, and it is especially common in some of the Gulf Coast states.

### *Tapinoma*

This is a worldwide genus with most species found in the Indo-Australian and Ethiopian regions. One or more of the tropical species are cosmopolitan and sometimes establish nests in buildings. Workers are 1.5–2.5 mm long and range in color from dark brown to black. The antennae are 12-segmented, and the petiole is small and often concealed by the first segment of the gaster. Anal glands are present and these ants produce a disagreeable odor.

**Ghost ant, *Tapinoma melanocephalum* (Fig. 9.8e)** Workers are 1.5–2 mm long. The head is brown, the thorax and gaster are yellowish brown, and the antennae and legs are pale brown. The prothorax is laterally compressed and broad anteriorly. Nests in natural habitats are in soil, beneath stones, rotten wood, dead trees, and under tree bark. Indoor nests may be in a variety of voids and cavities, such as in closets and in discarded clothing. It is occasionally found in greenhouses. Colonies usually contain several hundred workers and several functional queens. *T. melanocephalum* colonies are typically broken into subunits (satellites) that occupy different nest sites and exchange individuals along odor trails. New colonies are formed when one or more inseminated females migrate to a new site with numerous workers. Flights of winged reproductives are not known for this species. Natural foods includes honeydew, and live and dead insects; when indoors they forage primarily for sweets. This species may be carried to other regions in plant material and produce from the tropics. It is a common house-infesting ant in the tropics, and it occurs in heated buildings in Europe and the UK.

**Odorous house ant, *Tapinoma sessile* (Fig. 9.8f)** Workers are 2.4–3.2 mm long and the body color ranges from uniform brown to uniform black. The pedicel is small and hidden by the base of the gaster. Nests in natural habitats are in the ground, usually under stones or logs, but also under bark of trees, in the burrows of mammals, and in refuse piles. Nests in soil are indefinite in form and usually not permanent. Indoors the nests are numerous and scattered throughout the structure, but usually associated with moisture, such as in wall voids near water pipes, in bathrooms beneath toilets, and in termite-damaged wood.

Colony size varies according to age and can range from a few hundred to many thousand individuals. Each colony may have 200 functional queens; winged reproductives (males) are only produced in colonies that are 4–5 years old. The workers or other individuals from different colonies are not antagonistic to each other. Mating usually takes place in the nest, but flights of winged reproductives have been observed. New colonies are formed when one or more inseminated females migrate to a new site with numerous workers. In temperate regions, colonies are active and workers may forage throughout the year. Outdoors, colonies become active and workers begin foraging in March and may continue through November. Winged females and males appear in the colony in May and June, and sometimes into July. When a nest is disturbed, workers run rapidly, emitting an odor from their elevated abdomens. Nests are moved to new sites regularly during the warm season; they are often moved prior to rainfall. They forage day and night; foraging occurs at temperatures from 6 to 34 °C. Workers establish trails leading from the nest to the food, and follow guidelines, such as tree limbs, the edges of buildings, baseboards, and kitchen counter. Natural food includes honeydew, and live and dead insects, and dead animals. Indoors they feed on meat, fish, dairy products, vegetables, and fruit. This species is native to African or oriental regions, but it is nearly cosmopolitan in urban environments, both indoors and outdoors.

### *Technomyrmex*

This genus of relatively small ants is distributed in the world tropics, with most of the species occurring in Africa and the tropics of Asia. It is morphologically similar to *Tapinoma*, but can be distinguished by the presence of five visible tergites on the gaster; *Tapinoma* has four visible gastral tergites. The overhanging gaster often obscures the single segment of the pedicel.

**White-footed ant, *Technomyrmex albipes*** Workers are 3.5–4 mm long and the body is blackish brown to black, and the tarsi are pale yellow. The pedicel is one-segmented and the small and flattened node is partially obscured by the gaster. Nests in natural habitats are in the ground or in vegetation on the ground, in dead wood, around the base of trees, and under loose bark of trees. In peridomestic habitats nests may be around house foundations, and invading electrical fixtures and machinery such as air conditioners; they also nest in piles of lumber, and firewood. Indoors, they nest in wall voids, potted plants, and in household materials. They establish well-defined trails outside buildings, and indoors the trails are usually along the edges of baseboards and carpeting. Nest sites may consist of a main nest and several satellite nests; there is regular exchange of food and brood between all components. This type of nest structure results in supercolonies that can occupy a large area. Colonies may contain hundreds of thousands of workers and forms called intercasts, which have the characters of both queens and workers. These individuals have the ability to lay eggs and care for immatures. Natural food is primarily honeydew from aphids and other homopterans. Indoors they forage on a variety of household food, but prefer sweets. Oral exchange of food between workers is limited, and is replaced by workers and intercasts depositing trophic eggs. These are used as a food source in the colony and fed on by all adults. Winged females emerge from colonies in May and June, after they have mated with wingless males in the nest. New colonies are formed by young queens. After the nest is established, they are replaced by reproductive intercasts. This species is distributed in Florida, Hawaii, and in isolated areas in California. It may be spread to other warm southern regions of the USA on infested goods and plants.

*Tetramorium*

This is a large and widespread Old World genus, which contains several cosmopolitan tramp species. Workers are 2.5–4 mm long and the body color ranges from yellowish brown through brown to blackish brown; the legs are usually lighter in color. Antennae are 12-segmented and with a three-segmented club. The thorax is slender and the epinotum has spines; the pedicel is two-segmented. The femora are noticeably enlarged. Nests are in soil, often under stones, boards, sidewalks, or around building foundations. They are omnivorous, and feed on a variety of plant and animal material, including live and dead insects, honeydew, seeds, and plant sap.

**Pavement ant, *Tetramorium caespitum* (Fig. 9.8h)** Workers are 2.5–3 mm long and the body color ranges from light brown to brown to blackish brown; the legs are usually light brown. The dorsal surface of the head and thorax has distinct longitudinal grooves. The scape extends nearly to the posterior border of the head. The mesoepinotal suture forms a distinct constriction on the thorax. Epinotal spines are short. Nests in natural habitats are in exposed soil or under stones, pavement, and sidewalks. Indoors they may be in the masonry walls and along the foundation. Refuse expelled from the nest site is somewhat characteristic of this species, and includes fragments of seeds, dead insects, and fine wood fibers. Colonies are large and there is usually one functional queen. Winged forms emerge from the nest in June and July, but they may emerge at almost any time of the year. Queenless colonies with workers are known to occur, and they contain black, winged females and wingless males of a parasitic ant, *Anagates atratulus*. This parasitic species is known only by its association with *T. caespitum*. It is dependent on its host for food and care. Natural food for *T. caespitum* includes live and dead insects, honeydew, plant sap, and seeds. Indoors they forage for meat and grease. This species is native to Europe, but it is nearly cosmopolitan in urban areas.

**Guinea ant, *Tetramorium guineense* (Fig. 9.8g)** Workers are 3–3.5 mm long and the body color ranges from light brown to reddish brown, with a dark brown gaster. The head has longitudinal ridges and the distance between them widens posteriorly. The clypeus has three longitudinal ridges. The thorax is compressed on each side. Sculpturing on the cuticle is deep on the head, thorax, and pedicel. Nests in natural habitats are in exposed soil or under the cover of stones, or in logs and stumps. Indoor nests are located in cavities and voids in woodwork and other places. In northern portions of its range, this ant usually nests indoors. Natural food includes honeydew, and live and dead insects. Indoors they feed on meat, grease, fruits, and vegetables. This species is apparently native to Africa, but it is nearly cosmopolitan in urban environments. It occurs in northern Europe and the UK in heated greenhouses.

***Tetramorium simillimum*** Workers are about 2.5 mm long and yellowish brown to reddish brown. This cosmopolitan species occurs in heated glasshouses and buildings in Europe and the UK.

### *Wasmannia*

This is a neotropical genus with a species, the little fire ant, which is common in other regions of the world. Workers are 1.5–1.8 mm long and usually golden brown to light brown; the gaster is dark brown. Antennae are 11-segmented, with a distinct club. The head and thorax may have small punctures; the setae on the body are scarce. The thorax lacks a constriction and the epinotum has long spines; the pedicel is two-segmented.

**Little fire ant, *Wasmannia auropunctata* (= *Ochetomyrmex auropunctatus*) (Fig. 9.7g)** Workers are 1.5–2 mm long and yellowish brown to brown. Ridges on the front of the head are widely spaced, forming a partial groove for the antennal scape. Antennae have a three-segmented club, but the last two segments are greatly enlarged and give the impression of a two-segmented club. The scape does not reach the posterior border of the head. Epinotal spines are close together at base, long, and with an acute angle. Nests are in the ground and in soil between objects, in rotting wood, and in buildings and houses. Several nests may be interconnected by worker trails forming a large colony and a unicolonial social structure. This species is neotropical and it is sensitive to cold temperatures. Colonies are usually large and contain several functional queens. New colonies are probably formed by budding. Flights of reproductives have not been recorded. Natural food includes honeydew, live and dead insects and other arthropods, and small animals. Indoors they feed on meat, oil, seeds, milk, juice, and fruit. Workers are not aggressive, but have a painful sting. This species is native to neotropical regions, and it occurs in Central and South America, West Indies, Mexico, and in southern USA, especially Florida. This species is found in the Galapagos Islands and New Caledonia.

**Other Formicidae** The urban environment provides suitable habitats for a large number of ant species. They are usually unnoticed, because of their nest and foraging habits, and only occasionally come indoors. They are infrequent pests, and establish nests in peridomestic habitats. Most are noticed by the emergence of winged forms or the activity of foraging workers. The brown house ant, *Doleromyma darwiniana*, occurs in Australia. Workers are 2–3 mm long and uniformly brown, and the pedicel is one-segmented. It nests along paths, and in the joints of brick- and stonework, and workers forage indoors for high-protein foods. The black house ant, *Ochetellus glaber*, occurs in Australia. Workers are 2.5–3 mm long and black; the pedicel is one-segmented and the epinotum is concave. It nests outdoors along paths, but may establish nests indoors, including subfloor areas of houses. Indoors it forages for a variety of household foods, but prefers sweets. The greenheaded ant, *Rhytidoponera metallica*, occurs in Australia. Workers are 5–6 mm long, and they are black with a metallic green head. Nests are along paths, among rockeries, in bushy areas. It feeds on vegetable material. This species can inflict a painful sting.

*Anoplolepis custodiens* and *A. steingroeveri* are pests in South Africa. During late summer and fall in western Cape orchards and vineyards, *A. steingroeveri* workers attack farm laborers and hinder their work. *A. custodiens* workers are known to kill domestic chickens. *Dorylus helvolus* is a driver ant in South Africa that is common in gardens in most southern African regions. Although it is rarely seen because of its subterranean habits, workers occasionally come indoors through openings in outside walls. *Melissotarsus emeryi* is one of the South African ants that tunnel into live trees, including *Ficus sur* and *Schotia* spp. that grow along roadsides. Nests consist of a complex of tunnels and passages, mostly under the bark but also penetrating into the heartwood. Tunneling by these ants induces parts of the tree to die and this causes branches to fall, endangering pedestrians and property. This species occurs in northern and southern Africa.

## Halictidae

The sweat bees are 5–16 mm long and mostly black with metallic blue, green, brass, or copper iridescence on their body. A few species are partially red or yellow. Most nest in clay soil and often in steep banks, and for this habit they are called miner bees. However, some nest in dead wood. They are often communal in their nesting habits, and many nests may occur together. Females will build a common main entrance and long burrow, from which individual females (sometimes her offspring) construct side galleries for their brood cells. Adult sweat bees visit flowers and collect nectar and pollen. The females of some species hibernate and are active in early spring, and on through summer. *Dialictus* and *Lasioglossum* species are attracted to and sometimes sting perspiring (sweating) individuals. They will sting if pressed by clothing or the skin. Their nests are along banks of small streams, but also bare spots in residential turfgrass. The entrance to the burrow is often ringed by excavated soil. In severely infested sites there may be many holes rimmed with soil.

**Common miner bees, sweat bees, *Halictus* spp.** Adults are 7–10 mm long and have a black body; the thorax and abdomen may have pale white setae. They are gregarious and many nests may occur together in a small area. The female excavates a long burrow into the ground or a vertical surface, and along the sides of this main burrow are short branches, which lead to a brood cell. The sides of the cell are usually lined with a thin layer of clay, and in closed cells there is a small amount of pollen or nectar and a single egg. Several females may use the main burrow and build their brood cells separately. Large numbers of these bees nesting in urban habitats can be a nuisance, and their attraction to people can cause a stinging hazard. These bees are widely distributed in North America and Europe.

## Ichneumonidae

This large family has an estimated 60 000 described species. They are 2–60 mm long (excluding antennae and ovipositor), and have a front wing length 2–35 mm. The antennae are usually 30 segments or more, and with uniform diameter. The abdomen (metasoma) has five or more visible segments. The majority of ichneumonids are parasitoids of holometabolous insects and spiders. Those associated with insects normally attack the larval or pupal stage. A large number of species have economic importance in limiting populations of some agricultural insect pests. The scientific name of this group is derived from the name of the Egyptian ichneumon or Pharaoh's rat. It devours the eggs and young of the crocodile, and was held sacred by the ancient Egyptians. The parasitic habits of these wasps are linked to the habits of the Egyptian ichneumon.

Pest status of ichneumonids is based on their presence in domestic and peridomestic habitats. The size and habits of some species make them threatening; the females of some species are capable of inflicting a painful sting. Adults in some genera are nocturnal and attracted to lights at night. Several species of *Amblyteles*, *Dolichometus*, *Diphyus*, and *Ichneumon* are brightly colored with black and yellow bands, and some are common in urban and suburban gardens. *Ophion macrurum* and *O. bilineatus* are common species in peridomestic habitats in North America. *Venturia canescens* is a common parasite of the flour moth, *Ephestia kuehniella*, and large numbers may sometimes occur in flourmills. In Africa, *Eniscospilos* species are parasites of cutworms (Noctuidae), and they are attracted to lights at night. In the UK, *Netelia testacea*, which parasitizes noctuid moth caterpillars, comes to lights and windows at night. *Rhyssa* and *Megarhyssa* species are parasites of horntail (Siricidae) and wood wasp (Xiphydriidae) larvae. The larvae of *Rhyssa* and *Megarhyssa* feed externally on the host. In North America, *Rhyssa nitida* is a parasite of wood wasps. *Megarhyssa atrata*, *M. greenei*, and *M. lunator* attack horntails (*Tremex*).

***Gelis latrodectiphagus*** Adult females are about 6 mm long, including the ovipositor; they are wingless. Males are about 5 mm long and winged; the body is shiny black. Eggs are deposited in the egg sacs of black widow spiders, *Latrodectus mactans*. Development from egg to adult requires about 10 spider eggs; full-grown larvae of the parasite spin elongate cocoons inside the egg sac. Adults emerge through small holes cut in the surface of the egg sac. This ichneumonid occurs in Africa. Other *Gelis* species are parasites of spider eggs.

***Megarhyssa atrata*, *M. greenei*, *M. lunator*** Adults are about 25 mm long; the ovipositor is 40 mm long or longer. *M. atrata* has a black body and a yellowish-brown head; both *M. greenei* and *M. lunator* have brown abdomens with angular yellow stripes on the sides. The sheath of the female's ovipositor is used to locate the host larvae in the wood, perhaps by vibrations produced by the larva feeding. The ovipositor sheath is lifted away and the two halves of the ovipositor move rapidly backwards and forwards to drill into the wood. The ovipositor can penetrate 13 mm or more into wood. The egg is greatly deformed as it passes through the valvulae of the ovipositor but regains its shape after emerging into the gallery. Oviposition is on the host larva or in the gallery close to it. The female parasite may first puncture the host larva with her ovipositor; she then deposits an egg close to it. The full-grown *Megarhyssa* larva pupates in the gallery, and the adult ichneumonid chews through the wood to the surface.

***Rhyssa persuasoria*** Adults are about 25 mm long, excluding the ovipositor. The thorax and abdomen are blackish brown and marked with pale yellow to pale brown bands and stripes. Oviposition begins with the female palpating the bark of a tree with her antennae. When the larva of a siricid is detected, the ovipositor is inserted into the wood and the hind coxae form a guide to aid the insertion process. When a siricid larva is contacted, it is injected with a paralyzing fluid. The *Rhyssa* egg is about 12 mm long, and part of this length is a narrow pedicel. The egg substance flows into the pedicel during oviposition, then extrudes back again when it emerges from the ovipositor. The *Rhyssa* larva is an ectoparasite of the full-grown siricid larva. Development takes about 5 weeks, and the adult chews a exit hole to the surface.

## Megachilidae

These are leaf-cutting bees, *Megachile* spp., and mason bees, *Osmia* spp. and *Chalicodoma* spp. The adults are 2–18 mm long and their body is usually black or brown, but many are metallic green, blue, or purple; some have yellow markings. Nests are built in rotten wood or in holes in solid wood, in holes in soil, and other natural cavities. The nests of leaf-cutting bees are lined with oval or circular pieces cut from the leaves of plants. The pieces are fitted together to form tight brood cells in a small space. The nests of mason bees are lined with soil instead of leaves, and they utilize small openings and natural cavities for nest sites.

**Mason bee, *Chalicodoma muraria*** Adults are about 16 mm long and the body is densely setose; the female is black and the male is brown. This species builds small to large mud nests on walls or flat rock surfaces. The nest is built of soil particles mixed with a salivary secretion and with the addition of small pebbles. This mixture is applied to wall surfaces or on exposed surface of large stones. After eight or nine cells have been built, provisioned, and an egg deposited, the entire nest is covered with the same mixture of soil and salivary secretion to produce a dome that is 7–10 cm in diameter. Larvae complete development and emerge from the nest the following spring. Parasites of the developing larvae include species of *Leucospis* and *Stelis* (Bombyliidae). In *Chalicodoma*, there are some very large bees, including *C. pluto*, which may be 39 mm long, from eastern Indonesia.

***Chelostoma campanularum*** Adults are about 5 mm long and the body is black with pale yellow markings. Nests are constructed in the exit holes of wood-infesting beetles. This species collects pollen from *Campanula* spp. flowers. Pollen is carried on a pollen-brush on the underside of the abdomen, and the bee must enter the nest opening backwards to deposit the pollen mass. This species occurs in the UK.

**Leaf-cutting bees, *Megachile latimanus*, *Megachile* spp. (Fig. 9.9)** Adults are 1.5–12 mm long. The body is black, but sometimes appears yellowish white because of dense pale white setae. Females usually select small, natural cavities as nest sites; they may excavate tubular cavities in wood or other soft substrates. Females cut 2–3 oblong sections from green leaves and transfer these pieces to the nest site to form the side and the bottom of the brood cell. When these pieces are fitted in the cell, a supply of pollen and honey is positioned there and an egg is placed on this food supply. The female will then cut 1–3 circular pieces of leaves, the diameter of which is slightly greater than the diameter of the tubular nest. The pieces are gently forced into the opening to make a tight fit. Usually several cells of this kind are placed end to end in a burrow. Sometimes several females will build their nests together in the same site. A communal nest site of 30 nests is estimated to contain 1000 pieces of leaves.

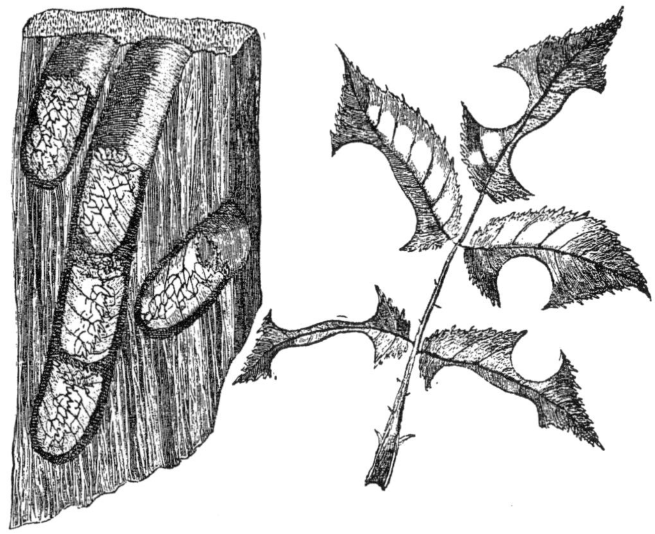

**Figure 9.9** Hymenoptera: Megachilidae. *Megachile latimanus* damage to plant leaves, and multiple nesting site.

**Common mason bees, *Osmia cornifrons*, *O. rufa*** Adults are 1.7–19 mm long and usually metallic blue or green. They construct small, earthen cells in burrows in the soil, or in decaying wood, beetle burrows in wood, in the brick veneer of structures, and other natural cavities. In spring, *O. rufa* females burrow into the soft mortar of stone- and brickwork, and line their nest cells with soil. This species may occur in large nesting aggregations and may do serious damage to the brick veneer of structures in the UK and parts of continental Europe. *O. cornifrons* occurs in eastern USA and parts of continental Europe.

## Mutillidae

Velvet ants are 3–23 mm long and their body is covered with long, hairlike setae; they are usually brightly colored with a banded abdomen. The females (Fig. 9.10b) are wingless and resemble ants, and their body coloring may be banded black, yellow, or red. The male (Fig. 9.10a) is winged and usually marked differently than the female; males are sometimes found on flowers. Females are well known for their painful (but not lethal) sting. The name cow-killers for some large North American species is from the myth that the sting is powerful enough

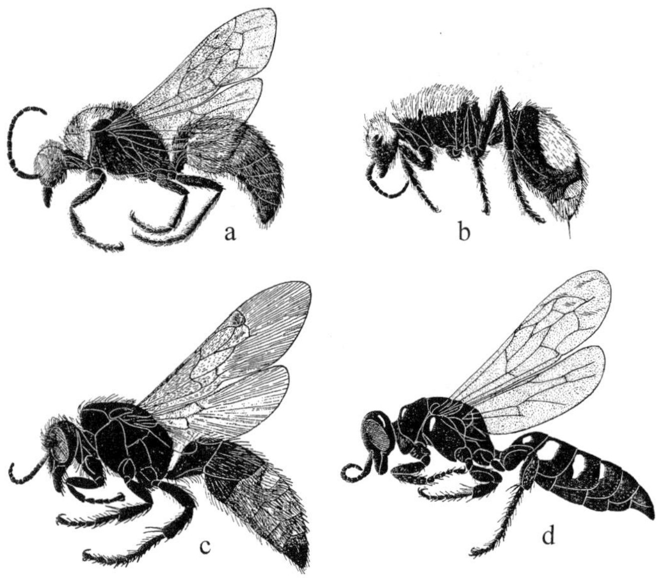


**Figure 9.10** Hymenoptera: Mutilidae, Scoliidae, Tiphiidae. (a) *Dasymutilla* sp. male; (b) *Dasymutilla* sp. female; (c) *Scolia* sp.; (d) *Tiphia* sp.

to kill livestock. Most species are external parasites of larvae of ground-nesting bees and mud dauber wasps. In the UK, *Mutilla europaea* is a parasite of *Bombus* spp. In Africa, *Dolichomutilla guineensis* is a parasite of the mud wasp, *Sceliphron spirifex*. Other hosts of mutilids include the puparia of flies, the pupae of moths and beetles, and the oothecae of cockroaches. Species of *Smicromyrme* have been reared from the puparia of tsetse flies (Glossinidae). Some species are believed to dig nests in the soil and provision them with insects.

Mutillids are active during the day or night. Diurnal species usually search for hosts during the coolest time of day. Females search the ground or vegetation for openings of host nests or insect cocoons. Species that attack wasp nests spend several days within the host nest laying eggs. The mutillid female opens the cell of the host to oviposit, and then the cell is resealed. The first-stage mutillid larva punctures the cuticle of the host with its sharp mandibles and feeds on the hemolymph. Development is rapid and there are about five larval stages. Pupation is in a cocoon within the host cell. Generally, only one mutillid develops per host. The males leave the nest of the host soon after they emerge, but the female may remain in the nest until the following spring. There is usually one generation per year.

Both sexes stridulate under stress or when grasped. They rub a finely striated medial area at the base of the third abdominal (metasomal) tergite against a transverse ridge on the underside of the second tergite. The sound is a hiss-like chirp.

**Common eastern velvet ant, cow-killer ant, *Dasymutilla occidentalis occidentalis*** The adult female is 16–18 mm long and the body is black and marked with red hairs, except on the apex of the metathorax, and abdominal segments 1 and 3. The male has red hairs on the pronotum and dorsal part of the thorax, and the abdomen has red hairs, except on the basal segment. This species is parasitic on the bumble bee, *Bombus fraternus*. It is distributed in eastern USA.

**Western velvet ant, *Dasymutilla sackeni*** The adult female is 12–14 mm long, and the black body is covered with pale yellow hairs, except for the legs and the underside, which are covered with black hairs. The posterior half of the male abdomen is covered with pale yellow hairs, while the anterior portion is black. This species is a parasite of the wasp *Bembix occidentalis*, which is a ground-nesting wasp that occurs in western USA.

***Dolichomutilla guineensis*** Adult females are about 12 mm long; they have a red thorax and a black abdomen with a white band. This species is a parasite in the nests of the black-and-yellow mud wasp, *Sceliphron spirifex*. It occurs in southern Africa.

**Velvet ant, ant-like fosser, *Mutilla europaea*** The adult female is about 14 mm long and has a black body and an orange-red thorax, which is lightly colored with black hairs. The abdomen of the female has three white bands, while the male has two white bands. The males may be found at flowers, such as bramble (*Rubus fruticosus*), and they are attracted to lights at night. Females are usually found in the nests of their host or foraging on the ground during the day. This species occurs in the UK and the range probably extends to Europe.

## Pompilidae

Spider wasps have a wing length of 2–50 mm and their body is black, bluish black to reddish brown. The abdomen is black with yellow markings, or yellow with black and/or red markings. The legs are long. The females have a potent sting but they are not aggressive. These wasps are rapid flyers and fast runners, and these behaviors may be threatening to people. Pompilids search for free-living spiders and, after locating, stinging, and paralyzing one, the female spider wasp carries it to a secluded place. She places it in the soil and then lays an egg on the spider. Many species excavate burrows in the ground, and some, such as *Auplopus*, construct elaborate

**Figure 9.11** Hymenoptera: Pompilidae, Sphecidae, Vespidae. (a) *Chalybion californicum* (Sphecidae); (b) *Sceliphron caementarium* (Sphecidae); (c) *Sphecius speciosus* (Vespidae); (d) *Eumenes fraterna* (Vespidae); (e) *Pepsis* sp. (Pompilidae).

mud nests above ground. Pompilids generally catch their prey before preparing the burrow or nest site; almost all sphecids (Sphecidae) build a burrow first then search for prey. One of the largest pompilids in Europe is *Cryptocheilus comparatus*. This species hunts wolf spiders (Lycosidae) and other large spiders. Among the largest wasps in this family are the tarantua-hawks of the genus *Pepsis* (Fig. 9.11e); they occur primarily in southwestern USA. These large wasps provision their nests with tarantulas.

**Orb-web spider wasp, *Agenioideus humilis*** Adults are frequently seen around buildings where they prey exclusively on large orb-web spiders (Araneidae). These spiders often build nests near outdoor lights. Female wasps excavate short burrows, often in cracks and crevices in walls of buildings. This species occurs in Puerto Rico. A related species, *Agenioideus birkmanni*, which is widespread in North and Central America, preys on gnaphosid spiders.

## Pteromalidae

These small wasps are 1–8 mm long, and most are metallic blue or green; some are black or yellow. Some species have large punctures in the cuticle of the thorax, and they have a small abdomen. Most are parasites and some are hyperparasites; some form galls on *Acacia* and *Eucalyptus* (Australia). The hosts of these wasps are spiders and insects. They attack eggs, larvae, pupae, and, rarely, adults.

Three cosmopolitan species are frequently found indoors and are parasites of beetle pests of stored grain: *Theocolax elegans*, *Anisopteromalus calandre*, and *Lariophagus distinguendus*; *Pteromalus cerealellae* parasitizes lepidopterous pests of grain. Several pteromalid species are associated with beetles infesting structural wood, including *Theocolax formiciformis* with anobiids, and *Cerocephala rufa* and *Cerocephala* spp. with Scolytidae, Anobiidae, Bostrichidae, and Lyctidae. The host of *Cratomus megacephalus* is unknown, but it has been found associated with wooden posts.

## Scoliidae

These wasps are 20–50 mm long, and are usually black with patterns of red, yellow, and white on the abdomen. The larvae are parasitic on the grubs of scarab beetles, particularly species of *Phyllophaga*. Courtship of some species, such as *Scolia dubia*, consists of large numbers of both sexes flying low over the ground; egg-laying begins soon after mating. Females have spiny legs adapted for burrowing into soil where they seek large Coleoptera larvae, usually Scarabaeidae. When the host larva is contacted, the female stings and paralyzes it. The venom of Scoliidae contains kinins, which apparently block portions of the insect nervous system. After the host is paralyzed, the female lays an egg on it. In some cases, the female drags the paralyzed host into a crude cell she made in the soil before ovipositing. Females lay 1–2 eggs per day for about 2 months. Larval development is completed in about 1 week and there are three larval stages. At the end of larval development, a cocoon is spun in or near the host, and the pupal period is about 3 weeks.

**Four-spotted grub parasite, *Campsomeris quadrimaculata*** Adults are 15–30 mm long and the body is black. Female abdominal segments 2 and 3 have two large orange spots. Some males have five abdominal segments with yellow bands. This species is distributed in the USA.

**Western scoliid, *Campsomeris tolteca*** Adults are about 25 mm long, and the head, thorax, and legs are black with pale white hairs. The abdomen is reddish brown with black markings; the wings are clear. The distribution of this species includes California, Arizona, Texas, Mexico, and Haiti. A related species in Central America, *C. dorsata*, attacks *Ligyrus nasutus* and *Phyllophaga* spp. (Scarabaeidae).

**White grub parasites, *Scolia dubia*, *Scolia* spp. (Fig. 9.10c)** Adults are about 21 mm long and black; the abdomen has the second basal segment black, and the remaining segments are reddish brown. Abdominal segment 3 usually has two large dorsal yellow spots. The female flies above the soil or turfgrass seeking a suitable site to search for grubs. She lands on the surface and digs into the soil or turf thatch with her mouthparts and legs. After locating a suitable host, she lays an egg on it. The wasp larva completes development on the beetle grub.

## Siricidae

Adult horntails or woodwasps are 20–35 mm long. Most species have a triangular or spearlike process at the apex of the female abdomen; this process is short in the male. Their body is cylindrical, and they have long slender antennae. Females are large and have a long ovipositor. Adults are mostly black or metallic dark blue, or combinations of black, red, and yellow. Eggs are deposited in the bark of trees and shrubs, and the larvae are wood borers. Larvae are cylindrical, yellowish white, and with a small spine at the posterior end.

Horntails attack both hardwood and softwood trees. A few species have been recorded attacking vigorous trees, but they prefer trees, or parts of trees that have been recently felled, dead, or that are weakened. Larvae may survive in wood that is air-dried and adults emerge after the wood is in use. Small adults emerge from wood that is dried during larval development. Kiln-drying lumber kills larvae. Larval tunnels are 4–8 mm diameter and circular, but they may be exposed longitudinally during sawing of infested logs and preparing finished lumber, resulting in an oval shape. The fibrous frass is tightly packed in the tunnels, but it may become loose after drying. Siricids are vectors of phytopathogenic fungi, and their activity can kill and damage trees. Qualitative losses of timber occur from trees that survive attack. An outbreak of *Sirex noctilio* in the *Pinus radiata* forests of New Zealand in late 1940s and establishment of this species in Tasmania and mainland Australia caused considerable economic losses.

The female uses her long ovipositor to penetrate the bark and insert eggs into the cambium to a depth of 10–20 mm. Eggs are laid in batches of about seven and egg-laying is completed in about 14 days. Hatching occurs in 21–28 days, but in *Tremex columba*, egg hatch is delayed until the following summer; fecundity is 300–400 eggs for large species. Early-stage larvae begin burrowing into the wood at right angles to the oviposition channel. They feed by excavating tunnels in the wood, and use the spine at the end of their abdomen to pack the frass and as support during feeding. After feeding in the outer sapwood, larvae tunnel into the center of the tree, then return to the surface. Cast skins from the three or four immature stages are incorporated into the frass. Pupation occurs at the enlarged end of a tunnel, about 2 cm from the wood surface; the pupal period is 5–6 weeks. The adult chews a circular hole to emerge. The minimum period for the life cycle is 1 year in warm climate regions, but it may be extended to 2 or 4 years in cold climate regions.

Adults generally emerge in late summer and early fall. Adults fly in sunshine, and make a buzz like bees. Mating takes place in treetops, which is where the male bees usually remain; females descend to low levels to lay eggs. Females are facultatively parthenogenetic, and many oviposit before mating. *T. columba* deposits about half of its eggs in the frass in its gallery before emergence. These unfertilized eggs produce males, some hatching in the fall, others the next spring. Mated females produce offspring of both sexes.

Horntail larvae are parasitized by several species of ichneumon wasps (Ichneumonidae) of the genera *Megarhyssa* and *Rhyssa*. *Megarhyssa atrata lineata*, *M. greenei greenei*, and *M. macurus lunato* have ovipositors that are 35–40 mm long and longer. The female wasp inserts the ovipositor through the bark and deep into the wood to deposit eggs on or near horntail larvae or pupae in their galleries. The female parasite first punctures the beetle larva with her ovipositor; she then deposits an egg close to the potential host. The larva of *Rhyssa* feeds externally on its host. The life cycle of the parasite is completed in 1 year.

Symbiotic fungi are utilized as food by siricid larvae. The larvae do not ingest wood, but they extract necessary nutrients from the fungal mycelium, which grows in the tunnels, and is digested by their salivary secretions. The fragments of wood removed by the larval mouthparts are passed along the outside of the body and packed in the tunnel behind them.

During oviposition, the female deposits a small amount of fungal spores with each egg, and the mycelia that develop from these spores provide food for the developing larva. The fungus associated with siricids is often species-dependent, and includes species of *Amylostereum* and *Daedalea*.

**Western horntail, *Sirex areolatus* (= *S. apicalis*)** Adult females are 24–35 mm long, dark metallic blue; the wings are dark. Males are 18–24 mm long, and the abdominal segments 3–7 and sometimes the front tibia and tarsi are dark red to reddish brown (and known as *S. apicalis*). This species attacks dead Monterey cypress, and other cypress trees, and it is known to infest redwood lumber in storage. This species occurs from New Mexico north to Washington in the USA, and British Columbia.

**Blue horntails, *Sirex californicus, S. cyaneus, S. edwardsii, S. juvencus, S. nigricornis, S. obsus*** Adults are 23–30 mm long, the body is usually dark metallic blue to black, and the wings are clear to completely dark. The legs, except the coxae, are red or yellow. The ovipositor extends 10–15 mm beyond the tip of the female abdomen. Larvae are 20–30 mm long and yellowish white; the head is dark yellow to yellowish brown, and the legs are small. The larvae of these species attack a variety of coniferous trees, including pine (pitch and shortleaf pine), fir, spruce, and Douglas fir. They occur in eastern Canada, and in northeastern and midwestern USA. *S. californicus* occurs in California, Oregon, Washington, and British Columbia, and has been reported from New Mexico.

**Long-tailed horntail, *Sirex longicauda*** Adult females are about 40 mm long, and the body is dark metallic blue. Males are about 25 mm; the abdominal segments 3–7, and the tibiae and tarsi of the front and midlegs are reddish brown. This species may be restricted to California; it is commonly found indoors.

**Sirex wasp, *Sirex noctilio*** Adult females are about 25 mm long and metallic blue. Males are about 18 mm long, metallic blue, and have a dark-orange abdomen. The emergence hole in the wood is 5–6 mm diameter. This species attacks *Pinus* spp. trees, and often causes the death of the tree. It prefers to oviposit in *Pinus radiata* trees that have wide annual rings; these trees typically have a small amount of heartwood. Larvae complete development in 1 or 2 years and have 7–8 larval instars. It is known from the Eurasian region, Canada, and South Africa. It was introduced into New Zealand and Australia.

**Pigeon tremex, *Tremex columba* (Fig. 9.1c)** The adult female is 37–50 mm long. The head, antennae, and thorax are reddish brown to black; the abdomen is black with yellow bands and spots along the sides. The wings are about 50 mm long, and light brown. Males are 18–37 mm long, and the body is reddish brown with some black markings. Antennae are slightly swollen in the middle, and are as long as the head and thorax combined. Eggs are laid singly at depths of 10–12 mm in the wood. Full-grown larvae are about 50 mm long, and yellowish white. The end of the abdomen has a sclerotized process that has two pairs of small teeth. This species attacks beech, maple, birch, elm, hickory, sycamore, and oak. It occurs throughout northern North America, but a subspecies has been reported from elm timber imported into New Zealand. A related species, *Tremex fuscicornis*, occurs in Europe and Asia.

**White-horned horntail, *Urocerus albicornis* (= *Sirex abdominalis*)** Adult females are 25–30 mm long, and the body is bluish black to black; wings are dusky. The middle of the antennae, the cheeks on the head, and the bases of the tibiae and tarsi are white; and white spots may occur on the abdomen. Males have abdominal segments 3–6 yellow, and the wings are clear. This species attacks many species of coniferous trees and occurs throughout northern North America. It may attack freshly sawn lumber.

**Yellow-horned horntail, *Urocerus flavicornis* (= *Sirex bizonatus, Urocerus riparius*)** Adults are 22–33 mm long, and the body is black. The antennae, tibiae, and abdominal segments 1, 2, 7, and 8 are yellow or reddish yellow. Wings are clear or slightly yellowish at the bases. The males have abdominal segments 3–6 yellow and the wings are clear. This species breeds in pine, fir, Douglas fir, and other coniferous trees. It occurs throughout northern North America.

**Wood wasp, *Urocerus gigas* (Fig. 9.1b)** Adults are 18–35 mm long, the body is bluish black, and there is a yellow patch below each eye. Larvae are about 30 mm long, yellowish white, and the spine at the posterior end is dark brown. This species attacks a variety of coniferous trees. It occurs in the UK and Europe. The ichneumonid, *Rhyssa persuasoria*, is a larval parasite of the larvae.

## Sphecidae

These are the digger and mud dauber wasps. The adults are 2–40 mm long and the body is black to reddish brown; some are metallic blue or green, and some are marked with yellow or reddish brown. The hind legs and antennae are long, and the body setae are sparse. They are very active and take short erratic flights. They usually nest in subterranean cells, which are provisioned with insects or spiders. This is a diverse group of wasps, and the prey used for provisioning brood cells includes homopterans, various species of wood-infesting beetles, winged ants, bees, and flies captured in flight. *Microbembix monodonata* is a large wasp with greenish-white markings; it is abundant along the Atlantic, Gulf, and Pacific coasts. It preys on dead or dying insects. Nests of most sphecids are usually in sandy areas, but many species utilize natural cavities above ground, including abandoned beetle emergence holes in wood, and a few species build mud cells.

**Thrips wasps, *Spilomena* spp.** Adults are about 9 mm long and black; the petiole is short. This wasp is abundant in the spring; thrips wasps capture thrips to provision their nests. The prey is stored in abandoned tunnels of the furniture beetle, *Anobium punctatum*. They may first clean the beetle tunnels, and create a pile of ejected frass below the holes, which gives the impression that the beetle larvae are active in the wood. *Spilomena* is a large genus and species are distributed around the world.

### Cockroach wasps

*Ampulex* wasps are 8–15 mm long with a shiny black to brownish black body. Their wings are clear and the antennae long. These wasps are widely distributed around the world. They are characterized by not making nests to rear their young, but drag their prey to a nearby cavity or hole in the ground. There are many species in the genus and cockroaches are the only known prey.

***Ampulex amoena*** This species occurs in Asia. It preys on small nymphs of *Periplaneta americana* and *P. australasiae*. The female cockroach wasp grasps the cockroach nymph and stings it several times in the thorax, and moves the nymph to a cavity, either indoors or outdoors. The wasp may amputate the cockroach antennae, and ingest hemolymph from the open end. One egg is deposited on the cockroach mesothorax. As many as three cockroaches may be attacked and stored each day. Eggs hatch in 2 days, and the first-stage larva feeds on the cockroach hemolymph for 1–2 days, then feeds on the internal body parts for about 2 days. Adults emerge from the pupal cocoon in about 30 days.

***Ampulex compressa*** This species occurs in Asia and Hawaii. It preys on adult *Periplaneta americana* and *P. australasiae* cockroaches. Females remove the legs, antennae, and wings of the cockroach, and then attach an egg to the mesothoracic coxa. These wasps will frequently enter houses in search of prey, and they may prey on 43–71 cockroaches.

### Mud dauber wasps

Adults of *Chalybion* and *Sceliphron* are 18–30 mm long, and the body may be black, purplish blue, metallic blue, or black with yellow markings. These wasps prey on spiders, grasshoppers, katydids, crickets, and caterpillars. Females build mud nests on the sides of buildings and in protected sites, including in attics and sheds. Cells of mud are constructed until they make a mass 4–8 cm wide. After provisioning each cell with prey and laying an egg, the cells are covered with mud to make a smooth outer surface. Larvae complete development in about 3 weeks and spin a cocoon, but they pupate the following spring. These wasps rarely sting, and do not defend their nest site.

**Sand wasps, *Bembix* spp.** Adults are 11–20 mm long; their body is black and the abdominal segments have yellow markings, which are similar to yellowjackets. They have relatively large heads and eyes. The abdomen has a broad base and tapers to a point posteriorly. They build nests in sunny and sandy locations; some species are common along the seashore. Numerous nests may occur in small areas, and the nests are often defended. They provision their nests with flies, often *Eristalis* (Syrphidae) and *Tabanus* (Tabanidae). Their nests consist of a tunnel in the soil about 1 m long with a horizontal terminal branch. Cells containing developing larvae are left unsealed, and females provision the cells daily. In *Bembix rostrata* a single female supports five or six larvae and each larva requires 50–80 flies during the 14–15-day development period. When the larva is nearly full-grown the female closes the brood cell and the larva spins a cocoon. Overwintering is spent in the prepupal stage, and the adult emerges in the spring.

***Chalybion californicum* (Fig. 9.11a)** Adults are 22–24 mm long and entirely metallic blue; the pedicel is about the same length as the remainder of the abdomen. They sometimes take over nests of other mud dauber species. Females oviposit on the provision of paralyzed spiders intended for the larvae of the true mud daubers (*Trypoxylon* spp. and *Sceliphron caementarium*) in mud cells under construction. Adult females often visit pools of water to gather water to soften the entrances of the cells in the nest. The female removes the spiders stored by the previous wasp, and replaces them with spiders she has collected, and seals the cell. This species is generally distributed in the USA.

**_Sceliphron caementarium_ (Fig. 9.11b)** Adults are 24–28 mm long and the body is black with yellow markings. This species builds mud nests on ceilings and walls of sheds, sides of buildings, and other peridomestic substrates. Females are often seen collecting water at the edge of pools and puddles. A related species, the black-and-yellow mud wasp, *Sceliphron spirifex*, occurs in Africa. It frequently builds large mud nests provisioned with spiders on the outside of buildings. The mutillid (Mutillidae), *Dolichomutilla guineensis*, and *Osprynchotus* spp. ichneumonids (Ichneumonidae) are parasites of this mud wasp.

**Cicada killer wasp, _Sphecius speciosus_ (Fig. 9.11c)** Adult females are about 39 mm long; males are about 29 mm long. The body is blackish brown with reddish-orange legs, and yellow markings on the thorax and abdominal segments 1–3. The prey is species of annual cicadas, *Tibicen* spp., and they nest in burrows in the soil. Nesting occurs in late summer and fall. Searching for prey begins early in the morning and continues throughout the day. The female flies around the tree or shrub in a circular pattern. When a cicada is located, it is stung between the abdominal segments. The wasp turns the cicada on its back, grasps it with its legs and flies back to the burrow in the soil. Burrows dug by the female may be 180 cm long, and contain about 15 cells provisioned with 1–3 cicadas each. Burrows are typically straight for half to a third of their total length; they then curve to form a J-shape. Larval development is completed in the fall and adults emerge the following spring. Females are not aggressive, but males will actively protect areas around burrows. Nests can be numerous in turfgrass. This species is widely distributed in North America, east of the Rocky Mountains.

**Horse guard, _Sticta carolina_** Adults are about 22 mm long and the body is black with yellow markings. It occurs in southern USA. Its habit of hunting for flies around horses is the basis of the common name.

**Pipe-organ wasp, _Trypoxylon politum_** Adults are about 14 mm long and the body is black, with pale white setae; the hind tarsi are pale. This species builds mud cells in parallel rows that resemble the pipes of an organ. The nests are provisioned with spiders, with 3–18 spiders per cell.

## Stephanidae

These wasps are 11–60 mm long, including the ovipositor. Front wing length is 6–15 mm. The body is elongate, cylindrical, and black to dark brown. The head is slightly spherical with a dorsal ring of blunt to acute teeth (*stephano*, Greek for crown). This is a cosmopolitan group, but the majority of species are in the eastern hemisphere tropics. The three genera occurring in the western hemisphere are *Foenatopus*, *Megischus*, and *Schlettereius*. They are solitary ectoparasites of wood-boring insects, including buprestids, bostrichids, and siricids. These wasps are most often encountered on or around dead standing trees, or around wooden buildings. The female stings and immobilizes the host larva prior to laying an egg on it. The egg hatches in 10–14 days, and the first-stage larva punctures the integument of the host to feed. Larval development is 6–7 weeks long and the host is completely consumed. The final instar pupates in the site where it finished feeding.

## Tenthredinidae

The adults are called sawflies; they are 5–20 mm long and the body is black or brownish black with some red and yellow markings on the thorax and abdomen. Antennae are slender, with 7–10 segments. Adults may be found on vegetation, some visit flowers; the adults of some species are predaceous on other insects. They can be numerous at certain times of the year and gather on the outside of houses and buildings. Larvae resemble Lepidoptera caterpillars, but they have seven pairs of prolegs and three pairs of thoracic legs. Most species feed on leaves, but a few are leaf miners or stem borers. There is usually one generation a year, and overwintering is in a pupal cell or cocoon in the soil or a protected site. Full-grown larvae of *Lophyrotoma* spp. move from their feeding site to a secure place to pupate. Wood in houses, fences, and other outside uses are often selected for pupation sites. These sawflies, which are characteristically pale green, occur in Australia. In peridomestic habitats in eastern USA, the common tenthredinids include *Dolerus nitens* and *Eutomostethus ephippium*, which feed on grasses, and *Ametastegia pallipes*, which feed on *Viola* spp.

**Pear slug, _Caliora cerasi_** Adults are 8–12 mm long, shiny black and with grayish-black wings. The full-grown larva is 10–13 mm long and usually dark green. The prothoracic area is much larger than the rest of the body; there are seven pairs of prolegs and three thoracic legs. Larvae feed on the epidermis of the leaves of a variety of fruit trees, including varieties of cherry and pear. Adults appear in early spring and again in late summer. Eggs are deposited in the leaves on the host tree; hatching is in 1–2 weeks, and the first-stage larvae feed on the upper surface of the leaves. They may remove the leaf surface, leaving only the veins. Females of the spring generation are parthenogenetic and produce a large generation in fall. Overwintering is in cells in the soil near the host trees. Adults can occur in large numbers in spring, collect around the outside of houses, and fly to lights at night. Their abundance and resemblance to subterranean termites (reproductives) often cause concern. This species occurs throughout North America, the UK, and Europe.

## Tiphiidae

These wasps are 6–26 mm long, and the body is black, sometimes black with yellow markings. Males are always winged, but females may be winged or apterous and ant-like in appearance. Most species of tiphiids are solitary, ectoparasites of large soil-dwelling insects. They are known to parasitize larvae of tiger beetles, and various ground-nesting bees and wasps. The majority attack beetle larvae, and show a preference for full-grown individuals. Some common species are parasites of white grubs, including June beetles, *Phyllophaga* species (Scarabaeidae), and the Japanese beetle, *Popillia japonica*. The single species in the subfamily Diamminae, *Dammia bicolor*, attacks mole crickets (Gryllotalpidae).

The *Tiphia* female burrows into the soil to gain access to a scarabaeid larva. The wasp stings the host and injects venom that causes temporary paralysis. After manipulating the beetle larva, she attaches an egg on its abdomen. There are five larval instars, and the beetle grub may not be killed until the last instar. The full-grown wasp larva forms a cocoon in the soil. Adult tiphiids feed on honeydew and nectar, often during the morning hours.

**Common grub parasites, *Myzinum maculatim*, *M. quinquecinctum*** Adults are 12–25 mm long, the body is black and the abdomen, legs, and sometimes the wings have yellow or orange markings. The male has an upturned spine at the tip of the abdomen (metasoma). These species are larger and slightly more slender than *Tiphia*. In some regions, they are the most common parasites of scarabs in turfgrass.

**White grub parasites, *Tiphia* spp. (Fig. 9.10d)** Adults are about 11 mm long, and generally black with yellow or orange marking on the abdomen. This genus contains several common parasites of scarabaeid larvae, and some species have been introduced to control turfgrass pests. Adults of *Tiphia intermedia* are common on flowers of wild carrot and other Umbelliferae during June to August, and they are parasitic on *Phyllophaga* grubs. *T. inornata* is a common species throughout the USA. Species introduced from Japan and Korea to control the Japanese beetle include *T. vernalis* and *T. popilliavora*. The Asiatic garden beetle, *Maladera castanea*, is parasitized by the introduced species, *T. asericae*.

## Vespidae

These are the well-known paper wasps, umbrella wasps, yellowjackets, hornets, mason wasps, and potter wasps. Most are black with yellow or white markings, or brown with yellow markings. These wasps usually fold their wings longitudinally at rest. Some species are considered eusocial, and the individuals in a colony are morphologically distinct queens, workers, and males. In other species, there is only a slight size difference between the queens and workers. Social vespids construct nests from papery material that consists of chewed wood or foliage. Colonies in temperate regions last one season, and the queens overwinter. Mason and potter wasps are solitary, and some use cavities in soil, twigs, or logs; others make a nest of mud or clay.

### Mason wasps, potter wasps

These wasps are predators, mostly of Lepidoptera larvae, and all construct nests involving soil they wet to produce mud. Some species excavate burrows in the ground and construct cells out of the soil. Nonburrowing species use mud in nest construction, and nests are built in pre-existing cavities, such as in plant stems, and beetle emergence holes. Species of *Pseudodynerus* and *Ancistrocerus* regularly use cavities in wood, such as beetle holes, and *Zethus histrionicus* is common in the holes found in houses (Costa Rica). Species of *Eumenes*, such as *E. fraterna* (Fig. 9.11d), and *Pachymenes* make mud cells in the shape of pots or jugs, which are widest in the middle and narrow to a small opening at the top. When the nest cells of mason and potter wasps are complete, but before provisioning begins, a single egg is attached to the upper surface of each cell. Most mason and potter wasps search for concealed prey, which is stung on the underside of the head or thorax. The prey is carried to the nest and placed in the cell.

Pest status of mason and potter wasps is based on their nesting habits. Nests are built in cavities or holes in wood siding of houses, or in logs of old or modern log houses and commercial buildings. Mud from the nest cavity can discolor surfaces, and wasp activity in nest construction can dislodge frass from old beetle holes and give the appearance of recent beetle emergence.

**Tube-dweller wasps, *Ancistrocerus antilope, A. parietum, A. parietinus, A. nigricornis*** Workers are about 15 mm long and black, with the head, thorax, and abdomen marked with yellow bands; the legs may be yellow. Nests may be in the twigs of plants, such as sumac, or in the holes and cavities in the old mortar between brickwork, and in household materials. These species occur in the UK.

**Chimney-maker wasps, *Odynerus spinipes*** Nests are built in vertical walls of clay banks, but also in cavities in mortar of brick walls and chimneys. Completed nests contain five or six cells and the chimney, which is made of pellets of mud, and is about 3 cm long. Saliva from the worker may help to loosen the sand in the mortar of brickwork. This species occurs in the UK.

**Keyhole potter wasps, *Pachodynerus astraeus, P. erynnis, P. nasidens*** Workers are 8–12 mm long, and black with white markings. These wasps nest in cavities in wood, and old mud dauber cells. Females of *P. erynnis* construct mud nests under siding boards of buildings, and they provision them with pyralid moth caterpillars collected from the surrounding turfgrass and landscape plants.

**Mason wasp, *Pison spinolae*** Adults are about 20 mm long. The body is black, but fine gray setae on the body mark the segments of the thorax and abdomen. The wings are slightly opaque and the tips are dark gray. Nests are in natural cavities in trees and under bark. In the urban environment this species will nest in keyholes, angles of walls, and the folds of stored clothing. This species occurs in New Zealand.

### Paper wasps, umbrella wasps

These wasps are long-legged and usually brown with few yellow markings; their middle tibia has two apical spurs. Nests for colonies founded by individual queens (*Polistes, Mischocyttarus*) are made of a single comb without an envelope, and attached to the substrate by a strong pedicle (giving the appearance of an upside-down umbrella). Queens apply a chemical secretion from a sternal gland on the abdomen to the pedicel and other nest surfaces. This chemical repels potential predators, including ants, parasitic flies, and yellowjackets. Nests for colonies founded by several queens have an envelope that extends over several combs. Adults partially feed on nectar and honeydew. Females hunt caterpillars, and often masticate the prey at the capture site. They carry a solid prey pellet back to the nest to feed other adults and larvae. In temperate regions, nests are used one season and queens overwinter. A new nest is often built next to the queen's natal nest, and the result may be a large number of new and used nests in one site. Colonies of the honey wasp, *Brachygastra mellifica*, are perennial. Tropical paper wasp colonies may last only 2–3 months, whereas the nests of others are perennial and may last up to 25 years.

Other wasps often visit the open nests of *Polistes* spp. In some cases, the visitors are homeless conspecifics whose nests were destroyed by predators, such as birds or people. In others the visitors are there to steal eggs to feed larvae in their own active colony. Before allowing visitors on to their nest, the wasps have to distinguish between orphaned kin, which will be helpers, and unrelated wasps, which are threats. Paper wasps make this distinction using chemical odors. Each wasp assimilates from its nest an odor specific to the insects that live there. This smell, which serves as a recognition cue, is locked into the wasp's epicuticle. The source of the compounds that make up the odor is the plant fibers used to construct the nest and secretions from the worker wasps in the colony. Because each colony uses a unique mixture of plants in nest construction, nonthreatening family members are more likely to share this environmentally acquired label.

*Polistes* is the most widespread of the social wasp genera, and its range encompasses that of nearly all others. There are about 200 species, and they are distributed almost worldwide in temperate and tropical regions. These wasps are often dominant or the only wasp genus present in many regions, and usually very successful. *Polistes* nests are relatively small and inconspicuous, and they can nest in a variety of above-ground locations. Their abundance and nesting habits often bring them into contact with people, especially in urban environments.

Pest status of paper wasps is based on their aggressive nest-defense behavior and their painful sting, which is life-threatening for some people. These wasps are pests when they nest near or on buildings, and workers respond aggressively toward movement near the nest. The cavity-nesting species, such as *Agelaia* in tropical regions, often use wall spaces or discarded domestic containers to build their nests. The aggressiveness of species in the South American genus *Stelopolybia* is variable; at least one species, *S. areata*, is considered non-aggressive. Species of *Synoeca* are considered dangerous because of the barbed sting that frequently remains after stinging. In Southeast Asia, colonies of *Polybioides* have been used in booby traps in warfare. Species of *Apoica* forage at night, and many are a stinging hazard to persons working in tropical forests at night. *Polistes* are serious pests in the urban and agricultural ecosystem. In southwestern USA and regions of Mexico, populations of *Polistes exclamans* threaten laborers picking citrus fruits. Nests per tree vary from 1 to 26 and the number of wasps per nest is 100–200. *P. apachus* and *P. fuscatus* have been a problem for agricultural workers.

**African paper wasp, *Belonogaster junecus*** Adults are about 35 mm long and dark brown. Nests consist of about 50 cells, but may sometimes be 100–200 cells. There are no sterile workers in these wasps, and the colony consists of one or more founding females, daughters, and males. Older females are involved in egg-laying and younger females forage. Males remain at the nest but they do not forage and are fed by females. When colonies reach a certain size, small groups of females leave to form new nests elsewhere. Nests may be built on the outside of buildings. This species occurs in most parts of southern Africa.

**Honey wasps, *Brachygastra* spp.** These are the honey-producing wasps, which are maintained in a semidomesticated state by native people in Mexico. The nests have several horizontal combs with the cells facing downwards. The combs are surrounded by a thin, paper envelope. Colonies may continue for many years with nests composed of several thousand workers.

**Petiolate paper wasp, *Mischocyttarus flavitarsis*** Workers are 16–20 mm long, and the body is black, with orange-yellow and reddish brown markings. The petiole is primarily black, and the abdomen (metasoma) is primarily yellow. It occurs in southwestern USA, including California. Species in this genus are distributed from northern Argentina to British Columbia in the west and through the West Indies to Florida in the east. Females overwinter under bark, and they have been found with *Polistes fuscatus aurifer* and *P. dorsalis californica*. They have also been found in an abandoned nest of *Vespula arenaria*. The nests of *M. flavitarsis* may be established under the eaves of houses and branches of trees, and are very similar to the nests of *Polistes*, but smaller. A related species, *M. drewseni*, in Guyana attacks honey bee hives and causes problems with beekeepers. *M. phthisicus* and *M. mexicanus* are neotropical, and extend into Texas and Florida; *M. flavitarsis* has been introduced into Hawaii.

**Common paper wasps, umbrella wasps, *Polistes* spp.** Workers are about 18 mm long, brown to blackish brown, and with yellow marking on the abdomen. These wasps occur in all regions of the world except the coldest part of continents. In Europe and North America, colonies of *Polistes* probably outnumber colonies of all other social wasps combined. In the UK, *P. dominulus* and *P. gallicus* occur in the southern parts of the country, and establish nests only during very warm summers. *P. gallicus* occurs in Europe and was introduced into eastern USA in the 1980s. *P. humilis* occurs in New Zealand; it was accidentally introduced from Australia. The Pacific Island hornet, *P. habraeus*, occurs on islands in the southern Pacific region, and sometimes in New Zealand. The hymenopterous parasite, *Elasmus polisti* (Elasmidae) is an ectoparasite of *Polistes* spp. (reported from *P. exclamans*). Species of *Sulcopolistes* are social parasites in the nests of *Polistes* in the UK and continental Europe.

Temperate-region *Polistes* species overwinter as fertilized queens overwinter. Single queens initiate nests in spring. However, she may be joined by other queens that have not constructed a nest of their own. These individuals establish a dominance hierarchy, with the auxiliary queens subordinate to the founding queen; they all cooperate in the care of the young. Nests with multiple founding queens are more likely to survive than nests with single queens. Multiple-queen colony foundation is common in *P. exclamans* and is the rule in *P. annularis*, *P. apachus*, and *P. carolinus*. Aggressive interaction frequently occurs between the nest-sharing queens until one attains a dominant status.

Tropical *Polistes* do not hibernate. Females may leave the nest at any time during the year and found new colonies. The nest is usually constructed by multiple queens. Nest architecture is variable, with some species making a single horizontal comb, while others make a vertical comb, or a series of cells, one cell below the other. The nest may have a single or multiple attaching pedicels. No *Polistes* species construct envelopes around the comb. Caterpillars are an important prey of these wasps, but they will capture a variety of other slow-moving insects. Adults feed on honeydew, and they are attracted to bruised fruits.

**Large paper wasp, *Polistes annularis* (Fig. 9.12b)** Workers are 20–25 mm long, the body is light brown, and there are coarse transverse ridges on the propodeum. The abdomen is dark brown, while the basal segment is distinctly light brown. This species builds combs in bushes and trees.

***Polistes apachus*** Workers are 16–20 mm long; the body is light brown to reddish brown with yellow markings. It has the habit of building a new nest close to that of the previous

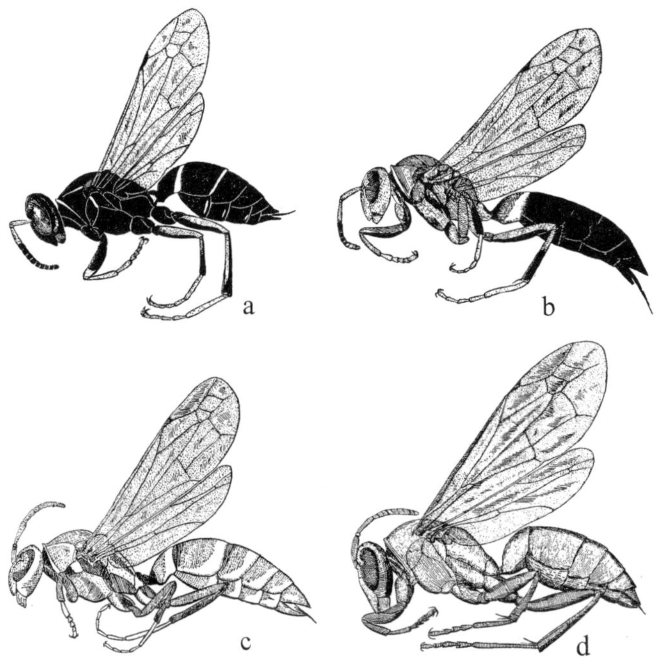

**Figure 9.12** Hymenoptera: Vespidae. (a) *Polistes fuscatus pallipes*; (b) *P. annularis*; (c) *P. exclamans*; (d) *P. rubiginosus*.

year. In one case, this habit was observed for 10 years. New colonies are typically founded by 2–6 queens. Nests are located in agricultural and urban environments. This species occurs in Mexico, Texas, New Mexico, and California.

**European paper wasp, *Polistes dominulus*** This species was accidentally introduced into the eastern coast of the USA in the 1970s and has been rapidly spreading westward, displacing *P. fuscatus*. The native species, *P. dominulus*, appears to have a number of advantages over *P. fuscatus*, including earlier production of workers, higher per capita foraging rates by queens and workers, higher queen survivorship, and lack of conspecific pressures. Nest site and prey availability may be limiting factors in the competition between the two species. This species typically has an Old World distribution.

***Polistes exclamans* (Fig. 9.12c)** Workers are about 18 mm long; the body is brown to dark brown, with yellow markings. This species is common in southeastern USA, especially along the Atlantic seaboard. Two or three cooperating queens form nests.

**Dark paper wasp, *Polistes fuscatus* complex** Workers are 9–15 mm long, the head and thorax are black with yellow markings; the abdomen is yellow with black bands on segments 1 and 2. Nests are found under the eaves of buildings and sometimes in attics. *P. fuscatus aurifer* is the most widespread subspecies of *P. fuscatus*. It ranges from British Columbia to southern California, where it merges with *P. fuscatus centralis*, and at the eastern part of its range, *P. fuscatus aurifer* merges with *P. fuscatus variatus* and *P. fuscatus utahensis*.

***Polistes fuscatus pallipes* ( = *P. metricus*) (Fig. 9.12a)** Workers are 17–21 mm long and black and yellow. The abdomen is black; the first segment has a yellow band posteriorly. This species is common in northeastern USA and ranges west to British Columbia.

**Orange paper wasp, *Polistes rubiginosus* (Fig. 9.12d)** Workers are 18–24 mm long; the head, thorax, and legs are uniformly orange brown. This species occurs in western North America, and nests in hollow trees or the walls of houses or other buildings.

**Asian umbrella wasps, *Polybiodes* spp.** These wasps occur in Southeast Asia and central Africa. They construct nests varying in structure from spirally produced combs to vertical combs surrounded by an envelope. Colonies may have 3000 workers, and nests of species in the genus have been used in booby traps during hostilities in Vietnam.

**Asian paper wasps, *Ropalidia* spp.** These wasps occur in Southeast Asia, India, and Australia, and in Africa and Madagascar. There are more than 100 species in *Ropalidia*. Nests are made of chewed wood fiber, and the type of nest constructed by the various species is diverse. There are simple nests comprised of only a few cells, and large nests with thousands of cells. The nests may be attached to the underside of tree branches, electric wires, and under the eaves of houses. The large nests and peridomestic nesting sites of some species may provide a stinging hazard for humans.

### Hornets

These vespid wasps are usually black and yellow. In the USA, the term hornet is often applied to the baldfaced hornet, *Dolichovespula maculata*, and European hornet, *Vespa crabro*, and sometimes to all aerial-nesting yellowjackets. Technically, only species in the genus *Vespa* are considered hornets. The genera of true hornets, *Provespa* and *Vespa*, occur throughout the northern hemisphere and in Southeast Asia. The three species of *Provespa* are only found in Southeast Asia and are nocturnal. *Vespa* contains 23 species, most of which are distributed throughout eastern Asia. The range of some species extends through the islands of the South Pacific to New Guinea. One of the largest species in the genus is the Himalyan *V. ducalis* whose queens are 40 mm long, with a wing expanse of 80 mm. The distribution of *V. crabro* and *V. orientalis* extends to Europe and North America. *V. germanica*, a European species, has been introduced into Australia, New Zealand, South Africa, Chile, and eastern USA. *V. vulgaris*, which occurs in North America and Europe, has been introduced into Australia.

Several species of vespid wasps are successful in a tropical environment, and their biology and habits differ from those in temperate regions. In the tropics, vespid colonies are not regulated by a cyclic climate or a season in which food is limited. During the cold winter period of temperate regions there is a shortage of food, and wasps usually adopt an annual life cycle. Colonies decline and die in fall, and inseminated queens overwinter and start new colonies in the following spring. In the tropics is it possible for vespids to start and dissolve colonies at any time of the year. Nest founding may be by one or more queens, or swarming from an existing colony. Colonies of most tropical *Vespa* species undergo an annual, colony life cycle. There is a slow-development phase, which lasts 4–6 weeks, in which the first workers of the colony emerge. This is followed by a rapid-development phase with an increase in the number of workers. When worker numbers reach a peak, males and new queens begin to emerge in the colony. Workers thereafter usually devote their time to taking care of reproduction in the colony, and then die. As males and new queens leave the nest, the colony declines and disintegrates.

Wasps in tropical regions usually make larger colonies than the same or related species in temperate regions. Large colonies in the tropics are supported by the abundant food supply and the availability of nesting sites. The populations of *V. affinis affinis* and *V. tropica pulchra* in Japan are considered small to medium-sized types. However, these species in Sumatra (represented by different subspecies) produce more than 10 times as many comb cells and new queens as the Japanese conspecifics. The initial nests of most *Vespa* species in the tropics do not differ in shape from those of temperate conspecifics; they are flask-shaped and consist of a single envelope sheet. However, with the emergence of workers and increase in nest size, thick shell-like scallops of envelope are added to the top of the nest. The large, conical roof may be an adaptation to climatic factors, particularly to protect the nest from daily heavy rains. The outer surface of the envelope is usually hardened by applying material with oral secretions, and this has a waterproof function. In temperate regions, the nest envelope may be thick and multilayered, consisting of 4–8 layers of air pockets. The thick and multilayered envelope may serve to maintain favorable temperatures for rearing the reproductive brood in late summer and fall.

Pest status of *Vespa* is based on the aggressive behavior and painful sting of some species, and to the damage foraging workers do to ripening fruit and stripping bark off twigs and branches of ornamental plants. *V. orientalis* in the Middle East is a serious pest to fruit crops and commercial honey bee colonies. *V. basalis*, *V. mongolica*, *V. mandarinia*, and *V. velutina* are aggressive species, and it is dangerous to be close to their nests, whether they are disturbed or not. In urban areas, the presence of *V. affinis*, *V. analis*, and *V. tropica* nests causes public concern, and there are cases of death from stings. Species of *Vespa* are also reported to cause damage to ripe fruits of papaya, guava, peach, fig, grapes, apricots, and apples. Crops such as tea may remain unpicked because of nests in the branches of the plant. *V. crabro* vigorously attacks several species of shrubs and trees to obtain plant sap. The plants commonly attacked include lilac, ash, birch, dogwood, rhododendron, and boxwood. In commercial nurseries, attacks from this hornet cause economic losses due to dieback on stock that has been attacked.

Several species of *Vespa* are problems to beekeepers; workers will invade hives and take bees, but may not take the honey. Individual workers of *V. orientalis* were observed to kill an average of 33 honey bees per hour, and this species may kill entire apiaries. Other species that damage or destroy beehives include *V. crabro*, *V. analis*, and *V. mongolica*. After killing a colony, worker hornets often remove the bee larvae and pupae.

Vespid wasps can be considered as both beneficial and pests insects. They are beneficial as predators of some agricultural and forest pest species (Coleoptera, Lepidoptera, Orthoptera, Hemiptera), and they are predators of species of synanthropic flies. The larvae and pupae of some vespids are utilized as food by people in some parts of the world, such as Asia. Vespine wasps usually inhabit natural or undisturbed environments, but several species now occur regularly in urban environments, such as residential areas, parks, and other recreational areas. Vespids have three major aspects as pests: they are nuisance insects in agricultural and urban environments, they are a medical threat to individuals who are extremely sensitive to their venom, and they are the natural enemy of honey bees.

Although the greatest threat from vespine and polistine wasps is being stung, overall the greatest economic and social importance is probably due to their nuisance. Most people are afraid or not willing to tolerate the presence of these wasps around food or any outdoor activity. The presence of foraging yellowjackets usually causes anxiety in most people, and stinging episodes reinforce the threat these insects have. The presence of nests around the perimeter of the home or in the walls is a health and structural threat. Nests in wall voids may be expanded to the interior of the building by the workers chewing through walls, and newly emerged queens and males may emerge inside. In Japan, *V. simillima* prefers buildings as nesting sites, and the most common locations are in wall voids, attics, and under the floor (crawl space). This nesting behavior increases the number of people stung by these wasps in urban environments.

Yellowjacket venom consists almost entirely of proteins. It is about 75% low-molecular-weight peptides, and 25% high-molecular-weight enzymes, and with a small proportion (by weight) of biogenic, low-molecular-weight amines. The three major enzymatic proteins in yellowjacket venom are: a hyaluronidase, a phospholipase, and a protein named antigen 5, which appears to be the major allergen. The substances responsible for the intense pain during a sting are probable amines, such as histamine, serotonin, and kinin (wasp kinins). There seems to be no allergenic or antigenic cross-reactivity between yellowjacket proteins and honey bee venom. There is cross-reactivity between venoms and individual allergens among yellowjacket species. A wasp sting with the injection of venom produces immediate pain and local inflammation, but the subsequent reaction is variable and dependent on an individual's immunological status. Nearly all fatalities resulting from a wasp or bee sting are the result of a hypersensitive reaction to the venom, but very few people have such a reaction.

**Common hornets, *Vespa* spp.** Workers are about 20 mm long, brown to blackish brown, and with yellow marking on the thorax and abdominal terga. These wasps occur in both temperate and tropical regions, and in both regions they have annual nesting cycles. In temperate regions, the new colony is founded by a queen, which copulated the preceding fall and successfully overwintered. The nest consists of a pedicel, a comb of about three cells, and an envelope. The comb consists of 40–60 cells by the time the first workers emerge. The colony develops rapidly after the first workers emerge, and the number of workers reaches its peak in late summer or fall. At that time males and new queens are produced. Soon after the production of the sexuals in the colony, the founding queen dies, the production of new workers stops and the colony declines. Males and new queens leave their nest and copulate, after which males die and the new queens construct individual hibernaculum underground or in natural cavities to spend a period of dormancy until spring.

The development from egg to adult of workers is variable by species, and may be influenced by several environmental factors, including temperature. In Japan, development for the common species is: *V. simillima* about 30.8 days; *V. crabro* about 32.4 days; *V. analis* about 32.4 days; *V. tropica* about 35.6 days; and *V. mandarinia* about 40.1 days. Survival of adult workers is also variable, and in Japan ranges from about 13 days for *V. simillima* to about 35 days for *V. tropica*. There is a general tendency for short adult longevity with increased colony size. The number of cells in a mature *Vespa* varies considerably among species. The largest nest of *V. tropica* in Japan may have 313 cells in three combs; a nest of *V. simillima* may have 14 272 cells in 10 combs. Colony size of *Vespa* species can be classified as: small, with up to 1000 cells in 3–4 combs, for *V. orientalis*, *V. tropica* and *V. analis*; intermediate, with 2000–4000 cells in 4–12 combs for *V. crabro*, *V. dybowski*, *V. mandarinia*, *V. simillima simillima*, and *V. affinis*; large, with 4000–10 000 cells in 6–12 combs for *V. simillima xanthoptera*.

The nesting period, from foundation of the nest to disintegration, varies among species and regions. Among the *Vespa* species that occur in Japan, *V. tropica pulchra* has a nesting period that lasts for about 4 months, and *V. simillima* has a nesting period that lasts about 7 months. Characteristics of the long-period nesting species are: the nesting period starts early and ends late in the year; the nest size and the number of males and queens are large; a large number of colony cells and workers are produced before sexuals are generated; and the feeding habit is that of generalist. In the short-period nesting species, every characteristic is opposite. However, it is often difficult to assign *Vespa* species to one of these two categories. Colonies of *V. simillima xanthoptera* and *V. crabro* in temperate regions relocate their nests when they become large and more space for nest expansion is required. The relocated nest sites are usually in open spaces, such as rock walls, under eaves of buildings, and in attics. In late summer in Japan, 88% of nests of *V. simillima* and 55% of the nests of *V. crabro* may be secondary or relocated nests. In the UK, 25% of the *V. crabro* nests collected during August and September are probably secondary nests.

Hornets gather nest materials from the xylem of decayed wood or from the dead parts of living trees, but sometimes they gnaw the bark of live trees. Nest sites preferred are above ground in an enclosed space, but below-ground nests are also established. In Japan, *V. luctosa*, *V. affinis*, and *V. analis* prefer sites above ground, such as under the eaves of houses and in rock walls. *V. crabro*, *V. dybowskii*, and *V. tropica* prefer enclosed sites, both above and below ground. In Europe and North America, *V. crabro* builds nests in hollow parts of trees, but other sites include thatched roofs, cavities in walls or buildings, and bird nest boxes. *V. orientalis* builds nests in similar locations.

*V. mandarinia* builds nests exclusively in underground cavities. In natural habitats, *V. simillima xanthoptera* nests in tree hollows, rock walls, or in underground cavities. In urban habitats, this species constructs nests under the eaves of houses, in attics, and in walls. In Japan, *V. mandarinia* may attack colonies of the Japanese native species, *Apis cerana japonica*, and colonies of the European honey bee, *Apis mellifera*.

Food gathered while foraging is both liquid carbohydrates and solid protein. The solid food is intended for developing larvae in the nest; liquids are distributed to adult nestmates. The main carbohydrate sources are tree sap, honeydew, flower nectar, ripe fruit, and discarded food material. The most important source of *Vespa* food in Japan is tree sap flowing from injuries to live trees by insects. In Europe and North America, *V. crabro* feeds on tree sap by gnawing the bark from young trees. In other regions, fruits with high sugar levels are fed on by *V. mandarinia* and *V. orientalis*. Insects and spiders are the main protein source for hornets. All species are solitary predators, except for *V. mandarinia*, which also attacks prey in groups. In Japan, *V. crabro* feeds primarily on cicada species, and in North America it has the same feeding habit. *V. mandarinia* preys mainly on coleopterans, including scarabs and longhorn beetles. *V. tropica* feeds on the larvae and pupae of wasps, such as species in *Polistes*, *Parapolybia*, and *Ropalidia*.

**Small hornet, *Vespa analis*** Workers are 24–37 mm long. It differs from other *Vespa* by having a small denticle medially on the anterior margin of the clypeus, between the lateral teeth. Nests are usually in bushes or shrubs, and rarely under eaves of houses. Nest combs are completely covered by a thick envelope. Mature colonies have 150–800 cells and 2–4 combs. Workers are relatively nonaggressive to humans. This species is generally distributed in Asia, from Malaysia to China, Korea, and Japan. Subspecies of *V. analis* (Kogata-suzumebachi) in Japan include *V. analis insularis*.

**European hornet, *Vespa crabro* (Fig. 9.13d, e)** Workers are 20–35 mm long. The body is brownish black with narrow to broad yellow bands on the abdomen (metasoma). Nests are usually built in trees, attics, walls of houses, and in other domestic and peridomestic sites. A brown envelope and carton and large cells distinguish this species from large *Vespula* species, such as *V. maculata* and *V. maculifrons*. A typical mature nest for *Vespa crabro* in North America has 1500–3000 cells in 6–9 combs, and the lower 2–4 combs contain queen cells; at its peak a nest may have about 1000 workers. Mature nests often have a foul odor, due to the large amounts of feces that accumulate beneath the combs. Sexual adults emerge from late August to November. Prey includes grasshoppers and other orthopterans, cicadas, flies, yellowjackets, and honey bees. Workers fly and hunt for prey at night, and they are attracted to outdoor lights, lighted windows in buildings, and can be serious pests as a result of this behavior. This species is widely distributed from Asia to Europe, and was introduced into the USA in the mid-1800s. It occurs along the eastern seaboard from New England, south to Georgia and Alabama, and west to the Mississippi river. However, there are records of this hornet in North and South Dakota and Louisiana. It is known to occur in southern England, and occasionally from Scotland; it is not known to occur in Ireland.

**Asian hornet, *Vespa mandarinia*** Workers are 40–45 mm long; the head is wholly orange yellow with no black markings. Abdominal segment 1 has an orange-yellow band at the base, and the last tergite is entirely yellow. This is the largest species in the genus. Nests are usually underground or in tree hollows. The nest has a thin envelope, which does not completely cover the lower combs. Mature colonies in Japan consist of 2000–5000 cells and 4–10 combs. Workers are very aggressive toward humans, and their sting causes a sharp pain. Prey includes slow-moving insects, such as adult scarab beetles and caterpillars. They also attack colonies of other social wasps and bees; it is a serious pest of cultivated and wild honey bee colonies in Japan. This hornet occurs in Asia, including the countries of Thailand, Laos, China, Korea, and Japan. The subspecies *V. mandarinia japonica* (Oo-suzumebachi) occurs in Japan.

**Yellow hornet, red wasp, *Vespa simillima*** Workers are 25–28 mm long; the area between the eyes on the vertex is black or with black markings. The pronotum has yellow or orange-yellow markings. Nests may be in open sites, such as on bushes and under eaves of houses, and in closed sites, such as in underground cavities, in hollow trees, and the walls of buildings. Nests in confined spaces may be moved during the season. The nest envelope is thick and covers the combs. Mature colonies in cold temperate regions of Japan have 500–2500 cells and 4–10 combs; in southern regions nesting activity may extend to November and have 4000–7000 cells and 7–10 combs. This species preys on a variety of insects, including flies and spiders, and occasionally visits colonies of *Apis* and *Vespula* to capture workers at the nest entrance. Workers are very aggressive and it is difficult to approach within 10 m radius of a nest during peak activity. This species occurs in Korea and Japan, and it

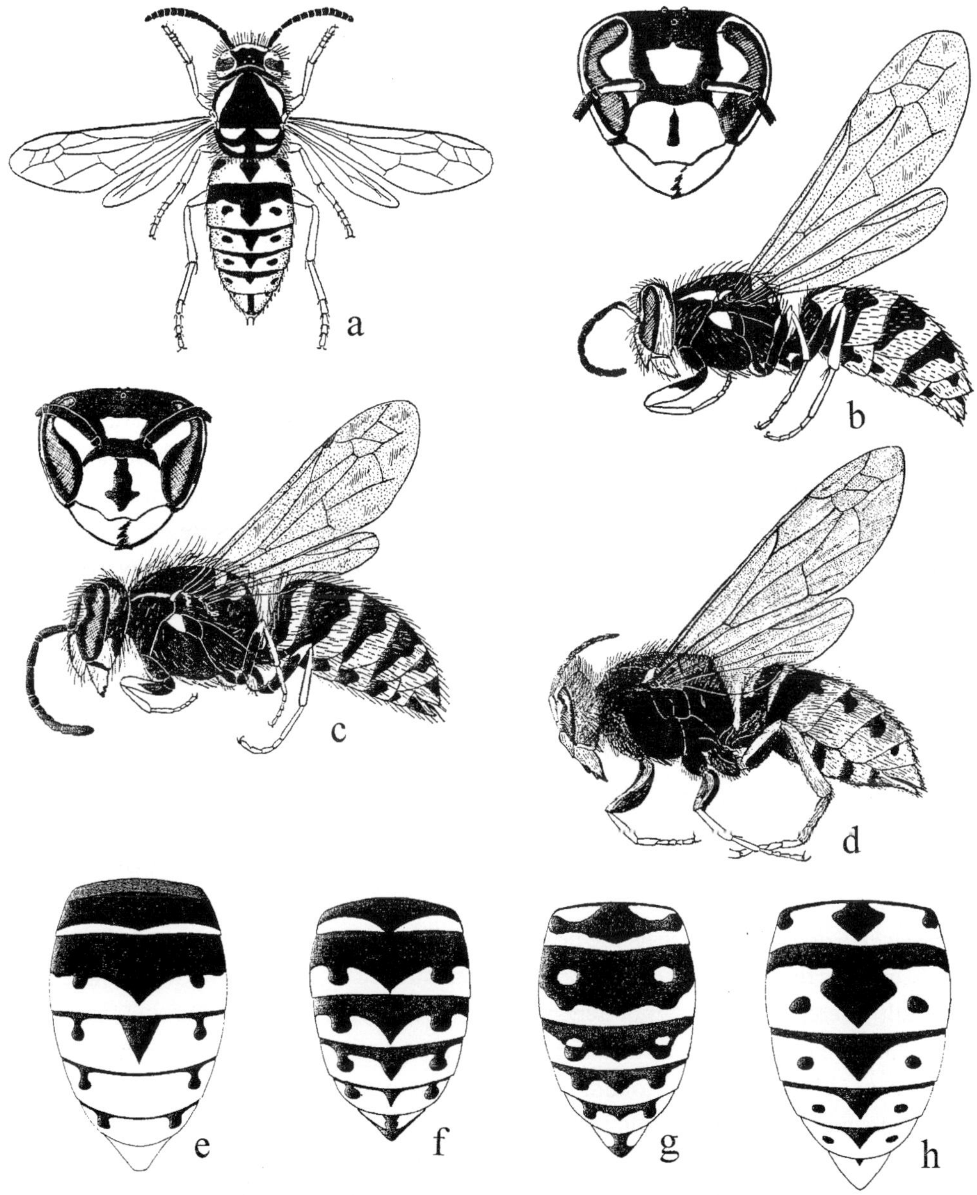


**Figure 9.13** Hymenoptera: Vespidae. (a) *Paravespula pensylvanica*; (b) *P. pensylvanica*, head anterior; lateral view of worker; (c) *P. vulgaris*, head anterior; worker; (d) *Vespa crabro*, worker; (e) *V. crabro*, gaster; (f) *Dolichovespula arenaria*; (g) *Vespula atripilosa*; (h) *P. germanica*.

is the most dominant hornet in Japan, and the most serious pest among the Japanese vespines. In rural areas *V. simillima* (Kiiro-suzumebachi) is often called kame-bachi (pot wasp) or aka-bachi (red wasp).

**Tropical hornet, *Vespa tropica*** Workers are 24–37 mm long; the area between the eyes on the vertex is yellow with slight brown markings. The last two segments of the gaster are usually black. Nests are in closed spaces, such as underground cavities, tree hollows, and house attics. The envelope is thin and does not cover the lower combs. Mature colonies in Japan have 200–300 cells, and 10–50 new queens are produced each year. Overwintered queens appear more than 1 month later than queens of other *Vespa* species. Workers are generally not aggressive toward humans. Prey includes other insects; *V. tropica* queens and workers may attack polistine nests for food to feed their larvae. This species occurs generally in Asia, from India to Japan. There are three subspecies of *V. tropica* (Hime-suzumebachi) in Japan.

## Yellowjackets

These black and yellow wasps belong to three distinct genera: ground-nesting *Vespula* and *Paravespula*, and aerial-nesting *Dolichovespula*. These genera are comprised of primarily north temperate species occurring throughout Europe, northern Africa, Asia, and North America.

Colonies are initiated by a single fertilized queen, usually after a period of hibernation. Queens feed on nectar, honeydew, and other sweet substances while they search for a suitable nest site. Once a nest site has been selected, the queen constructs a small number of cells in which she lays eggs. At this time, the queen performs all duties of foraging for nest materials and food for the developing larvae. The queen of a vigorous colony may lay 25 000–30 000 eggs during her lifetime. The same cell may be used two or three times for rearing larvae. Development time to complete the larval and pupal stage is about 30 days. The full-grown larva spins a silk cap over the cell, and then voids from its intestine a blackish mass of accumulated wastes, called meconium. This dries to become a hardened pellet, and the number of pellets at the bottom of the cell indicates the number of times it has been used. New adults cut their way out of the cocoon, and seek out larvae in the nest. These larvae provide a salivary liquid, which is the first food of the adult. Teneral adults remain in the nest for 2–3 days. If they are workers, they gradually take on the duties of nest building and brood care. Cocoons of the larvae that will become queens are larger than others, and extend farther out of the cells. Queens are larger than workers, and males have long antennae. For species that have large colonies, the queen maintains control of the colony with a queen pheromone. Yellowjackets do not store honey as do bees and some other vespids. They feed their larvae malaxated portions of arthropods, especially insects, and also nectar and honeydew. Adults feed on nectar, liquid from the larval food, and larval secretions. Trophallaxis or the exchange of alimentary liquids among colony members is a prominent activity in the colony.

The first workers to emerge in the colony assume all duties of maintenance and food gathering; the queen confines her activity to laying eggs and remains with the nest. There are three castes in each colony: the queen; males, which are produced from unfertilized eggs; and workers, which are infertile females. Two parasitic yellowjackets, *Dolichovespula artica* and *Vespula austriaca*, do not have a worker caste and utilize workers of the host species to care for their brood. Late in the season, workers build large reproductive cells in which males and queens are produced. At this time the colony enters a declining phase and workers remove some larvae from cells and feed them to other larvae or discard them. During this period workers are more aggressive and likely to sting, even when away from the nest. When new queens and males emerge they leave the nest and mate. Males die after mating, and fertilized queens enter a period of reproductive diapause, and in cold climates they overwinter. They hibernate in protected locations, such as under loose bark of trees, under boards and debris around buildings, and in other peridomestic locations.

Nest construction materials include plant fibers from decayed or weathered wood, the cortex of dead plants, and domestic debris, such as newspaper, cardboard, and paper bags. Workers chew the collected material and mix it with salivary secretions to make a doughy mass. The first structure to be built is a pedicel for the support of a small comb of cells. A paper covering is placed around this comb, and a small opening is at the bottom to permit entry and exit of the queen. As additional combs are built, secondary pedicels are built and the envelope is enlarged. Nest building is continuous until the colony declines. Most yellowjackets build annual nests. Perennial colonies occur in warm regions. In these, the new queens do not leave, but instead mate and rejoin the colony to lay eggs and rear young. Each queen has a region of influence over the workers in her section of the colonies. Perennial nests may have hundreds of queens and colony growth can be rapid.

Below-ground nests are usually made in abandoned animal burrows. The founding queen may excavate the burrow before preparing it for the first comb of cells. The roof of the hollow may be repeatedly wet by the queen; when dried, it becomes hard and firm. The base of the first pedicel may be expanded to form a broad area of attachment to the roof of the hollow. As the nest increases in size, many suspensoria are built from the envelope to the surface of the hollow, so the nest does not depend solely on the primary pedicel for support. Workers eventually clear away leaves, grass, and other material surrounding the entrance to the burrow. They also excavate the soil to enlarge the cavity for the nest, so that there is approximately a 12-mm clearance between the outer paper envelope of the nest and the wall of the cavity. Excavation is accomplished by workers regurgitating water on to the wall of the soil cavity, then the mud is scraped off in small pellets and carried away. Some of the excavated mud is used to fill in cavities leading out from the nest chamber. Sometimes there are two openings to an underground nest, and the wasps use both to enter and leave.

Nest associates of several species of yellowjackets include the ichneumonid wasp, *Sphecophaga vesparum burra*, and the pyralid moth, *Vitula edmandsae serratilineella*. The ichneumon is a pupal parasite and it frequently occurs in the nests of the *Vespula rufa* group species. It is less common and less abundant in *Dolichovespula* nests, and rarely found in the nests of the *Paravespula vulgaris* group species (with the exception of *P. vulgaris* nests). *Parcoblatta* cockroaches are common nest invaders of *V. squamosa*. Numerous insects are associated with subterranean colonies. The soil beneath the nests contains the organic waste from the colony, and is attractive to several saprophagous insects. Several Muscidae species are the most prevalent and abundant nest scavengers, including *Dendrophaonia querceti*, *Fannia canicularis*, and *F. scalaris*. Larvae of *Triphleba lugubris* (Phoridae) are common scavengers in colonies that are beginning to decline.

Pest status of *Vespula*, *Paravespula*, *and Dolichovespula* is based on their nesting and foraging habits, and the painful sting of some species. They do little damage to agricultural crops, except for the cases where their presence disrupts or prevents harvesting. Yellowjacket stings result in intense pain to most people, and can result in death from anaphylaxis in sensitive individuals. Ground nests of *Dolichovespula* and *Paravespula* create problems when they occur in peridomestic habitats or recreational areas. Below-ground nests are usually unnoticed until people and pets come near or into direct contact with them.

### Aerial yellowjackets

The nests of *Dolichovespula* are usually aerial, although species may nest below ground on occasion. *Dolichovespula* are not attracted to protein food and usually forage for live prey; they will occasionally scavenge flesh from animal carcasses. Colonies are usually initiated in early spring and the first workers appear in about 25 days. Cells for males and new queens are produced in the colony in late summer; reproductives emerge approximately 22 days following construction of the first reproductive cells. Colonies begin to decline after the production of reproductives. Typical nests have 300–1500 cells, but vigorous *D. arenaria* and *D. maculata* colonies may construct over 4300 cells. Mature nests usually have 2–6 combs and the peak number of workers ranges from 200 to 700 individuals. The paper covering of the nest consists of distinct laminar sheets, and the paper is resistant to water damage. Mature colonies of *D. arenaria* have 9–16 layers at one time and 35–40 over the life span of the nest. *D. maculata* nests may have 1–9 layers at one time. The outer diameter of the nest does not indicate colony size, because the number of layers and space between layers (usually 2–3 mm) vary greatly between nests. Construction material includes wood from willow, cottonwood, oak, cedar, birch, and weed stems. *Dolichovespula* build nest envelopes with laminar paper strips, in contrast to *Paravespula* species, which build nest envelopes that are scalloped. As the nest expands, layers are constructed on the outside as inner layers are removed. Unlike other yellowjacket species, *Dolichovespula* tend to incorporate obstructions such as leaves, twigs, and dead workers into the envelope layers.

**Parasitic yellowjacket, *Dolichovespula arctica*** This species is an obligatory social parasite in colonies of *D. arenaria* and *D. norvegicoides*. It relies on the host workers to care for its young. The queen invades a host nest before the first brood of workers has emerged, and after about 24 h it becomes established as a nestmate. The *D. arctica* queen may be aggressive and dominant with its hosts, and eventually kill the host queen before her worker production is completed. This shortens the life span of the colony and reduces the size of the mature colony. The host colony produces no new queens, but in some cases the workers rear a brood of males. This species is distributed throughout northern North America, from Alaska to Arizona and east to the northeastern seaboard.

**Aerial yellowjacket, *Dolichovespula arenaria* (Fig. 9.13f)** Workers are 11–13 mm long and black with yellow markings; the yellow genal band behind the eye is continuous, but sometimes deeply notched. The yellow band on gaster segment 1 and 2 is interrupted medially. Nests may be in shrubs or bushes, and in the tops of trees, but this species will readily nest on structures. Nests may be in the ground, and workers excavate the surrounding soil to expand the nest. Nests are initiated from April to June, colonies usually peak in mid-summer, and the nest declines by September–October. Variations occur with geographic region, and from year to year. In southern California, nest construction begins in March and many colonies are in decline by June and July; in mid-Atlantic states, colonies may be mature and producing reproductives by early July. Mature nest size is variable by region and year, and ranges from 644 to 4290 cells, and there may be as many as seven combs. Workers usually forage for only live prey, which includes grasshoppers, tree crickets, caterpillars, flies, and spiders. They usually do not scavenge for protein, but they are attracted to sugar in late summer. Foraging for flies may result in workers hovering around the head of people and large animals. Nests are often constructed on structures, and workers in large colonies can be a threat to people. This species is one of the most common yellowjackets in North America; it occurs from north-central Alaska to Arizona and New Mexico to northeastern and mid-Atlantic seaboard states.

**Baldfaced hornet, *Dolichovespula maculata* (Fig. 9.14d, h)** Workers are 14–18 mm long and black with white markings. The dorsal surfaces of gaster segments 1–3 are entirely black. Nests are usually made in forested or undisturbed areas. In the urban environment, nests are established in ornamental bushes and trees, electric power poles, houses, sheds, and other structures. Nearly all are constructed in exposed locations

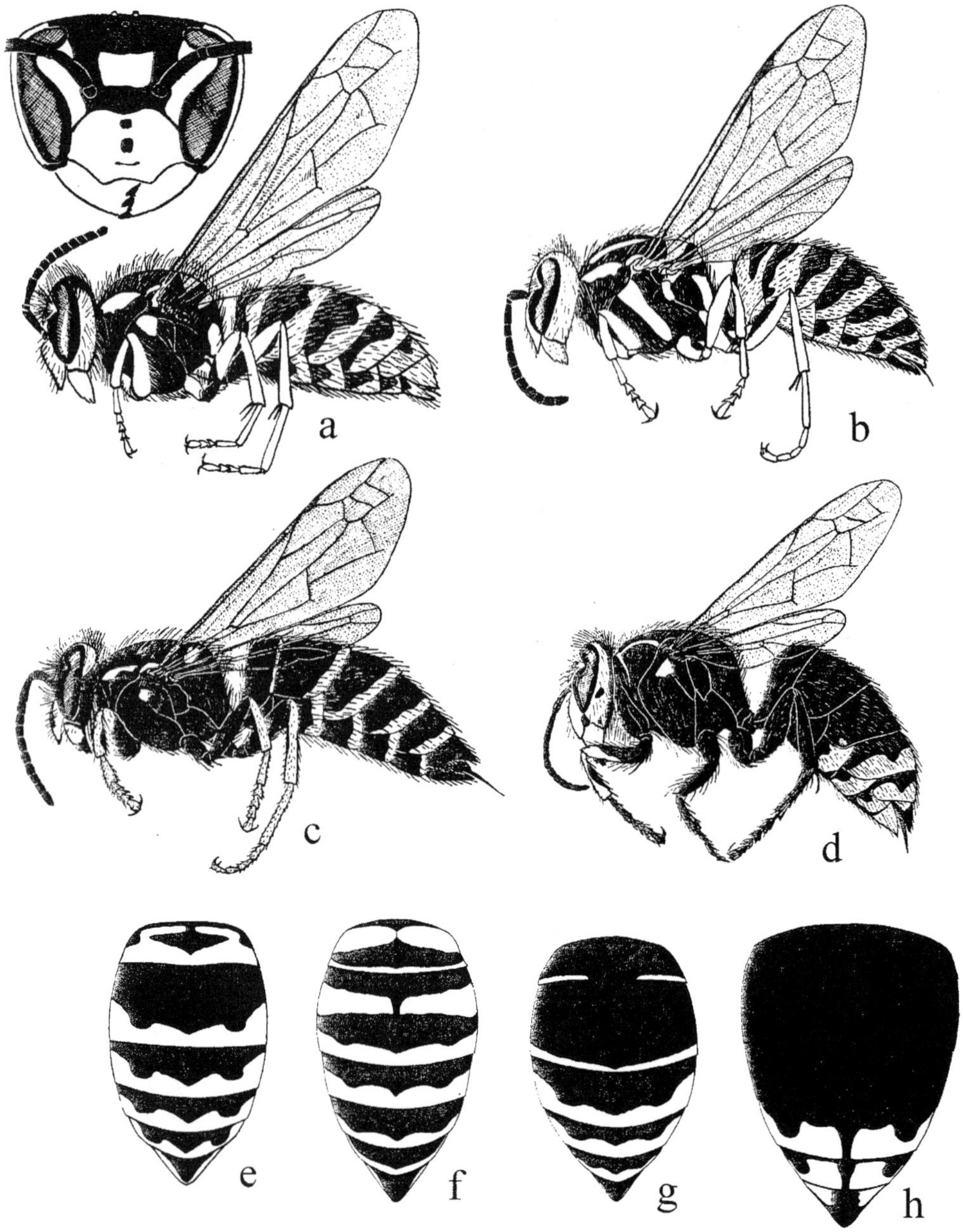


**Figure 9.14** Hymenoptera: Vespidae. (a) *Paravespula maculifrons*, head anterior; lateral view of worker; (b) *Vespula squamosa*, worker; (c) *V. consorbina*, worker; (d) *Dolichovespula maculata*, worker; (e) *P. maculifrons* gaster; (f) *V. squamosa*; (g) *V. consorbina*; (h) *Dolichovespula maculata*.

and initiated in April or May. Reproductives are produced in late July and August, and colonies decline in September. Large colonies construct 3500 cells in five combs, but the majority of nests contain fewer than 2000 cells in three or four combs. The peak number of workers is 100–400. One of the first nest envelopes constructed by the founding queen has a 6–11-cm-long tunnel extension to the nest entrance. New workers remove the tunnel soon after emerging. The primary duties of the queen after emergence of the first workers is oviposition; they obtain liquid food from late-stage larvae.

Workers are powerful, agile wasps and most forage for live prey. Flies are commonly taken, and in some areas other yellowjackets are an important prey. Workers locate prey by visiting locations such as animal carcasses, trashcans, and organic litter, or searching new areas by flying low over vegetation. Included in their searching behavior are peridomestic habitats, such as the eaves of houses and the screen doors and windows of structures. Peak flight activity for capturing prey may be immediately before sunrise. Workers ingest low to moderate amounts of honeydew and pollen. Foraging rates for colonies are generally linked to colony size. For a colony of 1090 workers, the daily foraging rate may be 19.2 trips per colony member, for an estimated 21 000 total trips per day. Nests of *D. maculata* are often large and this feature alone can be threatening. These wasps are not usually aggressive, and workers are not usually sensitive to the nest being disturbed. The abandoned nests of this species are commonly displayed in introductory biology classes and nature centers. This species occurs throughout Canada and eastern USA, and the Pacific coast states into Arizona and New Mexico. It is closely related to the Eurasian species, *D. media*, and the two may be considered allospecies of a superspecies.

***Dolichovespula media*** Workers are 19–22 mm long; body markings are yellow or orange yellow, the inner-orbital marking is wholly yellow. This large species is often misidentified as *Vespa*; it is the only *Dolichovespula* in Japan with yellow markings. Nests are usually in thick brush near the ground or in trees, but a small percentage may be on structures, such as on eaves and window frames. Colonies initiated in early to late spring produce workers by the end of June. Mature nests have 300–2000 cells, and 2–4 combs. Workers are aggressive towards humans. This species preys on insects, such as flies, and feeds on flower nectar and honeydew. This species occurs in central, northern, and southern Europe, and northern Asia, including Japan.

**Norwegian wasp, *Dolichovespula norwegica*** Workers are 9–15 mm. The body is black and yellow; the clypeus has a large central dark mark, and eyes do not extend to the base of the mandibles. Nests are constructed in a variety of habitats and substrates, including in low vegetation and high in trees, and in enclosed or subterranean sites. Mature colonies have about 500 workers. This species occurs in the UK, and it is the dominant vespid in northern Scotland.

***Dolichovespula saxonica*** Workers are 14–16 mm long; body markings are ivory-white, and the inner-orbital region has a pale white line in its lower half. Nests are constructed in open and enclosed sites, on the surface of the ground, under eaves, and in wall voids of structures. Mature colonies have 100–1500 cells and 3.7–6 combs. This species preys on insects, such as flies, and feeds on flower nectar and honeydew. The vespid, *D. adulterina*, is a social parasite of *D. saxonica* and *D. norvegicoides*. This species is distributed in North America.

**Tree wasp, *Dolichovespula sylvestris*** Workers are 9–15 mm long; the clypeus is entirely yellow or with a small central black spot; the gaster is not marked with brown. Nests are constructed in vegetation, underground and partially underground, and in structures. Colonies initiated in early to late spring produce workers in approximately 25 days, and reproductive cells are produced in less than 30 days. Reproductives emerge approximately 22 days following construction of the first reproductive cells. Mature colonies have about 329 workers. This species is generally common in the UK, and occurs from Scandinavia to North Africa.

### Ground-nesting yellowjackets

Nests of *Paravespula* and *Vespula* are usually constructed in soil cavities, or in decaying logs on the ground. Some species build above-ground nests. In the *V. rufa* species complex, there are two facultative parasites, *V. austriaca* and *V. squamosa*, of other yellowjackets, and their nesting habits are linked to the nests they parasitize. *V. squamosa* often nests in urban and recreational areas. *V. rufa* produce colonies that have 75–400 workers, and nests of 500–2500 cells; *Paravespula vulgaris* have colonies that have 500–5000 cells. These species forage for proteins and sweets. They are the cause of most of the wasp–human problems, especially at the end of the season when the colony begins to decline.

**Black yellowjacket, *Paravespula flaviceps*** Workers are 15–16 mm long. Body markings are pale white or yellow. The black bar on the clypeus rarely reaches the lower margin of the clypeus. Nests are in closed spaces, such as underground cavities, and the wall voids of buildings, and rarely under the eaves of houses. Nests have 8000–12 000 cells and 8–12 combs. In warm temperate regions the nest may remain active until early winter. Workers prey on insects and spiders, and collect flesh from dead animals (frogs, snakes, mammals); workers also forage for nectar and honeydew from aphids. This species occurs in northern India, China, Korea, and Siberia. In Japan, people excavate colonies to get larvae and pupae, which are used as food. Canned wasp larvae and pupae are sold in other regions in Japan.

**German yellowjacket, European wasp, *Paravespula germanica* (= *Vespula*) (Fig. 9.13h)** Workers are 12–14 mm long and brown with yellow marking on head, thorax, and bands on the gaster. Antennal segment 1 is usually entirely black; the yellow genal band behind the eye is interrupted with black, but continuous. Nests in Europe are usually below ground, but sometimes above ground in walls of buildings; nests in North America are in attics, and cavities of house walls. Colonies are usually annual, but nests of about 1 million cells found in New

Zealand and Tasmania, Australia are probably built by perennial colonies. Nests may contain as many as 2000 workers. This species preys on a variety of arthropods; they forage for protein and sweets in urban environments. Most workers forage within 200–300 m of the nest, but some will travel 1200 m. This species is native to Europe, but has been introduced to New Zealand, Australia, South Africa, Chile, and the USA. It is common in England, but absent from northwest Ireland and Scotland. *P. germanica* occurs in part of Israel and Northern India. It is a serious pest in eastern USA, and seems to be replacing *P. maculifrons* as the most abundant pest yellowjacket.

**Western yellowjacket, *Paravespula pensylvanica* (= *Vespula*) (Fig. 9.13 a–c)** Workers are 12–17 mm long. The body is black and yellow, the scutum is without stripes, and there is usually a continuous yellow ring around each eye. Antennal segment 1 is usually pale yellow anteriorly. Nests are usually built in subterranean cavities 10–15 cm below the soil surface; occasionally nests are built in walls of buildings. Nests may contain nearly 4000 workers. It preys on spiders, grasshoppers, flies, and hemipterans. This species occurs generally west of the Rocky Mountains from western Canada to Arizona and New Mexico, and in Hawaii. It is the primary pest yellowjacket in western USA, from Washington south to California. Periodic outbreaks of *P. pensylvanica* occur every 3–5 years, following a warm and relatively dry spring period in the spring. A large number of nests are established during these years and the opportunity for human exposure is increased.

**Eastern yellowjacket, *Paravespula maculifrons* (= *Vespula*) (Fig. 9.14a, e)** Workers are about 12 mm long. Antennal segment 1 is usually entirely black; the yellow genal band behind the eye is not interrupted with black, but is continuous. There is an anchor-shaped, median black mark on the tergum of gaster segment 1. Nests are subterranean along the banks of small streams in forested areas. In urban environments, nests are in the walls of houses and other buildings. In northern regions, colonies are initiated in May or June, and usually peak in August or September. Nests contain 2000–5000 workers. In most areas in which it occurs, *P. maculifrons* is the primary pest yellowjacket because of its foraging behavior, and habit of nesting around buildings, golf courses, and recreational areas. This species is widespread in eastern and midwestern USA, and southern Canada.

**Common yellowjacket, common wasp, English wasp, *Paravespula vulgaris* (= *Vespula*) (Fig. 9.13c)** Workers are 12–14 mm long. Antennal segment 1 is usually entirely black; the yellow genal band behind the eye is interrupted with black. Nests in Europe are subterranean, although some nests are constructed in walls of buildings or in aerial locations; nests in North America are usually built in decaying logs or stumps. The nest envelope is usually made of rotten wood fiber and is brittle. Colonies are founded in May or June, and they have 3–9 combs when the colony peaks in September; some colonies are active until October. Nest size is variable, and the number of workers can be nearly 3000. Prey includes caterpillars, small beetle larvae, flies, hemipterans, and homopterans. It is a scavenger over a wide area, and is attracted to nearly any protein or sugar source. This species is distributed in North America, Europe, and Asia; it also occurs in Mexico and Hawaii (Maui). *P. vulgaris* was introduced to New Zealand and Australia; in Australia it is established in Victoria. It is a nuisance in food-processing and food-dispensing facilities in many European countries and in the USA.

**Forest yellowjacket, *Vespula acadica*** Workers are 14–16 mm long. The body is black and yellow; the genal band is usually interrupted, rarely continuous. Antennal segment 1 is yellow ventrally; abdominal segment 2 has two yellow spots, but sometimes is faint or absent. Nests are sometimes above ground, but usually they are under the soil surface, and as deep as 15 cm. Nests may have one worker-producing comb, and 1–3 reproductive-producing combs. Workers prey on caterpillars, flies, and hemipterans. The nesting and foraging habits of this species rarely bring it into contact with people or urban environments. This species is restricted to the Canadian zone of North America.

**Prairie yellowjacket, *Vespula atripilosa* (Fig. 9.13g)** Workers are 12–14 mm long. The body is black and yellow; the yellow genal band is usually continuous. Antennal segment 1 is yellow ventrally. Nests are underground in grassland and prairie, but not in forested areas. Nests in peridomestic sites are in turfgrass around buildings, golf courses, and recreational areas. Most nests are found in abandoned rodent burrows, and located 10–15 cm below the soil surface. However, there are records of nests built above ground and in the walls of houses. This species occurs in western North America.

**Blackjacket, *Vespula consorbina* (Fig. 9.14c)** Workers are 9–15 mm long. The body is black and pale white, with lines and a spot on the head. The white band on the posterior margin of tergum of gaster segment 1 is usually interrupted medially. Nests are subterranean and usually in abandoned rodent burrows, but may be above ground in logs, rock cavities, and the walls of houses. Colonies are generally small, with about 100 workers. Nests consist of one worker-producing comb and 1–3 reproductive-producing combs, and fewer than 1000 cells. Aggressiveness varies with colony size; larger colonies that have more workers are usually more aggressive. This species is found in forested areas throughout Canada and the Boreal region of North America. In general, it is little threat to people in urban and suburban areas, but it is responsible for numerous stinging episodes among workers in commercial logging operations.

***Vespula rufa*** Workers are 16–17 mm long; the body markings are pale white; the posterior horizontal part of tergite 1 has a white basal band that is interrupted medially. Nests are in underground cavities and in the walls of buildings; they are usually not in open spaces. Mature colonies in Japan have 500–1000 cells and usually combs. Workers prefer to feed on flower nectar and honeydew.

**Southern yellowjacket, *Vespula squamosa* (Fig. 9.14b, f)** Workers are 14–17 mm long. The body has yellow and black markings; the scutum has two yellow stripes; the abdomen has narrow yellow bands. *V. squamosa* is a facultative social parasite of *Paravespula maculifrons* or *Vespa* species, depending on the region. The parasite queen takes over the nest from the host queen and assumes complete control of the colony. Most *V. squamosa* colonies are located in disturbed habitats, particularly in turfgrass surrounding houses and buildings, recreation areas, and roadsides. This species has an effective alarm pheromone that causes extremely vigorous attack and stinging behavior by workers. *V. squamosa* occurs as both polygyne and single-queen nests; single-queen nests usually have a 1-year life cycle, while polygyne nests may have a 2-year cycle. Colonies usually remain active into fall, with most producing reproductives from late August into November. At their peak, colonies contain 500–4000 workers; polygyne nests have more workers. Sometimes perennial colonies occur in subtropical areas of its distribution. This species occurs east of the Rocky Mountains in North America, and into southern Mexico and Guatemala. A related species, *V. sulphurea*, has the longitudinal stripes, but the yellow bands on the abdomen are broad, and it occurs west of the Rocky Mountains.

## Xiphydriidae

These are the wood wasps. The adults are 12–18 mm long, cylindrical, and the ovipositor sheath is seldom longer than the last abdominal tergite. They are reddish brown, black, and yellow or entirely black. Full-grown larvae are 12–15 mm long, subcylindrical and somewhat S-shaped. The larva head lacks ocelli and has a three- or four-segmented antenna; the thorax has three fleshy legs; the last segment of the abdomen has sclerotized, toothed processes. Larvae feed on sound to partly decayed hardwoods and are dependent on symbiotic fungi. They are not known to be active in structural timber, but may occur indoors in firewood.

Species in the USA and Canada include: *Xiphydria abdominalis*, which is found in basswood, maple, and elm from southern Canada to North Carolina and Iowa; *X. maculata* (Fig. 9.1d), which is found in maple in Canada and throughout the USA, except the Gulf states; *X. tibialis*, which is found in elm, birch, beech, and hawthorn in southeastern Canada and northeastern USA; *X. hicoriae*, which is found in hickory and elm in southeastern Canada and northeastern USA; and *X. millipes*, which is found in beech, birch, and alder trees across southern Canada and northern USA. The species that occur in the UK feed on birches and willows, and include *Xiphydria prolongata* and *X. camelus*. The aulacid (Aulacidae) wasp, *Aulacus striatus*, attacks the larvae of *X. camelus*.

## Bibliography


Akre, R. D. and H. G. Davis. Biology and pest-status of venomous wasps. *Annu. Rev. Entomol.*, **23** (1978), 19–42.

Barnard, J. H. Studies of 400 Hymenoptera deaths in the United States. *J. Allergy Clin. Immunol.*, **52** (1973), 259–64.

Barr, S. E. Allergy to Hymenoptera stings. Review of the world literature: 1953–1970. *Ann. Allergy*, **29** (1971), 49–66.

Breed, M. D., C. D. Michener, and H. E. Evans (eds). *The Biology of Social Insects*. Boulder, CO: Westview Press, 1982.

Edwards, R. *Social Wasps. Their Biology and Control*. East Grinstead, UK: Rentokil, 1980.

Evans, H. E. Predatory wasps. *Sci. Am.*, **208** (1963), 145–54.

The behavior patterns of solitary wasps. *Annu. Rev. Entomol.*, **11** (1966), 123–54.

The Bethylidae of America north of Mexico. *Mem. Am. Entomol. Inst.*, **27** (1978), 1–332.

Fowler, H. G. Human effects on nest survivorship of urban synanthropic wasps. *Urban Ecol.*, **7** (1983), 137–45.

Social insects as urban pests: an analysis of seasonality and human perception. *Mater. Organ.*, **18** (1983), 93–105.

Goulet, H. and J. Huber. Hymenoptera of the world: an identification guide to families. Agric. Can. Res. Branch publication 1894/E Ottawa, Canada: Agriculture of Canada Research Branch, 1993.

Hanson, P. E. and I. D. Gaud. *The Hymenoptera of Costa Rica*. New York: Oxford University Press, 1995.

Huddleston, T. and I. D. Gauld. Parasitic wasps (Ichneumonoidea) in British light-traps. *Entomologist*, **107** (1988), 134–54.

Jaisson, P. (ed.) *Social Insects in the Tropics*. vols. 1 and 2. Paris: Université de Paris-Nord, 1983.

LeBeck, L. M. A review of the hymenopterous natural enemies of cockroaches with emphasis on biological control. *Entomophaga*, **36** (1991), 335–52.

Michener, C. D. The Brazilian bee problem. *Annu. Rev. Entomol.*, **30** (1975), 399–416.

Naumann, I. D. Hymenoptera (wasps, bees, ants, sawflies), In Naumann, I. D., (ed.) *The Insects of Australia*, vol. II, pp. 916–1000. Melbourne: Melbourne University Press, 1991.

Payne, J. A. and W. R. M. Mason. Hymenoptera associated with pig carrion. *Proc. Entomol. Soc. Wash.*, **73** (1971), 132–41.

Richards, O. W. and M. J. Richards. Observations on the social wasps of South America (Hymenoptera, Vespidae). *Trans. R. Entomol. Soc. Lond.*, **102** (1951), 1–170.

Spradbery, J. P. *Wasps*. London: Sedgwick and Jackson, 1973.

Vinson, S. B. (ed.) *Economic Impact and Control of Social Insects*. New York: Praeger Special Studies, 1986.

Wille, A. Biology of the stingless bees. *Annu. Rev. Entomol.*, **28** (1983), 41–64.

Wilson, E. O. Jr. *The Insect Societies*. Cambridge, MA: Belknap Press, 1971.

Spradberry, J. P. *Wasps: An Account of the Biology and Natural History of Social and Solitary Wasps*. Seattle, WA: University of Washington Press, 1973.

### Anthophoridae

Anzenberger, G. Ethological study of African carpenter bees of the genus *Xylocopa* (Hymenoptera, Anthophoridae). *Z. Tierpsychol.*, **44** (1977), 337–74.

Balduf, W. V. Life of the carpenter bee, *Xylocopa virginica* (Linn.) (Xylocopidae, Hymenoptera). *Ann. Entomol. Soc. Am.*, **55** (1962), 263–71.

Barrows, E. M. Results of a survey of damage caused by the carpenter bee *Xylocopa virginica* (Hymenoptera, Anthophoridae). *Proc. Entomol. Soc. Wash.*, **82** (1980), 44–7.

Ben Mordechai, Y., R. Cohen, D. Gerling, and E. Muscovitz. The biology of *Xylocopa pubescens* (Spinola) (Hymenoptera: Anthophoridae) in Israel. *J. Entomol.*, **12** (1978), 107–21.

Camillio, E. and C. A. Garofalo. On the bionomics of *Xylocopa frontalis* (Oliver) and *Xylocopa grisescens* (Lepeletier) in southern Brazil. I. Nest construction and biological cycle. *Rev. Bras. Biol.*, **42** (1982), 571–82.

Camillio, E., C. A. Garofalo, and G. Muccillo. On the bionomics of *Xylocopa suspecta* (Moure) in southern Brazil: nest construction and biological cycle (Hymenoptera, Anthophoridae). *Rev. Bras. Biol.*, **46** (1986), 383–95.

Gerling, D. and H. R. Hermann. Biology and mating behavior of *Xylocopa virginica* L. (Hymenoptera, Anthophoridae). *Behav. Ecol. Sociobiol.*, **3** (1978), 99–111.

Hurd, P. D. Jr. Observations on the nesting habits of some New World carpenter bees with remarks on their importance in the problem of species formation (Hymenoptera, Apoidea). *Ann. Entomol. Soc. Am.*, **51** (1958), 365–75.

*An Annotated Catalog of the Carpenter Bees (Genus Xylocopa Latreille) of the Western Hemisphere (Hymenoptera: Anthophoridae)*. Washington, DC: Smithsonian Institute, 1978.

Kojima, J.-I. Observations on nest structure of the Japanese large carpenter bee, *Xylocopa appendiculata circumvolans* Smith (Hymenoptera, Anthophoridae). *New Entomol.*, **28** (1979), 33–8.

Malyshev, S. J. The life and instincts of the dwarf carpenter bee, *Xylocopa iris* Christ. *Izv. Akad. Nauk. SSSR Ser. Biol.*, **1** (1947), 53–77.

Sage, R. D. Observations on feeding, nesting, and territorial behavior of carpenter bees, genus *Xylocopa* in Costa Rica. *Ann. Entomol. Soc. Am.*, **61** (1968), 884–9.

Scott, H. Carpenter bees eating lead cable-covers. *Entomol. Lond.*, **68** (1932), 8.

Steen, Z. and M. P. Schwarz. Nesting and life cycle of the Australian green carpenter bees *Xylocopa (Lestis) aeratus* Smith and *Xylocopa (Lestis) bombylans* (Fabricius) (Hymenoptera: Apidae: Xylocopinae). *Aust. J. Entomol.*, **39** (2000), 291–300.

Watmough, R. H. Biology and behavior of carpenter bees in southern Africa. *J. Entomol. Soc. South Afr.*, **37** (1974), 261–81.

### Apidae

Alford, D. V. Bumblebees. London: Davis-Poynter, 1975.

Ayala, R., T. L. Griswold, and S. H. Bullock. The native bees of Mexico. In Ramamoorthy, T. P., R. Bye, A. Lot, and J. Fa (eds.) *Biological Diversity of Mexico: Origins and Distribution*, pp. 179–227. New York: Oxford University Press, 1993.

Cumber, R. A. The biology of bumble bees with special reference to production of worker caste. *Trans. R. Entomol. Soc. Lond.*, **100** (1949), 1–45.

Eickwort, G. C. and H. S. Ginsberg. Foraging and mating behavior in Apoidea. *Annu. Rev. Entomol.*, **25** (1980), 421–6.

Free, J. B. and C. G. Butler. *Bumblebees*. London: Collins, 1959.

Heinrich, B. *Bumblebee Economics*. Cambridge, MA: Harvard University Press, 1979.

Laidlaw, W. B. R. Notes on some bumble bees and wasps in Scotland. *Scott. Nat.*, **185** (1930), 135–6.

Sakagami, S. F. and R. Zucchi. Winterverhalten einer neotropischen Hummel, *Bombus atratus*, innerhalb des Beobachtungskastens. Ein Beitrag zur Biologie der Hummeln. *J. Faculty Sci. Hokkaido Univ. Ser. VI, Zool.*, **15** (1965), 712–62.

### Encyrtidae, Eulophidae, Eupelmidae, Evaniidae

Cameron, E. On the parasites and predators of the cockroach. I. *Tetrastichus hagenowii* (Ratz.). *Bull. Entomol. Res.*, **46** (1955), 137–47.

On the parasites and predators of the cockroach. II. *Evania appendigaster* (L.). *Bull. Entomol. Res.*, **48** (1957), 199–209.

Cros, A. *Blatta orientalis* et ses parasites. I. *Evania punctata* Brulle. II. *Eulopus* sp. etude biologique. *Eos, Madrid*, **18** (1942), 45–67.

Edmunds, L. R. Some notes on the habits and parasites of native wood-roaches in Ohio (Orthoptera: Blattidae). *Entomol. News*, **63** (1952), 141–5.

A study on the biology and life history of *Prosevania punctata* (Brulle) with notes on additional species (Hymenoptera: Evaniidae). *Ann. Entomol. Soc. Am.*, **47** (1954), 575–92.

Gordh, G. Biological investigations on *Comperia merceti* (Compere), an encyrtid parasite of the cockroach *Supella longipalpa* (Serville). *J. Entomol. A*, **47** (1973), 115–23.

Roth, L. M. and E. R. Willis. *Anastatus floridanus* (Hymenoptera: Eupelmidae) a new parasite on the eggs of the cockroach *Eurycotis floridana*. *Trans. Am. Entomol. Soc.*, **80** (1954), 29–41.

Thoms, E. M. and W. H. Robinson. Distribution, seasonal abundance, and pest status of the oriental cockroach (Orthoptera: Blattidae) and an evaniid wasp (Hymenoptera: Evaniidae) in urban apartments. *J. Econ. Entomol.*, **79** (1986), 431–6.

Vargas, M. and F. Fallas. Notes on the biology of *Tetrastichus hagenowii* (Hymenoptera: Eulophidae) a parasite of cockroach oothecae. *Entomol. News*, **85** (1974), 23–6.

#### Formicidae

Bhatkar, A. P. and W. H. Whitcomb. Artificial diet for rearing various species of ants. *Fla. Entomol.*, **53** (1970), 229–32.

Bolton, B. Apterous females and shift of dispersal strategy in the *Monomorium salomonis*-group (Hymenoptera: Formicidae). *J. Nat. Hist.*, **20** (1986), 267–72.

Bruder, K. W. and A. P. Gupta. Biology of the pavement ant, *Tetramorium caespitum* (Hymenoptera: Formicidae). *Ann. Entomol. Soc. Am.*, **65** (1972), 358–67.

Campbell, S. and H. Beatson. Ants: Formicidae, Hymenoptera. In Gorham, J. R. (ed.) *Ecology and Management of Food-Industry Pests*, pp. 207–15. FDA Technical Bulletin 4. Arlington, VA: Association of Analytical Chemists, 1991.

Delabie, J. H. C., I. C. Nacimento, P. Pacheco, and A. B. Casimiro. Community structure of house-infesting ants (Hymenoptera: Formicidae) in southern Bahia, Brazil. *Fla. Entomol.*, **78** (1995), 265–70.

Fowler, H. G. Spatial organization of foraging activity and colonization by colony emigration of the Pharaoh's ant, *Monomorium phararonis*. *Zool. J. Physiol.*, **97** (1993), 233–8.

Gauld, I. D. and B. Bolton. *The Hymenoptera*. Oxford: Oxford University Press and British Museum (Natural History).

Gotwald, W. H. Jr. Army ants. In Hermann, H. R. (ed.) *Social Insects* 4, pp. 157–254. New York: Academic Press, 1982.

Hantgartner, W. Trail laying in the subterranean ant, *Acanthomyops interjectus*. *J. Insect Physiol.*, **15** (1969), 1–4.

Hölldobler, B. Temperaturunabhängige rhythmische Erscheinungen bei Rossameisenkolonien (*Camponotus ligniperda* Latr. und *Camponotus herculeanus* L.) (Hym., Formicidae). *Z. Angew. Entomol.*, **49** (1961), 337–52.

Hölldobler, B. and E. O. Wilson. *The Ants*. Cambridge, MA: Harvard University Press, 1990.

Human, K. G., S. Weiss, A. Weiss, B. Sandler, and D. M. Gordon. Effects of abiotic factors on the distribution and activity of the invasive Argentine ant (Hymenoptera: Formicidae). *Environ. Entomol.*, **27** (1998), 822–33.

King, R. L. and R. M. Sallee. On the duration of nests of *Formica obscuripes* Forel. *Proc. Iowa Acad. Sci.*, **60** (1953), 656–9.

Klotz, J. H., J. R. Mangold, K. M. Vail, L. R. Davis, and R. S. Patterson. A survey of the urban pest ants (Hymenoptera: Formicidae) of Peninsular Florida. *Fla. Entomol.*, **78** (1995), 109–18.

Kutter, H. Hymenoptera: Formicidae. In Sauter, W. (ed.) *Insecta Helvetica: Fauna*, vol. 6a. Zurich: Erganzungsband Schweizerische Entomologische Gesellschaft.

Parry, K. and E. D. Morgan. Pheromones of ants: a review. *Physiol. Entomol.*, **4** (1979), 161–89.

Sanders, C. J. The biology of carpenter ants in New Brunswick. *Can. Entomol.*, **96** (1964), 894–909.

Schneirla, T. C. *Army Ants. A Study in Social Organization*. San Francisco, CA: W. H. Freeman, 1971.

Silva, E. J. and A. E. Loeck. Ocrrancia de formigas domicilares (Hymenoptera: Formicidae) em Pelotas, RS. *Rev. Bas. Agroc.*, **5** (1999), 220–4.

Sudd, J. H. and N. R. Franks. *The Behavioural Ecology of Ants*. Glasgow: Blackie, 1987.

Talbot, M. Local distribution and flight activities of four species of ants of the genus *Acanthomyops* Mayr. *Ecology*, **44** (1963), 549–57.

Traniello, J. F. A. Foraging strategies of ants. *Annu. Rev. Entomol.*, **34** (1989), 191–210.

Vander Meer, R. K., K. Jaffe, and A. Cedeno. *Applied Myrmecology, A World Perspective*. Boulder, CO: Westview Press, 1990.

Way, M. J., M. E. Cammell, M. R. Paiva, and C. A. Collingwood. Distribution and dynamics of the Argentine ant *Linepithema (Iridomyrmex) humile* (Mayr) in relation to vegetation, soil conditions, topography and native competitor ants in Portugal. *Insectes Soc.*, **44** (1997), 415–33.

#### Formicidae: Distribution and identification

Agosti, D. and C. A. Collingwood. A provisional list of the Balkan ants (Hym., Formicidae) with a key to the worker caste, II. Key to the worker caste, including the European species without the Iberian. *Mittel. Schweiz. Entomol. Gesell.*, **60** (1987), 261–93.

Alayo, P. D. *Introducción al Estudio de los Himenópteros de Cuba. Superfamilia Formicoidea*. Serie Biológica (Havana) no. 53. Havana: Academia de Ciencias de Cuba. Instituto de Zoologia, 1974.

Arnold, G. A monograph of the Formicidae of South Africa. *Ann. South Afr. Mus.*, **14** (1915–26), 1–766; **23**, 191–295.

Baroni Urbani, C. Studi sulla mirmecofauna d'Italia. 1. *Redia*, **47** (1962), 129–38.

Baroni Urbani, C. and C. A. Collingworth. The zoogeography of ants (Hymenoptera, Formicidae) in northern Europe. *Acta Zool. fenn.*, **152** (1977), 1–34.

Barrett, K. E. J. Provisional distribution maps of ants in the British Isles. In Brian, M. V. (eds) *Ants*, pp. 203–16. London: Collins, 1977.

Berndt, K. P. and W. Eichler. Die Pharaoameise, *Monomorium pharaonis* (L.) (Hym., Myrmicidae). *Mittel. Aus. Zool. Mus. Berlin*, **63** (1987), 1–188.

Bolton, B. The ant genera of West Africa: a synoptic synopsis with keys. *Bull. Br. Mus. (Nat. Hist.) (Entomol.)*, **27** (1973), 319–68.

A review of the *Solenopsis* genus-group and revision of Afrotropical *Monomorium* Mayr. *Bull. Br. Mus. (Nat. Hist.) Entomol.*, **54** (1987), 263–452.

Bolton, B. and C. A. Collingwood. *Hymenoptera: Formicidae. Handbooks for the Identification of British Insects.*, vol. 4, part 3 (c). London: Dramrite Printers, 1975.

Borgmeier, T. Die Wanderameisen der Neotropischen Region. *Studia Entomol.*, **3** (1955), 1–720.

Bourne, R. A. De Mierenfauna van Belgie (Hymenoptera: Formicidae). *Acta Zool. Pathol. Antv.*, **67** (1977), 1–191.

Brown, W. L. A review of the ants of New Zealand. *Acta Hymenopt.*, **1** (1958), 1–50.

Buren, W. F. A review of the species of *Crematogaster*, sensu stricto, in North America. 2. Descriptions of new species. *J. Georgia Entomol. Soc.*, **3** (1968), 91–121.

Chapman, J. W. and S. R. Capco. Check list of the ants of Asia. *Monogra. Inst. Sci. Technol. Manila*, **1** (1951).

Clark, J. *The Formicidae of Australia*, vol. 1: *Subfamily Mymeciinae*. Melbourne: CSIRO, 1951.

Collingwood, C. A. The identification and distribution of British ants (Hym., Formicidae), 1. A revised key to the species found in Britain. *Trans. Soc. Br. Entomol.*, **16** (1964), 93–114.

A synopsis of the Formicidae of North Europe. *Entomologist*, **104** (1971), 150–76.

The Formicidae of Fennoscandia and Denmark. *Fauna Entomol. Scand.*, **8** (1979), 1–174.

Hymenoptera: Fam. Formicidae of Saudi Arabia. *Fauna Saudi Arabia*, **7** (1985), 230–302.

Collingwood, C. A. and K. E. J. Barrett. The identification and distribution of British ants (Hym., Formicidae), 2. The vice-county distribution of indigenous ants of the British Isles. *Trans. Soc. Br. Entomol.*, **16** (1964), 114–21.

Creighton, W. S. The ants of North America. *Bull. Mus. Comp. Zool. Harvard Coll.* **104** (1950), 1–585.

Dlussky, G. M. Ants of the genus *Formica* (in Russian). Moscow: Izdatelstvo Nauka, 1967.

DuBois, M. B. A revision of the native New World species of the ant genus *Monomorium* (minimum group). *Univ. Kansas Sci. Bull.*, **53** (1986), 65–119.

Dumpert, K. *The Social Biology of Ants*, trans. C. Johnson. London: Pitman, 1978.

Durr, H. J. R. The Argentine ant *Iridomyrmex humilis* (Mayr). I. Its distribution, harmfulness and life cycle. *Farm. S. Afr.*, **27** (1952), 381–90.

Francoeur, A. Synopsis taxonomique et économique des fourmis du Québec. *Ann. Soc. Entomol. Québec*, **22** (1977), 205–12.

Les fourmis du Québec. 1. Introduction. 2. La famille des Formicidae. 3. La sous-famille des ponerinae. *Ann. Soc. Entomol. Québec*, **24** (1979), 48–64.

Greenslade, P. J. M. *A Guide to the Ants of South Australia*. Adelaide: South Australian Museum, 1979.

Gregg, R. E. Key to the species of *Pheidole* in the United States. *J. N.Y. Entomol. Soc.*, **66** (1958), 7–48.

*The Ants of Colorado*. Boulder, CO: University of Colorado Press, 1963.

Holgersen, H. The ants of Norway. *Nyt. Mag. Naturv.*, **84** (1944), 163–202.

Johnson, C. Species identification in the eastern *Crematogaster*. *J. Entomol. Sci.*, **23** (1988), 314–32.

Mann, W. M. The ants of the British Solomon Islands. *Bull. Mus. Comp. Zool. Harvard Coll.*, **63** (1919), 273–391.

The ants of the Fiji Islands. *Bull. Mus. Comp. Zool. Harvard Coll.*, **64** (1921), 401–99.

Morisita, M., M. Kubota, K. Onoyama *et al.*, *A Guide for the Identification of Japanese Ants. 1. Ponerinae, Cerapachyinae, Pseudomyrmecinae, Dorylinae and Leptanillinae* (in Japanese). Tokyo: Myrmecological Society of Japan, 1989.

Morisita, M., M. Kubota, K. Onoyama *et al. A Guide for the Identification of Japanese Ants. 2. Dolichoderine and Formicinae* (in Japanese). Tokyo: Myrmecological Society of Japan, 1991.

Morisita, M., M. Kubota, K. Onoyama *et al. A Guide for the Identification of Japanese Ants. 3. Myrmicinae and Supplement to Leptanillinae* (in Japanese). Tokyo: Myrmecological Society of Japan, 1992.

Nogueira, S. B. and M. R. Martinho. Leaf-cutting ants (*Atta* sp.), damage to and distribution along Brazilian roads. In Jaisson, P. (ed.) *Social Insects in the Tropics*, 2, pp. 181–6. Paris: Université de Paris-Nord, 1983.

Onoyama, K. An introduction to the ant fauna of Japan, with a check list. *Kontyû*, **48** (1980), 193–212.

Smith, M. R. House-infesting ants of the Eastern United States. *US Dept. Agric. Tech. Bull.*, **1326** (1965), 105.

Snelling, R. R. and J. H. Hunt. The ants of Chile. *Rev. Chilena Entomol.*, **9** (1975), 63–129.

Trager, J. C. A revision of the genus *Paratrechina* of the continental United States. *Sociobiology*, **9** (1984), 49–162.

Wheeler, W. M. The ants of the Belgian Congo. *Bull. Am. Mus. Nat. Hist.*, **45** (1922), 1–1139.

Wilson, E. O. and R. W. Taylor. The ants of Polynesia. *Pac. Insects Monogr.*, **14** (1967), 1–109.

Wing, W. M. Taxonomic revision of the Nearctic genus *Acanthomyops*. *Cornell Univ. Agricult. Exp. Sta. Mem.*, **405** (1968), 1–173.

Wu, J. Taxonomic studies on the genus *Formica* of China. *Forest Res.*, **3** (1990), 1–8.

Yano, M. On the ants of Japan. *Zool. Mag. (Tokyo)*, **22** (1910), 416–25.

### Formicidae: *Camponotus*

Arnold, G. A monograph of the Formicidae of South Africa, part 6. (Camponotinae). *Ann. South Afr. Mus.*, **14** (1924), 675–766.

Barlin, M. R., M. S. Blum, and J. M. Brand. Species-specificity studies on the trail pheromone of the carpenter ant, *Camponotus pennsylvanicus*. *J. Ga. Entomol. Soc.*, **11** (1976), 162–4.

Eidmann, H. Zur Kenntnis der Biologie der Rossameise (*Camponotus herculeanus* [L]). *Z. Angew. Entomol.*, **14** (1929), 229–53.

Fuchs, S. The response to vibrations of the substrate and reactions to the specific drumming in colonies of carpenter ants (*Camponotus*, Formicidae, Hymenoptera). *Behav. Ecol. Sociol.*, **1** (1976), 155–84.

Kusnezov, N. El genero *Camponotus* en la Argentina. *Acta Zool., Lillona*, **12** (1951), 2–88.

Pfeiffer, M. and K. E. Linsenmair. Contributions to the life history of the Malaysian giant ant *Camponotus gigas* (Hymenoptera, Formicidae). *Insectes Soc.*, **47** (2000), 123–32.

Pricer, J. L. The life history of the carpenter ant. *Biol. Bull.*, **14** (1908), 177–218.

Snelling, R. R. Taxonomic notes on Nearctic species of *Camponotus*, Subgenus *Myrmentoma*. In Trager, J. C. (ed.) *Advances in Myrmecology*. New York, 1988.

Smith, F. The nutritional requirements of *Camponotus* ants. *Ann. Entomol. Soc. Am.*, **38** (1944), 401–8.

Traniello, J. F. A. Recruitment behavior, orientation, and the organization of foraging in the carpenter ant, *Camponotus pennsylvanicus* (DeGeer). *Behav. Ecol. Sociobiol.*, **2** (1977), 61–79.

Wang, M., G. Xiao and J. Wu. Taxonomic studies on the genus *Camponotus* in China. *Forest Res.*, **2** (1989), 221–8.

Taxonomic studies on the genus *Camponotus* in China (conclusion). *Forest Res.*, **2** (1989), 321–8.

Yasumatsu, K. and W. L. Brown, Jr. Revisional notes on *Camponotus herculeanus* Linné and close relatives in the Palearctic regions. *J. Fac. Agric. Kyushu Univ.*, **10** (1951), 29–44.

A second look at the ants of the *Camponotus herculeanus* group in eastern Asia. *J. Fac. Agri., Kyushu Univ.*, **11** (1957), 45–51.


#### Formicidae: *Lasius*


Collingwood, C. A. The *Lasius (Chthonolasius) umbratus* (Hym., Formicidae) complex in North Europe. *Entomologist*, **96** (1963), 145–58.

Himalayan ants of the genus *Lasius*. *System. Entomol.*, **7** (1982), 283–96.

Seifert, B. A taxonomic revision of the Palearctic members of the ant subgenus *Lasius s. str. Abhand. Berichte Naturkunde-mus. Gorlitz*, **66** (1992), 1–67.

Wilson, E. O. A monographic revision of the ant genus *Lasius*. *Bull. Mus. Comp. Zool. Harvard Coll.* **113** (1955), 1–201.

Yamauchi, K. Taxonomic and ecological studies on the ant genus *Lasius* in Japan. 1. Taxonomy. *Sci. Rep. Fac. Educ. Gifu Univ.* (Nat. Sci.), **6** (1979), 147–81.


#### Formicidae: *Solenopsis*


Bruce, W. A., L. D. Cline, and G. L. LeCato. Imported fire ant infestation of buildings. *Fla Entomol.*, **61** (1978), 230.

Creighton, W. S. The New World species of the genus *Solenopsis*. *Proc. Am. Acad. Arts Sci.*, **66** (1930), 39–151.

Clemmer, D. I. and R. E. Serfling. The imported fire ant: dimensions of the urban problem. *South Med. J.*, **68** (1975), 1133–8.

Fletcher, D. J. C., M. S. Blum, T. V. Whitt, and N. Temple. Monogyny and polygyny in the fire ant, *Solenopsis invicta*. *Ann. Entomol. Soc. Am.*, **73** (1980), 658–61.

Lofgren, C. S., W. A. Banks, and B. M. Glancy. Biology and control of imported fire ants. *Annu. Rev. Entomol.*, **20** (1975), 1–30.

Markin, G. P., J. H. Diller, S. O. Hill, M. S. Blum, and H. R. Hermann. Nuptial flight and flight ranges of the imported fire ant, *Solenopsis saevissima richteri* (Hymenoptera: Formicidae). *J. G. Entomol. Soc.*, **6** (1971), 145–56.

Markin, G. P., H. L. Collins, and J. H. Diller. Colony founding by queens of the red imported fire ant, *Solenopsis invicta*. *Ann. Entomol. Soc. Am.*, **65** (1972), 1053–8.

Markin, G. P., J. O'Neal, J. H. Diller, and H. L. Collins. Regional variation in the seasonal activity of the imported fire ant, *Solenopsis saevissima richteri*. *Environ. Entomol.*, **3** (1974), 446–52.

Trager, J. C. A revision of the fire ants, *Solenopsis geminata* group. *J. N.Y. Entomol. Soc.*, **99** (1991), 141–98.

Vander Meer, R. K. Semiochemicals and the red imported fire ant (*Solenopsis invicta* Buren). *Fla. Entomol.*, **66** (1983), 139–61.

Wilson, N. L., J. H. Diller, and G. P. Markin. Foraging territories of imported fire ants. *Ann. Entomol. Soc. Am.*, **64** (1971), 660–5.


#### Mutillidae, Pompilidae, Pteromalidae


Becker, G. and W. Weber. *Theocolax formicformis* Westwood (Hymenoptera, Chalcidae) ein anobienparasit. *Z. Parasitenkunde*, **15** (1952), 339–56.

Brothers, D. J. Phylogeny and classification of the aculeate Hymenoptera with special reference to Mutillidae. *Univ. Kansas Sci. Bull.*, **50** (1975), 483–648.

Alternative life-history styles of mutillid wasps (Insecta, Hymenoptera). In Burton, M. N. (ed.) *Alternative Life-History Styles of Animals*, pp. 279–91. Dordrecht: Kluwer Academic, 1989.

Evans, H. E. and C. M. Yoshimoto. The ecology and nesting behavior of the Pompilidae (Hymenoptera) of the northeastern United States. *Misc. Publ. Entomol. Soc. Am.*, **3** (1962), 65–120.

Hurd, P. D. Jr. The California velvet ants of the genus *Dasymutilla* Ashmead (Hymenoptera: Mutillidae). *Bull. Calif. Insect Surv.*, **1** (1951), 89–114.

Richards, O. W. and A. H. Hamm. The biology of the British Pompilidae (Hymenoptera). *Trans. Soc. Br. Entomol.*, **6** (1939), 51–114.

Taylor, J. M. Studies on *Theocolax formiciformis* Westw. (Hymenoptera, Pteromalidae) a parasite of *Anobium punctatum* (Deg.), (Coleoptera, Anobiidae). *Bull. Entomol. Res.*, **54** (1964), 797–804.


#### Siricidae, Sphecidae, Stephanidae


Benson, R. B. Studies in Sircidae, especially of Europe and southern Asia. *Bull. Entomol. Res.*, **34** (1943), 27–51.

Bohart, R. M. and A. S. Menke. *Sphecid Wasps of the World: A Generic Revision*. Berkeley, CA: University of California Press, 1976.

Brockmann, H. J. Diversity in the nesting behavior of mud-daubers (*Trypoxylon politum* Say; Sphecidae). *Fla. Entomol.*, **63** (1979), 53–64.

Provisioning behavior of the great golden digger wasp, *Spex ichneumoneus* (L.) (Sphecidae). *J. Kansas Entomol. Soc.*, **58** (1985), 631–55.

Chrystal, R. N. and J. B. Myers. Natural enemies of of *Sirex cyaneus* (F.) in England and their life histories. *Bull. Entomol. Res.*, **19** (1928), 67–77.

Elliott, E. A. Monograph of the hymenopterous family Stephanidae. *Proc. Zool. Soc., Lond.*, **1922** (1922), 705–831.

Evans, H. E. Studies on the comparative ethology of digger wasps of the genus *Bembix*. Ithaca, NY: Comstock Press, 1957.

Gilbert, J. M. and L. W. Miller. An outbreak of *Sirex noctilio* (F.) in Tasmania. *Aust. Forestry*, **16** (1952), 63–9.

Hanson, H. S. Ecological notes on the *Sirex* woodwasps and their parasites. *Bull. Entomol. Res.*, **30** (1939), 27–65.

Maa, T. A. A synopsis of Asiatic Siricoidea with notes on certain exotic and fossil forms. *Notes Entomol. Chin.*, **13** (1949), 76–163.

Middlekauff, W. W. The siricid woodwasps of California. *Bull. Calif. Inst. Surv.*, **6** (1960), 59–77.

Morgan, F. D. Bionomics of Siricidae. *Annu. Rev. Entomol.*, **13** (1968), 239–56.

Morgan, F. D. and N. C. Stewart. The biology and behavior of the woodwasp *Sirex noctilio* (F.) in New Zealand. *Trans. R. Soc. N.Z.*, **7** (1966), 195–204.

Parkin, E. A. Symbiosis in larval Siricidae. *Nature*, **147** (1942), 329.

Parkin, E. A. Symbiosis and siricid woodwasps. *Ann. Appl. Biol.*, **29** 268–74.

Precupetu, A. and St. Negru. Contribution to the knowledge of the woodwasps of the Rumanian Peoples Republic fauna. *Rev. Biol. Bucharest*, **6** (1961), 81–9.

Rawlings, G. B. Recent observations on the *Sirex noctilio* populations in *Pinus radiata* forests in New Zealand. *N.Z. J. Forestry*, **5** (1948), 411–21.

Rodd, N. W. Some observations on the biology of Stephanidae and Megalyridae (Hymenoptera). *Aust. Zool.*, **11** (1951), 341–6.

Shafer, G. D. *The Ways of a Mud-Dauber*. Stanford, CA: Stanford University Press, 1949.

Spradbery, J. P. The oviposition biology of siricid woodwasps in Europe. *Ecol. Entomol.*, **2** (1977), 225–30.

Stillwell, M. A. The pigeon tremex, *Tremex columba* (Hymenoptera: Siricidae), in New Brunswick. *Can. Entomol.*, **99** (1967), 685–9.

Tobias, V. I. Family Stephanidae. In Zhelokhovtsev, A. N., V. I. Tobias, and M. A. Kozlov (eds.) *Key to the Identification of Insects of European USSR*, vol. 3, *Hymenoptera* 6, pp. 236–7, Moscow.

Tragardh, I. Survey of the wood-destroying insects in public buildings in Sweden. *Bull. Entomol. Res.*, **29** (1938), 57–62.

Tribe, G. D. The woodwasp *Sirex noctilio* Fabricius (Hymenoptera: Siricidae), a pest of *Pinus* species, now established in South Africa. *Afr. Entomol.*, **3** (1995), 215–17.

Williams, F. X. *Ampulex compressa* (Fabr.), a cockroach-hunting wasp introduced from New Caledonia into Hawaii. *Proc. Hawaiian Entomol. Soc.*, **11** (1942), 221–33.

### Vespidae

Carpenter, J. M. Biographic patterns in the Vespidae (Hymenoptera): two views of Africa and South America. In Goldblatt, P. (ed.). *Biological Relationships between Africa and South America*, pp. 139–55. New Haven, CT: Yale University Press, 1993.

Chan, K. L. The hornets of Singapore: their identification, biology and control. *Singapore Med. J.*, **13** (1972), 178–87.

Jeanne, R. L. Evolution of social behavior in the Vespidae. *Annu. Rev. Entomol.*, **25** (1980), 371–96.

Lord, W. D. and R. R. Roth. Density, distribution and reproductive success in *Vespula maculifrons* (Hymenoptera: Vespidae). *Am. Mid. Nat.*, **113** (1985), 353–60.

Madden, J. L. Factors influencing the abundance of the European wasp (*Paravespula germanica* [F]). *J. Aust. Entomol. Soc.*, **20** (1981), 59–66.

Makino, S., Sk. Yamane, T. Ban, and I. Kunou. The Japanese hornet *Vespa simillima* Smith, an important nuisance pest in urban areas (Hymenoptera: Vespidae). *Jpn. J. Sanit. Zool.*, **32** (1981), 203–13.

Matsura, M. Nesting habits of several species of the genus *Vespa* in Formosa. *Kontyû*, **41** (1973), 286–93.

Morse, R. A., G. C. Eichwort, and R. S. Jacobson. The economic status of an immigrant yellowjacket, *Vespula germanica* (Hymenoptera: Vespidae), in northeastern United States. *Environ. Entomol.*, **6** (1977), 109–10.

Pallett, M. J. Nest site selection and survivorship of *Dolichovespula arenaria* and *Dolichovespula maculata* (Hymenoptera: Vespidae). *Can. J. Zool.*, **62** (1984), 1268–72.

Parrish, M. D. and R. B. Roberts. Successful establishment of the German yellowjacket in urban New Jersey indicated by relative abundance of *Vespula germanica* and *V. maculifrons* (Hymenoptera: Vespidae). *J. Kansas. Entomol. Soc.*, **55** (1982), 272–6.

Ross, K. G. and R. W. Matthews (eds.) *The Social Biology of Wasps*. Ithaca, NY: Cornell University Press, 1991.

van der Vecht, J. The Indo-Australian species of the genus *Ropalidia* (= *Icaria*) Pt. 1. *Treubia*, **18** (1941), 103–91.

The Indo-Australian species of the genus *Ropalidia* (= *Icaria*) Pt. 2. *Zool. Meded. Rijks. Mus. Nat. Hist. Leiden*, **57** (1941), 1–72.

Wafa, A. K. Ecological investigations on the activity of the oriental hornet, *Vespa orientalis*. *Bull. Fac. Agr. Cairo Univ.*, **103** (1956), 1–35.

Willink, A. Sobre la presencia de *Vespula germanica* (Fabricius) en la Argentina (Hymenoptera, Vespidae). *Neotropica (La Plata)*, **26** (1980), 205–6.

Yamane, Sk. and S. Makino. Bionomics of *Vespa analis insularis* and *V. mandarinia latilineata* in Hokkaido, northern Japan, with notes on vespine embryo nests (Hymenoptera: Vespidae). *Insect Matsumurana N. S.*, **12** (1977), 1–33.

### Vespidae: *Polistes*

Bequart, J. An introductory study of *Polistes* in the United States and Canada with descriptions of some North and South American forms. *J. N.Y. Entomol. Soc.*, **48** (1940), 1–31.

Cervo, R., F. Zacchi, and S. Turillazzi. *Polistes dominulus* (Hymenoptera, Vespidae) invading North America: some hypotheses for its rapid spread. *Insectes Soc.*, **47** (2000), 155–7.

Corn, M. L. Notes on the biology of *Polistes carnifex* in Costa Rica and Columbia. *Psyche*, **79** (1972), 150–7.

Eberhard, M. J. The social biology of polistine wasps. *Misc. Publ. Mus. Zool. Univ. Mich.*, **140** (1969), 1–101.

Gamboa, G. J., E. I. Greig, and M. C. Thom. The comparative biology of two sympatric paper wasps, the native *Polistes fuscatus* and the invasive *Polistes dominulus* (Hymenoptera, Vespidae). *Insectes Soc.*, **49** (2002), 45–9.

Garcia, A. R. Observations of *Polistes peruvianus* in the environs of Lima. *Biota*, **10** (1974), 11–27.

Gillaspy, J. E. Behavioral observations on paper-wasps (genus *Polistes*). *Am. Midl. Nat.*, **90** (1973), 1–12.

The impact on man of *Polistes* wasps, with special reference to caterpillar suppression. In Breed, M. D., C. D. Michner, and H. E. Evans (eds.) *The Biology of Social Insects*, pp. 129–33. Boulder, CO: Westview Press, 1982.

*Polistes* wasps: biology and impact on man. In Vinson, S. B. (ed.) *Economic Impact and Control of Social Insects.*, pp. 332–52. New York: Praeger, 1986.

Miyano, S. Life tables of colonies and workers in a paper wasp, *Polistes chinensis antennalis*, in central Japan (Hymenoptera: Vespidae). *Res. Popul. Ecol.*, **22** (1980), 69–88.

Nelson, J. M. Parasites and symbionts of nests of *Polistes* wasps. *Ann. Entomol. Soc. Am.*, **61** (1968), 1528–39.

Rabb, R. L. Biological studies of *Polistes* in North Carolina. *Ann. Entomol. Soc. Am.*, **53** (1960), 111–21.

Rau, P. The nests and the adults of colonies of *Polistes* wasps. *Ann. Entomol. Soc. Am.*, **39** (1946), 11–27.

Soika, A. G. Vespidological notes XXXVII. New *Polistes* from the Australian continent. *Bol. Soc. Entomol. Ital.*, **107** (1975), 20–5.

West-Eberhard, M. J. The social biology of polistine wasps. *Misc. Publ. Mus. Zool. Univ. Michigan No.*, **140** (1969), 1–101.

Yamane, S. Preliminary observations of the life history of two polistine wasps, *Polistes snelleni* and *P. biglumis* in Sapporo, northern Japan. *J. Fac. Sci. Hokkaido Univ. Ser. 6*, **17** (1969), 78–105.

### *Vespidae:* Dolichovespula, Vespa, Vespula

Akre, R. D. The social wasps. In Hermann, H. R. (ed.) *Social Insects.*, vol. 4, New York: Academic Press, 1982.

Akre, R. D. and E. A. Myhre. Nesting biology and behavior of the baldfaced hornet, *Dolichovespula maculata* (L.) (Hymenoptera: Vespidae) in the Pacific Northwest. *Melandria*, **48** (1992), 1–33.

Akre, R. D. and H. C. Reed. Population cycles of yellowjackets (Hymenoptera: Vespinae) in the Pacific Northwest. *Environ. Entomol.*, **10** (1981), 267–74.

Akre, R. D., J. F. MacDonald, and W. B. Hill. Yellowjacket literature. *Melandria*, **18** (1974), 67–93.

Akre, R. D., W. B. Hill, J. F. MacDonald, A. Greene, and P. Landolt. Behavior and colony development of *Vespula pensylvanica* and *V. atropilosa*. *J. Kansas Entomol. Soc.*, **49** (1976), 63–84.

Akre, R. D., A. Greene, J. F. MacDonald, P. J. Landholt, and H. G. Davis. The Yellowjackets of America north of Mexico. USDA handbook, no. 552. Washington, DC: USDA, 1981.

Archer, M. E. The Euro-Asian species of the *Vespula rufa* group (Hymenoptera, Vespidae) with descriptions of two new species and one new subspecies. *Kontyû*, **49** (1981), 54–64.

Balduf, W. V. Observations on the white-faced wasp, *Dolichovespula maculata* (Linn.). *Ann. Entomol. Soc. Am.*, **47** (1954), 445–58.

Ban, T. and S. K. Yamane. Nesting sites of *Vespa simillima* in urban areas. *Jpn. J. Sanit. Zool.*, **31** (1980), 150.

Biegel, W. Zur Biologie und Okologie sozialer Wespen. *Sitzungsber. Phys. Med. Soc. Erlangen*, **76** (1953), 115–53.

Bluethgen, P. and J. Gusenleitner. Wasps from Iran. *Stuttg. Beitr. Naturkd.*, **223** (1970), 1–13.

Brian, M. V. and A. D. Brian. The wasp, *Vespula sylvestris* Scopoli: feeding, foraging and colony development. *Trans. R. Entomol. Soc. Lond.*, **103** (1952), 1–26.

Buckell, E. R. and G. J. Spencer. The social wasps of British Columbia. *Proc. Entomol. Soc. B. C.*, **46** (1950), 33–40.

Bunn, D. S. Notes on the nesting cycle of *Dolichovespula sylvestris* Scop. (Hym., Vespidae). *Entomol. Mon. Mag.*, **118** (1982), 213–18.

Chan, K. L. The hornets of Singapore: their identification, biology, and control. *Singapore Med. J.*, **13** (1972), 178–87.

Davis, H. G. Yellowjackets wasps in urban environments. In Frankie, G. W. and C. S. Koehler (eds.) *Perspectives in Urban Entomology*, pp. 163–5. New York: Academic Press, 1977.

Donovan, B. J. Occurrence of the common wasp, *Vespula vulgaris* (L.) (Hymenoptera: Vespidae) in New Zealand. *N.Z. J. Zool.*, **11** (1984), 417–27.

Duncan, C. D. A contribution to the biology of North American Vespine wasps. *Stanford Univ. Publ. Biol. Sci.*, **8** (1939), 1–282.

Eck, R. Zur Verbreitung und Varabilitatvon *Dolichovespula norwegica* (Hymenoptera, Vespidae). *Entomol. Abh. Mus. Tierkd. Dresden*, **44** (1981), 133–52.

Zur Verbreitung und Varabilitatvon *Dolichovespula saxonia* (Hymenoptera, Vespidae). *Entomol. Abh. Mus. Tierkd. Dresden*, **46** (1981), 151–76.

Edwards, R. *Social Wasps: Their Biology and Control*. UK: Rentokil, Sussex, 1980.

Fox-Wilson, G. Factors affecting populations of social wasps, *Vespula* species, in England. *Proc. R. Entomol. Soc. Lond. Ser. A.*, **21** (1946), 17–27.

Giganti, H. E. Una molesta avispa [*Vespula* (*Paravespula*) *germanica* (Fab.)] se ha introducido en Argentina. *Gaceta Agron.*, **3** (1983), 24–8.

Green, A., R. D. Akre, and P. J. Landolt. The aerial yellowjacket, *Dolichovespula arenaria* (Fab.): nesting biology, reproductive production, and behavior (Hymenoptera: Vespidae). *Melandria*, **26** (1976), 1–34.

Guiglia, D. Les guepes socials d'Europe Occidentale et Septentrionale. *Faune Eur. Bassin Méditerr.*, **6** (1972), 1–181.

Ikeda, F. and S. Kodomari. Biology of *Vespula lewisii* (in Japanese). *Sizuoka Pref. Agric. Exp. Stn.* (1982).

Ishay, J., H. Bytinski-Salz, and A. Shulov. Contributions to the bionomics of the oriental hornet (*Vespa orientalis* Fab.). *Isr. J. Entomol.*, **2** (1968), 45–106.

Ishikawa, R. A study of *Dolichovespula* media (Retzius) of Japan. *Bull. Nat. Sci. Mus. Tokyo*, **12** (1969), 179–82.

Jacobson, R. S., R. W. Matthews, and J. F. McDonald. A systematic study of the *Vespula vulgaris* group with a description of a new yellowjacket species in eastern North America (Hymenoptera: Vespidae). *Ann. Entomol. Soc. Am.*, **71** (1978), 299–312.

Jeanne, R. L. Evolution of the social behavior of the Vespidae. *Annu. Rev. Entomol.*, **25** (1980), 371–96.

Kemper, H. Nestunterschiede bei den sozialen Faltenwespen Deutschlands. *Z. Angew. Zool.*, **48** (1961), 31–85.

Kemper, H. and E. Döhring. *Die Sozialen Faltenwespen Mittleleuropas*. Berlin: Paul Parey, 1967.

Kojima, J. The genus *Vespa* in the Philippines (Hymenoptera, Vespidae). *Kontyû*, **50** (1982), 434–44.

Løken, A. Social wasps in Norway. *Nord. Entomol. Tidsskr.*, **12** (1964), 195–218.

MacDonald, J. F. and R. W. Matthews. Nesting biology of the eastern yellowjacket, *Vespula maculifrons* (Hymenoptera: Vespidae). *J. Kansas Entomol. Soc.*, **54** (1981), 433–57.

MacDonald, J. F., R. D. Akre, and R. Keyel. The German yellowjacket (*Vespula germanica*) problem in the United States. *Bull. Entomol. Soc. Am.*, **26** (1980), 436–42.

MacDonald, J. F., R. W. Matthews, and R. S. Jacobson. Nesting biology of the yellowjacket, *Vespula flavopilosa* (Hymenoptera: Vespidae). *J. Kansas Entomol. Soc.*, **53** (1980), 448–58.

Matsuura, M. Nesting sites of the Japanese *Vespa* species. *Kontyû*, **39** (1971), 43–54.

Nesting habits of several species of the genus *Vespa* in Formosa. *Kontyû*, **41** (1973), 286–93.

Miller, C. D. F. Taxonomy and distribution of Nearctic *Vespula*. *Can. Entomol.* (suppl. 22) (1961), 1–52.

Richards, O. W. The biology of the social wasps. *Biol. Rev.*, **46** (1971), 483–528.

*The Social Wasps of the Americas*. London: British Museum of Natural History, 1977.

Richards, O. W. and M. J. Richards. Observations on the social wasps of South America. *Trans. R. Entomol. Soc. Lond.*, **102** (1951), 1–169.

Smithers, C. N. and G. A. Holloway. Establishment of *Vespula germanica* (Fabricius) (Hymenoptera: Vespidae) in New South Wales. *Aust. Entomol. Mag.*, **5** (1978), 55–60.

Steffen, A. W. Uber Nestplatz und Nestbau von *Dolichovespula norwegica* (F.). *Z. Angew. Zool.*, **49** (1962), 383–92.

Thomas, C. R. The European wasp (*Vespula germanica* Fab.) in New Zealand. *N.Z. Dept. Sci. Ind. Res. Inf. Ser.*, **27** (1960), 1–74.

van der Vecht, J. Notes on Oriental Vespinae, including some species from China and Japan. *Zool. Meded. Rijks. Mus. Nat. Hist. Leiden*, **36** (1959), 205–32.

Visscher, P. K. and R. S. Vetter. Annual and multi-year nests of the western yellowjacket, *Vespula pensylvanica*, in California. *Insectes Soc.*, **50** (2003), 160–6.

Wagner, R. E. The genus *Dolichovespula* and an addition to its known species of North America (Hymenoptera: Vespidae). *Pan-Pac. Entomol.*, **54** (1978), 131–42.

Weyrauch, W. *Dolichovespula* und *Vespa*. Vergleichende Übersicht über swei wesentliche Lebenstypen bei sozialen Wespen., Mit Bezugnahme auf die Frage nach der Fortschrittlichkeit tierischer Organisation. *Biol. Zentralbl.*, **56** (1935), 287–301.

Whitehead, V. B. and A. J. Prins. The European wasp, *Vespula germanica* (F.) in the Cape Peninsula. *J. Entomol. Soc. South Afr.*, **38** (1975), 39–42.

Willink, A. Sobre la presencia de *Vespula germanica* (Fabricius) en la Argentina (Hymenoptera: Vespidae). *Neotropica*, **26** (1980), 205–6.

Yamane, S. On the genus *Vespa* from Nepal. *Kontyû*, **42** (1974), 29–39.

Yamane, Sk., R. E. Wegner, and S. Yamane. A tentative revision of the subgenus *Paravespula* of eastern Asia (Hymenoptera: Vespidae). *Insecta Matsumurana N. S.*, **19** (1980), 1–46.

### Tenthrendinidae, Tiphiidae, Xyphriidae

Allen, H. W. A revision of the Tiphiinae (Hymenoptera: Tiphiidae) of Eastern North America. *Trans. Am. Entomol. Soc.*, **92** (1966), 231–356.

A monograph study of the genus *Tiphia* (Hymenoptera: Tiphiidae) of Western North America. *Trans. Am. Entomol. Soc.*, **97** (1971), 201–359.

A monograph study of the subfamily Tiphiinae (Hymenoptera: Tiphiidae) of South America. *Smithsonian Contrib. Zool.*, **113** (1972), 1–76.

Benson, R. B. An introduction to the natural history of British sawflies. *Trans. Soc. Br. Entomol.*, **10** (1950), 45–142.

Clausen, C. P., H. A. Jaynes, and T. R. Gardner. Further investigations of the parasites of *Popillia japonica* in the Far East. *US Dept. Agric. Tech. Bull.*, **366** (1933), 1–58.

Smith, D. R. The xiphydriid woodwasps of North America (Hymenoptera: Xyphriidae). *Trans. Am. Entomol. Soc.*, **102** (1976), 101–31.

Smith, D. R. and E. M. Barrows. Sawflies (Hymenoptera: Symphyta) in urban environments in the Washington, DC area. *Proc. Entomol. Soc. Wash.*, **89** (1987), 147–56.

# 10 Isoptera

## Introduction

Winged termites vary in size, from the large *Macrotermes* with a body length of 19–20 mm and a wing span of about 90 mm, to *Afrosubulitermes*, which is 4–5 mm long and has a wing span of 9–10 mm. Their color ranges from light to dark brown, and their wings vary from transparent to dark brown. The head is round or slightly oval and flattened above and there are usually large mandibles. In the families Rhinotermitidae and Termitidae, there is a small pore, the fontanelle, medially on the frontal region, usually behind the posterior margin of the eyes. It is the external opening of the frontal gland. In the soldiers of some species, this external gland opening has been modified for the purposes of defense. The compound eyes are large and in all families except Termopsidae and Hodotermitidae there are ocelli situated near the upper border of the eyes. The antennae are long and there are 15–32 beadlike segments. Variations in the number of segments are sometimes related to the number of molts in the nymph stage; the two antennae may not always have the same number of segments.

The wings are equal-sized. In Mastotermitidae, the hind wings have an enlarged anal lobe, which provides a link with cockroaches and a cockroach-like common ancestor. The venation is simple, and the presence of cross-veins is limited to the primitive families. A feature unique to the Isoptera is a humeral suture near the base of the wing; this line of weakness allows the wings to be broken and shed after the mating flight. The abdomen consists of 10 segments; the dorsal plates (tergites) are similar in both sexes, and there is a ventral plate (sternite) on each segment except the first. The terminal segments differ in males and females, but there are no external genitalia. There are cerci on abdominal segment 10, and males have a pair of unsegmented anal plates on the posterior border of sternite 9. In some termites there are slight differences in the size between the sexes, and minor differences in the position of the ocelli. The male remains relatively unaltered throughout his life, but the female abdomen enlarges by the continuous growth of the intersegmental membrane. In the *Macrotermes*, there is a sixfold increase in length of the female abdomen, from 20 mm to over 120 mm, from the time she establishes the colony.

Food of termites is primarily cellulosic material, principally wood in stages of decomposition. Termites themselves are not capable of digesting cellulose. The phylogenetically placed lower termites rely on cellulolytic flagellate protists in the hindgut to break down cellulose. The flagellates are nutritional symbionts. They are in the hindgut of all termite castes that ingest or are fed wood particles, except newly molted individuals or those entering ecdysis. Termites lose their symbiotic flagellates before or during ecdysis, except in the molt to the winged adult. They are reinfected following ecdysis by proctodael or anal trophallaxis. Termite flagellates are somewhat host-specific, and the hindgut communities of these organisms are typically composed of 2–10 species. These communities are stable over space, time, and colony. Differentiation of protozoan fauna in *Coptotermes* has been used to trace geographic distribution and the relationships of sympatric families. The flagellates associated with alates, soldiers, and pseudergates of *Reticulitermes flavipes* are *Trichonympha agilis*, *Spirotrichonympha kofoidi*, *Holomastigotes elongatum*, *Pyrsonympha vertens*, *Dinenympha gracilis*, and *D. fimbriata*. Although the genus *Spirotrichonympha* is dominant in many termites, such as *Coptotermes* spp., the majority of cellulose digestion is by species of *Pseudotrichonympha* and *Holomastigotes*.

The food value of termites to humans has been determined for a few species that occur in southern Africa. The caloric value of fresh termites (species unspecified) is 619.6 J per 100 g (148 cal per 100 g); protein constitutes 10 g per 100 g of termites, and fat constitutes 12 g per 100 g of termites. The caloric value of *Macrotermes natalensis* is 523.35 J per 100 g of termites (125 cal per 100 g), with 25 g per 100 g of protein, and 2.6 g per 100 g of fat.

## Development and castes

Termites are social insects, but their incomplete metamorphosis, with egg, nymph, and adult phases, distinguishes them from other social insects, such as ants, bees, and wasps, which develop through egg, larva, pupa, and adult stages. Another difference is the presence of a functional male in the termite community. In other social insects, only the female survives after the nuptial flight, and total egg production is limited by the usually single mating. Termite eggs are small and laid singly, though sometimes in quick succession, as in the case of *Macrotermes* with an average of one every 2 s over long periods. In *M. darwiniensis* the eggs are laid in batches of 16–24 and cemented together in two rows, similar to oothecae of some cockroaches and grasshoppers. The nymph phase consists of a series of instars separated by molts, which may be recognized by the size of the individual (larva, worker, or nymph), the number of antennal segments, or the presence of wing pads. The reproductive organs become functional after the final molt, which may result in primary or secondary reproductives.

Colonies function with individuals that have particular roles and functions: defense and protection of the nest against predators, foraging for food, or reproduction to expand the existing colony or establish a new colony. Each of these functions is accomplished by a caste, such as soldiers, workers, and reproductives. A typical termite colony contains a functional male and female. This pair of primary reproductives (called king and queen) originally established the colony, and they were at one time winged alates. A colony may have wingless, secondary, or neotenic reproductives (males and females) that remain in the colony and produce eggs. The colony has numerous workers and soldiers; these forms are wingless, have large mandibles, and are sterile. The proportion of workers to soldiers in a colony depends on the species, and usually changes with the increasing age and size of the colony. The ratio of workers to soldiers differs between termite family and species, but colonies generally consist of 80–90% workers. In *Reticulitermes*, soldiers typically comprise about 2% of the colony individuals, but in *Coptotermes formosanus*, they comprise 10–15% of the colony.

Larvae hatching from eggs are essentially identical until the first molt. They have no eyes or wing buds, and they develop into other castes after one or more molts. In the primitive termites, nymphs develop from older larvae and they may become workers (called pseudergates) or soldiers with large heads and prominent mandibles. Older nymphs may also become reproductives, including winged males and females, and brachypterous or wingless neotenics. In the evolutionary advanced termites, workers develop from larvae after the second or third molt, and they may spend their entire 1–4-year life in this caste. However, a worker can molt twice (first to become a pre-soldier) to become a soldier; the soldier cast is a terminal caste. Workers that pass through the nymph stage can develop into winged reproductives. Nymphs have external wing buds and, as they progress toward the winged adult stage, they gradually develop eyes, full-size wings, and functional reproductive organs. Nymphs may undergo a regressive molt, lose their wing buds, and revert to the worker stage. Under certain conditions, secondary reproductives develop from nymphs or worker termites; they are called second- and third-form reproductives.

Workers are wingless individuals with well-sclerotized mandibles, lightly sclerotized heads, and, except for a few species, very lightly sclerotized bodies. In most cases, compound eyes and ocelli are lacking, and the number of antennal segments is fewer than the other castes. The size of the worker in the colony may vary, depending on their stage of development, and they are considered major and minor workers. Their task in the colony may be related to age and evident in morphological and physiological differences. The young workers tend to remain in the nest caring for the brood, and the old workers conduct nest repair and forage for food. In termite species that do not have a soldier caste, workers assume some nest defense responsibility. Workers in most of the African soldierless termites have a line of weakness in the cuticle behind the metanotum. When threatened by predators the abdominal muscles contract, the cuticle opens, and the intestines burst, scattering their contents on the intruder. The use of intestinal contents for defense has been observed in several genera, including *Skatititermes* and *Speculitermes*.

Soldiers differ from other castes in the colony by having modified heads and prominent mandibles. There are two types: mandibulate soldiers, which have a large head and mandibles, and nasute soldiers, in which the front of the head is elongated and the mandibles are reduced. Mandibulate soldiers occur in nearly all families. Defense against invaders of the nest is achieved by biting or a snapping action of the mandibles. The head in these individuals is usually elongated to accommodate the muscles necessary to move the large mandibles. The enlarged head is used by some species, such as *Cryptotermes*, to seal galleries and entrance holes against predators. Mandibulate soldiers use salivary secretions as a chemical defense. Saliva is an irritant discharged into wounds created by the mandibles, or functioning as a glue to entangle intruders. Nasute soldiers have reduced mandibles, and they rely entirely on chemicals to deter predators. Soldiers in *Nasutitermes* and other related genera have a large rounded head with a pointed snout, at the end of which the frontal gland opens. When disturbed, the frontal gland releases an irritating, sticky substance, which can be expelled up to 15 mm. They are able to disable intruders without physical contact. When soldiers of some species are disturbed, they exhibit considerable agitation. They frequently jerk the body and head back and forth, and may produce an audible knocking as the head strikes the sides of galleries. Such action by one soldier often stimulates others to similar agitated knocking.

Secondary reproductives or neotenics are individuals that develop functional reproductive organs without becoming alates. They are produced in colonies that have lost one or both of the primary reproductives; they are usually not produced in intact colonies. They may develop in portions of a colony that have become isolated or distant from the parent group. Neotenics develop from brachypterous workers and they are considered second-form reproductives. They may also develop from brachypterous nymphs and they are considered third-form reproductives. The primary reproductives in a colony may be replaced by a large number of secondary reproductives, and in this way the size and reproductive potential of the colony are increased. The original sexual pair in the colony is capable of chemically inhibiting development of the reproductive potential of all members of a large colony. The king and/or queen produce a pheromone that circulates through the colony and inhibits the formation of secondary reproductives. Several pheromones produced by the reproductives have been linked in the control of secondary reproductives. The inhibitory pheromone produced by the queen acts to prevent development of female secondary reproductives, but does not affect the development of alates. This substance is circulated through the larvae in the colony by anal trophallaxis. In *Kalotermes flavicollis*, several neotenics are developed following removal of the primary reproductives, but only one pair remains to serve the colony. Fights occur between neotenics of the same sex. Individuals that are severely wounded during these battles are eaten by other nestmates. In *Reticulitermes* and *Zootermopsis* the primary pair may be replaced by numerous secondary reproductives, and the colony may become very large with the added productivity of many queens.

## Colonies and life cycles

Dispersal of winged reproductives from established colonies is common to all termites and it is an important means of establishing new colonies. Winged forms or alates are usually produced by relatively mature colonies, and, depending on the species and resources, this occurs 3–5 years after the colony founding. In some of the primitive families, small alates are released over several months, but often there is a seasonal peak. In higher termites, large numbers of alates emerge at one time. The swarming of alates from these colonies is usually restricted to certain times of the year, and specific times of the day, evening, or night. Swarming is synchronized with regional and local weather conditions; it occurs during warm months in temperate regions and with the arrival of the rainy season in the tropics. The exact time of year and day that swarming takes place is species-specific, and dependent on the condition of the colony. Soldiers and workers are involved in excavating exit holes and protecting the emerging alates and the colony against predators. The number of winged reproductives produced ranges from 1 to 43% of the total individuals in the colony.

The flight of termites is weak and fluttering, and their flight direction and distance are often directed by winds. In general, the alates of primitive wood-dwelling termites fly further than do the highly evolved species. The latter tend to lose their wings after a short initial flight. They often lose one or more wings in flight, and they fly erratically to the ground; the process of flight weakens the wing at the basal suture and the wings are shed when they land. In the lower termites, the wings are stronger and a series of flights occur. To remove the wings the termites brace against some irregularity of the substrate and turn and twist to break the wings. Generally, alates are attracted to areas of greatest light intensity. Evening- and night-flying termites are often attracted to lights and windows. The emergence times for alates of different species in the same area are separated, and there is usually no overlap or more than one species flying at the same time. Typically, the flight of a single species occurs from many colonies over a wide area at the same time. Sporadic flights from isolated colonies are often encountered in urban areas, where the conditions suitable for flight exist only in small habitats. The emergence of alates from a colony may be interrupted before all the alates emerge. This results in a series of small and large flights from a colony during one season.

Alates are in the air for a short time, and predation by other insects and birds usually takes about 99% of the swarm. In the higher termites, the female lands first and assumes a position in which the abdomen is elevated to release a chemical sex attractant from either tergal or sternal glands. When a male approaches, and perhaps contacts the female, both sexes shed their wings and the female leads the male to a suitable nest site in moist soil. The dampwood termites seek cracks or holes in timbers above ground; subterranean termites select moist soil or decayed wood in tree stumps. Drywood termites fly to sites in buildings, or crawl up the outside to reach roofing timbers and begin excavating in cracks or regions softened by decay fungi. Mating usually does not occur until the male and female prepare and enter a small chamber and the entrance hole is sealed. This takes 1–2 days and involves displacing soil to make a cavity or chewing into a wood substrate. Establishment of a new colony by flight of alates is dependent on the survival of both reproductives until they are in a secure location and the female is inseminated. Continued development of the colony is dependent upon reinsemination of the female.

Egg-laying begins 1–2 weeks after mating. In lower termites, about 20 eggs are laid in the first batch. In higher termites, about 100 eggs are laid by young queens. Hatching takes 30–80 days, depending on species and temperature. After hatching, larvae are fed salivary and anal secretions from the male and female, but later in their development they care for the eggs produced by the queen, begin to expand the nest, and forage for food. At this early stage, the search for food is essential to the health and fertility of the male and female, because they do not leave the nest to feed. Chambers that are formed in wood are a ready source of food for the new workers and there is little need to leave the site, but subterranean termites must travel away from the nest site to find food. The number of workers and soldiers in the colony increases slowly. The colony enlarges with the addition of workers, and new galleries are excavated, and the queen and king are isolated in a cell at a central position in the nest. In most of the higher termites, the abdomen of the queen enlarges and, in this physogastric condition, she is nearly immobile. The king does not increase in size, and remains with the queen. Mating occurs throughout their life, which is up to 15 years. In lower termites, the king and queen are not isolated, and in some drywood termites, they move around the nest. Mating takes place at intervals and the queen's abdomen only slightly enlarges with the development of eggs. The rate of egg-laying in lower termites is several hundred to several thousand eggs per year.

The queen and king are groomed and fed by workers, which enter the royal cell through small openings in its hard protective wall. Queens lay eggs at a steady rate; workers carry the eggs to incubation chambers. The number of eggs produced by the queen varies depending on the species, and the age of the queen. In tropical regions, egg production is continuous throughout the year, although there are seasonal fluctuations. In termites living in temperate regions, egg production is often suspended during the cold months. After hatching, young larvae are taken to nursery chambers where they are fed and groomed by workers. They are moved to other chambers in the nest until their final molt into workers or soldiers. Subterranean termites use undigested and partially digested cellulose to build galleries and extend the nest. This material is mixed with excreted soil particles and moistened with saliva to make a pliable material that is formed into chamber walls.

Composition of the colony, in terms of young and mature forms, varies for different species and during different periods of the year. The cyclic presence in the colony of immature and terminal forms is more pronounced in termites in temperate regions than in those in the tropics. In north temperate regions colony activity and egg production are reduced during winter months. In mature colonies, the alates usually appear over a short time before emergence. They remain in the nest for an extended period if environmental conditions are not suitable for their release. The number of soldiers in the colony usually increases with the production of alates, and reaches a peak before swarming flights occur. The maximum number of soldiers in a colony coincides with the time the nest is opened to release alates, and when exposure to predators is highest.

Young colonies grow in relation to the rate eggs are laid during the first few years, and on the length of life of the colony members. Colony size is expanded by increased egg production and retention of workers to care for the young and forage for food. In lower termites, young colonies are distinguished by queens that feed on the available substrate and produce a small number of eggs for several years. These female alates emerge without fully developed eggs; development of eggs is completed when the young queen begins to feed. If feeding is delayed, egg-laying is also delayed. After the first group is deposited, egg production usually stops for several months; a second period of egg production may be followed by a period of rest and more feeding. As the young queen matures, a large number of eggs are laid during each egg-laying period, and the number of workers in the colony increases, along with the overall size of the colony. Mature colonies are usually defined as producing all castes and young normally produced by the species, including alates. The time required for a young colony to develop to a point where it will produce alates varies with species and local conditions, and the unique features of individual colonies.

### Distribution

Climatic factors, particularly temperature and rainfall, have the greatest influence on the distribution of termites around the world. They are restricted by arid areas and are rarely found at altitudes above 3000 m. The number of different species is greatest in the tropical rain forests close to the equator and the number decreases as the latitude increases. A few species extend into temperate regions; the limit of their distribution is approximately 45° N and S latitudes. Primitive forms living in what are now cold climates include species of *Archotermopsis* in the sub-Himalayas, *Zootermopsis* in British Columbia, and *Porotermes* in the tips of southern hemisphere continents. In the northern temperate regions, *Reticulitermes* is adapted to cold seasons, but the survival of some species is aided by favorable habitats provided by humans. Termites in the soil beneath and around buildings in the urban environment remain active throughout the year, which facilitates their success and spread in temperate regions.

Several species of termites have been distributed in agricultural and urban environments by indirect actions by humans and commerce. They are distributed in construction lumber, in wooden crates for merchandise, in manufactured articles such as furniture, in ship's ballast, and in the timbers of small boats. The termites potentially spread by these actions are few, and primarily limited to the drywood termites of Kalotermitidae, and the subterranean and wetwood species of Rhinotermitidae. In their native regions, termites occur in natural habitats, and secondarily in disturbed cultural habitats such as urban and suburban areas. When carried to new regions, introduced species occur primarily in artificial habitats, such as in seaports and urban buildings. Once established they usually spread widely in the adopted region, or they remain as discontinuous, small populations.

South American termite fauna of urban areas includes many native and some introduced, cosmopolitan species. A total of 77 species of termites have been reported as structural or agricultural pests, in South America. Among these, 40 species are reported as structural pests, 53 species as agricultural pests, and 15 species as both. However, only 18 species are considered major pests and the status of most of the others is uncertain. The most important structural pests are the introduced *Coptotermes havilandi* and *Cryptotermes brevis* and the native *Nasutitermes corniger*. In most South American countries, urbanization has increased dramatically in the last 40 years and the urban termite fauna is recent. Most species found in cities are native to the cerrado vegetation, a type of savanna that dominates the landscape in many regions. Surveys of urban termites reveal the presence of a large number of species, but the introduced species *C. brevis* and *Coptotermes havilandi* and the native species *Nasutitermes corniger* are major structural-wood pests. *C. testaceus*, *Heterotermes longiceps*, and *H. tenuis* are minor wood pests. *C. testaceus* can damage ornamental and shade trees. In Brazil, one forest species, *Syntermes dirus*, occurs in cities, and the mound-building species, *Cornitermes cumulans*, *Procornitermes araujoi*, and *Syntermes wheeleri*, cause aesthetic problems in lawns and gardens. *S. nanus*, a large, subterranean, litter-feeding termite, sometimes damages living plants. *Reticulitermes* was introduced into Chile and Uruguay, where it is a serious pest. Termites damage structural wood throughout Chile, but are most destructive in the north. In coastal areas the majority of the damage is by *Cryptotermes brevis* and *Neotermes chilensis*; *C. brevis* is the dominent species between Arica and Copiapo, a generally arid area. The degree of termite damage decreases towards the interior, and it rarely occurs in the pampas. *N. chilensis* occurs in areas where there is abundant vegetation, and it often infests trees.

Arabian peninsula termite fauna is predominantly subterranean, and is composed of about 16 endemic species, which are widely spread in the major ecosystems. Kalotermitidae is represented by one species, *Epicalotermes aethiopicus*, which is a widely spread species in the Ethiopian region. Hodotermitidae is represented by *Anacanthotermes ochracheus*, *A. ubachi*, and *A. vagans*. Rhinotermitidae is represented by *Heterotermes aethiopicus* and *Psammotermes hybostoma*. The sand termite *P. hybostoma* is adapted to the hot desert habitat; it occurs around sand dunes and oases where it feeds on plant residues and animal (camel and goat) waste. The family Termitidae is represented by 12 species, including *Microcerotermes diversus*, *M. buettikeri*, *M. gabrielis*, *M. parvulus*, *Amitermes messinne*, *A. stephensoni*, *A. villis*, *Microtermes najdensis*, and *M. yemenensis*. Species of *Microtermes* prefer the silt-type soils of valleys and mountainous areas where the temperature ranges from 26 to 30 °C. Species of *Anacanthotermes* and *Psammotermes* are in the periphery of oases and deserts, but not in areas with temperatures below 18 °C. *Anacanthotermes* species cause structural damage to buildings from the Middle East to Turkmenistan.

In Caucasia, in the region between the Black Sea and Caspian Sea and Turkmenistan, the most common pest species is *A. ahngerianus*. This subterranean species attacks timber-framed houses in the region, and the number of nests in non-saline soils may be 3–10 per hectare, and 7–30 per hectare in takyr soils. The pest status of *A. turkestanicus* is greater, and this species is widespread and infests various structures in the region.

Australia has a varied termite fauna. It is comprised of about 30 genera and about 250 species; but there are probably many undescribed species in this country. Approximately half of the genera in Australia are in the families Kalotermitidae, Mastotermitidae, Rhinotermitidae, and Termopsidae; the remainder is in the Termitidae. Only about 12 species account for most of the structural damage, and *Coptotermes acinaciformis* and *Mastotermes darwiniensis* are the most important. *M. darwiniensis* is the only surviving member of Mastotermitidae; it is one of the most destructive termites in Australia. In Kalotermitidae, *Kalotermes banksiae*, *Glyptotermes brevicornis*, and *G. tuberculatus* cause damage to structural wood. Rhinotermitidae are widely distributed with several pest species in *Schedorhinotermes*, *Heterotermes*, and *Coptotermes*. The common species, *Coptotermes acinaciformis*, is responsible for the most economically important damage in Australia. It has an extensive range and has adapted to urban environments. In Termitidae, there are several species in *Amitermes* and *Microcerotermes* that attack structural wood and paper products.

China has a large number of described species of termites, and probably more will be discovered. In this large country, termites are distributed in 25 provinces and autonomous regions, except for the provinces of Heilongjiang, Jilin, and Qinghai, and the autonomous regions of Inner Mongolia, Ningxia, and Xingjiang. Current estimates of the termite fauna include four families, 42 genera, and 379 species. The family Rhinotermitidae consists of eight Chinese genera and 169 species, including 33 *Coptotermes*. Current studies indicate that four *Coptotermes* occur in China: *C. curvignathus*, *C. formosanus*, *C. hainanensis*, and *C. emersoni*. In Hong Kong, nine species of *Coptotermes* are recognized, and *C. formosanus* is the most common pest. There is a large number of termite species in China and Hong Kong; however, estimates of building and household damage are based largely upon that attributable to *C. formosanus*.

Europe has a limited termite fauna. The genus *Reticulitermes* is the most abundant and naturally occurring subterranean termite. European species identified on the basis of morphological, chemical, and molecular features are: *R. santonensis* in western France; *R. grassei* in southwestern France, northwestern and southern Spain and Portugal; *R. banyulensis* in northeastern Spain, Iberian peninsula and southwestern France; *R. lucifugus* in Italy and southeastern France; and *R. balkanensis* in the Balkans. The above-ground, drywood termite in Europe is *Kalotermes flavicollis*. It occurs in the humid regions of the Mediterranean zone, including the coasts of Portugal, Spain, and southern France, all Italy south of the Alps, the Balkan and Black Sea coast, and south to the Middle East (Gaza).

India has a large number of domiciliary termites among the 260 species recorded from the country. *Cryptotermes dudleyi* damages houses along the coast and in eastern regions; it also attacks bamboo; *C. bengalensis* is a pest in Assam and Tipura; and *C. heimi* and *C. domesticus* are also a pest of structural wood. *Heterotermes indicola*, *Odontotermes feae*, and *O. obesus* are pests of structural wood in various regions of the country. These seven species cause about 90% of the damage to structural timber in India.

Japan has four families, 10 genera, and 17 termite species. Of these, *Incisitermes minor* and *Zootermopsis angusticollis* were introduced in imported timbers from North America. *Coptotermes formosanus* was probably introduced into Japan from China. In addition, *C. guangzhouensis* may be present in Fukuoka City and Yonakunijima Island. Most of the structural damage from termites in Japan is caused by the native species *Reticulitermes speratus*. *C. formosanus* accounts for less than 10% of the damage.

New Zealand has few native species of termites: *Kalotermes brouni*, *K. banksiae* (Kalotermitidae), and *Stolotermes ruficeps* (Hodotermitidae). Introduced into the country from Australia are three species belonging to the Rhinotermitidae, one Termitidae, and four Kalotermitidae; and there are three imported species of *Coptotermes*.

Philippine wood-infesting termites include species from several genera. Of the 54 species of termites reported from the Philippines, only six are relatively common and economically important. *Cryptotermes cynocephalus* and *C. dudleyi* are the most common and most serious pests of structural timber. *Microcerotermes los-banosensis* are often associated with dead and living trees, stumps, and posts in urban areas; *Macrotermes gilvus* is a pest in peridomestic sites, such as posts and logs, and building timbers. *Nasutitermes luzonicus* occasionally invades buildings and damages structural wood.

Peninsular Malaysia has three families, Kalotermitidae, Rhinotermitidae, and Termitidae, 42 genera, and 175 described species. Subterranean and drywood termites are economically important pests of forestry, agricultural crops, and urban structures. The most destructive wood pests are species of *Coptotermes*. Other structural pests are species of *Globitermes*, *Macrotermes*, *Odontotermes*, and *Schedorhinotermes*. The subterranean termites known to kill live trees in Malaysia are *Coptotermes curvignathus* and *Microcerotermes dubius*. The most common drywood species that infest wood in houses are *Cryptotermes domesticus* and *C. cynocephalus*.

South Africa has a large number of termites, and many are pests of agricultural land and structural timbers. *C. brevis* has been introduced into South Africa from the USA, and is a well-established pest of buildings in urban and rural areas. *Odontotermes badius* is a subterranean, fungus-growing termite that causes about 85% of all the termite damage to buildings in the country. The large termites, *Macrotermes bellicosus*, *M. natalensis*, and *M. goliath* (= *M. swaziae*), are common, and *M. natalensis* is the second most important pest of wood in South Africa. The genus *Ancistrotermes* is endemic to the Ethiopian region, and *A. latinotus* occurs in southern Africa, and infests structural wood in buildings.

The USA has four families, 17 genera, and 45 species. In North America, 95% of the damage to structural wood is caused by subterranean termites in the genera *Reticulitermes*, *Heterotermes*, and *Coptotermes*. The Formosan termite, *C. formosanus*, occurs in Hawaii, the southeastern USA and west to Texas, and one site in California. About 15 species are economically important structural pests. Subterranean termites occur nearly throughout the country, except Alaska. Pest species of drywood termites in the genera *Incisitermes*, *Cryptotermes*, and *Neotermes* occur primarily in the southern and southeastern coast states, Arizona, California, and Mexico.

## Pest status

Among the three main categories of termites, subterranean, drywood, and dampwood, the subterranean are the most common and most destructive. Termites are economically important pests in all zoogeographic regions. Yearly, about US$1920 million is spent on termite control in buildings in 29 major countries of the world. In Australia, termites cause about $200 million in damage every year, with one in five Australian homes being attacked: the costs for control each year are about $25 million. In the USA, termites cause about $1.1 billion in damage per year; the cost of termite damage is more than is caused by fires, floods, and storms. The Formosan termite alone causes about US$100 million annually in Hawaii. The damage from subterranean termites in China is extensive. Damage costs in some large cities may exceed US$3 million annually, and countrywide, control may exceed US$400 million. Damage in tropical and subtropical regions includes rural and urban structures; in Singapore treatment costs exceed US$11 million yearly, and are about US$20 million in Bangladesh and Pakistan.

## Classification

Isoptera is represented by six families: five families compromise the lower termites, and one the higher termites. The distinction between lower and higher is based primarily on their evolutionary development and social behavior, but also on the composition of the gut microflora. Lower termites have symbiotic bacteria and protozoa in their gut, which assist in the breakdown of cellulose food material. In higher termites, bacteria are absent and only protozoa and enzymes appear to be involved in food digestion. There are approximately 2000 species of termites: about 400 species are considered lower termites, and 1600 are higher. About 100 termite species are considered pests of structures or wood in use around the world.

Lower termites include:

1. Mastotermitidae. There is a single species *Mastotermes darwiniensis*, which occurs in northern Australia in subterranean nests.
2. Kalotermitidae. These occur as small colonies in dry wood and are widespread in the tropics and subtropics; common genera include *Cryptotermes*, *Kalotermes*, and *Marginitermes*.
3. Termopsidae. These occur in wood which has a high moisture content and is decayed by fungi. They occur in tropical and temperate regions; common genera are *Stolotermes* and *Porotermes* in the southern hemisphere.
4. Hodotermitidae. These harvester termites feed on plant litter. They occur in dry regions of Africa, Middle East, Australia, and India; common genera are *Hodotermes*, *Microhodotermes*, and *Anacanthotermes*.
5. Rhinotermitidae. These subterranean termites eat wood, and have a highly developed soldier caste. They occur throughout the world, and common genera include *Coptotermes*, *Reticulitermes*, *Heterotermes*, and *Schedorhinotermes*.

The category of higher termites is represented by the Termitidae. This is the largest family of termites; it includes species that feed on rotten wood and dry wood, and species that build nests on trees and poles, and those that construct large mounds. Common genera in this family are *Macrotermes*, *Microtermes*, and *Nasutitermes*.

The categories of drywood, dampwood, and subterranean can be used to classify the economically important genera of termites. Drywood termite nests are not in contact with the soil and they are dependent on wood moisture to provide suitable conditions for the colony. Distribution of these termites, such as *Cryptotermes* and *Kalotermes*, is greatly influenced by ambient temperature and humidity. They are commonly found in coastal regions, and they infest the upper parts of buildings. Their presence may be unnoticed because their feeding usually leaves a thin veneer of wood at the surface. However, their rounded fecal pellets are ejected from the galleries from small, round holes, and piles of these pellets are the only sign of infestation. Dampwood termites initially infest wet and decayed wood, but their feeding can extend to sound wood. Subterranean termites commonly occur in tropical regions, especially rain forests. However, some species occur in temperate regions of Europe,

North and South America, southern Africa, Australia, and parts of Asia. Their colonies are established in soil or in wood buried in soil, but they also forage above ground. Secondary nests constructed above ground are connected to the primary nest in the soil by shelter tubes. Wood attacked by these termites usually has a thin surface veneer and, during later stages of infestation, the wood below the surface is replaced with soil, fecal matter, and undigested wood, which is molded into a reinforcing material called carton.

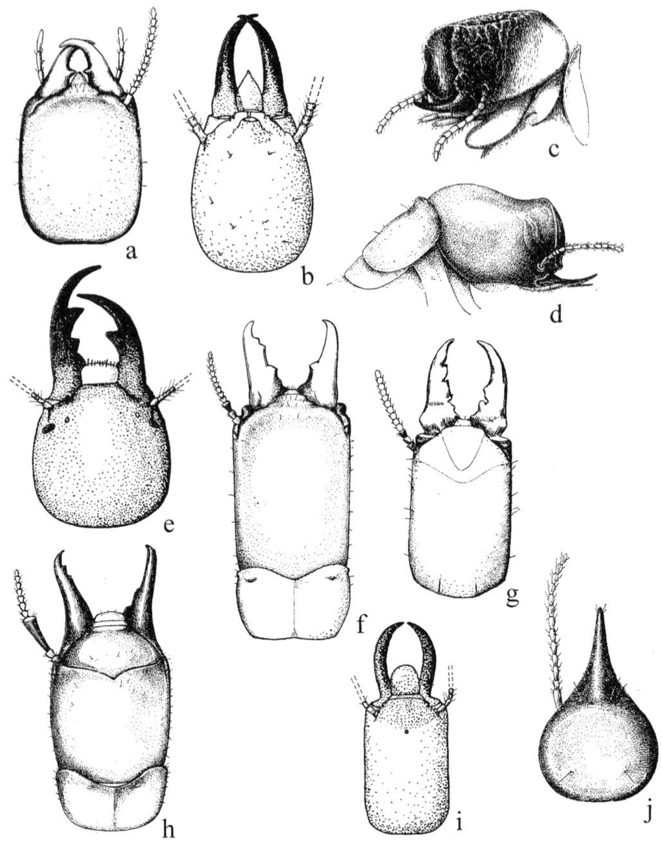

**Figure 10.1** Isoptera. (a) *Amitermes* sp.; (b) *Coptotermes* sp.; (c) *Cryptotermes brevis*; (d) *Hodotermes* sp.; (e) *Incisitermes snyderi*; (f) *Kalotermes approximatus*; (g) *Marginitermes hubbardi*; (h) *Microcerotermes biori*; (i) *Nasutitermes* sp.

## Hodotermitidae

These are the harvester termites. All castes, including workers, have functional compound eyes, but only rudimentary ocelli may be present. Alates have long and narrow wings, and the anterior wing scales never completely overlap the posterior scales. The pronotum is saddle-shaped. Soldiers are present in two forms and both are like workers with large legs. The mandibles are large with distinct marginal teeth (Fig. 10.1d), the antennae are composed of 23–31 segments. The groups included here are soil-dwelling and the (damp) wood-dwelling termites. *Hodotermes mossambicus* occurs from Ethiopia to South Africa, and *H. erithreensis* is limited to a northern distribution in Somalia and Eritrea. *Microhodotermes wasmani* occurs from Morocco to Tunisia and *M. viator* occurs over most of South Africa. *Anacanthotermes* occurs in the arid country stretching from Algeria through Egypt to Iraq and northeastern India. *Zootermopsis* occurs in northwestern North America, and some are pests of structures.

### *Anacanthotermes*

These are harvester termites, and they are distributed in the arid zone of desert and steppe that extends from the Atlantic coast of Morocco east to the mountains of northeastern Afghanistan. In natural habitats, the food of these termites is plant material, such as dry grass and straw from grain crops, debris from date palms, and animal dung. In urban habitats, they damage rural buildings constructed with mud bricks. They forage for the straw used in the bricks. Infestations also include the leaves and trunks of date palms used in building construction, and railroad ties. Workers forage at night for plant material that is deposited around the openings of nests; they also forage for buried animal droppings, and gnaw at plant material during the night. Nest sites may be mounds that are 20–30 cm high and 50 cm diameter at the base. Swarming occurs during the day following a period of rainfall. The major pest species include *A. ochraceus* in North Africa and the Arabian peninsula to the Persian Gulf; *A. ubachi* in southern Israel, Jordan, and northern Iraq; *A. vagans* in southern Iraq and nearly all of Iran; and *A. septentrionalis* in northeastern Iran and Afghanistan. Species from Turkmenistan include *A. turkestanicus* and *A. ahngerianus*, which is a pest of pastures with dense grass cover.

***Anacanthotermes ochraceus*** Nests may have external openings about 1 cm high made of loose sand or conical mounds 20–30 cm high. The nest is composed of an irregular assembly of 15–20-cm-diameter chambers at different levels and joined by galleries. The nest is about 30 cm from the surface and it may extend 1.5 m below ground. Upper chambers in the nest are used for storing small pieces of plant material; the colony usually occupies the chambers at intermediate depths. Swarming flights occur in the afternoon (15:30–17:30 h) at 12–16 °C, from small exit holes attended by workers and nymphs. Alates fly about 100 m before landing to shed their wings and form pairs. Within 8 months, the colony may have 146–182 individuals.

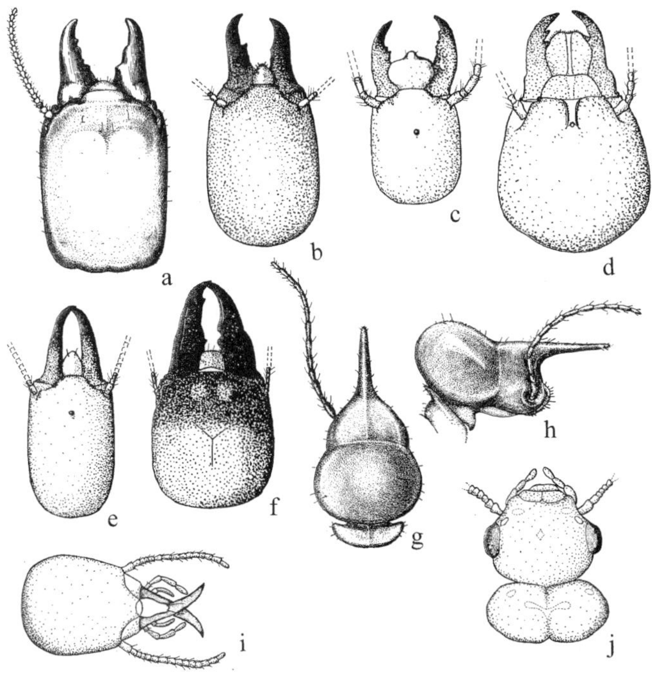


**Figure 10.2** Isoptera. (a) *Neotermes castaneus*; (b) *Odontotermes* sp.; (c) *Psammotermes hybostoma*; (d) *Schedorhinotermes* sp.; (e) *Reticulitermes lucifugus*; (f) *Zootermopsis angusticollis*; (g) *Tenuirostritermes tenuirostris*; (h) *T. tenuirostris*; (i) *Coptotermes havilandi*, soldier; (j) *C. havilandi*, worker.

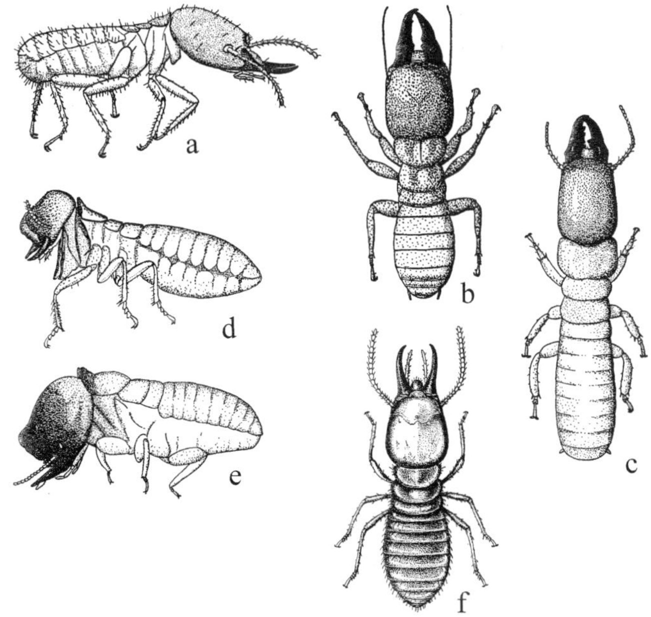


**Figure 10.3** Isoptera. (a) *Coptotermes niger*; (b) *Zootermopsis angusticollis*; (c) *Kalotermes flavicollis*; (d) *Cryptotermes brevis*; (e) *C. havilandi*; (f) *Prorhinotermes simplex*.

***Anacanthotermes septentrionalis*** Openings to the subterranean nests may be only small cones of sand 1–3 cm high. Mound nests are built in the eastern part of its geographic range; these are occupied only during summer; during winter the nest is 1–2 m below the surface. This species is distributed in northeastern Iran and Afghanistan.

## Zootermopsis

These are large termites; the alates may be 25 mm long. Alates usually have more than 23 antennal segments; the tarsi are five-segmented, and both alates and soldiers have one or more spines on the tibia of one or more pairs of legs. This genus is confined to North America, where there are three species. Two occur in Pacific northwest and Pacific coastal areas, and the third occurs in the southwestern desert. These are very large, wood-dwelling termites that generally require high-moisture habitats. *Zootermopsis nevadensis* and *Z. angusticollis* inhabit damp wood and can tolerate a range of temperatures. These termites are often characterized as inhabiting decayed wood, but this situation is encountered primarily in mature colonies. The colonizing pair usually seeks sound wood in down logs, standing dead trees, and dead portions of living trees. The fecal pellets of these termites are rounded and usually scattered through the nest galleries, or they may be expelled from the nest. They lack the longitudinal ridges that characterize drywood termite pellets. Although these dampwood termites do not cause damage on the same level as drywood or subterranean termites, they can cause economic damage when they are associated with urban structures.

***Zootermopsis angusticollis* (Fig. 10.2f; 10.3b)** Alates are about 25 mm long with the wings; body color ranges from yellowish brown to dark brown. Wings are 23–25 mm long and about twice the length of the entire body, and gray to dark gray. The body has only a few long setae, with few to none on the head; the anterior corners of the pronotum are rounded. The antennae are longer than the head and pronotum combined. Soldiers are 15–20 mm long, but the size varies with the age of the colony. The soldier head is 1.25 times longer than broad, and dark red and reddish black anteriorly; the mandibles are black. The sides of the head are concave and the head is somewhat narrower in front than behind. Antennae are not as long as the head and there are 25 segments. The fecal pellets are about 1 mm long, rounded, and usually the color of the wood infested. Swarming is at dusk, usually before sunset, and it occurs in May, and July to November. Natural infestations occur in dead tree trunks and branches (*Quercus*), and tree stumps (*Pinus*). In the urban environment the adults are attracted to lights at night. This species occurs along the northwestern coast of North America, from British Columbia south to Baja California.

***Zootermopsis laticeps*** Alates are about 22–24 mm long with the wings, and the abdomen and legs are yellowish brown. Wings are pale gray and with the costal area yellowish brown. The anterior corners of the pronotum are distinctly pointed. The antennae are longer than the head, with 23 or more segments in both alate and soldier. The soldier is 16–22 mm long, the abdomen and legs are yellowish brown, the head is brown, and the mandibles are blackish brown. The anterior margin of the pronotum is straight. Swarming occurs at night and during June and July. Although flights usually occur during the rainy season, they may not be directly associated with rainfall. This species inhabits dead wood. It occurs in tree stumps and large logs in southwestern USA, including New Mexico and Arizona.

***Zootermopsis nevadensis* (Fig. 10.4c)** Alates are about 20 mm long with the wings. The body color is dark brown to brownish black. The wings are 20 mm long or less, and dark gray. The body has short, fine setae, with a few on the head. Soldiers are 15–20 mm long, and the sides of the head are nearly parallel; the head is not narrowed in front. Swarming occurs at dusk and flights have been reported for January, July, August, and September. Winged adults are attracted to lights at night. Alates fly around trees to land in a crack, loose bark, or a decayed site in a tree. Then they shed their wings and burrow into the wood. The wingless adults are very active and able to jump. Colonies may contain the founding pair and secondary reproductives. This species occurs along the northern Pacific coastal area into Monterey county, California; and inland into the mountainous regions of the Sierra Nevada, Idaho, Montana, and northern Nevada.

## Kalotermitidae

This is the largest family of the lower termites, with 21 genera and 350 species. They are generally considered drywood termites. Individuals in this family lack a fontanelle, but usually have ocelli. Alates have oval heads with somewhat flattened sides, and the eyes are relatively small. The pronotum of the alate, soldier, and most other individuals in the colony is as wide as or wider than the head. The antennae of soldiers are usually shorter than the width of the head, and in many species segment 3 is enlarged. Soldiers usually have asymmetrical and toothed mandibles; there are three or more teeth on the left mandible and one or more on the right. Soldier antennae consist of 10–19 segments, and functional eyes are lacking. A worker caste is absent in the colonies in this family. Colonies are small, and the work in the colony is performed by nymphs. There is a range of nesting and feeding habits in this family. There are many pest species in the urban environment.

### *Calcaritermes*

These termites are distributed in the neotropics, and one species occurs in Florida. The alates are very small, and the front wing is about 7 mm long. There are 13 antennal segments. The soldiers have heads similar to those of *Coptotermes*.

***Calcaritermes nearcticus*** Alates are about 7 mm long with the wings. The wing membrane is punctate. There are 13 antennal segments. The soldiers' head is about as broad as long, and the front region of the head is high, and strongly concave (as in *Cryptotermes*). There is a distinct furrow from top to bottom of the front of the head. Swarming flights occur from January to April. Damage is to wood with high moisture content. This species occurs in Florida.

### *Cryptotermes*

These termites generally inhabit dry wood, and they are frequently referred to as powderpost termites. The genus is represented in nearly all of the zoogeographic regions. The head of the soldier is nearly as broad as it is long, and it is high in front; the steep frontal area of the head is distinctly concave. Mandibles are large and curved. Wings of the alates are iridescent and the antennae have 14–16 segments. Swarming flights often occur at night, and the alates are usually attracted to outdoor lights.

***Cryptotermes brevis* (Fig. 10.1c; 10.3d)** Alates are 10–11 mm long with the wings. The body is brown and the width of the head through the eyes is slightly more than 1 mm. Wings are translucent, and the veins are dark brown; the antennae have 16–18 segments. Soldiers are about 5 mm long and the head is brown to reddish brown. Fecal pellets of this species are small, round, and dry. Fecal pellets are usually expelled from the galleries, and they collect in piles below infested wood. This species infests dry wood and does not have a connection to the soil. In the USA, swarming flights occur in May and June. Damage is usually to floors, woodwork, furniture, and small wooden objects. This species has a wide distribution in tropical and neotropical regions, including Central and South America, Mexico, and the West Indies. It has been introduced into Hawaii, and continental USA. In the USA it is only known from structural infestations and not from natural habitats.

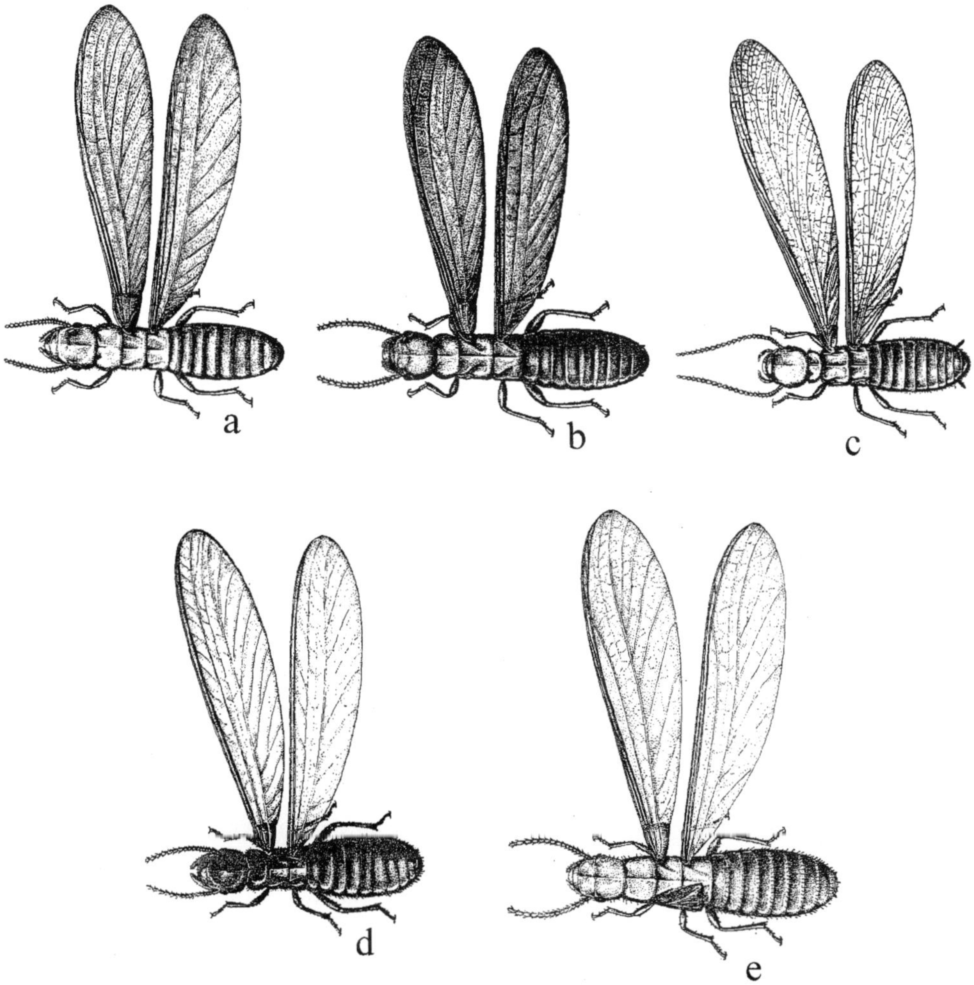


**Figure 10.4** Isoptera. (a) *Incisitermes schwarzi*, winged adult; (b) *I. minor*; (c) *Zootermopsis nevadensis*; (d) *Reticulitermes tibialis*; (e) *Cryptotermes cavifrons*.

*C. brevis* has been introduced into Australia several times, and has been detected in Brisbane, Maryborough (Queensland), and Sydney.

**Cryptotermes cavifrons (Fig. 10.4e)** Alates are 8.5–9 mm long, the body is pale brown, and the antennae and legs are very pale brown. The wings are clear, and the veins are brown. Mandibles are blackish brown. The width of the head through the eyes is less than 1 mm. The ocelli are large and subtriangular. Antennae are much longer than the head, and have about 15 segments. The soldier is 3.5–4.5 mm long. The head is blackish brown in front and reddish brown behind. The body is brownish yellow; legs are pale yellow. Antennae are yellowish white, and the mandibles are black. In the front of the head is a large cavity, and the sloped frontal area has a rough surface. The upper surface of the head is smooth and only slightly roughened in front toward the cavity. Antennae are nearly as long as the height of the head, and 13-segmented. Swarming flights occur from February to May. Natural infestations occur in the solid portions of decayed logs. This species is known to occur in Florida, from St. Johns county south.

**Cryptotermes cyanocephalus** Alates emerge from colonies in early morning from July to December. The alates leave the nest in low numbers and flights are repeated at intervals of 5–10 days; they are attracted to outdoor lights. They fly for short distances and land on the sides of buildings, then break off their wings and search for holes or cracks in the bare wood siding. Colony development is slow. This species is native to the Philippines and Indonesia, but has spread to other tropical and neotropical locations. Rarely it is found in natural habitats; it nests primarily in structures.

**Cryptotermes domesticus** This species nests in dry, seasoned wood at ground level. Swarming flights are usually during the rainy season, and small numbers of alates emerge at a time. This species is a household pest in southern Asia, but it also occurs in natural habitats. It is widely distributed in Asia, the south Pacific, and Central America. In Asia, it is found in India, Borneo, Sumatra, Malaysia, Thailand, Vietnam, Taiwan, and islands in the region.

***Cryptotermes dudleyi*** This species is primarily found infesting houses, with few nests found in natural habitats. Colonies are large, and several may live close to each other in the same piece of timber. It is widespread, and occurs in Africa, Asia, Australia, New Guinea, Central and South America, India, Indonesia, Malaysia, and Philippines.

***Cryptotermes havilandi* (Fig. 10.3e)** This species is probably native to east Africa where household infestations are rare. It has spread to other parts of Africa, southeastern Asia, and the Caribbean area, where it attacks structural wood as well as dead and living wood in trees in coastal and inland areas. In Africa, swarming occurs throughout the year, but is most common in January and February.

**Other *Cryptotermes*** There are several species in this genus that are known to utilize structural timber, as well as natural habitats, as nest sites. For some species, house infestations are more common outside their natural distribution. *C. declivis* is distributed in southern China and it is a pest of structural timber, such as door and window frames, ceiling beams, roof trusses, woodwork, and furniture.

### *Glyptotermes*

This genus occurs throughout the tropics, and in the subtropical regions of Australia. The soldiers are 4.5–9.7 mm long and they have long, narrow heads that are somewhat flattened; the mandibles are relatively short. The antennae have 10–12 segments. All species form small to moderate-sized nests. The species occurring in Australia usually establish nests in dead and decaying wood adjacent to sound wood of living trees, particularly eucalyptus. *G. brevicornis*, *G. tuberculatus*, and *G. iridipennis* nest in transmission poles made of eucalyptus, and infestations probably originate in the living tree.

### *Incisitermes*

Alates lack pigmented nodules on the membrane, and in most species in the USA the ocelli are somewhat elongated and set near the front margin of the eye. Antennal segment 3 is usually elongate. The species included here infest relatively dry wood, and they attack wood directly, rather than entering through the ground. The colonizing pairs may enter any exposed wood and use holes, cracks, or checks in the surface. Their colonies have swarming flights at dusk, and alates often collect at street lights at night. These termites occur in all the zoogeographic regions except the neotropics; many species cause economic damage to structures in southern USA.

***Incisitermes banksi*** Alates are about 11 mm long with the wings. The pronotum is not twice as broad as long, and the eyes are twice their diameter from the hind margin of the head. Soldiers are about 8.5 mm long, and the thorax, abdomen, and legs are yellowish brown. The head is brown to reddish brown in front and the mandibles are blackish brown. Antennae are pale brown, and segment 3 is dark brown. Antennae are 12-segmented, and extend to slightly past the mandibles; segment 3 is enlarged and as long as segments 4–6 together. The anterior margin of the pronotum is distinctly concave at the midline. Natural infestation sites include dry logs. Swarming flights are in June in Arizona, and winged adults are attracted to lights at night. This species occurs in Arizona and Texas.

***Incisitermes milleri*** Alates are 7.5–7.7 mm long with the wings. The body is dark brown; the wing membrane is clear and the veins are dark brown. Soldiers are 6.2–6.5 mm long and yellowish brown; the front of the head and mandibles are brown to blackish brown. There are 10–11 antennal segments, and segment 3 is elongate, but not as long as segments 4 and 5 together. The anterior margin of the pronotum is slightly concave at the midline, and the posterior margin is convex at the midline. Swarming flights occur during the day in spring. This species is one of the smallest of the drywood termites, and it occurs in the lower East Coast Keys of southern Florida.

**Western drywood termite, *Incisitermes minor* (Fig. 10.4b)** Alates are 11–12 mm long with the wings, the head is reddish brown, and the body is black to slightly bluish black. Wings are grayish black and the coastal veins are black. The ocelli are somewhat elongated; the eyes are round and relatively small. Antennae are 15-segmented; segment 3 is longer and darker than segment 2 or 4. Soldiers are 8–10 mm long, and the pronotum is twice as broad as long. Antennal segment 3 of the soldier is equal in length to the combined length of the next three or four segments. Abdomen lacks long setae in rows. Supplementary reproductives are about 10 mm long, light brown and with dark brown eyes. Soldiers are 8–10 mm long. The head is reddish brown and blackish brown on the anterior part; mandibles are black. The thorax and abdomen are pale brown to brown. Antennal segment 3 is elongated and enlarged at the tip, about twice as long as segments 4–6 combined. Mandibles are about as long as the width of the head.

Colonies live without contact with the soil. They consist of the founding reproductives, soldiers, and nymphs, there is no worker caste. Founding pairs bore directly into wood or they enter through cracks and other openings. Natural infestations occur in dead cottonwood, ash, walnut, and cypress trees, logs, and branches. In living trees, the galleries extend across the grain of the wood, and the sapwood and heartwood are excavated. In structures, the founding pair may enter houses through attic vents or shingled roofs, or the foundation vents. They will infest rafters, ridgepoles, and sheeting in the attic. In the living area, they infest window frames, and sills, doorframes, and doorsills, and in the substructure, they may infest floor joists. They also infest wooden furniture or other wooden materials in houses. Evidence of an established colony is usually piles of brown fecal pellets below small holes in the infested wood, particularly where the outer walls of the wood are thin. The pellets are about 0.8 mm long, rounded, and with six longitudinal ridges.

Swarming flights during the day generally occur from 11:00 to 15:00 h. In northern California, flights occur during June and July, and from September to November in southern California. In Arizona, flights occur in June and July. Large colonies may have about 3000 winged forms; typically, fewer than 20 alates emerge from the nest. New colonies are composed of about 20 nymphs and one soldier. After 2 years, the colony may consist of one soldier and 6–40 nymphs. Colonies at 3 years may have 70–700 individuals, and the ratio of soldiers to nymphs in a 3–4-year-old colony is 1:15–1:60. Winged adults are produced when a colony is about 4 years old, and well-established colonies have about 1000 individuals after 7 years. Optimum temperature for feeding is 26–32 °C, exposure to 37.8 °C is lethal in 2 weeks, and a 4-h exposure to 44.5 °C is lethal. In temperate climates, this species often infests upper floors and attics of buildings; in hot climates infestations are often close to the ground. At optimum temperatures, nymphs consume about 0.59 mg of wood per day; a large drywood termite colony containing about 1000 nymphs consumes about 0.25 kg of wood per day.

The natural distribution of this species in North America is from southern California east to the peripheral desert regions of Arizona and Utah. This species is easily transported in boxes, crates, and furniture. Infestations may occur in Ohio, New York, Pennsylvania, Wisconsin, and Virginia. This species is a structural pest in Japan.

***Incisitermes schwarzi* (Fig. 10.4a)** Alates are 15–16 mm long with the wings. The body is 8–9 mm long and yellowish brown; the pronotum is pale brown in front and dark brown to blackish brown on the posterior corners. Mandibles are black at their tips. There are 20 antennal segments; segment 3 is darker but not longer than the other segments. The head, pronotum, and abdomen have long, erect setae; on the abdomen the setae are in rows near the hind margin of each segment. Soldiers are 7–10 mm long and yellowish brown; the anterior margin of the pronotum is not concave. Mandibles are black. The femora are greatly swollen. There are two forms of soldiers: one with the head slightly longer than broad, and one with the head distinctly longer than broad. In both forms, antennal segment 3 is as long as segment 4 and 5 together, and darker than the adjacent segments. Soldiers usually have 20 antennal segments. Colonies may contain primary and secondary reproductives. Secondary queens are 6.5–10 mm long, and secondary males about 8 mm long. Secondary forms have 14–15 antennal segments. Swarming flights occur in April and May. Natural infestation sites include dead trees, tree stumps, and down logs. In the urban environment, damage is to woodwork in buildings and utility poles. This species is common in southern, coastal Florida.

***Incisitermes snyderi* (Fig. 10.1e)** Alates are 15–16 mm long with the wings. The body is yellow to yellowish brown. Soldiers are 7–10 mm long and yellowish brown, and the antennae are 12- or 13-segmented. Segment 3 of the antenna is elongate, but not as long as segment 4 and 5 together. The anterior margin of the pronotum is deeply concave at the midline. Natural infestation sites include dead trees, down logs, and branches. Secondary reproductive queens are about 9 mm long. Swarming flights occur in early evening after sunset in May and June. The small number of individuals that constitute a swarming flight may be attracted to lights at night. Damage is done to the woodwork in buildings, hardwood furniture (white ash), foundation timbers, and utility poles. This species is distributed from South Carolina to Florida and west to Texas, and into Mexico.

### *Kalotermes*

These drywood termites are widely distributed and have several pest species around the world. Alates of these termites have dark bodies and dark wings and there are nodules on the wings; the hind wing has distinct veins in the anal area. The anterior wing scales are large and cover those of the hind wing. The head is small and somewhat flattened and the eyes and ocelli are small; the pronotum is broader than the head. Soldiers have large heads with strongly developed mandibles.

**Kalotermes approximatus (Fig. 10.1f)** Alates are 9.5 mm long with the wings, the body is dark brown, and the wings are dark gray. The wing membrane has pigmented nodules. Soldiers are about 7.5 mm long. The head is yellowish brown and the mandibles are black, but reddish brown at the base. Antennae and legs are yellowish brown; segment 3 of the antenna is brown. The anterior margin of the pronotum is slightly concave at the midline; the posterior margin of the pronotum is roundly convex at the midline. There are 13 antennal segments, and segment 3 is nearly as long as segment 4 and 5 combined. Swarming flights occur during the day, from August to October. The mandibles are slender and toothed. Natural nest sites include dead tree stumps, dead trees, and branches. Damage may be done to structural wood. This species occurs in northern Florida, and it has been collected in Louisiana, coastal regions of Mississippi and Virginia, and western Texas.

**European drywood termite, *Kalotermes flavicollis* (Fig. 10.3c)** Alates are about 10 mm long and the body is yellowish brown. In some parts of its distribution the body color is dark brown. Soldiers are about 11 mm long; the head is yellowish brown and three times longer than the prothorax; mandibles are large. Colonies consist of reproductives, soldiers, and working nymphs. Activity in the colony is governed by the seasonal changes from summer to winter. Swarming flights occur during the day in July through October. A pair of adults excavate a chamber in decayed wood, and the female begins laying eggs in about 25 days. Egg-laying occurs from September to October; some hatch during this time and some overwinter. Eggs hatch in 50–60 days at 25 °C; the nymphs develop in 24–31 days. The first 1–2 batches of eggs produce 20–30 individuals and at least one soldier. In winter, foraging, egg-laying, and molting cease. The following spring eggs resume development, and winged adults develop from overwintered nymphs. Development to soldier takes about 85 days (laboratory colony), and the soldier caste is a small (2%) proportion of a mature colony. Soldiers can be developed from larvae and nymphs during instars 3–7. Nymphs can undergo regressive molting into pseudergates, and these large individuals function as workers. The founding pair of termites retains their symbiotic protozoa for several years, and during this time they forage and feed themselves. Colonies develop slowly and when mature there are 1000–1500 individuals.

In natural habitats, nests are established in trees and shrubs. It infests dead or diseased portions of a variety of trees and shrubs, including oak, poplar, fig, olive; it attacks vineyards in southern France. Along the Black Sea coast of the Caucasus it occurs in an oak forest zone within 2 km of the coast, and infests the dead parts of trees. Throughout its distribution, *K. flavicollis* attacks buildings, bridges, posts, and canal piles, and various household materials. This species is found in the humid regions of the Mediterranean. It occurs along the coasts of Portugal, Spain, and southern France, all of Italy south of the Alps, the Balkan and Black Sea coasts, and south to Gaza. Limited areas of North Africa provide suitable habitats, and these include Alexandria, Tripoli, Tunisia, and along the coast of Algeria and Morocco.

**Kalotermes immigran** Alates are about 12 mm long with the wings ; the body is dark brown. The ocelli are somewhat elongated and set obliquely at the front of the eyes. Antennal segment 3 is elongated and darker than the adjacent segments. The soldier has an enlarged antennal segment 3, which is longer than segments 4 and 5 combined. The eye of the soldier is not pigmented. This species occurs in Hawaii.

### *Marginitermes*

Soldiers in this genus have an extremely enlarged antennal segment 3. The alates are pale yellow with round ocelli, and normal-sized antennal segments.

**Desert drywood termite, *Marginitermes hubbardi* (Fig. 10.1g)** Alates are about 13 mm long with the wings. The body and wings are light yellow, and antennal segment 3 is not larger than segments 2 or 4. The abdomen is without long setae. The soldier is distinctive by having antennal segment 3 extremely enlarged (club-shaped) and elongated. Natural habitats include giant desert cacti and the branches of trees in canyons and along streambeds. Dry wood is attacked, and entrance holes may be in concealed cracks and crevices. It invades poles, posts, and structural timbers, and is frequently a serious pest in urban areas. Swarming flights occur at night, usually immediately after rain, from June through August. Alates often gather in large numbers at lights. This species is the most common drywood termite in the desert regions of Arizona and California; it occurs from the eastern border of California through central and southern Arizona to western Mexico.

### *Neotermes*

These are relatively large termites: the alates may be 15 mm long with the wings, and they range from yellowish brown to dark brown. Soldiers are about 12 mm long; the head is longer than wide and parallel-sided. Mandibles are shorter than the head, and they are usually asymmetrical and have several teeth. Nests are usually in wet logs and other wood with high moisture content, including wood in structures. These termites often nest in dead wood of living trees, but galleries may penetrate living areas and trees may be killed. *N. insularis*, the ring-ant termite of Australia, nests in the upper branches of eucalyptus trees, and may continue to feed in fallen logs. *N. luykxi* occurs in Broward and Monroe counties, Florida.

**Neotermes castaneus (Fig. 10.2a)** Alates are about 15 mm long with the wings, and they are pale brown. Antennae, wings, and legs are brown. Wings are slightly longer than the body. Antennae are longer than the width of the head and 16-segmented; segment 3 is not modified. The pronotum is much broader than the head, about 2.5 times as broad as long; sides are evenly rounded, and the anterior margin is slightly concave. Soldiers are 10–12 mm long; the head is light brown; the mandibles are blackish brown, and brown at the base. The head is 1.3 times as long as broad; it is broadest in front, behind the antennae. The antennae are 16-segmented, and segment 3 is not larger or darker than segment 4. The eyes of the soldier are not dark, as in *N. joutlei*. Natural sites for colonies include the decayed wood of trees, tree stumps, and branches. *N. castaneus* commonly infest living trees, including those that are grown as indoor plants for large buildings and shopping malls in the USA. Swarming flights occur in March, June, October, and November. This species occurs in West Indies, Central and South America, and North America (southern Florida, Lake county).

**Neotermes connexus** Alates are 17–18 mm long with the wings, and they are pale brown. Antennal segment 3 of the soldier is not as long as segments 4 and 5; the eyes are black and the pronotum is wide and projects beyond the margin of the head. Natural nest sites are dead wood, and dead wood in living trees. This species occurs in forested areas. It is native to Hawaii.

**Neotermes jouteli (= Kalotermes jouteli)** The alate body is about 10 mm long, yellowish brown, with very short setae on the head and pronotum. There is a black mark on each side, on the front edge of the head inward from the antennae. The eyes are large and round, and the ocelli are large, round, and nearly touching the eyes. Antennae are longer than the width of the head, and segment 3 is not longer than segment 2 and not distinctly darker. Soldiers are 9–13 mm long. The head is brown and the eyes are distinctly black; the mandibles are black. The thorax, abdomen, and legs are yellowish brown. The pronotum is twice as long as broad, and the antennae have 16–18 segments. Swarming flights occur in spring and fall. This species occurs along the east coast of Mexico, West Indies, and southern Florida (Fort Pierce south).

### *Paraneotermes*

In this genus, the median vein of the wing is heavily sclerotized, and the alate body is dark brown. The soldier has short mandibles, which are not as long as the width of the head. Antennal segment 3 is not enlarged, as it is in other Kalotermitidae.

**Paraneotermes simplicornis** Alates are about 12 mm long with the wings; the body and wings are dark brown. Soldiers are 8–9 mm long, yellowish brown to brown. The head is yellowish brown, and brown in front; the mandibles are blackish brown. The head is 1.5 times as long as broad. Colonies are located in wood partially or completely buried in the ground; it does not build covered galleries to enter wood above ground. Natural sites for nests and food include the dead roots and buried branches of desert trees and shrubs, and the roots of live trees, especially young citrus trees. Live trees may be cut horizontally across the base below ground by the foraging galleries, and the tree dies quickly. Damage is to poles, posts, and the woodwork and floors of structures. Swarming flights occur during the day in California from September to November; in Arizona, winged forms may be in colonies in June and July, and in Texas, flights occur in April. This species occurs in the desert regions of southwestern USA, including California, Arizona, Texas, and Idaho.

## Mastotermitidae

This family is represented by a single living species, *Mastotermes darwinensis*, from Australia. It occurs only in tropical northern Australia, in parts of western Australia, Northern Territory, and Queensland. Fossil termites in this genus include *M. anglicus*, *M. batheri*, and *M. bournemouthensis* from the UK, and several other species from Europe and North America. *M. darwinensis* alates are unique among living termites in possessing anal lobes on their hind wings. This feature links them to a cockroach-like common ancestor of termites, cockroaches, and mantids. Other primitive features of this termite include five-segmented tarsi, long and filamentous antennae, large eyes and ocelli. Eggs are laid in pods, similar to those of some cockroaches.

***Mastotermes darwinensis*** Alates are large and have a wing span of about 50 mm. Soldiers are 11.5–13 mm long and have a large head and mandibles. No mounds are built: the nests are at ground level in tree stumps, poles, or posts; workers may construct covered runways of soil when foraging above ground. The nest is composed of several tiers of horizontally arranged cells made of carton material. Nests may be 45 cm diameter and 1–1.5 m below ground. In natural areas, colonies may have tens of thousands of individuals; in urban environments the number of individuals in the colony may be 7 million when fully developed. Colonies usually contain several secondary queens, and primary queens are not common. In well-established colonies, soldiers are 1–2% of the total population. Foraging galleries may extend over 90 m, and be about 30 cm below the soil surface. Covered foraging galleries are on the ground, and over wood and metal surfaces. When attacking living trees and shrubs, roots, trunks, and branches may be excavated, and sometimes completely hollow. This habit of excavating small-diameter branches provides Australian Aborigines with the raw material for their traditional musical instrument, the didgeridoo. It is made from a length of hollow wood that has been eaten by *Mastotermes*.

Swarming flights usually begin in November and continue through the summer rainy season. This species is one of the most destructive termites in Australia and attacks a wide range of wood in use as well as living trees, shrubs, and vegetables. It is primarily tropical; the southern limit of its distribution approximates the Tropic of Capricorn, in both coastal and inland sites.

## Rhinotermitidae

These termites cause considerable damage to structures around the world. Most of the species nest in the ground, but many are capable of existing above ground when there is a water supply available. One genus, *Prorhinotermes*, nests in wood above ground, and *Psammotermes* are true desert termites. Most members of this family have ocelli and a fontanelle. The veins in the fore area of the wing are reduced. One large vein extends along the anterior edge of the wing; parallel to this is a sclerotized second vein (radial sector). The front-wing scale is large and overlaps the base of the hind-wing scale. Antennae consist of 14–22 segments and the cerci are two-segmented. Adult mandibles have on the left a group of one apical and two marginal teeth, and on the right a small tooth at the base of the first marginal tooth. The mandibles of the soldiers lack conspicuous teeth (Fig. 10.1b).

### *Coptotermes*

This genus is widespread around the world; there are 48 species, of which 23 are oriental. Alates are 12–15 mm long, and they are usually larger than those of other Rhinotermitidae. The ocelli are large but the fontanelle is usually small, and the wings are covered with setae. Several species are destructive pests where they occur, or have become established. *Coptotermes formosanus* has been introduced to many regions and is a serious structural pest wherever it occurs. *C. havilandi* has been introduced to Central and South America, and Madagascar. The mound-building species *C. acinaciformis* and *C. lacteus* are pests in Australia; *C. testaceus* and *C. brevis* are pests in Central and South America, and *C. heimi* is a pest in India.

***Coptotermes acinaciformis*** Alates are produced in colonies from October to May, flights are in early evening or at night, and they are often attracted to outdoor lights. Soldiers are about 6 mm long. It is one of the most destructive structural pests in Australia. Nests may be located in the root crown of a tree or in the main trunk that has been hollowed and filled with a damp, reddish-brown fecal material, called mud-guts, which is usually packed above the actual nest area. Trees frequently infested include the pepper tree (*Schinus areira*), English oak (*Quercus robur*), and eucalyptus. Structural infestations may occur from tree or mound nests, which can have a radius of 50 m. A single colony may attack as many as 15 neighboring trees. Underground tunnels connect the main colony with the infested sites. Soil contact is not required, and infestations may occur above ground and in isolated structures, such as boats and barges. It occurs nearly throughout Australia. In tropical Queensland and Northern Territory, it builds large mounds. In southwestern, subtropical, and temperate regions of Australia it nests in trees.

***Coptotermes frenchi*** Alates are produced in colonies from September to May. Flights from different colonies sometimes occur simultaneously over a wide area and involve large numbers of individuals. These flights are often before a rainfall, and start in the early afternoon and extend into the night; alates are strongly attracted to lights. Soldiers are about 4.6 mm long. This species occurs from north Queensland to Western Australia, and is usually a forest pest. It also attacks wood in use, and is a pest in Melbourne. Nest structure is similar to *C. acinaciformis*, with the formation of the mud-guts in hollowed trees. Colonies may contain 750 000 individuals.

**Formosan termite, *Coptotermes formosanus*** Alates are 12–15 mm long with the wings. The body is 6–7 mm long and the dorsal and ventral surfaces are uniformly brown; the mandibles are blackish brown. Wings are clear and have dark costal veins. The body and wings are densely covered with short, fine setae. The head is yellowish brown and without distinct marks anterior to each ocellus; there are two pairs of setae at the base of the fontanelle. Soldiers are about 6 mm long and have an oval-shaped, light-brown head and dark-brown mandibles; the body is yellowish brown. Workers are about 5 mm long and pale yellow; the mandibles lack teeth, and the fontanelle gland has a large opening at the front of the head. Primary queens are about 21 mm long; the abdomen is about 18 mm long and 6 mm wide. Secondary queens are about 12 mm long and brownish yellow, and without the chitinized, brown head of the founding queen. Nests may be established in the ground or above ground. When foraging for food and water below ground, workers damage materials such as asphalt, plaster, plastic, and thin sheets of metal. They may tunnel into electrical cables and cause power failure, and penetrate rigid board insulation in buildings.

Swarming flights are at dusk to about midnight, from March through June; the time varies with latitude and local conditions. Atmospheric pressure of 1001.2–1013.2 mb, temperatures of 21.7–29.4 °C, and 82–99% RH can stimulate swarming. *C. formosanus* swarms before the rainy season in some parts of its distribution. In China's Hainan province (20 °N), alates appear in nests at the end of March and early April; in Guandong province (23 °N), alates emerge in mid to the end of April. In Taiwan, swarming occurs at the end of May and early June, and in Japan at the end of June. Emergence flights last about 30 min, and there is a strong tendency for alates to fly towards electric lights. Large numbers of alates may collect at outdoor lights at night, and alates may fly as high as 90 m to establish colonies. Alates leave the nest through openings above the major portion of the colony, and through holes that are located at the edges. The number of alates that emerge from a colony depends on its size and age, and may be as many as 68 700 (field colony) or 45 969 (laboratory colony). The number of emergence flights in a season ranges from 2 to 6, but can be as many as 10. Alates are first produced from field colonies that are 5–6 years old. In laboratory colonies, the first alates are produced when they are 8–10 years old.

The founding pair establishes the initial nest in moist wood or soil, and the first eggs are produced in 6–7 days. Egg-laying lasts for several months, extending into November in some regions. In the first batch there may be 30 eggs; in the first 5 months about 46 eggs are deposited by the founding queen. Hatching occurs in 24–33 days. Development of the six larval instars takes about 20 days to complete. The first presoldier and soldier are produced in the colony 16–26 days after the first larvae appear. Following the first group of larvae and soldiers, colony growth rate is variable. The percentage of soldiers is 10–14% during year 1; the percentage then declines to 2–3% in mature colonies. In the laboratory, a *C. formosanus* colony at 2 years contains about 250 individuals, 1250 individuals at 3 years, and more than 50 000 at 4 years. Mature colonies are large and contain 1–4 million individuals. The life expectancy of a colony is variable, but estimates are at least 50 years. Colonies of *C. formosanus* are larger in size and number of individuals, and typically consume wood at a faster rate than colonies of *Reticulitermes* species.

Nest sites in natural habitats include below ground around tree stumps, hollow trees, and living trees. This species is known to attack more than 50 species of living trees and shrubs in Hawaii, and at least 20 tree species in Louisiana. In the urban environment, nest sites can be aerial, subterranean, and aerial–subterranean. *C. formosanus* often builds above-ground nests where there is a source of water, such as near plumbing leaks, condensation on water pipes, or from air-conditioners. Nests not close to water have distinct foraging tunnels (called water lines) established to connect the nest to a source of moisture; these tunnels are larger than typical foraging tunnels; they are moist, and usually contain a large number of workers. The moisture content of nest materials may be as high as 33%. In China, about 50% of the *C. formosanus* infestations in urban areas are aerial nests, and 15–20% of the infestations in rural areas is this type. Aerial or above-ground nests comprise about 25% of infestations in southeastern Florida, and in Hawaii about 50% of the high-rise buildings have above-ground infestations. Subterranean nests are common in rural areas, but are usually a small percentage of the nest sites in urban areas. They can extend to 0.5 acres or more, and tunnels may extend more than 3 m deep. Aerial–subterranean nests are primarily located in the ground, but have extensive galleries above ground to reach food. This type of nest is most common in rural areas of China. In urban areas, this type of nest is common in street trees.

The main nest is constructed with friable material called carton, which generally consists of a mixture of soil, chewed wood, saliva, and excrement. The size, which may be a cubic meter, and shape depend on the site and the conditions. The nest usually consists of a 3–5-cm outer layer of loose, wet soil, and an inner layer of overlapping pads arranged in concentric circles, with the thicker segments at the center. The chambers for the founding queen and male are at the center.

Nest location may change during the development of a colony. During the life of the colony, the main nest is often moved to a new and more suitable location, and satellite nests are established. There are tunnels connecting the main and satellite nests with the main colony. There may be 1–10 satellite nests, depending on the size of the colony and local conditions. The distance between the main colony and satellites varies from about 6 m to as far as 50 m. Satellites have nearly the same physical structure as the main nest and satellites, except they are smaller. There are no eggs and larvae in the satellite nests. Workers do not forage in winter months, but in areas of its southern distribution they are active throughout the year. Foraging and feeding do not occur at temperatures below 10 °C and activity is slowed at 10–20 °C. Survival and foraging are optimal at 20–36 °C. Lethal temperatures are −3 °C for 7 days, −1 °C for 9 days, and 1 °C for 14 days. The maximum foraging range is 60–100 m from the nest at about 24 °C. At several locations in the USA, *C. formosanus* colonies do not behave aggressively toward other colonies of the same species. Several colonies in the same location may intermingle to create a supercolony to exploit available food sources.

This species is native to China and is distributed in Taiwan, Japan, Philippines, Marshall Islands, Sri Lanka, South Africa, Hawaii, and southern USA. In China it is distributed in 14 southern provinces, with the northeastern limit in Jiangsu province (33° 28′ N) and extending through Anhui and Hubei provinces to a western limit in Sichuan province (104° 35′ E). It is abundant in the southern provinces of Fujian, Guandong, and Guangxi. It was introduced into the continental USA, and has spread to sites in several states, including Alabama, California (San Diego), Florida, Georgia, Louisiana, Mississippi, North Carolina, South Carolina, Tennessee, and Texas.

***Coptotermes gestroi*** This species is the most important subterranean termite in Thailand. It causes the most damage to wooden structures in urban environments in Thailand. It is also a serious pest of living trees, including rubber trees, in southeastern Asia. Subterranean tunnels are used for foraging, and covered runways of digested wood and soil particles are made above ground to reach a suitable food source. Colony structure includes a main nest, and subsidiary nests that contain only workers and soldiers. Nests in natural areas are in dead timber, logs, and old tree stumps. In the urban environment, structural wood above and below ground is infested. The number of foraging workers in mature colonies ranges from 1.3 to 2.7 million individuals.

***Coptotermes havilandi* (Fig. 10.2i, j)** Alates are about 6 mm long, and about 12 mm long with the wings. The body is long and slender; the dorsal surface of the head, thorax, and abdomen is brown to dark brown, and the ventral surface is yellowish brown. Wings are nearly translucent, and the costal margin is yellowish brown. The head has yellow marks (antennal spots) anterior to each ocellus; there is a single pair of setae at the base of the fontanelle. Neotenics are 5–6 mm long and yellowish brown; males have styli similar to those of male alates. The soldier has an oval-shaped head with a large fontanelle. Soldiers are very aggressive and will appear in large numbers in disturbed areas of the nest. Swarming flights occur at dusk, typically at temperatures from 16 to 22 °C. The central nest is subterranean or aerial, and the carton material is a mixture of soil, wood, saliva, and excrement. Galleries from the central nest can be 20 m long and connect to food or water resources, and to satellite or subsidiary nests. Large colonies of *C. havilandi* have a central nest where the primary reproductives are located, and satellite nests that may be located in a feeding area. Satellite nests are also made of carton and have workers, soldiers, and neotenics. The number of eggs laid by the founding queen is 8–26, and the first young hatch in 60–70 days after pairing. The primary male and female feed. The first pigmented worker typically appears in about 150 days and the first soldier appears in about 180 days.

Material attacked by this termite includes structural wood, paper, and living trees. The most severely damaged woods are softwoods, such as pine. When foraging, workers will damage material such as leather, electric and telephone cables, plastic, rubber, and metal. This species is native to Southeast Asia and Indonesia, but has been introduced to other regions. It occurs in the West Indies, including Antigua, Cayman Islands, Cuba, Jamaica, Montserrat, and Turks and Caicos Islands. It was introduced into North America in southern Florida (Miami), and may be established. This species is the most important subterranean pest termite in the urban environment in Brazil. Natural nest sites include the trunks of trees. In the urban environment, it attacks structural wood in buildings. In densely populated urban areas, colonies may become established on the upper floors of high-rise buildings. Populations of this termite have increased in the urban environment in Brazil, while the number of colonies in rural areas or natural sites remains stable.

***Coptotermes heimi*** Alates are produced in colonies from March to August. Flights usually occur during the monsoon (June–August) in the drier portion of the distribution. In natural habitats it attacks a variety of living plants, including bamboo, cactus, mango, palms, and *Ficus* spp. In the urban environment, it attacks structural wood, paper, and books. Galleries in wood are usually constructed in the direction of the wood grain: usually the soft portions of wood are consumed, and the outer layer, heartwood, and knots are not attacked. This species is an important pest throughout India and West Pakistan.

***Coptotermes niger* (Fig. 10.3a)** Alates are produced from colonies in April and May; flights are in mid-afternoon to dusk, and usually after rainfall. Natural nest sites include living plants and trees. In the urban environment, it attacks structural wood in buildings. This species constructs flat runways on trunks of trees that extend from the ground to about 1 m on the tree trunk, and 3–4 cm into the wood. It is distributed in Central America, and occurs in Guatemala, Costa Rica, Panama, Honduras, and Columbia. It is also found in Brazil, where it is a minor pest in the urban environment.

***Coptotermes raffrayi*** Alates are produced in colonies from October to January, and the flights are in mid-afternoon to dusk. Soldiers are about 5.5 mm long. This species occurs in southwestern Australia, where it is a serious pest of structures. It typically builds small mounds near living trees, particularly *Eucalyptus* spp.

***Coptotermes testaceus*** This species is widely spread through the neotropics, especially in the Antilles and northern South America; it also occurs in Brazil. Natural nest sites are in damp wood in forests. It damages trees, including rubber trees in managed forests. In the urban environment, it infests woodwork in buildings.

***Coptotermes traviansi*** Alates are produced in colonies in January, February, August, and October. Nests are usually in the trunks of trees and earthen galleries on the surface and under the bark of trees. In the urban environment, this species attacks fence posts, stored lumber, and above-ground wood kept moist by dripping pipes or roof leaks. It will also infest ships. This species occurs from east India through Pakistan to Malaysia, Singapore, Sumatra, and Borneo. A related species, *C. curvignathus*, has the same distribution and pest status. Colonies of these two species may have from 370 000 to 10 million individuals, with a foraging area of 13–88 m$^2$.

***Coptotermes vastator*** Alates are about 12 mm long with the wings, and the body length is 6–7 mm. Soldiers are 5–6 mm long. Nests are in the ground in tree stumps and down logs; in agricultural regions, this species attacks young sugarcane. In the urban environment, it is a pest of structural wood in buildings. Swarming flights occur from February to June, and large numbers of alates are released from colonies on days with rainfall. During this process, soldiers are at the openings, with their mandibles facing out. Emerging alates are strong flyers and are attracted to lights. This species occurs in the Philippines, where it is an important structural pest, and it has been introduced into Hawaii.

**Other *Coptotermes*** Species in this genus are structural pests in urban and rural areas, especially in tropical regions. *C. kalshoveni* infest houses in urban areas in Thailand, and *C. premrasmii* infestations are known from both rural and urban areas of the country.

### *Heterotermes*

This genus has nearly worldwide distribution; it occurs in the tropics, with some species occurring in arid or desert habitats. Nests are usually subterranean, and they attack wood below and above ground. Workers construct covered passageways above ground to reach wood. The soil tubes are generally smaller in diameter than the tubes of *Reticulitermes*. Species of *Heterotermes* and *Reticulitermes* may occur together in the same region. These two genera are distinguished from *Heterotermes* by having pale yellow alates and slender mandibles, and *Reticulitermes* having dark brown alates and broad mandibles. The protozoan gut fauna of *Heterotermes* species is much simpler than the fauna of *Reticulitermes* species.

***Heterotermes aureus*** Alates are 9–10.5 mm long and yellowish brown. The wings are translucent, and the margins are ciliate. Winged reproductives are produced in colonies in July (Arizona) through September (California); flights usually occur in late afternoon and evening after rainfall. This is the only *Heterotermes* species that occurs in North America. It occurs in the desert regions of southern Arizona and California, and

Mexico, where it accounts for considerable damage to wood in structures, fences, and utility poles.

**_Heterotermes ferox_** Alates are present in colonies from October to July according to locality; flights occur in late afternoon or early evening. Soldiers are about 4.74 mm long and have parallel-sided heads, and prominent dark mandibles. Colonies are small and nests are typically indistinct, with the young reared in galleries and chambers in the soil. The galleries and tunnels in damaged wood are lined with speckled fecal material. Nests are often located adjacent to other subterranean termite species. Workers and soldiers in the nest move slowly, and when confronted they have the habit of moving backwards. Damage is done primarily to softwoods; in hardwoods only the sapwood or weathered surfaces are consumed. This species is often encountered attacking posts, poles, weathered flooring, and timber decking. This species occurs in Australia, from southern Queensland to southeastern and southern areas, across to Western Australia.

**_Heterotermes indicola_** Alates are present in colonies from July to August, and flights occur at night between 21:00 and 24:00 h. Swarming flights have been observed to occur during rainfall. Nest sites in natural areas are in fallen trees and tree stumps in forests; in the urban environment, this species attacks structural timber in buildings, posts, and other wood in use. It is one of the most destructive termites in India. Almost all varieties of timber are attacked, above and below ground. When feeding in wood, the sapwood is removed first, and the annual rings and heartwood are usually left intact. The feeding galleries are filled with a gray carton made of fecal material and soil. Narrow galleries of soil and fecal material are built for moving above ground. A related species, *H. malabaricus*, occurs in India and it attacks structural timber. This species occurs throughout northern India, extending south to 15–20 °N; it occurs up to 2500 m in the Himalayas, and in West Pakistan and Afghanistan.

### Prorhinotermes

Species in this genus are primarily subtropical, and widely distributed. The absence of a median vein distinguishes the alate from other Rhinotermitidae. In the soldier, the fontanelle is very distinct; the mandibles are slender, curved, and lack teeth.

**_Prorhinotermes simplex_ (Fig. 10.3f)** Alates are about 9 mm long with the wings. The body is pale brown, the head is reddish brown, and the legs and antennae are yellowish brown. The wings are translucent and the veins yellowish brown. Eyes are circular and about twice their diameter from the hind margin of the head; antennae are longer than the head. The fontanelle is distinct and lies between the eyes. The ocelli are not visible from above; they lie at the front margin of the eyes and are obscured by a ridge that extends from the clypeus. The pronotum is twice as broad as long; the front margin is straight. The soldier is 5.5–6.5 mm long and yellowish brown. The head is 1.3 times longer than broad; mandibles are reddish brown and as long as the width of the head. Antennae are as long as the head and 17-segmented. The body, legs, and antennae have many long, pale white setae. Natural sites for nests include damp and dry wood in logs of tidal mangrove swamps, or in pine woods near the seacoast. It does not build subterranean nests, but will enter the soil beneath infested logs. Galleries are filled with frass pellets, and spotted with moist frass. In the urban environment, it infests wood around building foundations. The shelter tubes are typically soft and blackish brown, whereas shelter tubes of *Reticulitermes* are usually hard and light brown. Colonies are usually large, and supplementary reproductives are usually more common than primary pairs. Supplementary reproductives are 6–7.5 mm long and pale yellow; the eye is brown. Swarming flights occur at dusk from October through January, with a peak in late December. Alates are attracted to lights at night. This species occurs on the Florida coast; its northern limit of distribution is Fort Lauderdale.

### Psammotermes

This genus is distributed in the desert region of Africa and Madagascar. There are two distinct soldier castes; both have filiform antennae and mandibles with numerous marginal teeth. Alates have small wing scales and the head is broadly oval.

**Sand termite, _Psammotermes hybostoma_ (Fig. 10.2c)** This species is distributed in the deserts from Dakar to Sudan, and in Saudi Arabia; in Algeria it is common east and south of the Atlas mountains. In Egypt it is widespread outside the Delta, and is the main cause of damage to buildings in districts of Cairo and in towns and villages in the desert south to Sudan. The prevalence of this species in Egypt forced the Pharaohs to import cedar (*Cedrus libani*) wood from Syria and Lebanon for their coffins. The subterranean nest system consists of galleries that are about 40 cm below the sand surface, and chambers are filled with grass segments and fragments of subfossil trunks of *Tamarix* trees.

### *Reticulitermes*

This genus has primarily a north temperate distribution. It occurs in Europe, Asia, and North America, and includes 14 species. Species in this genus extend northward into continents, and some species occur in relatively cold climates. In the USA, *Reticulitermes* inhabits a variety of climatic zones, including coastal regions, high altitudes in the Rocky Mountains, and deserts. The color of alates ranges from yellowish brown to black. The wings have few, if any, setae, the antennae have 16 segments, and the ocelli are well separated from the eyes. In the soldiers, the sides of the head are nearly parallel; the head is longer than it is wide, and the mandibles lack teeth.

Cuticular hydrocarbons are useful in discriminating the *Reticulitermes* species in Europe and the USA. There may be 30–100 different hydrocarbons in the cuticle, including linear, monomethyl, or dimethyl alkanes or alkenes. The relative proportions of these molecules are controlled and different for each species. Evaluations indicate discrepancies between hydrocarbon-based species (phenotypes) and morphologically based phenotypes of *Reticulitermes*. In the USA, there are six morphologically described species. However, there are 26 species defined by distinct cuticular hydrocarbon patterns; these patterns in *Reticulitermes* are from a variety of geographic locations. There is growing evidence of the taxonomic value of this method of defining phenotypes. Analysis of the defense compounds from soldier frontal glands indicates that these chemicals can be used as a species diagnostic tool.

Species of this genus are referred to as subterranean termites, since their galleries are usually in or associated with soil. They have the ability to penetrate soil, to construct permanent passageways through it, and to build with particles of soil or other material. They attack wood buried in, or in direct contact with, the ground. They often extend their galleries to other wood, or they build covered galleries above ground to reach wood away from soil or to reach soil from infested wood. Galleries are usually constructed of soil and masticated wood. *Reticulitermes* require a humid environment. The moisture level in exposed galleries or tubes is generally equal to the moisture in the ambient air. The enclosed galleries primarily provide protection from air movement, and predators such as ants. Earthen tubes are built by workers using particles of soil, sand, or wood, which are coated with a glue-like substance. Fecal material may be used as cement. Tubes may be wide, flattened, and usually against a surface when used as a primary connection between the below-ground nest and food above-ground. Tubes may be built to explore for food sources, and these are usually free-standing and relatively small. Earthen tubes may be suspended from infested substrates above the ground, and they may connect to the ground. These tubes may be used by workers and soldiers to return to the soil from above-ground feeding sites. They are usually light-colored because they are constructed of masticated wood. Earthen tubes may be constructed to provide alates with an exit point from the nest. These tubes usually extend above ground 10–20 mm, but often remain hidden when colonies are indoors.

Localized sources of moisture, such as condensation in wall voids, leaking pipes, and plumbing fixtures, can provide suitable moisture for colonies. When above-ground moisture is sufficient, these termites may remain separated from the ground for an indefinite time. Colonies or portions of colonies can survive and flourish without ground contact, provided there is an adequate supply of moisture and food. *Reticulitermes* species do not require soil contact to survive. Soil is a substrate that can be tunneled to reach new sources of food; it serves as a reservoir for moisture, and as a building material for the above-ground galleries. If conditions of temperature, food, and moisture are satisfactory, colonies can become established without contact with the ground.

Founding pairs of *Reticulitermes* usually enter moist wood that is in contact with or in soil, but pairs may enter small openings in a variety of substrates. *Reticulitermes* alates may be carried by wind currents, and alight on above-ground surfaces, such as upper floors and roofs of buildings. Their survival depends on the subsequent conditions of moisture and temperature extremes. New colonies may be completely isolated from soil contact and capable of surviving and growing. Under laboratory conditions, *Reticulitermes* flourish as well without soil as with it, and when provided with both soil and wood, they will tunnel in wood. *Reticulitermes* nests typically do not have a permanent cell where the primary reproductives remain. The primary queen and secondary reproductives (brachypterous and apterous neotenic females) of this genus do not develop an enlarged abdomen. They are capable of moving around the colony to select favorable temperature and moisture conditions.

Mature colonies are generally considered dynamic, and mobile within a home range that can extend a linear distance of about 80 m. The entire colony, or only portions, may move to new sites as temperature, moisture, and food within this range are variable. Colonies are organized as a network of cells of various sizes and number of individuals. The cells are located in and around several food resources and they are connected to each other by a system of trails or passageways. Some connections have large numbers of workers and soldiers traveling them, while others have less activity, and some may be obsolete or simply not in use. The distribution of termites around food resources changes; when a new food is discovered and occupied, food in other locations may be exhausted or removed. A large number of individuals from the colony may remain for long periods at sites where there is favorable food, and move only occasionally to other portions of the colony. At one or more of the colony cells adjacent to food, there may be secondary reproductives. The egg production of several secondary reproductives may be equal to or exceed the primary queen, and the number of individuals in these cells can increase rapidly.

Swarming flights depend on the presence of alates in the colony, and on suitable conditions for emergence, which include temperature and humidity. Flights from the nest vary according to species and usually from region to region, and from year to year. Some variation in flight time can be expected from colonies that are located in buildings and colonies in natural habitats. Flights from colonies established in buildings usually occur before flights from colonies exposed to outdoor conditions. When large numbers of alates are present in the colony, they will be released when temperature, moisture, illumination, and possibly barometric pressure are suitable for the survival of the relatively fragile alates. Localized warming from sunlight on a portion of the nest, or from the conditions indoors, combined with moisture to prepare the substrate for the founding pair, and light (natural or artificial) will induce emergence.

An important reproductive pattern for *Reticulitermes* is colony splitting or budding. In general, this method usually begins with the isolation of a large group of individuals, such as those in a cell close to a sustained food source. There may be secondary reproductives within the group, or they may be developed later as the cell becomes independent of the pheromonal influence of the primary queen. Brachypterous neotenic females may travel above or below ground with a small number of soldiers and workers to a new location (or remain where they are) and begin to function as an independent group or new colony. Secondary reproductives with a group of workers and soldiers may travel to a new nest site through the many underground passageways, or actually move above ground. The connections of this group with the main colony can weaken or sever, and then they begin to function as a new colony. Colonies of other subterranean termite genera are known to divide this way, and one or more *Reticulitermes* species may use this method.

***Reticulitermes hageni*** Alates are about 8 mm long with the wings, and the body is yellowish brown. The wing membrane is nearly clear, and they are usually about 6 mm long and 1.7 mm wide. The ocelli are less than their diameter from the eye. The anterior border of the pronotum is slightly concave, while the posterior border is bilobed. The body is covered in pale setae. The soldier is about 7 mm long. The head is pale yellowish brown, and darker at the front; the mandibles are as long as the width of the head. The pronotum is narrower than the head. Swarming flights occur in late summer and fall, usually from July to August.

**Eastern subterranean termite, *Reticulitermes flavipes*** Alates are about 10 mm long with the wings, and the body is brownish black to black. The front wing membrane is gray to grayish brown, while the hind wing is lighter. The head is about as broad as long; ocelli are more than their diameter from the eye, and the antennae have 16–18 segments. The pronotum is about as broad as the head, and slightly concave posteriorly. The body is covered with yellowish-white setae. The soldier is about 6.7 mm long. The head is about 1.5 times longer than broad; the sides are nearly straight and tapering to the broadly rounded posterior. The ventral longitudinal edges of gena are convex medially. The antennae are 16-segmented. The pronotum is semicircular, and the front margin is slightly indented at the midline. Brachypterous females are about 12 mm long, and apterous females are 7–9 mm long. Natural nest sites include tree stumps, logs, and other wood in contact with the ground. In the urban environment, this species infests buildings, fence posts, and occasionally living trees and shrubs. Colonies are usually large, but develop slowly. During the early phase of colony growth in *R. flavipes*, colonies produce relatively small-bodied workers. Mature colonies have an average of 224 000 individuals, with about 2% of the population being soldiers. Primary reproductives may produce large number of eggs for 7–10 years. When there is a decline in egg-laying, or the isolation of a group of individuals from the primary queen, multiple secondary reproductives usually develop in the colony.

Swarming flights in northeastern USA occur about midday in April and May; in the southeast, flights occur from March to May, and February in Florida. Flights may occur at any time of year indoors, but rarely at night. Colonies may have several flights over a 1-month period; the first flight is usually the largest, and may consist of thousands of individuals. Flights are usually associated with rainfall, even with in-building colonies, flights are usually linked to rainfall. This species is the most common and widely distributed subterranean termite in North America. It occurs from Ontario, Canada south through eastern USA to northern Mexico and Guatemala. It does not occur in far western North America.

This species was introduced into northwest Germany on pine (*Pinus* spp.) trunks imported to Hamburg from the USA. In Hamburg, two parts of the city have active populations of *R. flavipes*. One is about 50 ha in area and consists primarily of old buildings, and infestations began here around 1937; the second area is about 2.5 km away and about 5 ha in area, and infestations here began about 1952. Infestations are generally in cellars and ground-floor levels. Swarming is in March.

Workers of *R. flavipes* will follow blue and red lines produced by some ballpoint pens. Gas chromatographic analysis indicates that the chemicals in the ink capable of producing this trail-following behavior are diethylene glycol monobutylether and diethylene glycol monoethylether. Other *Reticulitermes* species, including *R. lucifugus*, also respond to these glycol compounds, but some Kalotermitidae are repelled.

**Western subterranean termite, *Reticulitermes hesperus*** Alates are 8–9 mm long with the wings. The body is about 5 mm long and dark brown to brownish black; the anal area has a pale brown spot. Wings are gray to grayish brown. Ocelli are less than their diameter from the eyes. The soldier head is at least two times longer than broad, and yellowish brown. The pronotum is slightly concave at the midline. Natural nest sites include around the roots of living trees and plants. In the urban environment, woodwork in buildings, fence posts, and utility poles is attacked. Swarming flights occur during the day in November–January following rainfall; flights may occur from February to June. The founding pair excavates a small cell in soil or decaying wood. The first batch of about 10 eggs hatches in 30–100 days, and the first individuals to complete development are small workers and soldiers. Winged reproductives are not produced until colonies are 3–4 years old. Supplementary reproductives, males and females, occur in colonies in the absence of primaries. The egg-laying rate of these forms is higher than the rate for the single founding queen, and colonies rapidly increase in size. Shelter tubes are constructed of soil and excreted wood. Foraging galleries are excavated through soil or infested wood. In wood, galleries are longitudinal and usually follow the grain. Galleries are spotted with yellowish-brown fecal spots. This species is distributed in North America from British Columbia south to western Mexico, and east to Idaho and Nevada.

***Reticulitermes lucifugus* (Fig. 10.2e)** Alates are 10–12 mm long with the wings. The body is dark brown to black, the tibiae are pale, and the ocelli are not more than their diameter from the eyes. Colonies typically have a pair of primary reproductives, but some colonies have secondary reproductives. This species is primarily distributed in southeastern France and Italy. In Corsica, *R. lucifugus corsicus* is phenotypically close to *R. lucifugus*. An infestation of *R. lucifugus* was established in a rural, coastal setting in Staunton, North Devon, in southwestern UK. This is well outside its indigenous biogeographical range. The colony was in the walls of three buildings, and probably existed for 30 years.

***Reticulitermes santonensis*** This species produces large colonies that do not demonstrate intercolonial aggression. Colonies may be headed by primary reproductives or neotenic reproductives. In urban environments, colonies are founded by primary reproductives, and maintained by cycles of inbreeding among neotenics. In natural areas, *R. santonensis* may reproduce more frequently via sexual individuals. Colonies in urban areas are large and may have a foraging area of 1080 m$^2$. The lack of aggression may facilitate colony fusion and result in large urban colonies. In natural areas, colonies are generally small and may be sympatric with *R. grassei*, a closely related species. *R. santonensis* is distributed in forested areas in southwestern France, and in urban environments, including Paris and its environs, and in Hamburg, Germany. It was first reported infesting buildings in the 16th *arrondissement* in Paris in 1945; it is now widespread in the city, and also attacks street trees adjacent to infested buildings. *R. santonensis* could have originated from a North American population of *R. flavipes* that was introduced into Europe. In the wild, *R. santonensis* has an aggregated distribution related to human surroundings.

***Reticulitermes speratus*** This species is one of the most important pests in northern China; colonies survive at latitudes up to 41 °N. It is primarily distributed along the east coast of China. In Japan, *R. speratus* is a major pest species; it occurs as several subspecies in Japan, including *R. kyushuensis*, *R. leptolabralis*, *R. okinawanus*, and *R. yaeyamanus*.

**Arid-land subterranean termite, *Reticulitermes tibialis* (Fig. 10.4d; 10.5a–c; 10.6a, b)** Alates are 9–12 mm long with the wings, and the body is shiny, blackish brown. The tibiae are blackish brown; the anal area of abdomen has a light-brown spot. The wing membrane is nearly transparent;

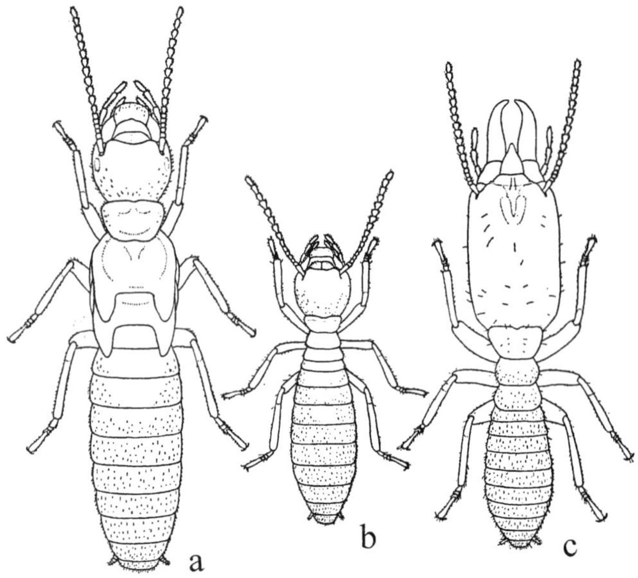


**Figure 10.5** Isoptera. (a) *Reticulitermes tibialis*, secondary queen; (b) *R. tibialis*, worker; (c) *R. tibialis*, soldier.

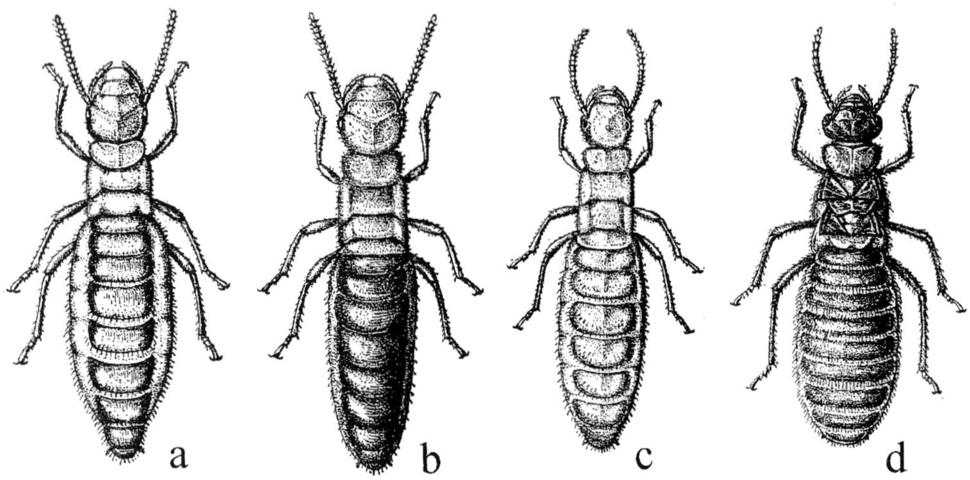


**Figure 10.6** Isoptera. (a) *Reticulitermes tibialis*, secondary queen; (b) *R. tibialis*, secondary king; (c) *R. virginicus*, secondary queen; (d) *Tenuirostritermes cinereus*, primary queen.

the wings are about 9 mm long and more than 2 mm wide. Ocelli are slightly less than their diameter from the eye. The pronotum is as wide as the space between the eyes, and concave at the midline anteriorly. The soldier head is brownish yellow; the head is about 1.5 times longer than broad, and the sides are nearly parallel. Mandibles are about as long as the width of the head. The pronotum is much narrower than the head. Ventral longitudinal edges of gena are nearly parallel. Swarming flights occur in spring and fall; in the central Rocky Mountains, alates may emerge in April and February. In this region, alates are present in colonies in late fall and early winter, and are apparently retained through the winter months to emerge in early spring. In Indiana, flights occur from March to May, and November; in Texas, flights occur in May, and October to December. This species is distributed in North America from Oregon and Montana south to western Mexico, and east to Missouri, Arkansas, and Texas. Where *R. hesperus* and *R. tibialis* overlap in distribution, *R. hesperus* prefers cool, shady, moist sites; *R. tibialis* prefers open, sunny, and dry locations.

***Reticulitermes virginicus*** **(Fig. 10.6c)** Alates are about 8 mm long with the wings, and the body is shiny, blackish brown. The wing membrane is nearly translucent; the wings are about 6 mm long. The head is slightly longer than broad, and the ocelli are much less than their diameter from the eye. The pronotum is much narrower than the head, and distinctly bilobed on the anterior border. The head is about as broad as long, and the antennae have 16 segments. Swarming flights occur in March and April. This species is distributed in eastern North America. In general, the northern limits of distribution are where the average annual minimum temperature does not fall below −23 °C; westward it extends through southern Illinois and Missouri, southward to the Gulf coast and to the southern tip of Florida. In the northeast, it extends to Pennsylvania and New York. Throughout most of its range *R. virginicus* overlaps the region occupied by *R. flavipes*, but it does not seem to extend into the northern regions occupied by *R. flavipes*. Swarming flights of *R. virginicus* tend to occur later than *R. flavipes*, although there is some overlap. Flights of *R. hageni*, whose range coincides with *R. virginicus*, tend to occur later.

**Other *Reticulitermes*** There are several species distributed in southern Europe. *R. grassei*, which may be a subspecies of *R. lucifugus*, and *R. santonensis* occur sympatrically in southwestern France, including Charente-Maritime, Lanes, and urban locations. Colonies are typically headed by a single pair of reproductives (46%), or primary and secondary reproductives (54%). *R. banyulensis* occurs primarily in Spain and southwestern France. Colonies of this species are typically headed by a single pair of primary reproductives (50–60%), or primary and secondary reproductives. *R. balkanensis* occurs in Greece and Albania.

## Termitidae

Genera in this family are referred to as the higher termites. This division is based on their lack of dependence on symbiotic protozoa for cellulose digestion. Over 80% of the genera and 74% of all the species of termites are placed in this family. It includes genera and species in all the zoogeographic regions with a wide range of habits, including species that are economically important.

### *Amitermes*

This is a large genus with species occurring in most parts of the tropics and subtropics, but also in some arid areas. The soldiers have rounded or pear-shaped heads with curved mandibles (Fig. 10.1a), which have a median tooth. Antennae have 13–17 segments. The alates are about 12 mm with the wings, and the body is dark brown to nearly black. They usually attack wood where it contacts the soil and often limit their feeding to wood buried in the ground. *Amitermes* spp. girdle and kill young citrus trees. Galleries are lined with brown fecal plaster, even in such objects as dry cattle dung (cow-chips). Swarming usually occurs during the day; some emerge at midday and others in the afternoon; in arid regions the flights may be at dusk and following rainfall. Many are mound-building termites, and some species are well-known for their nests. *A. meridionalis* constructs the so-called compass mounds in North Australia. These are wedge-shaped earthen mounds that are about 3.5 m high, and have their long axis always aligned north to south. Species in arid environments include *A. desertorum* in Algeria, *A. vilis* in Saudi Arabia, *A. capito* in Western Australia, and *A. wheeleri* in southwestern USA and Mexico.

***Amitermes coachellae*** Alates are about 10 mm long with the wings, and the body is dark brown to nearly black. In the soldier, the mandibles are slender, and the tooth is near the base; the right and left mandibles are nearly identical. Swarming is during the day and usually following rainfall. This species is known from California, Arizona, and Nevada.

***Amitermes emersoni*** Alates are about 12 mm long with the wings, and the body is dark brown to nearly black. The soldier is yellowish brown, the mandibles are long and slender, and the tooth is near the tip. Swarming occurs during the day in late January or early February. This species is known only from the Coachella valley in California, where it is one of the most common termites.

**Florida dark-winged, subterranean termite, *Amitermes floridensis*** Alates are 8.4–9.7 mm long with the wings; the wings are about 7.4 mm long; the body is about 4.3 mm long and dark brown to nearly black; the antennae and legs are pale to yellowish brown. Wings are brown, and the anterior veins dark brown. The soldier is yellowish brown; the head is longer than wide and light brown; the mandibles are blackish brown, long, slender, and curved. Workers are characterized by dark gut contents, which are visible through the abdominal wall. There is a single tooth on the inside of the mandibles. Swarming flights occur during the day, from 07:30–19:00 h, and from June to early August. Flights frequently occur after rainfall, and adults may begin to fly before the rain ceases. Nests are below ground, and above-ground foraging tubes are not constructed. This species occurs in and around the city of St. Petersburg, Pinellas county, Florida.

***Amitermes herbertensis*** Alates are produced in the colony from March to July. Natural nest sites include rotten logs and tree stumps in rain forest habitats. In the urban environment, it damages structural timber, the foundation piles of houses, fence posts, electrical transmission poles, and other timber in contact with the soil. The interior of round logs, trees, or structural lumber may be destroyed and replaced with earthen material, which contains galleries. House support beams and posts and the stems of woody shrubs may be covered with an earthen sheath. This species occurs primarily in northern Queensland, Australia.

**Amitermes minimus** Alates are 7.5–8.5 mm long with the wings, and the body is dark brown to nearly black. In the soldier, the mandibles are slender, and the tooth is near the base. The right and left mandibles are not identical, and the tooth on one may be large. Swarming is during the day. This species occurs in the desert region of southwestern USA, ranging from Texas and Arizona to Nevada and California. Damage is caused to poles and fence posts; it penetrates wood extensively and may construct hard, sand-covered galleries over the buried portion of the infested wood.

**Amitermes snyderi** Alates are about 9.5 mm long with the wings, and the body is dark brown. In the soldier, the mandibles are long and broad at the base, and the tooth is near the base; the right and left mandibles are nearly identical. Swarming is during the day. This species occurs in southwestern California in the Mojave and Colorado deserts. Damage is caused to redwood posts, pine boards partly buried in the soil, and the roots of desert trees.

**Amitermes wheeleri** Alates are 10–11.5 mm long with the wings, and the body is dark brown. The fontanelle is slit-like. The soldier is yellowish brown; the mandibles are strongly recurved and have a cone-like tooth on the basal half. Mandibles are about half as long as the width of the head. Swarming occurs in Arizona in late afternoon and at night in July, and they may be attracted to lights. In Texas, winged adults have been found in May to June and October. Flights occur in Arizona and Texas after rainfall in the afternoon. These termites are strong flyers. This is one of the most widespread and common *Amitermes* in the USA. Damage is caused to woodwork in buildings, windmill towers, fence posts, and utility poles. It is distributed from Mexico, southwestern Texas and Arizona to Nevada and California. It nests in dead trees and tree stumps, and sometimes in cactus and *Agave*, and in dry cattle dung (cow-chips).

**Other Amitermes** There are several other species of this genus in southwestern and western USA. *A. silvestrianus*, *A. pallidus*, and *A. parvulus* have soldiers in which the mandibular tooth is located about the midpoint of the curved mandible, and the alates are about 9.5 mm long with the wings. The alates of these three species are dark brown. *A. silvestrianus* and *A. parvulus* occur in California, and *A. parvulus* occurs in Arizona and Texas. The alates of *A. pallidus* are yellowish brown. This species is known from southern Arizona, and winged adults were found at lights from June to August. *A. dentosus* occurs in Western Australia through to Queensland, Australia. It has small colonies in underground galleries, and it damages wood close to or contacting the ground.

### *Anoplotermes*

Alates in this genus have a broad head, which is usually covered with fine setae, and they have small eyes and small ocelli. Antennae have 15 segments. Colonies of these species have no soldiers, and species identification is difficult. In the alates the mandibles are not much longer than broad. No soldier has been associated with colonies of this genus. Workers are dark gray and have an elongate, bag-like abdomen.

**Anoplotermes fumosus** Alates are about 7.5 mm long with the wings. The head is dark brown; the pronotum and abdominal segments are brown; legs are pale yellow. Wings are gray, and about twice as long as the body. No fontanelle is visible, and the ocelli are large. Antennae are as long as the head, and there are more than 14 segments; segments 3 and 4 are short. The body is covered with dense setae; setae on the head are dark brown, on the pronotum grayish, and on the abdomen yellowish brown. Colonies are in the ground in burrows, under buried logs, and in and under dry cattle dung (cow-chips). Swarming occurs in August. This species occurs in southwestern USA, in southwestern Texas, New Mexico, and northern Mexico.

### *Globitermes*

The few known species in this genus are found in southeastern Asia. Nests are small mounds, with parts above and below ground. They feed primarily on decayed wood. As with other species that occur primarily in forested or undisturbed areas, their pest status is limited to occasional infestations of rural houses and other isolated buildings.

**Globitermes sulphureus** Alates are present in the colony during the rainy season, and swarming flights occur in January and May. Nests are common in forested areas and colonies are usually large. The above-ground portion of the nest is hardened soil and may be 80 cm high and 60 cm diameter at the base; the covering is about 1 cm thick and apparently made of a mix of soil and saliva. Internally, the nest is divided into small chambers with thick walls. The central space is filled with a mass of carton, which holds the primary reproductives, the eggs, and developing larvae. The soldiers are active and readily attack other termites and ants. Intruders are killed with the large mandibles of the soldiers and covered with a yellowish-white salivary secretion. This species is distributed in Thailand, Myanmar, Vietnam, and Malaysia. It is a structural pest of houses in rural Thailand.

### *Gnathamitermes*

The long, and relatively straight mandibles of the soldier caste distinguish this genus. Mandibles are almost as long as the head is wide, and curve inward only at the tip. Alates are 13–16 mm long with the wings, and the body is dark brown. Members of this genus are primarily soil-dwelling, but they build extensive galleries above ground. Dry cattle dung (cow-chips), weathered vegetation, range grasses, posts, and other wooden structures are damaged. Most of the damage is the result of their galleries above ground, rather than actual consumption of materials.

***Gnathamitermes perplexus*** Alates are about 16 mm long, with the wings, and the body is brown to dark brown. Soldiers are variable and have considerable variability in the teeth on the mandibles; the teeth project from the face of the mandible and are located at about the midpoint. Often the teeth are not equal in size. Workers have a yellowish-brown head, and are slightly larger than the apterous nymphs. Alates develop from egg to adult in about 8 months. In California, soil tubes were built on the lower trunk of young trees, and the bark and wood were damaged. Swarming in Texas occurs in May to August, and it is usually in early morning after rain the previous night. In Arizona, flights take place in late June and July in the afternoon or evening during rainfall.

***Gnathamitermes tubiformans*** Alates are about 9 mm long with the wings; the body is brown and the legs are yellowish brown. Wings are nearly twice as long as the body, and brownish yellow, costal veins are dark brown. Antennae are longer than the head and 14-segmented. The body and legs have dense, short setae. Soldiers have mandibles that are not as long as the width of the head; the length of the head plus mandibles is about 2.4 mm. The abdomen is gray, the legs are yellowish white, and the mandibles are reddish brown; antennae are brown at the base and yellowish brown from the middle to the apex. The mandibular tooth is located on the apical portion of the mandible, and does not project from the mandible, but appears to be cut away from the bottom. Colonies are in the ground. Workers actively forage at night, usually after rainfall; earthen galleries may be in low vegetation and grass. Soil tubes are built 10–15 cm above ground and are usually attached to grass or bushes. Beneath the tubes are cells in the soil. Colonies may be very large and located deep in soil. Supplementary reproductive forms, with short wing pads and dark coloring, may head colonies. Swarming occurs during the day after rainfall in spring and summer. Damage is done to grass in grazing land, and truck crops. This species occurs in western Texas and possibly New Mexico.

### *Macrotermes*

This genus is widespread in Africa and southern and southeastern Asia. The soldiers and queens of several species are very large. These termites are mound builders and fungus growers. Their pest status in the urban environment is limited; they damage trees and young plants, but some are destructive to woodwork near the ground. *M. barneyi* occurs in central and southern China. The subterranean nest is compact and consists of a system of interconnecting chambers, which contain the fungus combs. This species is a structural pest in some regions. *M. gilvus* is a large species which is the most common mound-building termite of southeastern Asia, from Myanmar and Indonesia to the Philippines. The swarming of alates of this species in the Philippines occurs in the evening from May to September. The swarms are very large and composed mostly of males. This species, along with *M. pakistanicus* and *M. anadi*, occasionally infests houses in rural Thailand.

### *Microcerotermes*

This genus has more than 100 species and it occurs in most zoogeographical regions. Species are widely distributed in the tropics, neotropics, and in the dry steppe and desert fringes. They have subterranean or small-mound nests and the colonies are moderate-sized. Species in this genus build compact nests of hard carton with a honeycomb structure. Workers forage in below- and above-ground galleries made of tough carton material; they also forage within plant stems. Dead wood from trees such as *Acacia*, *Zizyphus*, and *Tamarix* is used, as well as plants growing beyond the limits of tree growth. The soldiers of this genus have long rectangular heads (Fig. 10.1h), which are dark brown. Their long mandibles have inward-curved tips, and the inner margins are serrated; the antennae have 12–15 segments.

Most species in this genus eat decayed or weathered wood. Some attack structural wood and have pest status. In Australia pest species include *Microcerotermes distinctus* and *M. implicadus*, which are widespread in the country. *M. nervosus* and *M. boreus* are restricted to Western Australia and the Northern Territory, and damage farm buildings. *M. biori* (Fig. 10.1h) is common on a large number of the islands of Melanesia and Central Polynesia. In the dry steppe and desert fringes, the common *Microcerotermes* species include *M. palearcticus* in Morocco and Algeria, *M. eugnathus* in Tunisia, *M. palestinensis* in Israel south of Tel Aviv, *M. diversus* from southern Iraq and adjacent Iran to the Arabian peninsula, and *M. gabrielis* in northern Iraq, Iran, and Afghanistan. Several species damage structural wood in southern Israel and southern Iraq, including *M. diversus*. *M. exiguus* and *M. arboreus* are structural pests in urban areas of Brazil and the West Indies. *M. strunckii* is a structural pest and damages paper products in northeastern Argentina. *M. crassus* is a very common pest of houses in rural areas of Thailand.

**Microcerotermes turneri** Alates are present in the nest from September to January. Soldiers are about 5.2 mm long. Nests are usually built on the ground, on the trunks of living or dead trees, on the tops of stumps or fence posts, or occasionally on the timber framing of buildings. The arboreal nests are rounded, and are 22–46 cm in diameter and 5 m above the ground. The arboreal nests are light brown, and not black. This species occurs north of Port Macquarie on the central coast of New South Wales and into coastal Queensland, Australia.

### *Nasutitermes*

Soldiers in this genus are usually smaller than the workers, and they have a large rounded head without a constriction, and there is a pointed snout (nasute) (Fig. 10.1i). The mandibles are small, the antennae are short, and the ocelli are small. Most species feed on grass, some infest decaying wood, and some attack structural timbers. The mound-building habit is pronounced in some species, but others build arboreal nests, and some are subterranean. Species that typically build arboreal nests may also infest structures, including boats. *Nasutitermes* is a widely distributed genus, with many species occurring in Australia, and Central and South America. *N. grandinasus* occurs in southeastern China, and it damages woodwork in houses that are within the foraging range of nests.

**Nasutitermes corniger** The alate body is about 7 mm long. The head and body are reddish brown; the mouthparts are yellowish brown. The head is broader than long and the eyes are small; the ocelli are rounded. The soldier head is 1.5 mm and dark brown. The thorax and dorsum of the abdomen are yellowish brown, while the legs are pale yellow. The worker head is dark brown, and the right mandible has three teeth. Nests may be arboreal and composed of brownish-black carton material; they are usually elongate and rounded. This species occurs in Central and South America, including Argentina and Brazil. In the USA, it is known from Texas and Florida.

**Nasutitermes costalis** Alates are present in the nest in spring (May), and swarming flights occur during the day, and sometimes during rainfall. Natural nest sites include dead wood in living trees. In the urban environment, nests may be in fences, and in the structural wood of buildings, and they nest around the base of residential trees. Ground nests may be mounds about 1 m diameter and are sometimes referred to as comejeneras. In Trinidad, this is the most abundant termite species. It is widespread in the Neotropics, occurring from Cuba to Bolivia, Puerto Rico and northern Brazil. It has been introduced into North America, and may be established as a structural pest in the region of Dania Beach, Florida.

**Nasutitermes exitiosus** Soldiers are about 4.2 mm long. This is a mound-building species; the mounds are 30–70 cm high and about 1 m diameter. In dry areas this species nests in tree stumps or below ground and often the soil is bare above the nest site. Colonies are usually established alongside a log, tree stump, or post, and it may take 20 years for a colony to reach mature size. Colonies usually have a single queen, and when mature the queen is often very large and capable of laying 2500–3000 eggs per day. Swarming occurs at night from September to January, but mostly in October and November. Colonies can produce 50 000 alates in a single season; they are attracted to light at night. Damage may be to fence posts, poles, structural timbers in bridges and buildings, and subflooring of houses when a nest beneath the house is undetected. It is not usually a pest in urban areas, because the nest mound is conspicuous. In rural areas, structures may be attacked, and sometimes from a nest that is located 45 m away. Wood attacked by this termite has large galleries and a rough surface. Galleries in wood are often filled with blackish-brown woody carton, and packed with soil. This species occurs in most areas of southern Australia, generally south of the New South Wales and Queensland border.

**Nasutitermes fumigatus, N. dixoni** Soldiers of *N. fumigatus* are about 3.5 mm long; soldiers of *N. dixoni* are about 3.8 mm long. These are subterranean species, and they attack decaying wood that is in contact with the ground. House infestations occur when there is an existing problem with decayed wood in flooring timbers, usually due to inadequate ventilation. These termites are secondary invaders in the decayed wood, and they do not infest sound wood. These two species occur along coastal Australia, from central Queensland to the Victoria and South Australia border. A related species, N. *occasus*, has similar habits; it occurs in southwestern Australia.

**Nasutitermes longipennis** Soldiers are about 4 mm long, and they have a reddish-brown head. This species constructs mounds similar to those of N. *exitiosus*, or the nests may be subterranean. It damages wood in contact with the soil, but it does not feed on sound, dry wood. This species occurs along coastal Queensland to Northern Territory, Australia.

**Nasutitermes walkeri** Soldiers are about 6 mm long, and these are among the largest in the genus. Colonies are usually started in the base or root crown of trees where there is some decay or basal fire damage. Years after a bush fire, the nests of this termite may be abundant in a comparatively small area. When colonies are well established at the base of the tree, an arboreal nest made of soil and carton material is constructed higher up in the tree. Damage is usually to fences, poles, and wood in the ground; attack of buildings is not common and usually is associated with decaying wood and high moisture. This species occurs in the coastal bushland of Sydney, Australia and further north. A related species, N. *graveolus*, has similar habits; the soldiers are about 4.3 mm long. It occurs in coastal northern Australia, from Townsville to Darwin.

**Other Nasutitermes** There are several species in this large genus that occur in urban areas and damage structures. N. *aquilinus* and N. *ehrhardti* are have been introduced into Brazil and Argentina, and often cause structural damage. N. *bivalens* constructs subterranean nests and is a pest in urban areas of eastern Brazil. N. *nigriceps* is widely distributed in Central and South America, Cayman Islands, and Jamaica, but it is known to infest boats and to be transported to other regions. N. *mexicanus* is common along the west coast of Mexico (Colima); nests are in the ground, and posts and other structures are attacked.

### *Odontotermes*

This large genus contains some of the most common termite species in Asia. Many of the species are mound-builders; they forage above ground in protected passageways. Soldiers in this genus have a marginal tooth on the left mandible (Fig. 10.2b), and sometimes on the right. The subterranean nests of some species, such as *Odontotermes bogoriensis*, are 3–4 m diameter and about 50 cm below ground. Above-ground nests may be 1 m high and capped with rounded domes. The pest status of the species in the urban environment is based on infestations of structural wood and the damage to banks of mountain ponds and reservoirs. People of the north Karnatak region of India have used soil from the mounds of O. *assmuthi* for plastering walls and making household cooking ovens. Analysis of the soil shows the presence of binding elements such as silica, iron, aluminum, calcium, and magnesium.

**Odontotermes feae** Alates are present in nests and swarming occurs in June, and September to November. Flights are usually after rainfall and begin after sunset, and they may last several hours or through the night. Alates emerge singly through small circular holes in the ground; workers are present around the holes, which they close when swarming concludes. Nests are large underground structures, consisting of interconnecting chambers that may cover an area of 1.5–2.5 m diameter. The fungus garden is usually in the center of the nest, and may be 75 cm below the surface. Foraging workers usually travel in rows in earthen tunnels and above-ground tubes. This species attacks dead wood, and structural timber in buildings. It is one of the most common termites damaging buildings in India, and frequently infests houses in rural Thailand. It is distributed throughout India, Sri Lanka, Myanmar, Thailand, and Vietnam.

**Odontotermes redemanni** Alates are present in nests and swarming occurs in September in eastern India, and in November and December in Sri Lanka. It builds a very large earthen mound that may be 2 m above ground, and 2–3 m diameter; the mound takes 10–12 years to develop fully. Inside and well below ground, there are large chambers that contain fungus combs. Nests usually have a large central chamber that contains a fungus and serves as an egg deposition site. Colonies are large and often contain several functional queens and kings. Foraging workers make covered passageways above ground. This species is common in Sri Lanka and damages woodwork in buildings close to nest sites. Distribution includes Sri Lanka and India.

**Other Odontotermes** Many of the subterranean species of this genus attack dead wood and infest buildings that are close to nest sites. O. *ceylonicus* is a structural pest in Sri Lanka. It often occurs in the nests of other termites, including O. *horni*, O. *redemanni*, and especially *Hypotermes obscuriceps*. The subterranean species, O. *formosanus*, attacks living trees and structural timber. It forages above ground under earthen passageways. This large species occurs in southeastern Asia, including eastern India, Myanmar, Thailand, and Vietnam. It is also a pest in southern China and Taiwan. *O. proformosanus* and *O. longignathus* infest houses in rural areas of Thailand. *O. wallonensis* builds low mounds which often have 15–25-cm-high chimneys extending above the surface. It is widespread in India, where it attacks structural wood in buildings.

### *Schedorhinotermes*

Species in this genus have a dimorphic soldier caste, and the minor (small) soldiers have mandibles with distinct teeth (Fig. 10.2d). Species occur in Africa, Asia, and Australia. All of the Australian members of the genus infest wood, and form small- to moderate-sized colonies in wood in the ground; none are mound builders. Most species produce swarming flights at night and the alates are attracted to lights. Several species are pests of structural wood. *S. intermedius* has several forms, and it may be a species complex.

***Schedorhinotermes actuosus*** Alates are present in colonies from November through May. Swarming is usually at night and there may be several flights from a colony during one season; flights are usually before or after rainfall. Natural nest sites include logs, trunks of trees, and hollow sections of eucalyptus, and in abandoned nests of other termite species. In the urban environment, this species is a pest of structural wood and buildings. This species occurs nearly throughout Australia, except Victoria.

***Schedorhinotermes intermedius*** Alates emerge from nests in the ground in November and December. Major soldiers are about 5.6 mm long, and minor soldiers are about 3.6 mm long. Colonies are usually large, and when only minor soldiers are found, it is not mature. Natural nest sites include trees, rotten logs, and tree stumps; in the urban environment, nests may be around building foundations and under patios. This species is distributed from New South Wales into southern Queensland, Australia. It is a pest of wooden buildings in the Sydney metropolitan area.

**Other *Schedorhinotermes*** Species in this genus are usually structural pests wherever they occur. *S. breinli*, *S. seclusus*, and *S. reticulatus* occur in Australia and the winged forms are present from January through May. They are found in trees, both hardwood and softwoods, but also attack structural timber in buildings. *S. malaccensis* occurs in Malaysia and parts of southern China. It nests in tree stumps and in the wounds of trees, including palms, but it also attacks woodwork in buildings. *S. medioobscurus* occasionally infests houses in urban and rural areas of Thailand.

### *Tenuirostritermes*

These termites are referred to as ant-like because of their rapid movement, dark color, and general behavior when the colony is disturbed. Soldiers in this genus have the front of their head produced into a snout (nasute). The head is longer than wide, and there is a constriction across the lower front, giving a distinct bilobed shape. The antennae are inserted ventrally on the head. The alate head is broader than long and the clypeus is large and bilobed. The colony contains small piles of vegetative material, primarily cuttings of weathered grass. Workers forage at night or on cloudy days, accompanied by an escort of soldiers.

***Tenuirostritermes cinereus* (Fig. 10.6d)** Alates are about 20 mm long with the wings, and the abdomen is brown and densely covered with pale yellow setae. The head of the alate is broader than long; the eyes and the ocelli are large; ocelli are less than half their diameter from the eyes. The head of the soldier is 1.3–1.4 mm long, primarily black; the snout is reddish brown. The antennae are 12-segmented. It is found in and under dry cattle dung (cow-chips) or under stones. Colonies are deep in the ground, and include workers, soldiers, and reproductives. Workers forage at night in columns and attack plants. Swarming flights are at night during rainfall in late summer (September). Damage has been to cedar posts. This species occurs in central and southwestern Texas, primarily in irrigated areas.

***Tenuirostritermes tenuirostris* (Fig. 10.2g, h)** Alates are 20–23 mm long with the wings; the abdomen is brown and densely covered with pale yellow setae. The head of the alate is broader than long; the eyes and the ocelli are large; ocelli are about half their diameter from the eyes. Wings are about three times longer than the abdomen, and often twisted at the tip; the wing membrane has minute, yellowish-brown setae. The head of the soldier is reddish brown, while the snout is blackish brown. The antennae are 13-segmented. Its habits are similar to those of *T. cinereus*. Swarming flights in Arizona occur from dusk to night during rainfall in June and July. This species occurs in Mexico, Texas, and in the mountains of Arizona.

## Termopsidae

This is a primitive family of dampwood termites. The alates lack ocelli and a fontanelle, and the pronotum is somewhat flat and not as broad as the head. The colonies are small and there is no worker caste. As in other species of lower termites, their intestines have flagellate protozoa. The genus *Stolotermes* occurs in eastern Australia, Tasmania, New Zealand, and Cape province in South Africa; the majority of species nest in decayed logs and stumps. *Porotermes adamsoni* occurs across southeastern Australia and Tasmania, *P. planiceps* is in Cape province of South Africa, and *P. quadricollis* occurs in Chile.

***Porotermes adamsoni*** Alates are present in the colony from December to February. Swarming flights occur in early evening, and all the alates appear to leave the nest at one time. Colonies frequently have primary and secondary reproductives active at the same time, especially when there are extensive gallery systems. This termite is primarily a pest of forest trees. It causes considerable damage to standing trees and it is found in down logs that are decaying. Attack is initiated where fungal decay is present in the standing tree, which may be in the vicinity of a wound or the weakened center of a hardwood tree. The economic importance of this pest is linked to infestations that extend into sound wood, along with the decay fungi. In the urban environment, this species will tunnel below ground to attack timber in contact with the ground, such as poles, fencing, and weatherboards and other structural wood in houses. It requires damp and decayed wood, and infestations are usually in buildings that have wood in contact with the ground, poor ventilation, or plumbing leaks.

## Bibliography

T. E. Snyder's annotated bibliography of literature pertaining to termites, extending from 1350 BC to AD 1965 (1956, *Smithsonian Inst. Misc. Collections* **130**; 1961, *Smithsonian Inst. Misc. Collections* **143**; 1968, *Smithsonian Inst. Misc. Collections* **152**) enabled scientists around the world to access the published scientific literature on this group of insects. E. Ernst of the Swiss Tropical Institute and R. L. Araujo of the São Paulo Museum of Zoology, in cooperation with the Tropical Development and Research Institute (London), continued this bibliographic work. They catalogued, abstracted, and annotated 3165 published documents on termites from 1966 to 1978 (Ernst and Araujo, 1986). Scientific publications on termites published from 1979 onwards are abstracted in the journal *Termite Abstracts*.


Ernst, E. and R. L. Araujo (eds.) *A Bibliography of Termite Literature 1966–1978*. Chichester: John Wiley, 1986.

Snyder, T. E. Annotated subject-heading bibliography of termites: 1350 BC to AD 1954. *Smithsonian Inst. Misc. Collections*, **130** (1956), 1–305.

Supplement to the annotated subject-heading bibliography of termites, 1955 to 1960. *Smithsonian Inst. Misc. Collections*, **143** (1961), 1–137.

Second supplement to the annotated subject-heading bibliography of termites 1961–1965. *Smithsonian Misc. Coll.*, **152** (1968), 1–88.


### General


Abe, T. *Termite Ecology*. (in Japanese) Tokyo: University of Tokyo Press, 1989.

Abe, T., D. E. Bignell, and M. Higashi. (eds.) *Termites: Evolution, Sociality, Symbioses, Ecology*. Dordrecht: Kluwer Academic, 2000.

Abushama, F. T. The role of chemical stimuli in the feeding behaviour of termites. *Proc. R. Entomol. Soc. Lond. A*, **42** (1967), 77–82.

Adamson, A. M. Termites and the fertility of soils. *Trop. Agr.*, **20** (1943), 107–12.

Ahmad, M. The phylogeny of termite genera based on the imago-worker mandibles. *Bull. Am. Mus. Nat. Hist.*, **95** (1950), 37–86.

Becker, G. and R. Mannesmann. Untersuchungen über das Verhalten von Termiten gegenüber einigen spurbildenen Stoffen. [Understanding the behavior of termites following a trail substance (ball-point pen substance).] *Z. Angew. Entomol.*, **62** (1968), 399–436.

Bodenheimer, F. S. Population problems of social insects. *Biol. Rev.*, **12** (1937), 393–430.

Breznak, J. A. and A. Brune. Role of microorganisms in the digestion of lignocellulose by termites. *Annu. Rev. Entomol.*, **39** (1994), 453–87.

Brian, M. V. *Social Insect Populations*. London: Academic Press, 1965.

(ed.) *Production Ecology of Ants and Termites*. London: Cambridge University Press, 1978.

*Social Insects: Ecology and Behavioral Biology*. London: Chapman and Hall, 1983.

Comparative aspects of caste differentiation in social insects. In Watson, J. A. L., B. M. Okot-Kotber, and C. Noirot (eds.) *Caste Differentiation in Social Insects.*, pp. 385–9. New York: Pergamon Press, 1985.

Buxton, R. D. Changes in the composition and activities of termite communities in relation to changing rainfall. *Oecologia*, **51** (1981), 371–8.

Chatterjee, P. N. and P. K. Sen-Sarma. Seasonal incidence of wood-destroying subterranean termites. *Indian Forest.*, **88** (1968), 139–41.

Constantino, R. The pest termites of South America: taxonomy, distribution, and status. *J. Appl. Entomol.*, **126** (2002), 355–65.

Crozier, R. H. and P. Pamilo. *Evolution of Social Insect Colonies*. New York: Oxford University Press, 1996.

Ebeling, W. *Termites: Identification, Biology, and Control of Termites Attacking Buildings*. University of California Agricultural Experimental Station Extended Service manual 38. Berkeley, CA: Agricultural Publications, 1968.

Eder, J. and H. Rembold (eds.) *Chemistry and Biology of Social Insects*. Munich: Verlag J. Peperny, 1987.

Emerson, A. E. The biogeography of termites. *Bull. Am. Mus. Nat. Hist.*, **99** (1952), 217–25.

Zoogeographical origins and dispersions of termite genera. *Fieldiana Zool.*, **37** (1955), 465–521.

Esenther, G. B., T. C. Allen, J. E. Cassida, and R. D. Schenefelt. Termite attractant from fungus-infected wood. *Science*, **134** (1961), 50.

Ettershank, G., J. A. Ettershank, and W. G. Whitford. Location of food sources by subterranean termites. *Environ. Entomol.*, **9** (1980), 645–8.

Fontes, L. R. Cupins em áreas urbanas. In Berti Filho, E. and L. R. Fontes (eds.) *Alguns Aspectos Atuais da Biologia e Controle de Cupins*, pp. 57–75, Piracicaba: FEALQ, 1995.

Forschler, B. T. and G. Henderson. Subterranean termite behavioral reaction to water and survival of inundation: implications for field populations. *Environ. Entomol.*, **24** (1995), 1592–7.

Fox, F. W. *Studies on the Chemical Composition of Foods Commonly Used in Southern Africa*. Johannesburg: South African Institute of Medical Research, 1996.

Gay, F. J. A world review of introduced species of termites. *Bull. Commonwealth Sci. Ind. Res. Organ. (CSIRO) Melbourne, Aust* **286** (1967), 1–88.

Harris, W. V. Termites attacking structural timber. *J. Inst. Wood Sci.*, **5** (1970), 9–15.

*Termites. Their Recognition and Control*, 2nd ed. London: Longman, 1971.

Haverty, M. I. The proportion of soldiers in termite colonies: a list and bibliography (Isoptera). *Sociobiology*, **2** (1977), 199–216.

Haverty, M. I. and W. L. Nutting. Natural wood consumption rates and survival of dry-wood and subterranean termite at constant temperatures. *Ann. Entomol. Soc. Am.*, **67** (1974), 153–7.

Hill, D. H., P. Hore, and I. W. B. Thornton. *Insects of Hong Kong*. Hong Kong: Hong Kong University Press, 1982.

Howard, R. W. and G. J. Blomquist. Chemical ecology and biochemistry of insect hydrocarbons. *Annu. Rev. Entomol.*, **27** (1982), 149–72.

Howse, P. E. *Termites: A Study in Social Behavior*. London: Hutchinson University Library, 1970.

Howse, P. E. and J.-L. Clément (eds.) *Biosystematics of Social Insects*. London: Academic Press, 1981.

Hungate, R. E. Studies on the nutrition of *Zootermopsis*. II. The relative importance of the termite and the protozoa in wood digestion. *Ecology*, **19** (1938), 1–25.

Kambhamputi, S., K. M. Kjer, and B. L. Thorne. Phylogenetic relationship among termite families based on DNA sequence of mitochondrial 16S ribosomal RNA gene. *Insect Mol. Biol.*, **5** (1996), 220–38.

King, E. J. and W. T. Spink. Laboratory studies on the biology of the Formosan subterranean termite with primary emphasis on young colony development. *Ann. Entomol. Soc. Am.*, **68** (1974), 355–8.

Kofoid, C. A. (ed.) *Termites and Termite Control*, 2nd edn. Berkeley, CA: University of California Press, 1946.

Climatic factors affecting the local occurrence of termites and their geographical distribution. In Kofoid, C. A. (ed.) *Termites and Termite Control*, 2nd edn. Berkeley, CA: University of California Press, 1946.

Koidzumi, M. Studies on the intestinal protozoa found in termites of Japan. *Parasitology*, **13** (1921), 235–309.

Krishna, K. A generic revision and phylogenetic study of the family Kalotermitidae (Isoptera). *Bull. Am. Mus. Nat. Hist.*, **122** (1961), 303–408.

LaFage, J. P., W. L. Nutting, and M. I. Haverty. Desert subterranean termites: a method for studying foraging behavior. *Environ. Entomol.*, **2** (1973), 954–6.

McKittrick, F. A. A contribution to the understanding of cockroach-termite affinities. *Ann. Entomol. Soc. Am.*, **58** (1965), 18–22.

Noirot, C. Les sexues de remplacement chez les termites superieurs. *Ann. Sci. Nat. (Zool.) Ser.*, **11** (1956), 400–595.

Remarques sur l'écologie des termites. *Ann. Soc. R. Zool. Belg.*, **89** (1959), 151–68.

Nutting, W. L. Termite flight periods: strategies for predator avoidance? *Sociobiology*, **4** (1979), 141–51.

Oster, G. and E. O. Wilson. *Caste and Ecology in Social Insects*. Princeton, NJ: Princeton University Press, 1978.

Parihar, D. R. Field observations on the nature and extent of damage by Indian desert termites and their control. *Ann. Arid Zone*, **17** (1978), 192–9.

Pearce, M. J. *Termites: Biology and Pest Management*. Wallingford: CAB International, 1997.

Prestwich, G. Defense mechanisms of termites. *Annu. Rev. Entomol.*, **29** (1984), 201–32.

Scheffrahn, R. H., N.-Y. Su, and J. R. Mangold. *Amitermes floridensis*, a new species and first record of a higher termite in the eastern United States (Isoptera, Termitidae, Termitinae). *Fla Entomol.*, **72** (1989), 618–25.

Smythe, R. V. and F. L. Carter. Feeding responses to sound wood by the Eastern subterranean termite, *Reticulitermes flavipes*. *Ann. Entomol. Soc. Am.*, **62** (1969), 335–7.

Snyder, T. E. Catalog of the termites (Isoptera) of the world. *Smithsonian Inst. Misc. Collect.*, **112** (1949), 1–490.

Spragg, W. T. and R. Paton. Tracing, trophallaxis and population measurement of colonies of subterranean termites using a radioactive tracer. *Ann. Entomol. Soc. Am.*, **73** (1980), 708–14.

Su, N.-Y. and R. H. Scheffrahn. Economically important termites in the United States and their control. *Sociobiology*, **17** (1990), 77–94.

Thorne, B. L. Termite terminology. *Sociobiology*, **28** (1996), 253–61.

Thorne, B. L. and M. Haverty. A review of intracolony, intraspecific, and interspecific agonism in termites. *Sociobiology*, **19** (1991), 115–54.

Thorne, B. L. and J. F. A. Traniello. Comparative social biology of basal taxa of ants and termites. *Annu. Rev. Entomol.*, **48** (2003), 283–306.

Traniello, J. F. A. and R. B. Rosengaus. Ecology, evolution and division of labour in social insects. *Anim. Behav.*, **53** (1996), 209–13.

Vander Meer, R. K., M. D. Breed, M. Winston, and K. Espelie (eds.) *Pheromone Communication in Social Insects: Ants, Wasps, Bees and Termites*. Colorado: Westview, 1998.

Vasconcellos, A. and A. G. Bandeira. Avaliação do consumo de madeira por espécies de *Nasutitermes* e *Microcerotermes* (Insecta, Isoptera, Termitidae). *Rev. Nord. Biol.*, **14** (2000), 17–24.

Veeresh, G. K., B. Mallik, and C. A. Viraktamath (eds.) *Social Insects and the Environment*. Oxford: IBH, 1990.

Wasserman, L. Termiten folgen Kugelschreiber-Strichen. [Termites follow ball-point pen lines. ] *Nat. Rundschau*, **20** (1967), 174.

Watson, J. A. L. and J. J. Sewell. The origin and evolution of the caste systems in termites. *Sociobiology*, **6** (1981), 101–18.

Watson, J. A. L., B. M. Okot-Kotber, and C. Noirot (eds.) *Caste Differentiation in Social Insects*. New York: Pergamon Press, 1985.

Williams, R. M. C. The ecology and physiology of structural wood-destroying Isoptera. *Mater. Org.*, **12** (1977), 111–40.

Wilson, E. O. *Insect Societies*. Cambridge, MA: Belknap Press, 1971.

Wood, T. J. and W. A. Sands. The role of termites in ecosystems. In Brian, M. V. (ed.) *Production Ecology of Ants and Termites*, pp. 245–325. Cambridge: Cambridge University Press, 1978.

Yamin, M. A. Flagellates of the orders Trichomonadida Kirby, Oxymonadida Grasse, and Hypermastigida Grassi & Foa reported from lower termites (Isoptera: families Mastotermitidae, Kalotermitidae, Hodotermitidae, Termopsidae, Rhinotermitidae,and Serritermitidae) and from the wood-eating roach *Cryptocercus* (Dictyoptera: Cryptocercidae). *Sociobiology*, **4** (1979), 1–119.

#### Geographic distribution

Adamson, A. M. Preliminary report on termites and termite damage in Trinidad, West Indies. *Trop. Agr. (Trinidad)*, **14** (1937), 141–9.

Notes on the termite fauna of the Lesser Antilles. *Trop. Agr. (Trinidad)*, **25** (1948), 53–5.

Ahmad, M. Termites of West Pakistan. *Biologia (Lahore)*, **1** (1955), 202–64.

Termites (Isoptera) of Thailand. *Bull. Am. Mus. Nat. Hist.*, **131** (1965), 1–113.

Akhtar, M. S. Zoogeography of the termites of Pakistan. *Pakistan J. Zool.*, **6** (1974), 85–104.

Taxonomy and zoogeography of the termites (Isoptera) of Bangladesh. *Bull. Dept. Zool. Univ. Punjab (New Series)* (1975).

Araujo, R. L. Contribuição á biologeografia dos térmitas de São Paulo, Brasil (Insecta, Isoptera). *Arq. Inst. Biol.*, (São Paulo), **25** (1958), 185–217.

Contribuição á biologeografia dos térmitas de Minas Gerais, Brasil (Insecta, Isoptera). *Arq. Inst. Biol. (São Paulo)*, **25** (1958), 219–236.

Assmuth, J. Indian wood-destroying white ants. *J. Bombay Nat. Hist. Soc.*, **23** (1915), 690–4.

Banks, N. The termites of Panama and British Guiana. *Bull. Am. Mus. Nat. Hist.*, **38** (1918), 659–67.

Banks, N. and T. E. Snyder. A revision of the Nearctic termites. *US Natl. Mus. Bull.*, **108** (1920), 1–228.

Becker, G. Einige Beobachtungen über hozzerstörende Insekten (Termiten und Kafer) in Guatemala. *Z. Angew. Entomol.*, **35** (1953), 330–73.

On termites in Central and South America (with Spanish summary). *Rev. Floresta*, **9** (1978), 71–5.

Bodot, P. Etude écologique des Termites savenes de Basse Côte d'Ivoire. *Insectes Soc.*, **14** (1967), 229–58.

Cachan, P. Les termites de Madagascar. *Mem. Inst. Sci. Madagascar A3* (1949), 177–275.

Cachan, P. Les termites de Madagascar. *Mem. Inst. Sci. Madagascar A5* (suppl. 1) (1951), 1–18.

Cai, B.-H. and N. S. Chen. *Economic Insect Fauna of China* (in Chinese). Fascicle 8. *Isoptera*. Beijing: Science Press, 1964.

Calaby, J. H. and F. J. Gay. Aspects of the distribution and ecology of Australian termites. *Monogr. Biol.*, **8** (1959), 211–23.

Chaudhry, G. H. and M. Ahmad. *Termites of Pakistan (Identity, Distribution and Ecological Relationships)*. Peshawar, India: Peshawar Forest Institute, 1969.

Clark, A. F. Termites in New Zealand. *N. Z. J. Forestry*, **4** (1938), 177–179.

Coaton, W. G. H. The hodotermitid harvester termites of South Africa. *Dept. Agr. Sci. Bull. S. Afr. Entomol. Ser.* 375, **43** (1958), 1–112.

Survey of the termites of the Kruger National Park. *Koedoe*, **5** (1962), 144–69.

Constantino, R. Catalog of the living termites of the New World (Insecta: Isoptera). *Arq. Zool.*, **35** (1998), 135–231.

Constantino, R. and E. C. Dianese. The urban termite fauna of Brasilia, Brazil (Isoptera). *Sociobiology*, **38** (2001), 323–6.

de Costa Lima, A. M. Insetos do Brasil. XV. Isoptera. *Compo (Rio de Janeiro)*, **7** (1936) (83), 8–17; (84): 10–17.

Creffeld, J. W. *Wood Destroying Insects. Wood Borers and Termites*. East Melbourne, Australia: CSIRO, 1991.

Dhanarajan, G. The termite fauna of Malaysia and its economic significance. *Malay Forest*, **17** (1969), 276–8.

Emerson, A. E. A review of the Mastotermitidae (Isoptera), including a new fossil genus from Brazil. *Am. Mus. Novitates*, **2236** (1965), 1–46.

Esenther, G. R. Termites in Wisconsin. *Ann. Entomol. Soc. Am.*, **62** (1969), 1274–84.

Faragalla, A. A. Termite problems in Saudi Arabian ecosystem. *Sociobiology*, **8** (1983), 119–25.

Fuller, C. The termites of South Africa. *S. Afr. J. Nat. Hist.*, **3** (1922), 70–131.

Gao, D.-R. and P. K. S. Lam. Notes on the termites (Isoptera) of Hong Kong, including description of a new species and checklist of Chinese species. *Mem. Hong Kong Nat. Hist. Soc.*, **17** (1986), 67–83.

Gao, D.-R., B.-Z Zhu, and X. Wang. Survey of termites in the regions of Jiangsu Province with description of two new species (in Chinese, English abstract). *Zool. Res.*, **3** (suppl.) (1982), 137–144.

Garcia, M. L. Philippine wood-destroying termites. *Forpride Digest*, **1** (1972), 40–1.

Gay, F. J. Isoptera. In *Insects of Australia*. Melbourne: Melbourne University Press, 1970.

The Termitinae (Isoptera) of temperate Australia. *Aust. J. Zool.* (Suppl. Ser. 3) (1971), 36.

Goetsch, W. Die chilenischen Termiten. *Zool. Jahrb. Abt. I. Systematik (Oekol.) Geograph. Biol.*, **64** (1933), 225–43.

Grassé, P.-P. Les termites en Afrique Occidentale francaise. Leur importance économique. Les moyens de lutte. *Rev. Pathol. Vegetale Entomol. Agric. France*, **23** (1936), 265–306.

Gusman, R. Termes citados para Chile. *Museo Nacl. Hist. Nat. Chile, Publ. Ocasion.*, **10** (1966), 1–11.

Hadington, P. *Australian Termites and other Common Timber Pests*. NSW, Australia: New South Wales University Press, 1987.

Hariri, G. The Economic Insects of Syria and Neighboring Countries (in Persian). Aleppo, Syria: University of Aleppo, 1972.

Harris, W. V. An introduction to Malayan termites. *Malay. Nature J.*, **12** (1957), 20–32.

*Isoptera*. British Museum (Natural History) Report Expedition South-West Arabia 1937–1938, No. **28** (1957), 421–433.

Termites in Europe. *New Scientist*, **13** (1962), 614–17.

The termites of Hong Kong. *Mem. Hong Kong Nat. Hist. Soc.*, **6** (1963), 1–9.

Termites from western Congo. *Rev. Zool. Botan. Afr.*, **71** (1965), 10–18.

Isoptera from Vietnam, Cambodia and Thailand. *Oposcula Entomol.*, **33** (1968), 143–54.

Huang, F.-S., G.-X. Li, and M. Zhu. *The Taxonomy and Biology of Chinese Termites – Isoptera* (in Chinese). Tianze Press, 1989.

Huang, F.-S., S.-M. Zhu, Z.-M. Ping *et al. Fauna Sinica. Insecta*, vol. 17. *Isoptera* (in Chinese). Beijing: Science Press, 2000.

Johnson, R. A. and T. G. Wood. Termites of the arid zones of Africa and the Asian Peninsula. *Sociobiology*, **5** (1980), 279–93.

Kemp, P. B. The termites of north-eastern Tanganyika: their distribution and biology. *Bull. Entomol. Res.*, **46** (1955), 113–36.

Krishna, K. Termites (Isoptera) of Burma. *Am. Mus. Novitates*, **2210** (1965), 1–34.

Li, G.-X., Z.-R. Dai, and D. Li. Chinese Termites and Control (in Chinese). Beijing: Science Press, 1989.

Li, G.-X., Z.-R. Dai, and B. Yang. Introduction to termite research in China. *J. Appl. Entomol.*, **117** (1994), 360–9.

Light, S. F. Notes on Philippine termites. III. *Philippine J. Sci.*, **40** (1929), 421–52.

Notes on Philippine termites. IV. *Philippine J. Sci.*, **42** (1930), 13–58.

The California species of the genus *Amitermes* Silvestri. *Univ. Calif. Pub. Entomol.*, **5** (1930), 173–215.

Liu, Y.-Z., Y. Jiang, X.-Y. Su *et al. Biology and Control of Termites in China* (in Chinese). Chengdu, China: Publishing House of Chengdu Science and Technology University 1998.

Mill, A. E. Populaçes de termites (Insecta: Isoptera) em quarto habitats no baixo Rio Negro. *Acta Amazônica*, **12** (1982), 53–60.

Generic keys to the soldier caste of New World Termitidae (Isoptera: Insecta). *System. Entomol.*, **8** (1983), 179–90.

Termites as structural pests in Amazônia, Brazil. *Sociobiology*, **19** (1991), 339–49.

Miller, D. The species of termites in New Zealand. *N. Z. J. Forestry*, **4** (1941), 333–4.

Morimoto, K. Biology of termites in the far east. *Rev. Plant Protect. Res.*, **8** (1975), 29–40.

Moutia, A. Termites in Mauritius. *Bull. Dept. Agric. Mauritius Sci. Ser.*, **21** (1936), 1–30.

Pangga, G. A. A biological study of some common Philippine termites. *Philippine Agric.*, **25** (1936), 233–65.

Perry, D. H., J. A. L. Watson, S. E. Bunn, and R. Black. Guide to the termites (Isoptera) from the extreme south-west of Western Australia. *J. R. Soc. Western Aust.*, **67** (1985), 66–78.

Peters, B. C. *Drywood Termites in Queensland*. Queensland Department of Forestry timber note no. 24.

Ratcliffe, F. M., F. J. Gay, and T. Greaves. *Australian Termites*. Melbourne: CSIRO, 1952.

Roonwal, M. L. and O. B. Chhotani. Indian-wood destroying termites. *J. Bombay Nat. Hist. Soc.*, **63** (1967), 354–64.

Roonwal, M. L. and P. K. Maiti. Termites from Indonesia including West Irian. *Treubia*, **27** (1966), 63–140.

Sands, W. A. Termite distribution in man-modified habitats in West Africa, with special reference to species segregation in the genus *Trinervitermes* (Isoptera, Termitidae, Nasutitermitinae). *J. Anim. Ecol.*, **34** (1965), 557–71.

*Termites in Saudi Arabia*. Report (COPR) to Ministry of Agriculture and Water, Saudi Arabia. 1971.

Scheffrahn, R. H. and N.-Y. Su. Keys to soldier and winged adult termites (Isoptera) of Florida. *Fla Entomol.*, **77** (1994), 460–74.

Scheffrahn, R. H., J. R. Mangold, and N.-Y. Su. A survey of structure-infesting termites of peninsular Florida. *Fla Entomol.*, **71** (1988), 615–30.

Sen-Sarma, P. K. Ecology and biogeography of termites of India. In Mani, M. S. (ed.) *Ecology and Biogeography of India*. The Hague: Junk. 1974.

Sewell, J. J. and J. A. L. Watson. Developmental pathways in Australian species of *Kalotermes hagen* (Isoptera). *Sociobiology*, **6** (1981), 243–322.

Snyder, T. E. Termites of West Indies, the Bahamas and Bermuda. *J. Agric. Univ. Puerto Rico*, **40** (1956), 189–202.

New termites from Venezuela, with keys and a list of the described species. *Am. Midl. Nat.*, **61** (1959), 313–21.

Swett, T. Y. and C. M. Ping. Studies on the faunal regions of Isoptera in China. *Acta Entomol. Sinica*, **13** (1964), 10–24.

Tang, C. and S. Li. Termites in Hangzhou (1) (in Chinese). *Entomol. Knowl.*, **9** (1959), 230–77.

Torales, G. J., E. R. Laffont, M. O. Arbino, and M. C. Gordy. Primeira lista faunistica de los Isopteros de la Argentina. *Rev. Soc. Entomol. Argentina*, **56** (1997), 47–53.

Vasconcellos, A., A. G. Bandeira, C. S. Miranda, and M. P. Silva. Termite (Isoptera) pests in buildings in João Pessoa, Brazil. *Sociobiology*, **40** (2002), 639–44.

Watson, J. A. L. and H. A. Abbey. *Atlas of Australian Termites*. Canberra, Australia: CSIRO Division of Entomology, 1993.

Weber, N. A. The insect fauna of an Iraq oasis, the city of Baghdad. *Entomol. News*, **65** (1954), 181–2.

Weidner, H. Termiten (Isoptera) aus dem Irak. *Entomol. Mitt. Zool. St. Inst. Zool. Mus. Hamburg*, **17** (1958), 4–12.

Die Termiten von Afghanistan, Iran and Irak (Isoptera) (Contribution a l'étude de la fauna d'Afghanistan 29). *Abhandl. Ver. Naturw. Ver. Hamburg (N. F.)*, **4** (1960), 43–70.

Weesner, F. M. *The Termites of the United States. A Handbook.* Elizabeth, NJ: National Pest Control Association, 1965.

Termites of the Nearctic region. In Weesner, F. M. and K. Krishna (eds.) *Biology of Termites*, pp. 477–525. New York: Academic Press, 1970.

Yaman, I. K. A. Insect pests of Saudi Arabia. *Z. Angew. Entomol.*, **58** (1966), 266–78.

Yule, R. A. and F. R. Wylie. 1987. *Subterranean Termites in Queensland.* Queensland Department of Forestry timber note no. 9.

### Coptotermes

Abe, T. On the distribution of the oriental termite *Coptotermes formosanus* Shiraki in Japan. *Sci. Rpt. Tohoku Univ. Ser.*, **4** (1937), 463–72.

Brown, W. V., J. A. L. Watson, F. L. Carter *et al.* The cuticular hydrocarbons of workers of three Australian *Coptotermes* species, *C. michaelseni*, *C. brunneus*, and *C. dreghorni* (Isoptera: Rhinotermitidae). *Sociobiology*, **23** (1994), 277–91.

Calaby, J. H. and F. J. Gay. The distribution and biology of the genus *Coptotermes* (Isoptera) in Western Australia. *Aust. J. Zool.*, **4** (1956), 19–39.

Chen, B.-Y., B.-Y. Ding, K.-H. Wang, and A.-S. Wang. A study on reproductive characteristics and proportion of various castes of *Coptotermes communis* Xia et He (in Chinese, English abstract). *Sci. Technol. Termites*, **6** (1989), 1–6.

Costa-Leonardo, A. M. and R. C. Barsotti. Swarming and incipient colonies of *Coptotermes havilandi*. *Sociobiology*, **31** (1998), 131–42.

Dai, Z.-R., X.-Y. Xie, and Z.-Y. Huang. Effect of worker : soldier ratio on feeding activities and *Coptotermes formosanus* Shiraki survival in laboratory kept colonies. *Insect Knowl.*, **17** (1985), 74–6.

Gay, F. J. and A. L. Watson. The genus *Cryptotermes* in Australia. *Aust. J. Zool.* (Suppl. Ser. 88) (1982), 64.

Greaves, T. Studies of foraging galleries and the invasion of living trees by *Coptotermes acinaciformis* and *C. brunneus* (Isoptera). *Aust. J. Zool.*, **10** (1962), 630–51.

Haverty, M. I., L. J. Nelson, and M. Page. Cuticular hydrocarbons of four populations of *Coptotermes formosanus* Shiraki in the United States: similarities and origins of introductions. *J. Chem. Ecol.*, **16** (1990), 1635–47.

Heather, N. W. The exotic drywood termite *Cryptotermes brevis* (Walker) and endemic Australian drywood termites in Queensland. *J. Aust. Entomol. Soc.*, **10** (1971), 134–41.

Hedlund, J. C. and G. Henderson. Effect of available food size on search tunnel formation by the Formosan subterranean termite (Isoptera: Rhinotermitidae). *J. Econ. Entomol.*, **92** (1999), 610–16.

Henderson, G. Alate production, flight phenology, and sex-ratio in *Coptotermes formosanus* Shiraki, an introduced subterranean termite in New Orleans, Louisiana. *Sociobiology*, **28** (1996), 319–26.

Henderson, G. and K. S. Delaplane. Formosan subterranean termite swarming behavior and alate sex-ratio (Isoptera: Rhinotermitidae). *Insectes Soc.*, **41** (1994), 19–28.

Huang, L.-W and L.-L. Chen. Biology and colony development of *Coptotermes formosanus* Shiraki (in Chinese, English abstract). *Acta Entomol. Sinica*, **27** (1984), 64–9.

Kalshoven, L. G. E. Biological notes on the *Coptotermes* species of Indonesia. *Acta Trop.*, **17** (1960), 263–72.

*Coptotermes curvignathus* as a cause of trouble in electric systems. *Symp. Genet. Biol. Ital.*, **11** (1962), 223–9.

Lai, P. Y., M. Yamashiro, J. R. Yates *et al.* Living plants in Hawaii attacked by *Coptotermes formosanus*. *Proc. Hawaii Entomol. Soc.*, **24** (1983), 283–6.

Lai, P.-Y., M. Tamashiro, and J. K. Fujii. Abundance and distribution of the three species of protozoa in the hindgut of *Coptotermes formosanus* Shiraki (Isoptera: Rhinotermitidae). *Proc. Hawaiian Entomol. Soc.*, **24** (1983), 271–6.

Lee, K. E. and T. G. Wood. *Termites and Soil.* New York: Academic Press, 1971.

Lenz, M. and R. A. Barret. Neotenic formation in field colonies of *Coptotermes lacteus* (Froggatt) in Australia, with comments on the roles of neotenics in the genus *Coptotermes* (Isoptera: Rhinotermitidae). *Sociobiology*, **11** (1982), 237–44.

Lenz, M., R. A. Barret, and L. R. Miller. Mechanisms of colony re-establishment after orphaning in *Coptotermes lacteus* (Froggatt) (Isoptera: Rhinotermitidae). *Sociobiology*, **14** (1988), 245–68.

Li, S.-Q. Formosan subterranean termite and its control in China (in Chinese). *Sci. Technol. Termites*, **3** (1986), 1–8.

Mannesmann, R. Comparison of 21 commercial wood species from North America in relation to feeding rates of the Formosan termite, *Coptotermes formosanus* Shiraki. *Mater. Org.*, **8** (1973), 107–20.

McMahan, E. A. Food transmission within the *Cryptotermes brevis* colony (Isoptera, Kalotermitidae). *Ann. Entomol. Soc. Am.*, **59** (1966), 1131–7.

Roonwal, M. L. Biology and ecology of oriental termites (Isoptera). No. 4. The drywood termite *Coptotermes heimi* (Wasm.) in India. *J. Bombay Nat. Hist. Soc.*, **56** (1959), 511–23.

Smythe, R. V. and F. L. Carter. Feeding responses to sound wood by *Coptotermes formosanus*, *Reticulitermes flavipes*, and *R. virginicus* (Isoptera: Rhinotermitidae). *Ann. Entomol. Soc. Am.*, **63** (1970), 841–846.

Sornnuwat, Y., K. Tsunoda, T. Yoshimura, M. Takahashi, C. Vongkaluang. Foraging populations of *Coptotermes gestroi* (Isoptera: Rhinotermitidae) in an urban area. *J. Econ. Entomol.*, **89** (1996), 1485–90.

Su, M.-Y. and J. P. LaFage. Comparison of laboratory methods for estimating wood consumption rates by *Coptotermes formosanus* (Isoptera: Rhinotermitidae). *Ann. Entomol. Soc. Am.*, **77** (1984), 125–9.

Su, N.-Y. and R. H. Scheffrahn. Alate production of a field colony of the Formosan subterranean termite (Isoptera: Rhinotermitidae). *Sociobiology*, **13** (1987), 167–72.

Foraging population and territory of the Formosan subterranean termite (Isoptera: Rhinotermitidae) in an urban environment. *Sociobiology*, **14** (1988), 353–359.

Su, N.-Y., M. Tamashiro, J. R. Yates, and M. I. Haverty. 1984. Foraging behavior of the Formosan subterranean termite (Isoptera: Rhinotermitidae). *Environ. Entomol.*, **13** (1984), 1466–70.

Tamashiro, M. and N.-Y. Su (eds.) *Biology and Control of the Formosan Subterranean Termite*. University of Hawaii, Honolulu: College of Tropical Agriculture and Human Resources, 1987.

Thompson, G. J., M. Lenz, and R. H. Crozier. Microsatellites in the subterranean, mound building termite *Coptotermes lacteus* (Isoptera: Rhinotermitidae). *Mol. Ecol.*, **9** (2000), 1932–4.

Wang, J. and K. Grace. Current status of *Coptotermes* Wasmann (Isoptera: Rhinotermitidae) in China, Japan, Australia and the American Pacific. *Sociobiology*, **33** (1999), 295–305.

Wang, C., J. E. Powell, and Y. Liu. A literature review of the biology and ecology of *Coptotermes formosanus* (Isoptera: Rhinotermitidae) in China. *Sociobiology*, **40** (2002), 343–63.

Wilkinson, W. Dispersal of alates and establishment of new colonies of *Cryptotermes havilandi* (Isoptera, Kalotermitidae). *Bull. Entomol. Res.*, **53** (1962), 265–86.

Williams, R. M. C. Factors limiting the distribution of building-damaging dry-wood termites (Isoptera: *Cryptotermes* spp.) in Africa. *Mater. Organ.*, (Suppl. 3) (1976), 393–406.

Yoshimura, T., T. Watanabe, K. Tsunoda, and M. Takahashi. Distribution of the symbiotic Protozoa in the hindgut of *Coptotermes formosanus* Shiraki (Isoptera: Rhinotermitidae). *Jpn. J. Environ. Entomol. Zool.*, **4** (1992), 116–19.

Yule, R. A. and A. L. Watson. Two further domestic species of *Cryptotermes* from the Australian mainland. *J. Aust. Entomol. Soc.*, **15** (1976), 349–51.

### *Mastotermes*

Goodisman, M. A. D. and R. H. Crozier. Colony and population genetic structure of the primitive termite, *Mastotermes darwiniensis*. *Evolution*, **56** (2002), 70–83

Watson, J. A. L. The development of 'workers' and reproductives in *Mastotermes darwiniensis* Froggatt (Isoptera). *Insectes Soc.*, **18** (1971), 173–6.

The development of soldiers in incipient colonies of *Mastotermes darwiniensis* Froggatt (Isoptera). *Insectes Soc.*, **21** (1974), 181–90.

Watson, J. A. L., R. A. Barrett, and J. M. Abbey. Caste ratios in a long-established, neotenic-headed laboratory colony of *Mastotermes darwiniensis* Froggatt (Isoptera). *J. Aust. Entomol. Soc.*, **16** (1977), 469–70.

### *Reticulitermes*

Aber, A. and L. R. Fontes. *Reticulitermes lucifugus* (Isoptera, Rhinotermitidae) a pest of wooden structures, is introduced into South American continent. *Sociobiology*, **21** (1993), 335–9.

Bagnères, A.-G, A. Killian, J.-L. Clément, and C. Lang. Interspecific recognition among termites of the genus *Reticulitermes*: evidence for a role for the cuticular hydrocarbons. *J. Chem. Ecol.*, **17** (1991), 2397–420.

Becker, G. *Reticulitermes* in Mittel und West Europa. *Z. Angew. Entomol.*, **65** (1970), 268–78.

Bulmer, M. S., E. S. Adams, and J. F. A. Traniello. Variation in colony structure in the subterranean termite *Reticulitermes flavipes*. *Behav. Ecol. Sociobiol.*, **49** (2001), 236–43.

Clément, J.-L. Ecologie des *Reticulitermes* français: position systématique des population. *Bull. Soc. Zool. Fr.*, **102** (1977), 169–85.

Spéciation des *Reticulitermes* européens, espéces et méchanimes d'isolements. *C.R. Soc. Biogéog.*, **58** (1982), 145–58.

Open and closed societies in termites of the genus *Reticulitermes*, geographic variations and seasonality. *Sociobiology*, **11** (1986), 311–23.

Clément, J.-L., A.-G. Bagneres, P. Uva *et al.* Biosystematics of *Reticulitermes* termites in Europe: morphological, chemical and molecular data. *Insectes Soc.*, **48** (2001), 202–15.

Dronnet, S., M. Ohresser, E. L. Vargo *et al.* Colony studies of the subterranean termite *Reticulitermes santonensis* (Isoptera: Rhinotermitidae), in the city of Paris. In Jones, S. C., J. Zhai, and W. Robinson (eds.) *Proceedings of the 4th International Conference on Urban Pests*, pp. 295–301. Blacksburg, VA: Pocahontas Press, 2002.

Evans, T. A., M. Lenz, and P. V. Gleeson. Estimating population and forager movement in a tropical subterranean termite (Isoptera: Rhinotermitidae). *Environ. Entomol.*, **28** (1999), 823–30.

Forschler, B. T. and T. M. Jenkins. Evaluation of subterranean termite biology using genetic, chemotaxonomic, and morphometric markers and ecological data: a testimonial for multi-disciplinary efforts. *Trends Entomol.*, **2** (1999), 71–80.

Subterranean termites in the urban landscape: understanding their social structure is the key to successfully implementing population management using bait technology. *Urban Ecosystems*, **4** (2000), 231–41.

Grace, J. K. Mark–recapture studies with *Reticulitermes flavipes* (Isoptera: Rhinotermitidae). *Sociobiology*, **16** (1990), 297–303.

Grace, J. K., A. Abdallay, and K. R. Farr. Eastern subterranean termite (Isoptera: Rhinotermitidae) foraging territories and populations in Toronto. *Can. Entomol.*, **121** (1989), 551–6.

Haverty, M. I., B. T. Forschler, and L. J. Nelson. An assessment of the taxonomy of *Reticulitermes* (Isoptera: Rhinotermitidae) from the southeastern United States based on cuticular hydrocarbons. *Sociobiology*, **28** (1966), 287–318.

Haverty, M. I., C. A. McDonald, and G. J. Blomquist. Cuticular hydrocarbons of the eastern subterranean termite, *Reticulitermes flavipes* (Isoptera: Rhinotermitidae). *J. Chem. Ecol.*, **4** (1978), 233–45.

Haverty, M. I., G. M. Getty, K. A. Copren, and V. R. Lewis. Size and dispersion of colonies of *Reticulitermes* spp. (Isoptera: Rhinotermitidae) in a wildland and a residential location in northern California. *Environ. Entomol.*, **29** (2000), 241–9.

Holway, R. T. Tube-building habits of the eastern subterranean termite. *J. Econ. Entomol.*, **34** (1941), 389–94.

Howard, R. and M. I. Haverty. Reproductives in mature colonies of *Reticulitermes flavipes*: abundance, sex-ratio, and association with soldiers. *Environ. Entomol.*, **9** (1980), 458–60.

Seasonal variation in caste proportions of field colonies of *Reticulitermes flavipes*. *Environ. Entomol.*, **10** (1981), 546–549.

Howard, R. W., S. C. Jones, J. K. Mauldin, and R. H. Beal. Abundance, distribution, and colony size estimates for *Reticulitermes* spp. (Isoptera: Rhinotermitidae) in southern Mississippi. *Environ. Entomol.*, **11** (1982), 1290–3.

Howard, R. W., C. A. McDaniel, D. R. Nelson *et al.* Cuticular hydrocarbons of *Reticulitermes virginicus* (Banks) and their role as potential species- and caste-recognition cues. *J. Chem. Ecol.*, **8** (1982), 1227–39.

Husby, W. D. Biological studies on *Reticulitermes flavipes* (Kollar) subterranean termite populations (Isoptera: Rhinotermitidae). *Pan-Pac. Entomol.*, **65** (1980), 381–4.

Jenkins, T., M. Haverty, C. J. Basten *et al.* Mitochondrial gene sequence questions *Reticulitermes* sp. social structure (Isoptera: Rhinotermitidae). *Sociobiology*, **34** (1999), 161–72.

Jenkins, T., M. Haverty, C. J. Basten, M. Page, and B. T. Forschler. Correlation of mitochondrial haplotypes with cuticular hydrocarbon phenotypes of sympatric *Reticulitermes* species from the Southeastern United States. *J. Chem. Ecol.*, **26** (2000), 1525–42.

Jenkins, T. M., M. L. Newman, and B. T. Forschler. Subterranean termite movements and relationships over time: a genetic characterization. In Jones, S. C., J. Zhai, and W. Robinson (eds.) *Proceedings of the 4th International Conference on Urban Pests*, pp. 95–102. Blacksburg, VA: Pocahontas Press, 2002.

Matsumura, F., A. Tai, and H. C. Coppel. Termite trail-following substance, isolation and purification from *Reticulitermes virginicus* and fungus-infected wood. *J. Econ. Entomol.*, **62** (1969), 599–603.

Pawson, B. M. and R. E. Gold. Caste differentiation and reproductive dynamics of three subterranean termites in the genus *Reticulitermes* (Isoptera: Rhinotermitidae). *Sociobiology*, **28** (1996), 241–51.

Snyder, T. E. Northward spread of *Reticulitermes* species in United States and Europe. *Proc. Entomol. Soc. Wash.*, **61** (1959), 40.

Su, N.-Y., P. M. Ban, and R. H. Scheffrahn. Foraging populations and territories of the eastern subterranean termite (Isoptera: Rhinotermitidae) in southeastern Florida. *Environ. Entomol.*, **22** (1993), 1113–17.

Tang, C. and S. Li. Forecasting the swarming of the yellow-thorax termite, *Reticulitermes flaviceps* Osh. in Hangchow. *Acta Entomol. Sinica*, **9** (1959), 477–82.

Thorne, B. L., N. L. Breish, and J. F. A. Traniello. Incipient colony development in the subterranean termite *Reticulitermes flavipes* (Isoptera: Rhinotermitidae). *Sociobiology*, **30** (1997), 145–59.

Thorne, B. L., J. F. A. Traniello, E. S. Adams, and M. S. Bulmer. Reproductive dynamics and colony structure of subterranean termites of the genus *Reticulitermes* (Isoptera: Rhinotermitidae): a review of the evidence from behavioral, ecological, and genetic studies. *Ethol. Ecol. Evolution*, **11** (1999), 149–69.

Tsunoda, K., H. Matsuoka, T. Yoshimura, and M. Tokoro. Foraging populations and territories of *Reticulitermes speratus* (Isoptera: Rhinotermitidae). *J. Econ. Entomol.*, **92** (1999), 604–9.

Vieau, F. Biologie comparée de *Reticulitermes santonensis* Feytaud et *Reticulitermes lucifugus* Rossi (Isoptera, Rhinotermitidae) en France: différences morphologiques entre les soldats, modes d'implantation urbaine et forestière, cycles reproductiers. *Actes Coll. Insectes Soc.*, **12** (1999), 151–8.

Comparison of the spatial distribution and reproductive cycle of *Reticulitermes santonensis* Feytaud and *Reticulitermes lucifugus grassei* Clément (Isoptera: Rhinotermitidae) suggests that they represent introduced and native species, respectively. *Insectes Soc.*, **48** (2001), 57–62.

Weidner, H. Weitere Mitteilungen über das Auftreten der Termite *Reticulitermes flavipes* in Hamburg. *Z. Hyg. Zool. Schädlingsbekämpf*, **39** (1951), 259–65.

Weesner, F. M. The biology of colony foundation in *Reticulitermes hesperus* Banks. *Univ. Calif. (Berekely) Pub. Zool.*, **61** (1956), 253–314.

# 11 Lepidoptera

## Introduction

Adults and caterpillars of butterflies, moths, and skippers are found in nearly all environments and are well known to everyone. Adults are characterized by having their wings and other parts of their body covered with a layer of short, flattened setae, or scales. The two pairs of wings are usually broad, subtriangular, and with the front pair larger. The mouthparts, when present, are a coiled proboscis or tube for siphoning liquid. The noctuid *Calyptra eustrigata* has a strong proboscis, which enables it to pierce the skin of mammals and suck blood. Moth antennae are usually thread-like or feather-like, while butterfly antennae are thread-like and clubbed at the tip. Butterflies fold their wings vertically above the body when at rest, while moths hold their wings rooflike or close around the body. They undergo complete metamorphosis.

Caterpillars are usually cylindrical, and besides the head there are three thoracic segments, and 10 abdominal segments. Thoracic segments bear a jointed leg, which terminates in a single claw. Abdominal segments bear unjointed, fleshy projections or prolegs. Typically there is one pair on segments 3–6, and 10, but some or all of the prolegs may be absent. Caterpillars have silk glands that open at the mouth; they use silk to make feeding shelters and to protect the pupal stage. Pupae are usually encased in a protective structure formed by the last caterpillar stage. This may be a silken cocoon, or detritus held together by silk secretions, a cell in the food substrate, a cavity in wood or other material adjacent to a food source, or an exposed chrysalid hanging from a tree branch. Lepidopterans feed primarily on flowering plants as caterpillars and adults. Most adults are capable of obtaining only fluid from flowers and water from pools; some are inactive and may not feed. Exceptions include geometrid moths in the genus *Eupithecia*, which are predaceous, clothes moth caterpillars which feed on wool, and adults of the noctuids *Loboscraspis griseifusa* and *Arcyophora sylvatica*, which scrape the skin of vertebrates to obtain blood.

Pest status for members of this group is based primarily on the damage they do to household stored foods and materials. Several species have adapted partly or completely to indoor habitats and to the food and fabric stored there. Most of them have been distributed around the world with commercial shipment of food and materials. In peridomestic habitats there are several species with urticating (poisonous or stinging) setae and they are capable of causing a skin rash or other physiological reaction in sensitive people. The few species that are known to feed as caterpillars in the wood of trees, such as the carpenterworms (Cossidae), are only minor pests of wood in use.

## Arctiidae

These moths have broad wings and large bodies. Many species are marked with brightly colored stripes and spots, which gives them the common name of tiger moths. All are night flyers and are generally attracted to outdoor lights. Eggs are laid on leaves and small branches of trees and are often covered with long setae from the ovipositing female. Caterpillars of most species are covered with dense clusters of setae and fine hairs, and usually there are some clusters of very long setae at the anterior and posterior end. The hairs of certain species cause skin irritation. Caterpillars feed on leaves of trees and shrubs, and are common on ornamental trees. Overwintering is usually in a cocoon covered with long hairs and located in ground litter, or protected places along walls or building perimeters. For most species there is one generation per year.

***Elima fuscodorsalis*** Adult wing span is 9–11 mm, the body and wings are light brown, and the front wings have two black spots at the anterior edge. The full-grown caterpillar is 20 mm long and slightly flattened. The head is black and the body is gray brown with green irregular markings. Full-grown caterpillars usually leave the tree in which they are feeding and search for a place to pupate, and during their movement they sometimes enter houses from under eaves. Dermatitis and skin lesions are caused by exposure to the caterpillars of this species. Caterpillar dermatitis occurs from April to October. Overwintering is in the pupal stage. There are 1–3 generations per year. This species is distributed in the main island (Honshu) of Japan, and China.

**Fall webworm, *Hypantria cunea*** Adult wing span is 40–60 mm, and the body and wings are uniformly white or with dark spots on the wings and abdomen; the bases of the front legs are orange or red. A full-grown caterpillar is about 25 mm long, pale yellow or greenish, with a brown stripe down the back and a yellowish-white stripe along each side. The body is covered with long gray hairs, which arise in tufts from orange-yellow or black tubercles. Eggs are white or golden yellow and are laid on the underside of leaves in masses of 200–300. They are covered with hairs from the female; hatching occurs in about 10 days. First-stage caterpillars spin a silk web over the leaves on which they are feeding; successive stages enlarge the web to enclose more foliage, and eventually several branches are enclosed. Small trees may be entirely enclosed in the webs of feeding caterpillars. Caterpillars feed within the protective web and it becomes littered with pellets of frass. Late-stage caterpillars are less gregarious than early stages and feed away from the web. Full-grown caterpillars leave the web to pupate on the ground or in protected places near the infested tree; pupation occurs in silken cocoons. Overwintering is in the pupal stage, and adults appear from May to July. There are one or two generations per year. This species occurs throughout the USA and southern Canada, and its hosts include more than 100 species of trees and shrubs. It may be a serious pest of ornamental trees, which can be completely defoliated, and the presence of the webs in the tree branches is unsightly. Full-grown caterpillars disperse from the webs to pupate and often occur in large numbers on the sides of nearby houses or buildings.

**Banded woollybear, *Pyrrharctia isabella*** Adult wing span is about 45 mm, the body and wings are orange-yellow, and there are dark spots on the wings and abdomen. A full-grown caterpillar is about 45 mm long, and covered with dense and uniform-length setae; it is brown in the middle and black at both ends. The length of the brown middle section is variable among individuals in a population, and this section of the body is often considered an indicator of winter temperatures to follow. In general, the relative amount of black (thus the length of the brown middle increases or decreases) at the ends of the caterpillar varies with its age and the wetness of the fall weather. When the weather has been wet there is more black at the ends, and the younger the caterpillar, the greater the extent of black. Full-grown caterpillars disperse in the fall to overwintering sites in protected places. They feed for a short time in the spring then make a cocoon covered with black and brown setae. Caterpillars feed on a variety of plants in peridomestic habitats, including plantain, grasses, dandelion, and burdock. They are not economically important, except that they are common in fall and attract attention. There are two generations per year. This species occurs in eastern North America.

## Cosmopterigidae

These small moths are 5–8 mm long, and have long and narrow wings, which are usually pointed at the apex. Some adults are brightly colored golden orange, yellow, and with silver spots or lines on the wings. Species of the genus *Cosmopterix* look similar the world over; they are brown with a golden band near the apex of each front wing. Most species in the family are leaf miners, but the range in habits includes parasites and predators. *Euclemensia bassettella* is an internal parasite of female coccids in the genus *Kermes*. In Australia, caterpillars of *Pyroderces terminella* invade deserted nests of *Polistes* wasps, and they have been found infesting the egg-sacs of the spider *Nephila edulis*.

**Pink scavenger moth, *Pyroderces rileyi* (= *Sathrobrota*)** Adult wing span is about 11 mm, the body is brown, and the antennae are banded. Front wings are banded and mottled with yellow, reddish-brown, and black; hind wings are pale gray, slender, and the edges have a long fringe. A full-grown caterpillar is 4–7 mm long, pink, and has a pale-brown head and thoracic shield. This insect infests maize in the field and in storage. Eggs are laid singly on the kernels, and the caterpillars feed on the kernel and the cob; infested grain may have a large amount of webbing and frass evident on the cobs. There is one generation per year. This insect is common in southern USA where it infests maize; infestation usually begins in the field and continues after the crop is in storage.

## Cossidae

Cossids are wood-boring in the immature stage. The adults have a wing span of 50–85 mm, and a large thorax and body; the front wings are usually spotted or mottled brownish white and black, which earns them the common name leopard moths. Males are about half the size of females and their hind wings are yellow or orange-red with dark lines. Adults are sometimes called goat moths, presumably because of their odor; the caterpillar of the carpenterworm also has an unpleasant odor. Full-grown caterpillars are 50–65 mm long, yellowish to pinkish white, and have a dark head and thoracic shield; there are brown spots at the base of the body setae. Life cycles of these moths require 1 or 2 years to complete, depending on the location and species. Infestations occur in a variety of living hardwood and pine trees.

**Carpenterworm, *Prionoxystus robiniae* (Fig. 11.1a)** Adult female wing span is 65–75 mm, and about 30 mm for the male. The body is dark brown to black, slightly mottled and with orange and black hind wings. The hind wings of the male are light brown, with a yellowish brown to orange spot with a black border. Full-grown caterpillars are 50–70 mm long, and greenish white; the head is shiny brown and there are large black mandibles. Eggs are laid in batches in spring; they are placed in bark crevices, near wounds, and under lichens or moss; fecundity is 450–800 eggs. Early-stage caterpillars feed on the inner bark, while late-stage caterpillars tunnel inward to the heartwood. Tunnels attain a diameter of 18 mm and may be 375 mm long, and they are usually lined with loose, brown silk. Development is completed in 1 year in southern USA, but extends to 3 or 4 years in northern regions and Canada. Pupation occurs at the surface of the tree, and the empty pupal case may be seen protruding from the emergence hole. Adults appear from April to July. This species is distributed throughout the USA and southern Canada, where it feeds on a variety of hardwoods, including oaks, particularly those in the red oak group, and also ash, locust, elm, maple, willow, and cottonwood. Trees are not usually killed, and the greatest damage results from the degraded lumber, and the persistence of holes and signs of damage in wood in use.

**Other Cossidae** The adult and caterpillar stages of the little carpenterworm, *Prionoxystus mcmurtrei*, are similar to *P. robiniae*, but slightly smaller in size. The caterpillars attack oak trees; the life cycle is usually 3 years. The adult pine carpenterworm, *Givira lotta*, is white and has a wing span of about 30 mm. The caterpillars mine the outer bark of yellow pine trees. *Acossus centerensis* and *A. populi* attack poplars and cottonwood trees. Wood damaged by the caterpillars may be found in old furniture and construction lumber.

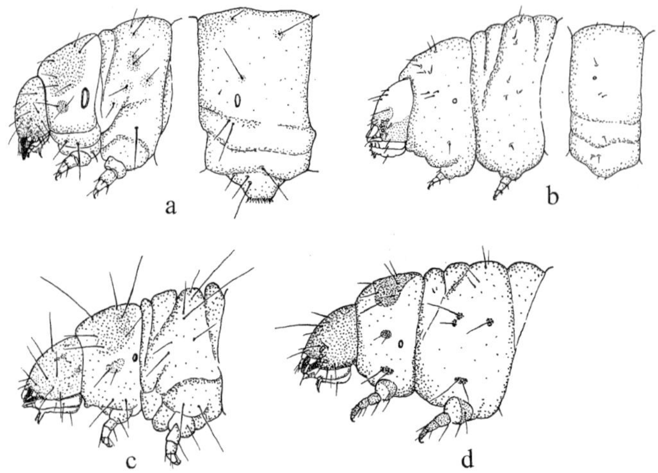


**Figure 11.1** Lepidoptera larvae. (a) *Prionoxystus robiniae* (Cossidae), head and thorax segments 1, 2, and abdomen segment 4; (b) *Sitotroga cerealella* (Gelechiidae), head and thorax segments 1, 2; (c) *Achroia grisella* (Pyralidae), head and thorax segments 1, 2; (d) *Galleria mellonella* (Pyralidae), head and thorax segments 1, 2.

## Gelechiidae

Adults are small, with a wing span of about 12 mm, and their wings are mottled gray and black. Hind wings are usually trapezoidal, pointed anteriorly, and the outer margin is sinuate. Caterpillars vary greatly in feeding habits; some are leaf miners, but many feed in rolled or spun-together leaves, or in stems or seed heads. There are some common species: *Gnorimoschema gallaesolidaginis*, which creates galls on the stems of goldenrod; the worldwide agricultural pest, pink bollworm, *Pectinophora gossypiella*; and the Angoumois grain moth. The potato tuber moth, *Phthorimaea operculella*, is a pest of stored potatoes, tobacco, tomato, and other Solanaceae. A related species, *Scrobipalpopsis solanivora*, damages stored tubers in Central America. There are several species that are serious pests of forest trees.

**Angoumois grain moth, *Sitotroga cerealella* (Fig. 11.1b; 11.2a)** Adult wing span is 13–17 mm, and the body is grayish to yellowish brown; the hind wings are narrow, pointed, and densely fringed. Full-grown caterpillars are about 5 mm long, white with a yellowish brown head and small legs and prolegs; the pupa is reddish brown. The head has six ocelli arranged in a slightly curved line. Eggs are laid directly on the surface of grain; fecundity is 80–200 eggs. Hatching occurs in 10 days at 20 °C, in 6–7 days at 25 °C, and 5–6 days at 30 °C. Caterpillars penetrate grain kernels, and cover the entry hole with silk. Two or three caterpillars can develop in a single kernel of maize. From other grains, such as wheat, only one adult is produced.

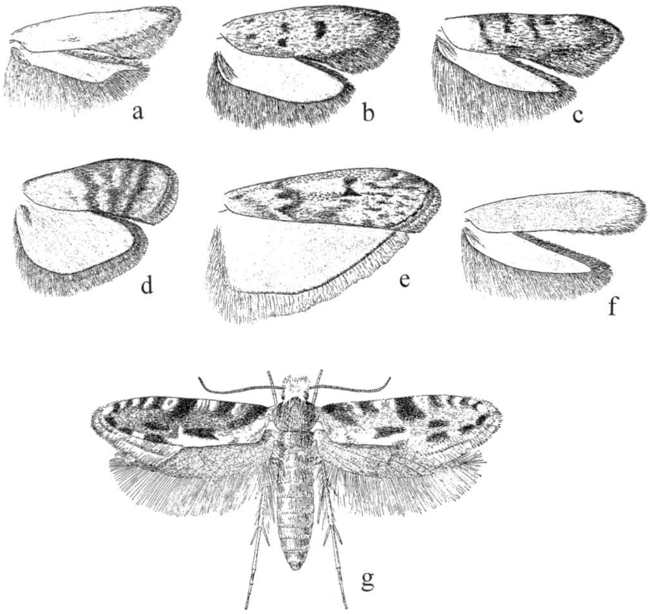


**Figure 11.2** Lepidoptera: Gelechiidae, Oecophoridae, Pyralidae, Tineidae. (a) *Sitotroga cerealella* (Gelechiidae); (b) *Hofmannophila pseudospretella* (Oecophoridae); (c) *Endrosis sarcitrella* (Oecophoridae); (d) *Plodia interpunctella* (Pyralidae); (e) *Anagasta kuehniella* (Pyralidae); (f) *Tineola bisselliella* (Tineidae); (g) *Nemapogon granella* (Tineidae).

Development is completed in about 21 days; although eggs hatch at 36 °C, but development is not completed above 34 °C. Minimum temperature for population increase is 16 °C; optimum relative humidity (RH) is 75–80%. The pupal stage lasts about 20 days at 20 °C, 10–12 days at 24–27 °C, and 8 days at 30 °C. Total development time at 70–90% RH is about 30 days at 30 °C and 40 days at 25 °C. There are four or five generations per year, but in heated buildings there may be 10–12 generations.

Caterpillars are a pest of stored whole grain or caked grain in containers, and they also infest grain in the field. Infestations occur in barley, rye, maize, oats, rice, and various seeds. The current distribution of this species is cosmopolitan, but it was first reported attacking grain in the French province of Angoumois. In the household habitat, ears of ornamental corn (Indian corn) may become infested, and also food and feed grain in storage. This insect is active at low temperatures and remains active during winter. It was introduced into North America before 1743.

## Lasiocampidae

Tent caterpillar adults have a wingspan of 37–50 mm; their body is brown or gray. The body, legs, and eyes are covered with hairs, and the antennae are somewhat feathery. Eggs are laid in masses on tree bark or small twigs, and they are held in place by a frothy substance produced by the female. Caterpillars vary in shape from nearly cylindrical to flattened, and are usually covered with fine setae or hairs. They feed on the leaves of a variety of trees. The tent-building species are gregarious and construct a silk retreat or tent on a branch or crotch in the tree. They do not feed from within the tent, like webworms (Arctiidae), but move out of the tent to the leaves of neighboring branches. Caterpillars return to the tent during periods when they are not feeding and at night. Species that do not build tents assemble in clusters during periods between feeding. Pupation is in a cocoon, and there is usually one generation per year.

**Eastern tent caterpillar, *Malacosoma americanum*** Adult wing span is 37–52 mm, and the body is yellowish to dark brown. Wings have some white scales; the fore wings have two oblique yellowish white lines, and the hind wings are uniformly brown with a faint white area. Full-grown caterpillars are about 50 mm long, with black head and with scattered long hairs; there is a dorsal white stripe, bordered with reddish-brown and black lines. Eggs are laid in masses of 150–250 that encircle small twigs on preferred host trees. Hatching occurs in the following spring, about the time new leaves appear on the tree. First-stage caterpillars feed on the egg chorion, then on buds and young leaves. A tent is usually constructed in the crotch of the tree, and it is expanded with the growth of the caterpillars. From this tent, caterpillars crawl out to feed on leaves, and return to the tent to rest. Development is 2–3 weeks, and full-grown caterpillars leave the tent and wander in search of a place to pupate. Pupation occurs in silken cocoons, covered with a yellowish powder. There is one generation per year. This species is generally distributed throughout eastern USA and southern Canada. The preferred hosts are wild cherry and apple, but it also attacks shade trees and fruit trees. The tents are unsightly on ornamental trees, and full-grown caterpillars often wander on to the sides of houses and buildings, and become a nuisance.

**The lackey, *Malacosoma neustria*** Adult wing span is 30–35 mm; the body is light brown. Wings are light brown with two pale brown stripes on the fore and hind wing. Full-grown caterpillars are about 50 mm long, and have a blue head and two black circles that look like eye-spots; the body is blue, with black, red, and orange stripes. Caterpillars live gregariously in a dense web on various trees; when full-grown, they move away to pupate and crawl on houses and buildings at this time. The pupa is dull black, with short reddish-brown hairs on the abdomen. It is enclosed in a double cocoon spun among leaves of the food plant. The pupal period is about 14 days. This species occurs in England, Wales, and Ireland.

## Limacodidae

The caterpillars in this family are short, fleshy, and appear somewhat like a slug. The thoracic legs are small and there are no prolegs on the abdominal segments; they move with a slow, creeping motion. Adults have a wing span of about 35 mm, and have a large and hairy body. Pupation occurs in oval, silken cocoons spun between leaves or attached to twigs. The cocoon is egg-shaped or nearly spherical, and often covered with urticating (poisonous or stinging) hairs; the cocoon has a hole covered by a cap at one end through which the adult emerges. Caterpillars of many species are unusually shaped, and distinctly marked. They have urticating hairs on the body; many caterpillars have body protrusions (scoli) with sharp spines, and often with urticating properties. Immatures of *Setora nitens* are nettle caterpillars, and feed on palms, tea, and coffee in southeastern Asia. Other limacodid caterpillars are sluglike and have reduced legs and a smooth body; they are called slug caterpillars or jelly grubs. These include *Thosea*, which feed on herbaceous and woody shrubs in Africa and tropical regions of Asia, and *Cheromettia*, which are polyphagous on plants in southeastern Asia. Caterpillars of *Niphadolepis* are the most common jelly grubs on coffee and tea in East Africa.

**Saddleback caterpillar, *Acharia stimulea* (= *Sibine*) (Fig. 11.3a)** Adult wing span is about 30 mm, and the wings are dark, velvety, reddish brown, with two white spots near the apex of the front wings. Full-grown caterpillars are about 25 mm long, and brown with a light-green patch over the back, and a purplish-brown spot in the middle. The body has along the sides several distinct tubercles (scoli) with urticating hairs, and there is a pair of spiny tubercles bearing poisonous hairs at each end. Caterpillars feed on a variety of ornamental trees and shrubs, but also agricultural crops, such as maize. Eggs are laid on the underside of leaves. Early-stage caterpillars feed on the lower leaf epidermis, while late-stage caterpillars consume the entire leaf. Development is completed in late summer, and pupation occurs in secluded sites on the ground or leaf litter. The cocoon is smooth and ovoid, and contains some of the urticating hairs of the caterpillars. Pupae or prepupae overwinter and adults emerge in June and July. Distribution is in eastern and southern USA.

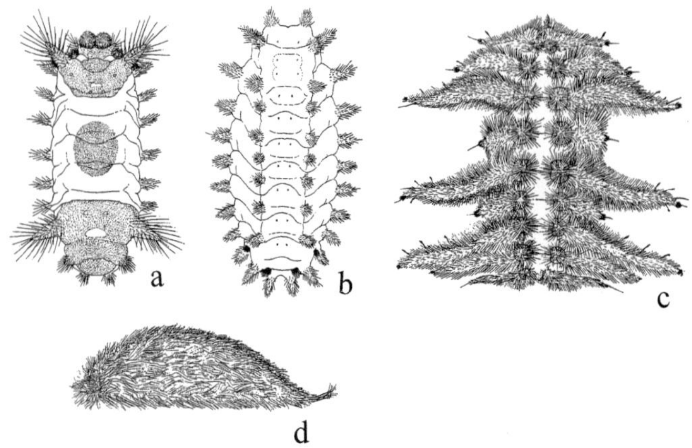


**Figure 11.3** Lepidoptera: Limacocidae, Megalopygidae. (a) *Acharia stimulea*; (b) *Euclea delphinii*; (c) *Phobetron pithecium*; (d) *Megalopyge opercularis*.

**Oriental moth, *Cnidocampa flavescens* (= *Monema*)** Adult wing span is 30–42 mm, and the body is yellowish brown. Full-grown caterpillars are about 22 mm long, and marked with yellow, blue, green, and purple. Caterpillars have tubercles with urticating hairs and setae along the outer edge of the body. This species feeds on a wide variety of trees, and is distributed in the USA.

**Spiny oak-slug, *Euclea delphinii* (Fig. 11.3b)** Adult wing span is about 28 mm, and the body is light brown with a variable number of bright-green spots on the fore wings. Full-grown caterpillars are about 20 mm long, and have large tubercles (scoli) bearing urticating setae and hairs on the perimeter and dorsum of the body. It feeds on oak, pear, willow, and other trees. Distribution is in eastern and southern USA.

**Asian slug moth, blue-striped nettle grub, *Latoia lepida* (= *Parasa*)** Adult wing span is 30–35 mm. Front wings are green, with a basal brown patch, and the hind wings are pale brown. Full-grown caterpillars are about 25 mm long and the body is green with blue longitudinal stripes, and flattened; the legs are indistinct. It has fleshy protuberences (scoli), each with spines. The spines on the scoli are urticating. Eggs are laid in batches of 20–30 on the underside of leaves of a variety of crop and ornamental plants. Development requires about 40 days, and pupation is in a round, hard cocoon on the bark of a tree or shrub. The pupal period is about 21 days. This species occurs in southeastern Asia, and in regions of East and West Africa. Related species include *Parasa hilarata*, *P. vivida*, and *P. latistriga*. The caterpillars of these species are similar in appearance to *L. lepida* and have urticating setae.

**Coffee jelly grub, *Niphadolepis alianta*** Adult wing span is about 40 mm. The body is yellowish brown and there are three or four dark brown patches on the front wings; the hind wings are uniformly yellowish white to pale brown. Full-grown caterpillars are about 35 mm long; the body is somewhat rounded oval, humped and without distinct features. This species occurs on coffee and tea in East Africa. The caterpillars have urticating qualities, and they are a threat to agricultural workers.

**Green slug moth, *Parasa chloris*** Adult wing span is 20–25 mm and the body is brown; the front wings are without a dark patch in the center. The caterpillar is bright purple, with four bluish-black lines along the back, with yellow tubercles (scoli) along both sides, and at the anterior and posterior ends. The cocoon is dark brown and egg-shaped, with a smooth surface. Adults emerge in June. This species occurs in northeastern and southeastern USA.

**Stinging rose caterpillar, *Parasa indetermina*** Adult wing span is 20–25 mm; the head and thorax are green above, and brown beneath. The front wings are green with a brown patch at the base and with a brown, irregular border. The caterpillar body is red or yellow, purple, and white. There are stinging spines on the second, fifth, and last segment of abdomen. This species occurs in north- and southeastern USA.

**Hag moth, *Phobetron pithecium* (Fig. 11.3c)** Adult wing span is about 20 mm. The body of the female is brown, and marked with yellow; the male is uniformly brown. Full-grown caterpillars are about 10 mm long, and have nine pairs of brown, lateral processes; the third, fifth, and seventh are the longest, curved and twisted, and are suggestive of the disheveled hair of a hag (witch). These processes bear stinging or urticating hairs. The thoracic legs are small, and on the ventral surface of the abdomen there is a series of disks, which serve as prolegs for locomotion. Caterpillars are general feeders and are found on the leaves of a variety of trees and shrubs. The cocoon is spherical, and has the tubercles and stinging hairs of the immature stages attached. Cocoons are attached to leaves and exposed. This species occurs in eastern and southern USA.

***Scopelodes nitens*** Adult wing span is about 30 mm, and the front wings are light brown and have one dark spot medially. Hind wings are yellow; the thorax is yellowish brown, and the abdomen is pale yellow. The apical segment of the labial palp is very long, and is expanded at the tip with long scales. This species occurs in Australia: Banks Island in Torres Strait, and at Iron Range, Cape York peninsula. The caterpillars defoliate cashew nut trees and have stinging hairs. A closely related species, *S. contracta*, occurs in Japan.

## Lymantriidae

The tussock moths have a wing span of 15–55 mm, and the thorax and abdomen are covered with long piliform scales. Females have specialized hairs on the abdomen that are attached to and cover their egg masses. Caterpillars are usually covered with dense and often secondary setae arranged on tubercles, frequently with dense dorsal tufts of setae. Caterpillars feed chiefly on trees, and the tussock, gypsy, and browntail moths are serious pests of forest and shade trees. The hairs of the caterpillars of some species often cause an irritating rash on human skin.

**Australian browntail moth, *Euproctis edwardsii*** Adult wing span is about 36 mm, and the wings are uniformly yellowish white. The thorax is dark yellow and the abdomen is dark brown, but dark yellow at the tip. This species occurs in Australia, and it is widespread from southern Queensland to southeastern South Australia. Caterpillars feed at night on mistletoes (*Amyema*) growing on *Eucalyptus*, and retreat during the day. Pupae are formed under loose bark on the trunk of the host tree. When populations are large, caterpillars move from the trees and pupate in other locations, including sites around the outside and inside of houses. Human contact with live caterpillars or with cast skins or cocoons can result in severe skin irritation and rashes. In susceptible people, wind-blown hairs can be sufficient to produce intense reactions.

**Browntail moth, *Euproctis chrysorrhoea*** Adults are 28–38 mm long, and the wings and thorax are white. The abdomen is brown with an apical tuft. Caterpillars feed on apple, cherry, oak, maple, and a variety of other trees. This species is distributed in northeastern USA.

**Japanese browntail moth, *Euproctis pseudoconspersa*** Adult wing span is 13–18 mm, and the body is yellowish brown. The front wings have black spots on the anterior edge. Full-grown caterpillars are about 25 mm long; the head is yellowish brown and the body is pale brown. The body has blackish-brown and yellowish-brown protrusions that have urticating setae. Contact dermatitis frequently occurs as a result of exposure to the adults and caterpillars. There are two generations per year, but human exposure and skin irritations usually occur in summer and fall. This species often infests and is brought indoors on camellia plants. It is distributed in Japan, Taiwan, Korea, and China. A related species, *E. subflava*, occurs in Japan and also causes skin dermatitis.

***Euproctis stenomorpha*** Adult wing span is about 32 mm. Front wings are patterned, with the apical edge with a pale border, and a large and small pale spot. Caterpillars of this species feed on *Eucalyptus tetardonta*, and they have caused skin irritations on schoolchildren in Darwin, Australia.

## Megalopygidae

Flannel moths have a wing span of about 30 mm. The wings have a dense covering of scales mixed with fine curly hairs, which gives them a somewhat woolly or flannel-like appearance. The body is large and covered with long hairs. Males are usually smaller than females; the body is usually brown to pale brown, with mottled wings. Caterpillars are covered with long hairs and stinging spines; in addition to the five pairs of prolegs, they have two additional pairs that are sucker-like and lack crochets. The cocoons are oval and usually covered with urticating hairs, and they are provided with a similar lid-like opening found in the cocoons of Limacodidae.

**Puss caterpillar, *Megalopyge opercularis* (Fig. 11.3d)** Adult female wing span is about 37 mm and the male wing span is about 25 mm. Wings have long, wavy white hairs, especially along the veins, with dark brown spots. The body is yellowish brown. Full-grown caterpillars are about 25 mm long, and slightly tapered posteriorly. The body is covered with long brown, yellowish-brown, and gray hairs; the hairs at the rear produce a tail-like appearance. Eggs are laid on leaves and are usually covered with hairs from the abdomen of the female. Early-stage caterpillars feed gregariously and skeletonize leaves; older caterpillars consume the entire leaf. Overwintering is in the pupa stage in an irregular-shaped cocoon. Cocoons are lightly covered with urticating hairs, and attached to small branches and twigs of the host tree. The operculum opening for the emergence of the adult is devoid of hairs. There are one or two generations per year. Distribution of this species is eastern and southern USA.

**Crinkled flannel moth, *Lagoa crispata*** Adult female wing span is about 40 mm and the male wing span is about 35 mm. The body is large, pale yellow, with black wavy lines and brown crinkled hairs on the fore wings. Full-grown caterpillars are about 25 mm long and covered with long, silky, brown hairs that meet in the form of a ridge along the back. Immersed in these hairs are venomous setae. Development is completed in late summer and winter is spent in the pupal stage in a silk cocoon. The cocoon is shaped like an urn, and has a flat, hinged lid, which is lifted when the adult emerges in the spring. Distribution of this species is throughout eastern USA.

## Oecophoridae

Adults have a wing span of about 20 mm, and many species are brightly colored or have patterned wings. Caterpillars are also marked with white or yellow bands, and most feed on plants, usually on exposed blossoms, leaves, or stems, or in folded or rolled leaves. The species that are household pests of stored food are scavengers as caterpillars, and have adapted to the food debris indoors. *Barea* species feed on fungi in decaying wood.

**Decay moths, *Barea* spp.** Adults are about 23 mm long and gray. Full-grown caterpillars are about 30 mm long, pale yellow, and have dark gray markings at the end of the abdomen. Caterpillars remain on the surface of decaying wood in silk retreats covered with feces. Caterpillars feed on the fungi in decaying wood.

**White-shouldered house moth, *Endrosis sarcitrella* (Fig. 11.2c)** Adult males are about 6 mm long and females are about 11 mm long. The prothorax and base of the wings are white, the fore wings are grayish white and marked with dark patches, while the hind wings are narrowed apically. Full-grown caterpillars are about 12 mm long, pale yellow to white, and have a brown head. Eggs are deposited deep into crevices by the long female ovipositor; hatching occurs in 42 days at 10 °C, 15 days at 15 °C, and 6–7 days at 25 °C. Development requires a minimum of 80% RH, and is 10–11 weeks at 15 °C, and 5–6 weeks at 25 °C. The pupal stage lasts 25 days at 15 °C, and 10–11 days at 25 °C. Development continues throughout winter in sheltered locations; there is no record of diapause in this species. Under favorable conditions, 2–4 generations are completed per year. Adults emerge in early spring, and the sex ratio may be 23:1 in favor of females. Caterpillars feed on a variety of plant material, including bird nests and thatched roofs, fungi on trees, corks in wine cellars, and dry seeds and vegetables, broken grain, such as maize and wheat, meal and flour. This species is cosmopolitan, but it is more common in temperate regions.

**Brown house moth, *Hofmannophila pseudospretella* (Fig. 11.2b)** Adult females are about 14 mm long and have a wing span of 18–25 mm; males are about 8 mm long and have a wing span of 17–19 mm. The body is bronze brown; the dark brown front wings have distinct black spots. Males fly more than do the females, who tend to seek a harborage when disturbed. Full-grown caterpillars are about 6 mm long, yellowish white, and with a light brown head. Eggs are laid singly or in small groups; fecundity is 400–500 eggs. Hatching is in 25 days at 15 °C and 10 days at 25 °C. Development is 70–80 days at 20–25 °C and 70–90% RH, but may be over 180 days at 10 °C. The immature stages produce only a little silk while feeding. Caterpillars prefer high humidity for development, and enter diapause when conditions are not suitable. The caterpillar overwinters in diapause, and in spring spins a cocoon that has attached to it debris from the feeding substrate. The pupal period lasts about 14 days at 10 °C, 7 weeks at 15 °C, and 15 days at 25 °C. Generally, there is one generation per year. This species is more common in households than in commercial storage facilities, and it infests dried and decaying organic matter in cool habitats. In food-processing plants it attacks bulk wheat and flour. In domestic habitats it damages wine bottle corks, book bindings, skins, and furs. Caterpillars are capable of digesting keratin, and feed on animal products. In house attics and roof spaces the caterpillars feed on animal droppings, dried corpses, and other organic material. This species is cosmopolitan but most abundant in temperate regions.

**Other Oecophoridae** Other species recorded in stored foods include: *Anchonoma xeraula*, grain moth, which occurs in grain storage facilities in Japan; and *Promalactis inonisema*, which occurs in stored cotton seed in Japan.

## Psychidae

Bagworm males have a wingspan of 4–60 mm, and are generally mottled brown and black; females of most species are wingless and resemble fly larvae (larviform). The caterpillar forms a portable case made of pieces of twigs and leaves of their host, and they carry the case as they feed on the leaves and needles of hardwood and coniferous trees. Many bagworms occur in the tropics, especially the large species, and some show a preference for palms as hosts. Infestations are sometimes large enough to defoliate the tree. Pests in Asia include species of *Acanthopsyche*, which attacks *Acacia* and *Pinus* in East Africa and India, and species of *Delonix* in Africa and Asia. Caterpillars of *Clania* and *Cryptothelea* are polyphagous and infest trees in Africa and Asia.

**Abbots bagworm, *Oiketicus abbotti*** Male wing span is about 20 mm. Wings are dark and the front wing has a narrow, irregular, white subapical band. Caterpillars place short twigs and leaves on the bag. This species has been recorded from southeastern USA (Florida).

**Bagworm, *Thyridopteryx ephemeraeformis*** Male wing span is about 25 mm. The wings are transparent and dusky, and the body is dark brown to black. Full-grown caterpillars are about 30 mm long and dark yellow to light brown; the anterior end is more pigmented than the posterior. The head, thorax, and anal plate have scattered dark-brown spots. The cuticle of abdominal segments is transversely wrinkled. Caterpillars attack evergreen and deciduous trees, and the most commonly infested trees are firs, juniper, pines, spruce, maple, sweet gum, and sycamore. Caterpillars complete their development in fall and leave the host tree or shrub and seek a place to pupate. They often move to the perimeter of houses and buildings, and attach their cocoons on sidings. They pupate in the case; males leave the case after development, but the larviform females remain in the case. In fall, females extend their abdomens to the outside of the bag and males are capable of mating. Eggs are laid inside the bag and hatch the following spring. First-stage caterpillars leave the bag and move out on to leaves to feed and form their own bag. This species probably originated in the West Indies, and is distributed in eastern and southeastern USA.

## Pyralidae

Moths in this large family are small and have a wing span of about 27 mm. Front wings are elongate or triangular, and the hind wings are usually broad. The proboscis has scales and the labial palps often project forward, giving this group the name snout moths. All species have abdominal tympanal organs, and many are active at night. This family includes species that are aquatic and have caterpillars that breathe by means of gills, species that are pests of ornamental and agricultural grains, and in foods in storage.

The origin of pyralid pests of stored foods is linked to their natural habitats and their ability to adapt to the indoor environment. Some species were originally associated with the habitats of other animals, such as the nests of bees, birds, and small rodents. Several species naturally infest ripe grains in the field, but readily adapted to indoor conditions to become pests of the harvested and stored grain. Many of the pests of dry nuts and fruits were associated with the natural products either on the tree or ripened and on the ground. The moths associated with stored grain are general scavengers on a variety of plant materials. Many of the species that occur indoors are also represented by populations in natural or undisturbed areas.

Caterpillars of some pyralid species secrete kairomones from their mandibular glands, and the quantity of these chemicals increases as caterpillar density increases. The concentration of these chemicals causes increased wandering by the late-stage caterpillar, and affects adult oviposition. Caterpillars

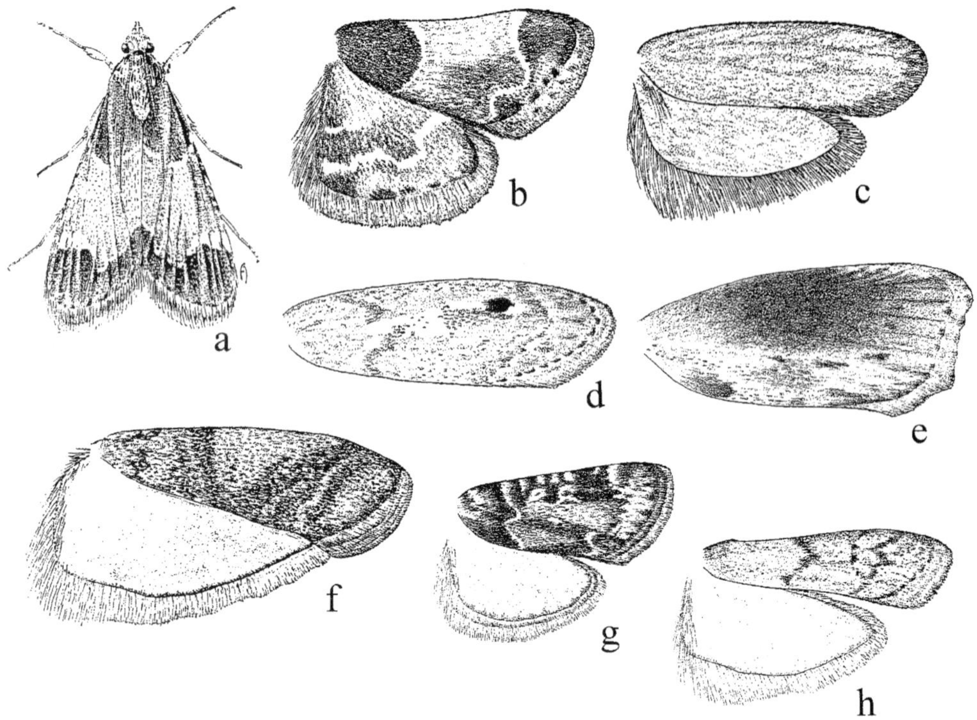

**Figure 11.4** Lepidoptera: Pyralidae. (a) *Pyralis farinalis* adult; (b) *P. farinalis* front and hind wings; (c) *Corcyra cephalonica*; (d) *Paralipsa gularis*; (e) *Galleria mellonella*; (f) *Ephestia elutella*; (g) *Aglossa caprealis*; (h) *Vitula edmandsii serratilinella*.

of most pyralids are negatively phototactic until they are about to pupate. They are also sensitive to crowding and this condition can delay development and result in small caterpillars and adults. Most of the stored-food pyralids produce sex pheromones. The chief component of the female pheromone of several species is *cis*-9, *trans*-12-tetradecadien-1-yl acetate. Within 24 h of emergence females assume the calling posture, in which the abdomen is lifted between the wings and the scent glands are extruded. Males locate females by flying against air currents and orient using the concentration of the pheromone in the air stream. Sex pheromones have been extracted from the wing glands of male *Ephestia elutella* and *Plodia interpunctella*. The rapid wing beating of males prior to mating may disperse these chemicals, which appear to be important in the final stage of courtship. The pheromones released from male *P. interpunctella* produce a turning response in receptive females.

**Lesser wax moth, *Achroia grisella* (Fig. 11.1c)** Adult wing span is 17–21 mm, the body is brown, and the head is yellowish brown. Front wings are mottled. The full-grown caterpillar is 15–18 mm long, slender, and yellowish white to gray. The head is brown, and the ocelli are absent. Thoracic segment 1 has a dark brown dorsal plate. Caterpillars feed on stored foods and dried fruits, and are also found in wild and commercial honey bee hives, especially in the debris at the bottom and, less often, in the combs. This species is cosmopolitan.

**Murky meal moth, fungus moth, *Aglossa caprealis* (Fig. 11.4g; 11.5a)** Adult female wing span is 9–10 mm; the male moth body is 13 mm long, and the female 16 mm long. The body is dark brown and the wings have light brown markings. Full-grown caterpillars are about 20 mm long, shiny dark brown to brownish black. The caterpillars are usually active in dark and damp locations where wood-decay fungi, such as *Poria incrassata*, develop. The surface of decay fungi is often covered with webbing produced by the caterpillar. When full-grown, caterpillars migrate before pupation, and are usually found away from their feeding site.

**Mediterranean flour moth, *Anagasta kuehniella* (= *Ephestia*) (Fig. 11.2e; 11.5b)** Adults are 7–12 mm long and have a wing span of about 22 mm; the front wings are pale gray with transverse, irregular bands; hind wings are grayish white. When at rest, the adult raises its front wings and elevates the anterior of the body. Full-grown caterpillars are about 15 mm long, yellowish white to pinkish white, and have a brown head. Eggs are laid singly; fecundity is 100–600 eggs. Oviposition can occur at 7.5 °C, but not at 5 °C. Hatching is between 10 and 31 °C, and is in 5 days when eggs are deposited at 25 °C, and 4 days when deposited at 30 °C. Eggs survive exposure to −10 °C for 5 days. Development is completed in 10 weeks at 25 °C and

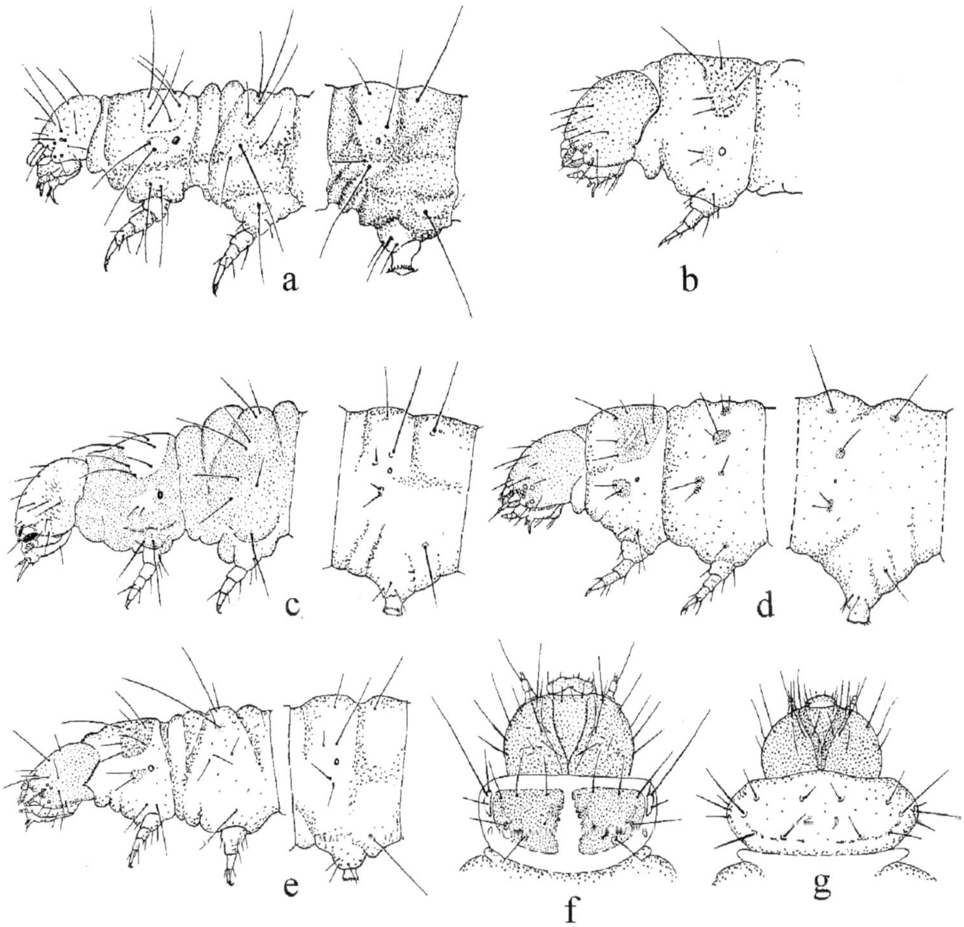


**Figure 11.5** Lepidoptera: Pyralidae larvae. (a) *Aglossa caprealis*, head and thorax 1, 2, and abdomen segment 4; (b) *Anagasta kuhniella*, head and prothorax; (c) *Pyralis farinalis*, head, thorax segments 1, 2, and abdomen segment 4; (d) *Ephestia elutella*, head, thorax segments 1, 2, and abdomen segment 4; (e) *Plodia interpunctella*, head and thorax segments 1, 2, and abdomen segment 4; (f) *Tinea pellionella*, dorsal view of head and prothorax; (g) *Tineola bisselliella*, dorsal view of head and prothorax.

75% RH on white flour. The lower limit for development is about 12 °C; it can be completed at 28 °C but not at 31 °C. Males become sterile when reared at 30 °C, and rearing in continuous light also induces male sterility. Pupae are in cocoons, or not when they are in cracks or crevices; the pupal period lasts 8–12 days. There are four or five generations per year. Food infested includes flour, nuts, seeds, beans, dried fruits, and chocolate. The silk produced by the caterpillars may be thick in infested areas; developing caterpillars usually remain in their silk feeding tubes. Full-grown caterpillars may be found far from the infested site.

**Wax moth, *Aphomia sociella*** This moth is a pest indoors in northern Europe (Denmark). The caterpillars develop in the nests of bumble bees, but they migrate before pupation and invade houses before they spin a cocoon.

**Almond moth, tropical warehouse moth, *Cadra cautella* (= *Ephestia*)** Adults are about 12 mm long, and have a wing span of about 14–22 mm. Front wings are reddish brown with indistinct white cross lines, while hind wings are uniformly pale gray with short fringe hair. The proboscis is rudimentary. Adults fly with rapid wing movements, and move quickly. Full-grown caterpillars are about 12 mm long, yellowish white, and with a uniformly brown head. Body segments have small, pigmented spots. Eggs are laid singly or in small batches; hatching occurs between 15 and 37.5 °C. Eggs hatch in about 17 days at 25 °C and 70% RH, 7–8 days at 15 °C, 4–5 days at 25 °C, and 3 days at 30 °C. Exposure to −10 °C for 9 h kills 95% of eggs. Development is longer at low RH; caterpillars survive above 0 °C. At 70% RH, the range of temperature for development is 15–36 °C; development is 30 days at 30–32 °C and 70–80% RH when immatures are reared on a mix of wheatfeed, wheat germ, and yeast. Diapause is influenced primarily by temperature and photoperiod. Caterpillars are more responsive to continuous darkness than to short photoperiods or continuous light.

Feeding at 25 °C rather than 20 °C limits diapause in most strains. Diapause in this species lasts about 4 months, even under constant conditions, and exposure to temperatures slightly below development threshold terminates diapause in about 30 days. The pupal stage lasts about 18 days at 20 °C, 9 days at 25 °C, and 7 days at 30 °C. Caterpillars feed on cereals, cocoa beans, dried fruits, flour, grain, peanuts, various seeds, and shelled nuts. This species has a worldwide distribution, especially in warm parts of the world.

Adults are mainly nocturnal, and become active in late afternoon and evening in response to changes in light intensity and temperature; they mate and lay eggs after dusk. They have a peak of activity in late evening and a second but smaller peak just before dawn. The visual system of *C. cautella* responds most strongly in the yellow-green and ultraviolet regions of the spectrum. The peak of spectral efficiency in the green region may be employed during motion detection or during the orientation towards yellow-green pigmented plants containing nectar.

**Raisin moth, *Cadra figulilella* (= *Ephestia*)** Adults are about 10 mm long and the wing span is 15–20 mm. The front wing is gray and the hind wing is shiny white; both wings have a long fringe of hairs at the margins. Full-grown caterpillars are about 13 mm long, yellowish white, and have four rows of purple spots along the back. Eggs are scattered over the surface of the host fruit; fecundity is about 350 eggs. Hatching is in about 4 days, but only 10% hatch at 15 °C, and none below. Development requires about 30 days; limits for complete development at 75% RH are at 15 °C and 36 °C. The low RH limit is 30–50% RH. Optimum conditions for survival and development are about 30° C and 70% RH. On wheatfeed, yeast, and glucose, development from egg to adult is about 36 days, and increases to 122 days at 17.5 °C. Development is about 34 days on ground carobs and 56 days on almonds at 30 °C and 70% RH. Diapause is in response to photoperiods of 13 h or less; at constant temperatures some caterpillars remain in diapause for about 1 year; in field populations most pupate after 3–5 months in diapause. This is a pest of ripening fruit in the orchard or partially dried fruit in storage; it is also known to feed on cereals and nuts. Damage to dried fruit, such as raisins, includes the caterpillars feeding on the surface of the fruit and the resulting fecal pellets and webbing. It has worldwide distribution, and is one of the most common pests of dried fruits.

**Rice moth, *Corcyra cephalonica* (Fig. 11.4c)** Adult wing span is 14–24 mm, and the body and wings are grayish brown; the wings are without distinct bands. The male has short and blunt labial palps, while the female has long and pointed palps. Adults fly at night. Full-grown caterpillars are 12–14 mm long, and yellowish white to bluish gray, and sometimes have a green tint. Caterpillars have long fine hairs on the body and a dark brown head and brown prothoracic shield. There is a seta above each spiracle, starting with abdominal segment 1; the posterior rim of the spiracles is thickened. Eggs are somewhat sticky and they are deposited directly on flour or other host food; fecundity is 100–300 eggs. Hatching is in 10 days at 20 °C and 4 days at 30 °C. Development from egg to adult is 26 days at 30–32.5 °C and 70% RH. Development on sorghum is 46–55 days at 28 °C and 70% RH, and on millet it is 31–41 days. Limits for development are 17 °C and 35 °C at 70% RH. The last stage spins a tough, double-layer cocoon covered by food particles and debris. This species feeds on stored rice, cocoa, nuts, and flour. Damage is done by direct feeding of the caterpillars, and by the tough silk webbing on the infested material. In hot and humid climates, it is a major pest in flourmills. Caterpillars produce dense webbing, and when feeding on grains they spin dense silken tubes and web the grain kernels into the walls of the tube. As its scientific name implies, this species was described from the Grecian archipelago, but it has spread from there. It is nearly cosmopolitan throughout the tropics, and it is abundant in Southeast Asia, and common in Africa.

**Grass moth, *Crambus flexuosellus*** Adults are about 12 mm long. The wings are pale yellow and usually wrapped around the body. Front wings have scattered brown markings and the front margin is bordered with a brown stripe and a broad white stripe. Full-grown caterpillars are about 12 mm long and brownish red. Caterpillars feed at night, and chew long narrow holes in the blades of grass; they sometimes burrow into the stem. Caterpillars also attack the roots of small plants. They remain in silken retreat in the soil during the day. Adults often fly to lights at night. This species occurs in New Zealand.

**Sod webworms, lawn moths, *Crambus* spp. and *Pediasia* spp.** Adults are 12–20 mm long and the body is gray to grayish white. The maxillary palps are long and give the head a distinct snout appearance. Full-grown caterpillars are 16–20 mm long and the body is brownish white. The head is dark brown and segments have dark brown sclerites. They remain in silk retreats at the soil or thatch layer of turfgrass, and other ground-cover vegetation. Adults remain inactive on low vegetation during the day, and become active at dusk and for 2–3 h after. They have an erratic flight pattern, stopping at numerous sites to rest and oviposit. Eggs are laid singly or in small batches while the female rests on blades of grass or plant stems. Eggs are small, somewhat barrel-shaped, and drop into the organic layer or soil. Fecundity is about 250 eggs. Hatching occurs in about 7 days, and eggs turn from white to reddish orange before the first stage emerges. Development is completed in about 3 weeks, and full-grown caterpillars pupate in a cavity formed in the thatch or soil. The pupal period is about 10 days. There are 1–3 generations per year.

**Carob moth, dried fruit moth, *Ephestia calidella*** Adults are about 12 mm long, and the body is grayish brown. The wing span is 19–24 mm; the front wing is grayish brown or brown with two cross-lines, while the hind wing is pale gray. Full-grown caterpillars are about 14 mm long, yellowish white, and have a dark head. At 30 °C and 70% RH development is complete in 27 days on wheatfeed, 41 days on carobs, and 59 days on almonds. This species is nearly cosmopolitan. In the UK the feeding period is September to May, and the adults fly in August and September.

**Tobacco moth, cocoa moth, warehouse moth, *Ephestia elutella* (Fig. 11.4f; 11.5d)** Adult are 5–9 mm long and the wing span is 14–20 mm; the body is grayish brown. Front wings have two pale regions. Full-grown caterpillars are 10–18 mm long and yellowish white; sometimes they are slightly brown or pink, and have a brown head. Eggs are deposited singly at dusk or night during the first week after emergence; fecundity is 150–200 eggs, but 327 and 500 eggs from one female have been reported. Eggs hatch in about 20 days at 15 °C, 10 days at 20 °C, 6–7 days at 25 °C, and 4–5 days at 30 °C. Fecundity and longevity are increased by about 30% when water is available. The temperature range for development is 10–30 °C. Development from egg to adult in wheatlings, yeast, and glycerol takes 6–7 weeks at 25 °C and 70% RH, and 11–12 weeks at 20 °C. Caterpillar diapause usually lasts 6–8 months and is stimulated by exposure of the recently molted last stage to photoperiods of 14 h or less, or to temperatures of 20 °C or below. Termination of diapause occurs when there is an extended period of temperatures from 5 to 10 °C, followed by a period when temperatures are above 20 °C. The pupal period is 6–7 weeks at 15 °C, 12–15 days at 25 °C, and 10–11 days at 30 °C. In warehouses, adults live up to 3 weeks. Caterpillars feed on a variety of dry fruits and vegetables, including tobacco, cereals, chocolate, cocoa beans, coffee, flour, nuts, seeds and spices. On most foods development is relatively slow. This species is nearly cosmopolitan, but is not common in the tropics. In the UK and Europe it survives outdoors and in unheated buildings.

**Greater wax moth, *Galleria mellonella* (Fig. 11.1d; 11.4e)** Adult wing span is 28–30 mm; the gray or pale brown body is marked with black tubercles on the fore wing, and the tips of the wings are gray. Males are smaller than females and lack maxillary palps. Full-grown caterpillars are 23–28 mm long, and the body color ranges from yellowish white to brown, with a dark brown to black prothoracic region and dorsum. The head is dark brown, with four pairs of ocelli. The thoracic spiracle has a yellow peritreme of uniform thickness. It infests abandoned or weakened bee nests, and can enter buildings from nests behind walls or in attics. It is an Old-World species, but is distributed in the USA and other regions where honey bees are kept. Eggs are laid at night on or near the comb, and hatching is in 10–12 days. The caterpillar burrows through the hive comb and feeds on wax, excrement, and exuviae of the bees; the feeding tunnels are lined with silk. Pupation occurs in white, thick cocoons, which are spun during the night along the edge of the comb, and in cracks and crevices of the hive. Development varies from 45 days for the first generation to 35 days for the second and third-generation; some of the third-generation caterpillars overwinter. Adults are active at night, but remain motionless in the hives during the day.

**Stored-nut moth, *Paralipsa gularis* (Fig. 11.4d)** Adult wing span is about 21 mm, and the body and legs are yellowish brown; the fore wings are light brown and the hind wings yellowish brown. Full-grown caterpillars are about 18 mm long, yellowish white, and with a brown head. They feed on seeds, flour, nuts, and dried fruits, peanuts, soybeans, and some cereals. Eggs are laid singly or in small batches; fecundity is 150–250 eggs; hatching occurs in 4–5 days at 30 °C. Development is completed in 12–15 weeks at 24 °C, and about 6 months at 18–20 °C. Full-grown caterpillars enter diapause in response to temperatures below 22 °C. Most diapausing caterpillars reared at 31 °C pupate after 3–4 months' exposure to 15–18 °C. It occurs principally in south Asia, including India, China, and Japan, but it has spread to other regions of the world.

**Indian meal moth, *Plodia interpunctella* (Fig. 11.2d; 11.5e)** Adult wing span is 6–20 mm, the wings are pale gray, and the outer portion of the fore wing is reddish brown. The body is 6–8.5 mm long; the labial palps extend forward. Full-grown caterpillars are 10–13 mm long and yellowish white, but may be greenish or pinkish white; the head is dark brown. Caterpillars are distinguished from *Ephestia* by not having dark spots at the bases of the setae (one spot per seta), and the rim of the spiracles is weakly sclerotized. Eggs are deposited at night, in batches of 39–275, and usually 3 days after females emerge; fecundity is 150–400 eggs. Hatching is in 7–8 days at 20 °C, 4–5 days at 25 °C, and 3–4 days at 30 °C; no eggs hatch at 15 °C. Development is at temperatures between 18 and 35 °C, and is completed in about 60 days at 20 °C, 30 days at 25 °C, and 25 days at 30 °C. Young caterpillars survive temperatures down to 10 °C. The number of caterpillar stages varies from 5 to 7. Diapause extends development periods. It is generally linked to photoperiods ranging from 12.5 to 13.5 h, but the intensity varies with individual strains. In some, diapause may be induced at about 20 °C, and last no longer than 2–4 months, while in other strains diapause is induced by temperatures above 25 °C, and can last for up to 9 months. Full-grown caterpillars move away from the infested site to pupate; they wander for many hours before stopping to make a cocoon. Diapausing, full-grown caterpillars wander for 36–48 h at about 26 °C; nondiapausing caterpillars wander for 12–24 h at 27 °C. The pupal period is 15–20 days at 20 °C, 8–11 days at 25 °C, and 7–8 days at 30 °C. Under favorable conditions there may be 18 generations per year. Adults may not remain close to the infested material. They are weak flyers, usually in a zigzag pattern, and remain at rest for long periods on walls, usually facing upwards.

This species is a major pest of packaged food in transit and storage. Starved fourth- and fifth-stage caterpillars are capable of making entry holes in bags made of 0.03-mm-thick polyethylene containing suitable food. First-stage caterpillars invade food containers through holes of 0.27 mm diameter or greater, and they can travel 38 cm in search of entry holes. Adults can pass through holes 3 mm diameter or greater, but will oviposit on the surface of containers holding suitable food, and generally avoid laying eggs on empty containers. Caterpillars generally prefer coarse grades of flour, wholewheat or graham flour, maize, and cornmeal. In household and commercial sites this insect infests broken grain and grain products, such as flour and meal, dried fruits and nuts, seeds (especially birdseed), powdered milk, chocolate, and dry pet food. These and adjacent products may be covered by extensive webbing. *P. interpunctella* has been reported infesting the body cavity of domestic cats and an Indian ring-necked parakeet. A live *P. interpunctella* caterpillar was recovered from the brain cavity of a male parakeet, and live caterpillars were excised from subcutaneous tissue in two domestic cats. This species has a cosmopolitan distribution in commercial and residential environments.

**Meal moth, *Pyralis farnalis* (Fig. 11.4a, b; 11.5c)** Adult wing span is about 25 mm, and the front wings are light brown in the middle and dark brown at the base and the tip. The wings are held slightly spread when the moth is at rest. Full-grown caterpillars are 20–25 mm long and the body is slightly gray; the head is black and the posterior is pale orange. Infested food has extensive silk matting over the surface. It feeds on flour, meal, damaged grain, seeds, sesame cake, peanuts, and vegetable refuse. They are common in flour-processing and storage facilities, and prefer damp and spoiled grain and flour. It occurs in damp locations where there is mold growth. Eggs are scattered on the food surface; fecundity is 200–400 eggs. Caterpillars feed from tubes of silk, which contain particles of food. Development is completed in about 2 months, while overwintering is in diapause. It has a wide distribution in temperate regions of Europe and North America.

**Other *Pyralis*** Two other species that are pests of dried and stored grains and seeds include *P. manihotalis* and *P. pictalis*. Adults of *P. manihotalis* are uniformly pale brown, and the wing span is 24–37 mm. This species is endemic to South America, but now occurs in Africa, Asia, and Pacific Islands. It feeds on dried foods, tubers, dried meats, and hides. *P. pictalis* feeds on stored grains. The front wing is basally black and distally reddish brown, while the hind wing is dark basally and pale distally. This species is nearly cosmopolitan.

**Dried fruit moth, *Vitula edmandsii serratilinella* (= *V. edmandsae*) (Fig. 11.4h)** Adult wing span is about 30 mm, and the body and wings are mottled gray. Full-grown caterpillars are about 18 mm long; the body is pale white or pinkish white and the head, thoracic shield, and anal plate are dark brown to black. This insect develops in dried fruits, especially those slightly fermented, and it has been recorded in rotting bee combs. It has been found in the cells of carpenter bees, *Xylocarpa tabaniformis orpifex*, where caterpillars were feeding on stored pollen.

**Other Pyralidae** Caterpillars of *Ectomyelois ceratoniae* feed on stored almonds, and other nuts, and also dried fruit, stored carob pods, and seeds in Europe, Middle East, and Africa. *Hypsopygia mauritialis* adults have a wing span of about 15 mm; the front wings are dark with two pale spots. This species is recorded from Africa, India, China, Taiwan, Japan, Malaysia, Australia, and Hawaii. In Hawaii the caterpillars feed in old nests of vespid wasps. *Corcyra cephalonica* is a common pest of stored food products in Europe, and it has been introduced into the USA. In Australia, caterpillars of *Sclerobia tritalis* damage couch grass growing in lawns; this species occurs as far north as Maryborough in Queensland; *Calamotropha cuneiferellus* damages lawns in the east and north of Australia.

## Sphingidae

The hawk and sphinx moths are large, with a wing span of 25–200 mm, and have a characteristic shape of a streamlined body, large front wings and small hind wings. Hawk moths in flight look similar to hummingbirds, and some species are called hummingbird moths. They visit flowers and are powerful flyers; they hover over open flowers much like birds. Many species are day flyers, some fly at dusk, and others fly after sunset. Caterpillars of many species have a large spine on the posterior end.

**Hummingbird moth, *Hemaris thysbe*** Adult wing span is 45–55 mm, and the wings are reddish brown with large clear areas in each wing. The thorax ventral surface is uniformly dark brown (other species of *Hemaris* have black or red bands). Caterpillars feed on a variety of trees and plants, including hawthorn, cherry, plum, and honeysuckle. The adult moths fly during the day and visit ornamental flowers.

## Tineidae

Most tineids are small moths; the adult wing span ranges from 12 to 30 mm. They are weak flyers and do not move far from the site of caterpillar feeding. These moths rarely fly to lights and are not active in lighted areas. Caterpillars are scavengers on a variety of dry plant material and fungi. Some tineids feed on animal material, including animal horns and woolen fabrics. Caterpillars of burrowing webworms, *Acrolophus* spp., make a tubular web in the ground and feed on the roots of grasses. *Acridotarsa conglomerata* occurs in eastern and northern Australia, and its immature stages are probably scavengers in the nests of the termites, *Neotermes insularis* and *Mastotermes*. Species in the genera *Dryadaula*, *Nemapogon*, and *Oinophila* are associated with fungi and corks in wine storage. Caterpillars of *Dryadaula pactolia* have been recorded as attacking the cork in wine bottles stored in cellars. *Haptotinea ditella* and *H. insectella* occur in Europe and western Asia; the immature stages of these moths feed on stored grains, rice, and groundnuts.

**Cork moth, *Nemapogon cloacella*** Adult wing span is 10–17 mm. Wings are mottled brown and white, and the hind wings are gray. Full-grown caterpillars are about 16 mm long, and yellowish white to greenish white. They feed on rotten wood, fences, and tree stumps, and they can also feed on corks in wine bottles in cellars. A closely related species, *Nemapogon personella*, is a pest of stored grain in Sweden.

**European grain moth, corn moth, cork moth, *Nemapogon granella* (Fig. 11.2g)** Adults are 6–7 mm long and their wing span is 8–16 mm. The body is light brown and mottled with silvery white spots. Wings are mottled with black and white, and the front and hind wings have a fringe of long hairs along the edge. Full-grown caterpillars are 11–16 mm long and yellowish white with a pale yellow to dark brown head. Eggs are laid singly or in small batches directly on food; fecundity is 50–200 eggs. Hatching is in 5–30 days, depending on temperature and humidity. Development takes 200–400 days at 18 °C and 80–90% RH. There are two generations per year. In wine cellars, caterpillars infest and feed on exposed corks. Females lay eggs in crevices on the surface of unprotected corks in wine bottles; damp corks are preferred over dry or very wet corks. Corks covered with mold are preferred, and it is uncertain whether *N. granella* caterpillars actually feed on the cork or the mold. Oviposition is usually on bottles containing red wine, but corks in other bottles are attacked. Caterpillars often leave small amounts of cork frass bound together by silk produced when feeding. Internal tunnels and cavities in the cork generally do not reach the cork bottom, but are restricted to the upper two-thirds. The quality of the wine in the bottle may not be adversely influenced by damage (caterpillars feeding) to the cork. This species feeds in grain and stored cereals, soybeans, garlic, dried fruits, dried mushrooms (*Psalliota* spp., *Boletus edulis*), the dry rot fungus, *Serpula lacrymans*, decayed wood, and pharmaceutical products. It is thought to have originated in the Palearctic region, but now distribution is nearly cosmopolitan in the cool regions of the world, and includes indoor and outdoor populations in Europe, North and South America, Japan, and parts of Africa. A closely related species, *N. variatella*, occurs on stored grain in northern Europe.

**Poultry-house moth, *Niditinea fuscipunctella*** Adults have a wing span of 10–16 mm. The front wings are yellowish gray and the hind wings are brownish gray. Caterpillars feed on animal material, including skins and bedding material. This species is common in commercial poultry houses, and also in bird nests, where it feeds on feces. In California, population peaks occur in spring and early summer, followed by mid- and late-summer peaks.

**Wine cellar moth, *Oinophila v-flava*** Adult wing span is 8.5–10 mm and the wings are slender. The front wings have two irregular, pale brown bands. Caterpillars are scavengers on a variety of plant and animal materials. This species often occurs in wine cellars where it feeds on fungi and wine corks.

**Plaster bagworm, *Phereoeca uterella* (= *P. walshingham*, *P. dubitatrix*)** Adult wing span is about 12 mm, and the body is dusky brown. Caterpillars remain in portable cases and are usually found crawling on walls, and in stored-food materials. This species is widely distributed in North America.

**Webbing clothes moth, *Tineola bisselliella* (Fig. 11.2f; 11.5g)** Adult wing span is 9–14 mm, and the body is dark yellow to reddish brown. Wings are uniformly gray, without dark spots. Full-grown caterpillars are about 12 mm long, shiny, and yellowish white. Eggs are deposited singly or in groups of two or more between threads on the cloth surface; they are held in place by a gelatinous material. Eggs are laid singly or in groups of about 25. Eggs are usually laid during a period of 2–3 weeks; fecundity is 40–100 eggs, and the maximum may be 221. Hatching occurs in 37 days at 13 °C and in 7 days at 33 °C; eggs are not capable of overwintering or delaying development. Development includes 5–45 stages (molts), and can last from 35 days to 2.5 years. Development on flannel is completed in 63 days at 23.5 °C and 50% RH, and 55 days at 31 °C and 66% RH. At 25 °C the optimum RH is 75%, but most caterpillars can complete development at 30% RH. They survive 67 days at −6.7 °C to −3.9 °C, and 21 days at −15 °C to −12.2 °C. Caterpillars make a cocoon for pupation. The pupal period is 8–10 days in warm weather, and 3–4 weeks in cold weather. Adults are active and can fly relatively long distances. Moths are capable of flying 400–800 m. Males live for 13–79 days, and females for 10–48 days at 20–25 °C. They are not attracted to lights, and gravid females are weak flyers. Clothes moth caterpillars feed on wool clothes, natural carpets, furs, stored wool, and piano felts; in the tropics they infest dried fishmeal and hides. Outdoors the immature stages feed on pollen, hair, feathers, wool, fur, dead insects, and dried animal remains in animal burrows. The caterpillar spins a feeding tunnel or a flat mat of silk and feeds below this webbing or in the temporary tube. This species is probably native to Africa, but it is now nearly cosmopolitan in distribution.

Adults uses sonic signals in addition to pheromonal signals for communication. Male *T. bisselliella* produce sounds of 27 db, with a base frequency of 40–50 Hz and a harmonic frequency of 80–100 Hz. Sound intensity increases to 55 db when calling males are within 2 cm of females or other males. There is no evidence that females produce sound and no evidence for ultrasonic sound production by either sex.

This species was first described in Russia by Arvid-David Hummel as *Tinea bisselliella* in 1823. His description was published in the journal *Essais Entomologiques*, which had only limited distribution. Zeller published a more extensive description in 1852, but dropped the second s in *bisselliella*. It was later moved to the genus *Tineola* and the misspelling of the name continued until about the mid-1940s.

**Case-making clothes moth, *Tinea pellionella* (Fig. 11.5f)** Adult wing span is 11–17 mm, and the body is grayish yellow. The front wings have three dark spots on the middle; the hind wings are yellowish brown and without spots. Males are active flyers; females are slow and relatively inactive. Full-grown caterpillars are about 10 mm long and yellowish white. They remain within a cylindrical retreat or case, which is open at both ends, and constructed of silk with particles of debris. Eggs are slightly sticky and have longitudinal ridges; they are laid singly or in small groups; fecundity is 37–48 eggs. Hatching occurs in 4–7 days. Development is complete in 68–87 days, and pupation takes place in the case after both ends are sealed. The pupal stage lasts 9–19 days. There are three or four generations per year from caterpillars that are fed on woolen fabric enhanced with 5% yeast at about 26 °C and about 82% RH. This species is nearly cosmopolitan.

**Tapestry moth, *Trichophaga tapetzella*** Adult wing span is 6–10 mm and the body is yellowish white or gray. The front wing has the basal third black, and the remainder grayish white. Caterpillars make a gallery composed of silk and mixed with fragments of cloth; the gallery may be long and winding.

**Other clothes moths** In Australia, *Tinea pallescentella* and *Niditinea fuscella* are usually pests of stored cereals, but will also attack animal fibers. Two case-making species, *T. translucens* and *T. dubiella*, are also pests of wool materials. Caterpillars construct portable flattened cases with an opening flap at each end; pupation occurs in the case. *T. vastella* is distributed throughout Africa, and the immatures infest dried animal material and dried fruits. In Australia, species of *Monopis* have case-bearing caterpillars that feed on animal fibers, feathers, bird droppings, and dead animal materials. Species of this genus, in both Australia and New Guinea, are ovoviviparous or viviparous. Fertilized eggs are retained in an enlarged genital chamber until the young are ready to hatch.

## Tortricidae

Species in this large family eat the foliage and fruits of a wide variety of trees. The caterpillars often roll or fold leaves or shoots together in which to rest and feed. The adults are small, with a wing span of 25 mm or less, and have wide, oblong, and fringed wings. The caterpillar is usually light or dark green and about 25 mm long. Current classification of this family incorporates species in the family Oleuthreutidae, which includes several species that feed on fruits and nuts.

**Uglynest caterpillar moth, *Archips cerasivoranus*** Adult wing span is 18–25 mm. The front wing is orange with purple iridescence, and it is spotted and speckled with dark red-brown spots. Caterpillars live on wild cherry trees in large nests of mixed silk and debris. When full-grown, they leave the silken nest to form cocoons, and they migrate to the perimeter of buildings and form cocoons on sidings and around foundations. Adults appear in late June and early July. This species is generally distributed in eastern USA.

**Filbertworm, *Melissopus latiferranus*** Adult wing span is about 13 mm and the body is pale or dusky bronze. The front wing has a complete, broad, coppery median band and a broken postmedian coppery band. Full-grown caterpillars are about 20 mm long, yellowish white, and have a light-brown head. This species infests the Catalina cherry tree (*Prunus lyonii*) in southern California, and the large, green galls of the blue oak (*Quercus douglasii*). It also infests walnuts, hazelnuts (filberts), and acorns. It is a pest only where the oak-apple gall is produced; the moths migrate from the oaks to walnuts and other trees. Caterpillars gain access to the walnuts when cracks appear in the husk at the time of maturity. Once nuts are infested, caterpillars and adults can occur indoors.

## Bibliography


Austin, A. D. A note on the life history of *Anatrachyntis terminella* (Walker) (Lepidoptera: Cosmopterigidae), whose larvae are predacious on the eggs of *Nephila edulis* (Koch) (Araneae: Araneidae). *J. Aust. Entomol. Soc.*, **16** (1977), 427–8.

Back, E. A. and R. T. Cotton. Insect pests of upholstered furniture. *J. Econ. Entomol.*, **23** (1930), 833–7.

Barlow, H. S. *An Introduction to the Moths of South East Asia*. Kuala Lumpur: Malayan Nature Society, 1982.

Benson, J. F. The biology of the Lepidoptera infesting stored products, with special reference to population dynamics. *Biol. Rev.*, **48** (1973), 1–26.

Capps, H. W. *Keys for the Identification of some Lepidopterous Larvae Frequently Intercepted at Quarantine*. ARS-33-20-1. Washington, DC: US Department of Agriculture, 1963.

Common, I. F. B. Lepidoptera (moths and butterflies) In *The Insects of Australia*, CSIRO, chapter 36. Melbourne: Melbourne University Press, 1970.

Corbet, A. S. and W. H. T. Tams. Keys for the identification of the Lepidoptera infesting stored food products. *Proc. Zool. Soc. Lond.*, **113** (1943), 55–148.

Cotton, R. T. *Pests of Stored Grain and Grain Products*. Minneapolis, MN: Burgess, 1956.

Covel, C. V. Jr. *Field Guide to the Moths of Eastern North America*. Boston, MA: Houghton Mifflin, 1984.

Fracker, S. B. The Classification of Lepidopterous Larvae, 2nd edn. *Contrib. Entomol. Lab. Univ. Ill.*, **43** (1930), 1–161.

Freeman, P. (ed.) *Common Insect Pests of Stored Food Products. A Guide to their Identification*. Economic Series No. 15. London: British Museum (Natural History), 1980.

Gorham, R. Filth in foods: implications for health. *J. Milk Food Technol.*, **38** (1975), 409–18.

The significance for human health of insects in food. *Annu. Rev. Entomol.*, **24** (1979), 209–24.

(ed.) *Ecology and Management of Food-Industry Pests*. Food and Drug Administration technical bulletin. 4. Arlington, VA: Association of Official Analytical Chemists, 1991.

Hickin, N. E. *Household Insect Pests*. London: Hutchinson, 1964.

Hinton, H. E. The larvae of the Lepidoptera associated with stored products. *Bull. Entomol. Res.*, **34** (1943), 171–8.

Holland, W. J. *The Moth Book: A Popular Guide to Knowledge of the Moths of North America*. New York: Dover Press, 1968.

Howe, R. W. A summary of estimates of optimal and minimal conditions for population increase of some stored products insects. *J. Stored Prod. Res.*, **1** (1965), 177–84.

Kino, T. and S. Oshima. Allergy to insects in Japan. I. The reaginic sensitivity to moth and butterfly in patients with bronchial asthma. *J. Allergy Clin. Immunol.*, **61** (1978), 10–16.

Kurtz, O. L. and K. L. Harris. Identification of insect fragments: relationship to the etiology of the contamination. *J. Assoc. Agric. Chem.*, **38** (1955), 1010–15.

Lamas, G. and J. E. Peres. Lepidopteros de importancia medica. *Diagnostico*, **20** (1987), 121–5.

Marsh, D., J. S. Kennedy, and A. R. Ludlow. An analysis of anemotactic zigzagging flight in male moths stimulated by pheromone. *Physiol. Entomol.*, **3** (1978), 221–40.

Mound, L. (ed.) *Common Insect Pests of Stored Products. A Guide to Their Identification*. London: British Museum (Natural History), 1989.

Okumura, G. T. Key to the lepidopterous larvae found in stored foods in California. *Sacramento State Coll. Nat. Hist. Ser.*, **5** (1951), 1–13.

Parkin, E. A. Stored product entomology. *Annu. Rev. Entomol.*, **1** (1956), 223–40.

Peterson, A. *Larvae of Insects. An Introduction of Nearctic Species. Part 1. Lepidoptera and Plant Infesting Hymenoptera*. Ann Arbor, MI: Edwards, 1948.

Scoble, M. J. *The Lepidoptera: Form, Function, and Diversity*. Oxford: Oxford University Press, 1992.

Sinha, R. N. Ecology of storage. *Annu. Technol. Agric.*, **22** (1973), 351–69.

Weisman, D. M. *Key to the Identification of Some Frequently Intercepted Lepidopteran Larvae*. APHIS/PPQ 81-47. Washington, DC: US Department of Agriculture, 1986.

Werner, K. *Die Larvalsystematik einiger Kleinschmetterlingsfamilien (Hyponomeutidae, Othoteliidae, Acrolepiidae, Tineidae, Incurvariidae und Adelidae)*. Berlin: Akademie-Verlag, 1958.

Wirtz, R. A. Allergic and toxic reactions to non-stinging arthropods. *Annu. Rev. Entomol.*, **29** (1984), 47–69.

Young, M. *The Natural History of Moths*. London: T. and A. D. Poyser, 1997.

Zimmerman, E. C. *Insects of Hawaii*, vol. 9, parts 1 and 2. *Microlepidoptera*. Honolulu: University of Hawaii, 1978.

### Gelechiidae

Boldt, P. E. Effects of temperature and humidity on development and oviposition of *Sitotroga cerealella*. *J. Kansas Entomol. Soc.*, **41** (1974), 30–5.

Hammad, S. M., M. G. Shenouda, and A. L. El-Deeb. Studies on the biology of *Sitotroga cerealella* Oliv. *Bull. Soc. Entomol. Egypte*, **51** (1967), 257–68.

Landry, B. Systematics of Nearctic Scythrididae (Lepidoptera: Gelechoidea): phylogeny and classification of supraspecific taxa, with a review of described species. *Ottawa Entomol. Soc. Can. Mem.*, **160** (1991).

Misra, C. P., C. M. Christensen, and A. C. Hodson. The Angoumois grain moth, *Sitotroga cerealella*, and storage fungi. *J. Econ. Entomol.*, **54** (1961), 1032–3.

### Lasiocampidae, Lymantriidae

de Jong, M. C. J. M. and E. Bleumink. Investigative studies of the dermatitis caused by the larva of the brown-tail moth, *Euproctis chrysorrhoea* L. (Lepidoptera, Lymantriidae) III. Chemical analysis of skin reactive substances. *Arch. Dermatol. Res.*, **259** (1977), 247–62.

Investigative studies of the dermatitis caused by the larva of the brown-tail moth, *Euproctis chrysorrhoea* L. (Lepidoptera, Lymantriidae) IV. Further characterization of skin reactive substances. *Arch. Dermatol. Res.*, **259** (1977), 263–81.

de Jong, M. C. J. M., E. Bleumink, and J. P. Nater. Investigative studies of the dermatitis caused by the larva of the brown-tail moth (*Euproctis chrysorrhoea* Linn.) I. Clinical and experimental findings. *Arch. Dermatol. Res.*, **253** (1975), 287–300.

de Jong, M. C. J. M., P. J. Hoedemaecker, W. L. Jongebloed, and J. P. Nater. Investigative studies of the dermatitis caused by the larva of the brown-tail moth (*Euproctis chrysorrhoea*) II. Histopathology of skin lesions and scanning electron microscopy of their causative setae. *Arch. Dermatol. Res.*, **255** (1976), 177–91.

Fitzgerald, T. D. *The Tent Caterpillars*. Ithaca, NY: Cornell University Press, 1995.

Kawamoto, F. Studies on the venomous spicules and spines of moth caterpillars II. Pharmacological and biochemical properties of the spicule venom of the oriental tussock moth caterpillar, *Euproctis subflava*. *Jpn. J. Sanit. Zool.*, **29** (1978), 175–83.

Stehr, F. W. and E. F. Cook. A revision of the genus *Malacosoma* Hübner in North America (Lepidoptera: Lasiocampidae): systematics, biology, immatures, and parasites. *Bull. US Natl. Mus.*, **276** (1968), 1–321.

### Noctuidae, Oecophoridae, Olethreutidae

Banziger, H. Skin-piercing blood-sucking moths. IV: biological studies on adults of 4 *Calyptra* species and 2 subspecies (Lep., Noctuidae). *Mitt. Schw. Entomol. Gesell.*, **59** (1986), 111–38.

Banziger, H. and W. Buttiker. Records of eye-frequenting Lepidoptera from man. *J. Med. Entomol.*, **6** (1969), 53–8.

Clarke, J. F. G. Revision of the North American moths of the family Oecophoridae, with descriptions of new genera and species. *Proc. US Natl. Mus.*, **90** (1941), 33–286.

Cole, J. H. *Hofmannophila pseudospretella* (Stnt.) (Lep., Oecophoridae), its status as a pest of woollen textiles, its laboratory culture and susceptibility to mothproofers. *Bull. Entomol. Res.*, **53** (1962), 83–9.

Hodges, R. W. *Gelechoidea: Oecophoridae* (in part) (fasc. 6.2). In Dominic R. B. (ed.) *The Moths of America North of Mexico*. Washington, DC: Wedge Entomological Research Foundation, 1974.

MacKay, M. R. Larvae of the North American Olethreutidae (Lepidoptera). *Can. Entomol.*, **91** (suppl. 10) (1959), 1–338.

Woodroffe, G. E. A life-history study of the brown house moth, *Hofmannophila pseudospretella*. *Bull. Entomol. Res.*, **41** (1951), 529–33.

### Psychidae, Pyralidae

Aitken, A. D. A key to the larvae of some species of Phycitinae (Lepidoptera, Pyralidae) associated with stored products, and some related species. *Bull. Entomol. Res.*, **54** (1963), 175–88.

Allyson, S. Last instar larvae of Pyraustini of America north of Mexico (Lepidoptera: Pyralidae). *Can. Entomol.*, **113** (1981), 416–518.

Bell, C. H. Factors influencing the duration and termination of diapause in the warehouse moth, *Ephestia elutella*. *Physiol. Entomol.*, **1** (1976), 169–78.

The regulation of development during diapause in *Ephestia elutella* (Hübner) by temperature and photoperiod. *J. Insect Physiol.*, **29** (1983), 485–90.

Burges, H. D. and K. P. F. Haskins. Life-cycle of the tropical warehouse moth, *Cadra cautella*, at controlled temperatures and humidities. *Bull. Entomol. Res.*, **55** (1965), 775–89.

Cox, P. D. The influence of temperature and humidity on the life-cycles of *Ephestia calidella* (Guenee) and *Ephestia figulilella* Gregson (Lepidoptera: Phyctidae). *J. Stored Prod. Res.*, **11** (1974), 75–85.

The suitability of dried fruits, almonds, and carobs for the development of *Ephestia figulilella*, *E. calidella* and *E. cautella*. *J. Stored Prod. Res.*, **11** (1975), 229–33.

Cold tolerance and factors affecting the duration of diapause in *Ephestia kuehniella* Zeller (Lepidoptera: Pyralidae). *J. Stored Prod. Res.*, **23** (1987), 163–8.

Davis, D. R. Bagworm moths of the western hemisphere (Lepidoptera: Psychidae). *Bull. US Natl. Mus.*, **244** (1964).

Fraenkel, G. and M. Blewett. The dietetics of the caterpillars of three *Ephestia* species, *E. kuehniella*, *E. elutella* and *E. cautella*, and of a closely related species, *Plodia interpunctella*. *J. Exp. Biol.*, **22** (1946), 162–71.

Hausenfuss, I. *Die Larvalsystematik der Zunster (Pyralidae)*. Berlin: Akademie-Verlag, 1960.

Jacob, T. A. and P. D. Cox. The influence of temperature and humidity on the life-cycle of *Ephestia kuehniella* Zeller (Lepidoptera: Pyralidae). *J. Stored Prod. Res.*, **13** (1976), 107–18.

Le Cato, L. G. Yield, development and weight of *Cadra cautella* and *Plodia interpunctella* on 21 diets derived from natural products. *J. Stored Prod. Res.*, **12** (1976), 43–7.

Lum, P. T. M. and B. R. Flaherty. Effect of mating with males reared in continuous light or in light–dark cycles on fecundity in *Plodia interpunctella* males. *Ann. Entomol. Soc. Am.*, **63** (1969), 1470–1.

Mathlein, R. Investigations on pests of stored products. II. The grain moths, *T. secalella* and *T. granella*. *Statens Vaxtskyddsanstalt Medd.*, **34** (1941), 1–56.

Nawrot, J. Effect of temperature and relative humidity on population parameters for almond moth (*Cadra cautella*) reared on natural products. *Prace Naukowe Insty. Ochrony Roslin*, **21** (1979), 41–52.

Population parameters for almond moth (*Cadra cautella*) reared on natural products. *Prace Naukowe Insty. Ochrony Roslin*, **21** (1979), 53–60.

Richards, O. W. and W. S. Thompson. A contribution to the study of the genera *Ephestia* Gn. (including *Stymax* Dyar), and *Plodia*, Gn. (Lepidoptera, Phyctidae), with notes on parasites of the larvae. *Trans. Entomol. Soc. Lond.*, **80** (1932), 169–250.

Richards, O. W. and N. Waloff. The study of a population of *Ephestia elutella* Hübner (Lep. Phycitidae) living on bulk grain. *Trans. R. Entomol. Soc. Lond.*, **97** (1946), 253–98.

Shinoda, K., S. Tanaka, T. Yoshida, and F. Nakasuji. Penetration of polyethylene film for packaging by larvae of Indian-meal moth, *Plodia interpunctella* (Hübner) (Lepidoptera: Pyralidae) and their dispersal. *Jpn. J. Environ. Entomol. Zool.*, **2** (1990), 128–32.

Tsuji, H. Experimental invasion of a food container by first-instar larvae of the Indian meal moth, *Plodia interpunctella* Hübner, through pinholes. *Med. Entomol. Zool.*, **49** (1998), 99–104.

Tzanakakis, M. E. An ecological study of the Indian meal moth, *Plodia interpunctella*, with emphasis on diapause. *Hilgardia*, **29** (1959), 205–46.

Ullyett, G. C. Oviposition by *Ephestia kuehniella* Zell. *J. Entomol. Soc. South Afr.*, **8** (1945), 53–9.

Vansell, G. H. The wax moth as a household pest. *J. Econ. Entomol.*, **36** (1943), 626–7.

Waloff, N., M. J. Morris, and E. C. Broadhead. Fecundity and longevity of *Ephestia elutella*. *Trans. R. Entomol. Soc.*, **99** (1948), 245–68.

### Tineidae

Austen, E. E., A. W. McKenny Hughes, and H. Stringer. Clothes moths and house moths. *Br. Mus. (Nat. Hist.) Econ. Ser.*, **14** (1935), 1–55.

Bender, E. Investigations of the biology and morphology of the Microlepidoptera that inhabit wine-cellars. *Rev. Appl. Entomol. A*, **35** (1941), 71–2.

Billings, S. C. Notes on clothes moths breeding. *J. Econ. Entomol.*, **29** (1936), 1014–16.

Cheema, P. S. Studies on the bionomics of the case-bearing clothes moths, *Tinea pellionella* (L.). *Bull. Entomol. Res.*, **47** (1956), 167–82.

Colman, W. Minimum size of openings through which clothes moth larvae can pass. *J. Econ. Entomol.*, **33** (1940), 582.

Geigy, R. and R. Zinkernagel. Beobachtungen beim Aufbau einer technischen Grosszucht der Kleidermotte (*Tineola bisselliella*). *Mitt. Schweiz Entomol. Ges.*, **18** (1941), 213–32.

Griswold, G. H. On the length of the adult life in the webbing clothes moth, *Tineola bisselliella* Hum. *Ann. Entomol. Soc. Am.*, **24** (1931), 761–4.

Griswold, G. H. and M. F. Crowell. The effect of humidity on the development of the webbing clothes moth (*Tineola bisselliella*). *Ecology*, **17** (1936), 241–50.

Hinton, H. E. The larvae of the species of Tineidae of economic importance. *Bull. Entomol. Res.*, **47** (1956), 251–346.

Legner, E. F. and R. E. Eastwood. Seasonal and spatial distribution of *Tinea fuscipunctella* on poultry ranches. *J. Econ. Entomol.*, **66** (1973), 685–7.

Madrid, F. J. and R. N. Sinha. *Haplotinea insectella* (Fabricius) (Lepidoptera: Tineidae), a potential pest of stored cereals in North America. *J. Stored Prod. Res.*, **20** (1984), 107–10.

Rawle, S. G. The effects of high temperature on the common clothes moth, *Tineola bisselliella*. *Bull. Entomol. Res.*, **42** (1951), 21–40.

Robinson, G. S. Clothes-moths of the *Tinea pellionella* complex: a revision of the world's species (Lepidoptera: Tineidae). *Bull. Br. Mus. Nat. Hist. (Entomol.)*, **38** (1979), 57–128.

Robinson, G. S. and E. S. Nielsen. The pest species of *Tinea* (clothes moths) occuring in Australia (Lepidoptera: Tineidae). *Gen. Appl. Entomol.*, **19** (1987), 45–8.

Takács, S. and G. Gries. Communication ecology of webbing clothes moth: evidence for male-produced aggregation signal(s). *Can. Entomol.*, **133** (2001), 725.

Takács S., M. Mistal, and G. Gries. Communication ecology of webbing clothes moth: attractiveness and characterization of male-produced sonic aggregation signals. *J. Appl. Entomol.*, **127** (2003), 127–33.

Titschack, E. Beitrage zu einer Monographie der Kleidermotte, *Tineola bisselliella*. *Z. Technol. Biol.*, **10** (1922), 1–168.

Yamada, Y. On the life-history of *Tineola bisselliella* Hamps (in Japanese). *Bochu Kagaku*, **2** (1938), 13–16.

### Zygaenoidea

Epstein, M. E. Revision and phylogeny of the Limacodid-group families, with evolutionary studies on slug caterpillars (Lepidoptera: Zygaenoidea). *Smithson. Contrib. Zool.*, **582** (1996), 1–102.

# 12 Mantodea, Neuroptera

## MANTODEA
## Introduction

Praying mantids are primarily tropical insects, and there are about 1800 species in eight families worldwide. They are characterized by an elongate prothorax and modified front legs. Front coxae are long and mobile, and the front femora and tibiae usually have strong spines, and flex for grasping prey. Legs are held in front of the head, simulating an attitude of prayer, which is the origin of the common name. They are predaceous on living insects. There may be an elaborate premating sequence, and the male usually approaches the female with caution. In some instances the female in copulation turns around and devours the participating male. Females of some species use a pheromone to attract males.

Eggs are laid in batches of 12–400 in an eggcase, which begins as a frothy liquid but hardens into a tough fibrous protective structure after the eggs have been deposited in it. The shape and placement of the eggcase are characteristic for each species. Most mantids spend the winter or tropical dry season in the egg stage; however, *Empusa* spp. overwinter as nymphs. There is usually one generation per year in north temperate regions, but several in the tropics. Adults and nymphs are found on trees and ornamental plants around buildings; some species fly to lights at night.

## Mantidae

Species in this family are exclusively carnivorous and occur in all warm regions of the world. Included in this group are species that commonly occur on ornamental trees and shrubs in urban and suburban areas. They are called praying mantids because of their habit of holding the front legs in an attitude of prayer. The South African name, hottentot god, seems to have been given for a similar reason. There are many mantid species in Africa, and some of the most common in peridomestic habitats include the large green mantid, *Spodromantis gastrica*, and the common brown mantid, *Tarachodes perloides*. Four species have been introduced into the USA: *Tenodera aridifolia sinensis*, found in Pennsylvania in 1896; *Mantis religiosa*, found in New York in 1899; *Tenodera angustapennis*, found in Maryland in 1926; and *Iris oratoria*, found in California in 1933. In the 1930s and 1940s, *M. religiosa* was released in Canada as a biological control agent; similarly *Tenodera aridifolia sinensis* was released in New York, Ohio, Illinois, and California. This may be the origin of the prevalent myth that mantids are protected by law, and killing one may result in a fine.

**Carolina mantid, *Stagmomantis carolina* (Fig. 12.1d)** Adults are 75–100 mm long. The male has grayish-brown wings and a green body and legs, while females are brown or greenish yellow, with light green front wings. Prey species for this mantid include orthopterans and other mantids. This is the most common species in southern USA, west to New Mexico. A related species, *S. floridensis*, occurs in Florida and is slightly longer than *S. carolina*.

**European mantid, *Mantis religiosa*** Adults are about 50 mm long; the body is pale green or brown. They readily fly in warm weather. This small species was introduced into the USA in 1899, and now occurs in many parts of the country, and is commercially available for school biology demonstrations and as a household pet. It also occurs in southern Europe.

**Mediterranean mantid, *Iris oratoria*** Adults are 45–55 mm long; the body is green with brown dorsal markings. Hind wings are often reddish brown with dark eye-spots. Prey for the nymphs and adults includes Hemiptera: Pentatomidae; Araneae: Salticidae; Hymenoptera: Vespidae; Coleoptera; Dermaptera; and Hymenoptera: Apidae. This species is attracted to lights at night. It occurs in southern California and Arizona.

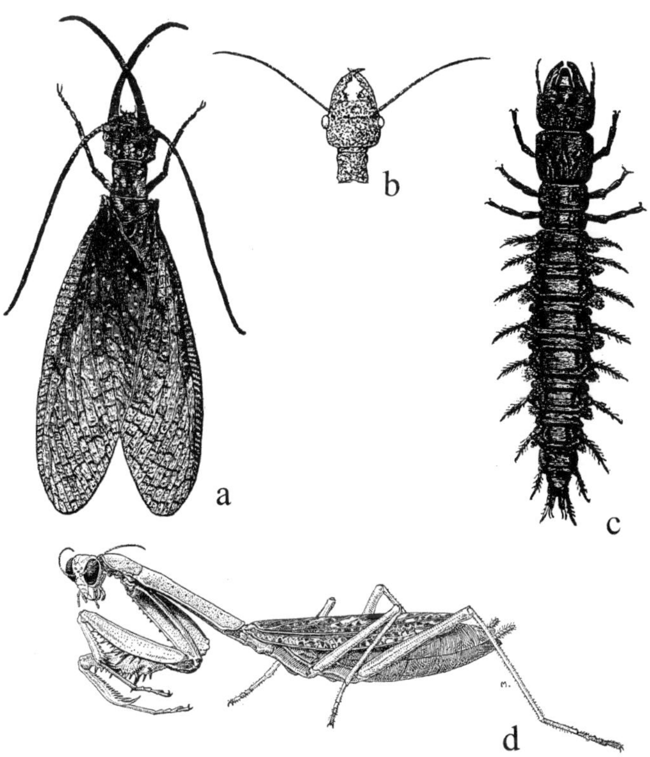


**Figure 12.1** Mantodea, Neuroptera. (a) *Corydalus cornutus* male; (b) *C. cornutus* female, head; (c) *C. cornutus* larva; (d) *Stagmomantis carolina* male, adult.

**Chinese mantid, *Tenodera aridifolia sinensis*** Adults are 75–100 mm long, and the body is green or sometimes brown. Adults of both sexes are winged, but only the males are normally observed flying. The females are probably too heavy during oogensis to fly, except for short distances. Under laboratory conditions, females can produce 4–6 oothecae, with an interval of 8–10 days between successive ovipositions. Egg masses are common on ornamental plants around buildings. This species was introduced into the USA in 1896 near Philadelphia, and is now widely distributed in eastern USA, Japan, and China.

**Narrow-winged mantid, *Tenodera angustipennis*** Adults are 70–105 mm long and green. The front margin of the front wings is narrow, and the hind wings are narrower than the front and marked with black. This species was introduced into the USA in 1926, and is now widespread throughout the country.

# NEUROPTERA
## Introduction

This order contains a variety of terrestrial and aquatic insects. Adults have two pairs of wings, which have numerous veins, and are held rooflike over the abdomen. Generally, they are weak flyers. Antennae are long, the tarsi are five-segmented, and there are no cerci. Adults have chewing mouthparts and are often predaceous on other insects. Larvae usually have well-developed mandibles and most are predaceous. They undergo complete metamorphosis. Pest status is limited to those that are attracted to lights at night, but there is at least one species that regularly occurs indoors. Common neuropterans at lights in the urban environment include green and brown lacewings, antlions, owlflies, and dobsonflies.

## Ascalaphidae

Owlfly adults are 40–80 mm long; the antennae are clubbed, and nearly as long as the body; the eyes are divided in some species. They are diurnal or crepuscular flyers, and usually remain on tree branches during the day. Larvae are in leaf litter and plants, and prey on other insects. This family is predominantly found in the western hemisphere. Adults of the owlfly, *Ululodes arizonensis*, are common at lights at night in southwestern USA.

## Chrysopidae

Lacewing adults are 10–25 mm; their body and wings are usually green, but sometimes brown. The wings are transparent and the eyes are usually golden. Adults and larvae of some species are predaceous, while others are nectar or pollen feeders.

**Golden-eye lacewing, *Chrysopa oculata*** Adults are 15–22 mm long. They are green with a black stripe on the face, and the eyes are golden. Wings are pale green. Eggs are laid on upright stalks and attached to leaves; larvae are predators. Adults are common on foliage, and fly from May to September. They fly to lights at night, and they are often found indoors. Adults of a related species, *C. californica*, are 9–14 mm long, and pale green, with a yellow stripe on the dorsum.

**Garden lacewing, *Chrysoperla carnea*** Adults are about 6 mm long and pale green. They become pale brown in fall. This lacewing is often found indoors. It is distributed in the UK.

## Corydalidae

Dobsonflies are usually over 25 mm long, with large wings and antennae. They are soft-bodied, weak flyers, usually found near water. Some species are attracted to lights at night and can be found far from their breeding site. Larvae are aquatic and occur under stones in streams; they are called helgramites and are used as bait for fishing. Dobsonfly larvae are collected by natives for food in the Peruvian Amazon region. In parts of Mexico, they are considered venomous and feared.

**Common alderfly, *Archichauliodes diversus*** Adults are about 30 mm long and orange with speckled wings. Larvae are known as black-creepers. This species occurs in New Zealand.

**Dobsonfly, *Corydalus cornutus* (Fig. 12.1a–c)** The adult male is 100–125 mm long and there are long and curving mandibles extending from the head. Females have small mandibles. This species occurs in North America, east of the Rocky Mountains. Eggs are laid in batches and attached with a white covering to objects such as bridges and tree branches above water. After hatching, nymphs drop to the water and crawl under stones, where they are predators on other insects. After about 2.5 years, they leave the water to pupate in the soil along the bank. Adults remain near streams and they are rarely seen, except when they fly to lights at night. The large mandibles on the males make them look more dangerous than they actually are. A related species, *C. cognata*, is about 60 mm long, and occurs in western North America.

**Fishfly, *Chauliodes rastricornis*** Adults are 30–50 mm long. The body is brown and the wings are grayish brown. Antennae are pectinate. Adults fly from May to July near lakes and large streams, and occur at lights at night.

## Hemerobiidae

Brown lacewings resemble the common lacewings (Chrysopidae), but are brown to pale brown and their wings are densely setose. Adults are 5–11 mm long and the wings are held rooflike over the body. Their wing venation is distinctly different: the chrysopids have a single radial sector, while the hemerobiids appear to have 2–4 veins that branch off the radial vein. These insects are generally found in wooded areas; the adults and larvae are predaceous on scale insects, mealybugs, aphids, and spider mites. They often occur in greenhouses, and adults occur indoors at lights. The large genus *Hemerobius* is represented by species that occur in North America, the UK, and continental Europe.

## Ithonidae

The mothlike netwing adults are 20–40 mm long; the body is brown to blackish brown. Larvae are subterranean, and grublike; they probably feed on roots in sandy soils.

**Moth lacewing, *Oliarces clara*** Adults are 21–35 mm long. The body is dark brown or blackish brown while wings are yellowish white, and shiny. The thorax has tufts of black setae. Larvae usually live around the roots of trees and shrubs. Adults swarm in large numbers during the day in April and May.

## Myrmeleontidae

Larvae of this family are called antlions (also doodlebugs), and are the most familiar form. The adults have a 40–80 mm wingspan, and resemble damselflies (Odonata), but have short, clubbed antennae. Larvae are predators and remain in sandy soil or in shallow, cone-shaped pits and wait in the bottom for ants (hence the common name) or other insects to become trapped.

The most common antlion around the perimeter of houses in northern USA is *Myrmeleon immaculatus*. Adults of the spotted-winged antlion, *Dendroleon obsoletus*, and *Vella* spp. are weak flyers, and they rest on trees or posts during the day. They are often attracted to lights at night.

**Antlion, *Weeleus acutus*** Adults are about 30 mm long. The body is pale brown to gray, and the wings are opaque with numerous dark markings and covered with fine setae. Full-grown nymphs are about 15 mm long and reddish gray, with dark spots. Eggs are laid in soil. The nymph constructs a shallow pit in loose soil during the night; it works backward and counterclockwise to eject the soil from the gradually narrowing pit. The pit is about 5 cm wide and 5–6 cm deep when the larva is full-grown. Larval food consists of insects and other arthropods that enter the pit and are captured by the antlion nymph at the bottom. Pupation occurs in the base of the pit. Development usually requires more than 1 year. Pits are constructed in loose soil in sunlight, but sheltered from rain, and they are located around buildings. This species occurs in New Zealand.

## Nemopteridae

This is a specialized family, in which adults have extremely long hind wings, and the head is prolonged into a specialized rostrum. Flight is an up-and-down motion with the long hind wings trailing behind.

**Katarru, zururituu, *Croce filipennis*** Adults are about 15 mm long; total length is about 35 mm (including wings). The body is black; fore wings have a dark spot near the tip, while hind wings are reduced to narrow bands ending in a wide fringe of hairs, and more than twice the length of the front wings. Larvae often occur indoors where they remain in secluded areas, covered with dust and debris, and prey on household arthropod pests. Eggs are laid singly in debris on the floor of houses. The full-grown larva has a large, quadrate head and long curved mandibles. The head is connected to the body with a distinct two-segmented extension of the thorax. The pupa is enclosed in a cocoon composed of silk and covered with sand and debris. This species occur in India and tropical regions of southern Asia.

## Bibliography


### MANTODEA

Breland, O. P. Notes on the biology of *Stagmomantis carolina* (Joh.) (Orthoptera, Mantidae). *Bull. Brooklyn Entomol. Soc.*, **36** (1941), 170–7.

Breland, O. P. and J. W. Dobson. Specificity of mantid othecae. *Ann. Entomol. Soc. Am.*, **40** (1947), 557–75.

Gurney, A. B. Praying mantids of the United States, native and introduced. *Smithsonian Inst. Rep.*, **1950** (1951), 339–62.

Further notes on *Iris oratoria* in California. *Pan-Pac. Entomol.*, **31** (1955), 67–72.

Harris, S. J. and M. D. Moran. Life history and population characteristics of the mantid *Stagmomantis carolina* (Mantodea: Mantidae). *Environ. Entomol.*, **29** (2000), 64–8.

Maxwell, M. R. and O. Eitan. Range expansion of an introduced mantid *Iris oratoria* and niche overlap with a native mantid, *Stagmomantis limbata* (Mantodea: Mantidae). *Ann. Entomol. Soc. Am.*, **91** (1998), 422–9.

Prete, F. R., H. Wells, P. H. Wells, and L. E. Hurd (eds.). *The Praying Mantids: Research Perspectives*. Baltimore, MD: Johns Hopkins University Press, 1998.

Robinson, M. By dawn's early light: matutinal mating and sex attractants in a Neotropical mantid. *Science*, **205** (1979), 25–7.

Tinkham, E. R. Western Orthoptera attracted to lights. *J. N.Y. Entomol. Soc.*, **46** (1938), 339–53.

Varley, G. C. Frightening attitudes and floral simulations in praying mantids. *Proc. R. Entomol. Soc. Lond.*, **14** (1939), 91–6.

### NEUROPTERA

Adams, P. A. New antlions from the southwestern United States (Neuroptera: Myrmeleontidae). *Psyche*, **63** (1956), 82–108.

Banks, N. Revision of Nearctic Myrmeleontidae. *Bull. Mus. Comp. Zool.*, **68** (1927), 1–84.

Bickley, W. E. and E. G. MacLeod. A synopsis of Nearctic Chrysopidae with a key to the genera (Neuroptera). *Proc. Entomol. Soc. Wash.*, **58** (1956), 177–202.

Brooks, S. J. and P. C. Barnard. The green lacewings of the world: a generic review (Neuroptera: Chrysopidae). *Bull. Bri. Mus. Hist. (Entomol.)*, **59** (1990), 117–286.

Carpenter, F. M. A revision of the Nearctic Hemerobiidae, Berothidae, Sisyridae, Polystoechotidae, and Dilaridae (Neuroptera). *Proc. Am. Acad. Arts Sci.*, **74** (1940), 193–280.

Faulkner, D. K. Current knowledge of the biology of the moth lacewing *Oliances clara* Banks (Insecta: Neuroptera: Itonidae). Advances in Neuropterology. Proceedings of the Third International Symposium. *Neuropterology*, 197–203.

Flint, Jr., O. S. Neuroptera. In Hurlbert S. H. (ed.) *Biota Acutica de Sudamerica Austral*. San Diego, CA: San Diego State University.

Froeschner, R. C. Notes and keys to the Neuroptera of Missouri. *Ann. Entomol. Soc. Am.*, **40** (1947), 123–36.

Glorioso, M. J. Systematics of the dobsonfly subfamily Cordalidae (Megaloptera: Cordalidae). *Syst. Entomol.*, **6** (1981), 253–90.

Gravely, F. H. and S. Maulik. Notes on the development of some Indian Ascalaphidae and Myrmeleonidae. *Rec. Ind. Mus.*, **6** (1911), 101–10.

Imms, A. D. On the life history of *Croce filipennis* Westw. *Trans. Linn. Soc. Lond.*, **11** (1911), 151–60.

Nakahara, W. Contribution to the knowledge of the Hemerobiidae of Western North America (Neuroptera). *Proc. US Nat. Mus.*, **116** (1965), 205–22.

Parfin, S. The Megaloptera and Neuroptera of Minnesota. *Am. Midl. Nat.*, **47** (1952), 421–34.

Penny, N. D. Neuroptera of the Amazon Basin. Pt. 7 Corydalidae. *Acta Amazonica*, **12** (1983), 825–37.

Tillyard, R. J. Studies in Australian Neuroptera. *Proc. Linn. Soc. N.S.W.*, **40** (1915), 734–752; **41** (1916), 41–70; **41**, 221–48; **41**, 269–332; **43** (1918), 116–22; **43** (1919), 750–86; **43**, 787–818; **44**, 414–37.

# 13 Orthoptera, Phasmatodea

## ORTHOPTERA

### Introduction

Members of this order include grasshoppers, crickets, and katydids. They are primarily plant feeders and distributed in all zoogeographic regions. They are characterized by having biting–chewing mouthparts, usually well-developed wings, with the fore wings enlarged or thickened; some are brachypterous, and some apterous. The body is elongate and the antennae are usually long. Development is gradual; the nymph stages resemble adults except for wings, when present. Egg and nymph stages survive dry seasons or overwinter.

Pest status is limited to a few domestic and peridomestic species in the urban environment. The house cricket, *Acheta domestica*, is the only species that occupies and reproduces indoors. However, other crickets frequently utilize household sites for harborage and foraging, or they are attracted to lights at night. The house cricket may damage materials, but other species are only a nuisance. Large numbers of field crickets can be a nuisance. In late summer large numbers may move to the perimeter of buildings during cool nights. The presence of these insects around and sometimes in the living space can contribute to respiratory asthma.

### Acrididae

These are typical grasshoppers. They have short, thick antennae, and the front of the head is prolonged forward a little beyond the eyes. The pronotum has a well-developed median ridge. The terminalia of the female are small. These grasshoppers are active during the day; they feed primarily on living plants. Males stridulate during the day, and the females of some species stridulate. Several species have wide variations in their body color pattern. These grasshoppers live in fields and meadows, and they infrequently occur in vegetation around buildings in rural and urban areas. In general, they are not attracted to lights at night.

Several species of grasshopper invade urban and suburban landscapes, gardens, and commercial nurseries during hot and dry summers in the southwestern USA. They consume foliage of many species of ornamental plants and turfgrass. The most common pest species include the differential grasshopper, *Melanopus differentalis*, the two-striped grasshopper, *M. bivittatus*, and the migratory grasshopper, *M. sanguinipes*. Only after they molt to adults are they pests in urban landscapes, but in rural areas severe damage may result from nymphs and adults. Problems with these grasshoppers are usually preceded by several years of hot, dry summers and warm falls. Dry weather increases the survival of adults and nymphs.

**Common field grasshopper, *Chorthippus brunneus*** Adult males are about 17.5 mm long, while females are about 22 mm long. Wings are well developed and the front wing extends beyond the bent hind legs. The underside of the thorax is densely covered with fine setae. It shows great variation in body color. The head, pronotum, and front wings may be blackish brown, brown, yellowish brown, gray, green, orange, orange-brown, pink, red, or reddish purple. The most common color varieties are striped, mottled, and semimottled with wedge-shaped marks. Eggs are laid in groups of about 14 in soil in August and September. Hatching occurs during May or June. Development is completed and adults emerge in late June or July. Adults live until October or November. Adults active in hot weather characteristically make short, hopping flights. The song of the male is a series of short chirps, and there are 6–10 chirps in each series. When two or more males are together, they often chirp alternately. Females chirp until they mate, and will resume chirping if mating does not occur frequently. The song of the female is similar to that of the male. It commonly occurs in agricultural fields, roadsides, and other artificial habitats, such as areas of concrete or asphalt, and around buildings. This species is distributed throughout Europe, North Africa, and temperate regions of Asia.

## Anostostomatidae

The wetas and king crickets are now grouped in this family. This family is represented by species from almost all parts of the world, and it is especially diverse in the southern hemisphere. These are large orthopterans, which live in humid forested habitats. Suburbanization in some regions of the world has increased their access to peridomestic habitats.

**Aboreal tree weta, *Hemideina thoracica, H. trewicki, H. crassidens, H. femorata*** Adults are 40–60 mm long. The body is smooth and shiny, and the abdomen has bands of contrasting brown and pale brown. Wings are absent and the upper side of the hind tibiae has large subapical spines. Femoroabdominal stridulatory apparatus is well developed; the abdominal stridulatory area has ridges. These weta normally live in galleries that have initially been excavated by cerambycid woodboring larvae (*Ochrocydus hutoni*) or hepialid lepidoptera larvae (*Aenetus virescens*); they also live in natural cavities in trees. Most favored plants are manuka (*Leptospermum scoparium*), kanuka (*Kunzea ericoides*), ngaio (*Myoporum laetum*), and mahoe (*Melicytus ramiflorus*). They are primarily herbivorous, but feed on living and recently dead invertebrates. Their galleries and foraging activity are usually above ground, but females oviposit in soil. These species are distributed in New Zealand, where they commonly occur in urban hedges and gardens. *H. thoracica* is widespread.

**King cricket, *Henicus monstrosus*** Adult male is about 25 mm long and dark brown to blackish brown. The head is large and the mandibles are large, broad, and curved, and they have four small teeth at the tip. The female is about 52 mm long and light brown. Wings are absent and the ovipositor is short and curved. This species lives in burrows in the ground during the day and forges at night. Adults stridulate by moving their mandibles and maxillae together. This species is found in domestic gardens and on sandy beaches along the southern African coast. Adults excavate burrows in gardens and turfgrass lawns.

**Parktown prawn, *Libanasidus vittatus*** Adult females are about 64 mm long; the ovipositor is 19 mm long and the entire insect may be 166 mm long. Males have very large mandibular tusks, which arise as anterior outgrowths of the mandibles. Males use the mandibles to dig their burrows, and in disputes with other males. These insects stridulate by rubbing the lateral abdominal pads against the modified inner surfaces of the femora. Tibial ears are visible on the front legs. Eggs are deposited singly in damp soil; the female inserts her ovipositor for each egg; fecundity is about 200 eggs. Oviposition is in the fall and eggs or nymphs overwinter. Adults and nymphs feed on slugs, snails, and other insects, including caterpillars such as cutworms. The natural habitat is in burrows under logs in forested areas in Mpumalanga and Northern province of South Africa, and probably Zimbabwe. This species is called the Parktown prawn, because adults accidentally occur in swimming pools in the (Parktown) suburbs of Johannesburg. They also enter houses and, when disturbed, they often jump on to people. They have foul-smelling feces. This species has generated several myths, mostly related to their sudden prominence in the 1960s. It is probable that they were originally present in the region in low numbers and environmental changes and urbanization (suburban spread) and gardens led to an increase in numbers and range.

## Gryllidae

These crickets appear somewhat flat dorsally because a portion of the front wing is held horizontal over the body, instead of being held vertically, as in grasshoppers. The female ovipositor is usually needle-like, and the cerci are long. Eggs are laid singly or in small batches in soil, or in crevices. Nymphs and adults feed primarily on vegetation. Males stridulate by rubbing together modified regions of the front wings. Crickets do not have a fixed song length. They will sing indefinitely as long as their surroundings are relatively warm; the chirps slow in cool temperatures. Several species regularly occur indoors or around the perimeter of buildings in rural and urban areas. The Indian house cricket, *Grylloides supplicans*, is a common indoor pest in Central America. The very large African cricket, *Macrogryllus consocius*, burrows in sandy areas along coastal southern Africa. These are social crickets; they have loud songs and can be heard from a long distance.

**Black field cricket, *Teleogryllus commodus*** Adults are about 34 mm long and black. Eggs are laid in the summer under objects on the ground or in crevices in the soil; hatching occurs the following spring (October). Large populations of this field cricket can develop; in the fall they move into urban environments and enter houses. This species occurs in New Zealand and Australia.

**House cricket, *Acheta domesticus* (Fig. 13.1a)** Adults are 16–26 mm long. The body is yellowish to mottled brown and pale brown; there are three dark cross-bands on the head. The ovipositor is long and the cerci are long in both sexes. Wings are well developed and these crickets readily fly. Eggs are 2.5 mm long, whitish yellow, and banana-shaped. They are deposited singly or in small batches in moist cracks and crevices; fecundity is 40–179 eggs, but 728 eggs were reported at 28 °C. Hatching occurs in 1–12 weeks, depending on temperature. Development in the laboratory is about 56 days for males and 53 days for females; under variable or field conditions development takes about 35 days at 32 °C and 60 days at 27 °C, with 9–11 nymph stages. Adults live about 90 days at 27 °C. Adults and nymphs feed on plant material outside, and on household foods inside, especially liquids. They overwinter in the egg stage and there is one generation per year. This species is native to northern Africa and Middle East, but it is distributed in eastern North America, Europe, and India. It has been introduced into Latin America.

This cricket occurs indoors and outdoors throughout its range, and it is established in buildings in Europe. References to the cricket of the hearth are common in English literature and refer to this species. They occur indoors in heated locations, and outdoors where temperatures are slightly above the surroundings. Adults are attracted to lights at night; they can climb rough-surfaced buildings. They can occur in large numbers in refuse dumps and urban landfills. These sites are usually kept warm throughout the year by fermentation of wet organic material. From landfills crickets can move to surrounding buildings. Large numbers of adults have been reported moving from breeding sites such as refuse dumps to nearby buildings. The normal song is short shrill bursts repeated regularly for long periods. They typically chirp at dusk and during the night, but they can be heard during the day in landfills during warm weather.

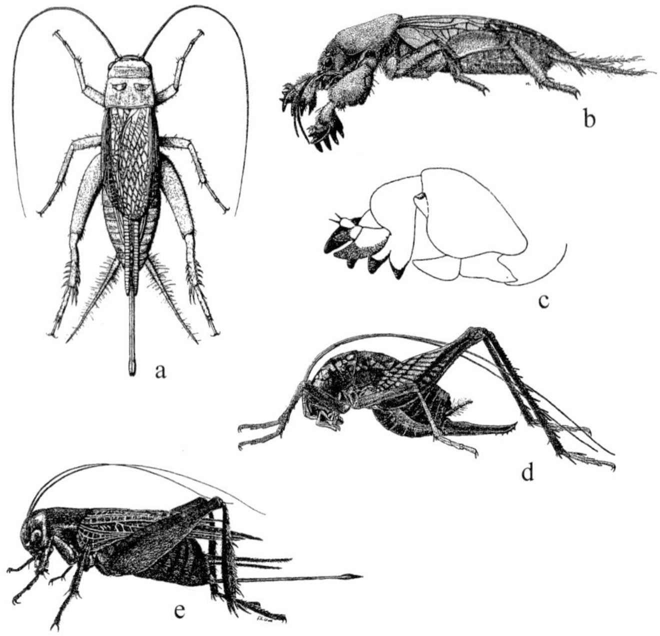


**Figure 13.1** Orthoptera. (a) *Achaeta domesticus* female (Gryllidae); (b) *Neocurtilla hexadactyla* female (Gryllotalpidae); (c) *Gryllotalpa gryllotalpa*, left front leg; (d) *Ceuthophilus maculatus* female (Raphidophoridae); (e) *Gryllus pennsylvanicus* (Gryllidae).

**Field crickets, *Gryllus assimilis*, *G. pennsylvanicus*, *G. campestris* (fig. 13.1e)** Adults are 13–28 mm long. The body ranges in color from black to yellowish brown. Front wings are shorter than the hind wings, and the front wings can have orange markings. Females have a long, slender ovipositor, and long cerci. Eggs are 2.8–3.2 mm, brownish yellow, and they are deposited singly in damp soil; 50 or more are placed within a small area; fecundity is 150–400 eggs. Development from egg to adult takes 78–93 days; there are 8–9 nymph stages. Adults live for about 2 months, and in the fall they are usually killed by frost. Adults and nymphs feed on plant material, but indoors they often damage plant and animal fabric, such as cotton, wool, silk, and fur. In north temperate regions there is usually one generation per year and they overwinter in the egg stage. In warm climates, there are two or more generations per year and they overwinter as eggs or large nymphs. These species have a wide distribution and they are found in North America and Europe. The several species of *Gryllus* are similar morphologically, but are separated based on habits, life history, and song.

Large numbers are known to occur around buildings in primarily rural areas. They are most abundant in the fall when adults gather at structures, perhaps attracted to lights or the sunlight heat retained by structures during the day, and radiated at night. Populations may persist in an area for several years, especially during drought conditions. In some desert valleys of southeastern California, large numbers of field crickets move to inhabited areas following summer thunderstorms.

***Gryllus bimaculatus*** Adults are about 28 mm long; the body and legs are shiny black. The thorax and base of the front wings have pale white markings. This species is widespread and common in Asia, Mediterranean Europe, and all of Africa. It occurs in peridomestic habitats, and occasionally enters houses and other buildings. A closely related species, *G. domesticus*, the black-headed cricket, occurs in India and is often a pest indoors when vegetation outdoors is scarce.

**Winter cricket, *Gryllodes sigillatus*** This cricket is legendary in Japan. Although of tropic origin, it was common in public baths and in households, primarily kitchens, in northern Japan. It feeds on greasy and fatty foods and is injurious to paper, cloth, and leather bookbinding. The modernization of household and introduction of gas cooking stoves and water heaters removed the niche occupied by this cricket. Limited populations exist in northern Japan.

**Temperature cricket, snowy tree cricket, *Oecanthus fultoni*** Adults are 12–15 mm long. The body is pale green with a yellow head and pale yellow antennal bases. This species is found on shrubs and small trees, where it feeds on aphids. It is called the temperature cricket because the air temperature can be estimated by counting the number of chirps per minute, dividing this number by four, and adding 40 to obtain the degrees Fahrenheit, or dividing it by six to obtain the degrees Celsius. This species occurs nearly throughout the USA.

## Gryllotalpidae

Mole crickets are brown to brownish black, and the body is usually covered with fine setae. Antennae are short and the front legs are very broad and have large spines (dactyls) for digging. These insects burrow in moist soil, usually along the edge of ponds and streams; they tunnel 150–200 mm below the surface. The males stridulate. They are capable flyers, and are attracted to bright outdoor lights at night. Mole crickets are pests of commercial turfgrass because they tunnel in soil, expose grass roots, and feed on grass and other plants. At night, mole crickets leave their underground tunnels to bite off stems and leaves of plants, which are dragged into their burrow to be eaten. Roots are eaten at any time from within the tunnels. In areas of severe damage, the surface 20–25 cm of soil is honeycombed with numerous galleries.

Three species of mole crickets, *Scapteriscus vicinus*, *S. borellii*, and *S. albibreviatus*, were introduced into southern USA from South America. *Scapteriscus vicinus* and *S. borellii* have spread throughout the coastal plain from southeastern Texas to southeastern North Carolina, with isolated populations of *S. borellii* in Arizona and California.

**African mole cricket, *Gryllotalpa africana*** Adults are about 25 mm long and the body is brown and covered with short setae to give it a velvety appearance. Wings do not cover the full length of the abdomen. Eggs are laid in chambers at the end of burrows 10–15 cm below the soil surface. Females construct several chambers and lay about 100 eggs; hatching occurs in 2–3 weeks. Eggs are usually laid during the rainy season and development takes about 12 months. Adults live for several months. They are attracted to lights at night. This species is common in warm regions of Africa and Asia, including India, China, and Japan.

**European mole cricket, fen cricket, *Gryllotalpa gryllotalpa* (Fig. 13.1c)** Adults are 35–40 mm long, and reddish brown to pale brown. Antennae are short, about the length of the pronotum. Front wings are about half as long as the abdomen, and the hind wings are fully developed. Front tibiae have four dactyls, and the hind tibiae have several spines on the inner margin. Eggs are laid in small chambers in the soil, usually close to the surface in moist soils. Fecundity is 100–300 eggs during 1–2 weeks, but 640 have been recorded. Females remain with the eggs and groom their surface until hatching. Mold fungi often damage eggs that are not cared for by females. Hatching occurs in about 2 weeks, and nymphs remain in the egg chamber for 2–3 weeks. Development is completed the following year or longer, depending on environmental conditions. Adults are capable flyers, and they are attracted to lights at night. A closely related species, *G. vineae*, occurs in southern Europe.

**Common mole cricket, *Neocurtilla hexadactyla* (Fig. 13.1b)** Adults are 25–30 mm long, brown with black markings. This species is native to North America, and it is common in eastern and southern USA.

**Southern mole cricket, *Scapteriscus borellii* (= *S. acletus*)** Adults are 26–36 mm long. The body is brown to pinkish brown and has a dark pattern on the pronotum. Tibial dactyls of the front legs have a deep, U-shaped separation. Eggs are laid in batches in underground chambers. Nymphs tunnel to the surface and feed in the upper layer of soil and litter. Juveniles and adults make and occupy extensive gallery and tunnel systems in the soil. Adults often fly to lights at night, and may be a pest in residential gardens. *S. borellii* is a predator of soil arthropods, and, while searching for prey, it digs tunnels in loose soil, and in commercial and residential turfgrass. Y-shaped tunnels are excavated in soil to facilitate feeding and for an escape route. Tunneling causes mechanical root damage and subsequent desiccation. It was introduced into the USA from South America. In South America it is found in Argentina, Uruguay, and southern Brazil. In the USA, it occurs from eastern Texas to Florida and north to South Carolina and North Carolina. It also occurs in Arizona and California.

**Tawny mole cricket, *Scapteriscus vicinus*** Adults are 26–36 mm long. The body is brown to dark brown. Tibial dactyls of the front legs have a short and V-shaped separation. They are distributed in southern USA. Adults fly to lights at night, and may be a pest in residential gardens and commercial turfgrass. *S. vicinus* is herbivorous and feeds on plant roots, and in the urban environment it feeds on the roots of turfgrass. Tunneling in turf also causes mechanical root damage and subsequent desiccation. Y-shaped tunnels are excavated in soil to facilitate feeding and for an escape route; the tunnels of this species are more extensive than tunnels of *S. borellii*. Along with *S. borellii*, this species was introduced into the USA from South America, where it is found in Argentina, Uruguay, and southern Brazil. In the USA, it occurs from eastern Texas to Florida and north to South Carolina and North Carolina.

## Pneumoridae

These are called bladder grasshoppers. The tracheae in the male abdomen are dilated and can hold a large amount of air. The sound produced by stridulating males is deep and very loud. Females are wingless and the hind legs are not adapted for jumping. This family is endemic to southern Africa, but some species occur as far north as Uganda and Tanzania.

**Bladder grasshoppers, *Physophorina* spp.** Adult males are about 50 mm long. The body is light green to green, and has longitudinal red stripes on the thorax and front wings.The legs are pale brown. The abdomen has lateral yellow and brown markings. Females are about 48 mm long, wingless or brachypterous, and pale green with various white markings. The hind legs are slender and adapted for walking and not jumping. Male bladder grasshoppers occur in two forms: in one, it develops wings, the large bladder-like abdomen, and the stridulatory apparatus. The other form is wingless, lacks the large abdomen, and is fertile. The role of the winged form is to attract females and wingless males to a common area where mating takes place. Females are fertilized by the wingless males. Males are attracted to lights at night and they sometimes occur indoors. This species is distributed in southern Africa.

## Rhaphidophoridae

These are wingless grasshoppers. Adults and nymphs are brown to dark gray, but may be mottled brown and black. The antennae are usually much longer than the body, and the hind legs are large; they are primarily terrestrial. The members of this family are nocturnal, and they are predominantly carnivorous. The peridomestic species with pest status include cave crickets and camel crickets.

**Cave crickets, camel crickets, stone crickets, *Ceuthophilus maculatus*, *C. pallidus*, *C. californicus*, and *C. pacificus* (Fig.13.1d)** Adults are 13–39 mm long and wingless. The body is light brown to dark brown, and many species have a mottled color pattern on the thorax, abdomen, and legs. Antennae are threadlike and longer than the body. Hind legs are long, and the femora are enlarged. Nymphs have a color pattern similar to the adults; the immature females lack an ovipositor. Camel crickets do not chirp.

Common names applied to these insects refer to their habits: cave cricket because they are found in dark habitats, camel cricket because of the high arched thorax of some species, and stone cricket for their habit of finding harborage under stones (*ceutho* = Greek, concealed). *C. pallidus* is widely distributed in North America, from southern Canada to Texas and northern Mexico. It has been reported eating fabric indoors. The spotted camel cricket *C. maculatus* was reported feeding on clothing hanging outdoors, at elevations of 2000–3000 m in New Mexico.

***Diestrammena japonica*, *D. apicalis*** Adults are 20–25 mm long and wingless. The body of *D. japonica* is yellowish brown with black spotted markings, while the body of *D. apicalis* is dark brown without black markings. These two species are found in domestic and peridomestic habitats throughout Japan and Taiwan.

**New Zealand cave crickets, *Pachyrhamma fascifer*, *P. acanthocera*** Adults are 44–50 mm long, excluding the antennae, which are extremely long (in *P. acanthocera* more than 550 segments). They occur in tunnels and other dark and moist peridomestic locations.

**Greenhouse stone cricket, greenhouse camel cricket, *Tachycines asynamorus*** Adult males are 11.3–11.9 mm long, while adult females are 12.5–14.6 mm long. The body is pale to dark brown; the legs are banded pale and dark brown. Eggs are laid singly or in groups of 50, and fecundity is about 1000 eggs. Hatching depends on temperature but is usually in 2–4 months. First-stage nymphs are about 3 mm long. Development takes 4–7 months through about 11 stages. Nymphs and adults prefer to eat animal material, and it will kill and eat other insects and small animals. It has been reported to move into buildings through sewer pipes. This species occurs in Asia. It is found in greenhouses, basements, and warm garages in other regions, including the USA and the UK.

## Stenopelmatidae

These are Jerusalem crickets or sand crickets. They have a large and robust head and body; they are usually brown to blackish brown, and have bands on the abdomen. They are found under stones or in loose soil. In the USA they are most common in western and Pacific coast states. There are 16 described species in the USA and Canada, but there may be as many as 60 undescribed species. Although they are found primarily in undisturbed areas, they occur around buildings during warm months.

**Jerusalem crickets, *Stenopelmatus fuscus, S. longispina, S. pictus*** Adults are 30–50 mm long. The head, thorax, and legs are shiny brown or yellowish brown. The abdomen is shiny brown with wide black bands dorsally, and ventrally pale brown. The apex of hind tibiae is ringed with large spines. Eggs are about 3 mm long, oval and yellowish white. They are laid in batches of 15–25 in holes or narrow chambers in soil below the frost line. Females use their head to excavate the holes. Hatching occurs in the fall or in the spring. Females that have overwintered lay eggs in spring. Development is through 9–11 molts and is completed in about 18 months in southern latitudes and longer in northern latitudes. Adults emerge in midsummer and mate within 4–14 days. Males produce a large, white sperm sac, which is removed and carried by the female for several hours after mating. After mating, the female occasionally consumes the male, who makes no effort to escape. There is one generation per year. Both sexes of some species stridulate by moving the abdomen past the hind femora or the hind femora past the stationary abdomen; the abdomen has short spines and the hind femora have raised, roughened areas. The sound produced is similar to rubbing sandpaper. Stridulation may be a defense mechanism. Food for the adults and nymphs is tubers and roots of plants, dead insects, and other animal matter.

The large, human-like head of the adult has created some superstitions around these crickets. In southwestern USA and Mexico they are called *niña de la tierra* or child of the earth. In other regions, the common name, potato bug, has been applied to this insect. This is apparently linked to records of their feeding on potatoes in the field. These crickets occur in states west of the Rocky Mountains, from British Columbia south to Mexico.

## Tettigoniidae

Longhorned grasshoppers and katydids have long, thread-like antennae, four-segmented tarsi, and a flattened, sword-like ovipositor. Most species stridulate; they are known for their loud, high-pitched song in trees during the afternoon and night. In the UK, the dark bush cricket, *Pholidoptera griseoaptera*, can be heard singing through the night. Most tettigoniid species are plant feeders, but some are predaceous on insects. They are occasionally attracted to lights at night, and accidentally come indoors. Several species are carried outside their normal range with the movement of commercial fruits and vegetables. The most common of the displaced species entering the UK include *Cosmoderus maculatus* on bananas from Cameroon, and *Jamaicana subguttata* and *Mastophyllum scabricolle* on bananas from West Indies.

**Edible grasshopper, *Homorocoryphus nitidulus vicinus*** Adults are about 6 cm from head to wing tips, and the body is pale green to brown. They are slender, long-horned grasshoppers that fly during the rainy seasons both day and night. They are often found in great numbers at streetlights. They are pests of a variety of cereal crops, especially rice. This species occurs in East Africa, including Uganda, Kenya, and Tanzania.

**Katydid, *Pterophylla camellifolia*** Adults are 45–55 mm long. The front wings are dark green and the body is pale green. The repeating nocturnal song of the males and females is the well-known katy-did, katy-did. They sing from midsummer until the first frost, and they are attracted to outdoor lights at night.

**Great green bush cricket, *Tettigonia viridissima*** Adult males are about 45.7 mm long and females are about 49.6 mm long. The body is green with a brown to blackish brown strip along the head and pronotum. Wings are well developed, and the wing span is 7 cm. The ovipositor is about 2 cm long, and slightly downward-curved. Eggs are laid in the fall in moist soil and suitable crevices in the ground. Hatching occurs during May and June. The nymphs are bright green with black or brown dorsal stripe. Nymph development is through 6–9 instars. This large bush cricket is found along hedgerows, gardens, and other locations in the urban environment where there are low shrubs and tall grass. Males sing loudly during the day and into the evening. Adults feed on a variety of plant materials, but may feed on other insects. This species occurs in the UK, continental Europe, North Africa, and regions of temperate Asia.

**Oak bush cricket, drumming katydid, *Meconema thalassinum*** Adults are about 16 mm long. The body is uniformly pale green; the antennae are inserted between the eyes. Wings are equal to the length of the abdomen; the ovipositor is narrow, and short. This species is commonly found at lights near trees. It is not restricted to feeding on oak trees. It does not stridulate, but the male produces a sound by striking (drumming) its hind tarsi on a leaf. This species is widely distributed in Europe, and has been introduced into northeastern USA.

**Pagago thread-legged katydid, *Arethaea gracilipes*** Adults are 30–35 mm long. The body is green and there is a colored strip on the side of the abdomen. They occur on grass and weeds along roadsides in southwestern USA. They are attracted to lights at night.

**Tizi, *Ephippiger ephippiger*** Adults are about 30 mm long. The body is green to brown, and has short, saddle-like wings (*ephippi* – Greek for saddle) extending to the base of the hind legs. Both males and females sing during the day, and produce a double chirp, from which it gets its common French name of tizi. This species is common on low bushes and shrubs around buildings in rural areas in southern Europe.

## PHASMATODEA
## Introduction

Walkingsticks and leaf insects have an elongate, cylindrical body, and the wings are either reduced or absent. Some tropical species are flattened laterally. All species have short cerci and small ovipositors. Nymphs are usually green and adults brown. They are plant feeders and found in trees and shrubs. Males are generally unknown and females are parthenogenetic. Females located on upper branches of trees lay eggs, and they drop through the canopy to the ground. Some species lay eggs in batches in small pits in the soil. Hatching occurs in 1 year in warm climates, and in 2 years in cold climates. There is usually one generation per year. Most of the nearly 2000 described species are found in the southern portion of the Indo-Malayan region. Species common in Europe include *Bacillus rossius*, *Clonopsis gallica*, and *Leptynia hispanica*.

## Heteronemiidae

Species in this family are long and slender and the most stick-like of all the walkingsticks. They have five-segmented tarsi, a mesothorax that is at least four times as long as the prothorax, and they are wingless. This family contains the longest insect in the USA, *Megaphasma denticrus*, which is 15–18 cm long. It occurs in southwestern states.

**Common walkingstick, *Diapheromera femorata*** Adults are 62–90 mm long. The body is brown to green, but may be mottled with gray, green, and red. First-stage nymphs are about 8 mm long and pale green. The egg is about 2 mm long, oval and shiny black or brown, and has a broad white or light green band. Winter is spent in the egg stage and hatching occurs in May and June; early-stage nymphs feed on ground vegetation, such as strawberry, blueberry, and Juneberry. Late-stage nymphs feed on tree leaves. Adults emerge in July and August and lay eggs until frost. In southern regions, eggs hatch the following spring; in northern regions eggs hatch the second spring following their deposition. This species occurs in southern Canada and throughout eastern USA. It is sometimes abundant and can defoliate trees, especially black oak, elm, black locust, wild cherry, and sometimes dogwood. They may occur in woodlots and residential areas. Other related species include *Diapheromera velii* and *D. blatchleyi*. Adults and nymphs of these species feed on grasses and tall shrubs.

**Prickly stick insect, *Acanthoxyla prasina*** The adult female is about 81 mm long. The body is greenish brown with spines on the head and thorax, and usually on the basal part of the abdomen. Cerci are large, flattened, and rounded at the tip. Eggs develop without fertilization, and males are unknown. This species is native to New Zealand, but was introduced into the UK on plants. It is known to occur in residential gardens and on ornamental plants. Closely related species, *A. geisovii* and *A. inermis*, are native to New Zealand, but populations are known from the Scilly Islands, UK.

**Smooth stick insect, *Clitarchus hookeri*** The adult female is about 86 mm long. The body is pale green to greenish brown. Cerci are large, flattened, and pointed at the tip. The thorax is black, with a longitudinal line along the middle of the dorsal surface. Eggs develop without fertilization in some regions. This species is native to New Zealand, but has been introduced to the UK. It occurs in cultivated gardens.

## Pseudophasmatidae

Species in this family have the tergum of the first abdominal segment at least as long as the mesothorax. Males are brownish yellow and the females are brown; both have a dark median and two lateral dorsal stripes. They have glands under the pronotum, which project a thick milky fluid at predators.

**Two-striped walkingstick, *Anisomorpha buprestoides*** Adults are 40–80 mm long. The male is uniform brown, and the female has a median dorsal and lateral stripe. Eggs are laid in batches of 8–10 in shallow pits in the soil. This species feeds on grass and low shrubs, and occurs in southern USA.

**Other walkingsticks** Species in the family Timematidae are short and stout. *Timema californica* is 12–25 mm long; the body is green, brown, or pinkish green, and found on shrubs and trees in California. The Indian stick insect, *Carusius morosus*, is native to Asia. It is sometimes kept as a house pet, and is established in some greenhouses outside Asia. Males are rare and females are parthenogenetic.

# Bibliography

## ORTHOPTERA

### General


Alexander, R. D. Sound production and associated behavior in insects. *Ohio J. Sci.*, **57** (1957), 101–13.

Alexander, R. D. and T. J. Walker. Two introduced field crickets new to eastern United States (Orthoptera: Gryllidae). *Ann. Entomol. Soc. Am.*, **55** (1962), 90–4.

Back, E. A. *Gryllus domesticus* and city dumps. *J. Econ. Entomol.*, **29** (1936), 189–202.

Bellman, H. *A Field Guide to the Grasshoppers and Crickets of Britain and Northern Europe*. Collins, 1988.

Boake, C. R. B. Natural history and acoustic behavior of a gregarious cricket. *Behaviour*, **89** (1984), 241–50.

Chapman, R. F. and A. Joern (eds.) *Biology of the Grasshoppers*. New York: Wiley, 1990.

Chopard, L. *La Biologie des Orthoptères*. Paris: Lechevalier, 1938.

Crampton, G. C. The cave cricket, *Ceuthophilus*, as a possible vector of parthenogenic organisms. *J. Econ. Entomol.*, **16** (1923), 460.

Forrest, T. G. Acoustic communication and baffling behaviors of crickets. *Fla Entomol.*, **65** (1982), 33–44.

Ghouri, A. S. K. and J. E. McFarlane. Observations on the development of crickets. *Can. Entomol.*, **90** (1958), 158–65.

Harz, K. *Die Orthopteren Europas*. vol. I. *Crickets and Bush-crickets*. The Hague: Junk, 1969.

Henne, D. C. and S. J. Johnson. Seasonal distribution and parasitism of *Scapteriscus* spp. (Orthoptera: Gryllotalpidae) in southeastern Louisiana. *Fla Entomol.*, **84** (2001), 209–11.

Hubbell, T. H. A monographic revision of the genus *Ceuthophilus*. *Univ. Fla. Publ. Biol. Sci. Ser.*, **2** (1936), 1–551.

Hutchins, R. T. and J. M. Langston. An unusual occurrence of the field cricket. *J. Econ. Entomol.*, **46** (1953), 169.

Janjua, N. A. A preliminary note of the bionomics and control of the black-headed cricket (*Gryllus domesticus*) in Usta Colony (Sidi District) of Baluchistan. *Rev. Appl. Entomol. A*, **28** (1940), 436.

Kemper, H. Observations on the biology of the house cricket. *Rev. Appl. Entomol. A*, **26** (1938), 37.

Laufer, B. Insect musicians and cricket champions of China. *Field Mus. Nat. Hist. Anthrop. Leaflet*, **22** (1927), 1–26.

McDonald, I. R. and S. A. Hanrahan. Aspects of foraging and behavior of the king cricket, *Libanasidus vittatus* (Kirby) (Orthoptera: Mimnermidae). *South Afr. J. Sci.*, **89** (1993), 354–8.

Mickle, D. A. and T. J. Walker. A morphological key to field crickets of southeastern United States (Orthoptera: Gryllidae: *Gryllus*). *Fla. Entomol.*, **57** (1974), 8–12.

Nagel, M. G. and W. H. Cade. On the role of pheromones in aggregation formation in camel crickets, *Ceutophilus secretus* (Orthoptera: Gryllacrididae). *Can. J. Zool.*, **61** (1983), 95–8.

Ordish, R. G. Aggregation and communication of the Wellington weta *H. crassidens* (Blanchard) (Orthoptera: Stenopelmatidae). *N.Z. Entomol.*, **15** (1992), 1–8.

Ragge, D. R. and W. J. Reynolds. *The Songs of the Grasshoppers and Crickets of Western Europe*. Colchester, Essex: Harley Books, 1998.

Stiedl, O. and K. Kalmring. The importance of song and vibratory signals in the behavior of the bushcricket, *Ephippiger ephippiger* Fiebig (Orthoptera, Tetigonidae): taxis by females. *Oecologia*, **80** (1989), 142–4.

Tinkham, E. R. Western Orthoptera attracted to lights. *J. N.Y. Entomol. Soc.*, **46** (1938), 339–53.

Tinkham, E. R. and R. D. C. Rentz. Notes on the bionomics and distribution of the genus *Stenopelmatus* in central California with the description of a new species (Orthoptera: Gryllacrididae). *Pan-Pac. Entomol.*, **45** (1969), 4–14.

Toms, R. B. Johannesburg's cricket invader the 'Parkmore prawn'. *Afr. Wildlife*, **39** (1985), 200–2.

Trewick, S. A. and M. Morgan-Richards. On the distribution of tree weta in the North Island, New Zealand. *J. R. Soc. N.Z.*, **25** (1995), 485–93.

Uvarov, B. P. Locusts and grasshoppers. London: *Imperial Bureau of Entomology*, 1928.

Vickery, V. R. and D. K. McE. Kevan. *A Monograph of the Orthopteroid Insects of Canada and Adjacent Regions*. *Lyman Entomological Museum* Memoir no. 13, vol. 1. Ottawa: McGill Univ, 1983.

*The Grasshoppers, Crickets, and Related Insects of Canada and Adjacent Regions: Ulonata, Dermaptera, Chleuthoptera, Notoptera, Dictyoptera, Grylloptera, and Orthoptera. The Insects and Arachnida of Canada*, part 14. Ottawa: Canadian Publishing Centre, 1985.

Walker, T. J. Factors responsible for intraspecific variation in the calling songs of crickets. *Evolution*, **16** (1962), 407–28.


### Geographic distribution


Alexander, R. D. The taxonomy of the field crickets of eastern United States (Orthoptera: Gryllidae, *Acheta*). *Ann. Entomol. Soc. Am.*, **50** (1957), 584–602.

Dirsh, V. M. *The African Genera of Acroidoidea*. Cambridge: Cambridge University Press, 1965.

Froeschner, R. C. The grasshoppers and other Orthoptera of Iowa. *Iowa State Coll. J. Sci.*, **29** (1954), 163–354.

Harz, K. *Die Orthopteren Europas*, vol. II. The Hague: W. Junk, 1975.

Hebard, M. The Dermaptera and Orthoptera of Illinois. *Ill. Nat. Hist. Surv. Bull.*, **20** (1934), 125–9.

Marshall, J. A. and E. C. M. Haes. *Grasshoppers and Allied Insects of Great Britain and Ireland*. Colchester, Essex: Harley Books, 1988.

Otte, D. *The Crickets of Hawaii*. Philadelphia, PA: Orthoptera Society, Academy of Natural Science, 1994.

Payne, R. M. The distribution of grasshoppers and allied insects in the London area. *Lond. Nat.*, **37** (1958), 102–15.

Ragge, D. R. *Grasshoppers, Crickets, and Cockroaches of the British Isles*. Warne, 1965.

Weismann, D. B., D. C. F. Rentz, R. D. Alexander, and W. Loher. Field crickets (*Gryllus* and *Acheta*) of California and Baja California, Mexico (Orthoptera: Gryllidae: Gryllinae). *Trans. Am. Entomol. Soc.*, **106** (1980), 327–56.

## PHASMATODEA

Bedford, G. O. Biology and ecology of the Phasmatodea. *Annu. Rev. Entomol.*, **23** (1978), 125–49.

Bradley, J. C. and B. S. Galil. The taxonomic arrangement of the Phasmatodea with keys to the subfamilies and tribes. *Proc. Entomol. Soc. Wash.*, **79** (1977), 176–208.

Caudel, A. N. The Phasmidae, or walking sticks of the U. S. *Proc. U.S. Nat. Mus.*, **26**, 863–85.

Clark, J. T. *Stick and Leaf Insects*. Winchester, Hants, UK: Sherlock, 1974.

Gueney, A. B. Praying mantids of the United States, native and introduced. *Smithson. Inst. Ann. Rep.*, **1949/1950** (1951), 339–62.

Manzzini, M. and V. Scali (eds.) 1987. *Stick Insects. Phylogeny and Reproduction*. Italy: University of Siena and Bologna.

# 14 Phthiraptera

## Introduction

Classification of the sucking and chewing lice has followed several schemes. Some place these insects in two orders: Anoplura, the sucking lice, and Mallophaga, the chewing lice. This classification considers the differences in the morphology and habits of the lice, and has a history of use and acceptance. Other entomologists recognize a single order, Phthiraptera, with four suborders: Amblycera, Ischnocera, Anoplura, and Rhynchophthirina. Lice in the suborders Amblycera and Ischnocera (both previously the Mallophaga) have biting mouthparts and are associated with birds and mammals. Anoplura and Rhynchophthirina have only placental mammals as hosts and possess sucking (anoplurans) and biting mouthparts. Classification using the single order of lice, Phthiraptera, has the utility of bringing together this monophyletic group of insects, without reducing the differences in morphology and habits.

Phthiraptera are wingless insects that are ectoparasites on mammals or birds, and spend all of their lives in the skin, fur, hair, or feathers of their host. They evolved from psocopteran-like ancestors that lived in the nests of mammals and birds, and in that habitat fed on the shed skin of the host and other detritus with biting–chewing mouthparts. Close association with animals at their nest site probably led to intimate and permanent association with the animal itself, and the development of blood-feeding mouthparts.

## Suborder: Amblycera (= Mallophaga)

These are the chewing lice. Adults are 1–6 mm long. They are somewhat flattened, brown to pale brown, and wingless. The head is distinct from the thorax, it is as wide as or wider than the prothorax, and has a pair of mandibles ventrally. Antennae are 3–5-segmented, and are clubbed or filiform; the eyes are reduced or absent. The abdomen is eight- or nine-segmented, with weak sclerotization. Legs have one-segmented tarsi. Immatures are similar to adults, except for reduced size, sclerotization, and chaetotaxy. Amblycera are permanent ectoparasites, primarily on birds, but some species feed on mammals. All stages are associated with the host animal. Eggs are deposited on to feathers or hairs or, in some cases, within the shaft of primary feathers; females lay 50–150 eggs in their lifetime. Hatching occurs in about 7 days, and there are three immature stages in most species. Food consists of bits of feathers or hair and debris on the skin; some species ingest blood. There are several generations per year.

### Boopidae

These lice are characterized by the presence of a maxillary palp. The antennae are five-segmented and strongly clubbed, and the legs are long and slender.

**Australian dog louse, *Heterodoxus spiniger*** Adults are about 2.5 mm long. The body is ovoid, and brown. The head is triangular, the mesothorax is quadrate, and the abdomen has eight apparent segments. A distinguishing character for this species is setae on prominent nodes on each side of the mesonotum. It is found on the domestic dogs throughout the world, and on coyote, and wolf. This louse probably shifted to dogs from the wild dingo, which frequently preys on kangaroo. The kangaroo is infested by the closely related but distinct species, *H. longitarsus*. The dog louse serves as a developmental host for the filarial nematode *Dipotalonema reconditum* that is parasitic in dogs. This louse has been taken from dogs in the San Francisco Bay region, California, and it is a pest of dogs in South Africa.

### Gyropidae

The species in this family are characterized by the presence of maxillary palps, clubbed antennae, and the tarsi have one or no claw. They are primarily distributed in Central and South America.

**Guinea pig louse, *Gliricola porcelli* (= *G. saviae*, *Gyropus gracilis*)** Adults are 1.0–1.5 mm long and yellowish white. The head is broad, angular, and dark yellow. The abdomen has eight or nine apparent segments. Legs are disproportionately long, 2 and 3 with 0–1 tarsal claws. This species is found on guinea pigs in North America.

**Oval guinea pig louse, *Gyropus ovalis*** Adults are about 1.5 mm long and pale brown. Legs have a single large, well-developed claw on each foot. It is found on guinea pigs in North and South America and Europe, wherever the host is domesticated.

## Suborder: Anoplura

These are the sucking lice. Adults are 0.5–3.0 mm long and have a somewhat flattened body; they are yellowish white to pale brown, and wingless. The head is conical, slender, and distinct from the thorax; the proboscis and sucking mouthparts project forward. The thorax is relatively small and fused; the legs are strongly developed, and have a one-segmented tarsus and a large claw. Immature stages are similar to the adults, but less sclerotized and with a reduced number of setae. These insects are obligate parasites of mammals, and remain on the host during their entire life. Sucking lice are generally host-specific, and many are restricted to one animal species. There are several generations per year. Eggs are firmly attached to the hairs of the host, or in the case of *Pediculus humanus*, to the clothing of the host. The number of eggs per female is about 30. Eggs hatch in about 10 days and larval development is completed in about 20 days, depending on environmental conditions. The three nymph stages maintain contact with the host. Food for the adults and immatures consists of blood of the host.

Legs end in a tibial thumb and tarsal claw, which adapt these insects to clinging on to hair or coarse fibers. The shape of the *Pediculus* claw has some epidemiological significance. In the USA the prevalence of head infestation of lice is 35 times higher among caucasians than among the black portion of the population. A major contributory factor is thought to be the adaptation of the claw to grasp to the predominant hair type in the host population. Caucasian hair is round in cross-section, whereas Afro-Caribbean hair is flattened oval. The claw of *Pediculus* may have difficulty grasping the hair of non-Caucasians. Coevolution of lice and their mammal hosts is evident by the distribution and host-specificity of current species. Chimpanzees, New World (western hemisphere) monkeys, gibbons, and great apes have a species of *Pediculus*, and gorillas have a separate species of pubic louse, *Phthirus gorillae*. It seems that the host preferences of these lice associated with primate hosts antedate the divergence of the ape and anthropoid stocks. The early domestication of the pig and its close association with humans is evident by the feeding habits of the sucking louse *Haematopinus suis*, which includes pig and humans.

## Linognathidae

This family includes species that feed on cattle, goats, sheep, foxes, coyotes, and dogs. Adult females are about 2.5 mm long and males are about 2 mm long. The head is without external evidence of eyes, antennae are five-segmented, and the abdominal spiracles are usually spherical, not elongated.

**Dog-sucking louse, *Linognathus setosus* (Fig. 14.1a, b)** Adult females are about 2 mm long and males are about 1.7 mm long. The head is short and somewhat broad, and the antennae are almost as long as the head. The thoracic dorsum has four long setae; the prothorax has two short setae on each side. Thoracic and abdominal spiracles are large. This species is found on domestic dogs worldwide. It also infests white or arctic fox (*Alopex lagopus*) and other carnivores.

## Pediculidae

There are two common species or one species with two subspecies in this family. These lice occur on New World monkeys, gibbons, great apes, which are infested by *Pediculus schaeffi*, and humans. Generally, the head and body louse of humans are considered as distinct subspecies: *Pediculus humanus humanus* and *Pediculus humanus capitis*. It is often impossible to distinguish a single specimen as either head louse or body louse, but populations of these two lice can be separated on a few morphological features. These two forms interbreed readily in the laboratory, but may not in natural habitats. However, there is evidence that subspecies rank may not be appropriate. Studies of a fragment of the cytochrome oxidase I (COI) gene of mitochondrial DNA in lice collected from nine countries indicate that the head and body lice belong to the same species (conspecific). The COI sequences from the head and body lice studied did not come from reciprocally monophyletic lineages. Instead, they shared three of the 10 haplotypes found, which is evidence for conspecificity. Head lice and body lice are considered here as one species, *Pediculus humanus*, with two distinguishable subspecies.

**Body louse, *Pediculus humanus humanus* (Fig. 14.2a, b)** The adult female is 2.4–3.6 mm long and the male is 2.3–3.0 mm long; the body is light to dark gray. The head

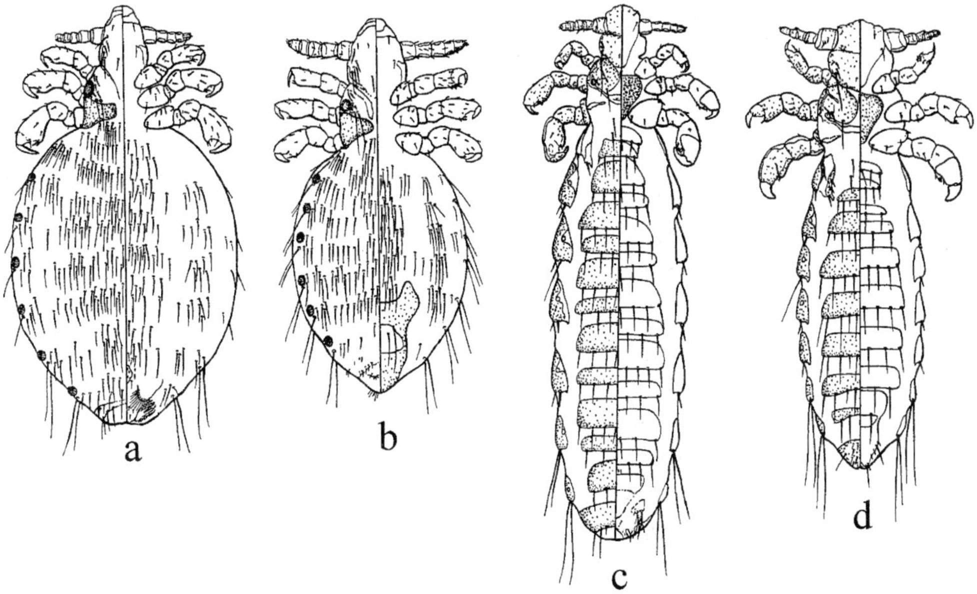

**Figure 14.1** Phthiraptera. (a) *Linognathus setosus* female (dorsal/ventral view); (b) *L. setosus* male (dorsal/ventral view); (c) *Polyplax serrata* female (dorsal/ventral view); (d) *P. serrata* male (dorsal/ventral view).

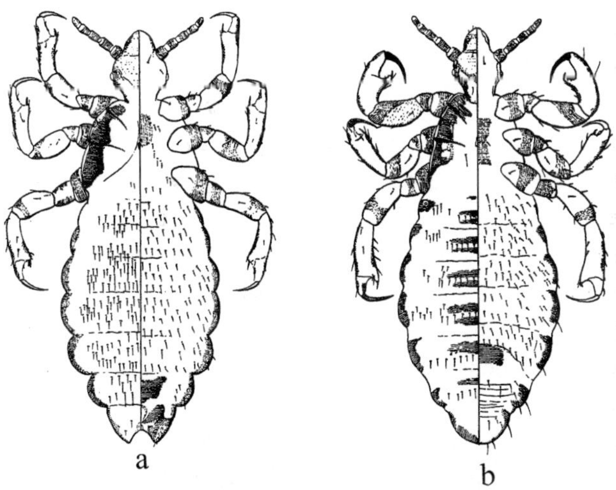

**Figure 14.2** Phthiraptera. (a) *Pediculus humanus humanus* female (dorsal/ventral view); (b) *P. humanus humanus* male (dorsal/ventral view).

is short, and constricted to create a short neck. Eyes are well developed; the antennae are five-segmented, and antennal segment 3 is usually longer than wide. The abdomen has lateral lobes, and there are sclerotized plates on lobes 3–8. Legs are subequal in shape and size, and with a claw. Nymphs resemble adults. All stages stay on clothing and in continuous contact with the body of the host, but make contact with the body while feeding. In severe infestations, some lice may remain on the host when clothing is removed. Eggs are deposited in clusters of three or four in the seams or other crevices of clothing, only rarely are they attached to coarse body hairs. Females deposit 3–11 eggs per day; hatching occurs in about 8 days at 30 °C; eggs do not hatch above 38 °C or below 23 °C. Fecundity is about 110 eggs; 91–94% hatch. Nymphs feed immediately after hatching, and frequently during the day and night throughout their development. Development is complete in 8–12 days; unfed lice die within 85 h at 23 °C and 45 h at 30 °C. Lice and eggs will die if clothing remains unworn for 17 days. Fecal material in spiral threads is extruded as the louse feeds, and the feces dry quickly in the air.

Location of lice in clothing is specific for life stage and sex of the adult. Females are more often found along the seams of clothing, while males are generally found over the surface of clothing. Adult lice move farther than nymphs, and the proportion of adults is higher in clothing farther from the skin. The body louse prefers wool clothing, but it survives in other fabrics. In warm climates where wool is not worn, body lice are few, regardless of personal hygiene. This louse feeds on skin close to clothing, and usually where the skin is soft or folded, as in the joints. While the mouthparts are inserted and the louse is feeding, it usually retains its hold on the adjoining fabric. Dissemination is through infested clothing and less by physical contact. Body lice leave the host when body temperature drops or increases, such as with a fever. Adults can travel a distance of about 23 cm in 1 min. The temperature of the human body is the optimum condition for this insect, and a rise of 4–5 °C is fatal to the louse within a few hours. Severe infestations of thousands of lice have been reported, but heavy infestations are not common. Most infected persons carry about a dozen lice. Body lice infestations often occur during times of war and civil unrest, when there is poor sanitation and crowding. Louse-borne typhus, *Rickettsia prowazekii*, and relapsing fever, *Borrelia recurrentis*, are diseases transmitted by body lice.

**Head louse, *Pediculus humanus capitis*** The adult female is 2.4–3.3 mm long and the male is 2.1–2.6 mm long. The body is gray to translucent, but usually resembles the hair color of the host. The head is short, constricted at the base and with a short neck. Antennal segment 3 is as long as it is wide. Eyes are distinct and set behind the base of the antennae; antennae are five-segmented. Legs are subequal in shape and size, and have a well-formed claw. The abdomen has lateral sclerites; the male abdomen is pointed at the tip, in the females the abdomen ends in two triangular projections. Nymphs resemble the adults, and range from 0.90 mm long for first stage to 2.70 mm long for third stage. Adults and nymphs live and feed on the body of the host; they are usually found on the neck and head, particularly behind the ears and on the back of the neck. Eggs are glued to hairs on the head and neck.

Eggs are about 0.8 mm long and yellowish white. Females attach eggs singly, about 1 mm from the base of a host hair. Hatching occurs in 7–10 days at 29–32 °C. The percentage of eggs hatching but not duration of egg stage is influenced by humidity, with the highest rate at 75% relative humidity. Females lay about seven eggs in 24 h, and a total of about 55 eggs in a lifetime. The maximum time for eggs to survive unhatched is 3–4 weeks. Scalp hair grows about 0.4 mm per day, and as it grows, the egg or nit is moved progressively farther and farther from the scalp. By the time the egg hatches, the empty chorion will be about 6 mm from the scalp. Immature development is completed in 8–9 days. Adults mate frequently throughout life. Males and females live about 10 days; adults and nymphs survive about 55 h at 23 °C and about 24 h at 30 °C away from the host without feeding. Dissemination is by physical contact, and usually by the exchange of clothing with stray hairs with eggs or lice attached, or by close and prolonged physical contact. Infestations occur under a range of sanitary conditions, particularly among schoolchildren. Obtaining lice in a school environment is common in developed and developing countries around the world, and is often wrongly associated with neglect or unclean conditions at home.

## Polyplacidae

This is a large family of lice that include species that infest numerous species of rats (*Rattus*) and mice (*Mus*), rabbits and hares, squirrels (*Spermophilus*), chipmunks (*Tamius*), and voles (*Microtus*). Adult males and females are usually less than 2 mm long.

**Mouse louse, *Polyplax serrata* (Fig. 14.1c, d)** The adult female is about 1.3 mm long and the male is about 0.9 mm long. The body is elongate and slender; the male head and thorax are broad. Antennae are as long as the head, and the thoracic sternal plate is broadly pointed posteriorly. This louse is commonly found on the house mouse (*Mus musculus*) worldwide; it also occurs on other mice.

**Spiny rat louse, *Polyplax spinulosa*** Adult females are about 1.3 mm long and males are about 1 mm long. The female body is elongate, while the male body is short and broad. Antennae are equal to the length of the head. The thoracic sternal plate has a blunt point and the lateral margins are nearly parallel. This species is associated with the Norway rat (*Rattus norvegicus*) and the roof rat (*R. rattus*), and it is distributed throughout the world wherever Norway rats occur.

## Pthiridae

There are two species in *Phthirus* in the world: a cosmopolitan species infesting humans, and the other, *Phthirus gorillae*, infesting the gorilla.

**Pubic louse, crab louse, *Phthirus pubis* (Fig. 14.3a–d)** Adults are 1.5–2 mm long. They are nearly as long as wide, and resemble a crab; the thorax is wide where it contacts with the abdomen. The first pair of legs is reduced and the claw is small; legs 2 and 3 are large and have a well-developed claw. The distance between two extended legs is approximately the distance apart of the coarse hairs on the parts of the body this louse infests. The abdomen has four pairs of lateral sclerotized tubercles. The first three pairs of abdominal spiracles are small and in a row; the remaining spiracles are large and located along the edge of the abdomen. Nymphs resemble adults. The third-stage male nymph is 1.3–1.4 mm long, and usually with two later tubercles. The third-stage female nymph is 1.0–1.5 mm long, and usually with four lateral tubercles. Adults and nymphs are found in the pubic region of the body, but they may occur in armpits, mustache, beard, and eyebrows. Eggs are yellowish white, about 1 mm long, and strongly attached to the hair or clothing. Several eggs may be attached to a single hair. Females lay about three eggs per day, and fecundity is 26–30 eggs. Hatching occurs in 6–8 days, and development takes 13–17 days at normal skin temperature. Adults live about 30 days, but away from the host they may live only 1 day. Adults

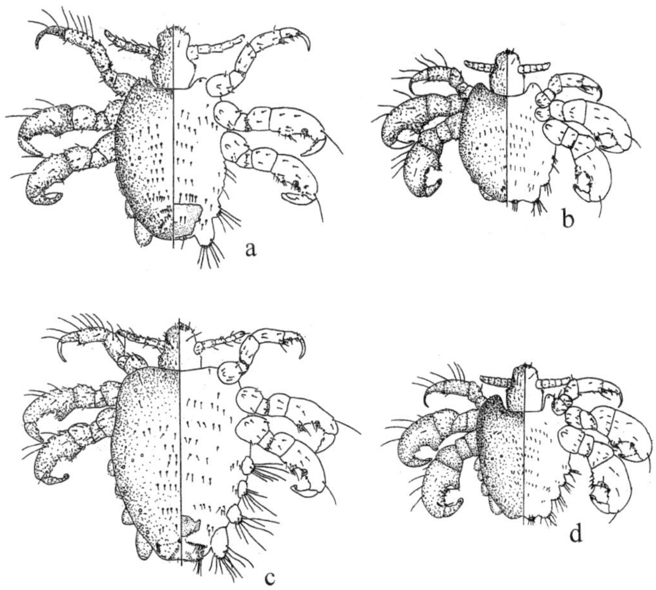


**Figure 14.3** Phthiraptera. (a) *Phthirus pubis* male (dorsal/ventral view); (b) *P. pubis* male, third-stage nymph (dorsal/ventral view); (c) *P. pubis* female (dorsal/ventral view); (d) *P. pubis* female, third-stage nymph (dorsal/ventral view).

are sedentary feeders and move only about 100 mm day. The tendency to take repeated blood meals in the same location may cause the formation of pale blue spots on the skin. Black spots of louse feces in underwear are a sign of *P. pubis* infestation. Dissemination is through physical contact, such as sexual activity or simply exposure in the same bed with an infected person. *P. pubis* may be spread on loose hairs transferred to towels and bedding. Dogs may carry this louse.

## Suborder: Ischnocera (= Mallophaga)

These are chewing lice and they are associated with birds and mammals. The body form and mouthparts of the suborders Ischnocera and Amblycera are similar in design to the mouthparts of Psocoptera. The mandibles are short, the head is broader than the thorax, and the antennae are less than half the length of the head.

### Menoponidae

Species in this family are ectoparasites of birds and are characterized by their large, wide head, which is broadly triangular and expanded behind the eyes. The slender antennae are inserted in grooves on the side of the head, and the tarsi have two claws. *Menopon gallinae* is the shaft louse of chickens, but also infests other birds, including turkeys, ducks, and guinea fowl, and may attack horses when stabled near infested fowl. *Menopus serini* is a pest of domestic canaries (*Serinus canarius*); lice occur under the wings and on the back of the birds.

### Philopteridae

Species in this family are characterized by having filiform or slender antennae, the absence of maxillary palps, and the tarsi have two claws.

**Pigeon louse, *Columbicola columbae* (= *Liperus baculus*)** Adults are about 2 mm long and slender. This species is nearly cosmopolitan on pigeons.

**European pigeon louse, *Goniocotes bidentatus* (= *Campanulotes*)** Adults are about 1 mm long. The body is pale or yellowish white, with brown margins. It is common on domestic pigeons in many parts of the world.

### Trichodectidae

This family contains species that are parasites of mammals instead of birds. Although some infest domestic animals, such as dogs, cats, and horses, they are of little medical importance.

**Dog-chewing louse, *Trichodectes canis*** Adults are about 1 mm long with a wide body. It is common on dogs around the world, and is often found on puppies.

**Cat louse, *Felicola subrostratus*** Adults are 2–4 mm long, broad to slender, but the head is narrowly rounded. Antennae are exposed and usually three-segmented; there is one tarsal claw on each leg. The cat louse is found on kittens and adult cats.

**Horse louse, *Trichodectes equi* (= *Werneckiella equi*)** Adults are about 2 mm long. The body is bright chestnut brown, with dark brown transverse bands on the abdomen. It occurs on horse, mule, and ass throughout North America and Europe.

## Suborder: Rhynchophthrina

There is only one family, Haematomyzidae, with two species in this suborder. The head is prolonged into a rostrum, with the mandibles at the end. Because these lice are blood-sucking, they are often considered a link between the Amblycera and Anoplura.

**Elephant louse, *Haematomyzus elephantis*** This species is found in natural locations in Africa, and in zoos in many parts of the world.

## Bibliography


Alexander, J. O'D. *Arthropods and Human Skin*. Berlin: Springer-Verlag, 1984.

Arakawa, Y. Studies on *Pthirus pubis*. *Insect World*, **37** (1933), 183–9.

Askew, R. R. *Parasitic Insects*. New York: American Elsevier, 1971.

Bacot, A. A contribution to the bionomics of *Pediculus humanus* (vestiment) and *Pediculus capitis*. *Parasitology*, **9** (1971), 228–58.

Busvine, J. R. Destruction of lice in clothing by hot and cold air. *Bull. Entomol. Res.*, **35** (1944), 115–25.

Evidence from double infestations for the specific status of human head lice and body lice (Anoplura). *System. Entomol.*, **3** (1978), 1–8.

Buxton, P. A. Studies on populations of headlice (*Pediculus humanus capitus*: Anoplura). I. *Parasitology*, **28** (1936), 92–7.

Studies on populations of head-lice (*Pediculus humanus capitus*: Anoplura). II. *Parasitology*, **30** (1938), 85–110.

*The Louse. An Account of the Lice which Infest Man, Their Medical Importance and Control*, 2nd edn. London: Edward Arnold, 1948.

Clay, T. The Amblycera (Phthiraptera: Insecta). *Bull. Br. Mus. Nat. Hist.*, **25** (1970), 75–98.

Cook, E. F. A study of the louse populations of the meadow vole and deer mouse. *Ecology*, **39** (1958), 645–59.

Combescot, C. Epidemiologie actualle de la pédiculose a *Pediculus capitus*. *Bull. Acad. Natl Med.*, **174** (1990), 231–7.

Cushing, E. C. *History of Entomology in World War II*. Smithsonian Institution Publication 4294. Washington, DC: Government Printing Office, 1957.

Durden, L. A. and G. G. Musser. The sucking lice (Insecta, Anoplura) of the world: a taxonomic checklist with records of mammalian hosts and geographical distributions. *Bull. Am. Mus. Nat. Hist.*, **218** (1994), 1–90.

Emerson, K. C. *Checklist of the Mallophaga of North America (North of Mexico)*. Part I, *Suborder Ischnocera*; part 2, *Suborder Amblycera*; part 3, *Mammal Host List*; part 4, *Bird Host List*. Dugway, Ut: Desert Test Center, Dugway Proving Ground, 1972.

Ewing, H. E. The taxonomy of the anopluran genus *Pediculus* Linnaeus. *Proc. Biol. Soc. Wash.*, **46** (1933), 167–74.

Ferris, G. F. The sucking lice. *Pac. Coast Entomol. Soc. Mem.*, **1** (1951), 1–320.

Fisher, I. and R. S. Morton. *Phthirus pubis* infestation. *Br. J. Venereal Dis.*, **46** (1970), 326–9.

Hase, A. Siphunculata; Anopleura; Aptera. Laüse. *Biol. Tiere Dtsch.*, **30** (1931), 1–58.

Hopkins, G. H. E. The host associations of the lice of mammals. *Proc. Zool. Soc. Lond.*, **119** (1949), 387–604.

The correct names of the body and head lice of man. *Entomologist*, **85** (1952), 90–1.

Ibarra, J. Head lice in schools. *School Health*, **4** (1989), 147–51.

Juranek, D. D. Epidemiology of lice. *J. School Health*, June (1977), 346–64.

Kim, K. C. and K. C. Emerson. Description of two species of Pediculidae (Anoplura) from Great Apes (Primates, Pongidae). *J. Parasitol.*, **54** (1968), 690–5.

Kim, K. C. and H. W. Ludwig. The family classification of the Anoplura. *System. Entomol.*, **3** (1978), 249–84.

Phylogenetic relationships of parasitic Psocodea and taxonomic position of Anoplura. *Ann. Entomol. Soc. Am.*, **71** (1978), 910–20.

Kim, K. C., H. D. Pratt, and C. J. Stojanovich. *The Sucking Lice of North America. An Illustrated Manual for Identification*. University Park, PA: State University Press, 1986.

Kraus, S. J. and L. H. Glassman. The crab louse: review of physiology and study of anatomy as seen by the scanning electron microscope. *J. Am. Venereal Dis. Assoc.*, **2** (1976), 12–18.

Ledger, J. D. *The Arthropod Parasites of Vertebrates in Africa South of the Sahara*, vol. IV. *Phthiraptera (Insecta)*. Johannesburg: South African Institute of Medical Research, 1979.

Maunder, J. W. Human lice – biology and control. *R. Soc. Health J.*, **97** (1977), 29–32.

Mellanby, K. The incidence of head lice in England. *Med. Officer*, **65** (1941), 39–43.

Natural populations of the head-louse (*Pediculus humanus capitus*: Anoplura) on infected children in England. *Parasitology*, **34** (1942), 180–4.

Relation between size of the family and the incidence of head lice. *Public Health*, **56** (1942), 31–2.

Nuttall, G. H. F. The biology of *Pediculus humanus*. *Parasitology*, **10** (1917), 180–5.

The biology of *Phthirus pubis*. *Parasitology*, **10** (1918), 383–405.

Orkin, M. and H. Maibach (eds.) *Cutaneous Infestation and Insect Bites*. New York: Marcel-Decker, 1985.

Piotrowski, F. Nimfy wszy lonowej *Phthirus pubis* L. (Anoplura). *Polskie Pismo Entomol.*, **31** (1961), 321–34.

Price, M. A. and O. H. Graham. Chewing and sucking lice as parasites of mammals and birds. *U.S. Dept. Agri. Tech. Bull.*, **1849** (1997).

Qui, M.-H and M.-L. Chen. Morphology of adult *Pthirus pubis* (L.) (in Chinese, English summary). *Endemic Dis. Bull.*, **8** (1993), 15–20.

Morphological studies on the human pubic louse, *Pthirus pubis* (L.) in post-embryonic development stages (in Chinese, English summary). *Endemic Dis. Bull.*, **8** (1993), 34–38.

Schaefer, C. W. Ecological separation of the human head lice and body lice (Anoplura: Pediculidae). *Trans. R. Soc. Trop. Med. Hyg.*, **72** (1978), 669–70.

Service, M. W. (ed.). *Biosystematics of Haematophagous Insects*. Oxford: Clarendon Press, 1988.

Sholdt, L. L., M. L. Holloway, and W. D. Fronk. *The Epidemiology of Human Pediculosis in Ethiopia*. Jacksonville, FL: Navy Disease Vector Ecology and Control Center, 1979.

Stojanovich, Ch. J. The head and mouth parts of the sucking lice (Insecta: Anopleura). *Microentomology*, **10** (1945), 1–46.

Traub, R. and H. Starcke (eds.) Fleas. In *Proceedings of the International Conference on Fleas*. Rotterdam: Balkema, 1980.

Trigger, D. S. Black fellows, white fellows and head lice. *Aust. Inst. Aboriginal Stud. Newslett.*, March (1981), 63–72.

Ubelaker, J. E. Scanning electron microscope of human pubic louse, *Pthirus pubis* (Linnaeus, 1758). *J. Parasitol.*, **59** (1973), 913–19.

# 15 Plecoptera, Psocoptera

## PLECOPTERA
### Introduction

Adult stoneflies are soft-bodied, 2–5 cm long, with a brown and gray body. They usually have two pairs of wings; they fly readily, but not well. Antennae are long, and the cerci are well developed. Mouthparts are the chewing type, but they are weak. Nymphs resemble the adults, except for the lack of wings; they have external filamentous gills on the thorax and legs. Most species live in cool, running streams, or in cold mountain lakes. Plecopterans are an important component in aquatic food chains, especially for fish. Adult stoneflies are imitated in the design of fishing fly used by anglers. These insects are physiologically fragile and intolerant of water conditions of low oxygen or pollution, and they are useful indicators of these and other stream conditions. Pest status of stoneflies is linked to their appearance in large numbers in areas adjacent to breeding sites, and sometimes a long distance from water. These incidents are usually associated with one or two species. Adults of *Perla* and *Pteronarcys* are attracted to lights at night.

Mating in stoneflies involves males attracting females by rubbing, tapping, or drumming a species-specific signal on an available substrate. Females respond to the male's vibrational signals with acoustic signals of their own. Male stonefly mating calls vary from a series of single beats to complex patterns of beats and intervals. These sounds are produced by a hammer-like structure on the underside of the abdomen. This morphological feature of the adult is struck or rubbed against the ground or other substrate. A few species in the Chlorperlidae produce sound by rapid vibration of their body, which is transferred to the substrate through the legs and tarsi. For most species, the males and females call back and forth, with some overlap (duetting) of the sounds. Males typically engage in a sequence of drumming and searching until a receptive female is located. Mating occurs immediately when male and female meet, and oviposition usually occurs soon after.

### Capniidae

Adults are about 10 mm long and grayish black to black. This is the largest family in the order, and adults emerge during the winter months. The wings are short and rudimentary in some species. Most of the species occur in eastern USA.

**Small winter stoneflies, *Allocapnia pygmaea*** Adults are 5–7 mm long, and dusky black. The immature stages are in creeks and lakes. Adults emerge in February through April, and may occur in large numbers on turfgrass, snow, the sides of buildings, and other objects in the peridomestic environment. This species occurs in northeastern USA; there are more than 30 species in the genus *Allocapnia*, and some of these occur in urban and suburban areas.

### Perlidae

Adults are 8–44 mm long, yellowish brown, and there are branched gill remnants at the posterior of the pro- and mesothoracic segments. Adults appear from April through September; they do not feed. Larvae are usually patterned in brown and yellow, somewhat flattened, and live in the bottom silt and mud as predators.

**Common stonefly, *Acroneuria evoluta* (Fig. 15.1a, b)** Adults are about 28 mm long to the wing tips; the body is uniformly brown, with dark brown marks on the head. Nymphs are 17–23 mm long, excluding tails, mottled brown and yellowish brown. The abdomen is distinctly banded brown and yellowish brown, with branched thoracic gills. Nymphs occur on rocks and gravel substrates of streams and rivers of central

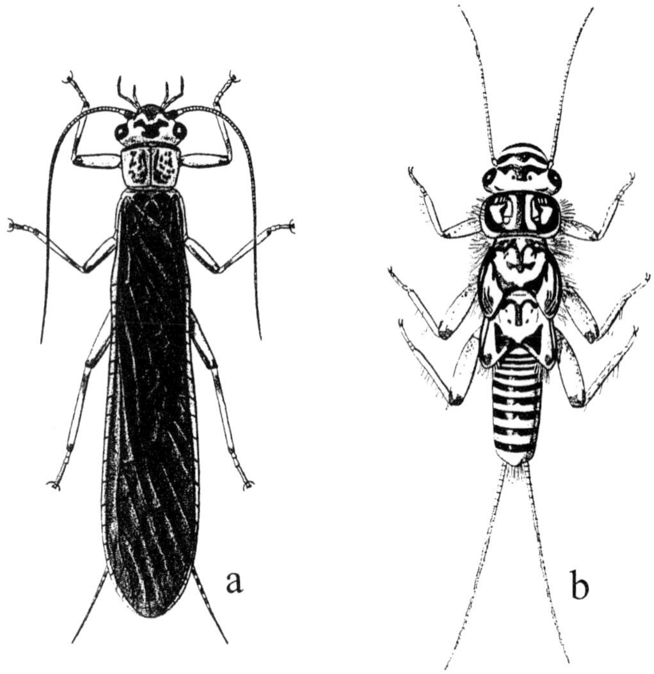

**Figure 15.1** Plecoptera. (a) *Acroneuria evoluta*, nymph; (b) *A. evoluta*, adult.

and southern North America. Adults emerge from May through July, and large numbers fly to lights at night.

## Pteronarcyidae

Adults are 38–63 mm long, and the body is brown or grayish brown. Larvae inhabit small streams and rivers, where they feed on plant material. Adults emerge at night from June through August; they do not feed.

**Giant stonefly, *Pteronarcys dorsata*** Adult wing span is 70–106 mm, and the body is yellowish brown. Adults are nocturnal and often fly to lights at night. This species is found in northeastern USA.

**The salmonfly, *Pteronarcys californica*** The adult male is 31–40 mm long, and wing span is 58–66 mm; the female is 40–46 mm long, with a wing span of 72–84 mm. The body is brownish gray, and the head is blackish brown with a tooth on the head plate above base of antennae; the prothorax has a dorsal red midline. The salmonfly is found in fast-flowing streams in western USA. Adults come to lights at night.

## Taeniopterygidae

Adults of most species are 10–55 mm long, with a dark brown or blackish brown body, short cerci, and long antennae. Larvae have divergent wing pads and live in rivers, feeding on plant material. Adults emerge from January through April; they may feed on flowers.

**Winter stonefly, *Strophopteryx fasciata*** Adults are 10–15 mm long, dark brown, and the cerci have 1–3 segments. Adults emerge from February to May and crawl on stones, fence posts, and the sides of buildings, especially if painted white. This species occurs in northeastern North America.

# PSOCOPTERA

## Introduction

Psocids are 1–6 mm long. They are somewhat flattened and delicate insects with a yellowish-brown or brown body, which is sometimes covered with scales. Individuals or generations of the winged species sometimes occur with the wings reduced. In other species, the female is consistently micropterous, and there are species in which both sexes are micropterous. Most Liposcelididae are wingless, and many of these are found indoors and called booklice. Wings may be clear or with color patterns, often with a dark spot at the anterior edge of the front wing, which is called the pterostigma. Winged individuals have membranous wings held rooflike over the abdomen. Antennae are long, eyes are present, and the mouthparts are mandibulate. Many psocids live gregariously, and clusters of individuals of various ages may be under a canopy of fine silken threads in protected places. Winged forms are often reluctant to fly; however, at times they fly in considerable numbers and drift through the air like aphids. Metamorphosis is simple, and immatures resemble adults. This is a large order of insects that are little known. There are nearly 2000 species in 226 genera, but there are probably many more undescribed species in the tropics. Most of the species associated with domestic and peridomestic habitats are known. Adults of *Lachesilla pedicularia* have occurred in buildings in large numbers.

Females lay 20–100 eggs singly or in a clutch of several over a span of 2–3 weeks. Eggs adhere to the substrate; a covering of silk usually protects egg clutches. Viviparous species produce young singly. Facultative parthenogenesis occurs, and males of some species are unknown. Nymphal development is 15–20 days, and there are 3–8 molts; for winged species the wing pads appear at the last molt. Immatures have a wide abdomen, narrow thorax, and a large head, with long antennae. Although these are generally considered soft-bodied insects, they are able to withstand a range of environmental conditions. Females of *Trogium pulsatorium* are capable of producing an audible sound by tapping their abdomen on a hard substrate.

The majority of psocids live on or under bark or foliage of trees and shrubs, in moist leaf litter, and in the nests of birds.

Some feed on lichens and algae, others on molds, pollen, and fragments of dead insects. Those that occur indoors feed on materials such as cereals and mold. Species living outdoors are sometimes called barklice; the species living indoors are occasionally found around books and are called booklice. Books are susceptible to mold growth in damp conditions because of the glue in the binding and sizing on the pages. The combination of mold and humid environment is often suitable for psocid infestations. Species of *Stimulopalpus* are found seasonally on moist exteriors of concrete buildings. Psocids do not live in isolation, and often interact with other structural pests in the household. *Trogium pulsatorium* and *Liposcelis bostrychophilus* are known to feed on the eggs of Angoumois grain moths and Indian meal moths.

Pest status of psocids is limited to a few species that occur indoors and are associated with household materials and stored cereal foods. About 3500 species of psocids have been described worldwide, but only about 50 of these are known to occur in domestic habitats, usually in food preparation and storage areas. Many species are sustained in outdoor populations and occasionally come indoors. Some species undergo parthenogenetic reproduction and indoor populations can become large when conditions are suitable. *Ectopsocus briggsi* and *Lachesilla pedicularia* occur indoors in the fall in the UK, but are not otherwise considered a domestic pest. Other species exist only as indoor pests, and are not represented by populations in natural habitats. The house psocid, *Liposcelis bostrychophilus*, is a tropical species, but it occurs primarily in households in temperate regions. This species is a common pest of food-storage facilities and retail food stores; in both locations infestations may occur on pallets, within packaging, and in the product. In retail stores this species is common in flour, cereal products, and sugar. The tropical origin of this species may be the reason for large populations in heated warehouses. *Lepinotus* spp. and *Trogium* spp. are adapted to low-temperature conditions.

## Amphientomidae

These are primarily tropical psocids. The wings and body are covered with scales, and the front femur usually has a row of spines on the anterior surface. One species, *Stimulopalpus japonicum*, is often found on concrete and stone surfaces. The distribution of this species includes eastern USA.

## Ectopsocidae

These psocids generally live in dry habitats, and many are found in leaf litter and similar sites. Adults have 13-segmented antennae, which lack secondary annulations. Species are macropterous or brachypterous. *Ectopsocus maindroni* is a widely distributed species. It is native to Africa. It is occasionally found in stored foods.

**Ectopsocus briggsi (Fig. 15.2n)** This species is macropterous; the front wings have light spots at the tips on longitudinal wing veins. It occurs indoors associated with stored foods, herbs, and fruits. It is probably cosmopolitan, and reported from houses along the Pacific coast of the USA, and in households in the UK.

**Large-winged psocid, *Ectopsocopsis cryptomeriae*** This species is macropterous, with the wings extending to the tip of the abdomen. It is nearly cosmopolitan, and has been recorded from stored foods in tropical and neotropical regions.

***Ectopsocus richardsi* (Fig. 15.2g)** This species is associated with stored grain and other dry food material. The adults are brachypterous; the head is brown, and the body is pale brown. The female subgenital plate has a single median lobe. It is nearly cosmopolitan. Related species, *E. pumilis* and *E. maindroni*, are known from tropical and subtropical regions, and are associated with stored food indoors. The adults are macropterous, and the body is pale yellow to light brown; the color pattern of the abdomen is not distinctly annulate.

***Ectopsocopsis vachoni*** This species is widely distributed in the subtropics. The wings of the female lack spots, and a pterostigma is absent; the body is reddish brown; the abdomen has reddish-brown segments on a pale brown background. The males are apterous.

## Lachesillidae

These psocids occur primarily in dry habitats, but some are found on the foliage of conifer trees. Adults have 13-segmented antennae, which lack secondary annulations. The wings extend to the tip of the abdomen or beyond; the wing-spot is constricted at the base, and rounded apically. *Lachesilla* is a large genus found throughout the world and contains a number of species that occur in habitats such as dead foliage, haystacks, and thatching. *L. quercus* is a rare species, but it is widely distributed and known from stored products in Europe. *L. greeni* occurs in outbuildings and may be a domestic species in the UK.

**Cosmopolitan grain psocid, *Lachesilla pedicularia* (Fig. 15.2o)** Adults are 1.5–2.1 mm long. The front wing of the macropterous form is about 1.8 mm long. The body is dark brown, and the

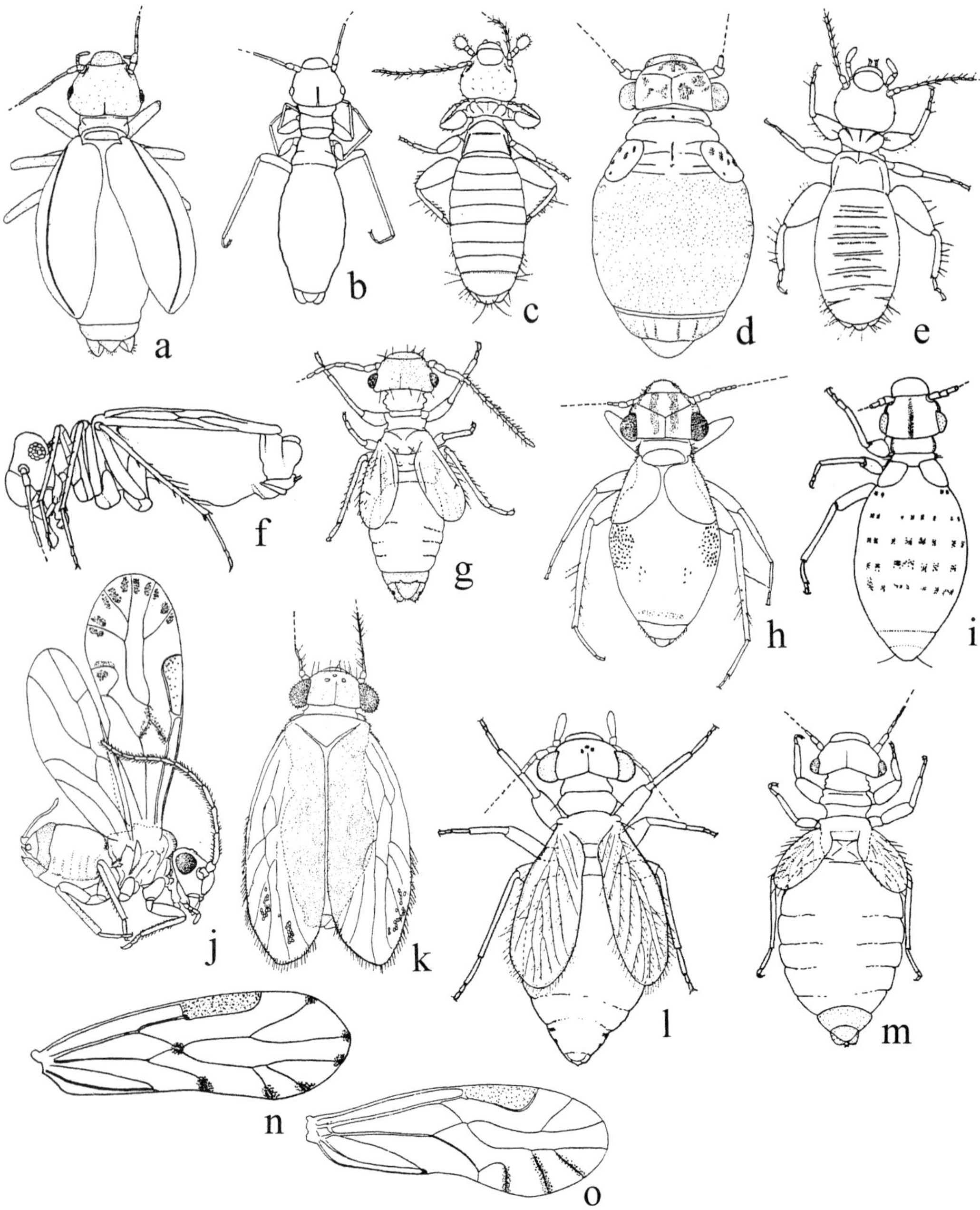

**Figure 15.2** Psocoptera. (a) *Badonnelia titei* female; (b) *B. titei* male; (c) *Blephotroctes ghesquierei*; (d) *Cerobasis annulata*; (e) *Embidopsocus oleaginous* female; (f) *Dorypteryx domestica* female; (g) *Ectopsocus richardsi*; (h) *Lepolepis bicolor*; (i) *Trogium pulsatorium*; (j) *Lachesilla rena* male; (k) *Soa flaviterminata*; (l) *Rhyopsocus bentonae* female; (m) *R. disparilis*; (n) *Ectopsocus briggsi* wing; (o) *L. pedicularia* wing.

abdomen is strongly annulated with reddish-brown segments. The front wing of the winged form is clear and without spots, while the veins at the posterior of the wing have light-brown borders. This species is distributed throughout Europe and North America outdoors, and is common indoors associated with stored grain. Closely related species, *L. rena* (Fig. 15.2j) and *L. nubilis*, are known primarily from outdoor habitats, but also occur with stored grains indoors. The fore wings of these species have distinct spots along the veins and at the apex.

## Lepidopsocidae

Species in this family are commonly found on trees, shrubs, and in dry ground habitats. Their head is very broad and the eyes are large and extend to the dorsal surface of the head. The antennae have more than 20 segments. They have pointed, slender wings; the wings and body are usually covered with scales, which is the basis for the family name.

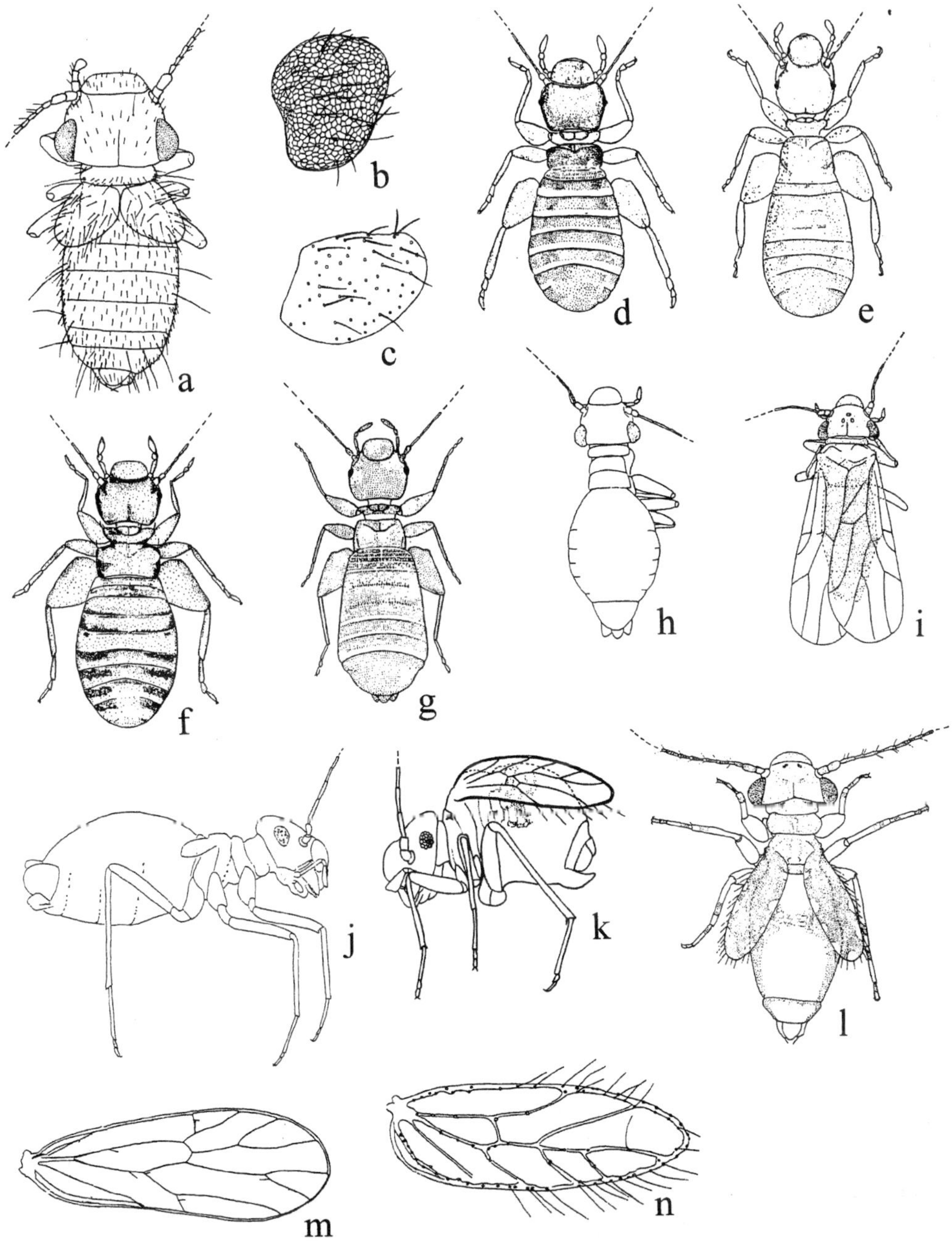

**Figure 15.3** Psocoptera. (a) *Lepinotus reticulatus* female; (b) *L. reticulatus*, female wing; (c) *L. patruelis*, female wing; (d) *Liposcelis bostrychophilus* female; (e) *L. corrodens* female; (f) *L. entomophilus* female; (g) *L. simulans*; (h) *Nanopsocus oceanicus*, apterous female; (i) *N. oceanicus*, winged female; (j) *Psyllipsocus ramburi*; (k) *Psocatropos micropos* male; (l) *Psoquilla marginepunctata* female; (m) *Psyllipsocus ramburi*, male wing; (n) *Psocatropos micropos*, male wing.

**Striped psocid, *Lepolepis bicolor* (Fig. 15.2h)** The adult has a pale yellow head with four longitudinal brown or blackish-red stripes. The fore wings are reduced to small, scale-like lobes and they are indistinct; the front wings extend to the posterior margin of abdominal segment 1, without distinct veins. It has been collected with peanuts, and is distributed from Africa to India; it has been intercepted at a port in the UK.

***Soa flaviterminata* (Fig. 15.3k)** Adults have long and broad wings, which are marked with scales apically, and extend beyond the tip of the abdomen. The antenna has fewer than 30 segments. The front wing is about 2.5 mm long and the membrane is blackish brown; the veins are dark brown; there is a row of yellow marginal setae. Ocelli are close together and the anterior of segment 3 is small. This species is found primarily in tropical regions, and from stored foods in West

Africa. It represents a genus with about five described species, which occur in leaf litter and dead foliage in the tropics.

## Liposcelididae

Most of the species in this family occur in dry locations, such as under tree bark and in dead leaves, but some are found in bird and mammal nests. The species that are found indoors are usually located in dusty places with high temperature and humidity. These psocids are wingless and they have enlarged hind femora (*lipo*, Greek for fat; *scelus*, Latin for leg). A large number of *Liposcelis* species have been recorded from stored foods and grain indoors. Males are usually identified to species by their association with females. *Liposcelis* species are characterized by the absence or presence of a dorsal posterior band on abdominal segments 3 and 4, and terga 3 and 4 are uniform in color.

***Blephotroctes ghesquierei* (Fig. 15.2c)** Adults are characterized by a dilated terminal segment of the maxillary palp, and the abdominal terga lack sclerotized transverse markings. This species has been recorded from Brazil, the USA (Florida), and from stored food in West Africa. A related species, *B. simulans*, has been found in stored food in West Africa.

***Embidopsocus oleaginous* (Fig. 15.2e)** Adults are about 1.7 mm long. The cuticle is composed of fine granulations. Lateral borders of the pronotum are extended in a rounded arc. The hind femur is rounded, without the small lateral anterior process. Abdominal terga have sclerotized transverse stripes, and the apical segment of the maxillary palp is not swollen. This species is known from stored foods in West Africa, Sri Lanka, Philippines, Taiwan, and Okinawa. A closely related species, *E. minor*, has been taken from African cacao in a warehouse in the UK, and from stored foods in West Africa.

**House psocid, *Liposcelis bostrychophilus* (= *L. granicola*) (Fig. 15.3d)** Adults are about 1 mm long, light brown, and wingless. This species is widely distributed in households, and it is probably dispersed with the movement of food materials. Although it undergoes obligate parthenogenetic reproduction, there is considerable variability in populations in domestic and commercial facilities. Morphologically the populations differ in body size and color.

**Cereal psocid, *Liposcelis simulans* (= *L. divinatorius*, *Termes divinatorium*) (Fig. 15.3g)** Adults are 1.1–1.3 mm long, with a brown body; the anterior margins of the abdominal terga 5–8 have sclerotized, transverse stripes. Eggs are laid in batches of 2–3 daily, but later this drops to about one per 7 days; fecundity is about 200 eggs. Females lay about 20 eggs from October to January at a temperature range of 10–30 °C; about 50 eggs are laid from June to August at 15–32 °C. Hatching occurs in about 11 days at 25 °C and 75% relative humidity (RH), but eggs may overwinter. Development is through four nymph stages over 24–65 days, but nymphs may overwinter. Death results when exposed to 0 °C for 3 h; 24-h exposure to 42.5% and 75% RH is lethal. This species does not develop at less than 55% RH at 25 °C or 65% RH at 35 °C. It commonly occurs in houses with stored food and grain products in central Europe, the UK, and USA; it is also known from bird nests.

**Stored food and grain psocids, *Liposcelis* spp.** There are about 50 wild species in this genus, and many are known as infesting stored food and grain. *L. albothoracicus* adults have the mesothorax, metathorax, and first abdominal segment white; other portions of the body are brown. This species is known from Turkish millet seed in a ship at a port in the UK, and from outdoor localities in southern India. *L. bouilloni* is known from outdoor habitats in Brazil and from stored coffee in West Africa. *L. decolor* has the anterior margins of abdominal terga 5–8 with prominent, dark, transverse stripes; the body is yellow. It is cosmopolitan and occurs with stored food in houses in Europe and USA. *L. entomophilus* (Fig. 15.3f) is pale yellow, and has lateral reddish-brown bands on abdominal terga 3–8. This cosmopolitan species infests stored grain in tropical countries. *L. rufus* adults are reddish brown; the compound eye of the female has 6–7 facets, and the male has five facets. It has been recorded outdoors and as an indoor pest in the USA and the UK. *L. kidderi*, *L. mendax*, *L. obscurus*, *L. paetulus*, *L. paetus*, and *L. rufus* females have eyes with 5–7 facets, and a brown to dark-brown body.

***Liposcelis transvaalensis*** Adults are uniformly brown, except for the apical half of the abdomen, which is dark brown. As the specific name implies, this species is known from South Africa, where it occurs in stored foods.

**Other domiciliary Liposcelidae** Grain storage facilities and stored food in households are infested with a variety of psocids. *Liposcelis bouilloni* adult females have eyes with only four facets. This species is known from stored foods in West Africa and from outdoor locations in Brazil. *L. brunnea* occurs in grain storage buildings in the UK and continental Europe. *L. exiguus varians* adult female is less than 1 mm long and the body color is pale brown to reddish brown. It has been reported from stored peanuts in Zaire. *L. kidderi* has a brown body; the eyes are small: the female has 5–6 facets, and the male 4–5 facets. It is known to occur in houses in Europe and North America. *L. lipara* is known from buildings in the UK, and is common outdoors in western USA and South Africa. Female adults of *L. mendax* are 1.2–1.4 mm long and the body color is light brown. It has been reported from houses in Europe and in stored products in Zaire. *L. obscurus* is known from wheat stored in bulk in a ship's hold in the UK. *L. paetus*, *L. minutus*, and *L. corrodens* (Fig. 15.3e) are nearly cosmopolitan, and known from stored grain. *L. pubescens* is known from stored grain in the UK and central Europe, Argentina, and New Zealand.

## Pachytroctidae

These psocids have wings with many wing veins, and the apterous forms with the meso- and metathorax are separated by a suture. *Nanopsocus oceanicus* (= *Tapinella africana*, *T. pallida*) (Fig. 15.3h, i) occurs in houses in southeastern USA and Japan; it is also known from West Africa and New Hebrides.

## Psoquillidae

Members of this family live in a variety of habitats, from dead leaves hanging on plants to leaves in ground litter. They may be fully winged or brachypterous, but the wings show distinct venation. *Psoquilla marginepunctata* (Fig. 15.3l) is circumtropical; it occurs primarily outdoors in tropical regions, but indoors in temperate regions. It has been recorded in houses in the USA, the UK, and continental Europe.

Species in the genus *Rhyopsocus* have front wings that are well-developed or reduced, but always with veins and uniform color. *Rhyopsocus peregrinus* has occurred in a food store in the UK; *R. disparilis* (Fig. 15.2m) has been taken from stored cacao from West Africa, rice from the Philippines and Japan, beans from West Indies, and peas from Guyana; *R. bentonae* (Fig. 15.2l) has been found in houses in southeastern USA.

## Psyllipsocidae

These psocids are found primarily in ground-level habitats, such as leaf litter, in caves, and in wine cellars. Long- and short-winged individuals occur within populations of most species. *Soa flaviterminata* is a long-winged species that has been recorded from stored foods in West Africa. *Psyllipsocus ramburi* (Fig. 15.3j, m) is a short-winged, cosmopolitan species, which occurs in damp and dark sites, such as cellars and caves. It has been recorded around the openings of wine and vinegar barrels.

**Dorypteryx domestica (Fig. 15.2f)** Adults are 1–2 mm long and the body is pale yellow to light brown. The front wing is long and narrow, and venation is reduced to two longitudinal veins. The hind wings are absent. This species is known from household habitats and stored-food facilities in the UK and continental Europe; it is probably cosmopolitan. A related species, *D. pallida*, has narrow front wings with reduced wing veins and with no closed cells. It is known from domestic habitats in eastern USA, and in central and southern Europe.

**Psocatropos spp.** These species have broad wings and there are several branches and closed cells on the posterior margin of the wing. *P. pilipennis* is known from domestic habitats in India, Madagascar, and the Seychelles; *P. micropos* (Fig. 15.3k, n) occurs in stored food in houses, and in outdoor habitats in the tropics.

## Sphaeropsocidae

These psocids have a 15-segmented antenna and their body is generally rounded and not flattened. *Badonnelia titei* (Fig. 15.2a, b) is known from stored food in domestic habitats in Europe and the UK. When present, wings are large, with few veins, and nearly cover the body; apterous forms have the meso- and metathorax fused.

## Trogiidae

These psocids are brachypterous, but none are wingless; the wings may be small and do not have veins. They are found in a variety of habitats, including with household materials, and commercial food facilities. *Lepinotus patruelis* occurs in low numbers in houses in the USA, the UK, and continental Europe throughout the year. *Cerobasis annulata* (Fig. 15.2d) occurs in houses in Europe and North America.

**Larger pale trogiid, *Trogium pulsatorium* (Fig. 15.2i)** Wings are uniform yellowish brown, and the frons has a dark longitudinal line. This species occurs in low numbers in houses in the USA, the UK, and Europe throughout the year.

**Reticulate-winged trogiid, *Lepinotus reticulatus* (Fig. 15.3a, b)** The small, rounded wings have a reticulated pattern when viewed under magnification; the head is uniform yellowish brown. This species is cosmopolitan and is usually found in outdoor habitats, but it is common in stored grain in the USA.

**Other Trogiidae** *Lepinotus inquilinus* is common in houses in Europe, North America, Africa, and Madagascar. *L. patruelis* (Fig. 15.3c) is common in houses in central Europe, and may have been introduced into the USA.

## Bibliography

### PLECOPTERA


Alba-Tercedor, J. and A. Sanchez-Ortega. *Overview and Strategies of Ephemeroptera and Plecoptera*. Gainesville, FL: Sandhill Crane Press, 1991.

Bauman, R. W., A. R. Gaufin, and R. F. Surdick. The stoneflies (Plecoptera) of the Rocky Mountains. *Mem. Am. Entomol. Soc.*, **31** (1977), 208.

Brinck, P. Studies on Swedish stoneflies. *Opusc. Entomol.* (Suppl. 11) (1949), 1–250.

Bueno-Soria, J. and S. Santiago Fragoso. Trichoptera. In Hurlbert, S. H. and A. Villalobos Figueroa (eds.) *Aquatic Biota of Mexico, Central America and the West Indies*. San Diego, CA: San Diego State University, 1982.

Campbell, I. C. (ed.) *Mayflies and Stoneflies: Life Histories and Biology*. Series entomologica 44. Dordrecht: Kluwer Academic, 1990.

Frison, T. H. The stoneflies, or Plecoptera, of Illinois. *Ill. Nat. Hist. Surv. Bull.*, **20** (1935), 281–471.

Gaufin, A. R., A. V. Nebeker, and J. Sessions. The stoneflies (Plecoptera) of Utah. *Univ. Utah Biol. Ser.*, **14** (1966), 9–89.

Hitchcock, S. W. Guide to the insects of Connecticut. Part VII. The Plecoptera or stoneflies of Connecticut. *Conn. State Geol. Nat. Hist. Surv. Bull.*, **107** (1974), 1–262.

Hynes, H. B. N. A Key to the Adults and Nymphs of British Stoneflies (Plecoptera) 2nd edn. *F. W. Biol. Assoc. Sci. Publ.*, **17** (1967), 1–86.

Biology of the Plecoptera. *Annu. Rev. Entomol.*, **21** (1976), 135–53.

Illies, J. Phylogeny and zoogeography of the Plecoptera. *Annu. Rev. Entomol.*, **10** (1965), 117–40.

*Katalog der Rezenten Plecoptera. Das Tierreich*, 82. Berlin: Walter de Gruyter, 1966.

Jewett, S. G., Jr. The stoneflies (Plecoptera) of the Pacific Northwest. *Ore. State Monogr.*, **3** (1959), 95.

Landholt, P. and M. Sartor (eds.). *Ephemeroptera and Plecoptera: Biology, Ecology, Systematics*. Fribourg: MTL Mauron, Tingley and Lachat., 1997.

Macan, T. T. *The Study of Stoneflies, Mayflies and Caddisflies*. London: Amateur Entomologist's Society, 1982.

Stark, B. P. and A. R. Gaufin. The stoneflies (Plecoptera) of Florida. *Trans. Am. Entomol. Soc.*, **104** (1979), 391–433.

Stewart, K. W. Vibrational communication in insects: epitome in the language of stoneflies. *Am. Entomol.*, **43** (1997), 81–91.

Surdick, R. F. and K. C. Kim. Stoneflies (Plecoptera) of Pennsylvania, a synopsis. *Bull. Penn. State Univ. Coll. Agr.*, **808** (1976), 1–73.

Zwick, P. Plecoptera (Sternfligen). In Beier, M. (ed.) *Handbuch der Zoologie*, vol. IV. *Insecta* **26**, pp. 1–115. Berlin: de Gruyter, 1980.


### PSOCOPTERA

The published information on Psocoptera is extensive, and, like the pest species, it is representative for all regions of the world. The bibliography by Smithers and Lienhard (1992) is helpful.


Smithers, C. N. and C. Lienhard. A revised bibliography of the Psocoptera (Arthropoda: Insecta). *Tech. Rep. Aust. Mu.*, **6** (1992), 1–86.


#### General


Back, E. A. Psocids in dwellings. *J. Econ. Entomol.*, **32** (1939), 419–23.

Badonnel, A. Order des Psocoptères. In Grass, P. P. (ed.) *Trait de Zoologie*, vol. 10, fasc, **2**, pp. 1310–40. Paris: Masson, 1951.

Baz, A. and V. J. Monserrat. Distribution of domestic Psocoptera in Madrid apartments. *Med. Vet. Entomol.*, **13** (1999), 259–64.

Broadhead, E. The life-history of *Embidopsocus enderleini* (Ribaga) (Corrodentia, Liposcelidae). *Entomol. Mon. Mag.*, **83** (1947), 200–3.

A new parthenogenetic psocid from stored products, with observations on parthenogenesis in other psocids. *Entomol. Mon. Mag.*, **90** (1954), 10–16.

The infestation of warehouses and ships' holds by psocids in Britain. *Entomol. Mon. Mag.*, **90** (1954), 103–5.

Broadhead, E. and B. M. Hobby. Studies on a species of *Liposcelis* (Correnditia, Liposcelidae) occurring in stored products in Britain. Part I. *Entomol. Mon. Mag.*, **80** (1944), 45–59.

Candura, G. S. Contributo alla conoscenza biologica del *Troctes divinatorius* (Müller). *Boll. Zool.*, **3** (1932), 177–84.

Finlayson, L. R. Some notes on the biology and life history of psocids. *Entomol. Soc. Ontario Annu. Rpt.*, **63** (1932), 56–8.

The life history and anatomy of *Lepinotus patruelis* (Psocoptera: Atropidae). *Proc. Zool. Soc. Lond.*, **119** (1949), 301–23.

Ghani, M. A. and H. L. Sweetman. Ecological studies of the booklouse, *Liposcelis divinatorius* (Mull.). *Ecology*, **32** (1951), 230–44.

Günter, K. K. Stäublause, Psocoptera. *Tierwelt. Dtsh.*, **61** (1974), 1–314.

Hawkins, J. Corrodentia pests of ground feed. *J. Econ. Entomol.*, **32** (1939), 467.

Knülle, W. and R. R. Spadafora. Water vapor sorption and humidity relationships in *Liposcelis* (Insecta: Psocoptera). *J. Stored Prod. Res.*, **5** (1969), 49–55.

Kučerová, Z. Faunistic records from Czechoslovakia. Psocoptera (Psylliposcidae): *Dorypteryx domestica*. *Acta Entomol. Bohemoslov.*, **89** (1992), 315.

Lienhard, C. Revision of the Western Palaearctic species of *Liposcelis* Motchulsky (Psocoptera: Liposcelididae). *Zool. Jb. (Syst.)*, **117** (1990), 117–74.

Linsey, E. G. Insect food caches as reservoirs and original sources of some food products pests. *J. Econ. Entomol.*, **35** (1942), 434–9.

Lovitt, A. E. and E. L. Sonderstrom. Predation of Indian meal moth eggs by *Liposcelis bostrychophilus*. *J. Econ. Entomol.*, **61** (1968), 1444–5.

Mockford, E. L. Notes on some eastern North American psocids with descriptions of two new species. *Am. Midl. Nat.*, **53** (1955), 436–41.

Life history studies on some Florida insects of the genus *Archiposocus* (Psocoptera). *Bull. Fl. State Mus. Biol. Sci.*, **1** (1957), 235–74.

Parthenogenesis in psocids (Insects: Psocoptera). *Am. Zool.*, **11** (1971), 327–39.

New, T. R. An introduction to the natural history of the British Psocoptera. *Entomologist*, **104** (1971), 59–97.

Psocoptera. In *Royal Entomological Society of London Handbook: Identification of British Insects*, vol. 1, pp. 1–102. London: Dramrite Printers, 1974.

Obr, S. Psocoptera of food processing plants and storages, dwellings and collections of natural objects in Czechoslovakia. *Acta Entomol. Boehm.*, **75** (1978), 226–42.

Pearman, J. V. On sound production in the Psocoptera and on a presumed stridulatory organ. *Entomol. Mon. Mag.*, **64** (1928), 179–86.

Biological observations on British Psocoptera. *Entomol. Mon. Mag.*, **64** (1928), 209–18, 239–43, 263–8.

More Psocoptera from warehouses. *Entomol. Mon. Mag.*, **67** (1931), 95–8.

Third note on Psocoptera from warehouses. *Entomol. Mon. Mag.*, **78** (1942), 289–92.

Scott, H. G., J. S. Wiseman, and C. J. Stojanovich. Collembola infesting man. *Ann. Entomol. Soc. Am.*, **55** (1962), 428–30.

Smithers, C. N. Keys to the families and genera of Psocoptera. *Tech. Rep. Aust. Mus.*, **2** (1990), 1–82.

Söfner, L. Zur Entwicklungsbiologie und Oekologie der einheimischen Psocopterenarten *Ectopsocus meridionalis* (Ribaga, 1904) und *Ectopsocus briggsi* McLach. 1899. *Zool. Jb. Syst.*, **74** (1941), 325–60.

Spieksma, F. Th. M. and C. Smits. Some ecological and biological aspects of the booklouse *Liposcelis bostrychophilus* Badonnel, 1931 (Psocoptera). *Netherlands J. Zool.*, **25** (1975), 219–30.

Turner, B. D. and N. Ali. Population variability in a domestic stored product pest, the parthenogenic psocid *Liposcelis bostrychophila*; implications for controls. In Wildey, K. W. and W. Robinson (eds.) *Proceedings of the International Conference on Insect Pests in the Urban Environment*, pp. 309–17. Cambridge, UK: St John's College, Cambridge University, 1993.

Turner, B. D. and H. Maude-Roxby. The prevalence of the booklouse *Liposcelis bostrychophilus* Badonnel (Liposcelidae, Psocoptera) in British domestic kitchens. *Int. Pest Control*, **31** (1989), 93–7.

Weber, H. Die Lebengeschichte von *Ectopsocus parvalus* (Kolbe). *Z. Wiss. Zool.*, **138** (1931), 457–86.

Welch, R. C. *Badonnelia titei* Peraman (Psocoptera, Sphaeropsocidae) from a second Huntingdonshire locality. *Entomol. Mon. Mag.*, **119** (1983), 236.

Wyniger, R. *Liposcelis pubescens*. Die Rolle der Staulabläuse in der Vorratshaltung. *Swiss Food*, **7** (1985), 11–16.

### Geographic distribution

Badonnel, A. Psocoptères nouveaux d'Afrique et d'Arabie. *Rev. Fr. Entomol.*, **2** (1935), 76–82.

Contribution à l'étude des Psocoptères de l'Atlantide. *Rev. Fr. Entomol.*, **11** (1944), 47–60.

Contribution à l'étude des Psocoptères du Maroc. Voyage de L. Berland et M. Vachon. *Rev. Fr. Entomol.*, **12** (1945), 31–50.

Chapman, P. J. Corrodentia of the United States of America. I. Suborder Isotecnomera. *J. N.Y. Entomol. Soc.*, **39** (1930), 54–65.

Hickman, V. V. A contribution to the study of Tasmanian Copeognatha. *Pap. Proc. R. Soc. Tasm.*, **1933** (1934), 77–89.

Lienhard, C. Revision of the Western Palaearctic species of *Liposcelis* Motchulsky (Psocoptera: Liposcelididae). *Zool. Jb. (Syst.)*, **117** (1990), 117–74.

Mockford, E. L. *North American Psocoptera. Fauna and Flora Handbook* No. 10. Gainesville, FL: Sandhill Crane Press, 1993.

Mockford, E. L. and A. B. Gurney. A review of the psocids, or book-lice and bark-lice, of Texas (Psocoptera). *J. Wash. Acad. Sci.*, **46** (1956), 353–68.

Roesler, R. Beitrage zur kenntnis der Copeognathenfauna Deutschlands. *Zool. Anz.*, **125** (1939), 157–76.

# 16 Siphonaptera

## Introduction

Fleas are 0.8–5 mm long, laterally compressed, and wingless insects. The body is well-sclerotized, yellowish brown to brownish black and usually with numerous setae and bristles. The head is small and the eyes are reduced or absent; the antennae are three-segmented, clubbed, and fit into a groove on the side of the head. Legs are short, and in most species the femur is enlarged. Mouthparts are piercing–sucking; the adult stage feeds on blood. Adult fleas are obligate parasites of warm-blooded vertebrates. In some species there is a comb of large spines, the genal ctenidium, located above the mouthparts. The thorax is compact, but consists of pro-, meso-, and metathorax. The pronotum lies immediately behind the head; on the posterior margin in some species there is a comb of large spines, known as the pronotal ctenidium.

There are about 2500 described species and subspecies in this order, and these are grouped in 239 genera and 15 families. The majority of species occur on mammals, and about 100 species are found on birds. Fleas have complete metamorphosis, with a distinct egg, larva, pupa, and adult stages. Fleas have evolved as parasites of animals that have dens, burrows, or nests. The adults remain on the body of the host, and the larvae feed and develop in the nest or burrow. Larvae are elongate and resemble fly maggots; they are distinguished by scattered, long setae on the body. Larvae have chewing mouthparts, and lack eyes and legs. Pupation is in a silken cocoon made by the third-stage larva.

Eggs are oval, smooth, and white. Several hundred eggs are produced at regular intervals during the life of the female. Females typically lay eggs on the host or in the nest or den of host. Eggs usually fall from the host to the ground; hatching occurs in about 5 days. Larvae feed on organic debris and dried fragments of host blood (produced by adult fleas) in the nest of the host animal. Larvae are sometimes found in the fur of the host, such as on dogs and cats, and in the feathers of nestling birds. However, larval development is usually off the host. Development depends primarily on temperature, and usually requires 2–3 weeks. There are three larval instars, which differ little except in size. Only the head and mouthparts are sclerotized; the remainder of the body is yellowish white to white. The body may be dark brown when there is blood or other food in the gut. The pupal period is 1–2 weeks. The adult flea may remain in the cocoon after emerging from the pupal case; emergence to the outside usually requires a mechanical stimulus or temperature change to induce it to emerge from the cocoon. Without a stimulus the adult can remain alive in the cocoon for long periods. Adult life span is variable and, depending on species and conditions, ranges from about 1 month to more than 12 months.

The coevolution of fleas and their vertebrate hosts shows several morphological adaptations, including the development and placement of combs of strong setae on the head, thorax, and abdomen of adult fleas. These large and posteriorly directed setae may have a hold-fast function. However, the primary structures used to hold adult fleas within the fur or feathers of the host are the long claws and mouthparts. The large setae and the setal combs (ctenidia) may serve to protect mobile joints and membranes. The combs and long setae on the body, which in general project posteriorly, may also protect and aid in the movement of the adult through the fur or feathers of the host. In general, fleas with well-developed combs have hosts that move by flying or gliding, or hosts that are both nocturnal and climbing, or have large home ranges.

Jumping is one of the most distinctive features of adult flea behavior. The rat flea, *Xenopsylla cheopis*, can jump 18–31 mm horizontally, while the human flea, *Pulex irritans*, can jump 20 mm horizontally, and about 13 mm vertically. The cat flea can achieve a jump of about 30 cm at up to 130 times the acceleration of gravity. Speeds average about 3.6 m/s, and range between 1.62 and 5.76 m/s – four times faster than other fleas.

Hungry fleas may jump hundreds of times an hour for several days to find a host. Jumping is by the hind legs set against the substrate, with the front and middle legs acting as supports. The femur of the hind leg is moved to an upright or vertical position, and the tibia and tarsus are pressed against the host. When jumping, the femur moves rapidly downward, transmitting its thrust to the substrate with the tibia. During the jump the flea may turn over and the legs project out, which may increase the probability of the flea holding on to a host when landing. Fleas regularly move from host to host, and in some species they can survive for more than a year without feeding.

Pest status of fleas is based on the transmission of diseases and the skin irritation caused by their bites to humans and pets. Fleas transmit protozoans, bacteria, rickettsiae, and viruses between mammals, but the main diseases transmitted to humans are bacterial and rickettsial. When a flea inserts its mouthparts into the skin of the host to obtain a blood meal, it first passes saliva down through the mouthparts and into the wound. The saliva contains an anticoagulant to keep the blood from clotting within the mouthparts while the flea is feeding. This saliva also contains a protein to which humans, and perhaps other animals, are allergic. The protein causes a swelling of the tissue and an itching sensation after the flea removes its mouthparts and leaves. A flea bite in human skin causes itching for several days and a patch of swollen and reddened skin. Frequently, humans scratch their flea bites, and this may result in disease organisms on the skin entering the wound caused by the mouthparts. Droplets of blood on the skin, which have passed through the flea's intestine, may contain pathogenic organisms and these may be introduced into the host body. This is one way in which murine typhus is transmitted. The dog and cat flea can be intermediate hosts of the dog and cat tapeworm, *Dipylidium caninum*, and the cat flea is primarily responsible for flea allergy dermatitis in dogs and cats. *C. felis* can transmit murine typhus, plague, and *Bartonella henselae*, which is the cause of cat scratch disease. Cats and dogs can groom themselves to remove fleas. The most efficient individual cat may remove only about 18% of the infesting fleas, and animals with the poorest grooming ability may remove only about 4%.

Plague is an acute infectious disease caused by *Yersinia pestis*. It is essentially a disease of rodents, and is usually transmitted by rodent fleas. However, this disease, under certain conditions, may cause serious and widespread epidemics among humans. The term bubonic plague is applied when an inflammation of the lymph glands results from the bacterial infection. The buboes or swollen lymph glands are 2–10 cm in diameter, and usually located in the neck, groin (femoral glands), and under the arms (axillary glands). When the course of the infection increases, the infection enters the blood stream and becomes septicemic plague. In cases of bubonic and septicemic plague, disease symptoms generally occur 2–6 days after infection. Pneumonic plague is the invasion of the respiratory system. This is the most dangerous form of the disease because it is fast-acting and is almost 100% fatal within 2–4 days of infection. Fleas are not the primary carrier of this form of plague; the bacteria can be transmitted by airborne droplets, such as during a cough. When a rodent dies of the disease, the infected fleas on its body, usually species of *Xenopsylla*, move to other hosts and at this time humans can be attacked. The bubonic form of the bacilli is transmitted while the flea is feeding on a host. The bacilli are contained in the digestive tract of the adult flea; the proventriculus usually becomes blocked by bacteria some time after infection, and bacilli are transmitted directly to the host by the mouthparts.

There is a reservoir of plague in wild rodent populations around the world, and plague outbreaks still occur in rural areas and towns. Urban outbreaks of plague have occurred since the first pandemic spread through Europe and North Africa in 542 AD. It killed many of the overcrowded and poorly housed people in cities and towns at this time. Plague epidemiology begins with the first signs of outbreak, which are large numbers of deaths in populations of the peridomestic brown rat, *Rattus norvegicus*. This is followed in about 10 days by deaths of domestic black rats, *R. rattus*. After about 2 weeks human cases appear. The explanation for this sequence is that infected rat fleas leave the dead bodies of brown rats and transfer to black rats, infecting them with plague. The rat flea, *Xenopsylla cheopis*, bites humans in the absence of live rats, and *X. cheopis* moves from *R. rattus* to humans. The recent status of the plague in Calcutta and Mumbai, India shows the relationship of rodent populations and fleas to the outbreak and spread of this disease. The lesser bandicoot rat (*Bendicota bengalensis*) is the dominant rodent in these cities. This rat is susceptible to plague, but carries fewer rat fleas than the black rat. *X. astia* is the primary flea on the bandicoot rat, but this species does not readily feed on humans. In Calcutta, the flea population is about 66% *X. astia*; in Mumbai *X. cheopis* is dominant, but *R. rattus* and *R. norvegicus* have developed some resistance to plague.

In the USA, plague has changed from a disease occurring in seaports (urban habitats) to one occurring in rural areas. This disease has been introduced repeatedly by shipping into major seaports on the Pacific and Gulf coasts. Plague in North America was first recognized in 1900 when it appeared in San

Francisco. It may have entered the city on infected rats escaping from a ship originating in Asia. The rats died and their infected fleas transmitted plague to other rats and wild rodents in the seaport. The San Francisco plague ended in 1904; there were 121 cases and 118 deaths. An outbreak of the urban type of zootic plague occurred in 1924 and 1925 in Los Angeles. There was a total of 40 cases, and 35 deaths from both pneumonic and bubonic cases. It is suspected that the human course of infection was domestic rats, which in turn were infected from California ground squirrels in the area. Cases of rural plague have been reported from Arizona, California, Colorado, Idaho, Nevada, New Mexico, Oregon, Texas, and Utah. These cases have been associated with wild rodents, rabbits, or hares.

Plague is maintained in populations of resistant rodent species that can harbor bacteria in their blood, but do not succumb to the infection. These reservoir populations are a source for disease outbreaks (epizootics) that occur in colonial rodent populations, such as ground squirrels and prairie dogs. The disease spreads rapidly among these animals and may kill more than 90% of infected populations. Areas that experience epizootics of plague range from suburbs to grassland and natural areas. Plague occurs in cycles, and may be linked to rainfall and food abundance, which in turn affect rodent populations.

Rapid suburban growth in western USA has resulted in increasing numbers of people living in or near areas where plague occurs in the rodent population. Urban sprawl has led to an increase in the suitable habitat and food sources for adaptable rodent species, such as the rock squirrel (*Spermophilus variegatus*), the California ground squirrel (*S. beecheyi*), and prairie dogs (*Cynops* spp.). These animals have increased in density in new suburban developments, and they are very susceptible to plague. Their fleas are efficient vectors of plague to humans and other animals. The number of human cases of plague in western USA has increased. This has followed an increase in the role of domestic cats as a source of human infection, including the pneumonic form of the disease. Between 1980 and 1997, 247 human plague cases were reported in the USA; of these, 37 (15%) were fatal.

## Ceratophyllidae

This is the largest family of fleas. Most species in this family are parasites of rodents; about 12% are parasites of birds. The squirrel flea, *Orchopeas howardii*, is found throughout the USA wherever the gray squirrel (*Sciurus griseus*) occurs. Adult fleas may disperse from active or abandoned squirrel nests in attics, and enter living spaces. The mouse flea, *Leptopsylla segnis*, is commonly found on domestic rats, and less commonly on the house mouse (*Mus musculus*). It is common in Europe and Asia, and in the USA.

Species of *Dasypsyllus*, *Mioctenopsylla*, and a few species of *Ceratophyllus* are parasites of birds. The important species include *C. gallinae*, which is a pest of poultry in North America and Europe, *C. gibsoni*, and *C. niger*, which are associated with chickens, and frequently with bird nests in buildings. *C. niger*, the western (USA) chicken flea, was originally described from specimens collected from humans and the Norway rat, *Rattus norvegicus*. All three of these bird species are known to bite people. In Japan, the sparrow flea, *C. gallinae dilatus*, is associated with six species of sympatric birds: tree sparrow (*Passer montanus saturatus*), gray starling (*Sturnus cineraceus*), violet-backed starling (*Sturnus philippensis*), great tit (*Parus major minor*), rock pigeon (*Columba livia domestica*), and the purple martin (*Delichon urbica*). Adult fleas from nests of these birds sometimes enter buildings and bite people. The pigeon flea, *C. columbae*, is associated with the common pigeon (*Columba livia*), and it will bite humans.

## Pulicidae

These fleas are parasites of a range of animals, including insectivores, carnivores, hyraxes, and rodents. Several species are cosmopolitan, and are often transported worldwide by humans. Many species are of medical or veterinary importance. There are fewer than 200 species in this family.

**Cat flea, *Ctenocephalides felis felis* (Fig. 16.1a, c)** Adults are about 2.5 mm long and brown to yellowish brown; males are slightly smaller than females. The head is twice as long as wide; there are eight pairs of spines on the genal and pronotal combs, and the last spine on the genal comb is small. It is distinguished from the dog flea by the first spine on the genal comb, which is equal in length to the second spine. Full-grown larvae are about 6 mm long and yellowish white, but may be reddish brown after feeding on dried blood in the substrate. Eggs are about 0.5 mm long, smooth, and translucent to slightly glistening. They are not attached to the host; about 70% fall from the animal within 8 h, usually when the host shakes or scratches. Most eggs are found at sites where the infested animal sleeps or rests. Hatching occurs in about 48 h at 26.7 °C, 1.5 days at 32 °C, and 6 days at 13–15.5 °C. Exposure to less than 50% relative humidity (RH) results in 20–60% decrease in egg hatch. At 16–27 °C and 50% RH, at least 70% of eggs hatch; exposure to 33% RH is lethal to eggs. Eggs are killed at 3 °C for 1 day, 8 °C for 10 days, and 2.7 °C for 5 days. Oviposition begins

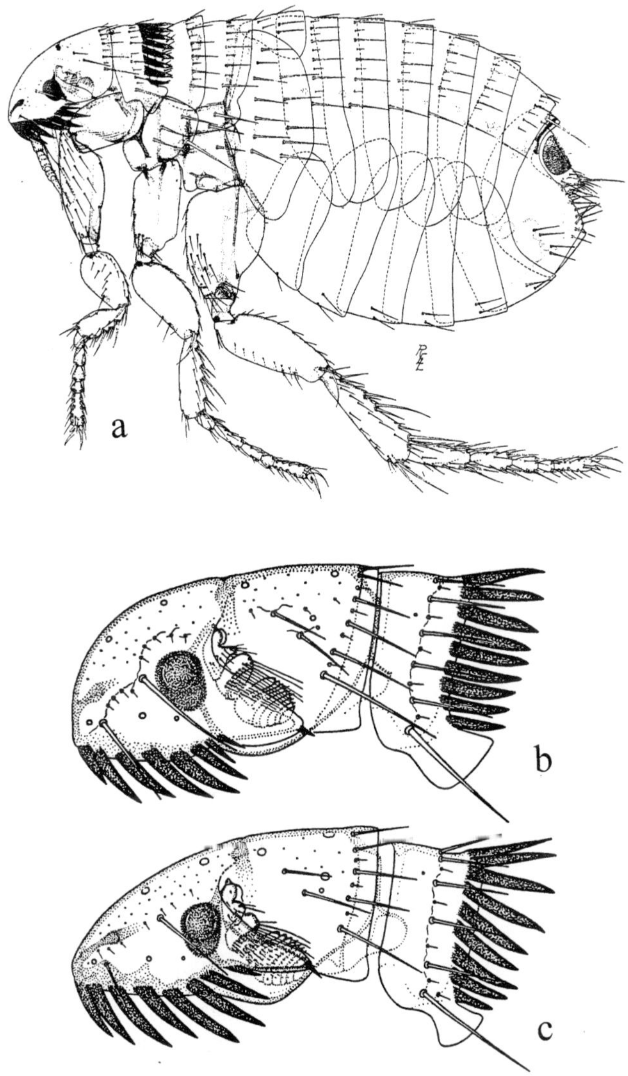


**Figure 16.1** Siphonaptera. (a) *Ctenocephalides felis felis* female; (b) *C. canis* female, head and pronotum; (c) *C. felis felis* female, head and pronotum.

24–36 h after the first blood meal; greatest egg production occurs 4–9 days after the first blood meal. Females consume an average of 13.6 μl of blood per day, which is equivalent to about 15 times their body weight. They lay 40–50 eggs per day during peak production; fecundity is 300–800 eggs.

Larval food is various organic matter and the dried blood feces of the adult flea. Development requires 6–36 days; at 50% RH it is in 10 days; at 90% RH it is 5 days. Exposure to less than 45% RH and greater than 95% RH is lethal; and exposure to 3 °C for 5 days and 8 °C for 20 days is lethal. Larvae survive temperatures up to 27 °C at 50% RH. Larvae disperse from the hatching site, and may move 46 cm in household carpeting. At 13 °C and 75% RH, about 50% of larvae complete development and pupate within 34 days of hatching. The pupal stage is in a silk cocoon made by the third-stage larva. Cocoons are usually established on a vertical surface. The cocoon surface is usually covered with pieces of debris from the surrounding substrate; the pupal period is 7–10 days. Emerged adults may survive for 20 days if a host is not available, and may live for about 113 days on a suitable host. Mechanical pressure and heat stimulate rapid emergence of fleas from the cocoon; vibrations may not be sufficient to stimulate emergence of adults. About 60% of adult fleas emerge in 155 days from cocoons held at 13 °C; adults emerge in 155 days at 15.5 °C. Exposure to 3 °C for 10 days and −1 °C for 5 days is lethal to pre-emerged adults. Females emerge 5–8 days after pupation, males 7–10 days at 26.6 °C. Unfed females die within 11 days at 27 °C. Once a host is acquired, feeding is initiated and mating occurs within 8–24 h. Overwintering in cold climates is by adult fleas on domestic and feral dogs and cats, and species of urban wildlife. Cold weather delays the development of immature stages in underground dens of wildlife and indoors.

Distribution of cat fleas indoors is linked to behavior of the infested pet dogs or cats. Sites where animals sleep or rest have accumulations of dried blood feces from adult flea feeding. These sites have large numbers of cat flea eggs and larvae, corresponding to where the dried blood (larval food) becomes dislodged and falls to the floor. Sites where pets do not rest or sleep, windows, or commonly traveled areas have fewer eggs and larvae.

This species is cosmopolitan and found primarily on domestic dogs and cats worldwide, but also feral mammals in urban areas, including opossums (*Didelphus* spp.), mongooses (*Herpestes* spp.), bobcat, civets and mustelids such as mink, ermine, and ferrets, skunk, and red and gray fox, and occasionally calves, lambs, rats, and chickens. It is an important pest of domestic dogs and cats worldwide. When they are indoors, recently emerged fleas will attempt to feed on humans. The dog and cat flea can be intermediate hosts of the dog and cat tapeworm, *Dipylidium caninum*, and the cat flea is primarily responsible for flea allergy dermatitis in dogs and cats. *C. felis* can transmit murine typhus, it can be a vector of plague, and it may transmit *Bartonella henselae*, which is the etiologic agent for cat scratch disease.

**Other cat fleas** There are three other subspecies of *Ctenocephalides felis*: *C. felis strongylus* and *C. felis darmarensis* are restricted to Africa. *C. felis orientis* occurs in Southeast Asia and East Indies. Other fleas associated with dogs and cats are *C. canis*, *Pulex simulans*, and *Echidnophaga gallinacea*.

**Dog flea, *Ctenocephalides canis* (Fig. 16.1b)** Adults are about 2.5 mm long, and brown to yellowish brown. This species closely resembles the cat flea. The differences include: the head is not twice as long as wide, head width and length are nearly equal; there are eight pairs of spines on the genal and pronotal combs, and the last two posterior spines are small. It can also be distinguished from the cat flea by the first spine on the genal comb, which is distinctly smaller than the second spine. Full-grown larvae are about 6 mm long and yellowish white; the gut may be reddish brown after feeding. Distribution is usually in elevated areas in the tropics and throughout temperate climates. It is typically found on dogs and rabbits, and rarely on cats.

**Oriental rat flea, *Xenopsylla cheopis*** Adults are about 2.5 mm long. The head is longer than wide, and without genal and pronotal combs. The front margin of the head is rounded, and the eyes are distinct. The ocular bristle is inserted medially in front of the eye, and not distinctly below the bottom margin of the eye. Egg hatch and development are optimal at 18–27 °C and 70% RH. Larval development is completed in 12–84 days, and the pupal period is 7–182 days. Unfed adults away from a host live about 38 days, and live for about 100 days on a host. This is a cosmopolitan species, and distribution is primarily from 35 °N–35 °S latitude. Its range closely follows the range of the black rat (*Rattus rattus*). It also infests brown rat (*R. norvegicus*), cotton rat (*Sigmodon hispidus*), house mouse, rabbits, and California ground squirrel.

**Other *Xenopsylla*** Several species of *Xenopsylla* are known to transmit plague. *X. brasiliensis* is a common species in central Africa, parts of South America, and India. It occurs more often in huts and primitive dwellings than in rural areas. *X. astia* occurs in the subcontinent Indo-Pakistan region. *X. vexbilis* is found on the Hawaiian field rat, *R. hawaiiensis*, and it occasionally occurs indoors.

**Human flea, *Pulex irritans* (Fig. 16.2a)** Adults are about 2 mm long, and brown to yellowish brown. The head is longer than wide, and without genal or pronotal combs. The ocular bristle is inserted distinctly below the eye. Eggs are about 0.5 m long, oval and white. Oviposition is 2–6 eggs per day, and fecundity is about 400 eggs. Hatching occurs in about 7 days. Larval development is completed in about 19 days at 23 °C, and about 105 days at 10 °C. Adult life span is 30–60 days, depending on temperature and humidity. Under high humidity and 7–10 °C, unfed adults live for 125 days; fed adults may live 513 days. This species is cosmopolitan and found primarily on humans, but it is also found on cats, dogs, pigs, and other domesticated animals. A large percentage of domesticated dogs may be infected with this flea. The range of hosts and variations in biology of different populations of *P. irritans* in various parts of the world indicate that this species may be best considered as a species complex.

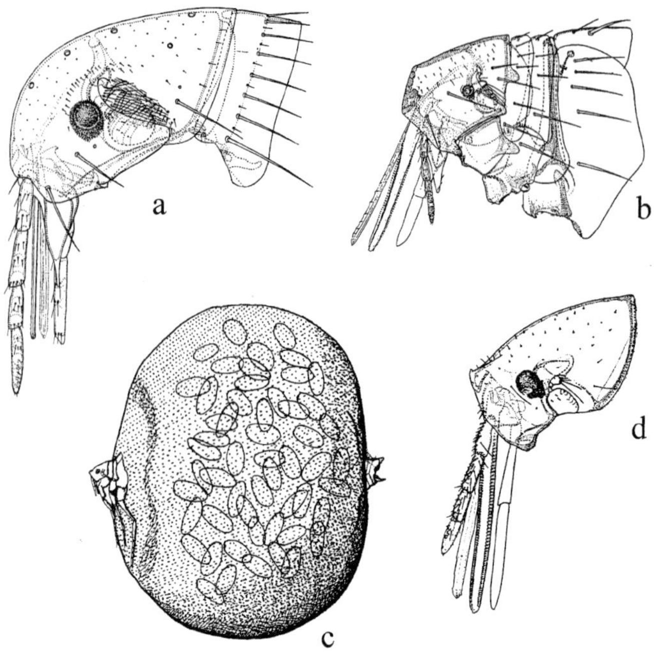


**Figure 16.2** Siphonaptera. (a) *Pulex irritans* male, head and pronotum; (b) *Echidnophaga gallinacea* female, head and pronotum; (c) *Tunga penetrans* female, enlarged with eggs; (d) *T. penetrans* female, head.

**Sticktight flea, *Echidnophaga gallinacea* (Fig. 16.2b)** Adult females are 1–1.5 mm long; males are less than 1 mm long, and yellowish brown to dark brown. The head is not longer than wide; it is distinctly angular on anterior margin, and the genal and pronotal combs are absent. The eyes are small and hidden; the thorax is reduced and contracted. Eggs are deposited in ulcers produced by females feeding in the skin of the host or dropped to the ground. Females deposit 5–20 eggs per day, depending on the timing of their blood meals. Hatching occurs in 6–8 days at 25 °C; larvae hatching in skin ulcers drop to the ground to feed on organic debris. Larval development requires about 14 days, and the pupal period lasts about 14 days. Adults overwinter in the cocoon and emerge in the spring, usually in response to a sharp rise in temperature. Adults may remain attached to the head of the host for several weeks. This species is distributed primarily in southern USA, tropical America, and warm climates of the world, including Australia and New Zealand. Adults feed on a variety of birds, including blackbird, house sparrow, bobwhite quail, and poultry. They will also feed on domesticated cats and dogs, rats, rabbits, ground squirrels, horses, and humans.

## Tungidae

These fleas are nearly permanent parasites in the adult stage. The adults are distinguished by having a small thorax and short legs, and the abdomen of the female is greatly enlarged and filled with eggs. They are mainly tropical, but some species occur widely outside the tropics. The smallest flea that attacks humans is the unfed male and female of *Tunga penetrans*. However, when mated and the female is enveloped into the skin, she becomes the largest flea known to infest humans.

### Chigoe, jigger, sand flea, *Tunga penetrans* (Fig. 16.2c, d)

Adults are about 1 mm long; the blood-engorged female is about 5 mm wide. The head is angular, the anterior margin straight, and the genal and pronotal combs are absent. Eyes are distinct. Eggs are deposited to the outside of the host, and hatching occurs in 3–4 days; fecundity is about 200 eggs. Larvae drop to the ground and develop in 10–17 days away from the host. Larvae develop in the soil in locations frequented by the host. There are only two instars and development from egg to adult takes about 3 weeks. The newly emerged adults actively search for a host; the females attach to the feet of mammals, including humans and pigs. The male chigoe is free-living. It feeds with its head and body up to five abdominal segments inserted into the host. The female does not actually burrow into the skin, but is enveloped by surrounding skin tissue, with the end of her abdomen remaining exposed. During the next 10 days, her abdomen distends and swells with eggs, which she ejects to the outside. Some eggs hatch within the skin cavity, but the larvae drop to the ground to develop. When the female dies, she remains in the skin of the host. The presence of an adult chigoe in the foot can cause a crippling injury, and the damage to the skin results in a secondary infection. Infestations are usually on bare feet, and the most common sites are between the toes and soles. This species occurs in hot and dry locations in neotropical regions of North and South America, West Indies, and Afrotropical regions, including Madagascar. It apparently originated in South America as a parasite of pigs, and was carried to Africa around 1872. This parasite is usually acquired by people traveling on dry roads, soiled by the excreta of domestic animals.

## Bibliography

The published information on the biology and distribution of fleas is extensive. The bibliographies by Jellison and Good (1942), Jellison *et al.* (1953), and Jellison and Glesne (1967), and the reviews by Jellison (1959), Rothschild (1975), and Rust and Dryden (1997) are helpful. *Flea News* is a biannual newsletter and bibliography devoted to Siphonaptera. It is compiled by Robert E. Lewis and distributed free at: www.ent.iastate.edu/fleaNews.


Jellison, W. L. Fleas and disease. *Annu. Rev. Entomol.*, **4** (1959), 398–414.

Jellison, W. L. and L. Glesne. *Index to the Literature of Siphonaptera of North America* (suppl. 2). Hamilton, Montana: Rocky Mountain Laboratory, 1967.

Jellison, W. L. and N. E. Good. *Index to the Literature of Siphonaptera of North America*. National Institute of Health Bulletin 178 US Public Health Service. Washington, DC: US Government Printing Office, 1942.

Jellison, W. L., B. Locher, and R. F. Bacon. *Index to the Literature of Siphonaptera of North America* (suppl. 1). Hamilton, Montana: Rocky Mountain Laboratory, 1953.

Rothschild, M. Recent advances in our knowledge of the order Siphonaptera. *Annu. Rev. Entomol.*, **20** (1975), 241–59.

Rust, M. K. and M. W. Dryden. The biology, ecology, and management of the cat flea. *Annu. Rev. Entomol.*, **42** (1997), 451–73.


#### General


Bacot, A. A study of the bionomics of the common rat fleas and other species associated with human habitation, with special reference to the influence of temperature and humidity at various periods of the life history of the insect. *J. Hyg.*, **13** (suppl.2) (1914), 447–652.

Bacot, A. W. and C. J. Martin. Observations on the mechanism of the transmission of plague by fleas. *J. Hyg.*, **13** (Suppl. 3), (1914), 423–39.

Bacot, A. W. and W. G. Ridgewood. Observations on the larvae of fleas. *Parasitology*, **7** (1914), 157–75.

Bates, J. K. Field studies on the behavior of bird fleas. I. Behavior of the adults of three species of bird fleas in the field. *Parasitology* **52** (1962), 113–32.

Bennet-Clark, H. C. and E. C. A. Lucey. The jump of the flea: a study of the energetics and a model of the system. *J. Exp. Biol.*, **47** (1967), 59–76.

Benton, A. H., M. Surman, and W. L. Krinsky. Observations on the feeding habits of some larval fleas (Siphonaptera). *J. Parasitol.*, **65** (1979), 671–2.

Beresford-Jones, W. P. Prevalence of fleas on dogs and cats in an area of London. *Small Anim. Rpt.*, **22** (1981), 27–9.

Berlinger, G. *Aphaniptera d'Italia*. Studio monografico. Rome: Il Pensiro Scientifico Editore, 1964.

Bibikova, V. A. Contemporary views of the interrelationships between fleas and the pathogens of human and animal diseases. *Annu. Rev. Entomol.*, **22** (1977), 23–32.

Bruce, W. N. Studies on the biological requirements of the cat flea. *Ann. Entomol. Soc. Am.*, **41** (1948), 346–52.

Byron, D. and W. H Robinson. Distribution of cat flea (*Ctenocephalides felis*) in a household environment. *Jpn. J. Environ. Entomol. Zool.*, **3** (1991), 70–4.

Chua, E. C. and K. T. Goh. A flea-borne outbreak of dermatitis. *Ann. Singapore Acad. Med.*, **16** (1987), 648–50.

Dryden, M. W. Host association, on host longevity and egg production of *Ctenocephalides felis felis*. *Vet. Parasitol.*, **34** (1989), 117–22.

Dryden, M. W. and J. C. Blakemore. A review of flea allergy dermatitis in the dog and cat. *Companion Anim. Pract.*, **19** (1989), 10–17.

Dryden, M. W. and S. M. Gafar. Blood consumption by the cat flea, *Ctenocephalides felis* (Siphonaptera: Pulicidae). *J. Med. Entomol.*, **28** (1991), 394–400.

Dryden, M. W. and M. K. Rust. The cat flea: biology, ecology, and control. *Vet. Parasitol*, **52** (1994), 1–19.

Edney, E. B. Laboratory studies on the bionomics of the rat fleas, *Xenopsylla brasiliensis* Baker, and *X. cheopis* Roths. *Bull. Entomol. Res.*, **38** (1945), 253–80.

Feingold, B. G. and E. Benjamini. Allergy to flea bites: clinical and experimental observations. *Ann. Allergy*, **19** (1961), 1275–89.

Guzman, R. F. *Cheyletiella blakei* (Acari: Cheyletiellidae) hyperparasitic on the cat flea *Ctenocephalides felis felis* (Siphonaptera: Pulicidae) in New Zealand. *N.Z. Entomol.*, **7** (1982), 322–3.

Haas, G. E. Cat flea–mongoose relationships in Hawaii. *J. Med. Entomol.*, **2** (1966), 321–6.

Hopkins, D. Ectoparasites of the Virginia opossum (*Didelphis virgiana*) in an urban environment. *Northwest Sci.*, **54** (1980), 199–201.

Hudson, B. W., B. F. Feingold, and L. Kartman. Allergy to flea bites. II. Investigations of the bite sensitivity to humans. *Exp. Parasitol.*, **9** (1960), 264–70.

Humphries, D. A. The function of combs on fleas. *Entomol. Mon. Mag.*, **102** (1966), 232–6.

The behavior of fleas (Siphonaptera) within the cocoon. *Proc. R. Entomol. Soc. Lond. A*, **42** (1967), 62–70.

Hunter, K. W., A. R. Campbell, and P. C. Sayles. Human infestation by cat fleas, *Ctenocephalides felis* (Siphonaptera: Pulicidae) from suburban raccoons. *J. Med. Entomol.*, **16** (1979), 547.

Joseph, S. A. On fleas of the genus *Ctenocephalides* in India. *Madras Vet. Coll. Ann.*, **28** (1970), 12–13.

Observations in the feeding habits of *Ctenocephalides felis orientis* Jordan, 1925 on human host. *Cheiron*, **5** (1976), 72–7.

Observations in the feeding habits of *Ctenocephalides felis orientis* (Jordan) 1925. *Cheiron*, **10** (1981), 275–80.

Karsten, H. Beitrag zur Kenntnis des *Rhynchoprion penetrans*. *Virchow's Arch. Pathol. Anat.*, **32** (1865), 269–92.

Kern, W. H. Jr., P. G. Koehler, and R. S. Patterson. Diel patterns of cat flea (Siphonaptera: Pulicidae) egg and fecal deposition. *J. Med. Entomol.*, **29** (1992), 203–6.

Kwochka, K. W. Fleas and related disease. *Vet. Clin. North Am. Small Anim. Pract.*, **17** (1987), 1235–62.

Lehane, B. *The Complete Flea*. New York: Viking Press, 1969.

Lyon, H. Notes on the cat flea (*Ctenocephalides felis* Bouche). *Psyche*, **22** (1915), 124–32.

Margalit, J. and A. S. Shulv. Effect of temperature on the development of prepupa and pupa of the rat flea, *Xenopsylla cheopis* Rothschild. *J. Med. Entomol.*, **9** (1972), 117–25.

Marshall, A. G. *The Ecology of Ectoparasitic Insects*. London: Academic Press, 1981.

Mellanby, K. The influence of temperature and humidity on the pupation of *Xenopsylla cheopis*. *Bull. Entomol. Res.*, **24** (1933), 197–203.

Metzger, M. E. and M. K. Rust. Egg production and emergence of adult cat fleas (Siphonaptera: Pulicidae) exposed to different photoperiods. *J. Med. Entomol.*, **33** (1996), 651–5.

Miyamoto, K. Cases of itchy eruption caused by the sparrow flea, *Ceratophyllus gallinae dilatus* (Siphonoptera: Ceratophyllidae) in Hokkaido, Japan. *Jpn. J. Sanit. Zool.*, **44** (1993), 97–9.

Osbrink, W. L. A. and M. K. Rust. Fecundity and longevity of the adult cat flea, *Ctenocephalides felis felis* (Siphonaptera: Pulicidae). *J. Med. Entomol.*, **21** (1984), 727–31.

Cat flea (Siphonaptera: Pulicidae) factors influencing host-finding behavior in the laboratory. *Ann. Entomol. Soc. Am.*, **78** (1985), 29–34.

Seasonal abundance of adult cat fleas, *Ctenocephalides felis* (Siphonaptera: Pulicidae), on domestic cats in southern California. *Bull. Soc. Vector Ecol.*, **10** (1985), 30–5.

Osbrink, W. L. A., M. K. Rust, and D. A. Reierson. Distribution and control of cat fleas in homes in southern California. *J. Econ. Entomol.*, **79** (1986), 135–40.

Pet'ko, B. Domestic cats as a source of human flea infestations in towns. *Cesko. Epidemiol. Mikrobiol. Immunol.*, **42** (1993), 190–1.

Razl, S. M., D. D. Cabral, and P. M. Linardi. Notes on the Siphonaptera (Pulicidae: Tungidae and Rhopalopsyllidae) from Brazilian domestic carnivores. *Rev. Bras. Entomol.*, **43** (2000), 95–7.

Rothschild, M. Fleas. *Sci. Am.*, **213** (1965), 44–53.

Rothschild, M. and T. Clay. *Fleas, Flukes and Cuckoos*. London: Arrow, 1957.

Rothschild, M., Y. Schlein, K. Parker, and S. Sternberg. Jump of the oriental rat flea, *Xenopsylla cheopis* (Roths.). *Nature, Lond.*, **239** (1972), 45–8.

Rust, M. K. Influence of photoperiod on egg production of cat fleas (Siphonaptera: Pulicidae) infesting cats. *J. Med. Entomol.*, **29** (1992), 103–5.

Interhost movement of adult cat fleas (Siphonaptera: Pulicidae). *J. Med. Entomol.*, **31** (1994), 486–9.

Schelvis, J. Caught between the teeth. A review of Dutch finds of archeological remains of ectoparasites in combs. *Proc. Sect. Exp. Appl. Entomol. Netherlands Entomol. Soc.*, **5** (1994), 131–2.

Silverman, J. and M. K. Rust. Some abiotic factors affecting the survival of the cat flea *Ctenocephalides felis* (Siphonaptera: Pulicidae). *Environ. Entomol.*, **12** (1983), 490–5.

Extended longevity of pre-emerged adult cat flea (Siphonaptera: Pulicidae) and factors stimulating emergence from the pupal cocoon. *Ann. Entomol. Soc. Am.*, **78** (1985), 763–8.

Silverman, J., M. K. Rust, and D. A. Reierson. Influence of temperature and humidity on survival and development of the cat flea, *Ctenocephalides felis* (Siphonaptera: Pulicidae). *J. Med. Entomol.*, **18** (1981), 78–83.

Strenger, A. Zur Ernahrungbiologie der Larve von *Ctenocephalides felis felis*. *B. Zool. Jb. Syst.*, **100** (1973), 64–80.

Thomas, R. E. Fleas and the agents they transmit. In Beaty, B. J. and W. C. Marquart (eds.) *The Biology of Disease Vectors*. Denver: University of Colorado Press, 1996.

Torgerson, P. and R. Breathnach. Flea dermatitis and flea hypersensitivity: the current situation in Ireland. *Irish Vet. J.*, **49** (1996), 426–34.

Traub, R. The relationship between the spines, combs and other skeletal features of fleas (Siphonaptera) and the vestiture, affinities and habits of their hosts. *J. Med. Entomol.*, **9** (1972), 601.

Traub, R. and H. Starcke (eds.) *Fleas*. Rotterdam: A. A. Balkema, 1980.

Yutuc, L. M. The cat flea hitherto unknown to sustain the larvae of *Dipylidium canium* (Linneaus, 1758) from the Philippines. *Philip. J. Sci.*, **97** (1968), 285–9.

### Geographic distribution

Amin, O. M. The fleas (Siphonaptera) of Egypt: distribution and seasonal dynamics of fleas infesting dogs in the Nile Valley and Delta. *J. Med. Entomol.*, **3** (1966), 292–8.

Beaucournu, J.-C. and H. Launay. Les puces (Siphonaptera) de France et du Bassin méditerraneén occidental. *Faune France Régi. Limitrophes*, **76** (1990), 1–548.

de Meillon, B., D. H. S. Davis, and F. Hardy. *Plague in Southern Africa*, vol. 1, *The Siphonaptera (Excluding Ischnopsylliae)*. Pretoria: Government Printer, 1961.

Dunnet, G. M. Siphonaptera. In Naumann, I. D. (ed.), *Insects of Australia*. Melbourne: CSIRO Division of Entomology, Melbourne University Press, 1970.

Dunnet, G. M. and D. K. Mardon. A monograph of Australian fleas (Siphonaptera). *Aust. J. Zool.* **30** (Suppl. Ser.) (1974), 1–273.

Fox, I. Fleas of eastern United States. New York: Hafner, 1968. (Facsimile edition of 1940 work published by Iowa State College Press, Ames, Iowa.)

Hastriter, M. W. and V. J. Tipton. Fleas (Siphonaptera) associated with small mammals of Morocco. *J. Egyp. Pub. Health Assoc.*, **50** (1975), 79–169.

Holland, G. P. The Siphonaptera of Canada. *Dominion Can. Dept. Agric. Tech. Bull.*, **70** (1949), 1–306.

Contribution towards a monograph of the fleas of New Guinea. *Mem. Entomol. Soc. Can.*, **61** (1969), 1–77.

The fleas of Canada, Alaska and Greenland (Siphonaptera). *Mem. Entomol. Soc. Can.*, **130** (1985), 1–630.

Hopkins, G. H. E. and M. Rothschild. *An Illustrated Catalogue of the Rothschild Collection of Fleas (Siphonaptera) in the British Museum (Natural History)*. London: British Museum, Natural History, 1953.

Hubbard, C. A. *Fleas of Western North America*. Ames, IA: Iowa State College Press, 1947.

Fleas and plague in Iraq and the Arab world, part 2. *Iraq Nat. Hist. Mus. Pub.*, **19** (1960), 1–143.

Iyengae, R. The Siphonaptera of the Indian subregion. *Oriental Insects* **3** (Suppl.) (1973), 1–102.

Lewis, R. E. The fleas (Siphonaptera) of Egypt. An illustrated and annotated key. *J. Parasitol.*, **53** (1967), 863–85.

Notes on the geographical distribution and host preferences in the order Siphonaptera. Part 1. Pulicidae. *J. Med. Entomol.*, **9** (1972), 511–20.

Siphonaptera collected during the 1965 Street Expedition in Afghanistan. *Fieldiana Zool.*, **64** (1973), 1–161.

Insects of Saudi Arabia, Siphonaptera. A review of the Siphonaptera of the Arabian peninsula. *Fauna Saudi Arabia*, **4** (1982), 450–64.

Lewis, R. E., J. H. Lewis and C. Maser. *The Fleas of the Pacific Northwest*. Corvallis, OR: Oregon University Press, 1988.

Linardi, P. M. and L. R. Guimaraes. *Sifonapteros do Brasil*. São Paulo Museu Zoologica USP/FAPESP, 2000.

Liu, Z. *Fauna Sinica: Insecta: Siphonaptera* (in Chinese). Beijing: Science Press, 1986.

Lumaret, R. Insectes Siphonaptères. *Faune Madagascar*, **15** (1962), 1–109.

Peus, F. Zur Kenntis der Flöhe Deutschlands (Schluss) (Insecta, Siphonaptera). IV. Faunistik und Ekologie der Saugertierflohe. *Zool. Jahr. (System.)*, **99** (1972), 408–504.

Rosický, B. Biechy-Aphaniptera. *Fauna ČSR*, **10** (1957), 1–439.

Sakaguti, K. and E. W. Jameson. The Siphonaptera of Japan. *Pac. Insects Mon.*, **3** (1962), 1–169.

Selim, M. K. Some studies on fleas infesting several mammals in U. A. R. *Vet. Med. J.*, **12** (1966), 391–6.

Skuratowicz, W. *Pchly-Siphonaptera (Aphaniptera). Klucze do Oznaczania Owadaw Polski* [Keys for the identification of Polish Insects], vol. 29, pp. 1–141 [in Polish].

Smit, F. G. A. M. Siphonaptera. *Handbooks Identi. Br. Insects*, **1** (1957), 1–94.

The fleas of New Zealand (Siphonaptera). *J. R. Soc. N.Z.*, **9** (1979), 143–232.

Tipton, V. J. and C. E. Machado-Allison. Fleas of Venezuela. *Brigham Young Univ. Sci. Bull. (Biol. Ser.)*, **17** (1972), 1–115.

Tipton, V. J., J. W. Southwick, H.-s. Ah, H.-s. Yu. Fleas of Korea. *Korean J. Parasitol.*, **10** (1972), 52–63.

Trembly, H. L. and F. C. Bishopp. Distribution and hosts of some fleas of economic importance. *J. Econ. Entomol.*, **33** (1940), 701–3.

Vater, V. G. and A. Vater. Flöhe (Siphonaptera) beim menschen. Befundanalyse 1961 im Bezirk Leipzig (DDR). Teil II. Räumliche und zeitliche Verteilung. *Angew. Parasitol.*, **26** (1985), 27–38.

Yurkina, V. I. Fleas (in Ukrainian). *Fauna Ukraine*, **17** (1961), 1–152.

# 17 Thysanoptera, Thysanura, Trichoptera

## THYSANOPTERA
### Introduction

Thrips are 0.75–2.0 mm long and have a slightly compressed or flattened body. They are usually pale yellow to dark brown, to black. Wings are long and narrow with few or no veins; there is an outer marginal row of long, delicate setae. Mouthparts are piercing–sucking. Most thrips feed on plant juices, and they are commonly found in flowers and leaves. Some feed on fungi, and a few species are predaceous on mites and other insects. Many of the species that feed on grasses and grains migrate in large numbers when the grain moisture content decreases, which is usually in the fall. Eggs are laid in the spring by females that overwintered in protected locations. Eggs are deposited on the surface of leaves or bark, or inserted into plant tissue; hatching occurs in 2–20 days, depending on temperature. Development from egg to adult is usually through four or five instars. The first two feeding instars are called larvae; the third instar does not feed and is called a prepupa; and the fourth instar is called a pupa. The pupa is sometimes enclosed in a cocoon. There are several generations per year.

Pest status is based on their biting people and being a nuisance during some seasons. Occasionally large numbers of thrips will gather around the outside of buildings, and move inside through doors and windows. Bites on the skin may be painful and result in a skin rash and an itching reaction. The itching produced by thrips bites is probably a result of their attempts to obtain water from the skin surface. Volatile components of sweat, including carponic and lactic acid, are attractive to some thrips, such as *Limothrips cerealium*, and *Haplothrips aculeatus*. Thrips that bite and provoke skin rashes include species that typically feed on cereals and grasses, such as *Chirothrips aculeatus*, and *L. cerealium*, the pear thrips, *Taeniothrips inconsequens*, and the onion thrips, *Thrips tabaci*. Species reported to cause skin rashes include *Gynaikothrips ficorum* in Algeria, *Caliothrips indicus* in Sudan, *T. imaginis* in Australia, and *L. denticornis* in Germany. A few are known to pierce the skin and suck blood. This habit has been reported for adults of *Karnyothrips flavipes*, a predator of scale insects in the Mediterranean region. Second instars of the plant-feeding thrips, *Thrips tabaci* and *Frankliniella moultoni*, have been reported to bite. The predatory species *Scolothrips sexmaculatus*, *Leptothrips mali*, *Aeolothrips fasciatus*, and *A. kuwanaii* also bite people.

### Thripidae

This is a large family and it contains many economically important species. The wings are pointed at the tips and the antennae are 6–9-segmented. Most species are plant feeders, and some are pests of agricultural crops.

**Greenhouse thrips, *Heliothrips haemorrhoidalis*** Adult females are about 1.2 mm long and dark brown to nearly black. The body is reticulated and the antennae are eight-segmented. Males are relatively unknown. Eggs are laid in plant tissue. The nymphs are pale brown. There are several generations per year. This species is nearly cosmopolitan and attacks a variety of ornamental plants indoors and outdoors. It is common in greenhouses, and brought into buildings on plants.

**Cereal thrips, thunder fly, *Limothrips cerealium*** Adults are about 1.5 mm long, and the body is black. They are common in cereal-growing areas, and often occur in large numbers in fall, prior to grain harvest. During periods when this species migrates, aerial densities of adults flying above wheat can exceed two thrips per cubic meter. Gravid females emerge from hibernation in early spring and fly to grasses and cereals to lay eggs. The next generation of adults emerge in summer, and continue to feed until the cereal begins to senesce. When the moisture content of the grain declines to about 45%, the adults no longer feed and begin to disperse. Females fly to hibernation sites in numbers comparable to migrating locusts and aphids. Migrating thrips can be carried by wind to buildings about 1 km from the host cereal. Large numbers can enter buildings and cause irritation to skin and eyes. Adults will move into small cracks and crevices that are about 0.3 mm wide, and sometimes this behavior activates electronic fire-detection systems. They often find suitable harborage in several types of fire detectors, including those based on ionization or optical scatter. In two-story buildings, they are usually found on the upper floors.

**Flower thrips, *Frankliniella tritici*** Adults are 1.2–1.3 mm long, slender, and yellow to yellowish orange. This species is common on ornamental flowers and brought indoors on cut flowers. It is known to bite people indoors.

# THYSANURA

## Introduction

Silverfish are 6–18 mm long and wingless. They are slender, dorsoventrally flattened, and tapered posteriorly, and they have three tail-like appendages at the end of the abdomen. Their body is usually covered with shiny, fish-like scales, which is the origin of their common name. Antennae are long and slender; the eyes are small and widely separated, and sometimes absent. Mouthparts are mandibulate, and considered to be chewing. Development progresses through distinct stages. At the third or fourth molt, the young develop scales that color their body. Thysanurans are long-lived insects that continue to molt after becoming sexually mature. They are mostly nocturnal and omnivorous scavengers. Natural populations occur in leaf litter or under bark, in caves, in the burrows of certain mammals, or in association with ant and termite nests.

Mating behaviors of *Lepisma saccharina* and *Thermobia domestica* have been documented. Courtship involves mutual contact with the antennae (antennation) and the male and female walking in tandem, usually in contact with a perpendicular wall. The male produces a number of silken threads that extend at an angle from the wall to the substrate, and, associated with the angular threads, there is a horizontal mat of silken threads. The male deposits a pear-shaped or oval spermatophore on the silk substrate. In *L. saccharina*, the female moves along the wall and her legs contact threads on the ground, which stimulates her to raise her cerci. When the cerci contact silk threads that are attached to the wall, the female stops and begins to search for the spermatophore with her ovipositor. In *T. domestica* the male indicates the presence of the spermatophore to the female by contacting her with his antennae. The silk threads on the substrate excite the female, delimiting the mating area.

Eggs are about 1 mm long, elliptical, and white to pale yellow when laid, but they turn brown before hatching. Females usually lay about 100 eggs in cracks and crevices; they are laid singly or in groups of about 45. Hatching occurs in 2 weeks to 2 months, depending on environmental conditions. Young closely resemble adults, except for size. Adults live for several years, and continue to molt every 2–3 weeks. These insects are primarily nocturnal and feed on plant and animal material, including insect cast skins and other insects.

Pest status is based on their presence indoors, but is limited to a few species. The majority of species live in moist wooded habitats, such as under bark of trees and logs, in mammal burrows; several species occur in ant and termite nests. Indoors, silverfish infest food and fabric storage sites where they feed on a starchy material, including old book binding, starched clothing, and starch-based wallpaper glue. They often occur in large numbers in attics of houses with wood-shingle roofs. Silverfish are trapped in sinks and bathtubs because they usually cannot climb smooth surfaces. Damage to fabric by silverfish is characterized by the presence of irregular feeding marks on individual fibers, small amounts of feces, and in the case of linen there may be pale yellow stains. Firebrats will attack either knitted or plain-weave fabric. The firebrat midgut has enzymes that digest starch, fat, and protein. The four-lined silverfish, *Ctenolepisma lineata*, has the enzyme cellulase that assists in the digestion of cellulose-material, and the crop has cellulose-digesting bacteria, and small amounts of fungal hyphae.

## Lepismatidae

This family includes the familiar silverfish and firebrats, which are common in both natural and domestic habitats. Outdoor species occur in caves, under stones and debris, and some species are associated with the nests of ants. The domestic species occupy a variety of habitats, and feed on starchy substances and plant material.

***Acrotelsia collaris*** Adults are 16–18 mm long, not including the terminal appendages, and about 5 mm wide. The dorsal scales are nearly black, but when lost, the insect is pale gray. The abdominal tergum 10 is triangular, and there is a tuft of bristles on the prosternum. This species occurs in the tropics and in northern Australia.

**Common silverfish, *Lepisma saccharina*** Adults are about 12 mm long, not including the long terminal appendages. They are silver-gray, with a metallic sheen. This species is cosmopolitan and a domiciliary pest in most regions of the world. Development and reproduction indoors require high temperatures and humidity. Eggs are laid on successive days or weeks, 2–3 eggs at a time; fecundity is about 100 eggs. Hatching occurs in about 19 days at 32 °C and 43 days at 22 °C; 90% RH is optimal for egg development. Development from nymph to adult takes 90–120 days at 27 °C. Nymphs are pale white for 1–3 months, then scales appear on the body at the third or fourth molt. The external genital appendages appear at the eighth molt. Adults live about 3.5 years at 27 °C, 2 years at 29 °C, and 1.5 years at 32 °C. This species may have originated in the tropics, but in Latin America it is only found in the cool highlands of Brazil, Bolivia, and Argentina.

Silverfish consume both carbohydrates and protein material; animal products are eaten readily and are important to the diet. In food-storage areas they feed on flour, meal, and other similar products. They damage paper products, especially paper that has a glaze or sizing, which consists of various amounts of starch, dextrin, casein, and gum. Paper with high cellulose content is preferred. Cotton and silk fabrics are also attacked.

**Four-lined silverfish, *Ctenolepisma lineata* (Fig. 17.1b)** Adults are about 15 mm long, not including the terminal segments. They are brownish gray, with four dark lines extending the length of the body. Nymphs are light brown and slightly pink until the fourth molt, which occurs about 4 months after hatching. This species can be found throughout infested structures, including basement, wall voids, and attic, where it often occurs in large numbers. Infestations are usually large in houses with roofs with wooden shingles. In California, it also occurs outdoors under bark. It is distributed in eastern USA, but is also reported from California.

**Gray silverfish, *Ctenolepisma longicaudata* (= *Ctenolepisma urbana*)** Adults attain a length of about 12 mm, not including the terminal appendages. They are uniformly light to dark gray, without a metallic sheen, as in *Lepisma saccharina*. This species occurs in southern USA, California, and Hawaii; it is a common household pest in Australia, and is the most common household silverfish in Latin America and South Africa. It is not restricted to humid sites indoors and may occur throughout a building. It is not known to occur outdoors.

**Other *Ctenolepisma*** Three other species that occur indoors include *C. ciliata*, known from California. *C. diversisquamis* is known from Florida, and *C. targionii* is known from South Carolina.

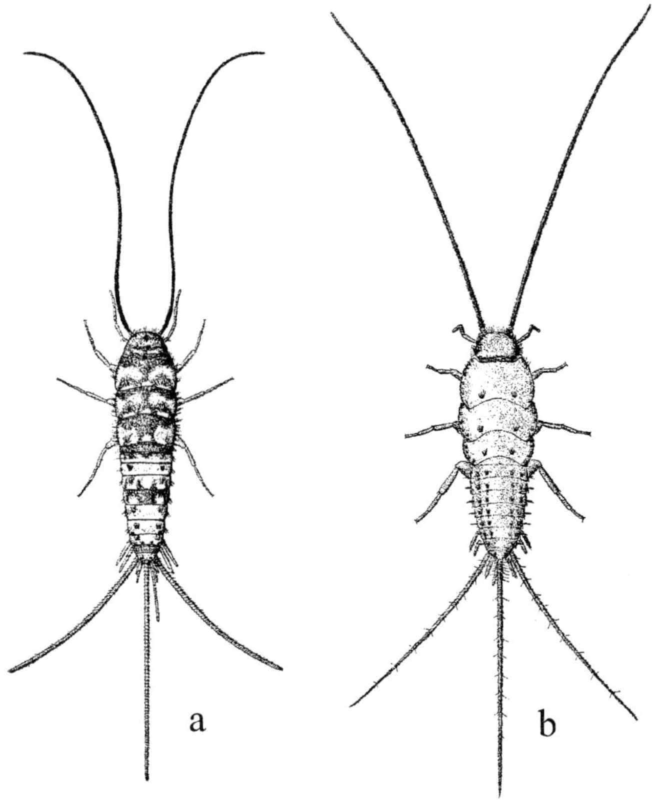


**Figure 17.1** Thysanura. (a) *Thermobia domestica* female; (b) *Ctenolepisma lineata* female.

**Firebrat, *Thermobia domestica* (Fig. 17.1a)** Adults are about 14 mm long, not including the terminal appendages. They are silver-gray, with a somewhat mottled gray appearance. Without the gray scales the underbody color is pale yellow. Firebrats prefer indoor locations with temperatures above 32 °C; optimum development occurs between 37 and 39 °C. Firebrats are pests in commercial locations that maintain high temperatures, such as food-processing plants, and equipment rooms. Eggs are elliptical, white, and about 1 mm long; hatching is in 9 days at 44 °C and 77 days at 25 °C. Optimum hatching occurs at 76–85% RH. Females begin ovipositing when 45–135 days old; fecundity is about 50 eggs. One clutch of eggs is produced between molts and fertilization is required following each molt. Development is about 1 day for first-stage young, 4 days for the second, about 6 days for third and fourth stage, and about 8 days for fifth to 10th stage. There are about 13 days between successive molts, and individuals have 45–60 molts in a lifetime. Adults live 2–2.5 years at 32 °C and 1–1.5 years at 37 °C. Molting and growth appear to be continuous during their life, but there is little change in size and shape of adults after the 35th molt. Females begin ovipositing when 45–135 days old, or about the 14th molt. This species is probably cosmopolitan.

Populations persist in hot and dry environments, but firebrats survive best when there is available water. Without access to water, development is slower than normal, and oviposition decreases. At room temperature, firebrats lose water from their body when the RH is below 45%; above this humidity level they absorb water.

**Other Thysanoptera** *Thermobia campbelli* is often associated with stored-food materials, and *Nicoletia meinerti* occurs in greenhouses.

# TRICHOPTERA
## Introduction

Caddisfly adults are 2–40 mm long. Their wings are covered in short, hairlike scales, and held rooflike over the body. Antennae are long and slender, the mouthparts are chewing, and the palps are well developed. Adults feed on liquids. Caddisflies undergo complete metamorphosis; the nymphs are caterpillar-like and aquatic. They have a well-developed head and thorax, and the abdominal segments have filamentous gills. Some larvae construct cases of leaves, twigs, and sand, while others construct nets, and a few are free-living. These insects occur in ponds, lakes, and streams. Pest status of caddisflies is linked to the emergence of large numbers of adults extending from spring to fall, and their attraction to lights at night. In some regions, large numbers of adults will gather on turfgrass and shrubs in peridomestic habitats after emerging from nearby ponds or streams. Allergic reactions and asthma (locally called sand fever) have been associated with swarms of caddisflies, the accumulation of bodies, and the odor of dead adults.

## Hydropsychidae

Adults are 7–18 mm long, brown to pale brown, and segment 5 of the maxillary palp is elongate. Larvae are filter feeders, but a few are predators. They live in fast-flowing streams and rivers.

**Common netspinner, *Potamyia flava*** Adults are about 10 mm long and pale brown. The wings are uniformly brown, with a short fringe of setae; ocelli are absent, and the antennae are long. Full-grown larvae are about 25 mm long, and the abdomen usually has a lateral fringe of short setae, but never with gills. Larval retreats consist of nets shaped like trumpets; they have a large opening and then taper to the end. This is one of the most common and abundant species in central and southeastern North America. Adult emergence occurs from May to September, sometimes in massive numbers.

**Other Trichoptera** Adults of *Hydropsyche bifida* and *H. colonica* swarm around trees and lights at night, and often enter houses. *H. bifida* occurs in western USA; *H. clonica* occurs in New Zealand.

# Bibliography


**THYSANOPTERA**

**General**

Ananthakrishnan, T. N. Bionomics of thrips. *Annu. Rev. Entomol.*, **38** (1993), 71–92.

Cuthbertson, D. R. *Limothrips cerealium*, an alarming insect. *Entomologist*, **108** (1989), 246–56.

Lewis, T. The weather and mass flights of Thysanoptera. *Ann. Appl. Biol.*, **53** (1964), 165–70.

*Thrips, Their Biology, Ecology and Economic Importance*. New York: Academic Press, 1973.

Mallmann, R. J. De. Observations sur le thigmotactisme de *Limothrips cerealium* (Thysanoptera). *Bull. Soc. Entomol., France*, **64** (1959), 151–7.

**Geographic distribution**

Bailey, S. F. The distribution of injurious thrips in the United States. *J. Econ. Entomol.*, **33** (1940), 133–6.

Dyadechko, N. P. *Thrips or Fringe-Winged Insects (Thysanoptera) of the European Part of the USSR*. New Delhi: Amerind, 1977.

Hinds, W. E. Contribution to a monograph of the insects in the order Thysanoptera inhabiting North America. *Proc. U.S. Natl. Mus.*, **26** (1902), 79–242.

Mound, L. A. and G. Kibby. *Thysanoptera: An Identification Guide*, 2nd edn. Wallingford: CAB International, 1998.

Priesner, H. *Die Thysanopteren Europas*. Vienna: Verlag Fritz Wagner, 1926–28.

Stannard, L. J. The thrips, or Thysanoptera, of Illinois. *Ill. Nat. Hist. Surv. Bull.*, **29** (1968), 215–552.

**THYSANURA**

**General**

Adams, J. A. Biological notes upon the firebrat, *Thermobia domestica* Pack. *J. N.Y. Entomol. Soc.*, **41** (1933), 557–62.

The early instars of the firebrat, *Thermobia domestica*. *Proc. Iowa Acad. Sci.*, **40** (1933), 217–19.

Broadhead, E. and B. M. Hobby. Studies on a species of *Liposcelis* occurring in stored products in Britain. *Entomol. Mon. Mag.*, **80** (1944), 45–9, 163–73.

Delaney, M. J. Life histories in the Thysanura. *Acta Zool. Cracov*, **2** (1957), 61–90.

Lasker, R. Silverfish, a paper-eating insect. *Sci. Monthly*, **84** (1957), 123–7.

Lindsay, E. The biology of the silverfish, *Ctenolepisma longicaudata* Esch. with particular reference to its feeding habits. *R. Soc. Victoria*, **52** (1940), 35–83.

Mallis, A. Preliminary experiments on the silverfish, *Ctenolepisma urbana* Slabaugh. *J. Econ. Entomol.*, **34** (1941), 787–91.

Sahrhage, D. Ekologische Untersuchungen an *Thermobia domestica* (Packard) und *Lepisma saccharina* L. Z., *Wiss. Zool.*, **157** (1952), 77–168.

Slabaugh, R. E. The silverfish in a new role. *Trans. Ill. Acad. Sci.*, **32** (1939), 227–8.

A new thysanuran, and a key to the domestic species of Lepismatidae (Thysanura) found in the United States. *Entomol. News* **51** (1940), 95–8.

Smith, E. L. Biology and structure of some California bristletails and silverfish. *Pan-Pac. Entomol.*, **46** (1970), 212–25.

Spencer, G. J. The firebrat, *Thermobia domestica* Packard (Lepismatidae) in Canada. *Can. Entomol.*, **62** (1930), 1–2.

Strum, H. Die Paarung beim Silberfischen *Lepisma saccharina*. *Z. Tierpsychol.*, **13** (1956), 1–12.

Strum, H. The mating behavior of *Thermobia domestica* (Packard) (Lepismatidae, Zygentoma, Insecta). *Braunschweiger Naturk. Schriften*, **2** (1988), 693–712.

Sweetman, H. L. Physical ecology of the firebrat, *Thermobia domestica* (Packard). *Ecol. Monogr.*, **8** (1938), 285–311.

Responses of the silverfish, *Lepisma saccharina* L. to its physical environment. *J. Econ. Entomol.*, **32** (1939), 698–700.

Sweetman, H. L. and W. M. Kulash. The distribution of *Ctenolepisma urbana* Slabaugh and certain other Lepismatidae. *J. Econ. Entomol.*, **37** (1944), 444.

Tys, K. Struktur des mannlichen Genitaltraktes und Bildung der Spermatophore bei *Thermobia domestica* (Packard) (Lepismatidae, Zygentoma, Insecta). *Zool. Jahrb. Anat.*, **119** (1989), 265–79.

Wall, W. J. Jr. and A. H. P. Swift. The digestive enzymes of the firebrat. *J. Econ. Entomol.*, **47** (1954), 87–188.

Watson, J. A. L. and C. S. Li. A further pest species of silverfish (Thysanura) from Australia, with a key to the domestic species. *J. Aust. Entomol. Soc.*, **6** (1967), 89–90.


### Geographic distribution


Delaney, M. J. Thysanura and Diplura. *R. Entomol. Soc. Handb. Ident. Br. Insects*, **1** (1954), 1–7.

Rijckaert, G., C. Theil, and E. Fuchs. Silberfischen und Staubause als Allergene. *Allergologie* (West Germany) **4** (1981), 80–6.

Roesler, R. Die gattung der Copengnathen. *Stn. Ento. Z.*, **105** (1944), 117–66.

Wygodzinsky, P. Thysanura and Machilidae of the Lesser Antilles and northern South America. *Stud. Fauna Curr. Carib. Is.*, **36** (1959), 28–49.

A review of the silverfish (Lepismatidae, Thysanura) of the United States and the Caribbean area. *Am. Mus. Novit.*, **2481** (1972), 26.


## TRICHOPTERA

### Geographic distribution


Barnard, K. H. South African Caddis-flies (Trichoptera). *Trans. R. Soc. S. Afr.*, **21** (1934), 291–394.

Betten, C. The caddis flies or Trichoptera of New York State. *N.Y. State Mus. Bull.*, **292** (1934), 570.

Denning, D. G. Trichoptera. In Usinger, R. L. (ed.) *Aquatic Insects of California*, pp. 237–70. Berkeley, CA: University of California Press, 1956.

Flint, O. S. Taxonomy and biology of Nearctic limnephilid larvae (Trichoptera), with special reference to species in eastern United States. *Entomol. Am. N. S.*, **40** (1960), 1–117.

Hickin, N. E. Larvae of the British Trichoptera. *Trans. R. Entomol. Soc. Lond.* **97** (1946), 187–212.

Mosley, M. E. and D. E. Kimmins. *The Trichoptera (Caddis-flies) of Australia and New Zealand*. London: British Museum Natural History, 1953.

Munroe, E. T. Pest Trichoptera of Fort Erie, Ontario. *Can. Entomol.*, **83** (1951), 69–72.

Ross, H. H. The caddis-flies, or Trichoptera of Illinois. *Ill. Nat. Hist. Surv. Bull.*, **23** (1944), 1–236.

*The Caddisflies or Trichoptera of Illinois*. Los Angeles, CA: Entomological Reprint Specialists, 1972.

Schmid, E. Genera des Trichopteres du Canada et des states adjacents. Les Insectes et Arachnides du Canada, Part 7. Agriculture of Canada Publication 1692. Ottawa: Agriculture of Canada.

Wiggins, G. B. Larvae of North American Caddisfly Genera (Trichoptera). Toronto: Toronto University Press, 1977.


### General


Bournaud, M. and H. Tachet (eds.) *Proceedings of the Fifth International Symposium on Trichoptera*. Series Entomologica 39. The Hague: Junk, 1987.

Flint, O. S. Taxonomy biology of Nearctic limnephilid larvae (Trichoptera), with special reference to species in eastern United States. *Entomol. Am.*, N.S., **40** (1960), 1–117.

Fremling, C. R. Biology and possible control of nuisance caddisflies of the upper Mississippi River. *Iowa State Univ. Res. Bull.*, **483** (1960).

Langolis, C., S. Schulman, and C. E. Arbesman. Immunological studies of caddis fly. *J. Allergy*, **34** (1963), 385–94.

Macan, T. T. *The Study of Stoneflies, Mayflies and Caddisflies*. London: Amateur Entomologist's Society, 1982.

Macay, R. J. and G. B. Wiggins. Ecological diversity in Trichoptera. *Annu. Rev. Entomol.*, **24** (1979), 185–208.

Merritt, R. W. and K. W. Cummins (eds.) *An Introduction to the Aquatic Insects of North America*. Dubuque, IA: Kendall/Hunt, 1984.

Morse, J. C. (ed.) *Proceedings of the Fourth International Symposium on Trichoptera*. Series Entomologica 30. The Hague: Junk, 1983.

Osgood, H. Allergy to caddis fly (Trichoptera). II. Clinical aspects. *J. Allergy*, **28** (1957), 292–300.

Pennak, R. W. *Fresh-Water Invertebrates of the United States*. New York: Wiley Interscience, 1978.

Peterson, D. G. Observations on the biology and control of pest Trichoptera at Fort Erie, Ontario. *Can. Entomol.*, **84** (1952), 103–7.

Part three

# Other arthropods in the urban environment

# 18 Arthropoda

## Introduction

Urban and suburban environments provide suitable habitats for a small number of scorpions, spiders, mites, millipedes, and centipedes, and other arthropods. Many of these live in stable populations in peridomestic habitats, such as turfgrass, flowers and shrubs, and ornamental trees. Individuals or groups of individuals from these outdoor populations occasionally enter structures. They rarely become established as pests indoors, because of the unsuitable environmental conditions and limited resources such as food and harborage. However, some arthropods have adapted to living indoors, in residential and commercial buildings. The populations and body size of these arthropods are usually small, and they usually remain unnoticed in basements, cellars, attics, and other secluded sites.

Pest status for arthropods that occur in peridomestic and domestic habitats is based primarily on their threat to the health and safety of people and pets. Spiders, scorpions, mites, and centipedes pose actual and perceived threats of painful bites and stings, and the medical consequences that follow. For some individuals, the bite or sting produces an immediate life-threatening anaphylactic or neurotoxic reaction or long-term illness by the transmission of bacterial or rickettsial disease organisms. Rocky Mountain spotted fever is an important tick-borne disease in suburban areas in southeastern USA. Human cases of this tick-borne disease have increased in areas where suburban development has expanded into farm- and woodland ecosystems. The increased incidence of this disease is linked to growing and expanding human populations, and conditions that have allowed for an increase in density of the American dog tick, *Dermacentor variabilis*. These conditions include a mixture of landscape vegetation and the prevalence of suitable hosts. The acceptance of domestic dogs as a host for immature and adult *D. variabilis* improves the opportunity for tick-to-human contact. A similar situation exists for the occurrence of boutonneuse fever (caused by *Rickettsia conori*) in South Africa. People living in Johannesburg suburbs where houses have extensive gardens may also visit natural areas where boutonneuse and its tick vector, *Amblyomma hebraeum*, are abundant. The disease has been introduced into South African suburban areas, where the tick survives in peridomestic gardens and landscaping, and utilizes rodents and dogs as hosts. The emergence of Lyme disease as an important disease in suburban areas in the USA may be linked to field mouse (*Peromyscus*) and white-tailed deer populations, and the deer-tick vector, *Ixodes scapularis* (= *Ixodes dammini*) of Lyme disease. The mixed-vegetation landscaping and gardens in suburban areas support populations of these animals.

Classification of the arthropod groups is based on various characters, including the appendages (mouthparts and legs) and body regions. There are different taxonomic arrangements in use, which are based on the emphasis of certain morphological features or phylogenetic relationships. The classification of the phylum Arthropoda followed here recognizes four subphyla: Trilobita, Chelicerata, Crustacea, and Atelocerata. In Chelicerata, the class Arachnida includes the orders Scorpiones (scorpions), Uropygi (whip-scorpions), Schizomida (short-tailed whip-scorpions), Amblypygi (tailless whip-scorpions), Araneae (spiders), Opiliones (daddy-long-legs, harvestmen), Acari (mites and ticks), Pseudoscorpiones (pseudoscorpions), and Solfugae (sun-spiders, wind-scorpions). In Crustacea, the class Malacostraca includes the orders Amphipoda (scuds) and Isopoda (sowbugs, pillbugs). Included in the Atelocerata are the classes Diplopoda (millipedes), Chilopoda (centipedes), and Insecta (insects).

## Acari

Mites and ticks are one of the largest orders of arthropods with about 20 000 described species, and probably many more undescribed. This is a group of small to minute arthropods that live in a variety of habitats. The body form is generally oval, with little differentiation of the two body regions. Newly hatched young are called larvae and they have three pairs of legs; after the first molt there are four pairs of legs. Acari are abundant in soil and water (fresh and salt), and many are parasitic on insects or vertebrates. Others are scavengers on plant and animal matter, and some feed on live plants. The majority of mites that are pests of stored-food products or otherwise occurring in the household environment are cosmopolitan, and not restricted to one geographical region.

The body consists of the gnathosoma or mouthparts and the idiosoma, on which is located the legs. The idiosoma is analogous to the thorax and parts of the head of insects. It is unsegmented, but often has sutures or cuticular sculpturing, and may be heavily sclerotized. Simple eyes (ocelli) or other photosensitive organs are present on the idiosoma of many species. Legs and spiracles (stigmata) are located on the prodosoma, and genital openings are on the region behind the legs, which is called the opisthosoma. Nymph and adult mites have four pairs of legs; the larvae have three pairs. The legs of mites are arranged in two sets; the anterior two pairs are usually separated from the posterior two pairs. Stigmata are the external opening of the complex tracheal system and their presence or absence (astigmatid mites), and their position provides diagnostic features for identification and classification of mites. The body has a variety of setae that function as mechanoreceptors or chemoreceptors; the location and arrangement of these organs are taxonomically useful.

Food of mites and ticks is principally liquids obtained from living animals and plants, or from decaying organic matter. Some mites feed on microorganisms, and a few ingest solid food. There are species entirely parasitic upon plants or other animals, including other arthropods, while others attach themselves on animals to feed during the larval stage and are free-living as an adult. Predatory species have chewing mouthparts. In parasitic species, the mouthparts are modified for piercing and sucking blood or other fluid from the host. Many families of mites and ticks have evolved to be ectoparasites of birds and mammals. The majority of the mite- and tick-borne diseases of humans probably developed first as diseases of animals closely associated with humans, such as domestic rats, mice, and poultry. These diseases were probably confined to a limited area at the start, but then were spread by movements of the original animal or infected humans to form new disease foci.

Eggs are deposited by nearly all species of mites and ticks. A few Oribatidae mites produce live young, and some mite species are parthenogenetic. Eggs hatch in 1–3 days; the six-legged larvae complete development in about 48 h and molt to eight-legged nymphs. Nymph development is completed in about 15 days. Females lay eggs soon after mating. Eggs formed by parthenogenetic females develop into males. In most parasitic mites the males do not feed on the host.

### Astigmata (= Sarcoptiformes)

The feeding habits of these mites is variable and ranges from feeding (filter-feeding) on microorganisms to the maceration and ingestion of solid food material. The important species are associated with grain, flour, dried meat and dried fruit, and the skin of vertebrates. They are small, 0.2–1.2 mm long, without a sclerotized dorsal shield, and yellowish white. Palps are two-segmented and the chelicerae are shaped like pincers. Stigmata and trachea are absent, and respiration is primarily through the thin cuticle. Sexual dimorphism is distinct and the males frequently have copulatory suckers on the tarsi or anal region.

This order includes species in about 10 families that are associated with humans, including stored products, and the household environment. One group is generally referred to as cheese mites; and another major group is parasitic and called itch mites. Metamorphosis in Astigmata includes the motile hypopus stage (deutonymph). Hypopi briefly attach themselves to flies and other insects for dissemination. These individuals lack chelicerae and a mouth, and attach only for the opportunity to disperse. Species in Acaridae and Glycyphagidae infest various stored foods, and handling these products can cause a contact dermatitis called grocer's itch. The Sarcoptidae are itch mites or scabies mites; they burrow into the skin of mammals and cause intense itching. The distinct forms of *Sarcoptes* are regarded as varieties of *S. scabiei*. They differ slightly from each other, and exchange hosts. *S. scabiei* var. *suis* is parasitic on swine and may attack humans, while *S. scabiei* var. *equi* temporarily moves from horse to human.

Dust fragments and dust-associated fungi are eaten by a number of mite species in the genera *Dermatophagoides*, *Euroglyphus*, and *Glycyphagus*; these mites are preyed upon by various species of *Cheyletus*. Other species of mites commonly found indoors associated with house dust include *Gymnoglyphus longior*, *Haplochthonius simplex*, *Hirstia domicola*, *H. passericola*, and *Cosmochthonius reticulatus*.

Mites commonly occur in animal feed mills and oilseed stores, and they generally infest cereal storage facilities. In the UK, the three most common species found at these sites were: *Acarus siro*, *Lepidoglyphus destructor*, *Tyrophagus longior*, *T. palmarum*, and *T. putrescentiae*.

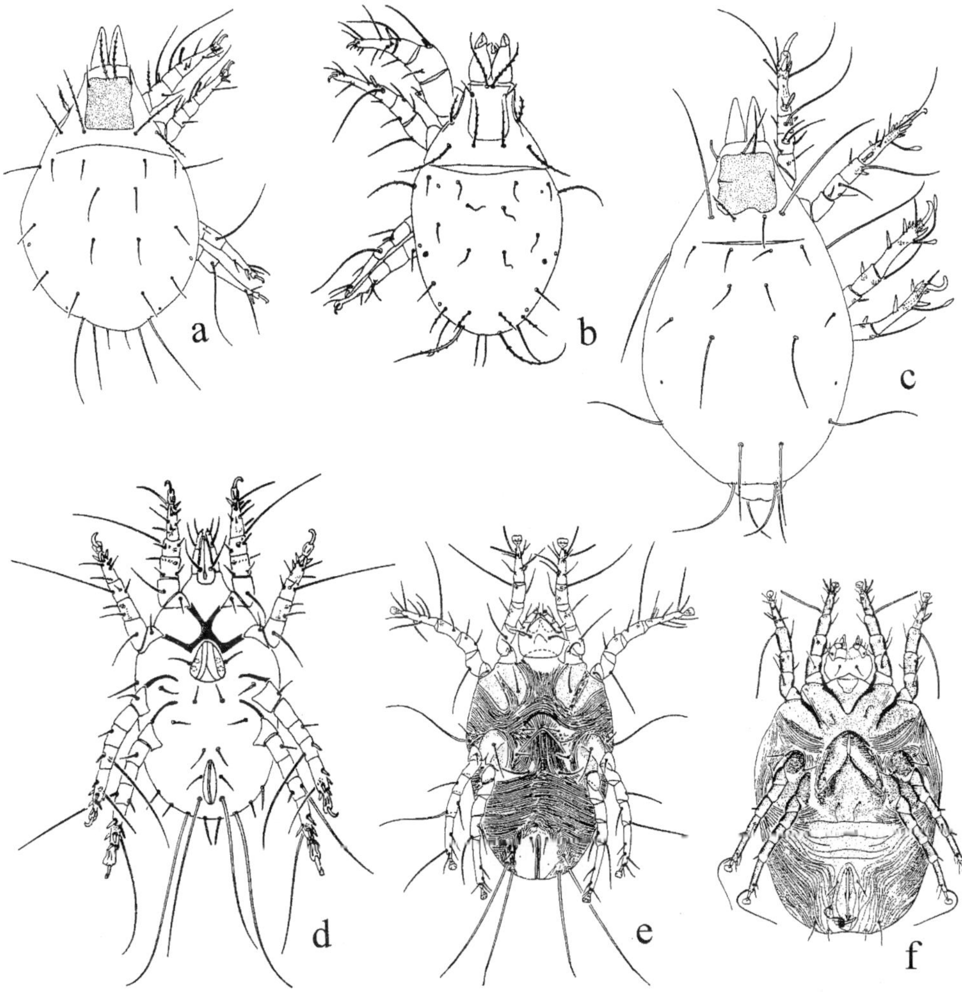


**Figure 18.1** Arthropoda: Acari. (a) *Acarus siro*; (b) *A. farris*; (c) *Caloglyphus berlesei*; (d) *Carpoglyphus lactis*; (e) *Dermatophagoides farinae*; (f) *Euroglyphus maynei*.

**Grain mite, cheese mite, *Acarus siro* (Fig. 18.1a)** Adults are about 0.4 mm long and opaque white. Infested materials include cereals, processed cereal products, cheese, medicinal herbs, and litter in poultry houses. This mite can infest a variety of food materials, and often lives on fungi. It occurs in house dust and is strongly allergenic to humans. Eggs are laid singly and females produce 1–24 eggs per day; fecundity is about 230 eggs, but is 670 with powdered milk or wheat germ as food, at 20 °C and 80% RH. Development from egg to adult is 78 days at 4 °C, and 9.2 days at 28 °C and 80% RH. Females reared on wheat germ live 42–51 days; males live a few days less. Grain moisture of at least 13.4% is required for survival; feeding and other activity stops at 0 °C. Optimal conditions are 20–25 °C and 75–80% RH. It is a cosmopolitan species, but is more abundant in temperate regions. Closely related species, *A. farris* (Fig. 18.1b) and *A. immobilis*, infest farm-stored grains and cheese.

This mite feeds on fungi associated with stored foods, including *Penicillium camemberti*, *Alternaria* spp., *Aspergillus flavus*, *A. candidus*, *A. repens*, *A. ruber*, *A. amstelodami*, *Trichoderma lignorum*, and *Fusarium moniliforme*. It is sensitive to antibiotics and antimicrobial agents, and small amounts of the preservative potassium sorbate, methyl paraben, and calcium propionate inhibit their development; high doses cause mortality. *A. siro* is killed by a 3-day exposure to 100% carbon dioxide.

***Caloglyphus berlesei* (Fig. 18.1c)** Infested food material includes damp and moldy commodities: wheat, copra, flaxseed, and peanuts. This mite also invades and damages insect, bacterial, and fungal cultures. It is a pest of stored food in Japan. Females lay about 27 eggs per day; fecundity is 588 eggs with a range of 60–1174. Oviposition period varies from 3–26 days. Development from egg to adult is 8 days at 27 °C and 80% RH. Optimal development is at 22–30 °C;

minimum is 16.5 °C. Fungi that support development of this mite include *Neospora crassa*, *Scopulariopsis brevicaulis*, *Aspergillus wentii*, and *Penicillium viridicatum*. A closely related mite species, *Caloglyphus anomalus*, occurs on decaying onions, fruits, and vegetables. It feeds on the semifluid materials associated with decay.

**Dried-fruit mite, *Carpoglyphus lactis* (Fig. 18.1d)** Adults are 0.4–0.5 mm long and pale white. Food infested includes milk products, dried fruits, honey, beer, wine, animal feeds, medicinal herbs, and beehives. It is a pest of miso in Japan. Hypopi have been found in the nests of birds, rodents, and ants. Females lay 20–56 eggs per day for about 14 days, and fecundity is about 278 eggs; hatching occurs in 2–4 days. Development from egg to adult is 9–11 days; thresholds for development are 3 °C and 35 °C and 60% RH. Adults live at 25 °C for 15 days at 60% RH and 29 days at 95% RH. Females live longer than males. Without food, adults can live 97–350 days at 0 °C and 85% RH, 10–57 days at 20 °C, and 3–7 days at 35 °C. The immature stages are less tolerant of starvation than adults. This species is cosmopolitan.

**American house dust mite, and European house dust mites, *Dermatophagoides farinae* (Fig. 18.1e) and *D. pteronyssinus*** Adult females are about 0.5 mm long and males are about 0.4 mm long. The body is striated and yellowish white, except for areas of sclerotized integument that are light brown. Setae on the ventroposterior margin are short in *D. farinae* and long in *D. pteronyssinus*. Infested materials and food include animal feeds, flour, and human skin scales.

***D. farinae*** Females lay 0–5 eggs per day for at least 31 days. Hatching occurs in about 38 days at 16 °C and 75% RH, 10 days at 23 °C, and 5 days at 30 °C and 35 °C. Fecundity is 31–100 eggs at 23 °C and 75% RH. Development time for males and females at 16 °C and 75% RH is about 140 days; at 23 °C it is about 36 days; and at 30 °C it is about 17 days. Most *D. farinae* do not develop to adult stage at 16 °C and 35 °C.

***D. pteronyssinus*** Females lay about 2.5 eggs per day at 23 °C, and 3.3 eggs per day at 35 °C. Hatching occurs in 26 days at 16 °C and 75% RH, 8 days at 23 °C, 5 days at 30 °C, and 4 days at 35 °C. Fecundity is about 65 eggs at 23 °C, and about 50 eggs at 35 °C. The percentage of eggs that successfully develop is 59% at 16 °C, 86% at 23 °C, 81% at 30 °C, and 87% at 35 °C. Development time for males and females at 16 °C and 75% RH is about 123 days, 34 days at 23 °C, 19 days at 30 °C, and about 15 days at 35 °C.

Development for *D. pteronyssinus* is optimal at 25 °C and 70–80% RH; populations do not grow well at humidities less than 60%. Optimal development for *D. farinae* is at 25–30 °C and 50–80% RH; critical RH for the female is 55–73%. Dust mites survive for long periods in buildings with humidities less than 50%. For these animals, the rate of water uptake exceeds the rate of water loss, and periods when the RH is high supplies mites with sufficient water to compensate for low RH levels. These two dust mite species are cosmopolitan. Their separation in geographic occurrence is correlated with their humidity requirements for survival. *D. farinae* occurs more often in dry inland sites, and *D. pteronyssinus* in regions with humid or coastal climates.

Natural habitats for these two *Dermatophagoides* include bird nests, on birds, and in mammal nests and burrows. In the urban environment these species occur predominantly indoors, in some stored food, on the floor in house dust, on clothing and upholstered furniture in commercial (including schools, hospitals, hotels) and residential settings, in automobiles, and other types of transportation. House dust is composed primarily of particles that are 0.01–1 mm in diameter, including human skin scales or flakes, cotton fibers, paper fibers, wool, synthetic fibers, mite feces, and mite exuviae. House dust samples range from 32 to 100% positive for species of *Dermatophagoides*, and they constitute 43–100% of the mite fauna indoors. The number of these mites in a 5-g dust sample may reach 4390. Distribution of mites indoors is centered in areas of human activity, and determining the number of mites per square meter is an accurate method of expressing the level of infestation. Populations fluctuate, and are usually higher in summer than in winter. These mites feed on cast skin flakes (from humans and indoor dog and cat pets), hair, and other detritus, along with the fungi that grow on skin flakes. Humans shed 0.5–1 g of skin scales per day. The fungus *Aspergillus repens*, which is carried on the body of dust mites, renders the skin flakes into a form that the mites can utilize as food. Allergens produced cause rhinitis, eczema, and asthma, affecting 50–100 million people worldwide. These allergens are contained primarily in dry mite feces, which are pellets 20–50 μm in diameter and covered with a peritrophic membrane (produced in the gut of the mite). Mites produce about 20 fecal pellets a day, which readily become airborne. Other allergens are present in the cast exuviae.

Extracts of *Dermatophagoides* mite bodies contain about 30 different protein molecules called allergens. The important allergens are called *Der p* I, *Der p* II, *Der p* III, *Der p* IV, and *Der f* I. The *Der p* and *Der f* are abbreviations of the scientific name of the mites (*D. pteronyssinus*, *D. farinae*). People allergic to house dust mites react to two or more of these proteins, and most people have a different combination of reactiveness. The ability of an allergen to sensitize an individual is related to its immunogenic properties, the susceptibility of the individual, and the degree of exposure. The proportion of individuals in an exposed population who become sensitized is probably under some genetic influence. While genetic influences and factors such as tobacco smoke and other pollutants are important, the level of the allergen in the environment plays the primary role. For the allergen *Der p* I, levels of 2 μg/g of dust may be the threshold for sensitization. The response occurs for the first time at any age, and infants are particularly susceptible. Seasonal changes and cultural habits, such as routine cleaning, cause fluctuations in mite populations, and these are accompanied by increasing or decreasing levels of mite allergens in house dust. Physiological actions such as enzymatic degradation of skin scales, and house dust-inhabiting fungi and bacteria also contribute to the decay of allergens, and influence allergen levels.

**Other *Dermatophagoides*** Other species in the genus have been isolated from natural habitats, such as bird nests, and from indoor habitats, such as house dust and stored food. These include *D. evansi*, *D. chelidonis*, *D. bakeri*, and *D. scheremetewski*.

***Euroglyphus maynei* (Fig. 18.1f)** Adults are about 0.4 mm long. The body is distinctly pointed anteriorly; the cuticle is sclerotized and it has striations dorsally and ventrally. Legs are cylindrical; the epistome terminates with two points. Male legs are of equal length, while female legs 3 are shorter than legs 4. Females lay 1.3 eggs per day over 25 days at 25 °C and 75% RH. Female nymphs emit a pheromone that induces male guarding behavior. This species is a common inhabitant of house dust, especially in temperate latitudes, and it has been shown to be highly allergenic. In dwellings where this species is present in large numbers, it may make a considerable contribution to the allergen pool. It occurs primarily in humid habitats, and is common in mattresses, which retain moisture for long periods. This species is distributed around the world.

**House mite, *Glycyphagus domesticus* (Fig. 18.2a)** Adults are 0.3–0.7 mm long; females are larger than males. The body has long and feathered setae posteriorly, while the legs have long setae. Food infested includes a variety of plant and animal materials, including flour, wheat, cheese, ham, and fish. It occurs in medicinal herbs and animal nests, such as the European house sparrow (*Hirundo rusticus*). It occurs in house dust and is strongly allergenic. Females lay 91.5 eggs per day at 15.6 °C and 98% RH, and 135.4 eggs at 20.6 °C; oviposition lasts 20–29 days. Development from egg to adult is 56.5 days at 11.7 °C and 95–100% RH, and 22.5 days at 23.3 °C and 98% RH. Fungi that support development of this mite are *Nigrospora sphaerica*, *Hormodendrum cladosporioides*, and *Scopulariopsis brevicaulis*.

Inert hypopi enclosed in the protonymphal cuticle often occur in house mite populations; 50% of the protonymphs develop through this stage. It is resistant to dry conditions and remains viable for several years. Hypopi occur at 25–30 °C and 75–95% RH. Development from hypopus to adult takes 4–5 weeks. A closely related species, *Glycyphagus cadaverum* (= *G. spinipes*), occurs in dry animal and plant material outdoors and indoors.

**Aquarium mites, *Histiostoma anguillarum*, *H. cyrtandrae*, *H. nigrelli*, *H. piscium*** Although there are many aquatic mites in natural habitats, only a few species are known from artificial aquatic habitats, such as swimming pools and aquaria. These *Histiostoma* have been reported from large aquaria in the USA (aquaria of New York Zoological Society) and Europe (aquaria in Antwerp).

***Histiostoma ocellatum*** Adults are distinguished by the copulatory orifice that leads to a bursa surrounded by a sclerotized structure. The propodonotal shield bears a faint pattern of lines, which become inconspicuous in the posterior half. This mite occurs in indoor swimming pools in Japan, and in some locations it is the dominant mite species. Adults and nymphs are found, indicating a successful population.

***Lardoglyphus konoi*** Infested foods include dried and salted fish in India, Japan, and European countries. Optimal conditions are 30 °C and 83% RH in fishmeal. Development from egg to adult takes 8.5 days at 35 °C, and 9–11 days at 23 °C and 87% RH. Motile hypopi appear at 20–30 °C and 76–94% RH. The hypopi are typically carried by insects, especially *Dermestes* and *Necrobia* beetles.

**Ear mite, *Otodectes cynotis*** Adults have long setae on legs 3 and 4; legs 1 and 2 have elongate claws. Leg 4 is the smallest in both sexes. This auricular mite lives in the ears of dogs, cats, ferrets, and foxes. Adult and immature stages feed on dead skin

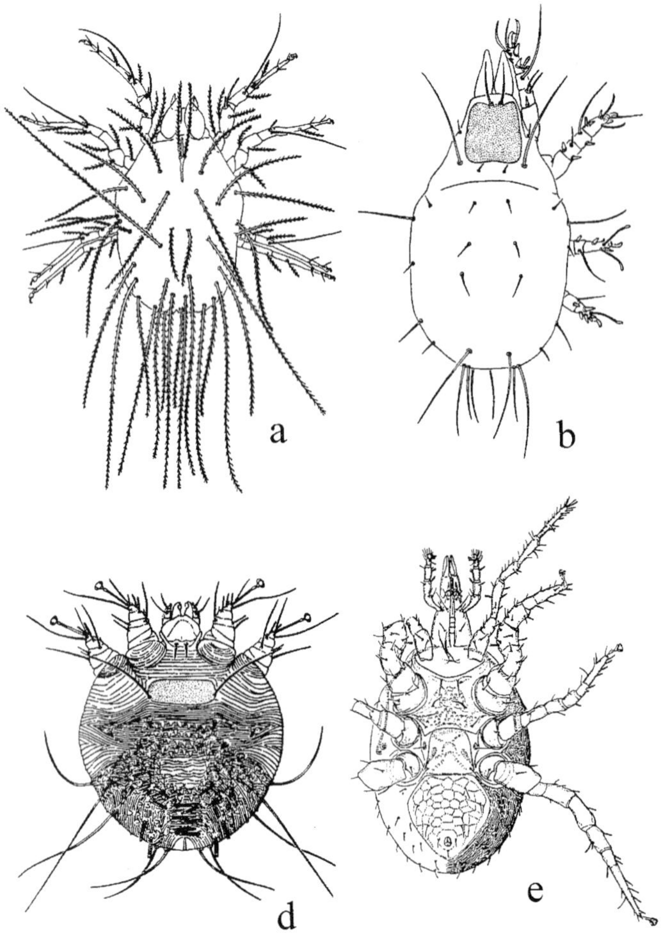


**Figure 18.2** Arthropoda: Acari. (a) *Glycyphagus domesticus*; (b) *Rhizoglyphus robini*; (c) *Thyreophagus entomophagus*; (d) *Sarcoptes scabiei*; (e) *Macrocheles muscaedomesticae*; (f) *Tyrophagus putrescentiae*.

tissue and complete their life cycle within the inner cavities of the ear. An infestation of these mites, otacariasis, causes irritation, liquid discharge from the ears, and rubbing, scratching, and head-shaking by the animal. Severe infestations can result in damage to the inner ear and loss of balance. This mite is common in domestic house pets around the world. It does not infect humans.

**Bulb mite, *Rhizoglyphus echinopus*** Adults are 0.5–0.7 mm long. The body is oval, usually with two brown spots laterally; the chelicerae and legs are brown. Immature stages have brown spots and brown legs. Females lay 2–26 eggs per day; fecundity is about 100 eggs when reared on onions and flower bulbs. Development from egg to adult (excluding the hypopal stage) is 8–14 days at 30 °C, 38–50 days at 10 °C, and 123 days at 5 °C. Optimal development is between 20 and 30 °C and 85% RH. Fungi that support development of *R. echinopus* include *Aspergillus repens*, *Botrytis cinera*, and *Trichothecium roseum*. These mites are usually found in large numbers, and because of their oval shape and slow movements they are often mistaken for insect eggs. Flower bulbs, primarily gladiolus, hyacinth, and onions, are attacked in the field and in storage. They cause irritation from contact with excretions as the mite walks on the skin. Other *Rhizoglyphus* associated with bulbs include *R. robini* (Fig. 18.2b).

**Scabies, itch mite, *Sarcoptes scabiei* (Fig. 18.2d)** Adult females are 0.3–0.4 mm long and 0.2–0.3 mm wide; males are about half as large. The body is rounded and the cuticle is striated, with specialized dorsal scales and setae. Legs are arranged in two groups: the anterior pair in both sexes end in stalked pulvilli, which assist in gripping the host; the posterior pair end in long setae, but in the male only the third pair ends in setae. Females move about 2.5 cm/min over the skin surface and select a site for making a burrow and oviposition. Preferred sites include between the fingers, at the bend of elbow and knee, and at other sites of soft skin. Eggs are deposited at 2–3-day intervals for about 2 months, and the tunnels in the skin may extend 3 cm. Hatching occurs in about 5 days, and newly hatched larvae move to the surface of the skin. Nymph development into adults takes 4–6 days. Females make a temporary burrow in the skin before mating. Males also burrow, but their burrows are no more than 1 mm long. Development from egg to adult requires 10–14 days. Peaks in the incidence of human scabies occur in 15–20 cycles, and this might be due to fluctuating levels of immunity in the human population. This species is widespread in the tropics and in temperate regions.

Scabies mites are usually transmitted from person to person, and infestations are common in dormitories, health- and mental-care facilities. These mites have acute sensory abilities and well-developed host-seeking behavior. They perceive specific host stimuli and will actively move from host to host. They respond independently to both thermal stimuli and host odor. In close proximity, both of these stimuli are effective; at a greater distance host odor is more effective. These mites die quickly away from the human body. *S. equi* is associated with horses and mules and may transfer to humans and cause a temporary skin irritation.

***Thyreophagus entomophagus*** **(Fig. 18.2c)** Adults are 0.3–0.6 mm long. The body is shiny white, and the mouthparts and legs are yellowish white to light brown. The dorsal suture is distinct; there are long setae on the legs and body. Materials commonly infested include flour, cereal products, animal feeds, spices, medicinal herbs, and insect collections. Female fecundity is about 76 eggs, and the maximum is about 170; the oviposition period lasts about 32 days. Development from egg to adult takes about 15.5 days. Growth and development occur between 3 and 32 °C and 75–100% RH. Optimum conditions are 15–25 °C and 85% RH. Adult females live about 46 days and males about 41 days; they can live 2 years at 3–5 °C, and 9 months at 6–10 °C. A closely related species, *T. carpio*, is associated with decaying hardwoods.

**Mold mite, *Tyrophagus putrescentiae* (Fig. 18.2f)** Adults are about 0.5 mm long. Chelicerae are enlarged; dorsally there are four long setae anteriorly and numerous long setae on the dorsoposterior margin, and four long setae on the ventroposterior margin. Food infested includes grain, wheat, flour, peanuts, cottonseed, rapeseed, sunflower seed, tobacco, dried fruits, dried eggs, cheese, and medicinal herbs. It occurs in commercial mushroom houses, and it often infests laboratory cultures of insects and microorganisms. These mites occur in large numbers in rice-straw mats (tatami) in houses, and they are a common allergen in house dust. Females lay about four eggs per day, but can deposit a maximum of 60 per day; fecundity is about 500 eggs for females that mate twice a month, depending on the food source. Females reared on wheat germ or yeast produce about 500 eggs, while those reared on rolled oats produce about 8. About 70% of the eggs are produced during the first 3 weeks of the female's life. Development from egg to adult takes 9.5 days at 25 °C and 85% RH, and 130 days at 8.5 °C and 85% RH. At 25 °C and 85% RH females reared on wheat germ live about 66 days, males about 59 days.

Medicinal herbs are often infested by *T. putrescentiae*. This mite responds favorably to many volatile oils and alkaloids, and to robinin and rutin in their food. It is generally unaffected by antibiotics and antioxidants added to food it infests. It can feed on insect eggs, but it is essentially a fungivore. Numerous fungi can support normal growth and development: *Polyporus* spp., *Poria* spp., *Stereum* spp., *Flammula* spp., *Alternaria tenuis*, *Aspergillus versicolor*, *A. candidus*, *A. amstelodami*, *Fusarium moniliforme*, *Helminthosporium sativum*, *Mucor racemosus sphaerosporus*, *Nigrospora sphaerica*, *Stemphylium botryosum*, *Trichothecium roseum*, *Botrytis cinerea*, *Penicillium purpureum*, and *P. chrysogenum*.

***Tyrophagus longior*** **(Fig. 18.3a)** Adults are about 0.5 mm long. The anterodorsal sclerotized plate has four plumose setae on the anterior margin, four short, plumose setae medially, and four long, plumose setae posteriorly. Materials infested include grains, cereal products, drugs, seeds, cheese, dried fruits, and prepared foods of animal and vegetable origin. It generally prefers substrates with a high moisture content, but not actually moldy or decomposing. Development from egg to adult takes 2–3 weeks at 23 °C and 87% RH. This species is probably native to Europe, but the current distribution is nearly worldwide.

### Cryptostigmata (= Oribatida)

This large group of mites exhibits considerable variation in form and habits. Adults are 0.2–1.3 mm long, shiny, and yellowish brown to brown. The body is convex oval, and some terrestrial species resemble small beetles (they are often called beetle mites). Some species have elaborate lateral projections of their body. Natural habitats include leaf litter and decaying vegetation, under bark, and in the soil. Many cryptostigmatid mites are aquatic and feed on algae. The majority of these mites occur in natural habitats, but some occur in artificial habitats, such as indoor swimming pools and aquaria. Two species that occur in swimming pools are *Hydronothrus crispus* and *Trimalaconothrus maniculatus*. The water supplied to some swimming pools in Japan originates from lakes, and species of

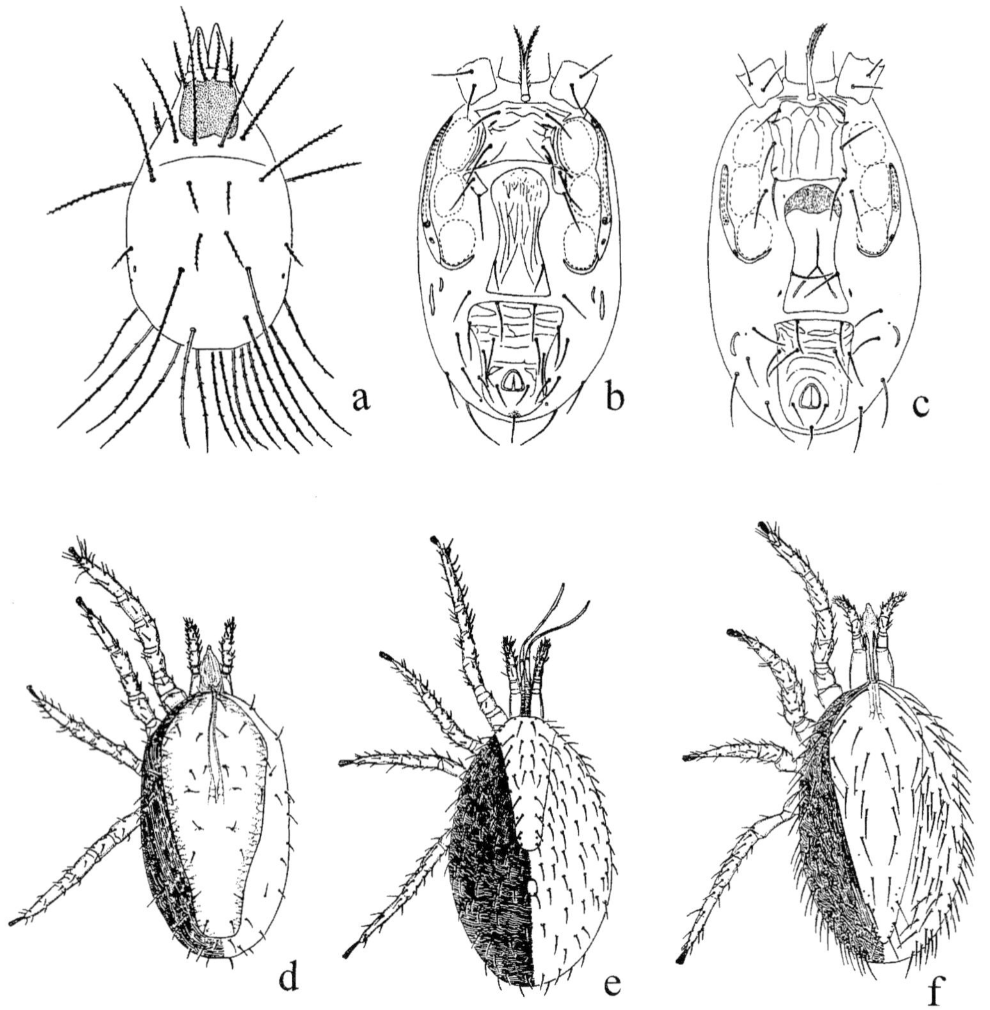


**Figure 18.3** Arthropoda: Acari. (a) *Tyrophagus longior*; (b) *Blattisocius dentriticus*; (c) *B. tarsalis*; (d) *Dermanyssus gallinae*; (e) *Liponyssoides sanguineus*; (f) *Ornithonyssus bacoti*.

*Hydronothrus* and *Trimalaconothrus* live in calm and shallow lake water, and may have access to water supplies.

**Swimming pool mite, *Hydronothrus crispus* (= *H. aquariorum*)** This species is characterized by the presence of curled dorsal setae, the absence of exobothridial setae, and an epimeral formula of 3–0–3–2. It occurs in indoor swimming pools in Japan. All life stages of these mites can be isolated from water in the pool, including eggs and larvae found in the abdominal cavity of the females, nymphs, and adults. It is parthenogenetic, so invasion of a single individual to a swimming pool may result in an infesting population. The occurrence is medically important because there are strong antigenic cross-reactivities between *H. crispus* and the common house dust mite, *Dermatophagoides farinae*. The swimming pool mite has allergenic components, which are common with those of the dust mite.

***Trimalaconothrus maniculatus*** The cuticle is smooth; tarsi 1 and 2 have unequal flattened spines bearing finger-like projections; the epimeral formula is 3–1–2–3. The rostral setae are smooth (in Japanese forms) or weakly barbed. In some swimming pools in Japan this species occurs as the dominant mite. The intestinal content of some *T. maniculatus* in swimming pools included *Cosmarium* sp. algae, which indicates that resources (food) and conditions in the pool water are suitable for this mite.

### Mesostigmata (= Gamasida)

This group of mites is highly diverse; most are free-living but others are ecto- and endoparasites of reptiles, birds, mammals, and a few insects. They are relatively large, 0.2–3 mm long, and usually heavily sclerotized and brown to reddish brown. Parasitic forms are usually colorless. The principal characteristics of the order relate to the stigmata, which are located near the middle (meso) of the body, next to the coxae of legs 2 and 4, or between legs 3 and 4. The gnathosoma or mouthparts are usually large and distinctly separated from the remainder of the body.

Several mesostigmatid families have free-living species associated with bark beetles, ants, and millipedes. Others are parasites of insects, including the Varroidea mites that infect and injure honey bees. There are species that are ectoparasites of snakes, and several that are internal parasites of domestic animals. The canary lung mite, *Sternostoma tracheacolum*, invades the lungs of canaries; *Pneumonyssoides caninum* occurs in the sinuses and nasal passages of dogs. Nymphs and adults of the house fly mite, *Macrocheles muscaedomesticae* (Fig. 18.2e), prey upon the eggs of *Fannia canicularis*, *Musca domestica* and *M. vetustissima*. The mesostigmatid *Fuscouropoda vegetans* is also a predator of the house fly and little house fly. The yellowish-brown adult attaches itself to the ventral surface of the base of the house-fly abdomen.

Mesostigmata associated with humans are in three families: Dermanyssidae, Macronyssidae, and Laelapidae. They are blood-feeders (hematophagous) and in natural habitats feed on birds and rodents. However, many are not host-specific, and if their primary host suddenly leaves, they will attack other animals, including humans. *Laelaps echidninus* and *L. nuttalli* commonly occur on the house rat, *Rattus rattus diardii*, in Malaysia.

**Predatory mites, *Blattisocius dentriticus, B. keegani, B. tarsalis* (Fig. 18.3b, c)** These predatory mites feed on several species of mites in the family Ascidae, and on eggs of some insects. They are generally found in flour, grain, and grain products, with various seeds, and in medicinal herbs. *B. dentriticus* females lay about 36 eggs at 70–90% RH. Development from egg to adult takes about 12 days at 20 °C and 70–100% RH. *B. keegani* occurs in stored food, and in nests of birds and rodents; this mite takes about six prey per day during development in these habitats. Female *B. keegani* lay a maximum of five eggs per day at 27 °C and 70–75% RH. Development from egg to adult is 6 days at 27 °C and 70–75% RH, and 8.7 days at 27 °C and 95–100% RH. Immature stages can live 2–7 days without food. Daily consumption of a single prey egg is sufficient for growth, maturation, and oviposition. Female fecundity is 11–61 eggs during 14 days of life at 27 °C and 70–75% RH. This mite preys on the eggs of the stored-product beetles *Cryptolestes*, *Tribolium*, *Trogoderma*, and *Oryzaephilius*; eggs and mobile stages of stored-product mites *Glycyphagus* and *Acarus* are attacked. *B. tarsalis* occurs in animal feeds infested with insect pests. The larvae and nymphs feed on the eggs and larvae of food-infesting pyralid moths. Female mites are carried on the bodies of moths, and they lay eggs on the silk webbing of the caterpillar. Development from egg to adult takes about 10 days at 27 °C; adults live about 10 days.

**Chicken mite, red chicken mite, roost mite, *Dermanyssus gallinae* (Fig. 18.3d)** Adults are 0.7 mm long and gray when unfed, and 1–1.5 mm long and red after a blood meal. The body is striated and the dorsal shield is oval and medially with two rows of setae. Eggs are deposited in cracks and in debris in poultry houses, and in crevices adjacent to bird nests in attics. Development from egg to adult can be completed in 7 days. Adults live for 4 months without feeding. Adults and nymphs remain in harborages during the day, and at night move to attack roosting birds. Infestations in buildings usually originate from active or recently abandoned bird nests in attics and under eaves. This species is cosmopolitan and it attacks domestic chickens and turkeys, pigeons, and wild birds such as sparrows, barn swallows, and starlings. When the normal hosts are not available, the mites will attack mammals, including humans. A related species, *D. hirundinis*, is also associated with poultry houses.

**House mouse mite, *Liponyssoides sanguineus* (= *Allodermanyssus*) (Fig. 18.3e)** Adults are 0.6–0.7 mm long when unfed; engorged females are about 1 mm long. The body is reddish black, and the idiosoma is unstriated and setose; the legs are slender and relatively long. The dorsal shield is divided: the anterior portion is slender and elongate, while the posterior portion is small and rounded. The chelicerae are long and whiplike. The house mouse (*Mus musculus*) is the preferred host, but the mite will feed on rats and other rodents, and will attack humans. Females feed several times, and each feeding is followed by oviposition. Development from egg to adult can be completed in 17–23 days. Unfed females can live 51 days. Engorged nymphs and adults are found indoors near rodent nests and runways. This mite is a vector of *Rickettsia akari*, which causes rickettsial pox in humans. It is distributed in the USA, Europe, Ukraine, South Africa, and South Korea.

**Tropical rat mite, *Ornithonyssus bacoti* (Fig. 18.3f)** Females are about 1.1 mm long when unfed, and 1.4 mm long when fed. The body is gray to yellowish gray, and it changes to red or reddish black when engorged with blood. They have long setae on their narrow dorsal shield. Females lay eggs in several batches, and hatching occurs in about 36 h; fecundity is about 100 eggs. Development from egg to adult takes 7–16 days at 21 °C. Adult females live about 62 days, while unfed nymphs can survive about 43 days. Unfertilized females reproduce parthenogenetically. Nymphs and adults do not remain on the host, but feed and drop off after every meal. The nymphs and adults are very active and within about 10 days will leave an empty nest or harborage of the host rodent. They can move to attack people in restaurants, warehouses, offices, and dwellings where there is an infestation of rodents, or that have been recently infested. Bites usually occur on the arms and ankles of people near the mites. The bite is painful as the mouthparts of the mite are inserted in the skin; the response is irritation, itching, and dermatitis. This species occurs in tropical and temperate regions in all continents and feeds on the blood of mice and rats. It readily bites humans in rat-infested buildings.

**Tropical foul mite, *Ornithonyssus bursa*** Adults are about 1 mm long and adult females have a broadly oval dorsal shield with two longitudinal rows of setae. The sternal plate has two pairs of setae. This species occurs in tropical and subtropical areas on all continents. It is a widespread parasite of domestic fowl and the English sparrow (*Passer domesticus*); other bird species serve as a host for this mite. Adults can live about 10 days away from an avian host, and mites move from infested nests on the outside (vents and exhaust ducts) or inside to bite humans nearby.

**Northern foul mite, *Ornithonyssus silvarium*** Adults are about 1 mm long and adult females have an oval dorsal shield with few and scattered setae. The sternal plate has three pairs of setae. Development from egg to adult can be completed in 7 days, and all stages remain on the host. Heavily infested birds appear gray; such birds lose weight and eventually die from loss of blood. Mites will move from dead birds and attack rodents and humans indoors. This species is a pest of poultry and wild birds in north temperate regions of Europe, North America, southern Australia, and South Africa.

Prostigmata (= Trombidiformes, Actinedida)

The feeding habits of these mites are variable and range from phytophagous and fungivorous to saprophagous; some are parasitic on vertebrates and invertebrates. Body length ranges from 0.1 to 10 mm, and the body surface may be striated. The integument ranges from colorless to bright red in the family Trombidiidae. Stigmata 1 or 2 are usually in the region of the mouthparts, usually near the chelicerae. Modifications of the legs result in one or more pairs missing, especially in the vertebrate parasites. Sexual dimorphism may be pronounced, and in some groups females give birth to adults.

***Cheyletus eruditus* (Fig. 18.4a)** Adults are about 0.3 mm long. Chelicerae are large and leg-like, and distinctly chelate. It is a predator of mites and is commonly found in grain storage facilities, stored foods, and in house dust. Females lay 5–13 eggs per day: most eggs are laid during the first 4 weeks of life in batches and attended by the female; fecundity is about 71 eggs. Development from egg to adult takes 84 days at 9.4 °C and 85% RH, and 10 days at 31.5 °C and 85% RH. Adults live 1–2 months. Growth and development occur between 8 and 31 °C and at least 52% RH; optimal conditions are about 20 °C and 80% RH. At 20.2 °C and 70% RH mortality is 42%, while at 31.5 °C mortality is 100%. This might explain why winter survival is low. All stages die at −2 °C and 40 °C. This mite prefers feeding on living stored-food mites, but will attack early-instar insect larvae. Females consume 1–3 adult acarid mites each day, or about 21 in a lifetime. Nymphs live 3 months without food. It also occurs in bird and mammal nests.

**Clover mite, goosberry mite, *Bryobia praetiosa* (Fig. 18.4e)** Females are 0.7– 0.8 mm long; this species is one of the largest plant-feeding mites. Body color varies from reddish brown to dark green, and dark red; immature stages may be red. Front legs are longer than the body and about twice the length of the other legs. Females lay about 70 eggs, which are bright red. Development from egg to adult takes about 30 days; optimum development is at 21 °C. Above 24 °C and below 4 °C the eggs become dormant or inactive; eggs laid in late fall hatch the following spring. Adults become dormant or inactive from May until September. In eastern USA this mite occurs indoors from October until May. Females are parthenogenetic; males are unknown in the USA and are rarely seen in other parts of the world. It is cosmopolitan and infests and feeds on about 200 species of plant, including grasses, shrubs, flowers, and agricultural crops.

Clover mites occur in large numbers on trees and shrubs, and turfgrass in the urban environment. When large populations develop, mites climb on the outside of buildings and enter through windows and doors. If crushed, they will leave a reddish stain. There are several species of phytophagous mites, including other *Bryobia*, that occur in peridomestic habitats, and invasions of mites in the spring or fall may be by one or more of these species. The brown wheat mite or obscure mite *Petrobia latens* and winter grain mite *Penthaleus major* invade houses in large numbers. *B. praetiosa* often occurs in recently established turfgrass, perhaps before populations of natural predators provide suppression.

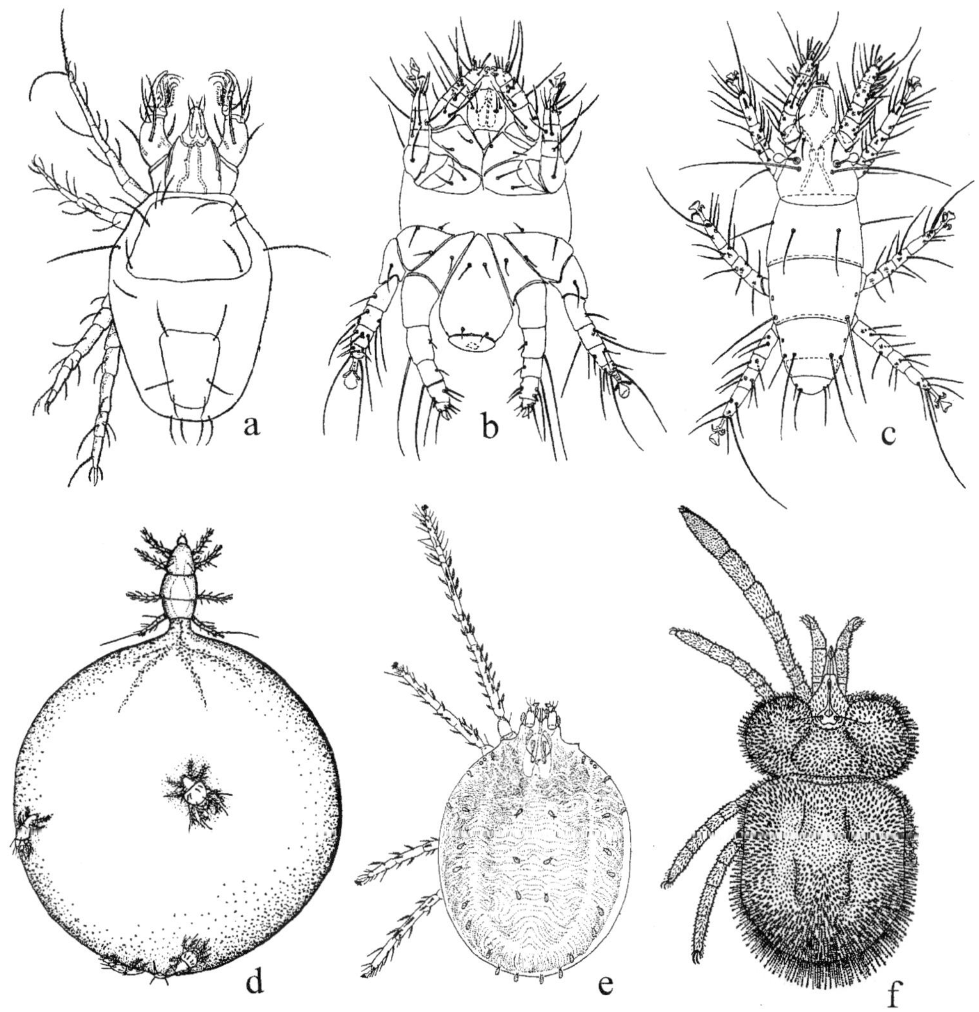


**Figure 18.4** Arthropoda: Acari. (a) *Cheyletus eruditus*; (b) *Pyemotes tritici* male; (c) *P. tritici* female; (d) *P. tritici*, enlarged female; (e) *Bryobia praetiosa* female; (f) *Eutrombicula alfreddugesi*.

**Chigger, redbug, *Eutrombicula alfreddugesi* (= *Trombicula*) (Fig. 18.4f)** Adults are 0.9–1.1 mm long; the cuticle is covered with densely pilose setae, giving the body a velvety appearance. The body is constricted behind the second pair of legs, giving the nymph and adult a figure of-eight shape; body color is often bright red. The larva is oval, without a dense covering of setae. Eggs are laid singly in soil; females lay about 7 eggs per day. Hatching occurs in 4–6 days; fecundity is about 200 eggs. The first stage is inactive and remains within the egg fragments; the six-legged larva emerges in about 7 days and crawls about rapidly in search of a host. Larvae live for about 14 days without feeding. The larva of *E. alfreddugesi* feeds for 1–3 days (sometimes up to 30 days) and takes only one meal. It then drops to the ground and burrows into the upper layers of the soil; after about 1 week, the eight-legged nymph emerges. It feeds for about 7 days on insect eggs and the early stages of insects and other arthropods in the soil. After about 7 days in an inactive stage the adult emerges. Development from egg to adult takes about 55 days. Adults are active, and females become inseminated when they walk over stalked spermatophores that males have deposited on the substrate. There are one or two generations per year; overwintering is in the adult stage in a small cell 2–3 cm deep in the soil. In the tropics the life cycle is short and there are two or three generations per year. This mite is parasitic in the larval stage (chigger) on a wide range of hosts, including mammals, birds, amphibians, and reptiles. Nymph and adult stages are red (redbug) and they are free-living. This species is distributed from Canada to South America and the West Indies.

Microclimate and vegetation can influence population size and activity patterns of adults and larvae. Larval populations along forest-edge habitats are greatest in areas of high RH, moderate temperature, and low-incident sunlight. The short- to tall-grass transition zone in a forest-edge habitat supports larger populations than a tree-canopy zone. Movement activity of larvae is late afternoon to early evening, and remains at low levels until sunrise.

Chigger dermatitis affects people in many parts of the world, particularly during the summer months in temperate climates. It results from the feeding of chigger larvae on exposed skin. Persons walking among tall grass, brambles, and low vegetation may be attacked and suffer intense itching 3–6 h after exposure. Severe dermatitis, consisting of pustules and wheals, develops on the ankles, knees, and waist. Larvae do not burrow into the skin (contrary to popular belief), but attach to the base of a hair or to the smooth surface of the skin. The chigger does not suck blood. When firmly attached, it injects a digestive fluid that causes disintegration of cells, and this cellular material is utilized as food by the chigger. The skin of the host becomes hardened and a tube is formed into the skin. The chigger lies in this tube and continues to feed, and when engorged it retreats and drops off. The action of the chigger digestive fluid probably causes the irritation and itching within a few hours.

**Harvest mite, *Neotrombicula autumnalis*** The larval stage is about 0.22 mm long, but when engorged it is about 0.6 mm long. Body color varies from yellowish white to reddish brown to orange. Eggs are laid in soil and larvae move upwards on the blades of grass and the leaves of low-growing shrubs. Larvae are parasitic; the normal hosts for this mite are small mammals and birds, but humans are bitten when walking or working in infested fields or the edges of woods. The larva feeds on fluids at a wound site on the skin, but does not penetrate below the skin surface; it completes feeding in 2–3 days. The eight-legged nymph stage and adult live on the soil surface and are primarily scavengers. This is the harvest mite of Europe.

**Other *Eutrombicula*** Several other species are known to attack humans, including *E. splendens* in eastern USA; *E. batatas* in Central and South America; *E. hirsti* and *E. wichmanni* in Asia, Australia, and the Pacific Islands; *E. lahillei* in Argentina; *E. samboni* and *E. sarcina* in Australia; and *E. akamushi* in Japan, China, and Indonesia.

**Straw itch mite, *Pyemotes tritici* (= *P. ventricosus*) (Fig. 18.4b–d)** Females are about 0.2 mm long and about 2 mm when gravid; the male is about 0.16 mm long. The body of the female is white to yellowish white; gravid females are distended behind the fourth pair of legs. Eggs are retained inside the female; they hatch internally and the young develop within the body of the female. Adult mites are extruded at the rate of about 50 per day; fecundity is 150–284 eggs. Males emerge first, and they usually remain on the distended abdomen of the female, clustered around the genital opening. Males mate with female mites when they emerge. Mated females immediately search for a host, while newly born males search for a female. Males remain on the female's abdomen and feed on her. Unmated females produce only male offspring. Broods are produced about every 14 days. Adults are active during warm months. Straw itch mite dermatitis often occurs when there are large populations of the mite's hosts present, such as grain moths, bean and pea weevil, and wood-infesting beetles. Infestations of the furniture beetle, *Anobium punctatum*, in floor joists of a house may result in mites moving from the beetle larvae in search of a host. This mite is a parasite of numerous species of insects, and some of these host insects infest straw, wheat, stored-food products, and wood. Humans can be attacked, resulting in dermatitis and itching.

### Ixodida (= Metastigmata)

Ticks are obligate hematophagous parasites of mammals, birds, reptiles, and amphibians. Their distinct morphology and large size distinguish them from other Acari. They are among the largest of the Acari, with a range of 2 mm to 3 cm. In ticks the cephalothorax and abdomen are fused into an oval body, which is flattened dorsoventrally; the epistome is absent and the mouthparts are formed into a denticulate hypostome. This toothed organ anchors the tick to its host. The stigmata or spiracles are located posterior to coxae 4 in hard ticks (Ixodidae), or anterior in soft ticks (Argasidae). Hard ticks have a sclerotized dorsal plate (scutum), and their mouthparts project anteriorly; soft ticks lack a dorsal plate and their mouthparts are located ventrally.

Ixodidae typically have a three-host life cycle. Eggs are laid on the ground; after hatching the larva (seed tick) feeds upon the first host, then drops off and molts to the nymph stage. The nymph feeds on a second host, and drops off to molt; the adult feeds and mates on a third host, and drops off to lay eggs. Each stage remains on the host for 1–3 days and takes one blood meal, but leaves to molt. The feeding habits of these ticks make them disease vectors to domestic animals and humans in the agricultural and urban environment. Argasidae typically have multiple hosts during their life cycle. Adults and immatures remain in harborages during the day and emerge at night to find and feed on a host. Each stage of development may take several blood meals. The larva stage of the soft tick, *Ornithodoros moubata*, emerges from the egg with food in its gut and does not feed. The larval mouthparts in this species are only used to escape from the egg.

Tick-borne diseases include Rocky Mountain spotted fever, Colorado tick fever, Lyme disease, and relapsing fever. Certain ticks, especially engorging females feeding on the neck or near the base of the skull, inject venom that produces a paralysis. Several species of ticks occur in peridomestic habitats and utilize domestic and feral mammals and birds as hosts. Urban and suburban expansion has resulted in increased human exposure to two diseases transmitted by ticks: Rocky Mountain spotted fever and Lyme disease. Rocky Mountain spotted fever, also known as Tobia fever and São Paulo fever, is widely distributed in the Nearctic and Neotropical regions. In spite of its name, this disease is most prevalent in eastern USA. Urban and suburban expansion into shrubby fields and woodlands, with a resident small- and large-animal population and various species of ticks, has increased human exposure to tick vectors of the disease. Transstadial and transovarial passage of the rickettsiae in the ticks maintains this disease in the vector population. Relapsing fever is caused by species of *Borrelia* and is transmitted by several species of *Ornithodoros* ticks, which occur in huts, outbuildings, seasonal houses, and sometimes in caves. The disease is endemic across many regions of the world, and *Ornithodoros* ticks serve as the disease reservoir.

Ticks are often associated with domestic dogs in the urban and rural environment. In Japan, *Haemaphysalis longicornis* is the species most frequently found on dogs, followed by *H. flava*, *Rhipicephalus sanguineus*, and *Ixodes ovatus*. Small numbers of *H. hystricis*, *H. campanulata*, *H. japonica*, *H. ias*, *I. persulcatus*, *I. nipponensis*, and *Amblyomma testudinarium* also occur. Dogs in rural areas of Japan frequently carry *H. longicornis*, *H. flava*, and *I. ovatus*, while *R. sanguineus* is more associated with the dogs in urban/suburban areas. Exposure of a domestic dog to a garden is associated with infestations of *R. sanguineus*, and exposure to woodland is associated with infestations of *H. flava* and *I. ovatus*.

Lyme disease, which is caused by the spirochete *Borrelia burgdorferi*, is perhaps the most important tick-borne disease in North America, where several hundred cases are reported every year. It is now known from the British Isles, Europe and parts of Russia, northeastern China, Japan, South Africa, and perhaps Australia. The vectors include *Ixodes* species, which feed as immatures on a variety of small animals. Mouse and vole populations in urban environments maintain and spread this disease. Transmission of Lyme disease is by tick bite, and attachment and feeding for 1–3 days are necessary to transmit

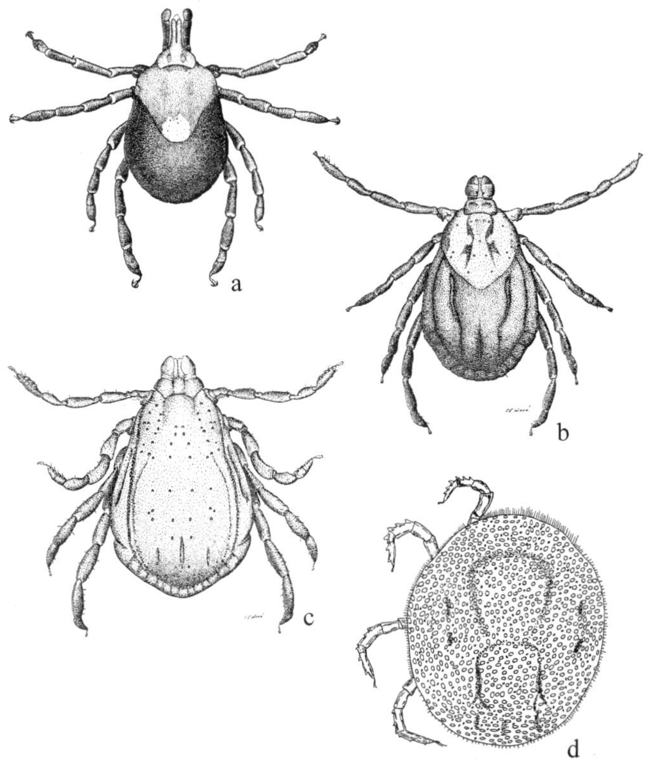

**Figure 18.5** Arthropoda: Ixodida. (a) *Amblyomma americanum* female; (b) *Dermacentor variabilis* female; (c) *Rhipicephalus sanguineus* male; (d) *Ornithodoros moubata* female.

the *Borrelia* pathogen. Transovarial transmission occurs at low rates in *Ixodes* ticks.

Boutonneuse fever produces button-like lesions on the skin at the tick attachment site. This rickettsial disease occurs in Africa, European and North African areas adjacent to the Mediterranean, the Middle East, and in Southeast Asia. Locally it is called Marseilles fever, South African or Kenya tick typhus, and Indian tick typhus. The primary tick vectors are *Amblyomma hebraeum* in the South African veldt, and the dog tick, *Rhipicephalus sanguineus*, in most other regions. Urban cases of this disease are associated with domestic dogs and the dog tick. Transmission may be directly through tick bite or by contact with skin and eyes after removing ticks from dogs. Transovarial passage of rickettsiae maintains the disease in the vector population.

**Lone star tick, *Amblyomma americanum* (Fig. 18.5a)** Adult males are about 5 mm long (including the long mouthparts), uniformly light brown to brown and with white markings at the margin of the dorsal plate. Females are about 10 mm long when engorged, and have a white spot at the posterior end of the dorsal plate. Females lay a mass of more than 5000 eggs, and hatching occurs in about 30 days. Larvae attach to host animals from the tops of grass stems and other vegetation. Nymphs and adults crawl to host animals in the immediate area. Overwintering is as a fed larva, as an unfed nymph, or as an unfed adult. Some fed nymphs develop to adults during winter. This species has a variety of hosts, including wild and domestic animals, birds, and humans. It attacks humans in the larval, nymph, and adult stage. The lone star tick is a vector of Rocky Mountain spotted fever and tularemia. This species is distributed in Mexico and southern USA, particularly Oklahoma, Missouri, Arkansas, Texas, and Louisiana.

**Other *Amblyomma*** Species in this genus have white markings; the mouthparts are long and inflict painful bites on humans and deep wounds on animals. *A. hebraeum*, South African bont tick, occurs from South Africa northwards to Zimbabwe and Mozambique. Adult males have two spots medially on the dorsal plate, festoons pale; the female dorsal plate is pale white with brown markings. The immatures feed on small and large mammals while the adults feed on large animals. In the South African veldt larvae and nymphs actively crawl towards and feed on humans. The Cayenne tick, *A. cajennense*, is distributed from South America and the Caribbean to southern Texas. Adult males and females have extensive white markings on the dorsal plate. It is a vector of Rocky Mountain spotted fever in Central America, Columbia, and Brazil; immatures readily feed on humans and produce intense itching and lesions.

**Pigeon tick, *Argas reflexus*** Adult females are about 8.5 mm long and 5 mm wide, but taper anteriorly; males are about 6.5 mm long and 4.5 mm wide. The margin of the body is composed of irregular striations. Palp segments are subequal, and segment 3 is the shortest. The capitulum has two long setae ventrally and directed forward. This species is associated with pigeons, although other avian and mammalian hosts, including humans, are known. This species is common indoors and attacks people in Israel. It is abundant in the Middle East, Europe, southwestern Russia, and Asia. A closely related species in the *A. reflexus* group, *A. latus*, occurs in large numbers indoors, and is associated with pigeons.

**American dog tick, *Dermacentor variabilis* (Fig. 18.5b)** Adult males are about 4.5 mm long and 2.5 mm wide. The dorsal plate is dark brown, covers the abdomen, and is marked with a variable pattern of white marks. Engorged females are about 13 mm long and about 10 mm wide, and the dorsal plate is small. Females lay a mass of 4000–6500 yellowish-brown eggs on the ground; hatching occurs in 36–57 days, depending on temperature and humidity. Larvae feed almost exclusively on small rodents such as mice and voles; engorgement takes 3–12 days. Nymphs usually feed on mice or voles and become engorged in 3–11 days; molting takes 3 weeks to 1–3 months. Adults usually attack dogs and other large animals, including humans; females engorge in 6–13 days. Mating takes place on the host. Unfed adults live for about 2 years. This species is distributed east of the Rocky Mountains and in California, Mexico, and Canada. It is particularly abundant along the east coast of the USA. It is a vector of Rocky Mountain spotted fever in eastern USA, where infested domestic and feral dogs bring infected ticks to peridomestic habitats.

**Other *Dermacentor*** Ticks of this genus are widely distributed; they occur in North and South America, Europe, Eurasia, and Asia. *D. andersoni*, the Rocky Mountain wood tick, is distributed in western North America. It occurs in shrubby or open areas, and the primary hosts are large animals. *D. albipictus* is widely distributed in North America and hosts include horse, moose, elk, and deer. Eurasian species include *D. marginatus*, *D. silvarum*, and *D. nuttalli*. Lowland forests and shrubby areas from Kazakhstan to central Europe are the habitats for *D. marginatus*; *D. silvarum* ranges from Kazakhstan to western Siberia, and *D. nuttalli* ranges from central Siberia to Mongolia, and China south to Tibet. *D. reticulatus* has been introduced into southern UK.

**Rabbit tick, *Haemaphysalis leporipalustris*** Females are 11.3 mm long and 7.5 mm wide when engorged. The body is narrow anteriorly and broad posteriorly, and coarsely punctate. Hosts are rabbits, birds, and occasionally domestic mammals such as dogs and cats. It rarely feeds on humans but it is sometimes found in houses, and is important in the spread of Rocky Mountain spotted fever. This tick occurs in Central and South America, continental USA, and Canada.

**Other *Haemaphysalis*** Members of this genus are usually small and the sexes are similar; they parasitize wild and domestic animals and are distributed worldwide. *H. parva* (=*H. otophila*) occurs on dogs in Jerusalem, Israel. *H. leachii*, the yellow dog tick, is common in Asia and Africa and is a vector of malignant jaundice in dogs. In South Africa, urban cases of boutonneuse fever caused by *Rickettsia conori* are associated with contamination of skin or eyes with infected ticks crushed while being removed by hand. *H. spinigera* is a vector of the arbovirus causing Kyasanur forest disease in monkey and humans in India. The immatures feed primarily on small forest rodents and monkeys. Humans are bitten during the dry pre-monsoon season when villagers enter the forest to gather firewood.

**English dog tick, hedgehog tick, *Ixodes hexagonus*** Males are 3.5–3.8 mm long; females are 3.5–4.0 mm long, and fed females are 13 mm long. Oviposition takes place at 3–30 °C, and about 1% of the eggs may develop parthenogenetically. Development from egg to adult in the laboratory at 22–23 °C takes about 60 days; under natural conditions development may last 2 years. This tick is a parasite of carnivores, such as foxes, dogs and cat, and the hedgehog (*Erinaceus europaeus*) in the urban environment. It is common in suburban locations in central Europe. As an endophilic tick, it spends most of its time inside the burrows or nests of its hosts, and does not occur in buildings. It is a vector of tick-borne encephalitis (TBE) virus and *Borrelia burgdorferi*; the hedgehog is a reservoir host for both these diseases. This species is widely distributed in Europe, North Africa, and southern Asia.

**Australian paralysis tick, *Ixodes holocyclus*** Males are 1.9–2.7 mm long and 1.6–2.0 wide. The body is yellowish brown, the dorsal plate is glossy and with numerous fine punctuations. Unfed females from Queensland and New South Wales are 2.6–3.8 mm long; those from Victoria are 4.3–4.6 mm long. Engorged females are 12 mm long and grayish green or greenish black. The female dorsal plate is small and covers about one-third of the abdomen; the male has a large plate that covers the abdomen. Females lay a mass of about 3000 eggs in damp locations on the ground; hatching occurs in 40–60 days. Larvae are about 1 mm long and feed on small, native mammals such as a bandicoot (*Parameles* spp.), kangaroo, or opossum; larvae feed for 4–6 days. Nymphs are about 2 mm long and feed on a variety of animals. They remain on the ground in moist vegetation for 14–40 days before finding a host; they feed for 4–7 days. Adult female ticks engorge for 6–20 days; males mate with females on the host but do not feed. This species is mainly distributed along the eastern coast of Australia, from Queensland to Victoria. It is also abundant on the central coastal plain, from Kempsey to Wollongong.

Symptoms of paralysis caused by *I. holocyclus* appear 3–7 days after the tick has attached and begun to feed. In humans the symptoms include severe headache, blurred vision, weakness in limbs, and increasing paralysis; after about 4 days, breathing becomes difficult. In domestic animals, the symptoms include loss of appetite, discharge from the eyes, difficulty walking, especially in the hind legs, vomiting, and dilated pupils. *I. holocyclus* is a vector of Queensland tick typhus, *Rickettsia australis*.

**European sheep tick, *Ixodes ricinus*** Adults are about 11 mm long. They are regularly oval, pale and reddish brown, with short, white hairs. It attacks a variety of vertebrate hosts. The free-living stages only survive in microclimates where the RH is high (80%), which includes leaf litter in temperate forested areas. In the urban environment, it occurs in naturally forested parts, gardens, and cemeteries. It occurs in some urban areas in Czech Republic, Germany, Austria, and the UK. Hedgehogs and foxes are the primary hosts for this tick in urban areas in central Europe. It is the most common vector tick in Europe and transmits a large number of human pathogenic organisms, including several genospecies of *Borrelia burgdorferi*. This tick is cosmopolitan, and occurs in North America, from Ireland and the UK to western and central Europe, Russia, and from Algeria and Morocco east to northern Iran.

**Black-legged tick, deer tick, bear tick, *Ixodes scapularis* (= *I. dammini*)** Females are about 3.5 mm long and males are about 2 mm long. The body is uniformly orange brown and it has a dark brown to black spot anterodorsally. The mouthparts are long and the dorsal plate of the male nearly covers the abdomen. This species is polyphagous and attacks a variety of hosts, including at least 27 species of birds, several small and large mammals, and humans. Immature stages commonly feed on the white-footed mouse (*Peromyscus leucopus*), and this mouse is the primary reservoir for Lyme disease in northeastern USA. Lyme disease is transmitted from mouse to mouse and mouse to human by larvae and nymphs of *I. scapularis*. White-tailed deer (*Odocoileus virginianus*) is the primary overwintering site and the main host for the adult tick. Adult ticks do not move from host to host and are not responsible for transmission of Lyme disease. The distribution range of this tick is expanding to urban and suburban areas, and this may be linked to the proliferation of deer in urban environments. It is distributed in northeastern USA and adjacent regions of Canada, and it is the primary vector of Lyme disease in this region. *Ixodes dammini* was at one time considered a distinct species and the main vector of Lyme disease in northeastern and northcentral USA.

**Other *Ixodes*** This genus is distributed worldwide, and several species occur in the urban and suburban environment along with their hosts. The Pacific coast black-legged tick, *I. pacificus*, is the primary vector of Lyme disease in western USA. The groundhog tick, *I. cookei*, is common in eastern USA, where it occurs from New England states south to Virginia. Nymphs and adults bite humans, and they occur in outbuildings or other structures frequented by rodents, such as groundhogs. *I. persulcatus* vectors Lyme disease in Russia. *I. redikorzevi* occurs in villages and small towns in Israel, and bites people.

**Relapsing fever tick, *Ornithodoros hermsi*** Adult females are 5–6 mm long and 3–4 mm wide and light brown. The engorged female is about 11 mm long and reddish brown to grayish blue. The integument of adult male and female is strongly tuberculated. Larvae survive 3 months and adults survive 7 months without food. Development from egg to adult is 202–314 days at 30 °C and 364–602 days at 21 °C. This species occurs in the mountains (1500 m) of California northward to British Columbia. It is a vector of relapsing fever in the mountainous area of California and in some other western states.

**Eyeless tampan, *Ornithodoros moubata* (Fig. 18.5d)** Adults are about 8 mm long and 6–7 mm wide; engorged females are 11 mm long. The integument of nymphs and adults is densely tuberculated and without spines. Mouthparts are small, and eyes are lacking. Females lay more than 400 eggs in six or seven batches; hatching occurs in about 8 days at 30 °C. Larvae emerge from the egg with food in their gut and probably do not feed. Nymphs feed four or five times. Adults feed six or seven times on different hosts. After each meal females deposit eggs; the preoviposition period is 5–25 days. This soft tick is active at night and it occurs as two forms. *O. moubata moubata* is found indoors, and feeds on humans and chickens; *O. moubata porcinus* occurs outdoors in warthog (*Phacohoerus* spp.) burrows and generally does not attack humans. *O. moubata moubata* is common in rest-houses on travel routes, and can be carried in household materials, such as mats and bed-rolls. Both subspecies have wild and domestic populations, and the domestic populations are linked to disease transmission to humans. This species is widely distributed in East, Central, and southern Africa; it is a vector of relapsing fever.

**Other *Ornithodoros*** This large genus is distributed in Africa, the Middle East, North and Central America. *O. coniceps* occurs in Israel where it has infested buildings and bitten people. *O. talaje* occurs in Central and South America and feeds on wild rodents, swine, cattle, and humans. It is a vector of relapsing fever. *O. rudis* occurs in Central and South America. It feeds primarily on humans, and regularly occurs indoors. The pajaroello, *O. coriaceus*, is a large tick (engorged females are about 20 mm long) that normally feeds on deer and cattle, but delivers a painful bite to humans. It occurs in the coastal areas of California, Mexico, and South America.

**Brown dog tick, kennel tick, *Rhipicephalus sanguineus* (Fig. 18.5c)** Males are about 3 mm long, flattened, uniformly reddish brown, and the body surface is punctate. Engorged females are about 12 mm long and the body is grayish blue to light green. Females lay a mass of 1000–3000 eggs in cracks and crevices above ground. Hatching occurs in 19–60 days, but cool and dry weather prolongs hatching. Larvae are about 2 mm wide (engorged) and light brown. They crawl on walls indoors, and attach themselves to domestic animals and people; they can survive about 8 months without food. Nymphs are about 3 mm diameter and dark gray. They feed for 4–9 days, and then drop off the host and molt to adults in 12–29 days. Adults attach to dogs or other animals, and engorge for 6–50 days; they live for 18 months before attachment and feeding. Development from egg to adult can be completed in about 63 days at 29 °C. This tick prefers warm and dry situations and seldom develops outdoors in continental USA. It is native to Africa, but it has successfully spread to much of the world. It survives in heated buildings in urban environments in countries between 50 °N and 35 °S. Populations of this tick have increased in urban areas in many parts of the world, along with increases in ownership of pet dogs and increases in feral dogs. *R. sanguineus* can complete its life cycle on dogs; it is well adapted to urban environments. It requires high temperatures for development, and in temperate regions it is successful indoors. This tick is a vector of Rocky Mountain spotted fever in southern USA, and with boutonneuse fever in the Mediterranean region of Europe.

**Other *Rhipicephalus*** Species of this genus occur in Eurasia and Africa, and from its homeland, *R. sanguineus* has spread to other parts of the world. *R. turanicus* occurs in shrub and tree habitats in some towns in Israel, along the coastal plain.

## Amblypygi

These are the tailless whip-scorpions. They are 8–45 mm long and somewhat flattened; they superficially resemble crab spiders. The carapace has a pair of median eyes, and three pairs of lateral eyes. Pedipalps are large and have strong spines, and they terminate in a movable hook. The first pair of legs is usually long and is used as a tactile sense organ. Ambulatory legs of some species have well-developed pulvilli, which enables them to climb vertical surfaces. The 12-segmented abdomen terminates in a short, rounded pygidum. These arthropods are distributed in the tropics and subtropics. They are predators of cockroaches, crickets, grasshoppers, termites, and other arthropods.

Several species are common indoors in tropical regions, including *Phrynichus ceylonicus* in India, *Paracharon caecus* in West Africa, and *Masicodamon allanteus* in Morocco. *Damon variegatus* occurs in buildings and unused outbuildings, and beneath stored furniture and packing cases in South Africa.

## Amphipoda

These crustaceans resemble shrimp and are variously called scuds, beach-fleas, or sand-fleas. Their body is elongate and laterally compressed. The thoracic segments are large and dominate the body; seven of these segments have leglike appendages. The abdominal segments are small and compressed. This group contains marine species, such as the sand-flea or beach-flea, *Orchestia agilis*, and a sea scud, *Gammarus annulatus*. The freshwater forms include *Dikerogammarus* spp. and *Hyalella* spp.

***Talitroides sylvaticus*** Mature individuals are about 8 mm long. Body color ranges from greenish brown, dark brown, to blackish brown, but it often turns pale red to coppery when dead (and it is dead specimens that are most often encountered). Populations of these small amphipods live in damp and shaded locations, and when disturbed they jump like fleas or springtails. Excess moisture may stimulate them to move to dry locations, which includes sidewalks, patios, and sometimes indoors (ground-level rooms). This species occurs in Hawaii, coastal California, and frequently in states along the Gulf coast.

## Araneae

Spiders are the dominant arachnids in most habitats. This group surpasses all others in number and variety of species, in complexity of behavior patterns, and in the range of habitats occupied. They are similar in morphology and behavior to some of the other arthropods, including the tailless whip-scorpions (Amblypygi) and pseudoscorpions (Pseudoscorpiones). The cephalothorax and abdomen are separated by a waist, which is formed by the constriction of the pregenital segment. The abdomen is composed of 11 segments. Two pairs of book-lungs open between the sternites of abdominal segments 1 and 2. The abdomen has specialized glands for producing and manipulating silk, which plays an important role in the foraging and dispersal of these arthropods. Spiders have poison glands in the chelicerae (mandibles); the second segment of the chelicerae forms a sharp, piercing fang with an opening at the tip to release the poison. Most spiders can give a poisonous bite, but few have mandibles strong enough to penetrate human skin, and when they do, the venom is usually harmless.

The food of spiders includes a variety of other arthropods, including sowbugs, millipedes, pseudoscorpions, harvestmen spiders, and other spiders. As predators their potential food is variable, but most species are adapted to a size and frequency of prey that occur in their habitat. Some attack large insects, while others capture small species. Some spiders capture diurnal insects, others hunt at night; some specialize in crawling insects, others in those that fly. The majority of spiders in the urban environment live around stones or leaf litter on the ground. Some excavate burrows in loose soil, while others find harborage under or around stones or logs, or in the moist layer of decaying leaves and plant litter. Some of the small-sized species are always wandering and may not construct a silken retreat. Large-sized species usually live permanently in a burrow and forage close to this site. Species that use large webs to trap their prey usually construct a silken retreat at the edge or close to the web. Vibration of the silk strands that connect the retreat with the web enables the spider to respond to a struggling prey. The daily life of web-building spiders centers on their ability to detect and respond quickly to vibrations in their web.

Digestion of food is initiated outside the body. After the prey has been immobilized by a poisonous bite or wrapped with silk, digestive fluid from the gut of the spider is applied to the victim. Within a short time, the spider sucks in the digestive fluid and the predigested body of the prey. This process is repeated until the prey has been consumed. Specific feeding behavior depends on the presence of cheliceral teeth. Theridiid and thomisid spiders lack cheliceral teeth and make only a small hole in their prey. Digestive fluid is passed in and out of this hole and the dissolved tissue is gradually consumed. After the meal the prey is an empty shell. Spiders with cheliceral teeth are capable of crushing their prey and consuming the predigested contents. After the meal the prey is an unrecognizable mass of small pieces of cuticle.

Spider venom is a mixture of substances, mostly of neurotoxic polypeptides with molecular weights of 5–13 kDa. Biogenic amines and proteolytic enzymes are also present. Poison glands are a feature of nearly all spiders, and they are all potentially harmful, at least to their prey. Only 20–30 of the 30 000 total species are dangerously poisonous to humans. The bite of the black widow spider is extremely painful at the site of the bite, but the physiological effects away from the site are potentially the most dangerous. These include accelerated heart beat, increased blood pressure, and paralysis of the diaphragm muscles, which often results in suffocation. The poison of the black widow spider consists of seven different proteins and is a neurotoxin that affects neuromuscular synapses. Bites of tarantulas (Theraphosidae) are relatively harmless, in spite of their reputation; generally, the bite is no worse than a wasp sting. The European tarantula (*Lycosa tarentula*) has a long history of causing severe and sometimes unusual reactions in people bitten, but the symptoms (including erratic dancing) of tarantism may be related to other causes or another spider. The bite of some tropical lycosid spiders, such as *Lycosa erythrognatha* of Brazil, can cause tissue lysis and necrosis at the site of the bite.

All spiders are capable of producing silk. They all possess spinning organs, which they use to form silk into egg-sacs, draglines when walking or hunting, and for building snares or webs to trap prey. Silk is a proteinaceous substance, fibroin; the amino acids alanine, glycine, and serine constitute 50–60% of the total fibroin. Silk of the orb weaver *Nephia* has a molecular weight of about 30 kDa as a liquid in the silk gland, and between 200 and 300 kDa as a solid silk thread. The tenacity of spider silk may be compared to that of nylon. The dragline of the garden spider, *Araneus spp.*, has a tenacity of 7.8 g/denier; the corresponding value for nylon is 8.7. Spider silk can be stretched by 31%, compared with 16% for nylon; the length at which it breaks under its own weight can be as much as 80 km. The silk from different spider species is slightly different, and the silk from different glands of the same spider is different. Silk from pseudoscorpions (*Neobisium maritimum*) differs from that of spiders, primarily by the smaller amount of glycine and alanine compared to spider silk.

Mating procedure and the premating behavior of males and females are elaborate, and generally species-specific. The mature male prepares a small pad of silk on which a small amount of sperm is deposited. This is taken up by the modified pedipalps, which are later inserted into the genital opening of the female to deposit the sperm. Males of each species have distinctly shaped palps, and a courtship that results in successful transfer of sperm to the female. Depending on the species, males and females mate more than once. Females lay their eggs in silken sacs, and often carry or remain with the eggs until they hatch. The number of eggs laid is highly variable, but the majority of species lay about 100 eggs. Those that produce more than one egg-sac usually lay a small number of eggs in each, and large-bodied species tend to lay more eggs than small species. After hatching, the spiderlings remain together for several days without feeding, then molt and disperse. The number of molts during development is variable. In general, small spiders molt four or five times; medium-sized spiders molt seven or eight times; and large spiders molt more than 20 times. In the majority of species the life cycle varies from 8 months to 4 years. Spiders in temperate regions usually live 1 year, while those in tropical regions usually live 2–3 years.

There have been several attempts at establishing a practical and perhaps phylogenetic classification of these spiders. However, with at least 30 000 described species and differing views on what internal or external characters should be used to show relationships, no system has sustained wide acceptance. Only family and species names are used here, and the alphabetical arrangement removes any phylogenetic implication.

### Actinopodidae

These spiders have an enlarged cephalothorax and the eyes are located at the front. The bases of the chelicerae are enlarged and project forward. The palps are long and leg-like.

**Redheaded mouse spider, *Missulena insignis*** Males are 12–18 mm long and females are 20–25 mm long. The body is black, but the chelicerae and anterior portion of cephalothorax are red. Mouse spiders occur throughout Australia. A related species, *M. bradleyi*, is often found in swimming pools on coastal New South Wales, Australia.

### Amaurobiidae

These spiders are large; some individuals are 15 mm long. In general, they occur in cool, moist, poorly lighted locations, such as in hollow logs, cracks in cliffs, and in cellar walls. The webs are irregular.

***Amaurobius ferox*** Females are about 15 mm long and the body is light brown to brown. At the base of the abdomen there are three yellowish-white longitudinal bands, and on the posterior of the abdomen there are four pairs of light-brown spots. Natural habitats include under logs and stones. Indoors it occurs in cellars, under floors of houses, and in outbuildings. This species is widely distributed in Europe and the USA.

### Amphinectidae

This family is represented by several genera that occur in Australia, New Zealand, and South America. Some genera in the family Amaurobiidae have been transferred to this family.

***Metaltella simoni*** Males are 7–8.5 mm long and females are 8–9 mm long. Male carapace and legs are yellow to yellowish orange; the carapace is dark anteriorly and the legs are dark distally. Female carapace and legs are brown. The abdomen of both sexes is mottled gray with four irregular white stripes on the venter. Natural habitats include under logs in forested areas, under bark, and under boards. In the urban environment, it occurs in outside crevices and corners of buildings. Males may be found wandering indoors. This species is native to Uruguay and Argentina. *M. simoni* was first recorded in the USA in Louisiana in 1994. It has spread to Alberta, Canada and southern California (Riverside, San Bernardino county, San Diego county).

### Angelenidae

These are the funnel-web spiders. They have three claws and the tarsi have a single row of erect sensory setae, trichobothria; the hind spinnerets are very long. The members of this family make sheet-like webs, which usually have a tubular retreat. These spiders often move over the surface of their web in an upright position.

The *Tegenaria* house spiders are closely associated with buildings, and several species are known to bite people indoors. In North America the bites of several species, especially *T. agrestis*, produce necrotic lesions and usually result in a scar. In the UK and continental Europe they bite humans, but without the adverse effects that occur in North America. There are no records of these spiders in Europe causing necrotic lesions. It is possible that the biotype of *T. agrestis* occurring in the Pacific Northwest is particularly venomous. The necrotic arachnidism syndrome that accompanies a *Tegenaria* bite includes severe headaches and joint pains that may last 7–10 days, low (less than 150 000/mm$^3$) blood platelet count, and a cratered scar.

**Grass spider, *Angelenopsis naevia*** Males are 13–18 mm long and females 16–20 mm long. The male palpus is greatly elongated, and the legs are distinctly banded. The abdomen is usually dark brown, and markings are obscured. Eggs are produced in fall and hatch in spring; adults die in fall. If a spider is not disturbed, it occupies the same web throughout the summer, extending it until it reaches 30 cm or more. The web is firmly attached to the grass, and there is an irregular network of threads above the sheet, supported by stalks of grass. The objective of the network is to impede the flight of insects, causing them to fall upon the sheet. In fall the males wander in search of females, and females leave their webs to seek a place to secure their egg-sacs. These spiders build their horizontal sheet webs on grass, bushes, stone fences, in the corners of buildings, and on elevated structures. This species occurs in the USA, from New England south to Florida and west to Kansas and Texas.

***Coras medicinalis* (Fig. 18.6d)** Males and females are about 12 mm long. The carapace is yellowish brown, dark brown anteriorly, and with gray bands. The abdomen is gray with dark markings. Legs have indistinct gray rings. Chelicerae are large. Natural habitats include hollow trees, tree stumps, and crevices among rocks. In the urban environment it occurs at the corners of buildings. This species is distributed throughout continental USA.

**Aggressive house spider, hobo spider, *Tegenaria agrestis*** Males and females are about 40 mm long. Anterior and posterior rows of eyes are in a straight line. Legs are uniformly brown, without bands. The abdomen is mottled brown and white, with four distinct pale-brown chevrons medially. Eggs are laid in fall and hatch in spring. Females produce egg-sacs in fall and often remain with them through the winter. Spiderlings molt at least once in the egg-sac before emerging. Spiderlings remain close to the egg-sac and build a communal web; they disperse in 2–5 days. Spiders molt 12–15 times before becoming an adult.

Habitats occupied in the urban environment include rock walls, discarded wood and firewood piles, and along house foundations. Adults often bite people without provocation, and for this behavior it is called the aggressive house spider. Males and females build webs, and both may be found on a single web in late summer and fall. Mature males leave their web in fall to search for females. Females enter houses during fall and winter after exposure to winter cold. Juveniles enter buildings in spring when they are searching for web sites. Males enter buildings in fall, but are usually restricted to basements and ground-floor rooms. They are poor climbers and are rarely found upstairs or on walls. Males are more venomous than females, and they are responsible for most human bites. *T. agrestis* venom is a necrotic type that causes tissue death and sloughing in the immediate area of the bite. The lesions that result from the bite take up to 6 months to heal. Pet dogs and cats are bitten on the head (face), legs and chest; death may occur. The current known range of this spider is from southern British Columbia to middle Montana, south through west-central Wyoming into northern Utah (east of Great Salt Lake) and into southern Oregon. All of Idaho and Washington is colonized.

**Domestic house spider, *Tegenaria domestica* (Fig. 18.6a)** Males are 6–9 mm long and females are 7.5–11.5 mm long. The carapace is pale yellow with two gray stripes. The abdomen has irregular gray marks. Legs are long and distinctly banded. The web is a horizontal sheet with a retreat at the edge. Eggs are laid through the warm season and spiderlings hatch in about

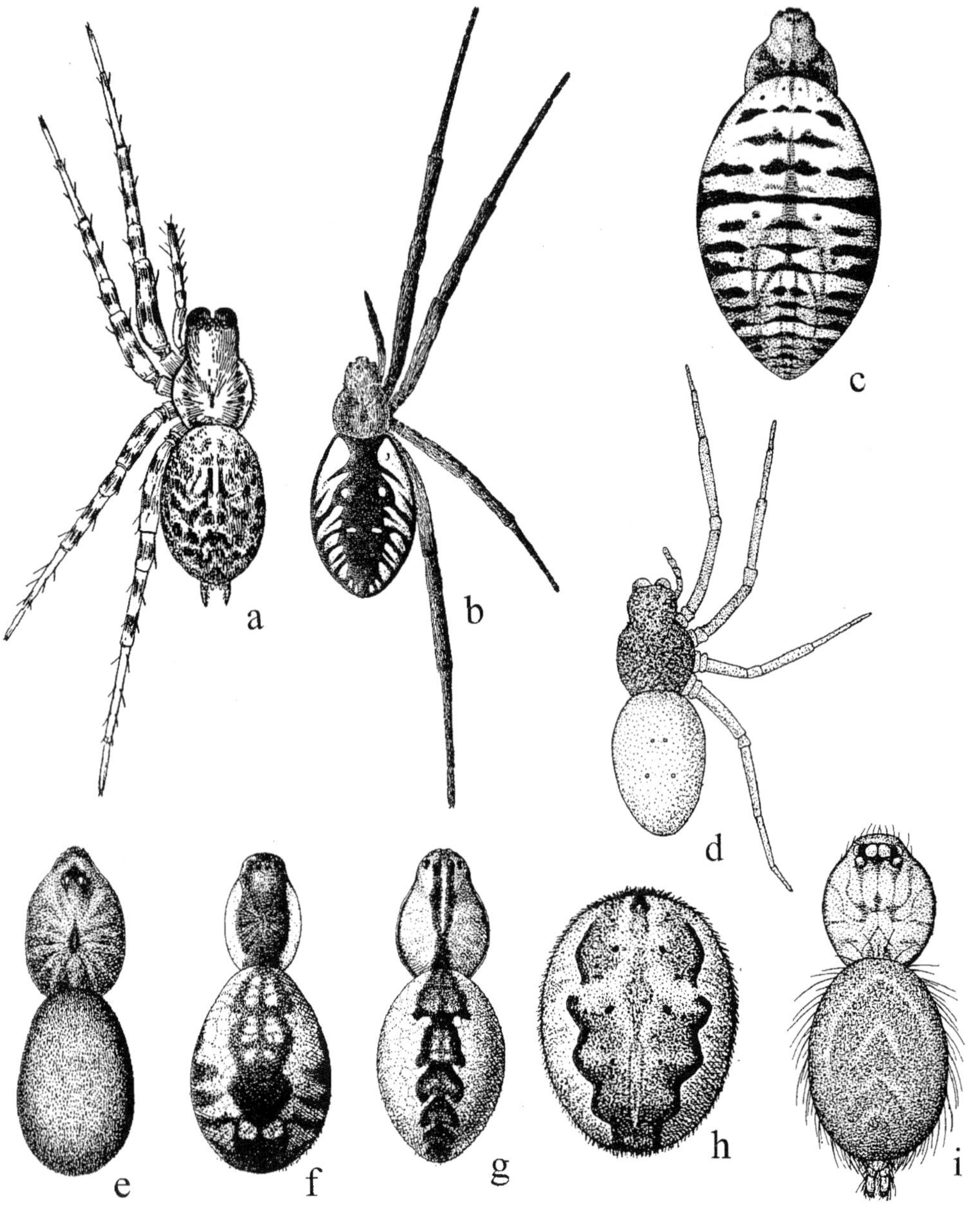


**Figure 18.6** Arthropoda: Aranea. (a) *Tegenaria domestica*; (b) *Argiope aurantia*; (c) *A. trifasciata*; (d) *Coras medicinalis*; (e) *Kukulcania hibernalis*; (f) *Linyphia marginata*; (g) *Pityohyphantes phrygianus*; (h) *Araneus cornutus*; (i) *Orchestina saltitans*.

39 days at 21–23 °C. Spiderlings molt at least once in the egg-sac before emerging. Spiderlings usually remain close to the egg-sac and build a communal web. Spiders molt after 1 week, and again in about 1 month. Numerous molts occur before becoming an adult. Natural habitats include under stones and in rock crevices; in the urban environment they occur in barns, cellars, and dark corners of rooms. Indoors, females usually remain at the funnel web, on the floor, and rarely on walls. Males wander in the house, searching for food and females. Adults live for several years, and males and females usually occur together on the same web. This species is distributed around the world, from the tropics to north temperate regions.

**Giant house spider, *Tegenaria gigantea*** Males are 11–14 mm long and females are 15–18.5 mm long. At rest the spider is nearly 50 mm across the legs. The carapace is yellowish brown with dark median stripes and a butterfly pattern at the posterior; the margin of the carapace is dark brown. The abdomen is dark brown with light pale markings; the chevrons are indistinct. Natural habitats include under stones and in rock crevices. Eggs are laid through the warm season and spiderlings hatch in about 39 days at 21–23 °C. Spiderlings molt at least once in the egg-sac before emerging. Spiders molt after 1 week, and again in about 1 month. Numerous molts occur before becoming an adult. This species is distributed in the Pacific Northwest of the USA and British Columbia.

**Cardinal spider, *Tegenaria parietina*** Adults are about 20 mm long; including the legs they are about 43 mm long. The cephalothorax has a reddish-brown stripe between yellowish-white spots with brown to brownish-black borders. This spider (named for Cardinal Wolsey, Lord Chancellor for Henry VII) occurs in southern England; there are a few records of this spider in Ireland.

**Other *Tegenaria*** Several species in this genus are common indoors in North America and Europe; there are 11 species in the UK and northern Europe, and seven of these occur in the UK and Ireland. *T. chiricahuae* is the only *Tegenaria* native to the USA, but it is not widely distributed. *T. atrica* adults are about 19 mm long and the cephalothorax has a brownish-green (khaki) stripe with lateral markings. It is common in houses and sheds in the UK, from the Scilly Islands northwards. *T. larva* is common in the UK and Ireland, and there are limited records for this species in Iceland.

### Araneidae

These are the orb weaver spiders: they spin webs in the form of an orb. Some adults remain at the center of the orb, while others build a retreat away from the web. In many species there is considerable sexual dimorphism; the males are often smaller and have a different-shaped abdomen than the female.

***Araneus cornutus* (Fig. 18.6h)** Males are 7–9 mm long and females are 9–12 mm long. The carapace is brown to gray. The abdomen has a distinct scalloped pattern, which is grayish brown, with pale areas within, and dark brown to black at the sides. Webs are built on low bushes, bridges, and around the outside of buildings. This species occurs in the USA, from New England south to North Carolina and west to Oklahoma and Nebraska.

**Black and yellow garden spider, *Argiope aurantia* (Fig. 18.6b)** Males are 5–8 mm long and females are 19–28 mm long. The carapace is usually covered with white or silvery-white setae. The abdomen is marked with black and bright yellow, or orange; it is slightly pointed posteriorly and curved at the sides to form a hump on each side. The front legs are entirely black; others have reddish-brown or yellow femora and the other segments black. Webs are constructed on vegetation, tall grass, and around houses (near outdoor lights). They usually occur in locations such as on porches or garden furniture. Webs often contain a zigzag stabilmentum of thick silk, which extends above and below the center of the web. The egg-sacs are light-brown spheres, 20–25 mm in diameter. This species occurs throughout USA.

**Banded garden spider, *Argiope trifasciata* (Fig. 18.6c)** Males are 4–5 mm long and females are 15–25 mm long. The carapace and abdomen are pale yellow with a thick covering of silver and white setae. Legs are pale yellow, and the femora and front legs may be entirely black. The abdomen is marked with black bands; the legs are spotted. The tip of the abdomen is somewhat pointed; the anterior margin lacks the notch and humps that characterize *A. argiope*. This species is generally distributed in the USA; adults are common in early September.

**St. Andrews cross spider, *Argiope keyserlingii*** Males are 5–6 mm long and females are 10–15 mm long. The cephalothorax is brown, and the abdomen is striped yellow and brown. The stabilmentum in its web is the shape of a cross. This species occurs around the outside of buildings in Australia.

***Metepeira globosa*** Females are about 8 mm long. The cephalothorax is brownish yellow with a dark longitudinal median line. The abdomen is oval with four pale yellow or pink spots that are partly united into a rectangular mark and surrounded by a black line. Legs are banded in both sexes. Webs are usually a composite of a vertical orb and an irregular and unstructured web that extends from the top of the orb, covering about half of it. There is a strong thread extending to the center of the orb. This species occurs in the USA.

***Neoscona domiciliorum*** Males are 6–8 mm long and females are 10–12 mm long. The abdomen is shaped slightly triangular; markings are yellow and brown, paired with black spots. Tibia 2 of the male is straight with strong setae. This species occurs in open woods, on fences, and around buildings. It is distributed throughout the USA.

***Neoscona nautica*** Males are 6–7 mm long and females are 8–12 mm long. The cephalothorax is greenish brown to yellowish brown. The abdomen is enlarged and slightly triangular; markings are yellowish brown and brown, paired with black spots. Legs are yellowish brown with dark brown tarsi. This species is distributed in Japan.

**Nook spider, *Zygiella x-notata* (= *Zilla litterata*)** Adults are about 10 mm long. The cephalothorax is yellowish brown and has a broad, yellow wedge-shaped stripe. Webs are often constructed in the corner of a window, doorframe, or post. The orb is incomplete and usually a small section remains open; through this section is a single thread that extends to the retreat. This species is distributed in continental USA and Europe. A closely related species, *Z. atrica*, occurs on the outside of houses.

### Clubionidae

These are two-clawed hunting spiders that are commonly found on the ground or on vegetation. They make flattened, tubular retreats in rolled leaves, under stones, and in leaf litter.

**Yellow sac spiders, yellow house spiders, *Cheiracanthium mildei*, *C. inclusum*** Males are about 6 mm long and females are about 9 mm long. The body of *C. inclusum* is light yellow, while the body of *C. mildei* is light green to yellowish white. The median strip on the abdomen is slightly darker than the remainder of the abdomen. Front legs are longer than the hind legs. Chelicerae are brown. The egg-sac is white and papery, and it is usually attached to the underside of objects. Females remain with the eggs until hatching. Juveniles overwinter and molt to adults in the spring; mating occurs in June and July. Natural habitats include leaf litter and around objects on the ground. In the urban environment, these spiders enter houses around doors and windows, heating ducts, and plumbing. They are most common indoors in fall and spring. Inside houses they are found on walls and in corners close to the ceiling; they drop from ceilings on silk threads. Most reports of yellow sac spiders indoors involve *Cheiracanthium mildei*, which was introduced into North American from Europe (and is distributed in Europe and the Middle East), and *C. inclusum*, which is native to North America. These two species are very similar in appearance and are generally known as sac spiders, but *C. inclusum* has also been called the running spider.

**European yellow sac spider, *Cheiracanthium punctorium*** Females are about 12 mm long. The cephalothorax is light brown to brown, while the legs are light brown. The abdomen is yellowish green and pale brown; there is a median brown stripe bordered by yellowish green. In the urban environment habitats include webs on walls, and in corners of rooms. This species is a common and medically important house spider in Europe. The bite of males and females produces symptoms such as intense local pain, itching, vomiting, enlargement of lymph nodes, and fever. This species occurs in Europe, primarily in the Mediterranean region; it is not known to occur in the UK. A closely related species, *C. japonicum*, occurs in Japan and is commonly found indoors.

***Trachelas tranquillus*** Males are 5–6 mm long and females are 7–10 mm long. The chelicerae and carapace are reddish brown and the cuticle has tiny punctures. The abdomen is pale yellow to light gray, slightly darker anteriorly and medially. Legs 1 and 2 are larger and darker than the other two pairs. Natural habitats include webs in rolled tree leaves; they invade buildings in the fall. It has been reported to bite people, and for mild to serious reactions to occur. This species occurs throughout the USA.

### Ctenidae

These spiders resemble wolf spiders (Lycosidae), and they are usually called running spiders or wandering spiders. The Ctenids are characterized by their eye arrangement: eight eyes in three rows, in a 2–4–2 pattern. In wolf spiders, the eyes are arranged in a 4–2–2 pattern.

**False wolf spider, *Ctenus captiosus*** Males are 10–14 mm long and females are 15–20 mm long. The male carapace is yellowish brown, with a wide pale median stripe that narrows posteriorly; the submarginal stripes on the carapace are gray. The abdomen is yellowish gray except for a pale gray median stripe consisting of a series of connected triangles. Female coloration is similar to male, except generally darker. This species wanders to catch prey and does not make a web. Natural habitats include wooded areas of oak and pine trees, in leaf litter, and in caves. In the urban environment, they enter houses in moist, woodland habitats. This species occurs in the USA, primarily Florida and southern Georgia.

***Phoneutria fera*** Males and females are about 40 mm long and brown to dark brown. Legs have some banding and strong spines. Chelicerae are reddish brown; the middle pair of eyes in the posterior row is larger than those in the front row of eyes. This spider does not make webs but wanders and hunts prey. Natural habitats include low vegetation and around stones and litter on the ground; they are common in banana plantations. It naturally occurs indoors. It is occasionally imported into countries outside its range, including the USA and the UK. This species is known to occur in Brazil, where there are cases of bites to young children, but fatalities from these bites are rare.

### Desidae

These large spiders have large chelicerae and an enlarged abdomen, and typically their legs are banded. They occur in leaf litter and under loose bark in natural habitats; a few species invade buildings.

**Black house spider, *Badumna insignis*** Males are 8–10 mm long and females are 15–18 mm long. The female is dark brown to black and the abdomen is mottled with light markings. Legs are black and banded at the segment joints. The web is a mat with a round entrance to the retreat. Natural habitats for this species include ground litter and under tree bark. In the urban environment they occur in sheds, outdoor toilets, under guttering and in various habitats indoors. This species occurs in urban and rural areas in Australia.

### Dictynidae

Species are distinguished by their long mandibles, and in males, the length may equal the length of the cephalothorax. Males are usually more distinctly marked than females. They construct irregular webs, with a characteristic zigzag pattern of silk threads. This family includes the large genus *Dictyna*, which has worldwide distribution. Webs of *Dictyna* are often found at the corners of windows or on walls, with the retreat hole near the center.

***Dictyna sublata*** Males are 2–2.5 mm long and females are 2.3–3.7 mm long. The female cephalothorax is yellow in front and brown posteriorly. The abdomen usually has a broad median yellow stripe, and is brown on the sides. The male lacks the distinct color differences present in the female cephalothorax. This species builds sheet-like webs on the sides of buildings, and the retreat may be at a corner of the web. It occurs throughout continental USA.

### Dipluridae

These are the funnel-web tarantulas. They are characterized by a generally flat cephalothorax, and there is a round pit medially in the center of the cephalothorax. The posterior spinnerets are very long and composed of three segments of about equal length.

**Sydney funnel-web spider, *Atrax robustus*** Males and females are about 35 mm long and the body is overall dark brown to black. The cephalothorax and legs are shiny, the abdomen is densely setose, and the surface is dull. The spinnerets are long, nearly half the length of the abdomen. Chelicerae are large and project forward. Natural habitats include under rocks, in loose soil, and among debris; it lives in a silken tube, which is not funnel-shaped. In the urban environment it lives in debris around the outside of buildings and indoors. This species occurs in Australia. It is common in the suburban area of Sydney and has spread into Queensland and south to Victoria. The venom of the male is considered more dangerous than that of the female. Symptoms following a bite of *Atrax* include pain at the site. It soon extends to other parts of the body; sweating, delirium, nausea, and vomiting are followed by respiratory failure. Closely related and morphologically similar species, *A. formidabilis* and *A. venematus*, are not common in urban environments.

### Filistatidae

These spiders have an oval cephalothorax, which is longer than broad. Eyes are in a group, and the anterior median eyes are dark and round; the others are opaque white and oval or angular. Chelicerae are fused together at the base. These spiders are distributed in the tropics and subtropics.

**Southern house spider, *Kukulcania hibernalis* (= *Filistata*) (Fig. 18.6e)** Males are 9–10 mm long and females are 13–19 mm long. Eyes are close together on a raised prominence. Leg 1 is about twice the length of the body. The male is yellowish brown, while the female is dark brown with irregular dark markings on the abdomen. The male palpus is long and nearly equal in length to the femur and tibia. In natural habitats this spider lives on the ground under stones; in the urban environment they build webs on the sides of buildings, especially around cracks and other narrow openings. The web is nearly circular in outline, and composed of zigzag bands of silk; in the center is the opening of the retreat of the spider. The web may accumulate household dust and be difficult to see. This species occurs in southern USA west to California.

### Gnaphosidae

In these spiders the cephalothorax and the abdomen are about the same size and somewhat flattened, and the legs are stout and nearly the same length. Many species are uniformly colored. They spin a small tubular web under stones in natural habitats.

***Herpyllus blackwalli*** Adults are about 11 mm long and uniformly gray to light brown; the cephalothorax is somewhat glossy. This species occurs indoors and is capable of surviving long periods in dry (household) conditions. It is nocturnal and finds harborage during the day in cracks and crevices in walls and behind picture frames. This species is distributed in the UK.

### Lamponidae

These spiders have an elongate oval-shaped and densely setose body. They occur in leaf litter and ground debris in natural areas.

**White-tailed spider, *Lampona cylindrata*** Males are 5–8 mm long and females are 12–15 mm long. The cephalothorax is broadly oval and uniformly dark gray to black. The abdomen is grayish black with a white mark at the end of the abdomen. This species is distributed in Australia; it often occurs indoors, typically in bathrooms.

### Linyphiidae

These spiders have long, thin legs, and usually have fine spines. They are morphologically similar to the Theridiidae. Most species construct a platform-like or dome web, and they are sometimes called canopy spiders and ceiling-net spiders. *Lessertia dentichelis* is found on the walls in coal mines, and building webs over the pebbles of sewage beds in the UK.

***Linyphia nebulosa*** Adults are about 4 mm long and light brown to yellowish brown with gray markings. The web is flat and can be 30 cm wide. This species occurs in the USA. A related species, *L. marginata* (Fig. 18.6f) occurs in cellars.

**Hammock spider, *Pityohyphantes phrygianus* (Fig. 18.6g)** Males are 4.5–6 mm and females are 5–7 mm long. The carapace is white or pale yellow and there is a thin black line at the margins; the carapace has a median forked mark. The abdomen has a herringbone pattern medially. Webs are built on fences, in garages and unused outbuildings. This species is distributed in the USA, from New England south to North Carolina and west to California.

### Loxoscelidae

The thoracic furrow of these spiders is longitudinal and the carapace is flat. The sternum is pointed posteriorly, and the tarsi have two claws. *Loxosceles* is a large genus in this family; there are over 50 species in the western hemisphere.

***Loxosceles laeta*** Males and females are 6–9 mm long and the body is uniformly brown to brownish black. The characteristic violin-shaped mark on the cephalothorax is not distinct. The palpal tarsus is longer than wide (in *L. reclusa* it is wider than long). Female leg 4 is longer than the other legs (in *L. reclusa* leg 2 is longest in both sexes). Indoors, their webs may be 30 cm in diameter and usually placed in corners; corner spider is a common name for this spider in South America. The egg-sac is loosely constructed and contains about 50 eggs. This species is distributed in South America, but isolated populations also occur in southern California. In South America this species is known as *arãna de rincones* (spider of corners) and *arãna de detrás de los cuadros* (spider of behind the pictures).

**Brown recluse spider, violin or fiddle-back spider, *Loxosceles reclusa* (Fig. 18.7c)** Males and females are 7–12 mm long and, except for marking on the cephalothorax, the body is uniformly brown. The cephalothorax has a dark brown violin-shaped mark in the middle, with the neck of the violin directed backwards. Eyes are arranged in three groups of two. The palpal tarsus is wider than long (in *L. leata* it is longer than wide). The abdomen is ovoid, and in the female it is dark brown; the abdomen of the male is light brown. The venter of both sexes is brown and the spinnerets are short. Legs are long, up to twice the body length, and they are not obviously setose. Mating occurs from February to October, but typically in June and July; a single mating is sufficient to fertilize several batches of eggs. Fecundity for one season is about five egg-sacs, with about 51 eggs each; total egg production varies from 31 to 300. Hatching occurs in about 33 days, and development to adult takes about 336 days. Adult males live 301–796 days and females 356–894 days. The preferred temperature range for this species is 23–27 °C. The web is an irregular maze of random threads, and there is often a silken tube or retreat where the spider rests. Adults feed on prey caught in their web and they hunt away from the web at night. Natural habitats include around rocks or other small cavities on the ground, sometimes on cliff faces. In the urban environment they occur around houses, sheds, outbuildings, and indoors. This species occurs widely in southeastern and central continental USA.

***Loxosceles rufescens*** Males are about 7 mm long and females about 7.5 mm long; the body markings are similar to *L. reclusa*. This species is known from Europe, primarily from the Mediterranean region, but it also occurs in urban areas in the USA. It has been reported in several cities, including Boston, New York, Philadelphia, Washington, DC, Los Angeles, and Spokane, Washington. Outside its normal range it usually remains in isolated populations in artificial habitats.

**_Loxosceles unicolor_** Males and females are 6–9 mm long; the body is uniformly light brown to yellowish brown. The violin-shaped mark on the cephalothorax is indistinct. Spiderlings hatch 15–19 days after formation of the egg-sac. Development of males takes 290–680 days and 276–562 days for females. The bite causes pain and skin necrosis that develop into a localized, skin ulceration. This species often occurs around houses in the urban environment. It is distributed in California, Arizona, New Mexico, Nevada, Utah, and Texas.

**Other _Loxosceles_** Males and females of *Loxosceles arizonica* are about 8 mm long. This species occurs in Arizona, New Mexico, western Texas, and southward into Mexico from Coahuila to Baja California. *L. devia* males and females are about 7.5 mm long. The violin mark on this species is not distinct or is sometimes lacking. It occurs in southern Texas and the adjacent states of eastern Mexico. In Brazil, the common medically important species are: *L. gaucho*, *L. similis*, *L. adelaida*, *L. intermedia*, *L. hirsuta*, and *L. amazonica*.

Loxoscelism is caused by the bite of *L. reclusa*, *L. laeta*, *L. rufescens*, and other *Loxosceles* species. The venom of these spiders produces mild to severe pain within 2–8 h, and a thick weal around the bite. At the site of the bite, the cells die and the necrotic area usually becomes dark and dry. An open ulcer develops in 7–14 days and persists for 2–3 weeks. As well as cutaneous lesions, the bite can result in systemic reactions, particularly in children. Bites by *L. laeta* produce similar reactions, but sometimes a deep-skin form of necrosis results, and healing is very slow.

### Lycosidae

These are the wolf spiders. They generally do not make webs but move about in search of their prey. The eyes are of unequal size; the anterior are the smallest and the second row is the largest. The chelicerae are relatively long, and the legs are long and often spinose. The egg-sac is globular with a median seam. Most wolf spider females carry the egg-sac attached to their spinnerets. When young emerge they climb on to the mother's abdomen, and they are carried there for some time. *Lycosa* is a very large genus and it includes most of the large-bodied wolf spiders. Some species build silken retreats under stones or in depressions on the ground, while others wander in search of prey. These spiders occur indoors during spring and fall, when adults are foraging or mating.

**Carolina wolf spider, _Lycosa carolinensis_ (Fig. 18.7a)** Males are 18–20 mm long and females are 22–35 mm long. The carapace is uniformly dark brown with gray setae but without distinct marking; it is light brown in males. Mandibles are brown with orange-yellow setae anteriorly. The abdomen is dark brown with a pale median longitudinal stripe. Legs are light gray with dark bands; leg 4 may be 4 mm long. The ventral surface is blackish brown, including the first and second segments of the legs. This is one of the largest spiders in North America. It occurs indoors in humid habitats, such as bathrooms and basements.

**Garden wolf spider, _Lycosa godeffroyi_** Males are 15–20 mm long and females are 20–25 mm long. This is a common species in coastal New South Wales, Australia.

### Mimetidae

These spiders are distinguished by having long spines on their legs. The chelicerae are long and slender, and may be fused at the base. These spiders are not known to build webs and are predators of other spiders.

**_Mimetus epeiroides_ (Fig. 18.7b)** Males are about 3.4 mm long and females are about 4.4–4.7 mm long. The body color is yellow with dark brown markings. There are two black spots on the front of each chelicera, near its base. The cephalothorax is 1.5 times longer than broad and yellowish white; there are four thin black lines extending from the abdomen to the front of the cephalothorax. Femora 1 and 2 have dark spots on the ventral surface. This species occurs in eastern USA, and may be found indoors in the webs of *Auchaearanea tepidarioum* (= *Theridium*). A closely related species, *M. puritanus*, has similar habits. There is a single black spot on each chelicera of *M. puritanus*.

### Oecobiidae

In this family the cephalothorax is broader than long, and the eyes are grouped in the center. The eyes are unequal in size, shape, and color; some are opaque white and some are elongate and angular. The anal tubercle has a long fringe of setae. In *Oecobius* the cephalothorax is convex and moderately elevated. The family Urocteidae of the Old World is sometimes placed within this family, and the genus *Uroctea* is included here.

**_Oecobius parietalis_ (Fig. 18.7f)** Males are about 2 mm long and females about 2.5 mm long. The cephalothorax is pale yellow with a marginal black line and some dark spots. The abdomen is brownish white or light brown with many dark markings; the venter is pale brown. This species constructs

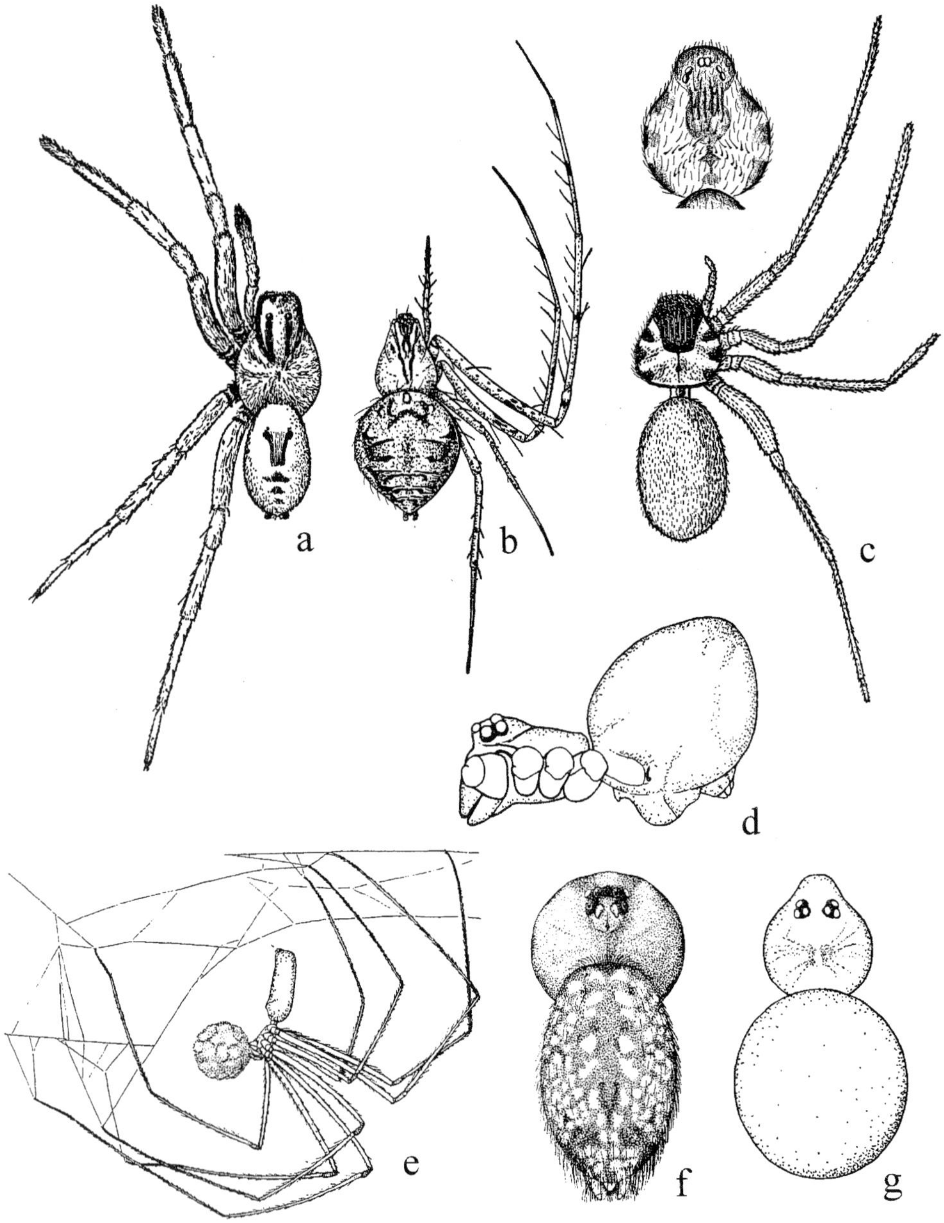


**Figure 18.7** Arthropoda: Aranea. (a) *Lycosa carolinensis*; (b) *Mimetus epeiroides*; (c) *Loxoceles reclusa*; male, female above; (d) *Physocyclus globosus*; (e) *Pholcus phalangiodes*; (f) *Oecobius parietalis*; (g) *Spermatophora meridionalis*.

small flat webs on windowsills, and over the cracks in walls of buildings. It occurs in the USA, from New England south to Florida and west to California.

***Uroctea compactilis*** Males and females are 8–10 mm long. The cephalothorax and legs are light brown to brown. The abdomen is grayish brown and has white markings at the base and laterally. This species builds dense sheet webs on flat surfaces and in corners indoors. It occurs in Japan.

### Oonopidae

These spiders have six eyes, which are usually close together in a compact group. In some members of the family the abdomen is relatively soft and covered with setae. The oonopids are very small spiders, and numerous species occur in the tropics and subtropics. Species of *Diblemma*, *Ischnothyreus*, and *Triaeris* often occur on ornamental plants, and they are shipped around the world; they often occur in greenhouses.

***Oonops domesticus*** Adults are about 2 mm long and the body is light red to pink. Tibiae of front legs have five pairs of long spines. This species occurs indoors, often on ceilings and walls. It is distributed in the UK and perhaps continental Europe.

**Orchestina saltitans (Fig. 18.6i)** Adults are 1–2 mm long. The cephalothorax is convex with the highest point in the middle; it is yellow with irregular purple markings and has a black line along the later margins. There is black around the eyes. Legs are yellowish orange, and the abdomen is purple with many long setae. Femur 4 is considerably larger than the other femora. These spiders are capable of jumping. Natural habitats include grass and bushes outdoors. In the urban environment, they occur inside buildings, on walls, and on the surface of tables and furniture. This species occurs in the USA, from New England south to Georgia to Missouri.

### Pholcidae

The quiver spiders have very long legs and build irregular or sheet-like webs in dark places. They hang from the webs with their back downward. When the web is vibrated the spider quivers up and down (thus the name). The eggs are wrapped by a few strands of silk and carried by the female in her mouthparts. The tarsi have three claws. The anterior median eyes are small or lacking, and the other eyes are arranged in two groups of three. Tropical species, including *Artema mauriciana* and *Smeringopus elongatus*, occur in houses and greenhouses in Florida, and *S. pullidus* is a peridomestic pest in Brazil.

**Long-bodied cellar spider, *Pholcus phalangiodes* (Fig. 18.7e)** Males are about 6 mm long and females are 7–9 mm long. The body is pale yellow, except for a gray mark in the center of the carapace. The abdomen is elongate, more than twice as long as wide, and twice as long as the cephalothorax. This is the most common cellar spider throughout continental USA, and it occurs in similar habitats in other regions of the world, including South America, Europe, and Japan.

**Physocyclus globosus (Fig. 18.7d)** Males are about 3.7 mm long and females about 4.7 mm long. The cephalothorax is dark yellow and the abdomen is dark gray. The abdomen is pale gray with dark spots on both sides of the midline. Legs are yellow with brownish-yellow rings at the distal ends of the femora and at both ends of the tibiae. This species occurs in houses in southern USA, from Florida to southern California. A closely related species, *P. simoni*, occurs in dry and warm house cellars and wine cellars. This species occurs in South America, the UK, and Europe.

**Short-bodied cellar spider, *Spermatophora meridionalis* (Fig. 18.7g)** Males are about 1.6 mm long and females are about 2 mm long. The body is pale yellow except for a pair of light gray spots on the carapace. Eyes are not elevated on a prominence. The abdomen is globose. This species occurs indoors in irregular webs in corners and on walls. It is distributed in eastern USA west to Missouri.

**Spermatophora senculata** Males and females are 2–3 mm long and the body is uniformly light brown. The cephalothorax, abdomen and legs are tan to light brown; eyes are brown and very distinct on the cephalothorax. This species occurs indoors in Japan.

### Plectreuridae

Spiders in this family have eight eyes in two rows, the labium is free, and there are three tarsal claws. Species occur primarily in southwestern USA, but some are known from Central America and Mexico. These spiders build a silken retreat in urban and natural habitats. Females and immatures usually remain at the web site, but males leave the web when mature and enter structures.

**Plectreurys tristis (Fig. 18.8a)** The adult male is about 12 mm long, and about 26 mm long with legs extended. The cephalothorax and legs are dark brown and unmarked. The carapace is granulated and without setae. The abdomen is yellowish gray, with gray setae. This species occurs in southwestern USA and California. It bites and produces a localized swelling and some numbness at the site of the bite.

### Prodidomidae

These are two-clawed, eight-eyed spiders; the dissimilar eyes are in three rows. They have large chelicerae. These are small spiders and they are not common; they are usually found under stones and in dry locations.

**Prodidomus rufus** Adults are about 2 mm long. The cephalothorax is pale yellow and the abdomen is red or yellow with some pink coloration. Anterior median eyes are larger than lateral eyes. It makes a tubular web or mesh of threads. Natural habitats include under stones and on the ground; indoors it occurs in cellars and dark closets. This species occurs in southern USA.

### Salticidae

These are the jumping spiders. They have a short body and stout legs with two tarsal claws. The eyes extend the entire length of the head region of the cephalothorax, and these spiders appear to have excellent vision. They use their eyes to hunt

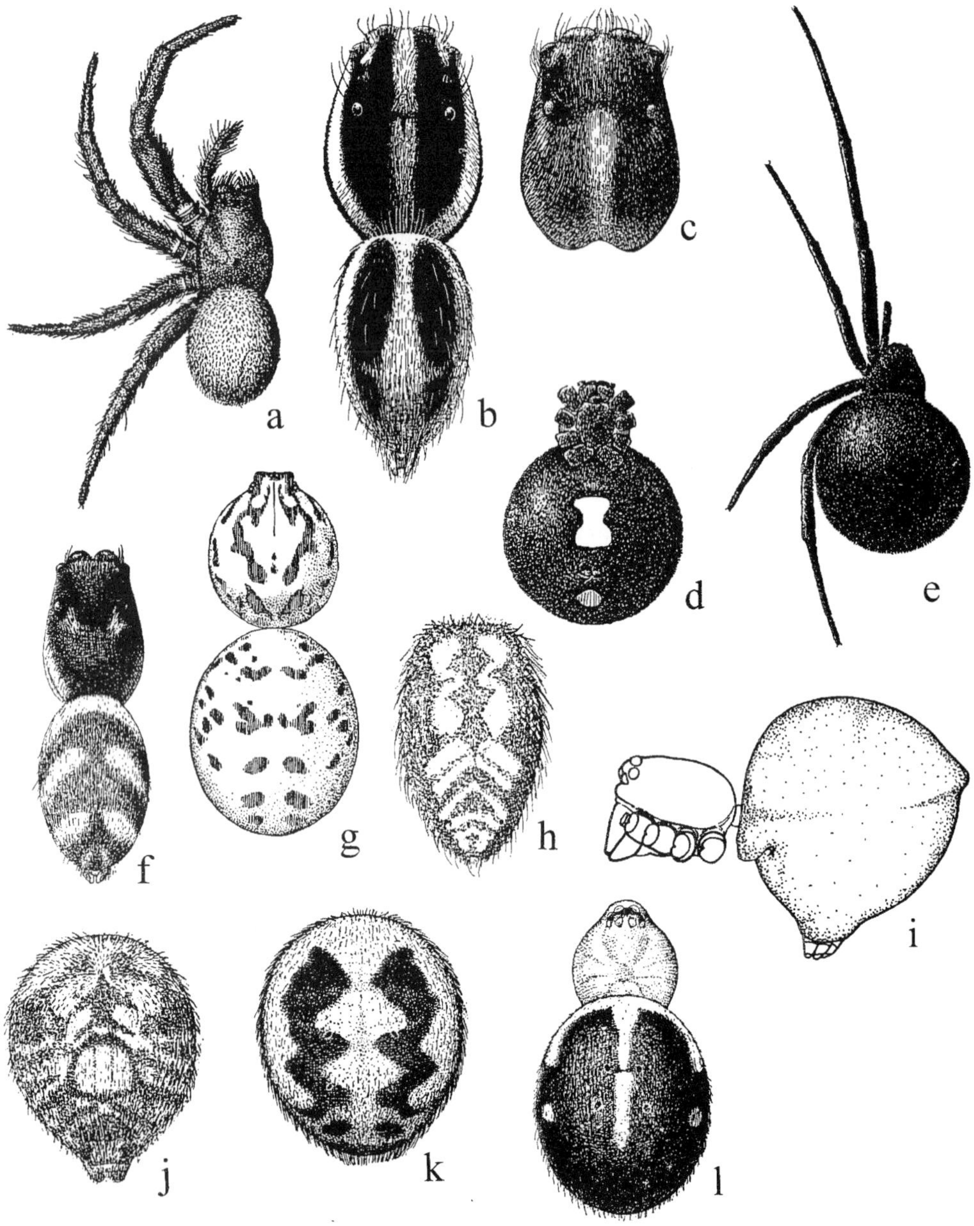


**Figure 18.8** Arthropoda: Aranea. (a) *Plectreurys tristis*; (b) *Plexippus paykulli* male; (c) *P. paykulli* female; (d) *Latrodectus mactans* female, thorax and abdomen of female; (e) *L. mactans* female, cephalothorax and abdomen venter; (f) *Salticus scenicus*; (g) *Scytodes thoracica*; (h) *Marpissa undata*; (i) *Theridion rupicola*; (j) *Achaearanea tepidariorum* female; (k) *Teutana triangulosa*; (l) *Steatoda borealis* female.

during the day, and they are common in sunny areas. Crawling on walls and ceiling is assisted by means of an adhesive tuft of setae, the scopula, on each of the tarsi. As the spider moves over surfaces it trails behind it a fine thread of silk, which is attached at frequent intervals so that it does not fall far if it slips from the surface. They slowly move to approach their prey and when close they suddenly jump to capture it. Before jumping, the front legs are extended forward to grasp the prey, and an anchor line of silk is secured to the substrate. There are several species that occur in peridomestic habitats, including *Hasarius adamsoni* and *Menemerus bivittatus* in Brazil.

***Marpissa undata* (Fig. 18.8h)** Males are 8.5–9.5 mm long and females are 10–13 mm long. The body is gray and covered with a mixture of white, gray, and reddish-brown setae. The abdomen has a pale yellow median stripe. Natural habitats include under loose bark, on fences; in the urban environment it occurs on the outside of buildings. This species is distributed in eastern USA, west to Utah.

***Plexippus paykulli* (Fig. 18.8b, c)** Males are about 9.5 mm long and females are about 11 mm long. The female cephalothorax is black anteriorly and brown posteriorly, and has a pale white median stripe; the male cephalothorax has lateral white stripes. The abdomen is black and the medial and lateral stripes are white. This species occurs primarily indoors; outdoor populations are not common. It feeds on a variety of arthropods, including the German cockroach (*Blattella germanica*) when infestations of this pest are large. This spider has a nearly worldwide distribution.

***Plexippus stipes*** Males and females are about 10 mm long, and the body is light brown. The cephalothorax has two dark brown to black stripes laterally; eyes are distinctly reddish brown. The abdomen has two dark brown to black stripes that are widely separated. Legs are brown with distinct setae. This species occurs indoors in Japan.

**Zebra spider, *Salticus scenicus* (Fig. 18.8f)** Males are 4–5.5 mm long and females are 4.3–6.4 mm long. Chelicerae of males are elongate and extended nearly horizontally forward. The body is gray with white markings; the abdomen has a white band at the base and two or three white bands on the sides or meeting in the middle to form a longitudinal stripe. This species occurs outdoors on fences, and on walls indoors. It is distributed throughout the USA, and probably other parts of the world.

### Scytodidae

These are the spitting spiders. They are closely related to the Pholcidae, and they are usually recognized by the presence of six eyes and short tarsi. They do not make webs, but instead they can eject a viscous material that ensnares their prey, which is the basis for the common name.

***Scytodes thoracica* (Fig. 18.8g)** Males are 3.5–4 mm and females are 4–5.5 mm long. The body is pale yellow with black markings on the cephalothorax and abdomen. Markings on the cephalothorax resemble a lyre. The legs are long and banded. Adults can eject a viscous secretion that fastens its prey to the substrate. This species does not build a web, and indoors it is found walking in shaded areas, dark corners, cellars, and closets. It is distributed in Japan, Europe, and eastern USA.

**Other *Scytodes*** *S. longipes* occurs in the Central American tropics, including Bermuda, where it is common indoors, and is known as the dust spider. Several species occur in peridomestic habitats in Brazil, including: *S. fusca*, which is cosmopolitan; *S. univittata*, introduced from Asia; *S. globula*; and *S. itapevi*.

### Sparassidae (= Heteropodidae)

These are the giant crab-spiders. They occur primarily in tropical and subtropical regions of Central America, but they are often brought to temperate regions on fruit and vegetable produce. The anterior median eyes are smaller than the anterior laterals. Tibia 1 has three or four pairs of ventral spines; the last pair of spines is distal and shorter than the others.

**Banana spider, *Heteropoda venatoria*** Males are about 20 mm long and females about 23 mm long. The carapace is yellow to brown, and it has black setae near the posterior of the carapace. Sometimes there is a white band on the carapace above the chelicerae. The abdomen is light brown with two or three indistinct longitudinal black lines; the marks are more distinct in males. It is not poisonous to humans. It feeds on other insects, and often on cockroaches. The egg-sac is flat and cushion-like, and it is carried by the female under her body. This species is distributed in tropical and subtropical regions of Central America, but occurs in tropical regions around the world. It occurs in southwestern USA, including Florida. Banana shipments to the UK and other parts of Europe may contain these spiders. It also occurs in fruit and vegetable produce shipped to Japan.

### Theraphosidae

These are the large spiders usually called tarantulas, although this name is also used for some *Lycosa*. Theraphosids are the largest spiders and their body size and thick covering of fine hair and setae often attract attention and fear. While some species in South America are very large and some have very toxic venom, the majority of tarantulas are not significantly harmful. Several species are known to travel to temperate regions of the world with shipments of fruits and vegetables from tropical countries, such as South America, West Indies and the Caribbean, and West Africa.

***Avicularia avicularia*** Males and females are 10–14 cm long and have a leg span of about 12 cm. The body is dark brown to black. Legs lack fringes of long setae, but they are densely setose; terminal segments of the tarsi are reddish pink. This species can deliver a painful bite, but the venom is not poisonous. It is native to South America and West Indies, but is sometimes shipped with fruit to other regions of the world, including the UK and continental Europe.

***Stromatopelma calceata*** Males and females are about 55 mm long and the body is light brown. Legs have long fringes of pale-brown setae along the sides and beneath. It is distributed in West Africa where it is common on banana plantations. It is transported to other regions with bananas and vegetables.

***Tapinauchenius santi-vincenti*** Males and females are 50–60 mm long. The body is reddish brown to brown, and densely covered with long, reddish-brown setae. The chelicerae are large. This species is distributed on islands in Lesser Antilles. It has been recorded on West Indian vegetable produce shipped into the UK.

### Theridiidae

These spiders build irregular webs and they remain in the web in an inverted position. The legs are moderately to very long and usually without spines, or have no spines on the tibiae and metatarsi. Several species in this family commonly occur in peridomestic and domestic habitats around the world.

**American house spider, domestic spider, *Achaearanea tepidariorum* (= *Theridium*) (Fig. 18.8j)** Males are 3.8–4.7 mm long and females are 5–6 mm long. The carapace is yellowish brown and the abdomen is grayish white to brown, with indistinct brown chevrons on the posterior half. Legs of the male are orange; female legs are yellow with brown bands at the ends of the segments. Natural habitats include under stones and boards on the ground. In the urban environment it occurs in barns, unused outbuildings, and houses. It makes webs in corners of rooms and frequently in the angles of windows. Adults are present year-round and some individuals live for 1–2 years. Egg-sacs are brown, oval or pear-shaped, 6–9 mm in diameter, and usually placed in the web. Eggs are laid 6–8 weeks after mating; hatching occurs in about 1 week. Females may produce as many as 17 egg-sacs, with a total of 3794 eggs. Female spiders complete development in six molts and males in seven molts. This species is distributed nearly throughout the world. In subtropical regions it occurs outdoors, while in temperate regions it occurs primarily indoors, including greenhouses. *Tepidariorum* is Latin for hot bath or hothouse. A related species, *A. frondeum*, occurs indoors in southern USA, from Florida to California.

**Florida red widow spider, *Latrodectus bishopi*** Males are about 5 mm and females about 10 mm long. The cephalothorax and legs are reddish orange; the abdomen may be black or have dorsal median red spots which have a yellow border. The venter of the abdomen may have one or two red spots. The species builds nests above ground in tree branches. The webs are large – typically a large-meshed network of thread above a convex sheet. This species occurs in southern USA, primarily southern Florida. *Argyrodes* spiders are associated with the webs of *L. bishopi*; the most common species are *A. baboquivari*, *A. elevatus*, *A. furcatus*, and *A. caudatus*.

**Brown widow spider, *Latrodectus geometricus*** This species varies in color from light gray to light brown, and sometimes nearly black. The dorsum of the abdomen has a pattern of black, white, red, and yellow markings. On the venter of the abdomen is an hourglass mark, which is orange or yellowish red. The egg-sac is covered with small spikes. This species is nearly cosmopolitan in peridomestic and domestic habitats. It has been introduced into southern USA and is very common around buildings in South America (Brazil) and South Africa.

**Redback widow spider, *Latrodectus hasseltii*** Males are 3–4 mm long and females are 12–15 mm long. The female body is black and has a red medial stripe on the posterior of the abdomen dorsum, and red spots anterior to the stripe. The venter has several red spots and a large, hourglass-like mark. This species occurs in urban habitats in India, Japan, and Australia. In Japan, it overwinters outdoors behind and beneath vending machines, inside telephone booths, and under benches.

***Latrodectus hesperus*** This species occurs in western USA and western Canada. Females are 14–15.5 mm long; males are 3.8–4.5 mm long and usually light brown. Egg-sacs contain about 196 eggs, they are light brown and about 11 mm in diameter. Hatching occurs in about 14 days; fecundity is about 21 egg-sacs. Adult males live about 196 days, females about 952 days. This species builds webs indoors in corners of rooms and outdoors in sheds and barns. Webs may be as high as 75 cm above the ground and have a long retreat.

**Southern black widow spider, button spider, *Latrodectus mactans* (Fig. 18.8d, e)** Males are about 6 mm long, while females are 10–15 mm long with a leg span of 30–35 mm. Females are shiny black and the abdomen is rounded; typically there is a red double-triangle or hourglass mark, or a similar red mark, on the venter. Red markings usually appear after the second molt but are sometimes absent, especially in immature forms. Immatures have complex patterns of red, white, and orange on the abdomen. The web may be 30 cm wide and nearly as high; it appears to be a random structure, but there is a common structural plan. The female hangs in an inverted position with legs extended. Females do not move far from their web. *Latrodectus* means robber or biter; *mactans* means murderous.

Mating occurs in April and May; a single mating is sufficient to fertilize several batches of eggs. Egg-sacs are gray, spherical or globular, about 9.5 mm diameter, and placed in the web. Fecundity for one season is about 10 egg-sacs, with 250–750 eggs per egg-sac; total egg production may exceed 2500. Hatching occurs in 14–30 days, and newly emerged young remain in the egg-sac until the first molt, and sometimes until the second molt. Young spiders are pale brown to reddish brown, and they have a pattern of stripes on the abdomen and alternating bands on the legs. Young spiderlings often use ballooning to disperse. Development is dependent on food and environmental conditions: for males it is 30–100 days and 4–7 instars, for females 60–120 days and 7–9 instars. Spiderlings usually overwinter and become adults the following year. Adult males live 28–40 days and females live 1–2 years.

Natural habitats for this species include around or under logs, tree stumps, and rocks. In the urban environment it occurs primarily in peridomestic habitats and rarely indoors. It occurs in downspouts, firewood piles, discarded household materials, electric and water-meter boxes, and near vents and doors in crawlspaces. This species is widely distributed throughout the USA, but it is more common in the southern than in northern states. It also occurs in Mexico, Central and South America, West Indies, and Hawaiian islands.

Prey for this spider is any insect or other arthropod that becomes entangled in the small but efficient web. The spider usually remains at the edge of the web, but responds quickly to any vibration of the strands. It usually approaches the trapped prey backwards while extending a strand of viscid silk, first tying down any moving legs or wings; viscous droplets of silk are ejected from the spinnerets to entangle the prey further. The spider delivers a lethal bite to the prey and it quickly dies. Immediately after the kill, the body fluids of the prey are sucked out by the spider. Once the prey is utilized, all points of attachment between its body and the web are cut and it drops from the web. The amount of food eaten varies with environmental conditions. Records for the life of one *L. mactans* include 250 house flies, 33 fruit flies (*Drosophila*), two crickets, and one small spider.

Mating habits of this spider are a sequence of movements that ensure the successful transfer of sperm to the female and safe departure of the male. After becoming an adult, the male leaves its web and seeks a female. When a male encounters the web of a female he moves his abdomen to cause the web to vibrate: the female usually responds with similar movements. The male approaches and strokes the female with his front legs. If the female is receptive the mating sequence proceeds; if not, the male may become prey. The male spins a small web to enclose the female, and then he transfers sperm from either palpus to the female genital opening. The female easily frees herself from the small web and may attack and feed on the male. However, if food is available to the female, the male is not attacked. Males mate several times, but females typically mate only once.

**Northern black widow spider, *Latrodectus variolus*** Males are 4.5–8.3 mm long and females are 7.4–13 mm long. The body is glossy black; the dorsum of the rounded abdomen and there is typically has a median longitudinal row of red spots; ventrally there are two transverse red bars (not an hourglass shape). Egg-sacs are gray and about 12 mm in diameter. Fecundity is about six egg-sacs; hatching occurs in about 13 days. Adult males live about 155 days, while females live about 822 days. This species builds webs in the branches of trees, 1–6 m above the ground; it infrequently occurs in peridomestic habitats. Adult males live about 155 days, while females live about 822 days. This species is distributed in northern USA and Canada, and its range in southern USA overlaps with *L. mactans*.

**Other *Latrodectus*** There are about 30 described species in this genus and species are distributed around the world. *L. cinctus*, the African widow spider, occurs from eastern to western Africa. *L. curacaviensis* occurs in Brazil, and is a medically important spider in the Amazon. *L. tredecimguttatus* occurs in Europe and North Africa.

Latrodectism is caused by the bite of *Latrocectus* species. The venom of these spiders is a neurotoxin and causes paralysis of voluntary muscle groups. The first symptom following a bite is a local swelling; this may develop into a large red area, with some localized skin sweating at the site. Severe pain occurs within 3 h, and this includes pain and rigidity; usually there is pain in the legs and abdomen. Other reactions are shock, fever, nausea, severe headache, elevated blood pressure, difficulty breathing because the diaphragm muscles are contracted, and sweating. In most cases, these symptoms subside in 2–3 days.

***Steatoda borealis* (Fig. 18.8l)** Males are 4.7–6 mm long and females are 6–7 mm long. The carapace is orange-brown with a covering of short setae. The abdomen is purplish brown to black with a yellow anterior margin and yellow median line on the anterior portion. Natural habitats include low vegetation, under bark and stones; in the urban environment they occur in corners of sheds and outbuildings. This species occurs throughout the northern USA.

**False black widow, *Steatoda paykulliana*** Males and females are about 7 mm long. Mature females closely resemble the black widow spider (*Latrodectus mactans*), but they lack red marks on the venter and the legs are brown with dark banding. The abdomen of the mature female has a red band anteriorly; the immature female has a yellowish-white chevron pattern dorsally, and there is a narrow white band around the anterior. The web is typically a random scaffolding of threads. Natural habitats include low vegetation and undisturbed areas. In urban environments they occur in unused sheds, outbuildings, and indoors. This spider is frequently carried on agricultural products (fruits), and it is imported into the UK. *S. paykulliana* is known to bite, and causes temporary illness in children. This species is distributed in the Mediterranean region.

**Other *Steatoda*** In the UK and Europe, *S. bipunctata* occurs indoor in attics and cellars, and on the outside of buildings. Natural habitats for this species include hollow trees and under loose bark. *Steatoda* are carried by commerce to regions outside their normal range. *S. nobilis* is native to the Canary Islands, but has been reported infesting buildings on the Isle of Wight, UK.

***Teutana triangulosa* (Fig. 18.8k)** Males are 3.5–4 mm long and females are 3.7–5.2 mm long. The cephalothorax is brownish orange and the legs are yellow with yellowish-brown bands at the ends of the segments. The abdomen has a pattern of purplish-brown markings on a yellow background. Indoors it is found in basements or in webs built at the corner of windows. This species occurs in the USA, from New England south to Alabama and west to Oklahoma and Colorado.

**Other *Teutana*** In Germany, *T. grossa* has been reported only in houses. *T. castanea* is yellow with dark stripes. It occurs outdoors under rocks and under bridges, and indoors it is found on ceilings and under the eaves. This species is distributed in the USA.

***Theridion rupicola* (Fig. 18.8i)** Males are about 2 mm long and females are 2.3–2.9 mm long. The body color is gray with dark gray and brown markings; the abdomen of the female is enlarged and oval; the spinnerets are located ventrally. The abdomen has a pointed tubercle at the posterior. Natural habitats include under stones and boards in woods and around houses. Females make a retreat camouflaged with debris, within which are placed the white to brown egg-sacs. This species occurs in the USA, from New England south to Alabama and west to California.

**Indian ornamental, *Poecilotheria regalis*** Males and females are about 12 cm long and have a leg span of 16–18 cm. The body is dark brown, and there are yellowish brown transverse bands on the legs. The ventral surface of the abdomen has a yellowish brown band. *Poecilotheria* spp. occur in southern and northeastern India and Sri Lanka. This species and others in the genus breed in captivity and are often kept as pets. *P. fasciata*, a xeric species from Sri Lanka, has been reported living indoors.

### Uloboridae

These spiders make geometric orb webs, or sections of orbs. The great length and size of the front legs are characteristic features. Spiders in this family have no poison glands and no venom.

**Triangle spider, *Hyptiotes cavatus*** Adult females are 3.5–4 mm long and males are about 2.4 mm long. The males and females are brown to brownish gray. The cephalothorax is angular and as wide in the middle as it is long. The posterior row of eyes is much longer than the anterior row. The abdomen of the female is broadly elliptical, and has a double row of tubercles. The palpal organ on the male is large and somewhat projecting. The web is triangular, resembling a 45° sector of a circular or orb web, and it is usually established on ornamental plants and small trees. This species occurs in eastern USA and southern Canada. A related species, *H. paradoxus*, occurs in Europe.

***Uloborus geniculatus*** Adults are about 7 mm long. The posterior row of eyes is nearly straight; the legs are distinctly banded. The abdomen has a single hump with a black transverse bar; in front of the bar the dorsum is gray, and behind the bar it is light brown. Indoors it may build webs on the ceiling. This tropical species is distributed in North and South America, and it extends into southern USA.

## Chilopoda

Centipedes are soft-bodied and dorsoventrally flattened arthropods characterized by having 15–181 pairs of legs, one leg per segment (always an odd number). Species in temperate regions are 1–10 cm long and yellowish brown to brown. Tropical species are 25 cm long and often bright red, orange, green, or violet. The head has a pair of antennae and three pairs of mouthparts. Behind the head is the first body segment (basilar segment), which contains the poison-claws that are used to capture prey. Some species have compound eyes, some have clusters of ocelli, and some lack eyes. These arthropods are nocturnal and are limited to moist habitats, since their cuticle lacks an impervious wax-layer (characteristic of insects). Spiracular openings to tubular tracheae are located above the insertion of the legs on the sides of the body. In most centipedes the spiracles are placed on alternate segments of the body. Legs have six segments: coxa, trochanter, prefemur, femur, tibia, and tarsus. In the Scutigermorphora the tarsus consists of a very large number of segments.

Centipedes are primarily carnivorous, and the majority feed on insects. Some species of Geophilomorpha will feed on plant tissues, and some of the large species of *Scolopendra* (Scolopendromorpha) have been known to kill and eat small birds, mice, snakes, and geckoes. The mouth is located ventrally on the head, and it is bounded by the clypeus, which is a plate anterior to the mouth. The appendages of the first three segments posterior to the mouth are modified to form mouthparts. The first pair is the mandibles, which have three or four teeth; the second pair of mouthparts (first maxillae) is located in front of the mandibles, and obscures them when the head is viewed ventrally. The third pair of mouthparts partly covers the second maxilla; they are leg-like in structure, and contain the poison gland.

### Geophilomorpha

These centipedes are subterranean, as their name implies. They are worm-like centipedes with slender bodies composed of 35–181 segments, and they have short legs. They live in loose soil in natural areas, but occur in suburban landscaping and gardens. They feed on soft-bodied insects and earthworms; their mouthparts are small and unable to bite humans. In a number of geophilomorphs the secretions of sternal glands are luminescent. In Europe, the luminescent species include *Geophilus simplex*, *G. electricus*, *G. carpophagus*, *Necrophloeophagus longicornis*, *Haplophilus subterraneus*, and *Strigamia crassipes*. In Micronesia, East Indies, and West Africa, the red secretion of *Orphaneus brevilabiatus* is strongly phosphorescent. In the USA, *G. vittatus* emits a faint blue-green glow. In general, the geophilomorphs lack eyes, which makes it unlikely that luminescence is used for recognition or sexual attraction. *G. carpophagus* is commonly found in houses and outbuildings in the UK.

***Necrophlaeophagus longicornis*** Adults are about 44 mm long and yellowish brown; the number of pairs of legs is 41–53 in males and 43–57 in females. Females deposit 25–45 eggs and remain with them until hatching. It is reported to be luminescent. This species occurs indoors in the UK.

### Lithobiomorpha

These are small, shiny centipedes with long legs. They are fast-moving and often occur under stones or in the top layer of soil. The body is usually limited to about 15 segments; dorsal tergites are not uniform in size, and usually a large segment alternates with a small one. Some of these species can produce a slimy secretion from pores on the terminal legs; this substance is used as a means of defense. The legs break at a specifically weakened joint near the base; the leg lost to predators is replaced at the next molt.

***Lithobius forficatus* (Fig. 18.9c)** Adults are 18–32 mm long, shiny and dark brown; there are 15 pairs of legs. This species is distributed in the UK, and it occurs indoors in sites such as under sinks and in other damp locations. A closely related species, *L. erythrocephalus*, is about 35 mm long and may occur in peridomestic habitats in Europe.

### Scolopendromorpha

These large centipedes live primarily in the tropics and subtropics, but several species occur in peridomestic habitats in temperate regions. There are about 100 species of *Cryptops* and 90 species worldwide in *Scolopendra* (Fig. 18.9d). These centipedes are distinguished by large bodies that consist of 25 segments, and by their enlarged terminal legs, which have large spines. They have large poison claws, and a history of biting people. They occur indoors and often find harborage in shoes or clothing. Numerous bites have been recorded from Hawaii and some of the Pacific Islands. There is some evidence that these centipedes have poison glands in their legs. In Nigeria, *Scolopendra morsitans* may inflict wounds by crawling over exposed skin. The large centipede, *S. heros*, irritates the surface of the skin when it crawls on the body. When alarmed, this centipede can make tiny incisions with its feet, and venom from the legs enters the punctures.

Many *Scolopendra* species are brightly colored. *Arthrorhabdus formosus* is common in the western Cape of South Africa. It is deep blue, with the head and last segment purplish red. *S. morsitans* is widespread throughout the African continent; it has a yellow body with dark-green cross bands; the head, antennae and last two segments are black, and the legs are bright

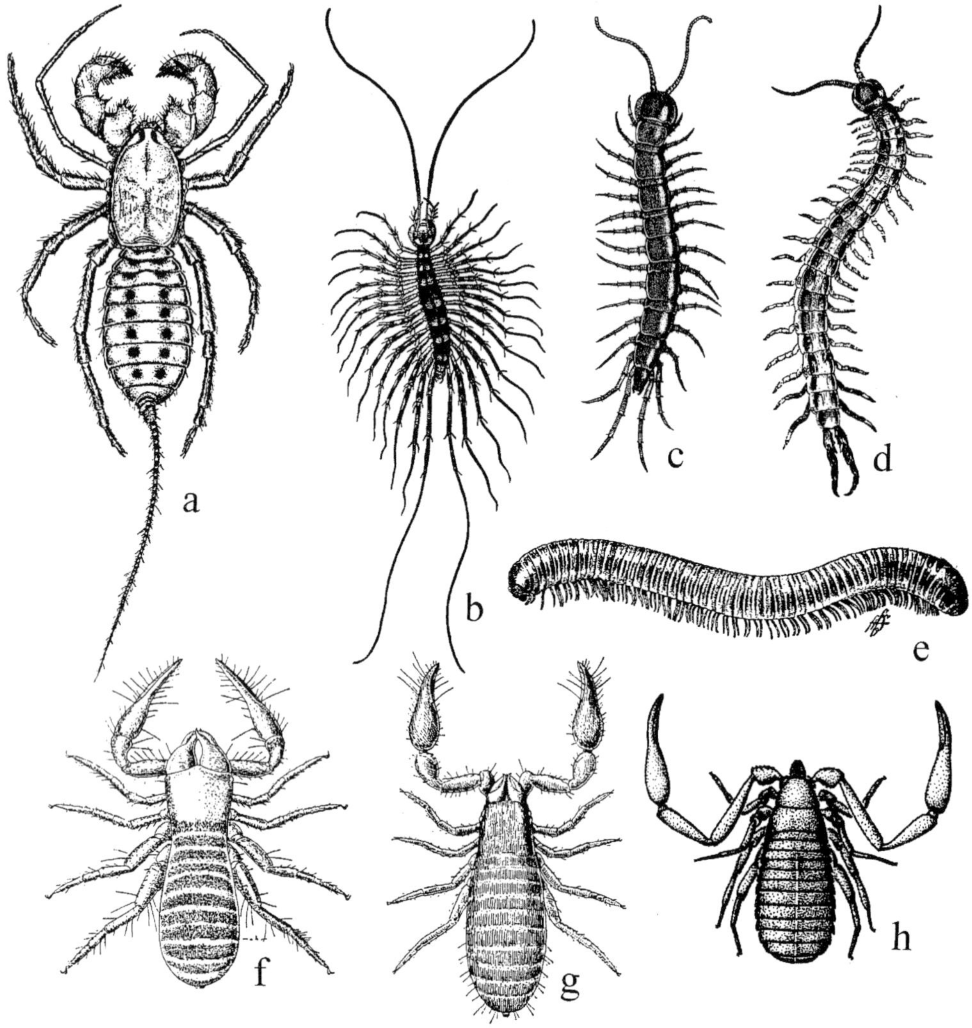


**Figure 18.9** Arthropoda: Uropygi, Chilopoda, Diplopoda, Pseudoscorpiones. (a) *Mastigoproctus giganteus*; (b) *Scutigera coleoptrata*; (c) *Lithobius forficatus*; (d) *Scolopendra* sp.; (e) *Spirobolus marginatus*; (f) *Apochthonius moestus*; (g) *Microbisium confusum*; (h) *Chelifer cancroides*.

orange. One of the largest species in the world, *S. gigantea*, occurs in Columbia and islands off the coast of Venezuela. It is about 26 cm long and 1.2 cm diameter. *S. obscura* is about 18 cm long and occurs in Europe, sometimes in peridomestic habitats. Nearly all the large forms like *Scolopendra* and *Ethmostigmus* live in India, Malaysia, Africa, and the tropics of South America.

***Cryptops hortnesis*** Adults are about 18 mm long and 1 mm wide. The first tergite overlaps the cephalic plate. The distribution of this species is usually urban habitats in North America, from south-central New York and southeastern Michigan to the Florida Keys, the Gulf Coast, west to Texas. It has been collected in Hawaii. This centipede is a European species that has been introduced into North America. A closely related species, *C. parisi*, has been recorded from greenhouses in Newfoundland.

**Australian giant centipede, *Ethmostigmus subripes*** Adults are 15–25 cm long and reddish brown; the legs are light brown. Natural habitats include under moist leaf litter and in damp areas. The bite of this centipede is painful and produces a local swelling. It occurs in Australia.

**Common eastern centipede *Hemiscolopendra marginata*** Adults are about 57 mm long and 4 mm wide. Legs, antennae, and terga are uniformly blue or gray with a blue tint (Virginia, North Carolina), dull gray and green (Florida), yellowish brown with blue to gray bands (central and western Texas), or uniformly blue (eastern Texas). This is a slow-moving centipede that preys on arthropods. Natural habitats for this species include under the bark of decaying pine logs and stumps. In the urban environment, this is the centipede most often encountered in buildings in southeastern USA. It frequently bites humans, often apparently without provocation. This species occurs generally across southeastern USA, with minimal occurrence in the southern Blue Ridge mountain region, and extending to Texas.

**Scolopendra heros** Adults are about 145 mm long and about 17 mm wide. Ultimate and penultimate palpal segments of second maxillae have distinct ventral ridges, usually ending in sharp teeth; antennal segments 2–6 (usually 3–5) are sparsely hirsute. This is one of the largest centipedes, and perhaps the largest terrestrial invertebrate. Body color is variable and ranges from uniformly yellowish brown, yellowish brown with blue bands along the caudal tergal margins, to orange, reddish brown, and black segments and legs. This centipede is active during the day and night; it is an aggressive predator of other arthropods and small animals. An individual was reported carrying a freshly killed juvenile long-nose snake, *Rhinochelius leconti*, across a road in Texas. The snake was over twice as long as the centipede, and was carried head-first under the centipede's body with its legs straddling the prey.

The bite of *S. heros* produces a sharp, temporary pain. It has been recorded in domestic and peridomestic habitats, including in swimming pools. This species is distributed in southwestern USA, from California east to New York and south to Florida. South of the USA it occurs from the Rio Grande and Mexico in general.

**Scolopendra polymorpha** Adults are about 111 mm long and 10 mm wide. The body is yellowish brown with blue to bluish gray bands on the metaterga. Maxillary palps are rounded ventrally and without ridges or teeth; antennal segments 1–3 are sparsely hirsute. This species is a voracious predator, and it actively searches for food at night. The venom has been variously reported as harmless to producing a sharp pain. This large centipede has been reported in gardens, indoors among furniture, and in a mouse trap. It is distributed in southwestern USA, Baja California, Mexico, and Central and South America, including Belize, Honduras, Venezuela, and Brazil.

**Common centipede, *Scolopocryptops sexispinosus*** Adults are about 69 mm long and 4 mm wide. The body is orange and without blue marks on the ventor and antennae. The cephalic plate is margined laterally; antennal segment 1 is nearly without fine setae, while segment 2 is densely hirsute dorsally, and slightly less hirsute than segment 3. This species is common in urban and natural habitats in eastern North America. In natural habitats it occurs under rotting logs and stones, and in moist leaf litter. It is native to North America, and occurs from Ontario, New York, and Michigan south to Florida and Texas.

### Scutigermorpha

These centipedes are relatively small and have 15 pairs of very long legs; the legs give them the appearance of being larger than they actually are. The antennae are also long and delicate, and are moved in a whip-like manner over the body. Eyes are large and faceted (about 100), and resemble the eyes of insects. Legs and the tarsi are very long, sometimes with as many as 400 segments, and the walking and running of this arthropod are very unusual. Instead of only the terminal tarsal segments contacting the surface, all the tarsal segments are applied to the surface. *Scutigera* spp. easily lose their legs, especially the last pair; they break at a weakened joint near the base of the leg. The mouthparts are strong and they have large mandibles. The respiratory system is a single tracheal opening in the middle of the back, one on each segment except the last. Scutigermorph centipedes are predatory hunters, and they feed primarily on insects. They run and leap upon their prey; the long legs form a cage to hold the insect before it is pierced with the mouthparts. Only the soft parts of the insect are eaten. They are active at night, but may be seen during the day.

**Australian house centipede, *Allothereua maculata*** Adults are 20–25 mm long. The body is pale brown with dark brown markings. Antennae and legs are long. This species occurs indoors in houses and greenhouses. It is distributed in Australia.

**House centipede, *Scutigera coleoptrata* (Fig. 18.9b)** Adults are 2.5–4 cm long; the body is grayish-yellow with three longitudinal dorsal stripes. The antennae and 15 pairs of legs are very long; the legs are banded with white. The compound eyes have about 100 facets. Newly hatched larvae have four pairs of legs; there are five larval stages, with 5, 7, 9, 11, and 13 pairs of legs, respectively. Eggs are placed singly or in groups of 7–10 into crevices in the soil or indoor substrates. In the laboratory, females lay about four eggs per day, and 20 is the maximum laid in 24 h; fecundity is about 63. In southern France, *S. coleoptrata* lays eggs from early May to late June. Hatching is in 30–38 days at 20–21 °C. The duration of the first three instars is 7–9, 12–13, and 13–17 days at 20 °C. Adults live for several years. Prey species for this species include house flies, cockroaches, moths, bees, wasps, and nearly any insect and spider that occurs in domestic and peridomestic habitats. This species is distributed around the world, and it occurs indoors and outdoors.

## Diplopoda

Millipedes are characterized by the presence of two pairs of legs on most of the body segments, and they have numerous body segments. Mature forms vary from 10 to 100 mm long. They have two body regions, and they range in color from reddish orange to dark brown and black. Their reproductive organs are located on the ventral side, at the front of the body near the head. The head has two pairs of unbranched seven-segmented antennae and at least two pairs of mandibles, and eyes. Some species lack eyes, but have a dermal light sense. Spiracles leading into tracheae open above the coxae; the spiracles lack a closing mechanism. Repugnatorial glands may be present on the middle and terminal body segments. A mixture of hydrocyanic acid, iodine, and quinone is released from these glands, and sometimes it is discharged as a spray. One pair of legs in the male is modified for mating. Two pairs of legs per segment enable millipedes to exert considerable forward thrust. The leverage provided by the legs, together with the calcified head capsule, enables millipedes to force their way into a variety of habitats. They are able to penetrate between the fibers of rotting wood, the spaces in closely packed soil particles, and the narrow openings around doors and windows. The power for their pushing ability is achieved by the backstroke of the legs; it is longer in duration than the front stroke, and this generates more energy for forward motion.

Millipedes typically occur in moist or wet habitats. Many species curl into a sphere when disturbed, and others form a compact spiral. These behaviors provide some protection from predators, but are also effective in reducing water loss when they are in dry habitats. Food for millipedes is a variety of soft or decomposing plant material. They have been recorded as eating dead worms, mollusks, insects, and vertebrates. Several species are pests of field crops and in greenhouses. Millipedes comprise a proportion of the food of other animals, including toads and birds. Europe starlings (*Sternus vulgaris*) are especially destructive to millipede populations.

Mass migrations of millipedes have been reported, and this is sometimes accompanied by large numbers of centipedes. Masses of individuals can be involved, and their path may intercept railroad tracks, buildings, and agricultural fields. In Japan, *Parafontaria laminata* has been reported in mass migrations that seem to occur at intervals of 7 or 8 years. Other large populations include those of *Gymnostreptus pyrrocephalus* in South Africa (Natal) and *Pseudopolydesmus serratus* in the USA (Ohio). Mass movements may be stimulated by mating activity or features of the habitat, such as moisture and overcrowding, or to temperature and humidity changes. Rainfall causes migration of *Unixenus nijobergi* in Western Australia. In India, large aggregations and migrations of *Streprogonopus phipsoni* occur. Millipedes move indoors from habitats in the mulch and ground cover around urban structures, or move in large numbers from adjacent natural areas. Some species, such as *Archiulus moreleti*, are positively phototactic and are attracted to lights at night.

Mating is essentially the same for all millipede species; there is relatively little courtship and pre-mating behavior. The male transfers spermatic fluid to the genital orifice of the female and fertilization is internal. The genital openings of male and females are situated on the third segment, behind or on the second pair of legs. During copulation the ventral surface of the terminal 13–14 body segments of the male is adjacent to the ventral surface of the anterior 10–11 body segments of the female. The seven anterior legs of the male are bent around the female. The region of the body segment 8–10 of the male is adjacent to the genital opening of the female, which is on body segment 3. Eggs are deposited singly in crevices in the soil or in a nest site and the female remains with the eggs for several days. The number of eggs deposited at one time and female fecundity vary among species. Some species lay about 20 eggs, while others lay up to 300; hatching occurs in 2–3 weeks. The immature stages develop through numerous molts, during which the number of legs and the postcephalic body segments increase. Many species reach sexual maturity in 2 years; some require 4 or 5 years and then will live several years more.

### Julida

In this order, the millipedes are small and cylindrical, and both pairs of legs on segment 7 of the male are modified as copulatory organs (gonopods). The first and sometimes the second pair of legs in the male may be modified to form secondary sexual structures. There are many species in Europe and western Asia, and several species have been introduced into North America. In the family Julidae, the male's first pair of legs is hooked. Several families in this order have species that occur in large numbers in or near urban environments. *Uroblaniulus jerseyi* is common in urban and natural habitats in eastern USA, and enters buildings in large numbers in fall. Two common species that enter buildings in eastern USA are *Oriulus venustus* (= *Prajulus*) and *O. impressus* (= *Prajulus*). They commonly occur in large numbers around the perimeter and inside of buildings.

**Portuguese millipede, *Archiulus moreleti* (= *Ommatoiulus*)** Adults are small and uniformly dark brown. This species is native to the Iberian peninsula, but has been introduced to other parts of the world. It is a pest in Australia and South Africa. In Australia it occurs in peridomestic habitats, where it destroys ornamental plants and vegetables, but it also enters structures. It is a problem during summer and early fall, which is the breeding season for this millipede. Invasion of structures also occurs in the spring, following rainfall. This species is attracted to lights at night and it enters buildings under doors and around windows.

***Brachyiulus lusitanus*** Adults have more than 30 body segments, and the body is long and slender. The head is without a distinct midfacial suture. Males lack short coxal lobes on the anterior legs. The dorsum has a broad, yellow, longitudinal stripe, which contains a medial black stripe. This millipede occurs under logs, rocks, and stones in urban gardens. This species is native to Europe, but has been introduced into northeastern North America. It is established in eastern USA and eastern Canada. In the USA, it is known from North Carolina and California (Contra Costa county). It is also known from urban areas in Mexico. A morphologically similar species, *B. pusillus*, was introduced to eastern North America from Europe.

***Ophyiulus pilosus*** Adults are about 4.5 cm long. The body has more than 30 segments and there are fine but distinct longitudinal striations on the segments. The dorsum is uniformly brown to grayish black, not striped. This species is common in urban habitats in Europe and eastern North America.

**Brown millipede, *Paraiulus venustus*** Adults are 25–40 mm long and about 8 mm in diameter; the body is uniformly brown. This species, and *P. impressus*, occurs around the perimeter and inside of buildings and other urban structures. It is distributed in eastern USA.

## Polydesmida

These are flat-backed millipedes; they have 18–21 segments, and are 3–13 mm long. They do not have eyes and they have completely fused sclerites; there are strong lateral projections on the posterior of each segment. The anterior pair of legs on segment 7 of the male is formed into gonopods. This is the largest order of millipedes, with more than 2700 species; there are about 250 species in North America, but fewer in Europe.

**Greenhouse millipede, *Orthomorpha gracilis* (= *Oxidus*)** Adult males are about 19 mm long and females are about 21 mm long. Body segments are dorsoventrally flattened, and they are dark brown to blackish brown. Postcephalic body segments have 30 or 31 pairs of legs in the males and females, respectively. Eggs are deposited in batches from 14 to 300 in small cavities in the soil, 7–15 mm deep. Development from egg to mature adult includes at least seven instars and takes 148–177 days (at indoor temperatures). Above 22 °C breeding occurs throughout the year. Natural habitats include moist leaf litter in undisturbed areas. In the urban environment it occurs around the perimeter of buildings and enters through doors and windows at ground level. This millipede can occur in large numbers in greenhouses, where densities can exceed 2500 per square meter.

**Orange and black millipede, *Sigmoria aberrans*** Adults are about 4 cm long and grayish black and bright orange along the margins of the segments. This species occurs in the mid-Atlantic states of Virginia and North Carolina. It is common in urban landscape plantings and natural habitats.

## Polyxenida

This order of small and bristly millipedes is considered soft-bodied and have 11–13 segments. The body wall is soft and not impregnated with calcium salts, and the body is covered with tufts of serrate bristles.

**Bristly millipede, *Polyxenida lagurus*** Adults are 3–4 mm long, somewhat flattened and covered with short setae or bristles. Each segment has two rows of serrated bristles, and bundles of detachable barbed bristles protrude from the tail end. Predators are entangled in these detachable bristles. This species occurs under the bark of trees, where it feeds on algae. Small to large numbers are known to occur indoors in England. *Polyxenida* species occur in northwestern USA and Europe; *P. lagurus* occurs in the UK.

## Spirobolida

In this order, there is only one pair of legs on segment 5; other millipedes have two pairs. There is a distinct suture extending vertically down the front of the head. The male copulatory organs are concealed in a pouch, and both pairs of legs on segment 7 of the male are modified into gonopods. These are generally tropical species, some of which are brightly colored.

**Narceus americanus annularis (= Narceus annularis)** Adults are about 10 cm long, but the length is variable; they have more than 30 body segments, and the body is robust. The head has a distinct, vertical, midfacial suture. Males have short coxal lobes on the anterior legs. This is an abundant and common millipede in eastern North America. It inhabits a variety of natural and urban habitats.

## Spirostrepida

This order includes the largest millipedes known. Some are up to 30 cm long and with 90 segments, and some are less than 6 mm long.

**Large millipede, *Spirobolus marginatus* (= *Arctobolus*) (Fig. 18.9e)** Adults are about 10 cm long and usually have 57 body segments. The body is reddish brown, each segment is marked with red, and the legs and antennae are red.

**Other millipedes** Several species have been reported from urban habitats around the world, and their frequency indoors depends on local populations and environmental conditions. *Orthomorpha coarctata* lives in the thatch layer of ornamental turfgrass (St. Augustine grass, *Stenotaphrum secundatum*) in southern USA (Florida), but it will frequently enter houses in large numbers. *Pseudopolydesmus serratus* is about 3 cm long, uniformly brown, and distributed in urban and natural habitats in eastern USA. The large species *Pachydesmus crassicutis* is about 7 cm long and brown but with orange laterally on the segments. It occurs in southeastern USA. In the UK, *Geophilus carpophagus* and *Tachypodoiulus niger* occur in urban populations and frequently enter buildings. *Proteroiulus fuscus* has also been recorded from buildings.

## Isopoda

These are woodlice, pillbugs, and sowbugs. They are familiar arthropods, and closely related to lobsters, crayfish, crabs, and shrimps. Isopods are dorsoventrally flattened; the last seven thoracic segments are distinct and have leg-like appendages. The abdominal segments are fused, and the thoracic segments have seven pairs of legs. The anterior abdominal appendages usually bear functional gills in aquatic forms. The terminal abdominal appendages are enlarged and elongate. In terrestrial species respiration is by means of tubelike air-channels or pseudotracheae. These open to the exterior by a single pore which lacks a spiracular closing device possessed by other arthropods. Some respiration takes place through the moist surface of the integument. The primitive pseudotracheae and the absence of a layer of cuticular wax expose isopods to desiccation. Isopods usually remain on or in damp soil or other moist habitats, and are active at night when humidity is high. They often occur in small to large numbers as a means of reducing evaporation and maintaining water balance.

Most isopods occur in marine or freshwater habitats, but some are terrestrial. Species in the family Limnoriidae are borers in marine timbers. The most common genera in this family are *Limnoria*, *Paralimnoria*, and *Phycolimnoria*. Species in the family Sphaeromatidae are marine borers in the tropics and subtropics, and the most common genera are *Sphaeroma*, *Exosphaeroma*, and *Cymdoce*. Isopods are cosmopolitan in urban and agricultural environments. They are suspected of being indigenous to Europe, and they have been introduced to the UK, the USA, and other regions of the world through commerce and shipping.

Pillbugs resemble sowbugs, but differ morphologically and in aspects of behavior. In pillbugs, the last abdominal appendages (uropods) are rounded posteriorly, whereas in sowbugs these appendages project from the posterior as a pair of pointed tails. The presence of rounded or pointed uropods easily distinguishes these two forms. When pillbugs are disturbed, they bend their body head to tail and form a compact ball; however, sowbugs are not capable of forming a compact ball.

Parasites of terrestrial isopods include species of flies in the subfamily Rhinophorinae of the Calliphoridae. Larvae of these flies are internal parasites of many of the common pillbugs, including *Porcellio scaber*, *Oniscus asellus*, *Armadillium vulgare*, and *Trachelipus rathkei*. The calliphorids *Parafeburia maculata* and *Styloneuria discrepans* are associated with these pillbugs. Successful parasitism begins with the access female flies have to harborages containing pillbugs. Females become negatively phototactic after mating, and they crawl into pillbug harborages and lay eggs on soil surfaces which have residues of secretions of the uropod glands. After hatching, larvae attach themselves to pillbugs and enter the body cavity through an intersegmental membrane. Within the pillbug body cavity, the maggot feeds on the hemolymph, the gonads, and the vital organs. The death of the host is followed by pupariation by the calliphorid maggot.

***Androniscus dentiger*** Adults are about 6 mm long and pink to light red. The surface is strongly tuberculate, and each tubercle bears a central spine. This species is common in gardens and disturbed areas; it frequently occurs in caves and cellars. It is widely distributed in the UK and Ireland.

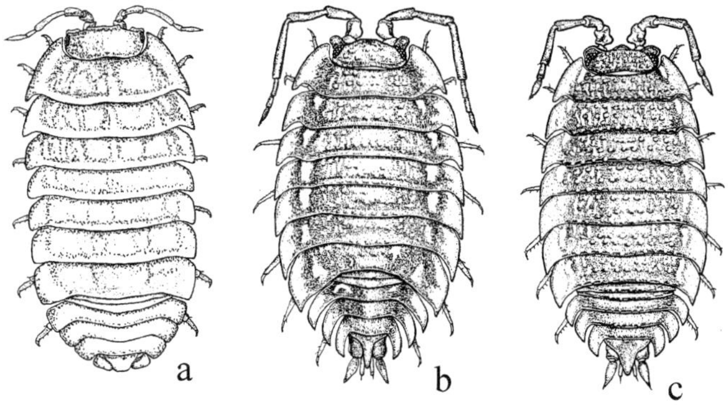

**Figure 18.10** Arthropoda: Isopoda. (a) *Armadillium vulgare*; (b) *Oniscus asellus*; (c) *Porcellio scaber*.

**Common pillbug, *Armadillium vulgare* (Fig. 18.10a)** Adults are about 14 mm long and shiny gray; red and variegated forms occur. They are capable of rolling into a compact ball when disturbed. Eggs are retained and young develop in a brood pouch (marsupium) on the underside of the female. In the USA, the number of young in each brood ranges from 20–29 (Texas) and 5–62 (Michigan); in France the range is 48–156. First-stage young are white and have six pairs of legs; the second stage occurs in about 24 h and has six pairs of legs; the third-stage molt occurs in 14–18 days and has seven pairs of legs. Molting depends on environmental conditions and food supply. Females reproduce when they are about 7 mm long. Adults live about 2 years; there are 1–3 generations per year. Adults and immatures feed on plant material, and this species is often a pest in gardens and greenhouses. It is distributed around the world.

**Gribble, *Limnoria lignorum, L. quadripunctata, L. tripunctata*** Mature individuals are 2–4 mm long and pale gray. These isopods tunnel at or slightly below the wood surface. In softwoods the earlywood is eaten, which leaves thin plates formed by the latewood (annual rings). They excavate long narrow tunnels that are 1–3 mm in diameter. Adults mate in the tunnels and fertilized eggs and developing juveniles remain in the tunnels. *Limnoria* can produce 1–3 clutches of eggs per year, depending on the water temperature. Young settle on timber with a roughened surface; in tropical and temperate sites they are active in wood at the tidal zone. In cool water, the damage to piers is done well below the tide line. *L. lignorum* occurs in coldwater regions, *L. quadripunctata* is found in temperate waters, and *L. tripunctata* occurs in both temperate and tropical waters.

Wood attacked by *Limnoria* has usually already been invaded by bacteria and fungi. Fungal decay of the surface layers of wood is not essential for the establishment of *Limnoria*, but the presence of fungi in the wood enhances feeding and reproduction. Feeding stages of *L. tripunctata* distinguish between fungus-infested wood and noninfested wood. The wood eaten by these isopods is broken down in their gut by cellulases. However, microorganisms in the wood probably assist in cellulose breakdown. *Limnoria* species are distributed around the world.

**Common woodlouse, *Oniscus asellus* (Fig. 18.10b)** Adults are about 16 mm long and gray with irregular light patches; yellow and orange forms are common near seacoasts. The head has distinct lateral lobes below the eyes; the telson ends in a long point. The body is glossy in adults, but often rough or granular in immatures. This species occurs in gardens, wasteland, and decaying wood in urban areas. It is common in the UK.

**Dooryard sowbug, *Porcellio laevis*** Adults are 15–18 mm long and about 8 mm wide. The body is brown and uniformly smooth and glossy; there is some gray to dark gray and there are two pale longitudinal lines. They have well-developed eyes and the antennae are nearly half as long as the body. Uropods are long and slender. Immatures are pale gray. Eggs are retained in a brood pouch (marsupium); hatching occurs in about 50 days. The number of young in each brood ranges from 24 to 88. Adults live about 2 years; there are 1–3 generations per year. All stages are nocturnal and feed on decaying plant material, but they will also attack tender plants. This species is nearly cosmopolitan, and it is usually associated with urban and suburban environments.

**Rough sowbug, *Porcellio scaber* (Fig. 18.10c)** Adults are 10–11 mm long and about 6 mm wide; the dorsal surface of the body is light gray with black spots, or the body is nearly black. Eggs are retained in a brood pouch, and hatching occurs in about 44 days; the number of young in each brood is about 28. Adults and immatures frequently attack live plant material and cause damage in agricultural and urban areas. This species is distributed around the world.

**Putty-bugs, pillbugs, *Sphaeroma terebrans, S. triste, S. quoyanum, S. annandalei*** Adults are 10–12 mm long and the body is gray to dark gray. They produce 5-mm-diameter tunnels across the grain of the wood, usually perpendicular to the wood surface. The mature adults excavate a single burrow, and the young are reared there. Severe infestations of *Sphaeroma* can lead to destruction and collapse of wood in use. Infestation of marine piling is usually in the middle and upper regions of the tidal range, so that the tunnels are exposed for several hours each day. The damage to marine piling often has an hourglass appearance, with the greatest loss of wood in the middle of the piling, at the tide line. The amphipod *Chelura terebrans* is associated with limnoriid infestations of maritime timbers worldwide. It lives in the galleries of *Limnoria*, and feeds on the feces and degrades the walls of the galleries.

**Other sowbugs** These crustaceans are present in a variety of natural and disturbed habitats. *Porcellio spinicornis* is commonly found in the urban environment. It is smaller than *P. laevis* and the dorsal surface of the body is covered with spine-like projections (thus the species name). The uropods are short and broad. *Alloniscus perconvesus* is about 16 mm long and the body is dull gray with black and white marks. Adults and immatures occur in the decaying vegetation debris along ocean beaches.

## Opiliones

The daddy-long-legs, harvestmen, or shepherd spiders are common arthropods. They have long, slender legs and a body shape in which the short abdomen is broadly joined to the cephalothorax (not by a narrow pedicel, as in spiders). In most species there are two eyes, which are on a prominent tubercle near the middle of the cephalothorax. Pedipalps are short and leg-like, but they are chiefly sensory and grasping organs used to contact objects close to the body. Males have large chelicerae and an extrusible penis; they engage in direct copulation. A pair of spiracles opens ventrally on abdominal segment 1; accessory spiracles occur in the tibia. Near the base of the first or second pair of legs is the opening for a pair of repugnatorial glands. The odorous liquid produced by these glands has been variously described as sweet or pungent, and lasting several minutes. A few species are able to stridulate.

Harvestmen are predators of snails, worms, other arachnids, and insects. Some are scavengers on recently dead animal tissue, and some eat fungi and bird droppings. They ingest solid food, and not simply liquid, as do spiders and scorpions. When eating, the chelicerae, pedipalps, and legs are used to hold and tear apart food. Water is important to survival and most species do not live long without it. They are usually active at night and early morning, and during the day they may be active following a rain shower. Mating occurs frequently and there is little courtship or pre-mating behavior. Frequent mating is necessary because eggs develop a few at a time and are deposited as they mature. The female has a long, extensible ovipositor and she deposits eggs in cracks and crevices in soil, under stones, in decaying wood, and at other moist sites. Eggs are white, without a sculptured chorion, and adhere loosely together. Hatching takes about 20 days, but the eggs of some species overwinter. Newly hatched young are about 1 mm long. Development is completed in 6–9 months and there are six or seven nymph stages.

***Phalangium opilio*** Adults are 3.5–9 mm long and the body is gray. Some individuals have a pattern of gray and black medially. Front legs are very long. This species has a nearly cosmopolitan distribution; it occurs in northern USA, continental Europe, and the UK. The name *opilio* means a shepherd in Latin. The name shepherd spiders for the Opiliones comes from the belief that fields that have many of these arthropods are good for grazing sheep. *P. opilio* is commonly found in fields at harvest time in the UK.

***Phalangium parientinus*** Adults are 5–7.5 mm long. The last segment of the pedipalps is longer than the preceding segment; the pedipalps of the female are nearly as long as the body. The base of the pedipalps lacks prominent spines. This species occurs in northern USA, and is commonly found in sheds, outbuildings, and wood and lumber piles.

## Pseudoscorpiones

These unusual arthropods are 1–8 mm long and have large pedipalps. They are yellowish brown to dark brown and resemble scorpions, but are much smaller and lack the enlarged tail. The carapace is large and on it are the eyes, and the six pairs of appendages. Chelicerae are situated in front of the mouth and they are composed of two segments: one movable and one stationary. Ducts of the silk glands open at the tip of the movable segment, and salivary glands are located in the chelicerae. Pedipalps serve as prehensile organs to capture and kill prey; the coxae are usually large and serve to crush food close to the mouth. The ambulatory legs are long and setose. Pseudoscorpions are carnivorous and feed on collembolans, psocids, thysanurans, and other small arthropods. Adults eat only once or twice a month, but the nymphs eat more frequently. Prey is grasped by pincers on the pedipalps and brought to the chelicerae; these organs pierce the cuticle and inject enzymatic saliva into the body. The liquefied internal contents of the host body are taken into the mouth.

Most of the nearly 2000 species live in leaf litter, under bark, and in other decaying vegetation. A large number of species live in the nests of ants and termites, and the females of at least 25 species are reported to cling to the legs of flies and other arthropods as a means of dispersal (phoresy). In Europe, a common phoretic species is *Lamprochernes nodosus*, which is found on the legs of house flies and syrphid flies. Some pseudoscorpions regularly occur or live indoors, including *Chelifer cancroides*, *Allochernes italicus*, and *Cheiridium museorum*. A number of species are associated with stored-food products in warehouses and occur in the nests of birds. These species include *Withius subruber*, *Cheiridium museorum*, and *Toxochernes panzeri*.

### Cheliferidae

Several species in this genus occur indoors, including *Chelifer panzeri* in attics, breweries, and stables; *C. depressus* occurs in houses in India; and *C. scorpioides* under wallpaper, and in hay and straw.

**Booklouse, *Chelifer cancroides* (Fig. 18.9h)** Adults are 4–6 mm long and brown to dark brown. The abdomen has 12 distinct segments (which appear as 11 segments) and is broadly attached to the cephalothorax. Pedipalps are long and enlarged apically. Adults and nymphs can move rapidly forwards, backwards, and sideways. This species occurs indoors where it preys on mites, flies, bed bugs, caterpillars and other arthropods. They are most active from March to May.

### Chthoniidae

Species in this family are recognized by the single row of long, slender setae across each of the abdominal tergites. The body and palps are weakly granular.

**_Apochthonius moestus_ (Fig. 18.9f)** Adults are about 2 mm long and the body is brown to dark brown; coxa I has three spines. This species is widely distributed over eastern USA, east of the Great Plains region. It may occur in peridomestic habitats, including under the bark of decaying logs and in the nests of mammals.

**_Microbisium confusum_ (Fig. 18.9g)** Adults are about 2 mm long and dark brown. This species is common and widespread in eastern USA. It occurs in forested areas, and in decayed logs and stumps. It is active nearly throughout the year. Closely related species with similar distribution include *M. brunneum* and *M. parvulum*.

## Schizomida

These are the short-tailed whip-scorpions and they are 5–7 mm long. The carapace is divided into three segments; median eyes are lacking but a pair of lateral eyes may be present which are reduced to small, pale areas of cuticle in some species. The pedipalps lack claws, and often have numerous spines and setae. The first pair of legs is long, slender, and used as tactile organs. The abdomen is 12-segmented; the last three segments are small, annular, and form a pygidium. The pygidium bears a short extension made up of 1–4 segments; in males it is often fused into a rounded or elongate knob. There are three genera in the order: *Schizomus* and *Trithyreus* have a wide tropical and equatorial distribution, and some species have been introduced into greenhouses in temperate regions; *Stenochrus* contains a single species from Puerto Rico. In Sri Lanka, *Schizomus crassicaudatus* and *S. vittatus* occur under dry leaves.

They are nocturnal and prey on small arthropods; they remain in dark and moist harborages during the day. These arthropods are very sensitive to light and to ground vibration, and move quickly when disturbed. The front legs are carried above the ground as the animal searches for food, and for defense they have a pair of anal glands that produce and expel an irritating liquid (acetic acid). In *S. crassicaudatus*, the female excavates a cavity in the soil, about 15 mm from the surface, and lines it with soil cemented together. Eggs are laid singly, and attached as a mass in the region of the genital opening. In *Trithyreus sturmi* the male cements a spermatophore to the ground and the female positions over it to transfer sperm into her genital ducts.

**_Trithyreus pentapeltis_** Adults are 4.5–7.5 mm long. The males are dark reddish brown, and the females are bright yellow. They occur under leaves and plant litter, and they are distributed in the semiarid and desert regions of southern California. A related species, *T. paradenigensis*, occurs among the fungal hyphae under wet leaves.

## Scorpiones

Scorpions are well-known and easily distinguished arthropods. The carapace is unsegmented and has a pair of median eyes and 3–5 pairs of lateral simple eyes. In front and above the base of the pedipalps are large, paired chelicerae that are positioned horizontally and used for crushing and shredding food. Scorpions have large pedipalps with large claws. The abdomen is divided into two portions: a broad preabdomen (mesosoma) consisting of seven segments that are as wide as the cephalothorax, and a tail-like postabdomen (metasoma). There are four pairs of book-lungs, which open ventrally through spiracles (stigmata), on the abdominal segments. At the end of the post-abdomen is a curved and pointed sting (telson). Two venom glands are contained in the telson, and each gland discharges it contents through a pore in the curved sting. On the ventral side of the abdomen, behind the genital opening, is a pair of comb-like organs known as pectines. The walking legs consist of seven segments and terminate in two lateral claws and a median claw.

Distribution of scorpions is primarily in hot and tropical regions. In Europe several species are found in Greece, Italy, Spain, France (five species), and southern Germany. In southern USA they extend from coast to coast, and into Canada they are recorded from British Columbia, Alberta, and Saskatchewan. In Central and South America they are widespread, but seem to be absent from Patagonia. They are also absent from New Zealand and the oceanic islands.

The order Scorpiones is divided into six families that are distributed in the New and Old World. None of the Bothriuridae or Chactidae is medically important, and in the Diplocentridae and Scorpionidae there are one or two species of minor medical importance in the Middle East. In the family Vaejovidae, species of *Vaejovis* and *Hadrurus* can deliver painful stings, but they are not common in peridomestic or domestic habitats. In Australia, the sting of the common brown scorpion, *Liocheles waigiensis*, is painful and produces a local swelling. The most important family is Buthidae, with more than 600 species in the New and Old World. The most dangerous genera in this family are *Androctonus*, *Buthus*, *Centruroides*, *Hottentotta*, *Leiurus*, *Mesobuthus*, *Parabuthus*, and *Tityus*. In general, scorpions with large claws (Scorpionidae, Chactidae, Diplocentridae, Bothriuridae) do not deliver a harmful sting; the dangerous species (Buthidae, Vaejovidae) usually have slender claws.

All scorpions possess venom glands, but fewer than 25 of the 1400 known species are considered dangerous and capable of inflicting human death. Of the 22 species identified in the Arabian peninsula, 14 belonging to the Buthidae and Scorpionidae have been indentified in Saudi Arabia. *Leiurus quinquestriatus* (Fig. 18.11h), *Androctonus crassicauda* (Fig. 18.11c), and *Apistobuthus pterygocercus* are the species of medical importance in Saudi Arabia. These species are also the most common cause of human stings throughout the Middle East. Scorpion stings in Saudi Arabia numbered 72 165 cases during a 5-year period from 1993 to 1997, with an average of nine cases per 10 000 inhabitants annually. Most of the stings were mild and all evolved to cure, except for one death. Deadly scorpion species reported in other countries include: *Tityus stigmurus*, *T. serrulatus*, and *T. brazilae* in Brazil; *Centruroides suffusus* in Mexico; *Hemiscorpion lepturus* in Iran; *Centruroides exilcauda* in the USA; *Mesobuthus tamulus* in India; and *Androctonus australis* (Fig. 18.11b) and *A. mauretanicus* in Algeria and Morocco, respectively.

Scorpion stings cause more morbidity in Mexico than any other country, leading to about 100 deaths annually. In 1999, the state of Morelos reported nine deaths among 30 663 cases of scorpion sting. The two species found in and around Mexican houses are *Centruroides limpidus limpidus* and *Vaejovis mexicanus smithi*.

Scorpion venom produces localized symptoms that last a few hours to a day, or systemic symptoms that affect bodily functions within a few hours and last for several days. The symptoms following stings by poisonous scorpions throughout the world are very similar. A local reaction includes pain and swelling at the site of the sting, followed by generalized symptoms of restlessness and excitement, fever, convulsions, and respiratory difficulties. Long-term reactions to stings, such as the muscular palsy and permanent limb weakness that may accompany the sting of *Tityus serrulatus*, are not common. Mortality from a scorpion sting is not common, and is highest among children. The death rate in Brazil to *T. serrulatus* ranges from 0.8 to 1.4% for adults, 3–5% for school-aged children, and 15–20% for young children. Death from a scorpion may be due to respiratory paralysis, peripheral vascular failure, or myocarditis.

Food consists primarily of insects and other arthropods. Some of the large scorpions are known to feed on frogs, small snakes, lizards, and small vertebrates. Prey is grasped by the pedipalp claws and moved to the chelicerae for maceration. Active or large prey is stung several times before being moved to the chelicerae for feeding. The sting is delivered by arching the postabdomen over the top of the scorpion so the stinger contacts the prey held by the claws. Once subdued, the prey is torn into small pieces by the chelicerae, and these pieces are held between the enlarged and often spined coxae of the pedipalps. Digestive fluid from the mouth of the scorpion liquefies the tissue held by the coxae, and the mixture is drawn into the mouth. When all the fluid is removed from the food, the remaining dry pellets are discarded. Scorpions forage at night, and their eyes play only a small role in location of prey. This is accomplished by the pectines, some specialized innervated setae called trichobothria, and other chemosensory setae and organs on the body.

Mating involves elaborate courtship behavior and the transfer of a spermatophore. The male initiates mating by grasping the claws of the female. He then begins to vibrate his body and the vibration moves the joined claws of the male and female. The two scorpions move back and forth, during

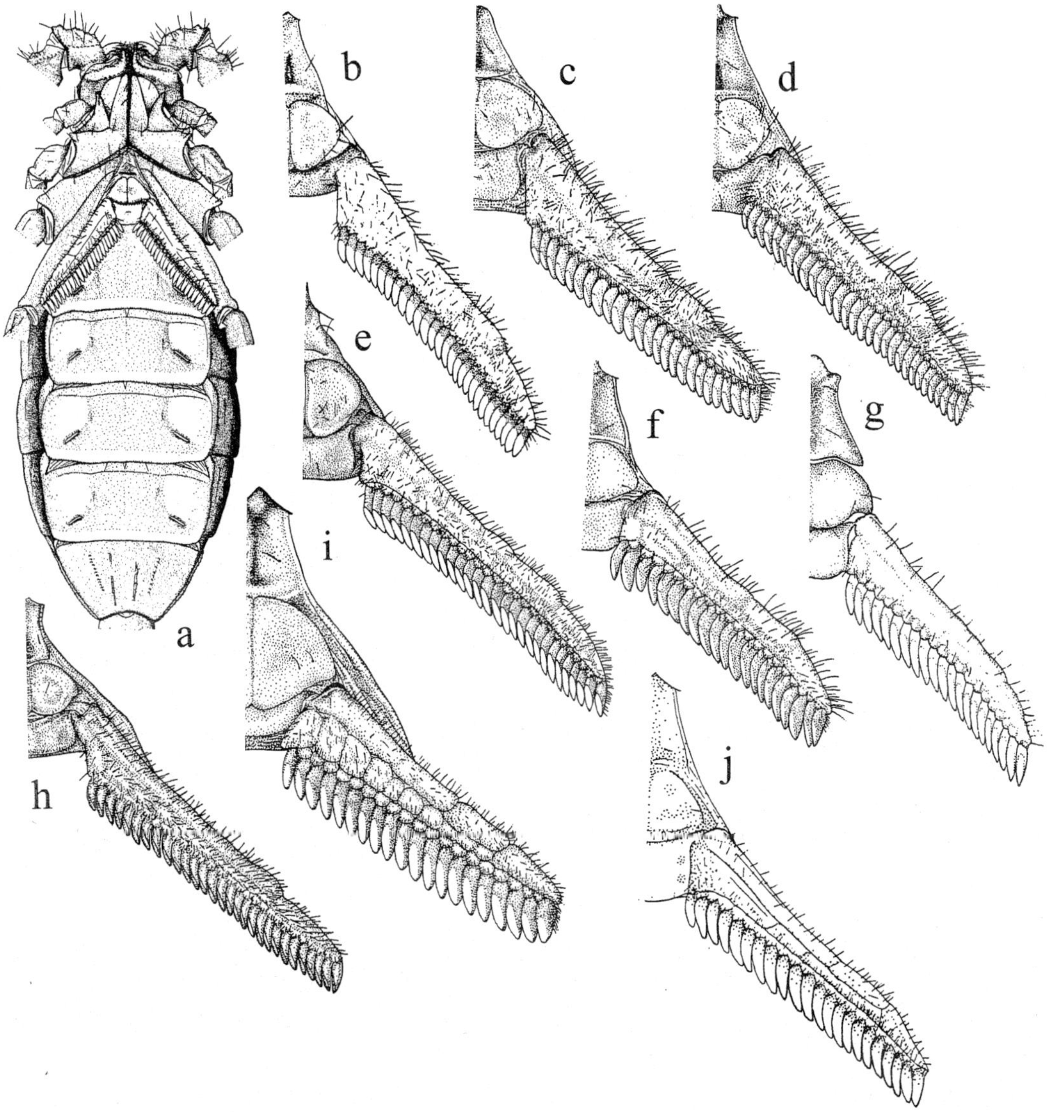


**Figure 18.11** Arthropoda: Scorpiones. (a) *Centruroides vittatus*, venter of cephalothorax and abdomen; (b) *Androctonus australis*, left half of pectines; (c) *A. crassicauda*; (d) *Buthotus tamulus*; (e) *Buthus occitanus*; (f) *C. elegans*; (g) *C. exilcauda*; (h) *Leiurus quinquestriatus*; (i) *Tityus bahiensis*; (j) *C. limpidus*. (Adapted from Keegan, 1980.)

which the pectines are swept over the substrate to locate a suitable site for depositing the stalked spermatophore. When the spermatophore has been deposited, the male moves the female over it, and once the sperm has been transferred to her genital atrium he releases the grip on her claws and they separate. Scorpions produce live young, and their behavior at the delivery time is variable. In some species the female positions one or more of the anterior legs below the genital opening to retain the emerging young. Young scorpions climb up one of the legs and on to the back of the mother. Newly born scorpions are yellowish white and do not achieve adult coloration for 2–3 months. They remain on the back of the mother until the first molt, which occurs 1–2 weeks after birth.

Fluorescence under ultraviolet light is a characteristic of most species of scorpions, tailless whip-scorpions (Amblypygi), sun-spiders or wind-scorpions (Sulfugae), and spiders (Araneae). The dorsal and ventral sclerites usually show the fluorescence response, while in some the effect is confined to the intersegmental membranes.

***Androctonus*** These are the fat-tailed scorpions (named because of their enlarged postabdominal segments), and several species are medically important. They occur in Northwest Africa and the Middle East from Turkey to India and Pakistan. *A. australis* (Fig. 18.11b) is responsible for many deaths in North Africa, but it is primarily distributed in arid mountainous regions and not in urban areas. Adults often reach a length of 10 cm and their body color varies from yellow to brown in different geographic areas. *A. crassicauda* (Fig.c) occurs over much of the Middle East, and it is considered the most dangerous scorpion in Turkey.

**Buthotus** Species occur throughout Africa, the Middle East, and Central Asia. Adults are 4–10 cm long and have lyre-shaped markings on the carapace. Of the approximately 20 species and subspecies in this genus, only a few have medical importance. *B. jayakari* occurs in peridomestic habitats on Kish Island (Iran) in the Persian Gulf. *B. tamulus* (Fig. 18.11d), an Indian scorpion, has significant medical importance. It is a dangerously venomous species, especially among children. It is not typically a house-infesting scorpion.

**Buthus** These are distributed in a variety of habitats from southern France and Spain through the Middle East into North Africa, and at altitudes from sea level to mountains. Adults are 4–11 cm long and have a characteristic H-shaped mark on the cephalothorax. *B. occitanus* (Fig. 18.11e) is the most widely distributed species, and the only one of medical importance. It occurs in urban and rural areas of France and Algeria.

**Centruroides** This genus contains some of the most dangerous scorpions. Species occur in Mexico, the USA, West Indies, and Central and South America. They are called bark scorpions because they are frequently found in harborages above ground (rather than in burrows), including under loose bark of trees, and in crevices of dead trees and logs. In the urban environment, *C. exilcauda* (= *C. sculpturatus*) (Fig. 18.11g) occurs in peridomestic habitats, such as lumber piles, firewood, and other debris. It frequently occurs indoors, where it is responsible for numerous stings. *C. exilcauda* is the only dangerous scorpion in the USA. In Mexico, *C. elegans* (Fig. 18.11f), *C. limpidus*, *C. infamatus*, and *C. suffusus* are dangerous species. In some regions, large numbers of *C. limpidus* (Fig. 18.11j) use harborages in adobe walls of village houses.

**Striped scorpion, *Centruroides vittatus* (Fig. 18.11a)** Adults are 5–6 cm long and the cephalothorax and abdomen are light brown to yellowish brown. The abdomen has two longitudinal, brown to reddish-brown stripes. There is a dark, triangular mark on the anterior portion of the carapace. This is the most widely distributed and common scorpion in the USA, from Florida to New Mexico, including Colorado, Kansas, Missouri, Oklahoma, and southern Illinois. It also occurs in Mexico. Adults and immatures are found under rocks and other ground debris, and the striped scorpion occurs indoors. The sting of *C. vittatus* causes temporary local pain, and sometimes light swelling. *C. pantheriensis* and *C. chisosarius* are color phases of *C. vittatus*.

**Hemiscorpion** This is an Old World genus in the large family Scorpionidae. These scorpions are large and have a threatening behavior when disturbed. The Middle Eastern species *H. lepturus* is dangerous and common in urban areas of Iran and Iraq.

**Hottentotta** This genus has about 20 species and they occur throughout Africa and the Middle East. *H. minax* is found under the bark of acacia trees in Sudan, and in dwellings in Chad.

**Mesobuthus** There are about 20 species in this genus and they occur throughout Asia, from Iran and Turkey to Korea. *M. eupeus* is a dangerous species in urban areas of Iran, and *M. tamulus* is a peridomestic pest in villages in India.

**Odontobuthus** In urban areas of Iran *O. doriae* is common; it is very active at night searching for food and harborage.

**Scorpion** This genus has species distributed in southern Europe, the Middle East, and Northern Africa. *S. maurus* is represented by several subspecies in the Middle East and North Africa. Fatalities due to this species have been reported from Algeria, and it is a common pest in urban areas of Iran. It is not known to be dangerous in southern France.

**Tityus** Of the approximately 37 species in this South American genus, several are dangerously venomous, and they often occur in urban environments. These are: *T. bahiensis* (Fig. 18.11i) in Brazil and Argentina; *T. cambridgei* in Guyana; *T. serrulatus*, *T. trivittatus charrenyroni*, and *T. stigmurus* in Brazil; and *T. trivittatus* in Argentina, Brazil, Paraguay, and Uruguay. The pest status of these species is based on the toxicity of their venom, plus their common occurrence in domestic and peridomestic habitats. *T. trivittatus charrenyroni* is the most frequently encountered scorpion in the Distrito Federal, Brazil. *T. trinitatis* is venomous, but this species does not enter houses. *T. caripitensis* occurs in Caripito, Bolívar municipality of Monagas state, Venezuela. *T. isabelceciliae* lives on the northern central slope of the Cordillera de la Costa, Venezuela. It belongs to the *T. discrepans* group and is dangerous to humans due to its high number, aggressive behavior, domiciliary habits, and high toxicity of its venom.

***Tityus serrulatus*** Adults are about 7 cm long and the cephalothorax and all but the last segment of the preabdomen are yellowish brown to blackish brown; the remainder of the body is usually yellow; the claws and the stinger are reddish brown. This species is parthenogenetic. *T. serrulatus* originally occupied a restricted area in the state of Minas Gerais, Brazil; today it is widely distributed throughout the southeast region of the country. The geographical expansion is related to human colonization, which started about 300 years ago from the Atlantic coastal region westward. Typically, newly erected towns are invaded within a few years after their foundation, although the surrounding natural areas are virtually devoid of this scorpion. It is generally considered to be the most dangerous scorpion in Brazil. It occurs in the states of Sergipe, Bahia, Minas Gerais, Espirito Santo, Rio de Janeiro, Goiás, and São Paulo.

## Solifugae

These arachnids are known as sun-spiders and wind-scorpions, and they are among the most formidable of the terrestrial arthropods. They are feared in many regions, but they are not poisonous. The large cephalothorax is divided into six segments, and the first three segments resemble a head. The segmented abdomen is constricted at the base, which gives the appearance of three distinct body regions, and the basis for the reference to them as spiders. The first pair of appendages, the chelicerae, are jaw-like, and the second pair, the pedipalps, are long and leg-like. The first pair of legs are primarily tactile, and the last three pairs are ambulatory. The body is densely covered with fine setae. These arthropods are generally active at night, and they move very quickly as predators of other arthropods, small lizards, and small rodents. Prey are crushed by the large jaws and reduced to pulp, and pressed against the mouth opening where food is absorbed in a semiliquid condition.

Solifugae occur in tropical and subtropical regions, and sometimes in hot, dry, desert areas. In Europe six species occur in southeast Spain, Greece, the Balkans, and in the vicinity of the Black Sea. *Gluvia dorsalis* occurs in urban habitats in Spain, and *Mummucia varigata* and *Pseudocleobis moricans* are common during the day on the streets of Santiago, Chile. In the USA, there are about 10 described species, and these occur in western states. Several species, such as *Solpuga sericea* and *S. lineata*, burrow into the ground to feed on termites. A California species of *Eremobates* kills bees, and a small nocturnal species, *E. pallipes*, is known to prey on bed bugs. In eastern USA the only solfugid is *Ammotrechella stimpsoni*, which occurs in southern Florida and the West Indies. Solifugae are particularly abundant in Africa, but none occurs in Madagascar, and they are absent from Australia and New Zealand. In Egypt and surrounding regions, *Galeodes arabs* is common, and it is said to screech when confronting danger. *Gylippus rickmersi* inhabits the Pamir plateau, north of the Hindu Kush mountains in central Asia.

After mating and fertilization, the female increases food consumption for the next 2–3 weeks. Females lay their eggs at night in a deep burrow in the soil. The eggs are round and white or with dark markings, and adhere to one another in compact groups. Females deposit 1–5 batches of eggs; hatching occurs within a few days or in 3–4 weeks. In *Galeodes*, the young larvae emerge 1–2 days after they are laid. At first, they are teneral, white, and unable to move. During their first 2–3 weeks and before the first molt, the young remain with the female. Following this molt, the young are more active and the integument hardens.

***Eremobates pallipes*** Adults are 24–26 mm long, pale brown, and have a slightly darker brown dorsal band. The body is moderately setose. The cephalothorax is dark brown, and joints of the pedipalps and legs have dark brown marks. This common species occurs in western USA, including California, Colorado, Utah, Wyoming, Nevada, Arizona, New Mexico, and Kansas.

**Other *Eremobates*** Common species in this genus include *E. californica*, which is about 25 mm long and yellowish brown. It occurs in southern California, Arizona, and northern Mexico. *E. durangonus* is about 28 mm long and the cephalothorax is light brown. *E. formidabilis* is about 27 mm long and reddish brown, and it has a broad, dark, dorsal longitudinal band on the abdomen. This species occurs in California and Arizona. *E. magna* has the tips of the pedipalps black, and it occurs in California, Arizona, and Texas. It has been reported feeding on ants in Texas.

## Uropygi

These are the whip-scorpions and vinegaroons. They are 25–120 mm long (including the caudal extension) and have an unsegmented cephalothorax, which bears eight eyes, and a flattened abdomen of 11 or 12 segments. The chelicerae are simple and two-segmented; the pedipalps are strong and six-segmented. In many species the last two segments of the pedipalps are modified to form pincers, which are used to seize prey. The first pair of legs is used only as tactile organs and they have modified tarsi. The organs for respiration are two pairs of book-lungs on the venter of abdominal segment 2 and 3. The last three segments of the abdomen are small and annular, forming a pygidium, which bears a long, slender telson from which the name whip-scorpion is derived. There are two glands that open one each side of the anus and discharge a fluid when the animal is disturbed. This fluid has the odor of vinegar (acetic acid), from which the name vinegaroon is derived (from the Spanish word *vinegare*). These non-poisonous arthropods are nocturnal and prey on insects such as cockroaches, crickets, grasshoppers, termites, and other arthropods. They live primarily in damp habitats in the tropics of the world, although a few species are found in arid regions. The genus *Hypoctonus* occurs in Malaysia, *Typopeltis* in Japan and northern China, *Thelyphonus* in southern Asia and Indonesia, and *Mastigoproctus* in southern USA and perhaps northern Mexico.

**Vinegaroon, whip-scorpion, *Mastigoproctus giganteus* (Fig. 18.9a)** Adults are about 65 mm long and brown to reddish brown. The caudal extension may be twice the length of the body proper. This species occurs in southern USA, nearly from coast to coast. It hunts at night, and moves slowly with its large pedipalps extended and open, and the first pair of legs touching objects in front and to the side. Prey is carried to a burrow for further mastication and eating. These arthropods are greatly feared because of their supposedly venomous bite, but they have no poison glands and they are not aggressive.

## Bibliography

### General

Baker, E. W. and G. W. Wharton. *An Introduction to Acarology*. New York: Macmillan, 1952.

Bettini, S. (ed.) *Arthropod Venoms*. Berlin: Springer Verlag, 1978.

Blum, M. S. *Chemical Defenses of Arthropods*. New York: Academic Press, 1981.

Bücherl, W. and E. E. Buckley (eds.) *Venomous Animals and Their Venoms*, vol. 3, *Venomous Invertebrates*. New York: Academic Press, 1972.

Clausen, C. P. Phoresy among Entomophagous insects. *Annu. Rev. Entomol.*, **21** (1976), 343–68.

Cloudsley-Thompson, J. L. *Spiders, Scorpions, Centipedes, and Mites*. London: Pergamon, 1958.

Evans, G. A. *Principles of Acarology*. Wallingford: CAB International, 1992.

Fager, E. W. The community of invertebrates in decaying oak wood. *J. Anim. Ecol.*, **37** (1968), 121–42.

Kekan, H. L. and W. V. MacFarlane (eds.) *Venomous and Poisonous Animals and Noxious Plants of the Pacific Region*. Oxford: Pergamon Press, 1963.

Lawrence, R. F. Fluorescence in arthropoda. *J. Entomol. Soc. South Afr.*, **17** (1954), 167–70.

Manton, S. M. Habits of life and evolution of body design in Arthropoda. *J. Linn. Soc. (Zool.)*, **44** (1958), 58–72.

*The Arthropoda. Habits, Functional Morphology and Evolution*. Oxford: Clarendon Press, 1977.

Mitchell, W. F., G. W. Wharton, and D. G. Larson. House dust, mites and insects. *Ann. Allergy*, **27** (1969), 93–9.

Savory, T. H. *Arachnida*. New York: Academic Press, 1977.

Tu, A. T. *Venoms: Chemistry and Molecular Biology*. New York: John Wiley, 1977.

(ed.) *Handbook of Natural Toxins: vol. 2, Insect Poisons, Allergens and other Invertebrate Venoms*. New York: Marcel Dekker, 1984.

Zumpt, F. (ed.) *The Arthropod Parasites of Vertebrates in Africa South of the Sahara*, vol. I. *Chelicerata*. Johannesburg: South African Institute for Medial Research, 1961.

### Acari

Arthur, D. R. *Ticks and Disease*. London: Pergamon, 1962.

*British Ticks*. London: Cambridge University Press, 1963.

Baker, E. W. A review of the mites of the family Cheyletidae in the United States National Museum. *Proc. U.S. Nat. Mus.*, **99** (1949), 267–320.

Baker, E. W., M. D. Delfinado, and M. J. Abbatiello. Terrestrial mites of New York. II. Mites in birds' nests. *J. N.Y. Entomol. Soc.*, **84** (1976), 48–66.

Barker, P. S. Bionomics of *Glycyphagus domesticus* (de Geer) (Acarina: Glyphagidae), a pest of stored grain. *Can. J. Zool.*, **46** (1968), 89–92.

Brennan, J. M. and M. L. Goff. Keys to the genera of chiggers of the western hemisphere (Acarina: Trombiculidae). *J. Parasitol.*, **63** (1977), 554–66.

Brennan, J. M. and E. K. Jones. Keys to the chiggers of North America with synonymic notes and descriptions of two new genera (Acarina: Trombiculidae). *Ann. Entomol. Soc. Am.*, **52** (1959), 7–16.

Burgdorfer, W. A review of Rocky Mountain spotted fever (tick-borne typhus), its agent and its tick vectors in the United States. *J. Med. Entomol.*, **12** (1975), 269–78.

Cooley, R. A. and G. M. Kohls. The Argasidae of North America, Central America and Cuba. *Am. Midl. Nat. Monogr.*, **1** (1944), 1–152.

Cunningham, A. M. Physical limits for complete development of the grain mite, *Acarus siro* L. (Acarina: Acaridae) in relation to its world distribution. *J. Appl. Ecol.* **2** (1965), 295–306.

Cusack, P. D., G. O. Evans, and P. A. Brennan. A survey of the mites of stored grain and grain products in the Republic of Ireland. *Sci. Proc. R. Dublin Soc. B3*, **20** (1975), 272–329.

Czajkowska, B. Influence of active substances of medicinal herbs on stored products mites. *Zeszyty Problemowe Postepow nauk Rolniczych*, **129** (1972), 197–232.

Evans, G. O. An introduction to the British Mesostigmata (Acarina) with keys to the families and genera. *Linn. Soc. J. Zool.*, **43** (1957), 203–59.

Evans, G. O. and E. Browning. Some British mites of economic importance. London: British Museum (Natural History), 1955.

Evans, G. O. and W. M. Till. Mesostigmatic mites of Britain and Ireland (Chelicerata: Acari-Parasitiformes). *Trans. Zool. Soc. Lond.*, **35** (1979), 139–270.

Evans, G. O., J. H. Sheals, and D. Macfarlane. *The Terrestrial Acari of the British Isles: An Introduction to their Morphology, Biology, and Classification*. London: British Museum (Natural History), 1961.

Fain, A. Notes sur les Acarines des familles Cheyletidae et Harpyrhynchidae producteurs de gale chez les oiseaux ou les mamifères. *Acta Zool. Pathol. Antverp*, **56** (1972), 37–60.

Observations on Cheyletid mites parasitic on mammals (Acari: Cheyletidae and Cheyletiellidae). *Acarologia*, **21** (1979), 408–22.

Griffiths, D. A. A revision of the genus *Acarus* L., 1758 (Acaridae, Acarina), with a key to species. *Bull. Br. Mus. (Nat. Hist.) (Zool.)*, **11** (1964), 415–64.

Griffiths, D. A., A. C. Hodson, and C. M. Christensen. Grain storage fungi associated with mites. *J. Econ. Entomol.*, **52** (1959), 514–18.

Ho, C.-C. Studies on the Biology of *Macrocheles muscadomesticae* (Scopoli) (Acarina: Macrochelidae). *Chin. J. Entomol.*, **3** (1989), 181–7.

Hoogstral, H. Ticks in relation to human diseases caused by viruses. *Annu. Rev. Entomol.*, **11** (1966), 261–308.

Ticks in relation to human diseases caused by *Rickettsia* species. *Annu. Rev. Entomol.*, **12** (1967), 377–420.

Hughes, A. M. The mites of stored food and houses. Ministry of Agriculture, Fisheries and Food, technical bulletin 9. London: Her Majesty's Stationery Office, 1976.

Needham, G. R. Evaluation of five popular methods for tick removal. *Pediatrics*, **75** (1985), 997–1002.

O'Connor, B. M. Evolutionary ecology of astigmatid mites. *Annu. Rev. Entomol.*, **27** (1982), 385–409.

Proctor, H. C. Feather mites (Acari: Astigmata): ecology, behavior, and evolution. *Annu. Rev. Entomol.*, **48** (2003), 185–209.

Rose, I. A review of tick paralysis. *J. Can. Med. Assoc.*, **70** (1954), 175–6.

Sasa, M. Biology of chiggers. *Annu. Rev. Entomol.*, **6** (1961), 221–44.

Sauer, J. R. and J. A. Hair. (eds.) *Morphology, Physiology and Behavioral Biology of Ticks*. Chichester: Ellis Horwood, 1986.

Schuster, R. and P. W. Murphy (eds.) *The Acari: Reproduction, Development, and Life History Strategies*. London: Chapman and Hall, 1991.

Shimada, Y., T. Beppu, H. Inokuma, M. Okuda, and T. Onishi. Ixodid tick species recovered from domestic dogs in Japan. *Med. Vet. Entomol.*, **17** (2003), 38–45.

Smiley, R. L. A review of the family Cheyletidae (Acarina). *Ann. Entomol. Soc. Am.*, **63** (1970), 1056–78.

Snodgrass, R. E. The feeding of Arachnida, including mites and ticks. *Smithsonian Misc. Coll.*, **110** (1948), 1–93.

Solomon, M. E. Ecology of the flour mite, *Acarus siro* L. ( = *Tyroglyphus farinae* DeG). *Ann. Appl. Biol.*, **50** (1962), 178–84.

Sonenshine, D. E. *Biology of Ticks*, vol. 1. New York: Oxford University Press, 1991.

Southcott, R. V. Arachnidism and allied syndromes in the Australian region. *Rec. S. Aust. Children's Hosp.*, **1** (1976), 97–186.

Summers, F. M. and D. W. Price. Review of the mite family Cheyletidae. *Univ. Calif. Publ. Entomol.*, **61** (1970), 1–153.

Tagami, K., T. Ishihara, J. Hosokawa, M. Ito, and K. Fukuyama. Occurrence of aquatic oribatid and astigmatid mites in swimming pools. *Water Res.*, **26** (1992), 1549–54.

Thomas, C. M. and R. J. Dicke. Response of the grain mite, *Acarus siro* (Acarina: Acaridae), to fungi associated with stored-food commodities. *Ann. Entomol. Soc. Am.*, **64** (1971), 63–8.

Woodroffe, G. E. An ecological study of the insects and mites in the nests of certain birds in Britain. *Bull. Entomol. Res.*, **44** (1953), 739–72.

Woodroffe, G. E. and B. J. Southgate. Birds' nests as a source of domestic pests. *Proc. Zool. Soc. Lond.*, **121** (1951), 55–62.

Woolley, T. A. *Acarology: Mites and Human Welfare*. New York: Wiley, 1988.

Zhang, Z.-Q. Biology and ecology of trombidiid mites (Acari: Trombidoidea). *Exp. Appl. Acarol.*, **22** (1988), 139–55.

### Acari: *Amblyomma*, *Dermacentor*, *Ixodes*

Anastos, G. The scutate ticks, or Ixodidae of Indonesia. *Entomol. Am.* (N.S.), **30** (1950), 1–144.

Arthur, D. R. *Monograph of the Ixodoidea, Part V. The Genera* Dermacentor, Anocentor, Cosmiomma, Boophilus, *and* Margaropus. Cambridge, UK: Cambridge University Press, 1959.

Arthur, D. R. *Ticks of the Genus* Ixodes *in Africa*. London: University of London, 1965.

Childs, J. E. and C. D. Paddock. The ascendancy of *Amblyomma americanum* as a vector of pathogens affecting humans in the United States. *Annu. Rev. Entomol.*, **48** (2003), 307–37.

Cooley, R. A. and G. M. Kohls. The genus *Amblyomma* (Ixodidae) in the United States. *J. Parasitol.*, **30** (1944), 77–111.

The genus *Ixodes* in North America. *Bull. Nat. Inst. Health*, **184** (1945), 1–246.

Duffy, D. C., D. D. Clark, S. R. Campbell *et al.* Landscape patterns of abundance of *Ixodes scapularis* (Acari: Ixodidae) on Shelter Island, New York. *J. Med. Entomol.*, **31** (1994), 875–9.

Elbl, A. and G. Anastos. Ixodid ticks (Acarina, Ixodidae) of Central Africa. *Mus. Rep. Afr. Cent. Tervuren. Belg. Ann. Ser.*, **8** (1966), 145–8.

Koch, H. G. Seasonal incidence and attachment sites of ticks (Acari: Ixodidae) on domestic dogs in southeastern Oklahoma and northwestern Arkansas, USA. *J. Med. Entomol.*, **19** (1982), 293–8.

Labruna, M. B., V. S. F. Homem, M. B. Heinemann, and J. S. Ferreira Neto. Ticks (Acari: Ixodidae) associated with rural dogs in Uruara, eastern Amazon-Brazil. *J. Med. Entomol.*, **37** (2000), 774–6.

Ogden, N. H., P. Cripps, C. C. Davison *et al.* The ixodid tick species attaching to domestic dogs and cats in Great Britain and Ireland. *Med. Vet. Entomol.*, **14** (2000), 245–51.

Papazahariadou, M. G., M. N. Saridomichelakis, A. F. Koutinas, E. G. Papadopoulos, and L. Leontides. Tick infestation of dogs in Thessaloniki, northern Greece. *Med. Vet. Entomol.*, **17** (2003), 110–13.

Rich, G. B. Disease transmission by the Rocky Mountain wood tick, *Dermacentor andersoni* Stiles, with particular reference to tick paralysis in Canada. *Vet. Med. Rev.*, **1971** (1971), 1–27.

Walker, J. B. *The Ixodid Ticks of Kenya*. London: Commonwealth Institute of Entomology, 1974.

Yeoman, G. H. and J. B. Walker. *The Ixodid Ticks of Tanzania*. London: Commonwealth Institute of Entomology, 1967.

### Acari: *Dermatophagoides*, *Sarcoptes*

Arlian, L. G. Biology and ecology of house dust mites, *Dermatophagoides* spp. and *Euroglyphus* spp. *Immun. Allergy. Clin. North Am.*, **9** (1989), 339–56.

Biology, host relations, and epidemiology of *Sarcoptes scabei*. *Annu. Rev. Entomol.*, **34** (1989), 139–61.

Arlian, L. G., R. A. Runyan, L. B. Sorlie, and S. A. Estes. Host-seeking behavior of *Sarcoptes scabei*. *J. Am. Acad. Dermatol.*, **11** (1984), 594–8.

Haarlov, N. and M. Almi. House-dust mites (*Dermatophagoides pteronyssinus* (Trt.), *D. farinae* Hughes, *Euroglyphus mayeni* (Cooreman) in Denmark (Acarina). *Entomol. Scand.*, **1** (1970), 301–6.

Maunsell, K., D. G. Wraith, and A. M. Cunnington. Mites and house-dust allergy in bronchial asthma. *Lancet*, **7555** (1968), 1267–70.

Miyamota, T., S. Oshima, T. Ishizaki, and S. Sato. Allergic identity between the common floor mite (*Dermatophagoides farinae* Hughes) and house dust as a causative antigen in bronchial asthma. *J. Allergy*, **42** (1968), 14–28.

Mumcuoglu, K. Y., Z. Gat, T. Horowitz *et al.* Abundance of house dust mites in relation to climate in contrasting agricultural settlements in Israel. *Med. Vet. Entomol.*, **12** (1999), 252–8.

Rosen, S., I. Yeruham, and Y. Braverman. Dermatitis in humans associated with the mites *Pyemotes tritici*, *Dermanyssus gallinae*, *Ornithonyssus bacoti* and *Androlaelaps casalis* in Israel. *Med. Vet. Entomol.*, **16** (2002), 442–4.

Samsinak, K., F. Dusbabek, and E. Vobrtazkova. Note on the house dust mites in Czechoslovakia. *Folia Parasitol. (Praha)*, **19** (1972), 383–4.

Spieksma, F. Th. M. and M. I. A. Spieksma-Boezema. The mite fauna of house dust with particular reference to the house-dust mite *Dermatophagoides pteronyssinus* (Trouessart, 1897) (Psorpotidae: Sarcoptiformes). *Acaralogia*, **9** (1967), 226–41.

van Bronswijk, J. E. M. H. *Dermatophagoides pteronyssinus* (Trouessart, 1897), in mattress and floor dust in a temperate climate (Acari: Pyroglyphidae). *J. Med. Entomol.*, **10** (1973), 63–70.

House-dust as an ecosystem. In Rodriguez, J. G. (ed.) *Recent Advances in Acarology II*. New York: Academic Press, 1979.

van Bronswijk, J. E. M. H. and E. J. de Kareek. *Cheyletiella* (Acari: Cheyletiellidae) of dog, cat and domesticated rabbit, a review. *J. Med. Entomol.*, **13** (1976), 315–27.

van Bronswijk, J. E. M. H. and R. N. Sinha. Pyroglyphid mites (Acari) and house dust allergy. *J. Allergy*, **47** (1971), 31–52.

Wharton, G. W. Mites and commercial extracts of house dust. *Science*, **167** (1970), 1382–3.

House dust mites. *J. Med. Entomol.*, **12** (1976), 577–621.

### Amphipoda

Friend, J. A. and A. M. M. Richardson. Biology of terrestrial amphipods. *Annu. Rev. Entomol.*, **31** (1986), 25–48.

### Aranae

Anderson, S. O. Aminoacid composition of spiders' silk. *Comp. Biol. Physiol.*, **35** (1971), 705–11.

Baerg, W. J. The brown widow spiders in Jamaica (Araneae, Therididae). *Ann. Entomol. Soc. Am.*, **47** (1954), 52–60.

Baldwin, G. A., D. F. Smith, and S. D. Fike. Loxocelism in Canada. *Can. Med. Assoc. J.*, **138** (1988), 521–2.

Bristowe, W. S. *The Comity of Spiders*. London: Adlard, 1939.

The distribution of harvestmen in Great Britain and Ireland. *J. Anim. Ecol.*, **18** (1949), 100–14.

Clyne, D. *A Guide to Australian Spiders*. Sydney: Nelson, 1969.

D'Amour, F. E., F. E. Becker, and W. van Riper. The black widow spider. *Q. Rev. Biol.*, **11** (1936), 123–60.

Dondale, C. D. and J. H. Redner. The sac spiders of Canada and Alaska (Aranae: Clumbionidae and Anyphaenidae). *The Insects and Arachnids of Canada*, Part 9. Ottawa: Canada Government Centre, 1982.

Efrati, P. Bites by *Loxoceles* spiders in Israel. *Toxicom.*, **6** (1969), 239–41.

Exline, H. *Tegenaria agrestis* (Walckenaer), a European agelenid spider introduced into Washington state. *Ann. Entomol. Soc. Am.*, **44** (1951), 308–10.

Foelix, R. F. *Biology of Spiders*. Cambridge, MA: Harvard University Press, 1982.

Forester, R. R. and L. M. Forester. *New Zealand Spiders*. Auckland: 1973.

Gertsch, W. J. *American Spiders*. New York: Van Nostrand, 1949.

The spider genus *Loxoceles* in North America, Central America, and the West Indies. *Am. Mus. Novitates*, **1907** (1958), 1–46.

Gorham, J. R. The geographic distribution of the brown recluse spider *Loxosceles reclusa* (Araneae, Scytodidae) and related species in the United States. *Coop. Econ. Insect. Rpt USDA*, **18** (1968), 171–5.

Guarisco, H. House spiders of Kansas. *J. Arachnol.*, **27** (1999), 217–21.

Habermehl, G. Die biologische Bedeutung tierischer Gifte. *Naturwissenschaften*, **62** (1975), 15–21.

Herms, W. B., S. F. Bailey, and B. McIvor. The black widow spider. *Science*, **82** (1935), 395–6.

Hickman, V. V. *Spiders of Tasmania*. Hobart: Tasman Museum, 1967.

Hillyard, P. *The Book of Spiders*. London: Hutchinson, 1994.

Ingram, W. W. and A. Musgrave. Spider bite (arachnidism): a survey of its occurrence in Australia, with case histories. *Med. J. Aust.*, **2** (1933), 10–15.

Jarlsfors, U., D. S. Smith, and F. E. Russell. Innervation of the venom-secreting cells in the black widow spider (*Latrodectus mactans*). In DeVries, A. and E. Kochva (eds.) *Toxins of Animal and Plant Origin*, vol. 1. London: Gordon and Breach, 1971.

Jones, D. *The Country Life Guide to Spiders of Britain and Northern Europe*. Middlesex Country Life, 1983.

Kaston, B. J. Notes on a new variety of black widow spider from southern Florida. *Fla. Entomol.*, **21** (1938), 60–2.

The spiders of Connecticut. *Conn. State Geol. Nat. Hist. Surv. Bull.*, **70** (1948), 1–874.

Comparative biology of American black widow spiders. *Trans. San Diego Soc. Nat. Hist.*, **16** (1970), 33–82.

*How to Know the Spiders*. Dubuque, IA: Wm. C. Brown, 1978.

Lamoral, B. H. On the nest and web structure of *Latrodectus* in South Africa. *Ann. Natal Mus.*, **20** (1968), 1–14.

Levi, H. W. The spider genus *Latrodectus (Araneae, Theridiidae)*. *Trans. Am. Microsc. Soc.*, **78** (1959), 7–42.

Levi, H. W. and A. Spielman. The biology and control of the South American brown spider, *Loxoceles laeta* (Nicolet), in a North American focus. *Am. J. Trop. Med. Hyg.*, **13** (1964), 132–6.

Levi, H. V., L. R. Levi, and H. S. Zim. *Spiders and their Kin*. New York: Golden Press, 1968.

Lucas, F. Spiders and their silks. *Discovery*, **25** (1964), 1–12.

Macchiavello, A. La *Loxoceles laeta*, causa del aracnoidismo cutáneo o mancha gangrenosa, de Chile. *Rev. Chil. Hist. Nat.*, **41** (1937), 11–19.

Main, B. Y. *The Spiders of Australia: A Guide to their Identification with Brief Notes on the Natural History of Common Forms*. Brisbane: Jacaranda, 1967.

Maretic, Z. Spider venoms and their effects. In Newting, W. (ed.) *Ecophysiology of Spiders*. Berlin: Springer-Verlag, 1987.

Maretic, Z. and F. E. Russell. A case of necrotic arachnidism in Yugoslavia. *Toxicon*, **17** (1978), 412–23.

Marples, B. J. and M. J. Marples. Notes on the behavior of spiders in the genus *Zygiella*. *Bull. Br. Arach. Soc.*, **2** (1971), 16–17.

McCrone, J. D. and H. W. Levi. North American widow spiders of the *Latrodectus curacaviensis* group (Araneae: Theriidae). *Psyche*, **71** (1964), 12–27.

Parrott, A. W. The banana spider (*Heteropoda venatoria* Linn.) recorded from New Zealand. *N.Z. Sci. Rev.*, **10** (1952), 129–30.

Petrunkevitch, A. The spiders of Porto Rico. *Conn. Acad. Arts Sci.*, **30, 31** (1930), 159–355; 1–191.

Preston-Mafham, R. A. and K. G. Preston-Mafham. *Spiders of the World*. London: Blandford Press, 1984.

Roberts, M. J. *The Spiders of Great Britain and Ireland*. vol. 1. *Atypidae to Theridiosomatidae*. Colchester, Essex: Harley Books, 1985.

Savory, T. H. *The Biology of Spiders*. London: Sidgwick and Jackson, 1928.

Schenone, H., A. Rojas, H. Reyes, F. Villaroel, and G. Suarez. Prevalence of *Loxoceles laeta* in houses in central Chile. *Am. J. Trop. Med. Hyg.*, **19** (1970), 564–7.

Schmidt, G. Giftspinnen – auch ein Problem des Ferntourismus. *Munch. Med. Wschr.*, **115** (1973), 2237–42.

Smith, A. M. *The Tarantula. Classification and Identification Guide*. London: Fitzgerald, (1987).

*Baboon Spiders: 1. Tarantulas of Africa and the Middle East*. London: Fitzgerald, 1990.

Vetter, R. S., A. H. Roe, R. G. Bennett *et al.* Distribution of the medically-implicated hobo spider (Araneae: Agelenidae) and its harmless congener, *Tegenaria duellica*, in the United States and Canada. *J. Med. Entomol.*, **40** (2003), 159–64.

## Chilopoda

Attems, C. The Myripoda of South Africa. *Ann. South Afr. Mus.*, **26** (1930), 1–431.

Auerbach, S. I. The centipedes of the Chicago area with special reference to their ecology. *Ecol. Monogr.*, **21** (1951), 97–124.

A key to the centipedes of the Chicago area. *Bull. Chicago Acad. Sci.* **9** (1951), 109–14.

Branson, B. A. and D. L. Batch. Valley centipedes (Chilopoda: Symphyla) from northern Kentucky. *Trans. Kentucky Acad. Sci.*, **28** (1967), 77–89.

Bücherl, W. Die Scolopendromorpha der Neotropischen Region. *Symp. Zool. Soc. Lond.*, **32** (1974), 99–133.

Chamberlin, R. V. The Chilopoda and Diplopoda of the West Indies. *Bull. Mus. Comp. Zool.*, **62** (1918), 151–262.

The myriapod fauna of the Bermuda Islands with notes on variation in *Scuterga*. *Ann. Entomol. Soc. Am.*, **13** (1920), 271–303.

On 10 new centipedes from Mexico and Venezuela. *Proc. Biol. Soc. Wash.*, **55** (1942), 17–24.

On Mexican centipedes. *Bull. Univ. Utah*, **33** (Biol. Ser. 7, no. 3) (1943), 1–55.

Chamberlin, R. V. and Y. M. Wang. Some records and descriptions of chilopods from Japan and other oriental areas. *Proc. Biol. Soc. Wash.*, **65** (1952), 177–88.

Cloudsley-Thompson, J. L. Some aspects of the biology of centipedes and scorpions. *Naturalist*, **1955** (1955), 147–53.

Eason, E. H. *Centipedes of the British Isles*. London: Hutchinson, 1965.

Easterla, D. A. Giant desert centipede preys upon a snake. *Southwestern Nat.*, **20** (1975), 411.

Hoffman, R. L. and R. M. Shelley. The identity of *Scolopendra marginata* Say (Chilopoda: Scolopendromorpha: Scolopendridae). *Myriapodologica*, **4** (1996), 35–42.

Jangi, B. S. On the ecology of the centipede *Scolopenda morsitans* L. (Scolopendridae) in Nagpur. *Entomol. Mon. Mag.*, **91** (1955), 211–13.

Jones, T. H., W. E. Conner, J. Meinwald, H. E. Eisner, and T. Eisner. Benzol cyanide and mandelonitrile in the cyanogenetic secretion of a centipede. *J. Chem. Ecol.*, **2** (1976), 421–9.

Lawrence, R. F. *Myripodes Chilopodes. Faune de Madagascar*, vol. 12. Tananarive: Institut de Recherche Scientifique, 1960.

Lewis, J. G. E. The taxonomy and biology of the centipede *Scolopendra amazonica* in the Sudan. *J. Zool. Soc. Lond.*, **149** (1966), 188–203.

*The Biology of Centipedes*. Cambridge: Cambridge University Press, 1981.

Shelley, R. M. The centipede order Scolopendromorpha in the Hawaiian Islands (Chilopoda). *Bishop Mus. Occ. Pap.*, **64** (2000), 39–48.

A synopsis of the North American centipedes of the order Scolopendromorpha (Chilopoda). *Va. Mus. Nat. Hist. Mem.*, **5** (2002), 1–108.

Summers, G. An illustrated key to the chilopods of the north-central region of the United States. *J. Kansas Entomol. Soc.*, **52** (1979), 690–700.

Verhoeff, K. W. Zur Biologie der *Scutigera coleoptrata*, und über die jungeren Larvenstadien. *Z. Wiss. Zool.*, **150** (1937), 262–82.

Wang, Y. M. The Myriapoda of the Philippine Islands. *Serica*, **1** (1951), 1–80.

## Decapoda, Diplopoda

Appel, A. G. Water relations and desiccation tolerance of migrating garden millipedes (Diplopoda: Paradoxosomatidae). *Environ. Entomol.*, **17** (1988), 463–6.

Baker, A. N. Some aspects of the economic importance of millipedes. *Symp. Zool. Soc. Lond.*, **32** (1974), 621–8.

Baker, G. H. The distribution and dispersal of the introduced millipede, *Ommatoiulus moreleti* (Diploda: Iulidae) in Australia. *J. Zool.*, **185** (1978), 1–11.

Barlow, C. A. Distributional and seasonal activity of three species of diplopods. *Arch. Neerland. Zool.*, **13** (1960), 108–33.

Bennett, D. R. and S. H. Kerr. Millipedes in and around structures in Florida. *Fla. Entomol.*, **56** (1973), 43–8.

Bernardi, M. de, G. Mellerio, G. Vidari *et al.* Quinones in the defense secretions of African millipedes. *Naturwissenschaften*, **69** (1982), 601–2.

Blower, G. British millipedes with special reference to Yorkshire species. *Naturalist*, **1952** (1952), 145–57.

Brolemann, H. W. Myrapodes Diplopodes (Chilognathes) I. *Faune France*, **29** (1935), 1–368.

Camatini, M. *Myriapod Biology*. London: Academic Press, 1979.

Causey, N. B. Studies on the life history and the ecology of the hothouse millipede, *Orthomorpha gracilis* (C. L. Koch, 1847). *Am. Midl. Nat.*, **29** (1943), 670–82.

Chamberlin, R. V. On Mexican millipedes. *Bull. Univ. Utah*, **34** (1943), 1–103.

Cloudsley-Thompson, J. L. The significance of migration in Myriapods. *Ann. Mag. Nat. Hist.*, **2** (1949), 947–62.

Fryer, G. Observations on some African millipedes. *Ann. Mag. Nat. Hist.* **10** (1957), 47–51.

Ingle, R. *Crayfish, Lobsters and Crabs of Europe*. London: Chapman & Hall, 1977.

Lewis, J. G. E. The ecology of centipedes and millipedes in northern Nigeria. *Symp. Zool. Soc. Lond.*, **32** (1974), 423–31.

Loomis, H. F. and R. Schmitt. The ecology, distribution and taxonomy of the millipedes of Montana west of the continental divide. *Northwest Sci.*, **45** (1971), 107–31.

Meinwald, Y. C., J. Meinwald, and T. Eisner. 1,2-dialkyl-4(3H)-quinazolinones in the defensive secretion of a millipede (*Glomeris marginata*). *Science*, **154** (1966), 390–1.

Minelli, A. *Proceedings of the Seventh International Congress of Myriapodology*. Leiden: E. J. Brill, 1990.

Mitra, T. R. Millipedes entering houses. *Entomol. Mon. Mag.*, **112** (1976), 44.

Niijima, K. and K. Shinohara. Outbreaks of the *Parafontaria laminata* group (Diplopoda: Xystodesmidae). *J. Ecol.*, **38** (1988), 257–68.

P. O'Neill, R. V. and D. E. Reichle. Urban infestation by the millipede, *Oxidus gracilis*. *J. Tennessee Acad. Sci.*, **45** (1970), 114–15.

Palmen, E. The Diplopoda of Eastern Fennoscandia. *Ann. Soc. Zool. Fenn., Vanamo*, **13** (1949), 1–54.

Peitsalmi, M. Population structure and seasonal changes in activity of *Proteroiulus fuscus* (AmStein) (Diplopoda, Blaninulidae). *Acta Zool. Fennica*, **161** (1981), 1–66.

Ramsey, J. M. Vast migrating armies of the millipede *Pseudopolydemus serratus* (Say) in the Dayton region. *Ohio J. Sci.*, **66** (1966), 339.

Read, H. J. The life histories of millipedes: a review of those found in British species of the order Julida and comments on endemic Madeiran *Cylindroiulus* species. *Rev. Ecol. Biol. Sol.*, **25** (1988), 451–67.

Saki, E. Diplopoda obstructive to railway traffic. *Botany Zool., (Tokyo)*, **2** (1934), 821–33.

Schubart, O. Diplopoda. *Tierwelt Dtsch.*, **28** (1934), 1–318.

Scott, H. Migrant millipedes and centipedes entering houses 1953–1957. *Entomol. Mon. Mag.*, **94** (1958), 73–7.

Shinohara, K. On sprawling of myriapod injuries in city. *Jpn. J. Sanit. Zool.*, **32** (1981), 249–50.

Toye, S. A. Studies on the locomotory activity of three species of Nigerian millipedes. *Ent. Exp. Appl.*, **9** (1966), 369–77.

Verhoeff, K. W. Zur Geographie, Ökologie und Systematik sudalpenlandischer Chilognathen (116. Diplopoden – Aufstaz). *Z. Morph. Okol. Tiere.*, **18** (1930), 575–668.


### Isopoda


Brereton, J. Le G. The distribution of woodland Isopods. *Oikos*, **8** (1957), 85–106.

Cloudsley-Thompson, J. L. Diurnal rhythms in woodlice. *J. Exp. Biol.*, **29** (1952), 295–303.

The biology of woodlice. *Discovery*, **16** (1955), 248–51.

Edney, E. B. The woodlice of Great Britain and Ireland. A concise systematic monograph. *Proc. Linn. Soc. Lond.*, **164** (1953), 49–98.

Woodlice and the land habitat. *Biol. Rev.*, **29** (1954), 185–219.

Harding, P. T. Notes on the biology and distribution of *Armadillidium album* Dollfus (Crustacea: Isopoda, Oniscoidea) in the British Isles. *Entomol. Mon. Mag.*, **104** (1968), 269–72.

Hatchett, S. P. Biology of the Isopoda of Michigan. *Ecol. Monogr.*, **17** (1947), 47–79.

Heeley, W. Observations on the life-histories of some terrestrial Isopods. *Proc. Zool. Soc. Lond. B.*, **111** (1941), 79–149.

Paris, O. H. The ecology of *Armadillidium vulgare* (Isopoda: Oniscoidea) in California grassland: food, enemies and weather. *Ecol. Monogr.*, **33** (1963), 1–22.

Thompson, W. R. The tachinid parasites of woodlice. *Parasitology*, **26** (1934), 378–448.

Vandel, A. Sur le mode de répartition des sexes chez l'isopode terrestre, *Armadillidium vulgare* (Latr.). *C. R. Acad. Sci. Paris*, **208** (1939), 1050–2.


### Opiliones


Bishop, S. C. The Phalangida (Opiliones) of New York. *Proc. Rochester Acad. Sci.*, **9** (1949), 159–235.

Bristowe, W. S. The distribution of harvestmen (Phalangida) in Great Britain and Ireland, with notes on their names, enemies and food. *J. Anim. Ecol.*, **18** (1949), 100–14.

Forster, R. R. The New Zealand harvestmen. *Canterbury Mus. Bull.*, **2** (1954), 1–329.

Gueutal, J. La ponte chez un Opilion: *Phalangium opilio* Linne. *Rev. Fr. Entomol. Paris*, **11** (1944), 6–9.

Holm, A. On the development of *Opilio parietinus* Deg. *Zool. Bidr. Uppsala*, **25** (1947), 409–22.

Kastner, A. Biologische Beobachtungen an Phalangiiden. *Zool. Anz.*, **95** (1931), 293–302.

Klee, G. E. and J. W. Butcheer. Laboratory rearing of *Phalangium opilio*. *Mich. Entomol.*, **1** (1968), 275–8.

Lawrence, R. F. The harvest-spiders of South Africa. *Ann. S. Afr. Mus.*, **29** (1931), 341–508.

Phillipson, J. The seasonal occurrence, life histories and fecundity of harvest-spiders in the neighborhood of Durham city. *Entomol. Mon. Mag.*, **95** (1959), 134–8.

Sankey, J. H. P. British harvest-spiders. *Essex Nat.*, **28** (1949), 181–91.

Savory, T. H. *Synopses of British Fauna*, no. 1. *Opiliones (Arachnida) or Harvestmen*, 2nd edn. London: 1948.

Spoek, G. I. The Opilionida of the Netherlands. *Zool. Verh. Leiden*, **62** (1963), 1–70.

Todd, V. Key to the determination of the British harvestmen (Arachnida, Opiliones). *Entomol. Mon. Mag.*, **84** (1948), 109–13.

The habits and ecology of the British harvestmen (Arachnida, Opiliones), with special reference to those of the Oxford district. *J. Anim. Ecol.*, **18** (1949), 209–16.

### Pseudoscorpiones

Beier, M. Pseudoscorpionidae. *Tierreich*, **57** (1932), 1–258 **58**, 1–294.

Gilbert, O. Observations on the feeding of some British false-scorpions. *Proc. Zool. Soc. Lond.*, **121** (1951), 547–55.

Hoff, C. C. The pseudoscorpions of Illinois. *Ill. Nat. Hist. Surv. Bull.*, **24** (1949), 411–98.

Kew, H. W. A synopsis of the false-scorpions of Britain and Ireland. *Proc. R. Irish Acad. B.*, **29** (1911), 38–64.

Levi, H. W. Notes on the life history of the Pseudoscorpion *Chelifer cancroides* (Linn) (Chelonethida). *Trans. Am. Micr. Soc.*, **67** (1948), 290–8.

Muchmore, W. B. On phoresy in pseudoscorpions. *Bull. Br. Arachnol. Soc.*, **2** (1971), 38.

Vachon, M. Recherches sur la biologie des Pseudoscorpionides. *Bull. Sci. Bourgogne*, **2** (1932), 21–6.

La reproduction et le développement des pseudoscorpions. *Ann. Sci. Nat. Paris*, **19** (1938), 84–7.

Weygoldt, P. *The Biology of Pseudoscorpions*. Cambridge: Harvard University Press.

### Scorpiones

Alexander, A. J. On the stridulation of scorpions. *Behaviour*, **12** (1958), 339–52.

Al-Sadoon, M. K. and S. A. Al-Faraj. *Scorpions of Saudi Arabia*. Riyadh: Al-Mehmas Press, 2001.

Al-Sadoon, M. K. and B. M. Jarrar. Epidemiological study of scorpion stings in Saudi Arabia between 1993 and 1997. *J. Venom. Anim. Toxins*, **9** (2003), 54–64.

Bücherl, W. Escorpiões e escorpionismo no Brasil. X. Catálogo da coleção escorpiônica do Instituto Butantan. *Mem. Inst. Butantan*, **29** (1959), 255–75.

Brazilian scorpions and spiders: 1. Biology of scorpions and effects of their venoms. *Rev. Bras. Pesq. Med. Biol.*, **1** (1968), 181–90.

Escorpionismo no Brasil. *Mem. Inst. Butantan*, **34** (1969), 9–24.

Byalynitskii-Birylya, A. A. *Arthrogastric Arachnids of Caucasia*. Part I. *Scorpions*. Jerusalem: Israel Programme for Scientific Translations, 1964.

Cloudsley-Thompson, J. L. On the function of the pectines of scorpions. *Ann. Mag. Nat. Hist.*, **8** (1955), 556–60.

Dittrich K., A. P. Power, and N. A. Smith. Scorpion sting syndrome, a 10 year experience. *Ann. Saudi Med.*, **15** (1995), 148–55.

El-Hennawy, H. K. A catalogue of the scorpions described from the Arab countries (1758–1990). Serket, **2** (1990), 95–153.

Ewing, H. E. The scorpions of the western part of the United States, with notes on those occurring in northern Mexico. *Proc. U.S. Natl Mus.*, **73** (1928), 1–24.

Francke, O. F. Systematic revision of diplocentrid scorpions (Diplocentridae) from circum-Caribbean lands. *Special Publ. Mus. Texas Tech. Univ.*, **14** (1978), 1–92.

Gertsch, W. J. and M. Soleglad. Studies of North American scorpions of the genera *Uroctonus* and *Vejovis* (Scorpionida, Vejovidae). *Bull. Am. Mus. Nat. Hist.*, **148** (1972), 547–608.

Glauert, L. Scorpions. *Aust. Mus. Mag.*, **9** (1946), 93–8.

Gonzalez-Sponga, M. A. G. D'Suze, and C. Sevick. Venezuelan arachnids: two new species of the *Tityus* genus (Scorpionida: Buthidae) and the chromatographic profile of venom as a possible taxonomical tool. *J. Venom. Anim. Toxins*, **7** (2001), 219–39.

Johnson, J. D. and D. M. Allred. *Scorpions of Utah*. Great Basin Naturalist, **32** (1972), 157–70.

Keegan, H. L. Scorpions of Medical Importance. Jacksonville, MS: University Press of Mississippi, 1980.

Lamoral, B. H. and S. C. Reynders. A catalog of the scorpions described from the Ethiopian faunal region up to December 1973. *Ann. Natal Mus.*, **22** (1975), 489–576.

Lourenço, W. R. Sur l'écologie du scorpion Buthidae: *Tityus trivittatus fasciolatus* Pessôa. *Vie Milieu*, **31** (1981), 71–6.

The scorpion families and their geographical distribution. *J. Venom. Anim. Toxins*, **7** (2001), 3–23.

Lourenço, W. R. and O. Cuellar. Scorpions, scorpionism, life history strategies and parthenogenesis. *J. Venom. Anim. Toxins*, **1** (1995), 51–62.

Mathew, A. P. Mating of scorpions. *J. Bombay Nat. Hist. Soc.*, **54** (1957), 853–7.

Mello-Leitão, C.de. Escorpiones Sul-Americanos. *Arq. Mus. Nac. Bras.*, **40** (1945), 1–468.

Polis, G. A. (ed.) *The Biology of Scorpions*. Stanford, CA: Stanford University Press, 1990.

Pringle, G. Notes on the scorpions of Iraq. *Bull. Endem. Dis.*, **3** (1960), 73–87.

Probst, P. J. A review of the scorpions of East Africa with special regard to Kenya and Tanzania. *Acta Trop.*, **30** (1973), 312–35.

Serfat, A. and M. Vachon. Quelques remarques sur la biologie d'un scorpion de l'Afghanistan: *Buthotus alticola* (Pocock). *Bull. Mus. Hist. Nat. Paris*, **22** (1950), 215–18.

Shulov, A. On some Israeli scorpions. *Dapin Refuiim (Folia Med.)*, **21** (1962), 3–14.

Stahnke, H. L. A review of *Hadrurus* scorpions (Vejovidae). *Entomol. News*, **80** (1969), 57–65.

Some observations of the genus *Centruroides* Marx (Buthidae: Scorpionida) and *C. sculpturatus* Ewing. *Entomol. News*, **82** (1971), 281–307.

A key to the genera of Buthidae (Scorpionida). *Entomol. News*, **81** (1972), 297–316.

Stahnke, H. L. and M. Carlos. A key to the species of the genus *Centruroides* Marx (Scorpionidae: Buthidae). *Entomol. News*, **88** (1977), 111–20.

Stockwell, A. A. Systematic observations on North American Scorpionida with a key and checklist of the families and genera. *J. Med. Entomol.*, **29** (1992), 407–22.

Tirgari, S. and J. Zargan. Scorpions in urban areas in Iran and recent progress of laboratory research (Scorpionida: Scorpionidae, Buthidae). In Jones, S. C., J. Zhai, and W. Robinson (eds.) *Proceedings of the 4th International Conference on Urban Pests*, pp. 399–400. Blacksburg, VA: Pocahontas Press, 2002.

Vachon, M. *Etudes sur les Scorpions*. Algiers: Institut Pasteur d'Algerie, 1952.

The biology of scorpions. *Endeavour*, **12** (1953), 80–9.

Arachnids of Saudi Arabia: Scorpions. *Fauna Saudi Arabia*, **1** (1979), 30–66.

Waterman, J. A. Scorpions in the West Indies with special reference to *Tityus trinitatis*. *Caribb. Med. J.*, **12** (1950), 167–77.

### Solifuge

Cloudsley-Thompson, J. L. Observations on the natural history of the camel-spiders *Galeodes arabs* C. L. Koch (Solfugae: Galeodidae) in the Sudan. *Entomol. Mon. Mag.*, **97** (1961), 145–52.

Cooke, J. A. L. and M. U. Shadab. Whipscorpions from Africa. *Am. Mus. Novit.*, **2526** (1973), 1–11.

Muma, M. H. The arachnid order Solpugida in the United States. *Bull. Am. Mus. Nat. Hist.*, **97** (1951), 31–142.

Burrowing habits of North American Solpugida. *Fla. Entomol.*, **49** (1966), 199–216.

The life cycle of *Eremobates durangonus* (Arachnida: Solpugida). *Fla. Entomol.*, **49** (1966), 233–42.

Basic behavior of North American Solpugida. *Fla. Entomol.*, **50** (1967), 115–23.

# Family, genus, species index

# Index